Solar System Plasma Physics

Geophysical Monograph Series

Including

IUGG Volumes

Maurice Ewing Volumes

Mineral Physics Volumes

GEOPHYSICAL MONOGRAPH SERIES

Geophysical Monograph Volumes

1 **Antarctica in the International Geophysical Year** *A. P. Crary, L. M. Gould, E. O. Hulburt, Hugh Odishaw, and Waldo E. Smith (Eds.)*
2 **Geophysics and the IGY** *Hugh Odishaw and Stanley Ruttenberg (Eds.)*
3 **Atmospheric Chemistry of Chlorine and Sulfur Compounds** *James P. Lodge, Jr. (Ed.)*
4 **Contemporary Geodesy** *Charles A. Whitten and Kenneth H. Drummond (Eds.)*
5 **Physics of Precipitation** *Helmut Weickmann (Ed.)*
6 **The Crust of the Pacific Basin** *Gordon A. Macdonald and Hisashi Kuno (Eds.)*
7 **Antarctica Research: The Matthew Fontaine Maury Memorial Symposium** *H. Wexler, M. J. Rubin, and J. E. Caskey, Jr. (Eds.)*
8 **Terrestrial Heat Flow** *William H. K. Lee (Ed.)*
9 **Gravity Anomalies: Unsurveyed Areas** *Hyman Orlin (Ed.)*
10 **The Earth Beneath the Continents: A Volume of Geophysical Studies in Honor of Merle A. Tuve** *John S. Steinhart and T. Jefferson Smith (Eds.)*
11 **Isotope Techniques in the Hydrologic Cycle** *Glenn E. Stout (Ed.)*
12 **The Crust and Upper Mantle of the Pacific Area** *Leon Knopoff, Charles L. Drake, and Pembroke J. Hart (Eds.)*
13 **The Earth's Crust and Upper Mantle** *Pembroke J. Hart (Ed.)*
14 **The Structure and Physical Properties of the Earth's Crust** *John G. Heacock (Ed.)*
15 **The Use of Artificial Satellites for Geodesy** *Soren W. Henricksen, Armando Mancini, and Bernard H. Chovitz (Eds.)*
16 **Flow and Fracture of Rocks** *H. C. Heard, I. Y. Borg, N. L. Carter, and C. B. Raleigh (Eds.)*
17 **Man-Made Lakes: Their Problems and Environmental Effects** *William C. Ackermann, Gilbert F. White, and E. B. Worthington (Eds.)*
18 **The Upper Atmosphere in Motion: A Selection of Papers With Annotation** *C. O. Hines and Colleagues*
19 **The Geophysics of the Pacific Ocean Basin and Its Margin: A Volume in Honor of George P. Woollard** *George H. Sutton, Murli H. Manghnani, and Ralph Moberly (Eds.)*
20 **The Earth's Crust: Its Nature and Physical Properties** *John G. Heacock (Ed.)*
21 **Quantitative Modeling of Magnetospheric Processes** *W. P. Olson (Ed.)*
22 **Derivation, Meaning, and Use of Geomagnetic Indices** *P. N. Mayaud*
23 **The Tectonic and Geologic Evolution of Southeast Asian Seas and Islands** *Dennis E. Hayes (Ed.)*
24 **Mechanical Behavior of Crustal Rocks: The Handin Volume** *N. L. Carter, M. Friedman, J. M. Logan, and D. W. Stearns (Eds.)*
25 **Physics of Auroral Arc Formation** *S.-I. Akasofu and J. R. Kan (Eds.)*
26 **Heterogeneous Atmospheric Chemistry** *David R. Schryer (Ed.)*
27 **The Tectonic and Geologic Evolution of Southeast Asian Seas and Islands: Part 2** *Dennis E. Hayes (Ed.)*
28 **Magnetospheric Currents** *Thomas A. Potemra (Ed.)*
29 **Climate Processes and Climate Sensitivity (Maurice Ewing Volume 5)** *James E. Hansen and Taro Takahashi (Eds.)*
30 **Magnetic Reconnection in Space and Laboratory Plasmas** *Edward W. Hones, Jr. (Ed.)*
31 **Point Defects in Minerals (Mineral Physics Volume 1)** *Robert N. Schock (Ed.)*
32 **The Carbon Cycle and Atmospheric CO_2: Natural Variations Archean to Present** *E. T. Sundquist and W. S. Broecker (Eds.)*
33 **Greenland Ice Core: Geophysics, Geochemistry, and the Environment** *C. C. Langway, Jr., H. Oeschger, and W. Dansgaard (Eds.)*
34 **Collisionless Shocks in the Heliosphere: A Tutorial Review** *Robert G. Stone and Bruce T. Tsurutani (Eds.)*
35 **Collisionless Shocks in the Heliosphere: Reviews of Current Research** *Bruce T. Tsurutani and Robert G. Stone (Eds.)*
36 **Mineral and Rock Deformation: Laboratory Studies—The Paterson Volume** *B. E. Hobbs and H. C. Heard (Eds.)*
37 **Earthquake Source Mechanics (Maurice Ewing Volume 6)** *Shamita Das, John Boatwright, and Christopher H. Scholz (Eds.)*

38 Ion Acceleration in the Magnetosphere and Ionosphere *Tom Chang (Ed.)*

39 High Pressure Research in Mineral Physics (Mineral Physics Volume 2) *Murli H. Manghnani and Yasuhiko Syono (Eds.)*

40 Gondwana Six: Structure, Tectonics, and Geophysics *Gary D. McKenzie (Ed.)*

41 Gondwana Six: Stratigraphy, Sedimentology, and Paleontoloty *Garry D. McKenzie (Ed.)*

42 Flow and Transport Through Unsaturated Fractured Rock *Daniel D. Evans and Thomas J. Nicholson (Eds.)*

43 Seamounts, Islands, and Atolls *Barbara H. Keating, Patricia Fryer, Rodey Batiza, and George W. Boehlert (Eds.)*

44 Modeling Magnetospheric Plasma *T. E. Moore, J. H. Waite, Jr. (Eds.)*

45 Perovskite: A Structure of Great Interest to Geophysics and Materials Science *Alexandra Navrotsky and Donald J. Weidner (Eds.)*

46 Structure and Dynamics of Earth's Deep Interior (IUGG Volume 1) *D. E. Smylie and Raymond Hide (Eds.)*

47 Hydrological Regimes and Their Subsurface Thermal Effects (IUGG Volume 2) *Alan E. Beck, Grant Garvin and Lajos Stegena (Eds.)*

48 Origin and Evolution of Sedimentary Basins and Their Energy and Mineral Resources (IUGG Volume 3) *Raymond A. Price (Ed.)*

49 Slow Deformation and Transmission of Stress in the Earth (IUGG Volume 4) *Steven C. Cohen and Petr Vaníček (Eds.)*

50 Deep Structure and Past Kinematics of Accreted Terranes (IUGG Volume 5) *John W. Hillhouse (Ed.)*

51 Properties and Processes of Earth's Lower Crust (IUGG Volume 6) *Robert F. Merev, Stephan Mueller and David M. Fountain (Eds.)*

52 Understanding Climate Change (IUGG Volume 7) *Andre L. Berger, Robert E. Dickinson and J. Kidson (Eds.)*

53 Plasma Waves and Istabilities at Comets and in Magnetospheres *Bruce T. Tsurutani and Hiroshi Oya (Eds.)*

IUGG Volumes

1 Structure and Dynamics of Earth's Deep Interior *D. E. Smylie and Raymond Hide (Eds.)*

2 Hydrological Regimes and Their Subsurface Thermal Effects *Alan E. Beck, Grant Garvin and Lajos Stegena (Eds.)*

3 Origin and Evolution of Sedimentary Basins and Their Energy and Mineral Resources *Raymond A. Price (Ed.)*

4 Slow Deformation and Transmission of Stress in the Earth *Steven C. Cohen and Petr Vaníček (Eds.)*

5 Deep Structure and Past Kinematics of Accreted Terranes *John W. Hillhouse (Ed.)*

6 Properties and Processes of Earth's Lower Crust *Robert F. Merev, Stephan Mueller and David M. Fountain (Eds.)*

7 Understanding Climate Change *Andre L. Berger, Robert E. Dickinson and J. Kidson (Eds.)*

Maurice Ewing Volumes

1 Island Arcs, Deep Sea Trenches, and Back-Arc Basins *Manik Talwani and Walter C. Pitman III (Eds.)*

2 Deep Drilling Results in the Atlantic Ocean: Ocean Crust *Manik Talwani, Christopher G. Harrison, and Dennis E. Hayes (Eds.)*

3 Deep Drilling Results in the Atlantic Ocean: Continental Margins and Paleoenvironment *Manik Talwani, William Hay, and William B. F. Ryan (Eds.)*

4 Earthquake Prediction—An International Review *David W. Simpson and Paul G. Richards (Eds.)*

5 Climate Processes and Climate Sensitivity *James E. Hansen and Taro Takahashi (Eds.)*

6 Earthquake Source Mechanics *Shamita Das, John Boatwright, and Christopher H. Scholz (Eds.)*

Mineral Physics Volumes

1 Point Defects in Minerals *Robert N. Schock (Ed.)*

2 High Pressure Research in Mineral Physics *Murli H. Manghnani and Yasuhiko Syono (Eds.)*

Geophysical Monograph 54

T.M

Solar System Plasma Physics

J. H. Waite, Jr.,
J. L. Burch,
and R. L. Moore
Editors

American Geophysical Union

Associate Editor T. W. Moorehead

Published under the aegis of AGU Books Board.

Library of Congress Cataloging-in-Publication Data

Solar system plasma physics.
(Geophysical monograph ; 54)
Includes index.
1. Space plasmas. 2. Sun. 3. Magnetosphere.
4. Astrophysics. I. Waite, J. H. (John H.) II. Burch,
J. L., 1942– . III. Moore, R. L., 1942–
IV. Series.
QC809.P5S65 1989 523.01 89-15177
ISBN 0-87590-074-7

Printed in the United States of America.

CONTENTS

PREFACE

Science involves a well orchestrated interplay between theory and experiment. Past unexplained observations suggest new questions to be asked, and answering these questions many times requires new observational techniques or at least new applications of old techniques. Solar system plasma physics is a classic example of the scientific process at work and has benefited from the rapid technological exploration of our near space environment over the last 35 years. This book is a 1988 snapshot of the scientific process in solar system plasma physics. It is structured by a series of scientific questions. Under each of these headings are theoretical papers which review the pertinent science and properly formulate the question in specific observational terms. These are followed by experimental papers which address the present state of observational techniques which can be used to investigate these outstanding problems. In addition, two introductory papers offer overviews of the fields of solar physics and magnetospheric physics; each addresses itself, as a kind of short course in its respective discipline, to scientists from the other discipline.

The outstanding problems in solar system plasma physics which are addressed in the book are divided into two major topics: (1) three-dimensional structure and turbulence, and (2) mass, momentum, and energy release and transfer. These broad topics are further divided into a series of questions.

In the area of three-dimensional structure and turbulence, the questions addressed are: How is free magnetic energy built up and held in the solar atmosphere? What new insight can be gained by understanding the three-dimensional structure and temporal variations of macroscale and microscale plasma and electromagnetic phenomena in the solar wind and planetary magnetospheres? What is the role of turbulence in the transfer of energy in solar-wind and magnetospheric plasmas? What determines the composition and charge state of solar-wind and magnetospheric ions?

The questions that relate to the second topic (mass, momentum, and energy release and transfer in solar system plasmas) were again concerned with solar and planetary plasma processes; e.g.: What is the magnetic energy conversion process in flares? What do solar radio bursts tell us about particle beams and wave-particle interactions in flares? How are coronal mass ejections driven? What macroscopic and microscopic processes are responsible for particle acceleration in the solar wind and in planetary magnetospheres? What are the critical phenomena associated with solar wind interactions with magnetic and non-magnetic bodies? What are the important ring, moon, and dust interactions in planetary and cometary magnetosphere-ionosphere systems?

In order to address these questions from both theoretical and experimental points of view, the 1988 Yosemite Conference on Outstanding Problems in Solar System Plasma Physics: Theory and Instrumentation was held at Yosemite National Park, California, February 2-5, 1988. The conference attracted over 90 scientists from the United States and Europe. The format of the conference was based on invited review talks with extensive periods of time reserved for open discussion of issues of interest. Evening sessions were devoted to poster presentations of contributed papers and general discussion among the conference participants.

The conference continued a tradition of bringing together scientific communities in related areas of space physics to focus on their common interests. The first Yosemite Conference on Magnetosphere-Ionosphere Coupling in 1974 was organized to foster communication between magnetospheric physicists and those scientists interested in the ionosphere and thermosphere, who were just beginning to realize the strong interactions that occur among these regions. Subsequent meetings in the even years from 1976 through 1982 continued with the same general theme. In 1984 a new community was included under the broader topic of Comparative Planetary Magnetospheres, and again in 1988 a new community, the solar physics community, was included with pleasing results.

Yosemite meetings have traditionally been meetings in which scientific exchange continues throughout the day and night in a cloistered and focused environment. We thought that a meeting which allowed for the presentation of new experimental techniques was long overdue and realized that the new directions in experimental space physics are being set by ongoing work in theory and simulation. By addition of the solar physics discipline to those groups previously associated with the Yosemite meeting, the important theoretical and experimental aspects of the field of solar system plasma physics were now addressed.

This philosophy for the meeting was enthusiastically accepted by the initial gathering of the program committee:

S.-I. Akasofu, University of Alaska
A. L. Broadfoot, University of Arizona
J. L. Burch, Southwest Research Institute
J. T. Clarke, University of Michigan
G. A. Dulk, University of Colorado
J. Geiss, Universität Bern
C. K. Goertz, University of Iowa
W. B. Hanson, University of Texas at Dallas
E. Hildner, NOAA/Space Environment Laboratory
R. L. Moore, NASA/Marshall Space Flight Center
M. Neugebauer, NASA/Jet Propulsion Laboratory
C. T. Russell, University of California, Los Angeles

G. L. Siscoe, University of California, Los Angeles
R. S. Steinolfson, University of Texas at Austin
J. H. Waite, NASA/Marshall Space Flight Center
D. J. Williams, Applied Physics Laboratory
D. T. Young, Southwest Research Institute.

However, the committee felt that something was still missing. The meeting needed to draw out and present some of the outstanding problems in the field and focus on solving them. Thus, the philosophy of the meeting took final form: present some key outstanding issues in space plasma physics and then discuss ways of finding solutions to these problems using new experimental techniques or new applications of old experimental techniques. The book follows the conference closely in format and content.

We are grateful to the members of the program committee for their work in putting together the program and for serving as session chairpersons. We wish to acknowledge the organizational efforts of Rose Mary Bryant and Leah Roberson, who skillfully arranged for the smooth running of the conference. Most importantly, we thank the participants of the conference, without whom none of this would have been possible.

The conference was sponsored by the NASA/Marshall Space Flight Center and the Southwest Research Institute. We wish to acknowledge financial support from the National Aeronautics and Space Administration, the National Science Foundation, and the sponsoring agencies of all the participants.

J. L. Burch
R. L. Moore
J. H. Waite, Jr.

A SKETCH OF SOLAR PHYSICS

Ronald L. Moore

Space Science Laboratory, NASA/Marshall Space Flight Center, Huntsville, AL 35812

The Scope of Solar Physics

Solar physics is an important, exciting branch of science in three ways. To begin with, solar phenomena and the physics that governs them are fascinating and worthy of study in their own right. The length scales, temperatures, densities, magnetic fields, gravity, and rotation of the Sun yield an array of magnetohydrodynamic (MHD) phenomena that are captivating to observe, confounding to explain, and impossible to truely replicate in the laboratory. In another way, solar physics is important because the Sun is the nearest star. This makes our knowledge and understanding of the Sun, i.e., solar physics, a key to stellar astrophysics. Moreover, as in the MHD and plasma phenomena of the Sun, magnetized plasma is an essential ingredient of most cosmic systems of stellar and galactic scale, including the violent objects prominent in modern astrophysics (e.g., collapsed stellar objects with accretion disks, active galactic nuclei, and stellar and galactic jets). Because of this and the nearness of the Sun, solar physics guides and tests our understanding of processes that are important for much of astrophysics beyond that of normal stars. Finally, solar physics is important because the Sun, through its direct radiation and the solar wind, is the origin or driver of phenomena central to space physics: the interplanetary medium and the magnetospheres, ionospheres, and atmospheres of the planets. This domain includes solar-terrestrial effects of great practical importance: the Sun sustains life on Earth, drives our weather, and regulates our climate. As in astrophysics, plasma processes that govern solar phenomena also pervade space physics. For example, this is evident in the strong similarity between solar flares and magnetospheric substorms (e.g., see Svestka, 1976). So, in broad perspective, solar physics is a worthy endeavor because solar physics by itself is a challenging and rewarding science, because solar physics (together with space physics) is a key to much of astrophysics, and because of the preeminence of the Sun in space physics phenomena and the terrestrial environment.

The cutaway view of the Sun in Figure 1 depicts the major components of the Sun; these set the scope of solar physics. The Sun has three main parts, spherically nested and distinguished by different modes of energy transport: the radiative interior, the convective envelope, and the directly observable atmosphere. The Sun is powered by nuclear fusion in the core of the radiative interior, wherein the temperature is high enough ($>10^7$ K) for fusion collisions between nuclei. Throughout the radiative interior, the opacity is small enough that the heat flowing out from the fusion core can be carried by radiative conduction; the radiative interior is thereby convectively stable and static to first order. In the outer few tenths of the Sun's radius, because the temperature is lower, the opacity is great enough that the same heat flow drives free convection. In this convective outer envelope, practically all of the escaping heat is carried by convection instead of radiation. Just below the visible surface, the mass density becomes so low that the opacity again becomes small enough for the energy transport to switch back to radiation. The solar atmosphere above the surface is largely transparent to the Sun's light and receives little of its energy radiatively. Instead, it is heated by some yet undetermined nonthermal means of energy transfer from the convective interior. This nonthermal energy transfer is strongly controlled by the magnetic field.

The essential function of the core is the release of nuclear energy, the ultimate source of energy that sustains the Sun's luminosity and powers all of the phenomena in the convective envelope and solar atmosphere. The state of the core (central values and radial change of temperature, density, composition, and rotation, and the form and magnitude of meridional circulation) determines the rate of nuclear burning and the rate at which neutrinos are produced by the nuclear reactions. The expected flux of solar neutrinos at the Earth, calculated from standard stellar structure and evolution theory, is a factor of 3 greater than observed (Davis and Evans, 1978). This now famous problem of the Sun's missing neutrinos is a measure of the uncertainty with which the state of the deep interior of the Sun and similar stars is known, and illustrates the value of the Sun for guiding and testing astrophysical theory.

Obviously, the state of the Sun's deep interior is a major quandry of solar physics, but it

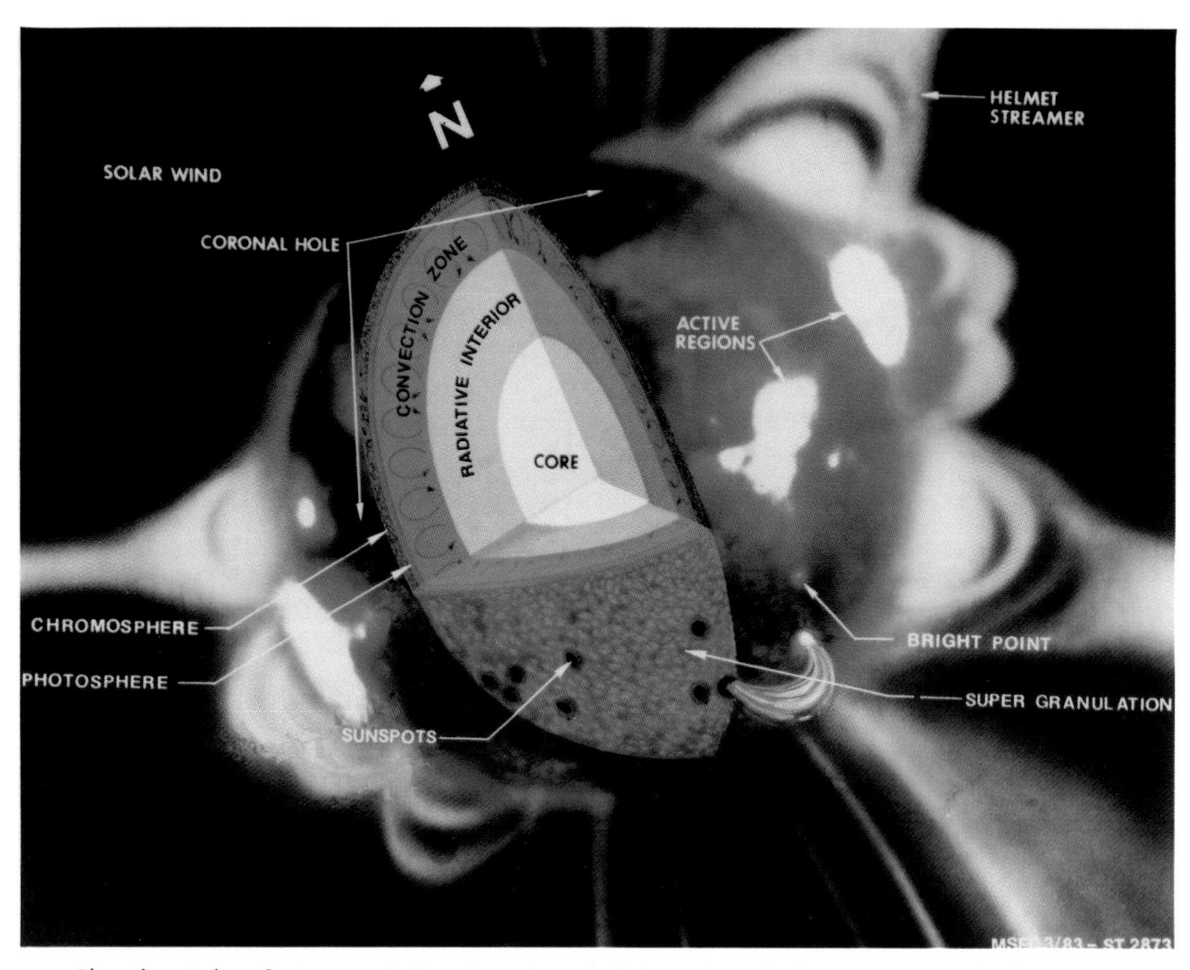

Fig. 1. Major features of the atmosphere and interior of the Sun. The physical processes governing the Sun are fundamentally different above, within, and below the convection zone. Dynamo action in the convection zone generates the magnetic field that controls the structure, heating, and dynamics of the solar atmosphere above the photosphere. The magnetic field is necessary for the very existence of sunspots, flares, the corona, and the solar wind.

does not dominate the scope of solar physics. It is only one of several major problems of comparable interest in solar research and of comparable broad importance for astrophysics and space physics. The other major topics are: the convective envelope and magnetic cycle, the heating of the chromosphere and corona, flares and coronal mass ejections, and the generation of the solar wind. It is these topics that encompass the bulk of solar physics.

In the convective envelope, the combination of the convection and the Sun's rotation is believed to yield the dynamo process that generates the solar magnetic activity cycle (the 11-year sunspot cycle). In any case, the convective envelope is the immediate source of the magnetic field that erupts through the photosphere and gives the solar atmosphere its intriguing magnetic structure and activity, including sunspots, the magnetic network, spicules, filaments, coronal loops, coronal holes, flares, coronal mass ejections, and the solar wind. Observed detailed correspondence of the magnetic field with enhanced heating shows that the field is intimately involved and likely essential in the heating process that results in the 10^6 K temperature climb from the low chromosphere to the low corona. The magnetic field is certainly essential in the buildup and explosive release of energy in the form of particle acceleration, heating, and bulk mass motion in solar flares and coronal mass ejections. The solar wind is driven away from the Sun by some combination of the gas pressure in the corona and energy and momentum deposition mediated by the magnetic field. The global structure of the Sun's magnetic field strongly

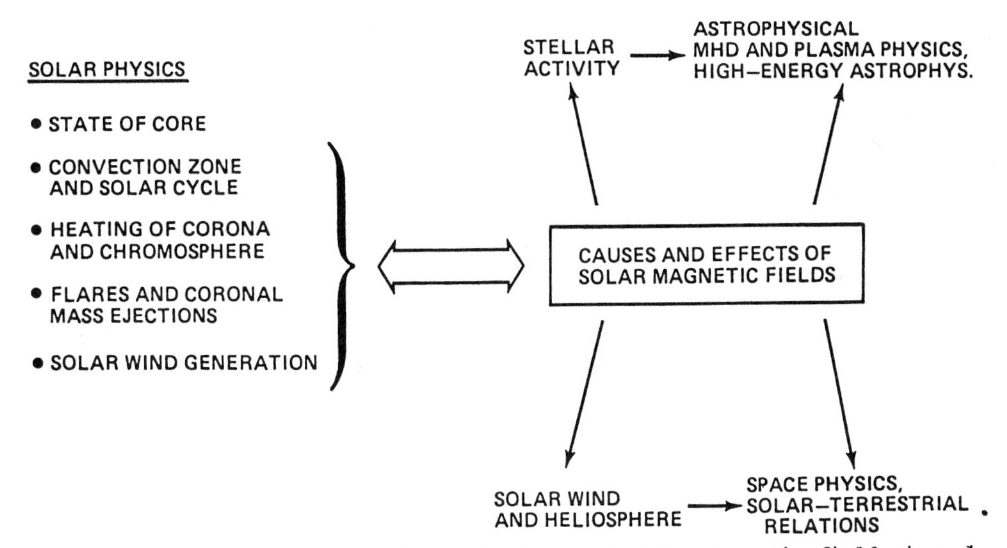

Fig. 2. Diagramatic summary of the central role of solar magnetic fields in solar physics and the resulting importance of solar physics for astrophysics and space physics.

modulates the interplanetary medium through the formation of coronal holes, coronal streamers, and the resulting fast and slow streams in the solar wind. All known variations in solar radiation and corpuscular emission are caused by the Sun's magnetic activity.

From the above, we see that the general problem of solar physics may be broadly labeled "the causes and effects of solar magnetic fields." This phrase covers all of the major problems of solar physics, except (probably) the state of the interior below the convective envelope, and is central to most of the connections of solar physics with astrophysics and space physics. This is the point of the diagram in Figure 2.

Each of the solar papers in this book bears on some aspect of solar magnetic fields, their origin, or their effects in the solar atmosphere.

Outstanding Problems of the Sun's Magnetosphere

The Pervasive Problem: Fine-Scale Magnetic Structure

Images of the Sun in visible to X-ray radiation with the best resolution achieved so far indicate that the magnetic field at all levels of the solar atmosphere is strongly structured on scales of 1000 km and less, and that this structure is fundamental for the evolution and action of the field throughout the solar atmosphere (Withbroe and Noyes, 1977; Vaiana and Rosner, 1978; Giovanelli, 1982; Feldman 1983; Moore and Rabin, 1985; Dowdy et al., 1986; Porter et al., 1987; Parker, 1987). Hence, a problem that is central to most aspects of the causes and effects of solar magnetic fields is that of determining the fine-scale structure of the field and understanding the formation and consequences of this structure in the Sun's magnetic activity. The present perception of the fine-scale character of the magnetic structure in and around active regions that have magnetic fields great enough to make sunspots is illustrated in Figure 3; the fine structure shown in the transition region and corona is largely a guess and is probably much oversimplified.

Direct observation of the Sun's magnetic structure and activity on the scale of individual photospheric granules (1000 km) and less requires sub-arc-second resolution. The required X-ray, EUV, and UV observations of the corona and chromosphere-corona transition region can only be made from space. The required high-resolution observations of the photosphere and chromosphere in visible light probably can only be achieved from above the atmosphere as well. Impressive steps in this direction have recently been achieved or soon will be achieved by new instruments (see the papers by the following authors in these proceedings: Title and Tarbell, 1989; Simon et al., 1989; Canfield and Mickey,

1989; Kohl et al., 1989; Bruner et al., 1989; Crannell, 1989; Wagner, 1989).

Specific Problems

Magnetic flux removal. An unsettled observational question basic to the solar cycle is that of the removal of magnetic flux from the solar atmosphere, the process that over the course of each cycle removes all of the flux that emerges from the interior. Analysis of full-disk magnetograms over a cycle has shown that the removal rate nearly matches the supply rate in all phases of the cycle and that nearly all of the removal happens in the sunspot belts, i.e., in the latitudes where the flux emerges (Howard and LaBonte, 1981). This result has been confirmed by studies of the decay of magnetic flux in active regions; these indicate that most of the flux vanishes within the active region rather than by diffusing into surrounding quiet regions (Wallenhorst and Howard, 1982; Wallenhorst and Topka, 1982; Liggett and Zirin, 1983; Gaizauskas et al., 1983; Rabin et al., 1984). This suggests that flux is mainly removed by submergence within the active region in which it originally emerged (Moore and Rabin, 1985; Zwann, 1987).

Sequences of magnetograms having nearly the best resolution that can be obtained from the ground (about 1 arc sec) have revealed examples of fine-scale merging and cancellation of positive and negative clumps of flux. H-alpha movies show that such flux cancellation is accompanied by microflare activity in the chromosphere above (Martin et al., 1985). These findings are consistent with flux removal by a combination of reconnection and submergence, but the present observations do not have enough resolution to prove this. We need sub-arc-second resolution of the photosphere, chromosphere, transition region, and corona to see how the flux is removed or destroyed. That is, sub-arc-second imaging simultaneously in visible, UV, and X-ray radiation is required to show the form of the associated fine-scale activity in the chromosphere, transtion region, and corona and its role in the flux cancellation in the photosphere.

Heating of the solar atmosphere. Images of the transition region and corona from rockets, Skylab, and the Solar Maximum Mission with spatial resolution in the range 1-10 arc sec, along with photospheric magnetograms and chromospheric images from the ground, have demonstrated that the heating of the corona, transition region, and chromosphere is directly tied to the magnetic field. In active regions, studies of these data have shown that the heating is strongest where the field is most nonpotential, e.g., as evidenced by shear in the field across polarity inversion lines (Orrall, 1981; Webb and Zirin, 1981; deLoach et al., 1984; Gary et al., 1987). This suggests that the heating is accomplished by some form of current dissipation. In any case, all fields in active regions are strongly inhomogeneous on sub-arc-second scales (Zwann, 1978, 1987; Moore and Rabin, 1985), so it is likely that the basic character of the heating is hidden below the resolution of present observations. Apparently, simultaneous sub-arc-second resolution of the photosphere, chromosphere, transition region, and corona is required to establish how the substructure in field configurations with large-scale shear is involved in the heating.

In quiet regions, the heating is concentrated in the magnetic network, the home of chromospheric spicules (spicules along the network are illustrated in Figure 3). It is plausible to suppose that spicules are basically smaller versions of macrospicules, the surge-like eruptions known to be generated in microflares in the small magnetic bipoles that appear as bright points in soft X-ray images of a few arc seconds resolution (Moore et al., 1977). Mircroflaring in UV emission from the transition region has also been observed in tiny bipoles throughout the magnetic network (Porter et al., 1987). These observations raise the question of the role of microflares and spicules in the heating of the chromosphere, transtion region, and corona (Rabin and Moore, 1980; Parker, 1983a,b; Porter and Moore, 1988; Parker, 1988a,b). It will take sub-arc-second resolution in visible light to see the structure of chromospheric spicules and how they are generated. It will take similar resolution in UV and X-ray radiation to see if and how microflares are important for coronal heating.

Flares, coronal mass ejections, and generation of the solar wind. (Most of the solar papers in these proceedings are in this area because the tropical questions chosen for the solar review papers mainly concern these solar phenomena, which strongly affect the interplanetary medium and the Earth's magnetosphere.) It has been recognized for many years that chromospheric filaments mark sheared magnetic fields that are likely sites for flares. The filament often erupts in step with the onset of the impulsive phase of the flare, indicating that the overall instability driving the flare energy release involves a strong transient change in the field configuration, a convulsion in which the field in and around the filament is expelled into the corona (Hagyard et al., 1984; Kahler et al., 1988; Moore, 1988; Sturrock, 1989). The subsecond time structure observed in the hard X-ray and microwave emission in the impulsive phase of flares may reflect sub-arc-second fibrous sturucture in the magnetic field (Sturrock et al., 1984). In any case, observations to date do not show in detail how large-scale magnetic shear leads to flares (Hagyard and Rabin, 1986). Because of the inherent fine-scale substructure of the field and the great range of temperature

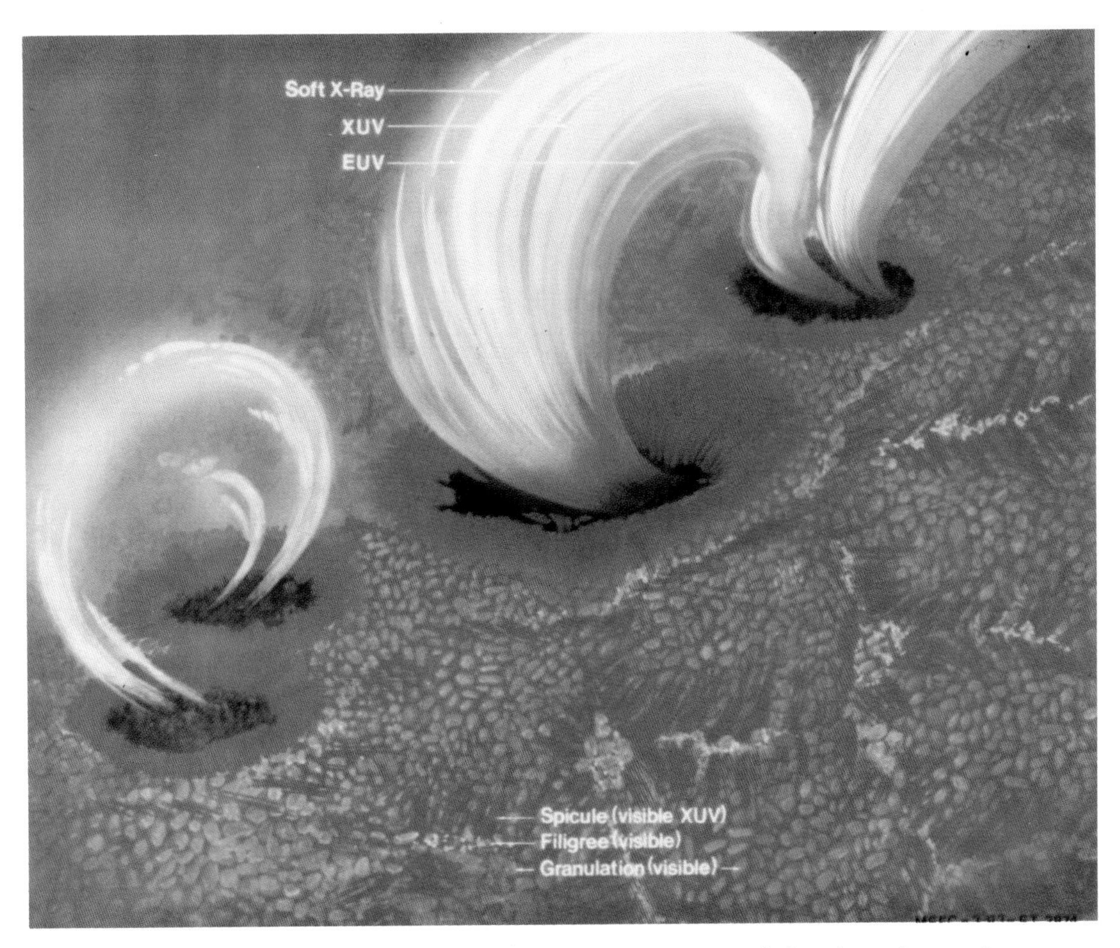

Fig. 3. Depiction of fine-scale magnetic structure and activity in and around a sunspot region in the absence of flaring. The diameter of the largest sunspot and the diameters of the magnetic network cells (bordered with spicules) are about 30,000 km; individual photospheric granules are 1000-2000 km across. Because the magnetic structure everywhere in the photosphere, chromosphere, and transition region is observed to be strongly fibrous on scales of order 1000 km and less, similar fine-scale structure is expected in the corona. The most highly resolved coronal soft X-ray images obtained so far do show magnetic loops and striations down to the limit of resolution of about 1000 km (Davis et al., 1979).

(10^4-10^8 K) of the plasma threaded by the field before and during the flare, it is likely that sub-arc-second resolution and simultaneous coverage of visible, UV, EUV, soft X-ray, and hard X-ray radiation are necessary to discern how the shear develops and how flares arise in sheared fields. The required telescopes must have larger collecting areas that previous space-borne solar telescopes in order to obtain the necessary subsecond time resolution of events on sub-arc-second scales in the impulsive phase of flares (Walker et al., 1986).

Filament-ejection flares are a source of coronal mass ejections (Rust et al., 1980; Wagner, 1984). When we have determined the configuration and change in configuration of the magnetic field in such flares, we will have taken a big step toward determining the field configuration of the mass ejection in the corona. This will clarify whether and how the magnetic field propels the ejection. A combination of coronagraph images of the high corona taken simultaneously with high-resolution visible, UV, and X-ray images of the chromosphere, transition region, and low corona are needed to reveal the form of full-blown mass ejections in the high corona and how they are launched from the low corona by flares and filament eruptions.

Coronal mass ejections supply some small fraction of the solar wind mass flux (Hildner, 1977; Wagner, 1984). It has been proposed that much of the solar wind is launched from the transition

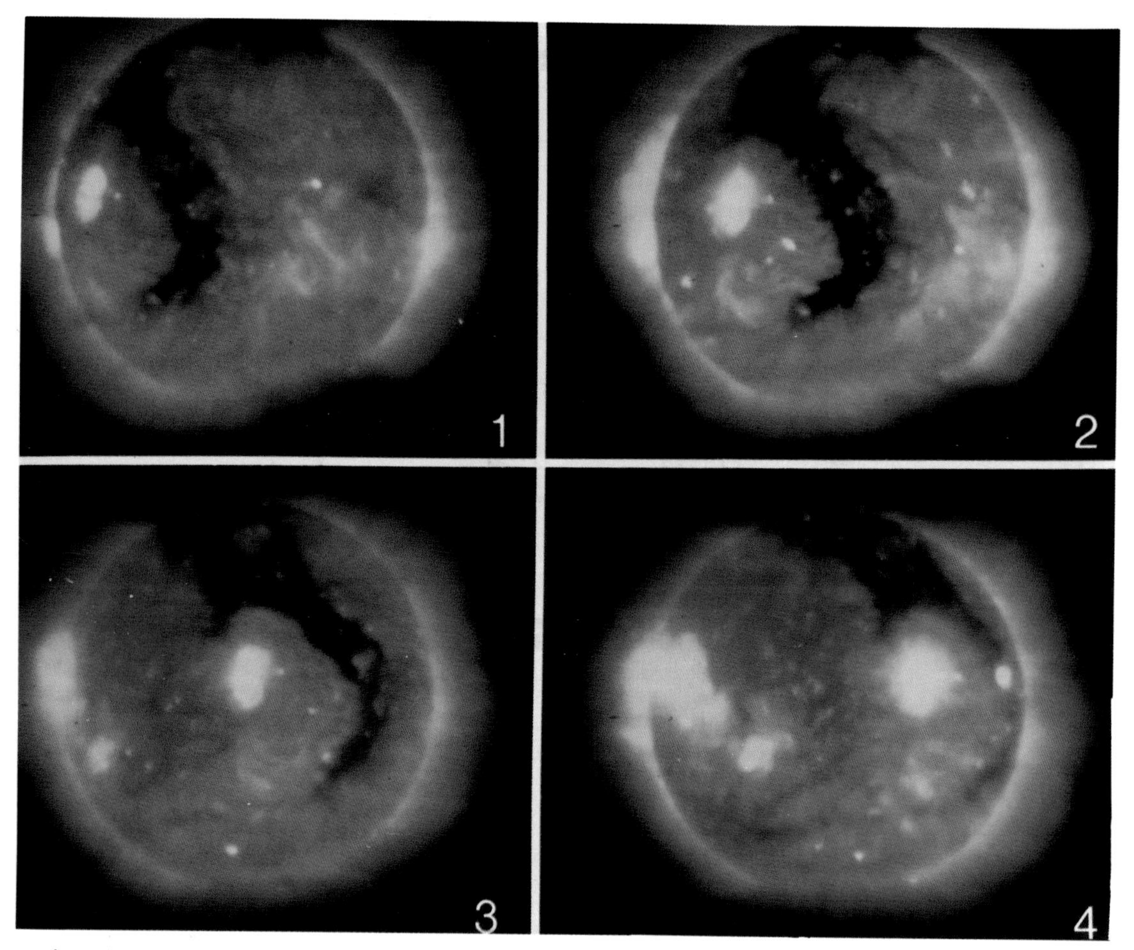

Fig. 4. Sequence of soft X-ray filtergrams showing the magnetic structure, rotation, and evolution of the corona over the course of a week. These photographs were taken on August 19, 20, 23, and 25, 1973, with the American Science & Engineering X-ray telescope on Skylab. These and hundreds of other such photographs from this telescope conclusively showed that the corona is completely controlled by the magnetic field and is strongly structured on all scales resolved (down to 2 arc sec in the originals); coronal holes, active regions, and bright points in the soft X-ray images all match magnetic features in photospheric magnetograms. Filter passband: 2-32 Å plus 44-54 Å; exposure time: 64 s.

region in the form of observed impulsive mass ejections ("jets") that are spatially much smaller than the events commonly called coronal mass ejections (Brueckner and Bartoe, 1983). Some of these jets appear to be erupting loops, as do many macrospicules (Moore et al., 1977). This suggests that the jets and macrospicules may be tiny versions of coronal mass ejections. Both high-resolution coronagraph measurements of the beginnings of the solar wind outflow in the corona and high-resolution visible, UV, and X-ray observations of the jets, macrospicules, and spicules in the transition region and low corona are needed to test whether a substantial part of the solar wind is generated from many tiny mass ejections.

Closing

Observations from space and from the ground have shown that magnetic fields control the solar atmosphere on all scales from the global reach of coronal holes, through the intermediate scales of active regions, down to the sub-arc-second scales of spicules and elementary field bundles. The truth of this in the corona is vividly demonstrated by the soft X-ray images from Skylab, such as those in Figure 4. The larger structures are rife with substructure down to the limit of resolution, about 2 arc sec in the originals of the photographs in Figure 4. From such observations of the corona, along with other observations of the transition region and chromosphere at arc second resolution, and their comparison with photospheric magnetograms, it is clear that the magnetic field basically causes all aspects of the atmosphere above the photosphere. In the future, a complement of solar telescopes in space, each with sub-arc-second resolution and together covering the spectrum from visible light to X-rays, will reveal "how" the field does it.

Acknowledgments. This work was supported by NASA through the Solar Physics Branch of its Space Physics Division and by the Air Force Geophysical Laboratory through the Solar Research Branch of its Space Physics Division. The present paper is based on reports from the NASA Advanced Solar Observatory Science Definition Study (Walker et al., 1986; Moore and Bohlin, 1986).

References

Brueckner, G. E., and J.-D. F. Bartoe, Observations of high-energy jets in the corona above the quiet Sun, the heating of the corona, and the acceleration of the solar wind, Ap. J., 272, 329, 1983.

Bruner, M. E., L. W. Acton, W. A. Brown, R. A. Stern, T. Hirayama, S. Tsuneta, T. Watanabe, and Y. Ogawara, The soft x-ray telescope for the Solar A Mission, these proceedings, 1989.

Canfield, R. C., and D. L. Mickey, An imaging vector magnetograph for the next solar maximum, these proceedings, 1989.

Crannell, C. J., Imaging solar flares in hard x-rays and gamma rays from balloon-borne platforms, these proceedings, 1989.

Davis, R., Jr., and J. C. Evans, Jr., Neutrinos from the Sun, in The New Solar Physics, edited by J. A. Eddy, p. 35, Westview, Boulder, Colorado, 1978.

Davis, J. M., A. S. Krieger, J. K. Silk, and R. C. Chase, Quest for ultrahigh resolution in X-ray optics, SPIE, 184, 96, 1979.

deLoach, A. C., M. J. Hagyard, D. Rabin, R. L. Moore, J. B. Smith, Jr., E. A. West, and E. Tandberg-Hanssen, Photospheric electric current and transition region brightness within an active region, Solar Phys., 91, 239, 1984.

Dowdy, J. F., Jr., D. Rabin, and R. L. Moore, On the magnetic structure of the quiet transition region, Solar Phys., 105, 35, 1986.

Feldman, U., On the unresolved fine structures of the solar atmosphere in the 3×10^4-2×10^5 K temperature region, Ap. J., 275, 367, 1983.

Gaizauskas, V., K. L. Harvey, J. W. Harvey, and C. Zwann, Large-scale patterns formed by solar active regions during the ascending phase of cycle 21, Ap. J., 265, 1056, 1983.

Gary, G. A., R. L. Moore, M. J. Hagyard, and B. M. Haisch, Nonpotential features observed in the magnetic field of an active region, Ap. J., 314, 782, 1987.

Giovanelli, R. G., On the relative roles of unipolar and mixed-polarity fields, Solar Phys., 77, 27, 1982.

Hagyard, M. J., and D. M. Rabin, Measurement and interpretation of magnetic shear in solar active regions, Adv. Space Res., 6(6), 7, 1986.

Hagyard, M. J., R. L. Moore, and A. G. Emslie, The role of magnetic field shear in solar flares, Adv. Space Res., 4(7), 71, 1984.

Hildner, E., Mass ejections from the solar corona into interplanetary space, in Study of Travelling Interplanetary Phenomena, edited by M. A. Shea et al; p. 3, Reidel, Dordrecht, Holland, 1977.

Howard, R., and B. J. LaBonte, Surface magnetic fields during the solar activity cycle, Solar Phys., 74, 131, 1981.

Kahler, S. W., R. L. Moore, S. R. Kane, and H. Zirin, Filament eruptions and the impulsive phase of solar flares, Ap. J., 328, 824, 1988.

Kohl, J. L., H. Weizer, and S. Livi, Spectroscopic measurements of solar wind parameters near the Sun, these proceedings, 1989.

Liggett, M., and H. Zirin, Naked sunspots, Solar Phys., 84, 3, 1983.

Martin, S. F., S. H. B. Livi, and J. Wang, The cancellation of magnetic flux. II. In a decaying active region, Aust. J. Phys., 38, 855, 1985.

Moore, R. L., Evidence that magnetic energy shedding in solar filament eruptions is the drive in accompanying flares and coronal mass ejections, Ap. J., 324, 1132, 1988.

Moore, R., and D. Bohlin, Anticipated scientific return of the advanced solar observatory, NASA/MSFC Space Science Laboratory Preprint Series No. 86-107, NASA/Marshall Space Flight Center, Alabama, 1986.

Moore, R., and D. Rabin, Sunspots, Ann. Rev. Astron. Astrophys., 23, 239, 1985.

Moore, R. L., F. Tang, J. D. Bohlin, and L. Golub, Hα macrospicules: Identification with EUV macrospicules and with flares in X-ray bright points, Ap. J., 218, 286, 1977.

Orrall, F. Q. (editor), Solar Active Regions, 350 pages, Colorado Associated University Press, Boulder, Colorado, 1981.

Parker, E. N., Magnetic neutral sheets in evolving fields. I. General theory, Ap. J., 264, 635, 1983a.

Parker, E. N., Magnetic neutral sheets in evolving fields. II. Formation of the solar corona, Astrophys. J., 264, 642, 1983b.

Parker, E. N., Why do stars emit X-rays?, Physics Today, 40(7), 36, 1987.

Parker, E. N., The origins of the stellar corona, in Solar and Stellar Coronal Structure and Dynamics, edited by R. C. Altrock, National Solar Observatory, Sacramento Peak, Sunspot, New Mexico, in press, 1988a.

Parker, E. N., Nanoflares and the solar X-ray corona, Ap. J., 330, 474, 1988b.

Porter, J. G., R. L. Moore, E. J. Reichmann, O. Engvold, and K. L. Harvey, Microflares in the solar magnetic network, Ap. J., 323, 380, 1987.

Porter, J. G., and R. L. Moore, Coronal heating by microflares, in Solar and Stellar Coronal Structure and Dynamics, edited by R. C. Altrock, National Solar Observatory, Sacramento Peak, Sunspot, New Mexico, in press, 1988.

Rabin, D., and R. L. Moore, Coronal holes, the height of the chromosphere, and the origin of spicules, Ap. J., 241, 394, 1980.

Rabin, D., R. Moore, and M. J. Hagyard, A case for submergence of magnetic flux in a solar active region, Ap. J., 287, 404, 1984.

Rust, D. M., et al., Mass ejections, in Solar Flares, edited by P. A. Sturrock, p. 273, Colorado Associated University Press, Boulder, Colorado, 1980.

Simon, G. W., et al., Magnetoconvection on the solar surface, these proceedings, 1989.

Sturrock, P. A., Energy conversion in solar flares, these proceedings, 1989.

Sturrock, P. A., P. Kaufmann, R. L. Moore, and D. F. Smith, Energy release in solar flares, Solar Phys., 94, 341, 1984.

Svestka, Z. (editor), Flare build-up study (proceedings of the study workshop held at Falmouth, Cape Cod, Massachusetts, September 8-11, 1975), Solar Phys., 47, 432 pp., 1976.

Title, A., and T. Tarbell, Optical disk processing of solar images, these proceedings, 1989.

Vaiana, G. S., and R. Rosner, Recent advances in coronal physics, Ann. Rev. Astron. Astrophys., 16, 393, 1978.

Wagner, W. J., Coronal mass ejections, Ann. Rev. Astron. Astrophys., 22, 267, 1984.

Wagner, W. J., CME and solar wind studies using GOES solar x-ray imagers and SOHO remote sensing, these proceedings, 1989.

Wallenhorst, S. G., and R. Howard, On the dissolution of sunspot groups, Solar Phys., 76, 203, 1982.

Wallenhorst, S. G., and K. P. Topka, On the disappearance of a small sunspot group, Solar Phys., 81, 33, 1982.

Walker, A. B. C., Jr., R. Moore, and W. Roberts, The Advanced Solar Observatory, report of the Advanced Solar Observatory Sciece Working Group, NASA/Marshall Space Flight Center, Alabama, 1986.

Webb, D., and H. Zirin, Coronal loops and active region structure, Solar Phys., 69, 99, 1981.

Withbroe, G. L., and R. W. Noyes, Mass and energy flow in the solar chromosphere and corona, Ann. Rev. Astron. Astrophys., 15, 363, 1977.

Zwann, C., On the appearance of magnetic flux in the solar photosphere, Solar Phys., 60, 213, 1978.

Zwann, C., Elements and patterns in the solar magnetic field, Ann. Rev. Astron. Astrophys., 25, 83, 1987.

TERRESTRIAL AND PLANETARY MAGNETOSPHERES

J. L. Burch

Southwest Research Institute
San Antonio, TX 78284

Introduction

The field of space plasma physics began with experimental studies of the Earth's magnetosphere. This magnetosphere, unlike the others, has been explored from the inside out, beginning with ground-based observations and progressing outward with the addition of balloon, sounding rocket, and spacecraft experiments. All of these techniques are still very much in use and each plays an important role in magnetospheric physics. With the exception of the induced magnetosphere of Venus, which has been explored in a long-term orbital mission (Pioneer Venus Orbiter), all of the planetary and cometary magnetospheres that have been explored to date have been sampled only by flyby missions. Future orbital planetary science missions are imminent (Galileo to Jupiter) or planned (Comet Rendezvous Asteroid Flyby and the Cassini mission to Saturn). In the case of the Earth, the International Solar-Terrestrial Physics program (ISTP), which is just now getting underway, will see approximately a dozen well-instrumented spacecraft placed in various strategic orbits to study the large-scale transport of energy throughout the geospace region (the Earth's magnetosphere and nearby solar wind).

The ISTP program, like the various planned planetary programs, has a strong theoretical component. In addition, NASA has instituted several theory programs such as the Solar-Terrestrial Theory Program, which have improved the balance between theory and experiment. The NSF is also planning to start a coordinated program of Geospace Environment Modeling (GEM), which will conduct theoretical studies and analysis in support of magnetospheric physics objectives.

This introductory paper is meant to provide an overview of magnetospheric physics for the nonspecialist and to introduce many of the topics that are covered in much greater detail by other papers in this volume. Since a large number of solar physicists participated in the conference, this material was prepared especially for that community in mind.

Solar Wind/Magnetosphere Interactions

A feature of many planetary and cometary bodies that are immersed in the solar-wind flow is the formation of a collisionless shock wave or bow shock. Bow shocks, which slow down and heat the solar-wind plasma, form when an object that the solar wind encounters has either a significant intrinsic magnetic field or a significant atmosphere or both. For example, comets and the planet Venus have atmospheres, which are partially ionized by solar ultraviolet radiation and x rays (see the paper by Cravens, this volume). Such ionospheres mass-load the solar wind and interact with its magnetic and electric fields to produce a bow shock and an induced magnetosphere, which is confined to a cavity by the flow of the solar wind around the obstacle. The only magnetic field within such magnetospheres is the entrained solar-wind magnetic field. Planets with strong intrinsic magnetic fields, on the other hand, can divert the solar-wind flow with magnetic forces alone so that their magnetospheres are dominated by the planetary magnetic field. In the case of these planets, the mixture of planetary and solar-wind plasmas occurs only in localized regions, such as the Earth's polar cusps and boundary layers, or hardly at all, as appears to be the case in the Jovian and Saturnian magnetospheres.

The essential question of solar wind/magnetosphere interactions is the means by which the magnetosphere of a rotating body is formed and energized by the solar wind. Planetary rotation is itself a significant source of plasma energization and is the primary energization mechanism in the giant rotating magnetospheres of Saturn and Jupiter. In the terrestrial magnetosphere the rotation of the Earth leads to the confinement of a vast region of corotating thermal plasma extending to equatorial altitudes of up to about five Earth radii (the plasmasphere) and competes with the solar wind flow in determining the global patterns of plasma within the magnetosphere.

The question of how the solar-wind flow is coupled into magnetospheres has proven to be very difficult to answer. Only a few per cent of the solar-wind particles that reach the Earth's magnetospheric boundary (the magnetopause) actually enter the magnetosphere. Efficient entry occurs only through the polar cusps, although weaker, diffusive entry is possible over larger regions of the magnetopause. However, the coupling of solar-wind momentum into the magnetosphere is much larger than this rather low plasma entry efficiency might suggest. For example, the solar-wind electric field, which produces a potential difference across the magnetospheric cross-section of about 250 kV, is coupled into the magnetosphere by a poorly understood process with about a 20% efficiency, producing an average potential difference across the polar caps of about 50 kV. More efficient coupling, leading to magnetospheric substorms and associated geomagnetic activity, occurs when the interplanetary magnetic

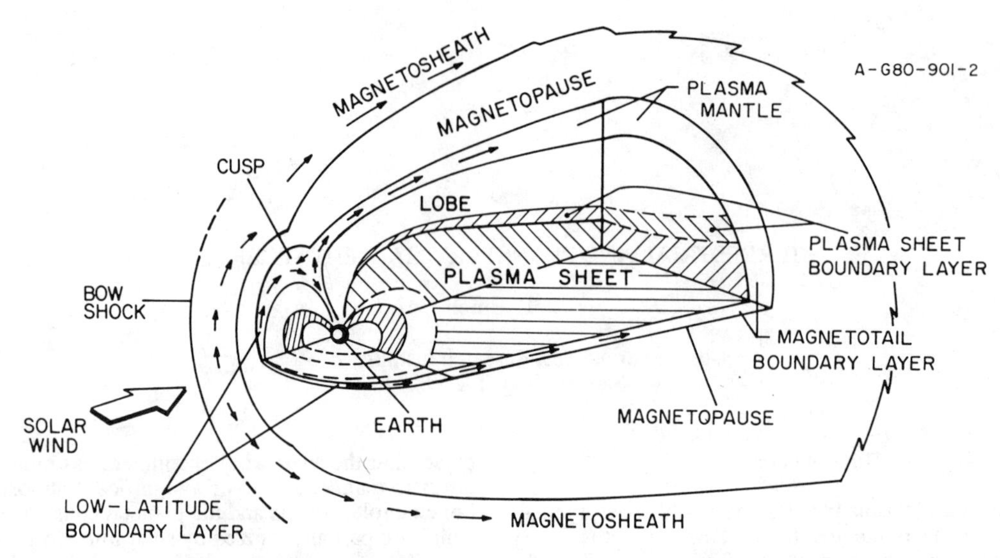

Fig. 1. Three-dimensional view of the Earth's magnetosphere. Illustrated in this sketch are the magnetosheath, the cusp, the lobe, the plasma sheet boundary layer, and the low-latitude boundary layer. (Figure courtesy of T. E. Eastman)

field (IMF) becomes more southward. This IMF dependence suggests that merging of the IMF with the magnetospheric magnetic field is an important process. Viscous interaction and impulsive plasma penetration are other proposed processes that may act in combination with merging to energize the magnetosphere.

If merging at the magnetopause is an important process, it almost certainly proceeds sporadically and in a rather localized manner. Recent spacecraft such as ISEE and AMPTE may have detected directly the combined solar-wind and magnetospheric flux tubes that should result from the merging process. These so-called flux transfer events (FTEs) appear preferentially when the IMF has a southward component and contain the magnetic-field, field-aligned current, and plasma-distribution signatures to be expected in and around a merged magnetic flux tube (see the paper by Kan, this volume). Much present effort is being expended in identifying the ionospheric signature of flux transfer events.

Studies of solar-wind interactions with the terrestrial magnetosphere provide a comprehensive and unique framework with which to plan and implement studies of the interactions with cometary and other planetary magnetospheres. Of the other planets, Venus has by far the best-studied solar-wind interaction. Because its internal magnetic field is very weak or nonexistent, the solar wind interacts directly with the Venusian ionosphere. A bow shock is formed, much as in the terrestrial case. However, instead of a magnetopause as at the Earth, there is an ionopause whose altitude is a strong function of the solar-wind dynamic pressure. A unique aspect of the dayside Venusian ionosphere is that it is magnetized by the formation of magnetic flux ropes. These structures consist of twisted solar-wind magnetic flux tubes that sink through the ionosphere as they are draped around the planet in the solar-wind flow and produce a long magnetic tail on the night side of the planet.

The brief flybys of the other planets and two comets have produced tantalizing glimpses of solar-wind interactions, which in some instances represent either limiting or very strange cases indeed. An extreme case of the nonmagnetized body with an atmosphere is represented by comets (see the papers by Gombosi and Korosmezey and by Cravens, this volume). Comets exhibit the most complex interactions imaginable among gas, radiation, plasma, and dust. The dust-laden neutral gas and plasma from comets mix with the solar-wind flow, producing counterstreaming interactions. These lead to significant mass-loading of the solar wind, to the formation of a bow shock, a cometopause, and a long ion tail that exhibits unstable behavior resulting in occasional ion tail disconnections, and to various wave emissions associated with unstable plasmas. If one can consider the disturbed plasma environment around a comet as a magnetosphere, then cometary magnetospheres are certainly among the most active magnetospheres available for study.

Magnetospheric Structure and Processes

The solar-wind ions and electrons that gain entry to the Earth's magnetosphere first populate the magnetospheric boundary layer, which lies just inside the magnetopause. The plasma population of this thin layer is similar to magnetosheath plasma, although it is somewhat less dense and has a lower flow velocity. Three different, but topologically connected, boundary layers have been discovered. The low-latitude boundary layer and the plasma mantle together constitute the magnetospheric boundary layer mentioned above. The plasma sheet boundary layer, as the name implies, lies along the northern and southern edges of the plasma sheet on the night side of the Earth. The plasma sheet itself is the great reservoir of hot plasma for the magnetosphere. It is populated by solar-wind plasma, which enters through the various boundary layers, and by thermal plasma that escapes from the ionosphere along magnetic field lines. These plasma regions are illustrated schematically in Figure 1.

The boundary layers are the sites of most of the high-speed plasma flows that are observed in the outer terrestrial magnetosphere. Transport of plasma is, therefore,

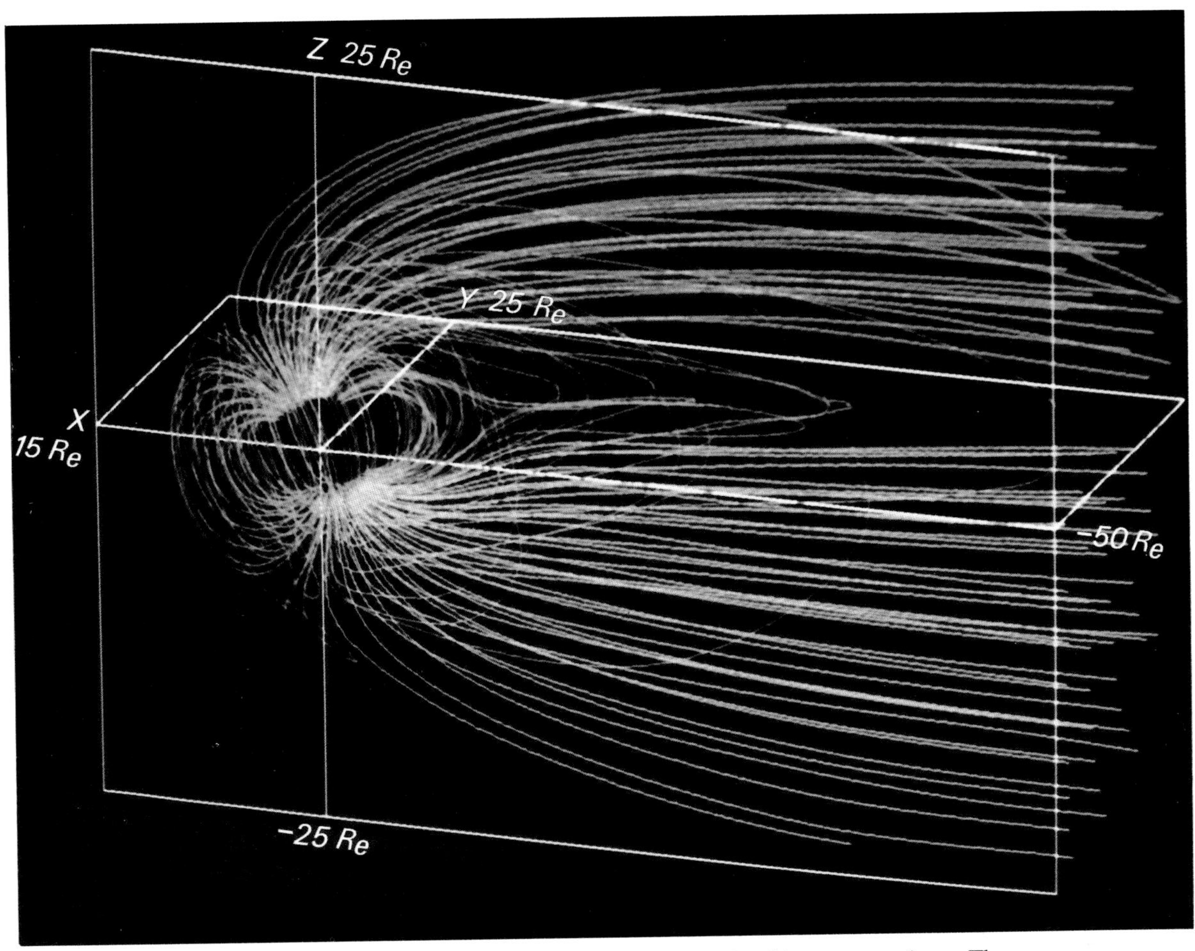

Fig. 2. Computer-generated three-dimensional image of the Earth's magnetosphere. The severe tilt angle (30°) corresponds to conditions near summer solstice and near midday at North American longitudes. Closed field lines are shown in green, open field lines from the northern polar cap in blue, and open field lines from the southern hemisphere in yellow. (Figure courtesy of R. J. Walker, UCLA)

concentrated in these regions. While this transport occurs mostly along magnetic field lines in the plasma mantle and plasma sheet boundary layer, there is a significant cross-field component in the low-latitude boundary layer. Such flows lead to a separation of positive and negative charges such as occurs in an electrical generator. This generation region is connected by conductive magnetic field lines to the resistive ionosphere, which acts as a load. The currents that flow between this boundary layer and the ionosphere are observed routinely by spacecraft. Although they enter and leave the ionosphere through narrow regions, these currents feed a large-scale system of horizontal currents that fill the high-latitude region. Because of the electrical resistivity of the ionosphere, electric fields are produced by these currents, and these electric fields are mapped throughout the magnetosphere by the magnetic field lines that thread the ionosphere. In this way the low-latitude boundary layer becomes responsible for a significant part of the energization of the magnetosphere.

Historically, the magnetic configuration of the magnetosphere and its various plasma populations have been illustrated by scale drawings based on numerous individual spacecraft measurements, as shown in Figure 1. In the future we may well have the capability of "photographing" the magnetosphere by imaging the energetic neutral atoms produced by the charge exchange of magnetospheric ions (see the paper by McEntire and Mitchell, this volume) as well as by imaging the ultraviolet photons produced by the resonant scattering of sunlight by the thermal helium ions of the plasmasphere. Work on both of these techniques is underway with some flight data having been obtained on the energetic neutral atom technique [*Roelof,* 1987].

Another technique that has become available for imaging magnetospheric plasma processes is large-scale computer simulation. Several classes of simulations are being used depending upon the phenomena being studied, the size of the region of interest, and the relevant time scales. The classes of simulation can be categorized as particle, fluid (or MHD), and hybrid. Of course, particle simulations would be ideal for all studies of plasma phenomena if the size and speed of computers would allow it. At present, however, particle

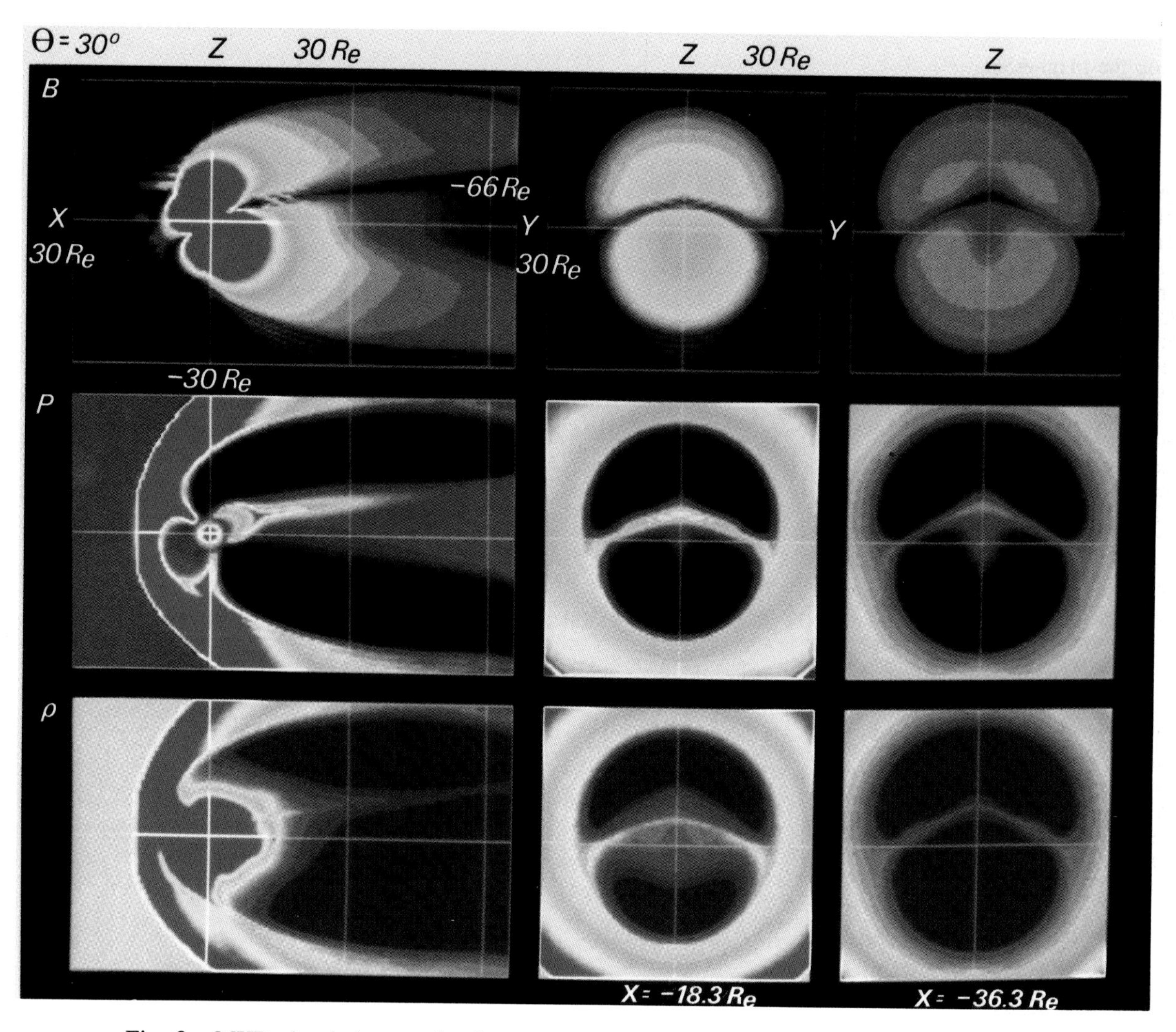

Fig. 3. MHD simulation results for magnetic field and plasma properties for the same conditions as in Fig. 1. The three rows contain, from top to bottom, the magnetic field magnitude (B), the plasma pressure (P), and the plasma density (ρ). The left-most column contains the values in the noon-midnight meridian. The values are color-coded with red being the largest, then yellow, light blue, and dark blue. The plasma sheet can be seen in the magnetotail. The middle and right-hand columns contain the profiles in the y-z plane at x= −18 and −36 Earth radii in the tail. The bending of the tail as a result of the large tilt angle is evident. The boundary layers can be seen inside the magnetopause and along the edge of the plasma sheet. (Figure courtesy of R. J. Walker, UCLA)

simulation of large-scale phenomena exceed the capacity of even the most advanced supercomputers. Thus the particle approach is reserved for studies of local phenomena such as beam-plasma interactions and wave-particle interactions. In fact, even for these phenomena so-called Debye spheres are often used to represent groups of particles that interact as one. For larger-scale phenomena, such as the magnetopause and bow shock, hybrid simulations are often used. In these simulations the ions are treated as individual particles or Debye spheres, while the electrons, which have much smaller gyroradii, are treated as a fluid. Finally, simulations of large plasma regions or of the entire magnetosphere are tractable only with MHD simulations. Figures 2 and 3 are examples of how the entire magnetosphere can be "imaged" by MHD simulation as it interacts with the solar wind (see, e. g., the paper by Walker et al., this volume). These cases apply to a rather severe magnetic tilt angle of 30°, corresponding to conditions near summer solstice and near midday at North American longitudes. Figure 2 shows the three-dimensional configuration of magnetic field lines. Simulation results for magnetic field and plasma properties for the same conditions are shown in Figure 3. The plasma sheet can be seen in the magnetotail, and the bending of the tail as a result of the large tilt angle is evident. The boundary layers can also be seen inside the magnetopause and along the edge of the plasma sheet.

Magnetospheric Substorms

The configuration of the magnetosphere that is shown in Figures 1 through 3 is subject to rapid and significant changes during disturbances that are called magnetospheric substorms because of their phenomenological relationship to the longer duration (days vs. hours) magnetic storms. Substorm activity increases when southward IMF components appear in the solar wind near the Earth. The auroral electrojet index (AE), which is the most commonly used measure of substorm intensity, shows both an immediate response to southward turnings of the IMF and a larger delayed response. The immediate response results from processes that are directly driven by the solar wind and that consist primarily of a large-scale reconfiguration of the magnetosphere and the enhanced currents associated with it.

The immediate effects of southward IMF components on the magnetosphere include a transfer of magnetic flux from the day side to the night side, presumably resulting from merging processes, and the formation of flux transfer events. Associated with the buildup of magnetic flux in the magnetotail is a thinning of the plasma sheet in the north-south direction. In the ionosphere, these effects lead to a gradual enlargement of the auroral oval, with the locus of discrete auroral forms moving toward lower latitudes.

The delayed reaction occurs through a large-scale instability of the magnetotail neutral sheet, which results in a diversion of the current that maintains it along magnetic field lines into the ionosphere. Most of the energy flow into the ionosphere during substorms is channeled through the plasma-sheet boundary layer. The specific instability leading to disruption of the neutral-sheet current is not known for sure, although most contemporary models involve some form of tearing-mode instability through which the neutral sheet is broken up into one or more islands of plasma with entrained magnetic fields. These plasma islands, or plasmoids, then move rapidly down the magnetotail as the substorm progresses. The plasmoid model is by no means universally accepted; however, it is widely enough accepted to provide a reasonable context for further study of substorm phenomena.

The concept of a magnetospheric substorm has not yet been applied to any significant degree to other planetary magnetospheres. We mentioned above that the ionosphere of Venus becomes more highly magnetized with the intrusion of magnetic flux ropes when the solar wind intensifies. The lack of an intrinsic magnetic field may render the Venusian magnetosphere incapable of storing and releasing the large amounts of magnetic energy that the Earth's magnetosphere stores and releases. On the other hand, the directly driven processes should be much stronger at Venus than at Earth, and auroral-type emissions have been observed in the nightside upper atmosphere of Venus.

Mercury, with a small intrinsic magnetic field and little or no atmosphere, may exhibit a substorm-like phenomenon. Data from the first flythrough of Mercury's magnetosphere by the Mariner 10 spacecraft suggest the occurrence of a substorm.

Jupiter, Saturn, and Uranus have all been observed to produce auroral emissions in the ultraviolet. The Jovian aurora seems to be related to the Io plasma torus and is probably powered by the rotational energy of the planet. The most intense emissions at Saturn are located near 80° magnetic latitude as would be expected from a magnetotail process, but they are strongly biased toward the day side and to particular longitude sectors, suggesting processes quite different from those at Earth. On the other hand, Uranus' auroras show a local-time and latitude dependence consistent with the typical terrestrial substorm; in situ magnetospheric data also indicate plasma injection processes similar to those seen at Earth [*Behannon et al.*, 1987].

Plasma Sources

There are two major sources of plasma in the Earth's magnetosphere: the ionosphere and the solar wind. The only minor source of any note is the Jovian magnetosphere and then only for very high energy particles. Overall the solar wind and the ionosphere are thought to provide roughly equal amounts of plasma to the magnetosphere, although individual estimates vary widely. The most commonly-used tracer of plasma source regions is the composition of the ionic components of the plasma. Complicating matters greatly is the fact that both the solar wind and the outermost ionosphere are dominated by hydrogen ions. The minor species associated with each are different, however. Singly-charged helium and oxygen characterize the ionospheric source, while doubly-charged helium and the higher charge states of oxygen indicate a solar-wind source. In addition, the characteristic energy of a plasma is a strong indicator of its source. For the lower energy ions (below a few hundred eV) the ionosphere is generally considered to be the source, while the very high energy ions (above several hundred keV) are normally attributed to the solar wind. The intermediate-energy ions, which make up the bulk of the plasma in the Earth's magnetosphere, particularly in the plasma sheet, could have either source. Further complicating the matter is the apparent occurrence of acceleration by magnetic-field-aligned potential differences in the auroral regions. Parallel electric fields would accelerate the electrons of the auroral source plasma downward into the atmosphere, while excluding the ions. At the same time, ions from the topside ionosphere would be accelerated upward by the same fields. If this type of acceleration is, in fact, the dominant one for auroral particles, it may be nearly impossible to determine what the ultimate source of the auroral plasma is.

In the case of comets and, to a lesser extent, the planet Venus, the atmosphere/ionosphere is the primary source of plasma to the magnetosphere and is in fact a more important source of plasma to the solar wind than the solar wind is to the magnetosphere. In these cases cometary (and Venusian) ions are picked up by the solar wind, adding mass to the solar wind as it slows down and is diverted around the planetary or cometary obstacle. The outer planets, on the other hand, are populated to a large extent by ions that are outgassed or sputtered from the surfaces of satellites and rings in addition to the more standard solar wind and ionospheric sources (see the paper by Goertz, this volume). One such extreme example is the Io plasma torus of the Jovian system, which contains highly charged sulfur and oxygen ions.

Plasma Acceleration

Plasmas in all planetary magnetospheres are accelerated by electric fields of various origins. For example, the terrestrial polar wind is produced by a very small ambipolar electric field generated by the different gravitational forces on ionospheric ions of various masses and electrons, leading to supersonic upward flow. In the auroral regions of the Earth's magnetosphere, waves with frequencies near the local ionic gyrofrequencies can create a resonance between the wave electric field and the gyromotion of ions, accelerating the ions in the plane perpendicular to the local magnetic vector. This type of resonant acceleration, combined with the magnetic mirror force, can produce conical ion distributions of the type observed frequently by numerous spacecraft. These distributions are called ion conics because the ion velocities remain roughly along the surface of a cone centered on the magnetic field line with its apex pointing downward toward the Earth. There is some controversy about the specific mechanisms responsible for ion conics. Most, but not all, candidate processes do, however, involve resonant acceleration by wave electric fields.

Waves near the ion gyrofrequencies can also resonate with the field-aligned motion of electrons. This type of wave-induced acceleration has been suggested as the source of the strongly field-aligned electron distributions known as suprathermal bursts. These bursts, which are found adjacent to and between the primary auroral fluxes within inverted-V events, contain electron energies in the few hundred eV to few keV range.

As mentioned above, the acceleration mechanism for primary auroral electrons has many of the characteristics of a field-aligned potential difference. The parallel electric fields seem to occur preferentially in regions of strong field-aligned currents associated with convective ion-velocity shear reversals in the magnetosphere and ionosphere. Whether the parallel electric fields are localized in altitude as double layers or are greatly extended in altitude is not known. The major portion of the inferred auroral potential differences is located at altitudes near and slightly above one Earth radius, although there are indications of smaller potentials at somewhat higher altitudes.

Another important class of magnetospheric particle acceleration phenomena is associated with the large-scale dawn-to-dusk electric field that exists across the Earth's magnetotail. The gradient and curvature drifts of electrons and ions cause them to drift eastward and westward, respectively. Hence the dawn-dusk electric field will accelerate ions and electrons while also causing them to drift closer to the Earth. There is another feature of the magnetic topology of the geomagnetic tail that causes ion acceleration as well. The magnetic field has a strong earthward component in the northern half of the tail and a strong anti-earthward component in the southern half. The magnetic field reversal that occurs near the center of the plasma sheet is supported by a thin sheet of current (the neutral sheet) that flows from dawn to dusk. Ions that approach the neutral sheet can move in meandering orbits that take them successively above and below it while moving westward. As the ions move along the electric-field direction, they are accelerated. Detailed analyses of such trajectories have shown that some ions are subsequently ejected out of the neutral-sheet region along magnetic field lines toward the Earth. Ions that mirror above the atmosphere return to the plasma sheet but arrive nearer to the Earth than they left it because of the earthward flow caused by the dawn-dusk electric field.

During substorm disruptions of the neutral sheet, ions are accelerated to very high energies in the MeV range. Some of these energetic ions escape the magnetosphere and travel into interplanetary space where they become mixed with other ion populations such as those of Jovian or bow-shock origin. Since the total potential difference across the magnetotail is generally less than 100 kV, it is is not easily understood how such high-energy ions could be produced so rapidly. Once they are understood, these very energetic ions will no doubt help greatly in determining the nature of the reconnection phenomena that are thought to disrupt the neutral sheet during substorms.

Magnetospheric Wave Phenomena

Typical magnetospheric plasma distributions contain significant amounts of free energy in the form of beams, loss cones, and other types of velocity-space density gradients. This free energy leads to wave growth and wave-particle interactions, which result in the acceleration of electrons and ions and in their precipitation into the atmosphere. At Earth, new data have recently been obtained on the processes involved in the generation of auroral kilometric radiation (AKR) and auroral hiss by spacecraft that have probed the auroral acceleration regions directly. AKR appears to be generated within plasma density cavities with energy derived from electron loss-cone distributions. A relativistic effect known as the cyclotron maser instability is generally accepted as the generation mechanism. While there is not yet universal agreement on the source of auroral hiss emissions, it is believed that they also might be generated by electron velocity-space density gradients, possibly associated with the electron "hole" distributions produced by a loss-cone distribution in the presence of a parallel electric field. Other important wave modes, such as ion cyclotron and lower hybrid waves, with frequencies near the ion gyrofrequencies have been invoked for the transverse acceleration of ionospheric ions (ion conics) and the parallel acceleration of electrons (suprathermal bursts) as noted before.

Many other important wave modes exist throughout the Earth's magnetosphere, and the reviews by *Gurnett and Inan* [1988] and *Fukunishi* [1987] discuss them comprehensively. Interesting, too, is the fact that many of the same types of waves and, apparently the same physical processes involved in their generation, occur in other planetary magnetospheres. The Jovian decametric radiation was, of course, known long before its terrestrial counterpart AKR was discovered. Saturnian kilometric radiation is a very interesting comparative phenomenon in that it seems to emanate from the day side of the planet rather than from the night side as is the case with kilometric radiation at the Earth. Some current theoretical thinking suggests that the cyclotron maser interaction may also be responsible for kilometric radiation at Saturn (SKR) and the recently detected UKR emissions from the night side of Uranus.

Summary

As a brief overview of some of the contemporary topics of interest to terrestrial and planetary magnetospheric physicists, the discussion in the preceding paragraphs was not meant to be comprehensive or even to cover all of the important outstanding questions in the field. It does, however, introduce many of the topics that are discussed in detail in the papers presented at Yosemite and contained in these proceedings.

The intent of the conference was to focus on theory, modeling, and instrumentation aspects of solar-system plasma physics in order to achieve a forward-looking view of research in this broad field.

A number of new experimental techniques for investigations of magnetospheric phenomena were presented for the first time at Yosemite and are discussed in papers contained in this volume. These new techniques included instrumentation for the detection of electrons injected as test particles for the measurement of electric fields (McIlwain); suprathermal ion mass spectrographs and electron spectrometers for the acquisition of three-dimensional distribution functions from three-axis stabilized spacecraft (Bame et al.; McIlwain); advanced time-of-flight ion mass spectrometers (Young et al.); energetic neutral atom imaging systems (McEntire and Mitchell; Hsieh and Curtis; Keath et al.); and instrumentation for the complete specification of plasma waves (Kintner).

This volume also contains theory and modeling papers on many important problems related to terrestrial and planetary magnetospheres and their interaction with the solar wind. Especially interesting from the theoretical point of view is the significant overlap between solar physics and magnetospheric physics, which clearly was not well appreciated before the conference. Hopefully, future interdisciplinary meetings will continue the fruitful interaction between these two groups of space physicists.

References

Behannon, K. W., R. R. Lepping, E. C. Sittler, Jr., N. F. Ness, B. H. Mauk, S. M. Krimigis, and R. L. McNutt, Jr., The magnetotail of Uranus, *J. Geophys. Res.*, *92*, 15, 354-366, 1987

Fukunishi, H., Plasma wave phenomena in the Earth's magnetosphere, in *The Solar Wind and the Earth*, ed. by S.-I. Akasofu and Y. Kamide, 183-212, Terra Scientific Publishing Co., Tokyo, 1987.

Gurnett, D. A., and U. S. Inan, Plasma wave observations with the Dynamics Explorer 1 spacecraft, *Rev. Geophys.*, *26*, 285-316, 1988

Roelof, E. C., Energetic neutral atom image of a storm-time ring current, *Geophys. Res. Lett.*, *14*, 652-655, 1987.

I. Three-Dimensional Structure and Turbulence in Solar System Plasmas

Half Dome, Merced River, Winter, Yosemite National Park. c. 1938.

How is free magnetic energy built up and held in the solar atmosphere?

MAGNETIC FREE-ENERGY IN THE SOLAR ATMOSPHERE

B. C. Low

High Altitude Observatory, National Center for Atmospheric Research,* Boulder, CO 80307

Abstract. The solar atmosphere is an excellent electrical conductor and the electric current flowing in it is a source of stored energy which can, under the appropriate circumstance, fuel a variety of plasma and hydromagnetic instabilities in such spectacular phenomena as the flares, prominence eruptions, and coronal mass ejections. These electric currents are associated with magnetic fields which permeate the solar atmosphere. The available spectroscopic methods of detecting solar magnetic fields allow only the fields in the photosphere and chromosphere to be measured with useful spatial resolutions down to about an arc sec. The line-of-sight component of the field can be measured and interpreted with considerable confidence while the measurement of the transverse field component involves a different piece of physics, straining existing techniques in both instrumentation and data interpretation. It has long been recognized that measurement of the line-of-sight component of the magnetic field in the lower solar atmosphere by itself tells us nothing about the electric current in the solar atmosphere. Whereas, with measurement of all three field components, the prospect exists for extrapolating for the magnetic field and its electric current in the upper atmosphere. For long-lived structures in the solar active region, a reasonable assumption takes the field to be force-free. A program for extrapolating magnetic field in the force-free approximation has much to teach us about solar magnetic fields and this review surveys what can be done and the kind of pitfalls that beset the program. Of particular note is the possibility of detecting the magnetic free energy associated with the electric currents in the solar atmosphere directly from the vector field measurements alone. Many of the ideas described are not new but interest in these ideas has heightened with the current effort to build new magnetographs and to improve on the existing vector magnetographs.

*The National Center for Atmospheric Research is sponsored by the National Science Foundation.

1. Introduction

The magnetic field plays a central role in the physics of the solar atmosphere. The magnetic field inhibits the diffusion of plasma across the field lines as a consequence of the high electrical conductivity and dominates the energetics of the atmosphere through its influence on energy transport and the addition of heat from the dissipation of electric currents. It is believed that the magnetic field has its origin in dynamo processes operating below the visible solar surface. In the course of an 11-year activity cycle, new magnetic flux emerges to spawn a rich variety of plasma structures in the atmosphere and to replace the old flux leading to a global reversal of the large-scale polarity. At any time, the solar atmosphere is observed to be populated by structures from the largest scale down to the current instrumental limit of arc sec spatial resolution. These structures are never completely steady, evolving over time scales ranging from months and days for the larger structures to sub-sec of the highly time-dependent processes of the spectacular flare.

The study of these structures and their dynamics is what makes solar physics an exciting subject from the point of view of plasma physics and hydromagnetics. The subject is made very difficult by the fact that it is not technically feasible to measure solar magnetic fields directly with any useful spatial resolution, except at the lower levels of the solar atmosphere. Even in that case, only the line-of-sight component of the magnetic field can be measured with confidence. The measurement of the transverse field component is much more involved, pushing present day efforts to the technical limits of both instrumentation and data interpretation (e.g., Harvey, 1977; Stenflo, 1978; Skumanich and Lites, 1987). Based on the knowledge of the magnetic field in the lower atmosphere, the structure of the magnetic field higher up must be inferred by the use of suitable theoretical models. Such an approach is generally regarded to be useful for long-lived structures so that, as a first approximation, the magnetic field may be taken to be in static equilibrium. There has been a recent revival of interest to measure all three components of the vector magnetic field in the lower regions of the solar atmosphere (see Hagyard, 1985). Vector field measurements, even if confined only to the lower regions of the atmosphere, are useful constraints for pinning down the appropriate choice of theoretical models for inferring solar magnetic fields. In this review,

I shall discuss several issues in this kind of work from the theoretical point of view. It is widely believed that the electric currents associated with long-lived magnetic structures constitute a store of free energy which, under the appropriate circumstance, can be liberated explosively such as in a flare, prominence eruption, or mass ejection. With quality data, it may be possible in the near future to directly measure this magnetic free energy and this prospect is the main motivation of this review. I shall concentrate on the active region magnetic fields which are of intensity 100 Gauss or more so that the plasma pressure is negligible compared to the magnetic pressure. In the quiescent static state, these magnetic fields may be approximated to be force-free; i.e., the electric current density is everywhere parallel to the magnetic field (Gold, 1964). Section 2 describes the general procedure of modeling force-free magnetic fields with observational input. Section 3 discusses various pitfalls and challenges of the procedure and section 4 gives a brief conclusion.

2. Modeling Solar Magnetic Fields

Let us take the solar atmosphere to be the infinite half space $z > 0$, using the usual Cartesian coordinates, with the plane $z = 0$ taken to be the solar photosphere. We shall adopt an idealization in which we neglect the chromosphere and the transition region and take the photosphere to be an infinitely thin rigid surface. The active region magnetic fields we are interested in are of the order of 100 Gauss or greater. For such a field $\mathbf{B}$ in a quiescent state, we may neglect plasma pressure as a first approximation and describe the field to be force-free, governed by the equation

$$\left(\nabla\times\mathbf{B}\right)\times\mathbf{B} = 0 , \tag{1}$$

supplemented by Maxwell's equation

$$\nabla.\mathbf{B} = 0 . \tag{2}$$

These extreme assumptions are commonly made in order to keep the model simple and tractable but later, in section 3, we will have occasion to again address some of these assumptions as nontrivial limitations of the model. The problem posed by solar magnetic fields calls for the construction of a magnetic field in $z > 0$ satisfying equations (1) and (2) subject to given information on $\mathbf{B}$ at $z = 0$ such as might be provided by direct measurement. For application, we may consider a magnetic structure spatially isolated in z in the sense that $\mathbf{B} \rightarrow 0$ at large distances from the structure. This condition, together with the boundary value of $\mathbf{B}$ at $z = 0$, constitutes a boundary value problem.

Different Formulations of the Boundary-Value Problem

Whether the boundary value problem we have posed makes mathematical sense depends on the manner in which the boundary condition at $z = 0$ is prescribed. For instance, if we take the z axis to be the line of sight and we prescribe only B_z at $z = 0$, an infinity of magnetic fields satisfies the boundary value problem. These magnetic fields correspond to different equilibrium states having the same normal magnetic flux distributions on z but different magnetic field topologies (or electric current distributions) in $z > 0$. In the early attempts to model solar magnetic fields, an ad hoc assumption was popular, which takes the magnetic field to be potential in $z > 0$:

$$\mathbf{B} = \nabla\Phi , \tag{3}$$

so that the electric current vanishes with $\nabla\times\mathbf{B} = 0$ and equation (2) gives the Laplace equation

$$\nabla^2\Phi = 0 . \tag{4}$$

The boundary value problem then has a unique solution (e.g., Schmidt, 1964; Altschuler and Newkirk, 1969; Levine, 1975; Sakurai, 1982). The solar atmosphere is an excellent electrical conductor in which significant electric currents can be readily induced by the motions in the photosphere. The potential model is thus not expected to be valid in general.

If we are able to measure all three components of $\mathbf{B}$ on $z = 0$, it has been pointed out that the boundary value problem may be meaningful in the following sense (Schmidt, 1968). Equation (1) can be recast in the form

$$\nabla\times\mathbf{B} = \alpha\mathbf{B} , \tag{5}$$

where α is a scalar function of space such that, in consequence of equation (2)

$$\mathbf{B}.\nabla\alpha = 0 , \tag{6}$$

i.e., α is a constant along magnetic lines of force. The distribution of $\mathbf{B}$ on $z = 0$ determines the spatial distribution of α in the sense that the latter is fixed at $z = 0$ by the z component of equation (5)

$$\alpha = \frac{\dfrac{\partial B_y}{\partial x} - \dfrac{\partial B_x}{\partial y}}{B_z} \tag{7}$$

and these values are carried as constants along the computed magnetic field lines into the region $z > 0$. It is an attractive idea but the precise conditions under which this boundary value problem is mathematically well posed have yet to be carefully worked out. The problem is certainly not well posed for an arbitrary prescription of the vector field at the boundary, a point we shall discuss in the next section.

If we are limited to prescribing only the normal field component at $z = 0$, electric currents may be introduced into the model by prescribing α in some manner. The function α may be interpreted to characterize the twist of the magnetic lines of force. An obvious choice is to set α uniform in space so that equation (6) is trivially satisfied and equation (5) is transformed into the familiar vector Helmholtz equation

$$\nabla^2\mathbf{B} + \alpha^2\mathbf{B} = 0 \tag{8}$$

which is tractable for many situations of interest (Chandrasekhar and Kendall, 1957). This was pursued with some enthusiasm in the early seventies to model observed solar structures (Nakagawa and Raadu, 1972). In particular, it was suggested that a pre-flare magnetic field may be modeled by a force-free field undergoing energy buildup with a constant α that progressively increases with time (Tanaka and Nakagawa, 1973). The choice of a constant α is difficult to justify physically. Moreover, it was pointed out that for a prescribed constant α, the boundary value problem may not have a unique solution (Chiu and Hilton, 1977). It has recently been suggested that a constant α force-free field may be the end state of a Taylor-type, resistive relaxation process, originally developed in the laboratory context (Woltjer, 1958; Taylor, 1974, 1986; Norman and Heyvaerts, 1983). The nonuniqueness of solution may not be such a bother if this suggestion can be put on a tenable basis; see the discussion in Low (1985). According to this theory of relaxation, the end state is a minimum energy state conserving the total magnetic helicity (Taylor, 1974). This conservation law serves to determine the value of the constant α and the selection of the minimum energy state breaks the degeneracy arising from the nonuniqueness of solutions (Woltjer, 1958; Berger and Field, 1984; Berger, 1985). In this theory, a constant α force-free field is more appropriately interpreted as a relaxed post-flare state rather than a pre-flare state as envisaged by Tanaka and Nakagawa (1973).

If α is not a constant in space, equations (5) and (6) are nonlinear, leaving very little useful that can be said about the circumstances under which the boundary value problem has meaning (Courant and Hilbert, 1963). Even with the use of numerical techniques, the treatment has to be largely empirical. This factor as well as the need to treat fully three-dimensional geometries for realistic structures makes the nonconstant α models generally unattainable. Leaving these technical obstacles aside, the physics of the highly conducting medium calls for the following formulation of the force-free field model which may be physically easier to interpret than the above boundary value problem.

Consideration of Magnetic Topology

The starting point is that the solar atmosphere may be approximated to have an infinite electrical conductivity so that the magnetic field may be taken to be frozen into the plasma as described by the induction equation

$$\frac{\partial \mathbf{B}}{\partial t} = \nabla \times \left(\mathbf{v} \times \mathbf{B}\right) \tag{9}$$

for the transport of the magnetic field in a velocity field $\mathbf{v}$. Suppose the force-free magnetic field we wish to construct is known to have evolved from some known initial state, say, a given potential state. The photosphere is so massive that we may regard the magnetic field in this layer of the atmosphere to be passively transported in a prescribed motion which is slow compared to the typical Alfvén speed of the tenuous upper atmosphere. In response to this slow boundary motion, the magnetic field in the upper atmosphere adjusts at the Alfvén speed through a sequence of stable force-free states. This is an assumption that must be valid for those solar structures observed to be evolving quasi-statically. The actual dynamical evolution involves the full set of hydromagnetic equations and is extremely complex. However, tracking this complex dynamical evolution is quite unnecessary for the determination of the end state as long as it can be assumed that the evolution leads to a stable force-free field rather than a highly time-dependent dynamical state. The end state force-free field is related to the initial magnetic field in terms of field topology and the magnetic flux distribution at $z = 0$ as accounted for by equation (9). The field topology and surface flux distribution of the force-free end state do not depend on the details of the motion of the medium in $z > 0$, under the assumed frozen-in condition, but is uniquely determined by equation (9) from the initial state and the known motion of the magnetic footpoint motion on $z = 0$. The end state we seek is then a solution to equations (1) and (2) subject to the constraints of these two pieces of information (Gold, 1964; Sturrock and Woodbury, 1967; Low, 1982a). This end state defined in terms of a static problem may not be unique and we shall return to discuss this point later in section 3. The formulation of the static problem can be made explicit by the use of variational calculus (e.g., Woltjer, 1958; Roberts, 1967; Low, 1982a; Sakurai, 1979). It suffices for our purpose to point out that the solution we seek minimizes the total magnetic energy subject to a prescribed normal flux distribution at $z = 0$ and a prescribed field topology. Actually, the force-free field merely extremizes the total magnetic energy but we seek a minimum in the total magnetic energy to ensure stability of the field. This variational problem is not readily tractable, again for technical reasons, but it is logically and physically more appealing. The magnetic field we construct is determined by its evolutionary history and the constructed field predicts a specific distribution of the transverse field component on $z = 0$ which we can compare with actual measurement. Where agreement is obtained by some appropriate experimental criterion, a basis may be said to have been established for believing that the constructed magnetic field represents the solar magnetic field in $z > 0$. Quite apart from the technical obstacles, it probably will not be practical to carry out such a program for many more years to come because it is unclear at the present how the photospheric footpoint motions can be observed to a quantitative detail adequate for fixing the magnetic field topology. On the other hand, there is much to learn about the properties of force-free fields from the above variational problem and we will discuss this point again in section 3.

An Application of the Virial Theorem

The extrapolation for the force-free field from its measurements in the lower solar atmosphere is a formidable undertaking. From the preceding discussion, it seems that solving the nonlinear boundary value problem with prescribed boundary distributions of the normal flux and the function α, the latter defined by the transverse field component, is the obvious goal to set for future model development. If this is disheartening, observers should be encouraged by the fact that certain important properties of force-free magnetic fields can be obtained directly from boundary vector measurements without having to compute the full magnetic field in $z > 0$. A simple application of the virial theorem shows that the total energy of a force-free magnetic field is determined completely by the boundary distribution of the magnetic field

$$\int_V dV\, \frac{B^2}{8\pi} = \frac{1}{4\pi}\int_{\partial V} \left[\frac{B^2\mathbf{r}}{2}.d\,\mathbf{S} - (\mathbf{B}.\mathbf{r})(\mathbf{B}.d\,\mathbf{S})\right], \quad (10)$$

where $\mathbf{r}$ is the position vector, V denotes the volume occupied by the force-free field, and $d\,\mathbf{S}$ is the normal surface element on the boundary ∂V (Chandrasekhar, 1961). For the idealized geometry considered above, we have

$$\int_{z>0} dV\, \frac{B^2}{8\pi} = \frac{1}{4\pi}\int_{z=0} dxdy \left(xB_x + yB_y\right)B_z\,. \quad (11)$$

This is a well-known result but its significance for solar physics has been recognized only recently (Molodensky, 1974; Low, 1982b). For most flares, the energy liberation is believed to be confined to the chromosphere and corona, with little detectable photospheric disturbances. If we identify the flare energy to be due to the dissipation of the electric currents in $z > 0$ with no dissipation or motion at $z = 0$, the maximum energy that can be so liberated must result in a potential magnetic field with the same normal flux distribution on $z = 0$. This potential field $\mathbf{B}_{pot}$ is uniquely determined by solving equation (4). Its total energy can then be calculated or else simply determined from its field distribution on $z = 0$ by the virial theorem

$$\int_{z>0} dV\, \frac{B_{pot}^2}{8\pi} = \frac{1}{4\pi}\int_{z=0} dxdy \left(xB_{pot\,x} + yB_{pot\,y}\right)B_z\,. \quad (12)$$

The maximum amount of energy that can be liberated is then the difference of the two total magnetic energies

$$\Delta E = \frac{1}{4\pi}\int_{z=0} dxdy \left(x\left[B_x - B_{pot\,x}\right] + y\left[B_y - B_{pot\,y}\right]\right)B_z\,. \quad (13)$$

The difference in energy between the force-free field and its associated potential field is reflected in the orientation and magnitude of the transverse field components at the boundary. More interesting will be to determine the total energies for the magnetic fields shortly before and after a flare by the use of the virial theorem and to compare the energy difference with the estimated flare energy release. This interesting part of the physics can be done without the need to extrapolate the force-free field into the region $z > 0$.

3. Some Pitfalls and Challenges

There are three pitfalls confronting the modeling effort which seem worth discussing. These pitfalls arise from the facts that (i) solar structures in the active region are usually not spatially isolated, (ii) the force-free condition breaks down at the photosphere, and (iii) it is necessary in general to account for electric current sheets in three-dimensional magnetic fields. Let us examine each in the order I have raised them.

Adequate Spatial Resolution Versus Large Field-of-View

Any useful magnetograph must be built upon an optimum compromise between the need to sample the small scale structures with an adequate spatial resolution and the desire to cover a large region of the solar surface. Typically, one needs to resolve structures as small as 1 arc second and cover a region with linear dimensions of the order of 10^5 km characteristic of the active region. A compromise between the two demands is to favor the need for fine spatial resolution and raster scan the large region of interest, a strategy adopted by the High Altitude Observatory for its new Stokes Polarimeter to measure the vector magnetic field through direct determination of absorption line profiles (Lites, 1988, private communication). This instrument is expected to scan an active region once in about 10 min. The full set of field measurements obtained this way must assume that the magnetic field is reasonably stationary over the time taken for each scan. This assumption is not severe, given the low speed of 0.5 km s^{-1} typical of large-scale photospheric motions and our interest in modeling quasi-static structures. Other goals have been set for other future vector field measurements with instruments that can cover an active region in a matter of seconds, using filtergrams with large fields of view (see, e.g., Canfield and Mickey, 1989). These instruments are aimed at studying rapid time evolution of magnetic structures expected to accompany a flare eruption. While such rapid evolutions are of great interest in their own rights, the vector field measurements are more difficult to model as the simplifying assumption of a static equilibrium can no longer be made.

Most active region structures are complex with several identifiable structures in close proximity so that no one structure can be approximated as spatially localized. This raises the question of the applicability of the boundary condition for an infinite half-space, an assumption essential to the use of the virial theorem to determine the total magnetic energy in terms of only the boundary vector field. If the entire active region is taken as a single structure, to estimate the magnetic energy would require the coverage of a region even larger, probably pushing the scanning time for the larger region beyond acceptability. The successful application of the virial theorem for estimating the magnetic energy will depend on the judicious choice of particular structures for study which are sufficiently isolated and simple so that sampling over an active region size is adequate to account for the surface integral in equation (11). Theoretical modeling suggests that an isolated sunspot with an penumbral radius of 3×10^4 km treated as a potential field requires covering a surrounding region with a linear dimension of 1.5×10^5 km in order to account for the total energy by the use of equation (11) to an accuracy of 10%. If the field is twisted, there is a tendency for the force-free field to be inflated relative to the potential field with the same boundary flux distribution (Yang et al., 1986). For a sunspot with a twisted field, then, the use of equation (11) to estimate the total energy at the same 10% accuracy level may require covering an even more extensive surrounding region, depending on the degree of twist.

Testing for Force-Free Conditions

The need for a spectroscopic signal usable with current methods of magnetic field measurement means that only the magnetic field in the photosphere and, with greater difficulty, in the chromosphere can be measured. The use of photospheric measurements as inputs for the force-free model should be avoided. The photosphere is actually a highly structured layer with the magnetic field not diffusely distributed but believed to be in fibril structures of diameter of about 150 km and field strength of the order of 1500 Gauss or larger (e.g., Livingston and Harvey, 1969; Stenflo, 1973; Chapman, 1973). This is a highly non-force-free state in which plasma flow and radiative transfer processes may play important roles (e.g., Parker, 1979; Deinzer et al., 1984; Ribes et al., 1985). The above model of taking the force-free condition all the way down to the boundary $z = 0$ is clearly inadequate if the photospheric measurements were used as boundary values. The density scale height of the photosphere is small, about 150 km, so that the low-plasma-β condition prevails just above the photosphere somewhere in the chromosphere. For a meaningful construction of the force-free field, we therefore require magnetograph measurements at these higher levels. A particular property of force-free fields actually allows one to test whether a given measurement of the vector field serves as an appropriate boundary condition for equations (1) and (2). The nature of the Lorentz force is such that the total force exerted on a volume of plasma is expressible in terms of the surface integral of the Maxwell stresses on the boundary. For a force-free field, this total force is of course zero. Setting the three components of the integrated Maxwell stresses to zero, we have the conditions for the magnetic field spatially isolated in $z > 0$

$$\int_{z=0} dxdy\ B_x B_z = \int_{z=0} dxdy\ B_y B_z = \int_{z=0} dxdy\ \left(B_z^2 - B_x^2 - B_y^2\right) = 0 . \qquad (14)$$

Equation (14) therefore provides a ready test for any vector field measurements taken on a given level to see if a net force is exerted on the medium above it. Given that we believe the photospheric field is not force-free, it will be interesting to verify this belief with direct measurements of the photospheric field. It will also be interesting to see if the condition is met more satisfactorily for field measurements taken at higher layers where the plasma β is expected to be small. Moreover, if magnetic field measurements can be made with time resolutions of a few seconds for the study of flares, it will be very interesting to observe whether chromospheric fields obey equation (14) reasonably well except during the time of flares and related eruptions. What is an appropriate experimental criterion for meeting the conditions expressed in equation (14)? The following is an obvious choice. For a magnetic field of intensity B, the force it can exert on a surface area S has a typical strength of the order of $S\ (B^2 / 8\pi)$. Thus, we can calculate from the measured magnetic field the hypothetical total force

$$F = \int_{z=0} dxdy\ B^2 , \qquad (15)$$

integrated over the field of view, as a measure of the magnitude of the total force the magnetic field is capable of exerting on the solar atmosphere. The conditions in equation (14) may then be regarded to be reasonably met if each of the integrals on the left sides of equation (14) integrated over the field of view is demonstrated to be small compared to F.

It was pointed out earlier that an arbitrarily prescribed distribution of the vector field on $z = 0$ cannot be assured of being consistent with a solution of equations (1) and (2). One reason is now apparent: any arbitrary boundary field distribution which violates equation (14) is inherently incompatible with a force-free field in $z = 0$. Such a boundary field distribution cannot lead to a meaningful solution of the boundary value problem for the force-free field. This property is useful as a test for any numerical algorithm that claims to have the capability to compute force-free fields in a volume in terms of their boundary distribution. It should be pointed out that equation (14) is a necessary but not sufficient condition for the existence of a force-free field in $z > 0$.

Some Theoretical Questions

In our consideration so far, we have implicitly assumed that the magnetic field we are interested in is everywhere smooth. There is a fundamental difference between magnetic fields which are geometrically simple, such as an axisymmetric field, and those which are fully three dimensional with no obvious symmetry. In a medium approximated with an infinite electrical conductivity, the magnetic field is frozen into the medium. As the magnetic field deforms and becomes twisted by footpoint motion in some arbitrary manner, as happens in the solar atmosphere, the field topologies are never expected to be simple and in this circumstance, the force-free states we will encounter are not likely to be everywhere smooth but will contain thin electric current sheets, a remarkable property first pointed out by Parker (1972, 1979, 1986a,b). The inevitable formation of the electric current sheets is due to the combined effects of force-balance and the frozen-in condition of the infinitely conducting medium. This is a fundamental property of ideal hydromagnetics and I refer the reader to the literature for the basic theory and demonstration (Grad, 1967; Sweet, 1969; Parker, 1972, 1979, 1986a,b, 1987, 1988; Arnol'd, 1974; Hu and Low, 1982; Tsinganos et al., 1984; Priest and Raadu, 1975; Syrovatskii, 1981; Moffatt, 1985, 1987; Low, 1985, 1986, 1987; Low and Wolfson, 1988; Aly, 1988). In terms of the solutions to equations (1) and (2) in a boundary value problem, singular solutions with current sheet structures may be unavoidable if these solutions are required to have a prescribed field topology. The prescribed topology in these situations overconstrains equations (1) and (2) so that these equations do not have everywhere smooth solutions in the mathematical sense. It should also be pointed out that this is a subtle property not universally accepted and there had been attempts to refute its existence. These attempts have not been successful, in my opinion. For example, Antiochos (1987) presented an analysis to show that if equations (1) and (2) are cast in terms of Euler potentials, a smooth solution (without electric current sheets) can always be constructed for a magnetic field in any prescribed topology in which there are no magnetic neutral points. This construction has been shown to fail, under the same conditions assumed by Antiochos, in several explicit examples (Parker, 1987; Low, 1987; Low and Wolfson, 1988; Moffatt, 1987; Aly, 1988).

The point is best seen in the variational formulation of the force-free field described in section 2. We recall that the problem seeks a field of minimum magnetic energy subject to a prescribed field topology and a prescribed normal flux distribution at the boundary. Physically, the prescribed quantities are properties endowed by the induction equation (9) as a result of a particular evolutionary history of the magnetic field. The details of the actual evolution are not relevant. The topology of the magnetic field is completely fixed in terms of the topology of the initial field and the footpoint motions at $z = 0$ that account for the change in topology through equation (9). What makes this problem formidable is the unavailability of a convenient mathematical description of the topology of an arbitrary three-dimensional magnetic field. Just describing the mapping of boundary points as pairs of footpoints of lines of force is not sufficient to specify the magnetic topology which involves also the winding pattern of the continuum field lines in the interior region. While the problem is difficult to solve, the hypothetically possible outcomes of such a problem have the following physical significance. Let us perform an extremization of the total magnetic energy subject to the two prescribed constraints of field topology and boundary flux distribution. It is in the nature of some variational problems that there may be no solution to a particular problem, or a solution may exist and in the latter case, the solution may or may not be unique. Take the case where solutions exist and allow for more than one solution to exist for given prescribed constraints. There are then more than one force-free states available to the magnetic field being considered. Different sets of evolutionary paths having identical surface motions take the same initial field, under the frozen-in condition, to these different possible force-free states. Physically, only the linearly stable members of this set of possible force-free states are of interest; the unstable ones can only be realized along very specialized evolutionary paths which do not excite any of the unstable modes of the particular force-free state. We suggest that the stable force-free states are all realizable with a probability measured by the density of possible evolutionary paths (in phase space) leading to these states. It is clear that the state with the lowest energy has the highest likelihood of realization. It is in principle possible that the force-free states, in a particular case, are all unstable and hence all physically not realizable. There is also the possibility that in a spectrum of admissible force-free states, the lowest-energy state is unstable. These interesting possible cases have important physical implications to which we shall return presently. It is instructive to first consider the special situations where no solution exists for a particular problem.

If there is no solution, the magnetic field cannot find an equilibrium state and a Lorentz force must always act to drive the medium. Let us consider such a dynamical state within the following artificial construction first employed in building plasma equilibrium states in the laboratory (Chodura and Schlüter, 1981; Parker, 1979; Moffatt, 1985; Yang et al., 1986). This construction is not intended here to imply an actual dynamical evolution; it merely serves the purpose of demonstrating the inevitability of the formation of electric current sheets in a given situation. Suppose the medium is taken to be moving with some artificial velocity and exerting no other forces than a viscous drag proportional to the velocity such that the drag everywhere balances the non-vanishing Lorentz force. Such a system must evolve with a monotonically decreasing magnetic energy due to work expended by the Lorentz force. We have the frozen-in condition and the motion is set to vanish at the boundary so that neither the field topology nor the boundary flux distribution change with time in the course of this artificial evolution. Now the magnetic energy cannot decrease to zero in the situation we consider here, where the magnetic field threads through the boundary. The magnetic energy is actually bounded below by the energy of a potential field having the prescribed boundary flux distribution. It follows that the above system must eventually approach a minimum energy state and to be consistent with the assumption that there are no smooth solutions, this minimum energy state must be force-free almost everywhere except at surfaces of singularity which we interpret to be electric current sheets. In variational calculus, this is not an unfamiliar situation for the minimizing sequence of trial functions taken from the set of admissible trial functions in a variational problem is often found not to be compact in the sense that the limit trial function does not belong to the admissible set (Courant and Hilbert, 1963). The above construction also leads to a similar conclusion if applied to the cases where the variational problem has solutions but the lowest energy state is unstable. By perturbing the unstable state of the lowest magnetic energy in the artificial viscous medium, the magnetic field cannot reach any of the other available higher-energy states and must again reach some even lower energy state with some form of current sheet singularity.

How certain are we about the above hypothetically possible outcomes of the variational problem? In the absence of a general treatment, we only have particular examples to guide us. Here theoretical studies by slow progress have taught us some very interesting properties of force-free fields. The first obvious question to ask is whether stable force-free fields in fact exist. Laboratory plasma physics has shown the existence of many unstable force-free fields (e.g., Friedberg, 1982). It has been suggested that in contrast to the laboratory circumstances, solar magnetic fields may be stable because the field lines are anchored almost rigidly by the dense photosphere (e.g., Raadu, 1972; Hood and Priest, 1979; Cargill et al., 1986). This is not easy to demonstrate theoretically as stability analyses are difficult to carry out to completeness for proof of stability, even for idealized simple fields. Recently, some exact examples of stable force-free fields have been found (Low, 1982b, 1988a,b). Included among the examples are a set of geometrically realistic models for the three-dimensional configurations associated with a pair of opposite polarity photospheric regions intruding one into the other. Examples have also been found of the situation in which the prescribed boundary flux distribution and field topology are not compatible with any smooth force-free fields and the inevitable formation of the current sheets was demonstrated (Parker, 1987; Low, 1987; Moffatt, 1987; Low and Wolfson, 1988; Aly, 1988). As far as I am aware, it has not been possible to demonstrate the possibility of a situation where multiple solutions exist. An explicit example of such a situation is likely to prove conceptually very instructive. What seems clear is that the computation of a three-dimensional force-free field in general is a difficult undertaking, due in large part to the need to account for electric current sheets, and there is still much we need to learn about their theoretical properties. In reality, the solar atmosphere has a large but finite electric conductivity. If current sheets form, they eventually must be dissipated away by electrical resistivity resulting in reconnection of magnetic field lines and a change in field topology. We no longer have an ideal system, the physical system becomes unwieldingly complex, and it is unclear how the topology as a constraint to determine the force-free field is to be introduced. It is interesting to note that the energy integral in equation (10) adequately accounts for the existence of electric current sheets in the interior volume. Electric current sheets are discontinuities in the derivatives of the magnetic fields across which a tangential magnetic field rotates abruptly with no change in the magnetic intensity. To avoid these surfaces in the application of Gauss theorem, we can partition the volume into sub-volumes bounded by surfaces containing the tangential discontinuities. It can then be shown that a subset of the various surface integrals arising from the partition of the volume would mutually cancel by virtue of the properties of the tangential discontinuities, leaving the same formula in equation (10). In other words, the energy associated with the presence of the electric current sheets is built into the volume distribution of the magnetic field and is, in fact, reflected in the distribution of the transverse component of the boundary magnetic field.

4. Conclusion

During the past three decades, the idea that the magnetic free energy drives various activities in the solar atmosphere has become common and popular in our attempts to understand solar phenomena. With the new observational instruments being planned, we can for the first time hope to detect directly this free energy through the measurement of the solar vector magnetic field. Even with the existing instruments, there has been some moderate success (see, e.g., Gary et al., 1987). Fortunately, the properties of force-free fields are such that the magnetic free energy can be obtained directly from the field measurements by the simple integral formula in equation (11). As pointed out in section 3, the procedure is not straightforward but the pitfalls are not insurmountable. Measurements of the vector field will also give direct, if only qualitative, information on various other physical quantities of interest that until now have been lacking in pinning down theoretical ideas. We can test whether the field in the upper atmosphere is force-free through equation (14) and whether there are substantial electric currents flowing across the height of field measurement by the direct use of Ampere's law. Actually, the morphology of the electric currents in the atmosphere can also be made out by the application of Ampere's law (see, e.g., Hagyard et al., 1981). Finally, it will be interesting to test the hypothesis that a flare eruption is a Taylor-type relaxation taking a stressed magnetic field into a constant α force-free field. This can be done by looking for the constancy of the boundary values of α derived in terms of the measured vector field at the base of a post-flare magnetic field. The prospect of doing all this exciting science should be sufficient motivation to solve the technical problems in instrumentation and data interpretation that make quality measurement of the vector field a nontrivial matter.

To make full use of the vector field data, a workable means of inferring the long-lived force-free magnetic structures in the upper atmosphere from the data is desirable. Without quantitative information on magnetic field structure and morphology, our understanding of the physics of the solar atmosphere will remain incomplete and superficial. In the long term, the interaction between the magnetic field and plasma (both in static situations and in dynamical evolution) which is neglected in the force-free approximation must be addressed. The development of such a program should be a high priority in future modeling efforts. It seems desirable in the immediate future to pursue a two-step program in which the boundary field is first shown to satisfy the necessary conditions in equation (14) for the existence of the force-free state and then with the boundary distribution of α defined by equation (7), the field above is computed as a boundary value problem. The development of this program is not simply a matter of devising the necessary numerical tools. There is much about force-free fields we have to learn, their stability, availability of equilibrium states, and current sheet formation, and, the role of the magnetic morphology in determining these properties. Only when adequate progress has been made in theoretical understanding can we hope to devise computational schemes that we understand physically and, in the application to observational data, relate observations to quantitative hydromagnetic principles.

Acknowledgments. I thank Andy Skumanich and Bruce Lites for discussion, and, Mike Knölker and Peter Sturrock for reading and commenting on the manuscript.

References

Aly, J. J., On the existence of a topological heating mechanism for a plasma in a sheared magnetic field, Astron. Astrophys., in press, 1988.

Antiochos, S., The topology of force-free magnetic fields and its implications for coronal activity, Ap. J., 312, 886, 1987.

Arnol'd, V., The asymptotic Hopf invariant and its applications, in Proc. Summer School in Differential Equations, Armenian SSR Acad. Sci.: Erevan (in Russian), 1974.

Altschuler, M. D., and G. Newkirk, Jr., Magnetic fields and the structure of the solar corona. I. Methods of calculating coronal fields, Solar Phys., 9, 131, 1969.

Berger, M. A., Structure and stability of constant-α force-free fields, Ap. J. Suppl., 59, 433, 1985.

Berger, M. A., and G. B. Field, The topological properties of magnetic helicity, J. Fluid Mech., 147, 133, 1984.

Canfield, R. C., and D. L. Mickey, An imaging vector magnetograph for the next solar maximum, this volume, 1989.

Cargill, P. J., A. W. Hood, and S. Migliuolo, The magnetohydrodynamic stability of coronal arcades. II. Sheared equilibrium fields, Ap. J., 309, 402, 1986.

Chandrasekhar, S., Hydrodynamic and Hydromagnetic Stability, Oxford University Press, Oxford, 1961.

Chandrasekhar, S., and P. C. Kendall, On force-free magnetic fields, Ap. J., 126, 457, 1957.

Chapman, G., On the nature of the small-scale solar magnetic field, Ap. J., 191, 255, 1973.

Chiu, Y. T., and H. H. Hilton, Exact Green's function method of solar force-free magnetic-field computation with constant α. I. Theory and basic test cases, Ap. J., 212, 873, 1977.

Chodura, R., and A. Schlüter, A 3D code for MHD equilibrium and stability, J. Comput. Phys., 41, 68, 1981.

Courant, R., and D. Hilbert, Methods of Mathematical Physics, Vol. II, Interscience, New York, 1963.

Deinzer, W., G. Hensler, M. Schlüssler, and E. Weisshaar, Model calculations of magnetic flux tubes. II. Stationary results for solar magnetic elements, Astron. Astrophys., 139, 435, 1984.

Friedberg, J. P., Ideal magnetohydrodynamic theory of magnetic fusion system, Rev. Mod. Phys., 54, 801, 1982.

Gary, G. A., R. L. Moore, M. J. Hagyard, and B. M. Haisch, Nonpotential features observed in the magnetic field of an active region, Ap. J., 314, 782, 1987.

Gold, T., Magnetic shedding in the solar atmosphere, in AAS-NASA Symposium on the Physics of Solar Flares, NASA SP-50, edited by W. N. Hess, p. 389, NASA, Washington, D.C., 1964.

Grad, H., Toroidal containment of a plasma, Phys. Fluids, 10, 137, 1967.

Hagyard, M., B. C. Low, and E. Tandberg-Hanssen, On the presence of electric currents in the solar atmosphere. I. A theoretical framework, Solar Phys., 73, 257, 1981.

Hagyard, M. J. (editor), Measurements of Solar Vector Magnetic Fields, NASA Conference Publication 237, Washington, D.C., 1985.

Harvey, J. W., Observation of small-scale photospheric magnetic fields, in Highlights of Astronomy, 4, Part II, edited by E. A. Müller, p. 233, Reidel, Dordrecht, 1977.

Hood, A. W., and E. R. Priest, Kink instability of solar coronal loops as the cause of solar flares, Solar Phys., 64, 303, 1979.

Hu, Y. Q., and B. C. Low, The energy of electric current sheets. I. Models of moving magnetic dipoles, Solar Phys., 81, 107, 1982.

Levine, R. H., The representation of magnetic field lines from magnetograph data, Solar Phys., 44, 365, 1975.

Livingston, W., and J. W. Harvey, Observational evidence for quantization in photospheric magnetic flux, Solar Phys., 10, 294, 1969.

Low, B. C., Nonlinear force-free magnetic fields, Rev. Geophys. Space Phys., 20, 145, 1982a.

Low, B. C., Magnetic field configurations associated with polarity intrusion in a solar active region. I. The force-free fields, Solar Phys., 77, 43, 1982b.

Low, B. C., Some recent developments in the theoretical dynamics of solar magnetic fields, Solar Phys., 100, 309, 1985.

Low, B. C., Blow-up of force-free magnetic fields in the infinite region of space, Ap. J., 307, 305, 1986.

Low, B. C., Electric current sheet formation in a magnetic field induced by continuous magnetic footpoint displacements, Ap. J., 323, 358, 1987.

Low, B. C., On the hydromagnetic stability of a class of laminated force-free magnetic fields, Ap. J., in press, 1988a.

Low, B. C., Magnetic field configurations associated with polarity intrusion in a solar active region. II. Linear hydromagnetic stability, Solar Phys., in press, 1988b.

Low, B. C., and R. Wolfson, Spontaneous formation of electric current sheets and the origin of solar flares, Ap. J., 324, 574-581, 1988.

Moffatt, H. K., Magnetostatic equilibria and analogous Euler flows of arbitrary complex topology. Part I. Fundamentals, J. Fluid Mech., 159, 359, 1985.

Moffatt, H. K., Geophysical and astrophysical turbulence, in Advances in Turbulence, edited by G. Comte-Bellot and J. Methieu, p. 228, Springer-Verlag, 1987.

Molodensky, M. M., Equilibrium and stability of force-free magnetic fields, Solar Phys., 39, 393, 1974.

Nakagawa, Y., and M. A. Raadu, On practical representation of magnetic fields, Solar Phys., 25, 127, 1972.

Norman, C. A., and J. Heyvaerts, The final state of a solar flare, Astron. Astrophys., 124, L1, 1983.

Parker, E. N., Topological dissipation and the small-scale fields in turbulent gases, Ap. J., 174, 499, 1972.

Parker, E. N., Cosmical Magnetic Fields, Oxford University Press, Oxford, 1979.

Parker, E. N., Equilibrium of magnetic fields with arbitrary interweaving of the lines of force. I. Discontinuities in the torsion, Geophys. Ap. Fluid Dyn., 34, 243, 1986a.

Parker, E. N., Equilibrium of magnetic fields with arbitrary interweaving of the lines of force. II. Discontinuities in the field, Geophys. Ap. Fluid Dyn., 35, 277, 1986b.

Parker, E. N., Magnetic reorientation and the spontaneous formation of tangential discontinuities in deformed magnetic fields, Ap. J., 318, 376, 1987.

Parker, E. N., Tangential discontinuities and the optical analogy for stationary fields. I. General principles, Geophys. Ap. Fluid Dyn., submitted, 1988.

Priest, E. R., and M. A. Raadu, Preflare current sheets in the solar atmosphere, Solar Phys., 47, 41, 1975.

Raadu, M. A., Suppression of the kink instability for magnetic flux ropes in the chromosphere, Solar Phys., 22, 425, 1972.

Ribes, E., D. Rees, and C. Fang, Observational diagnostics for models of magnetic flux tubes, Ap. J., 296, 268, 1985.

Roberts, P. H., An Introduction to Magnetohydrodynamics, American Elsevier Publ. Co., New York, 1967.

Sakurai, T., A new approach to the force-free field and its application to the magnetic field of solar active regions, Publ. Astron. Soc. Japan, 31, 209, 1979.

Sakurai, T., Green's function methods for potential fields, Solar Phys., 76, 301, 1982.

Schmidt, H., On the observable effects of magnetic energy storage and release connected with solar flares, in AAS-NASA Symposium on the Physics of Solar Flares, NASA SP-50, edited by W. N. Hess, p. 107, NASA, Washington, D.C., 1964.

Schmidt, H., Magnetohydrodynamics of an active region, in Structure and Development of Solar Active Regions, IAU Symp. No. 35, edited by K. O. Kiepenheuer, p. 95, Reidel, Dordrecht, 1968.

Skumanich, A., and B. W. Lites, The polarization properties of model sunspots: The broad-band polarization signature of the Schluter-Temsvary representation, Ap. J., 322, 483, 1987.

Stenflo, J. O., Magnetic field structure of the photospheric network, Solar Phys., 32, 41, 1973.

Stenflo, J. O., The measurement of solar magnetic fields, Rep. Prog. Phys., 41, 865, 1978.

Sturrock, P. A., and E. T. Woodbury, Force-free magnetic fields and solar filaments, in Plasma Astrophysics, edited by P. A. Sturrock, p. 155, Academic Press, New York, 1967.

Sweet, P. A., Mechanisms of solar flares, Ann. Rev. Astr. Ap., 7, 147, 1969.

Syrovatskii, S. I., Pinch sheets and reconnection in astrophysics, Ann. Rev. Astr. Ap., 19, 163, 1981.

Tanaka, K., and Y. Nakagawa, Force-free magnetic fields and flares of August 1972, Solar Phys., 33, 187, 1973.

Taylor, J. B., Relaxation of toroidal plasma and generation of reverse magnetic fields, Phys. Rev. Letters, 33, 1139, 1974.

Taylor, J. B., Relaxation and magnetic reconnection in plasma, Rev. Mod. Phys., 58, 741, 1986.

Tsinganos, K., J. Distler, and R. Rosner, On the topological stability of magnetostatic equilibria, Ap. J., 278, 409, 1984.

Woltjer, L., A theorem on force-free magnetic fields, Proc. Nat. Acad. Sci. USA, 44, 489, 1958.

Yang, W. H., P. A. Sturrock, and S. K. Antiochos, Force-free magnetic fields: The magneto-frictional method, Ap. J., 309, 383, 1986.

OPTICAL DISK PROCESSING OF SOLAR IMAGES

Alan Title and Theodore Tarbell

Lockheed Palo Alto Research Laboratory, 3251 Hanover Street, Palo Alto, CA 94304

Abstract. The current generation of space and ground-based experiments in solar physics produces many megabyte-sized image data arrays. Optical disk technology is the leading candidate for convenient analysis, distribution, and archiving of these data. We have been developing data analysis procedures which use both analog and digital optical disks for the study of solar phenomena. The analog disks allowed us to view movies much more effectively than video tape. In particular, it has been possible to develop a PC-based movie control program for interactive analysis of movies. The digital disks make it easy to have many, very large image data bases on the computer without loading numerous tapes. Our experience indicates that both the system design and software must be developed based on studying real data. Our basic approach to learning about managing large blocks of image data is to: (1) produce, analyze, and publish the high resolution movies from observing runs at Sacramento Peak Observatory with a tunable filter and a charge coupled device (CCD) camera; (2) collaborate with the Swedish Solar Observatory to obtain high resolution movies in the solar continuum; and (3) study three-dimensional Fourier filtering techniques for hydrodynamic studies and suppression of atmospheric seeing.

Introduction

Solar phenomena modify and control the Earth's atmospheric and magnetospheric environments. Solar flares, magnetic storms, and solar particle events affect the radiation belts, ionospheric communications, and electromagnetic background levels (gamma ray through radio). Most of these effects come from high-energy radiations (UV, X, gamma rays, and fast particles) which originate in the Sun's corona. The sources of coronal activity and instability lie in magnetic and hydrodynamic processes that are lower in the solar atmosphere and can be studied by optical techniques. For many decades, astronomers have taken movies and spectra which show solar activity of bewildering variety. The physical processes responsible for this activity are not well understood and in some cases, such as coronal heating, the process has not even been identified.

A serious problem in observational solar physics and other types of remote sensing is the great volume of data produced in a short observation time. In the past, movies on film and video tape have been used to examine and discover the phenomena of solar activity. These media are recordings of images in a fixed order. New orderings are possible only by editing and recopying the original data. They do not easily allow quantitative comparison of phenomena in several layers of the atmosphere, which requires exploration of different sequences for different processes.

The new generation of imaging solar experiments use solid state digital imaging detectors. These can produce data arrays as large as 2048 x 2048 pixels, with 8 to 16 bits per pixel. Filter camera systems are now being constructed which can generate data in excess of 5 Mbyte s^{-1}. A few minutes of imagery would fill a large magnetic disk. Although tape recording can solve the storage problem, analysis of the images from computer tape is so slow that it is of little value and sufficient magnetic disk storage is expensive. Even more important, loading tapes to magnetic disk is time consuming. To efficiently work with image sets and exploit their digital precision, techniques must be developed to select, review, and compare images. The user must be able to extract specific measurements of interest from gigabytes of potentially relevant data.

Optical Disk Processing

We have been exploring the use of optical disk memory to store and analyze large data bases of imagery. The random access capability of disks and the high volume of storage available (24,000 frames on one removable analog platter) are exciting new features of this medium. In particular, we are developing the ability to interactively study multi-spectral movies using both visual and digital inspection of the data.

In 1985, we began to intensively use analog video disk recorders and players to study digital movies of the Sun obtained from the solar optical universal polarimeter on Spacelab 2 (Title et al., 1986). These devices were integrated with image processors on a VAX 780 for recording movies, and with personal computers (PCs) for off-line study of movies previously recorded. Because the analog laser disk player can rapidly index to any track (image) on the disk, it is simple to observe image sequences in any order, rate, or direction by a simple command sequence from the control computer. Lists of commands can be stored for creation of "movies" which illustrate particular phenomena.

In addition to the analog disks, three Optimem 1000 digital optical disk drives, with 1.0 Gbyte capacity on each side of each removable platter, are dedicated to movie storage on the VAX system. The digital disks serve two important functions. First, they provide 3 Gbytes of storage on the VAX at relatively low cost. Second, and more important, the images have to be loaded from tape to disk only once. (A disk holds data from a dozen 2400 foot, 6250 bpi tapes on each side.)

Both the analog and digital systems have indices for simple data access. An interactive image data processing language (ANA) optimized for studies of movies is available. The optical disk drives have reduced the movie storage problem considerably, but the VAX 780 (shared with many other projects) is not fast enough for some of the computationally intensive image processing described below. This situation should improve with the installation of a dedicated Vax station 3200 with an array processor. We hope in the near future to have a clear concept of the requirements on scientific work stations for movie analysis.

In mid-1987, we carried out two observing runs using the breadboard tunable filter (TF) system for the Coordinated Instrument Package for the High Resolution Solar Observatory (HRSO) at the Tower Telescope. The breadboard TF consisted of: a 1024 x 1024 CCD camera (although the data system only stored 512 x 512 arrays), a universal tunable birefringent filter with 50 mA bandpass, spectral and polarizing prefilters, an image stabilizer using a fast tilt mirror and sunspot tracker, and a PC controller that allowed complex observing sequences to be run automatically. In 1988, we expect to obtain more observations with this system at the Swedish Solar Observatory (SSO) (Scharmer et al., 1985) telescope in the Canary Islands.

The National Solar Observatory (NSO) observing run yielded 85 magnetic tapes (containing roughly 12,000 images). The first processing step was to transfer all of the images to analog video disk. This allowed us to view the entire set, find interesting phenomena, and select the best frames and movies for further study. Without the video disk, we would still be unaware of most of the dynamic phenomena in the movies, since they are not perceptible from static displays of a small number of frames.

Processing steps on the digital images include flat field corrections, photometric conversion, compensation for atmospheric image wander and distortion, exposure balancing, flaw removal, and differencing and ratioing among wavelengths or polarizations. The results are new movies of longitudinal magnetic fields, Doppler shifts, and intensities at different heights in the atmosphere. These are also recorded on video disk. In addition, transverse velocities can be measured by correlation tracking (November et al., 1987), following the gradual displacement of fine structures advected by larger-scale flows. This procedure produces movies of velocity vectors, divergence, and vorticity. Any of the movies may also be displayed in more exotic fashion, as in space-time slice movies or color or graphic overlays of one movie on another. Features or correlations which are recognized by watching these video disk movies can be then studied quantitatively in the original digital data that are on the VAX.

Example Results

We have concentrated recent numerical efforts on an 82-min observing sequence in an active region and some surrounding quiet Sun obtained on August 6, 1987. This set of images had an average image quality of better than 1 arc sec. The real time stabilized field-of-view was 90 x 90 arc sec. The run consisted of 60 repeats of a 13-frame sequence collecting data at several heights in the atmosphere in lines that were both sensitive and insensitive to the magnetic field. Figure 1 shows four images from one time step: (a) continuum near 5576 Å, (b) H-alpha line center, (c) Fe I 5576 blue wing at -30 mA, and (d) H-alpha red wing at 600 mA. Tic marks are at 2 arc sec intervals.

All frames have been sorted by wavelength, spatially registered, and recorded as movies on video disk. The H-alpha movies show a small flare, with changes in the fibril structure that strongly suggest reconnection across the neutral line into a simpler magnetic topology. Figure 2 shows: (a) 5576 line center intensity, (b) H-alpha blue wing (-600 mA), (c) Doppler velocity in the magnetically insensitive line 5576, and (d) a longitudinal magnetogram (made in the blue wing of Fe I 6302). Currently, a magnetic movie is in production, and the Doppler and line center movies are being Fourier analyzed to make power spectra and filtered movies, separating waves and convective flows. With these, we plan to study the following topics: the relative phases and amplitudes of f (gravity) and p (pressure) mode waves of short spatial wavelength (large l) in velocity, line center, and continuum; changes in

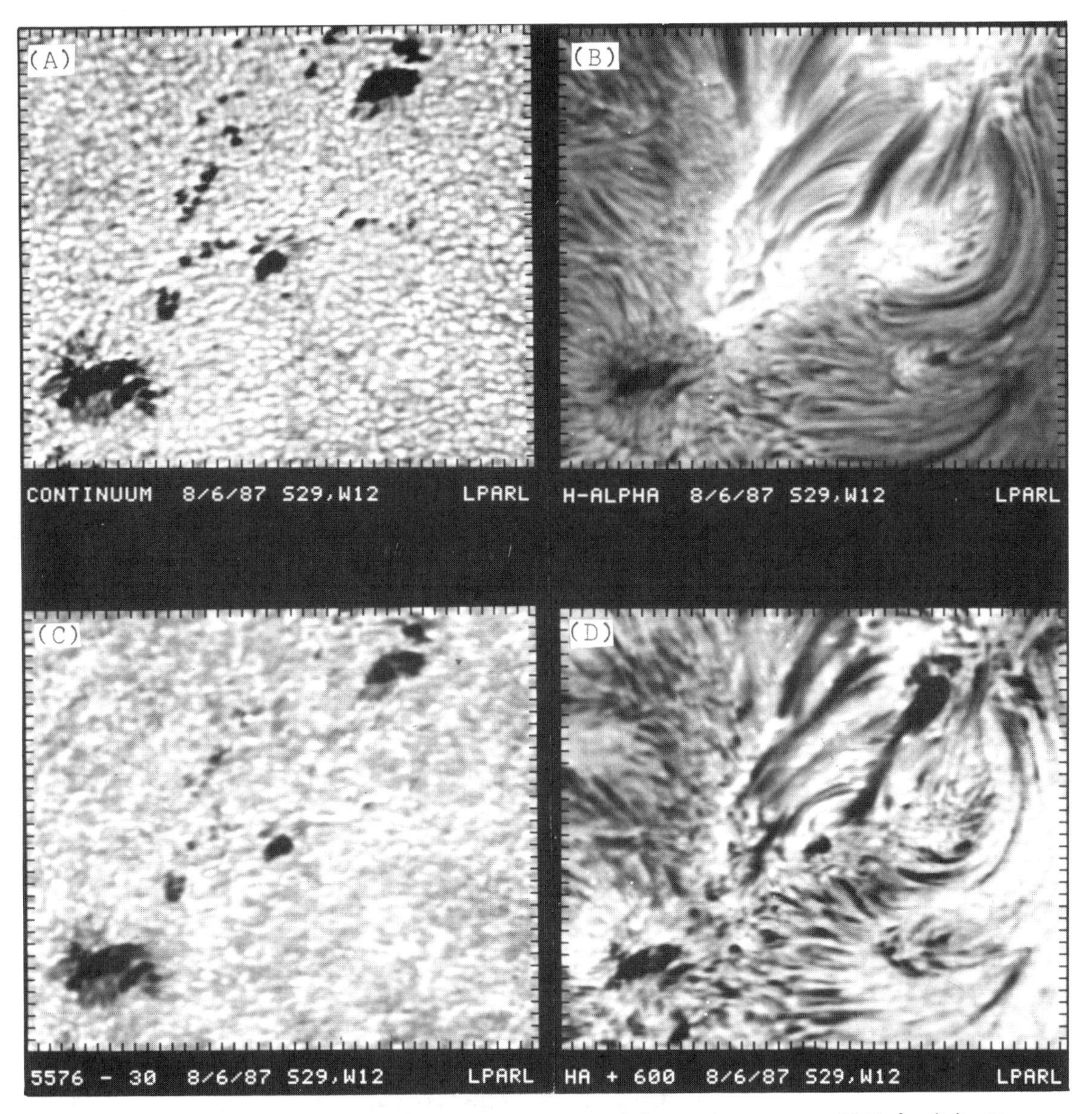

Fig. 1. Four CCD filtergrams from one time step: (a) continuum near 5576 Å, (b) H-alpha line center, (c) Fe I 5576 blue wing at -30 mA, and (d) H-alpha red wing at 600 mA. The tic marks are at 2 arc sec intervals.

wave power spectra between quiet Sun and an active region; systematic flows in small flux tubes and pores; and flux emergence events and other changes and their relation (if any) with the flare.

Three-dimensional Fourier analysis routines compute transforms from x, y, and t for movies up to approximately 256 x 256 x 200 pixels. A conventional power spectrum (k-ω or l-ν diagram) can be derived by suitable averaging over one spatial dimension. The three-dimensional transforms are also used to produce filtered movies. For example, a "subsonic" filtered movie is made by zeroing all Fourier components whose phase speeds (ω/k) exceed a chosen fraction of the sound speed and then transforming back to x, y, and t space. Alternatively, a movie of a particular wave mode can be made by zeroing all components which do not lie on the dispersion relation predicted for that mode. These filters, developed originally for Spacelab 2 data (Title et al., 1987), have proven very effective at separating different physical phenomena, such as f and p mode waves from turbulent convection. Unprocessed movies which are extremely chaotic because of the superposition of different effects often become comprehensible after properly designed filtering.

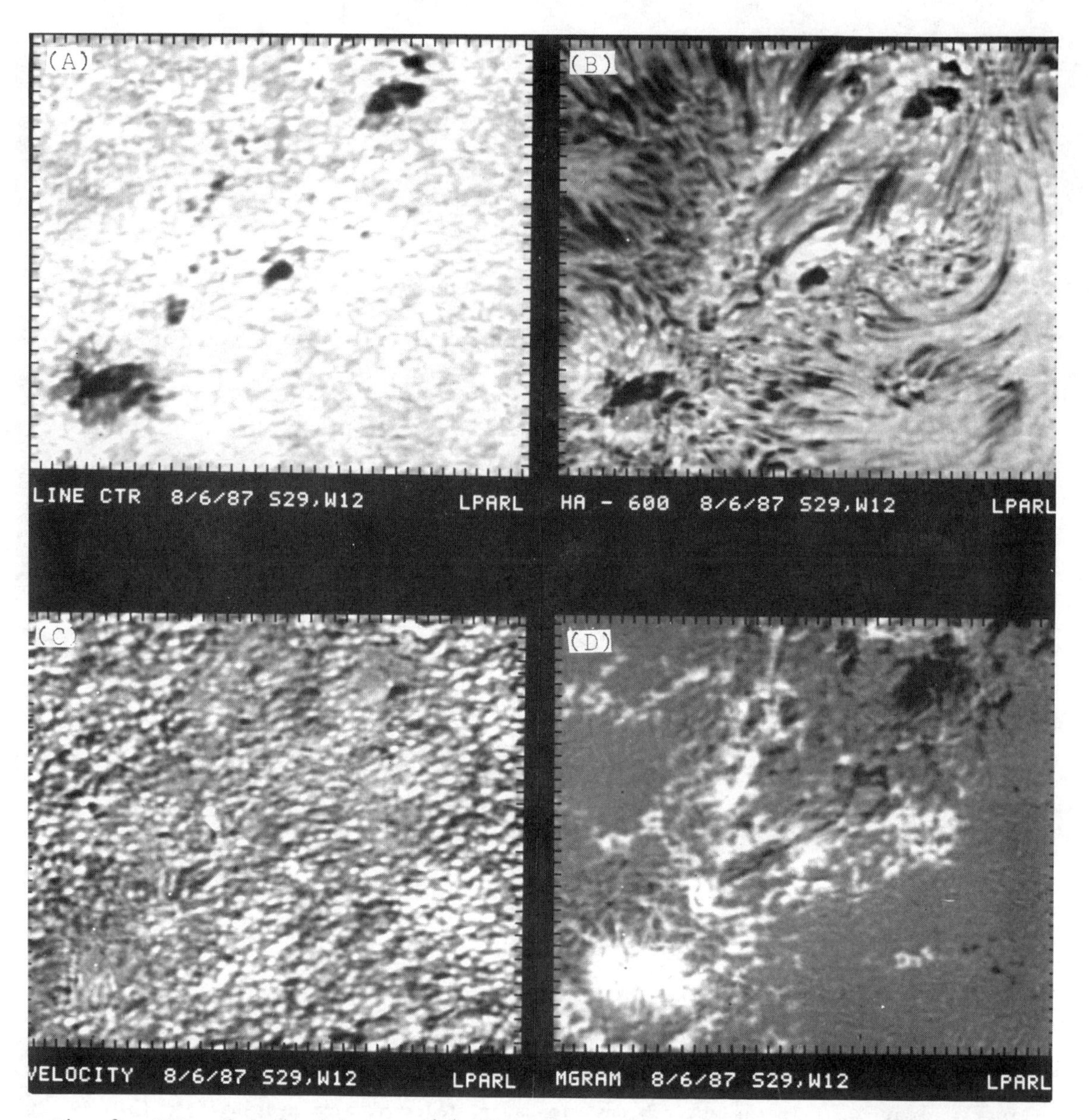

Fig. 2. CCD and derived images: (a) 5576 line center intensity, (b) H-alpha blue wing (-600 mA), (c) Doppler velocity in the magnetically insensitive line 5576, and (d) a longitudinal magnetogram (made in the blue wing of Fe I 6302). The tic marks are at 2 arc sec intervals.

Initial Fourier processing of the velocity data has shown that the 1-ν diagrams differ significantly in regions that are relatively field free and those with significant magnetic fields. Figure 3 shows 1-ν diagrams of: (a) the entire 90 x 90 arc sec region, (b) a 45 x 45 arc sec region which is relatively field free in the lower right-hand corner, and (c) a similar sized area in the center of the region which is full of magnetic field. It is clear from the figure that in the magnetic field regions all the wave amplitudes are strongly suppressed, and that in particular the f mode is almost completely absent.

Another major data analysis effort used a very high resolution continuum movie from the SSO. SSO collects its data using a real time, high-speed image sharpness monitor to control a commercial Sony CCD camera and image processor. The output of the system is a time sequence of digital images on tape. Working with Drs. Brandt, Kiepenheurer Insitut, and Scharmer, Royal Swedish Academy of Sciences, we processed 388 frames from raw data to finished movies on digital and video disk in a few weeks. The processing included flat field correction, rigid registration to remove telescope guiding errors, "destretching"

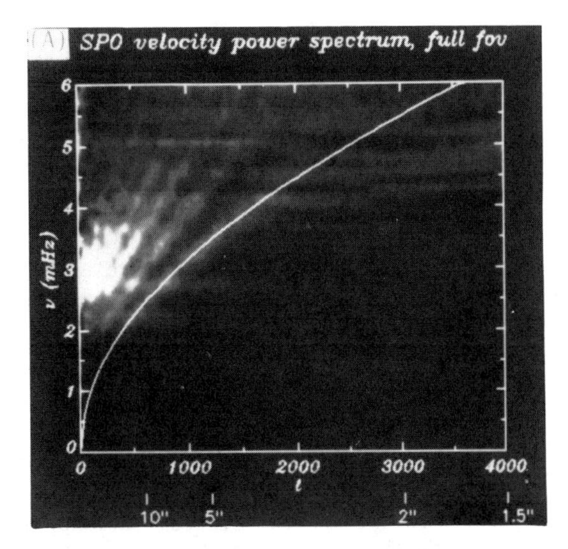

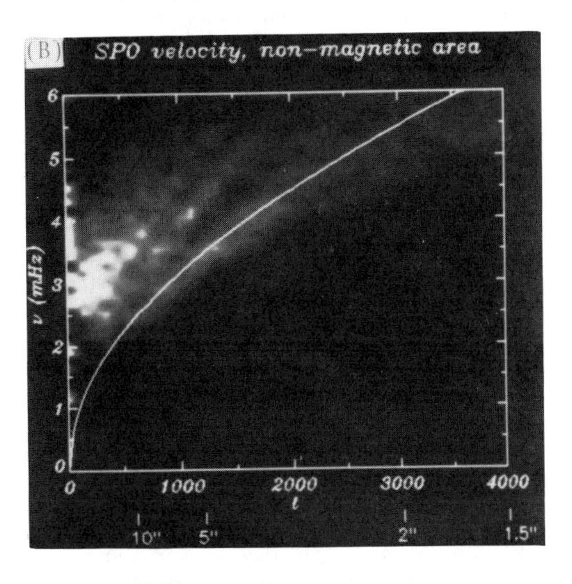

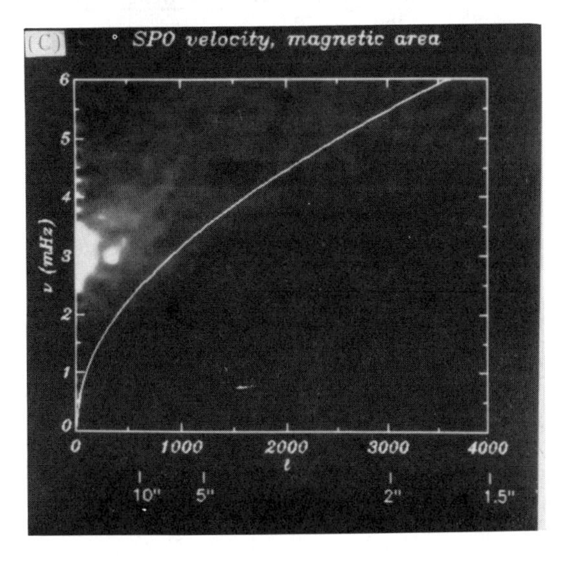

to remove distortion caused by seeing, editing of bad frames, and recording. The result is the sharpest solar movie in existence, with 0.3 arc sec resolution on the best frames, and few frames worse than 0.5 arc sec. Root-mean-square granulation intensity contrast ranges from 8.5 to 10.6%. The movie covers 79 min and has a field-of-view near disk center of 14.2 x 12.2 arc sec. Since mid-October, we have been using the software developed in previous years to study granulation, transverse flows, and waves on this movie.

Figure 4 shows (a) one frame of the movie and (b) a map of the average transverse velocity. The transverse velocity was measured by correlation tracking and was averaged over the entire 79-min interval for this display. Figure 4c shows a gray-scale image of the divergence of the velocity field, which is roughly proportional to the vertical velocity. Upflows (sources) are light and downflows (sinks) are dark. A long-lived, ordered flow field of the mesogranulation scale (5-7 arc sec) is evident. Speeds in the average velocity map range from 0.2 to 1.2 km s^{-1}; single frames of the velocity movie show speeds up to several kilometers per second in the flow of individual granules. The vortex in the upper left is a remarkable feature (Brandt et al., 1988), evident in the raw movies and stable over the entire time interval. Granules are seen to spiral around and disappear in the vortex center, resembling a solar "maelstrom." This is the clearest observation of such vortex motion in photospheric levels, although numerical simulations of granular convection have predicted similar features. The effect of such a flow on magnetic field lines may have interesting implications for coronal energy storage and heating.

Conclusions

Our current complement of digital and analog optical disks together with control software and a relatively small computer with an image display has been a remarkably effective tool for the analysis of a wide range and vast amount of solar data. We have also been able to transport the system to other institutions, both national and international. Hopefully, in the next few years we will be able to considerably increase the capabilities for analysis by incorporating special purpose processors into the computer system, which allow for both real time processing of image data and the production of multiple movie displays. We are convinced that the method of developing software and hardware for specific scientific tasks which we have been following is

Fig. 3. l-ν diagrams of: (a) the entire 90 x 90 arc sec region, (b) a 45 x 45 arc sec region which is relatively field free in the lower right-hand corner, and (c) a similar sized area in the center of the region which is full of magnetic field.

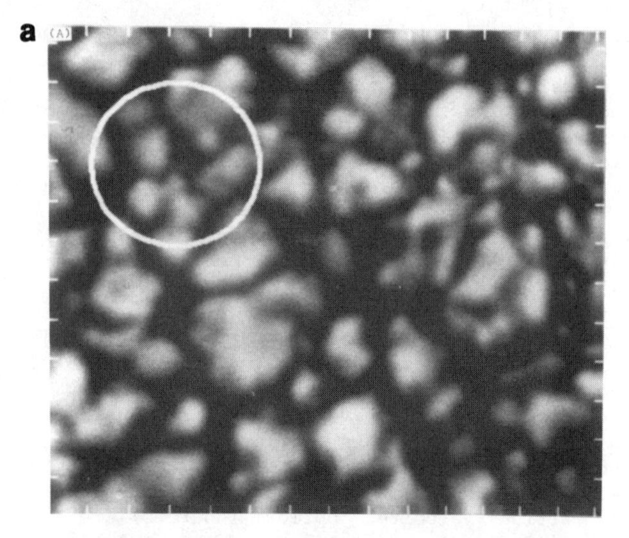

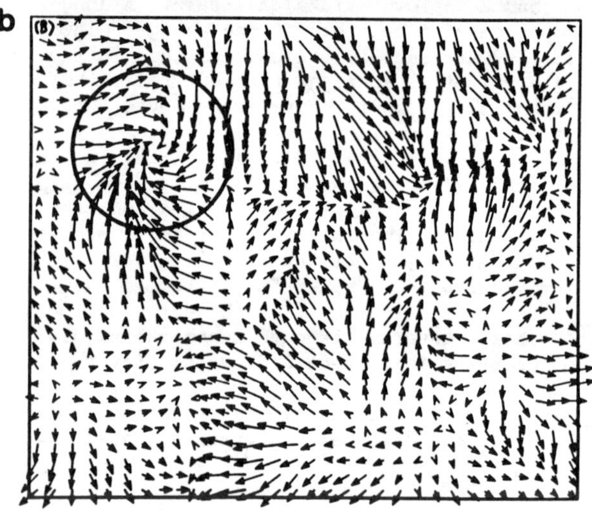

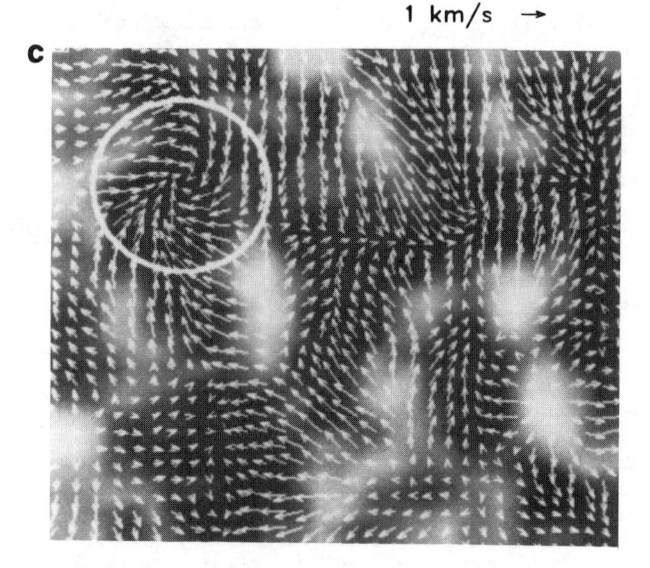

Fig. 4. Swedish Observatory data: (a) one frame of the movie, (b) a map of the average transverse velocity, and (c) a gray-scale image of the divergence of the velocity field, which is roughly proportional to the vertical velocity.

efficient. The original system was developed to handle data from Spacelab 2, but with relatively simple extensions it has been used to handle TF data and data from other ground-based observatories.

Acknowledgments. The development of optical disk image processing is supported by Lockheed Independent Research Funds. The data analysis procedures and science data analysis have been supported by NASA contracts NAS8-32805 and NAS5-26813. The scientific data were collected by the Tower Telescope of the National Solar Observatory (Sacramento Peak) and at the Swedish Solar Observatory at the Observatory del Roque de los Muchachos in the Canary Islands.

References

Brandt, P. N., G. B. Scharmer, S. Ferguson, R. A. Shine, T. D. Tarbell, and A. M. Title, Vortex flow in the solar photosphere, Nature, in press, 1988.

November, L. J., G. W. Simon, T. D. Tarbell, A. M. Title, and S. H. Ferguson, Large-scale horizontal flows from SOUP observations of solar granulation, in Theoretical Problems in High Resolution Solar Physics II, NASA Conf. Publ. 2483, edited by G. Athay and D. S. Spicer, p. 121, 1987.

Scharmer, G. B., D. S. Brown, L. Pettersson, and J. Rehn, Concepts for the Swedish 50-cm vacuum solar telescope, Appl. Opt., 24, 2558, 1985.

Title, A. M., T. D. Tarbell, G. W. Simon, and the SOUP Team, White light movies for the SOUP instrument of Spacelab 2, Adv. Space Res., 6(8), 253, 1986.

Title, A., T. Tarbell, and the SOUP Team, First results on quiet and magnetic granulation from SOUP, in Theoretical Problems in High Resolution Solar Physics II, NASA Conf. Publ. 2483, edited by G. Athay and D. S. Spicer, p. 55, 1987.

AN IMAGING VECTOR MAGNETOGRAPH FOR THE NEXT SOLAR MAXIMUM

Richard C. Canfield and Donald L. Mickey

Institute for Astronomy, University of Hawaii, Honolulu, HI 96822

Abstract. Measurements of the vector magnetic field in the sun's atmosphere with high spatial and temporal resolution over a large field-of-view are critical to understanding the nature and evolution of currents in active regions. Such measurements, when combined with the thermal and nonthermal x-ray images from the upcoming Solar-A mission, will reveal the large-scale relationship between these currents and sites of heating and particle acceleration in flaring coronal magnetic flux tubes. We describe the conceptual design of a new imaging vector magnetograph that combines a modest solar telescope with a rotating quarter-wave plate, an acousto-optic tunable prefilter as a blocker for a servo-controlled Fabry-Perot etalon, charge-coupled device cameras, and a rapid digital tape recorder. Its high spatial resolution (0.5 arc sec pixel size) over a large field-of-view (4 by 5 arc min) is expected to be sufficient to significantly measure, for the first time, the magnetic energy dissipated in major solar flares. Its millisecond tunability and wide spectral range (5000-7000 Å) enable nearly simultaneous vector magnetic field measurements in the gas-pressure-dominated photosphere and magnetically-dominated, chromosphere, as well as effective co-alignment with Solar-A's x-ray images.

Objectives

Flare Research at the Next Maximum

In the Max '91 report, Flare Research at the Next Solar Maximum, six fundamental questions identify the highest priority flare research during solar cycle 22:

1. How and where is flare energy stored?
2. What causes flare energy release?
3. What are the mechanisms of flare energy release?
4. How are energetic particles accelerated?
5. What are the mechanisms of flare energy transport?
6. How do flare effects propagate to Earth?

None of these questions can be answered in a satisfactory manner without vector magnetic field measurements. Such observations are as important to understanding solar flares as are γ-ray, x-ray, or microwave observations. The collective research experience during the last two solar cycles has shown that only through coordinated multi-spectral sets of observations can we hope to understand how and why solar flares occur.

Magnetic Structures, X-Ray Structures, and Particle Acceleration

Although large energetic flares did not occur during the Skylab mission, Skylab observations and the subsequent analysis provided important clues regarding the origin of flares (Sturrock, 1980). The soft x-ray telescope observations clearly showed the existence of magnetic loops in the corona and their importance for solar flares. Canfield et al. (1974) and Heyvaerts et al. (1977) identified flare kernels with emerging flux regions (EFRs), and suggested that flares begin in neutral sheets between EFRs and larger loop structures. Spicer (1976) called attention to the importance of mode coupling and the possibility that flares occur in a single current-carrying sheared loop. Many other equally plausible models, too numerous to mention, have been put forth, and are reviewed in Sturrock (1980) and Kundu and Woodgate (1986).

It is presently unclear what magnetic field conditions are both necessary and sufficient for flare occurrence. The emerging flux model, for example, has received considerable observational support. However, Martin, et al. (1985), presented contradictory evidence that clearly indicates a spatial and temporal relationship between canceling (not emerging) magnetic flux and solar flares. During the Flare Buildup Study of the Solar Maximum Analysis, Hagyard and colleagues at

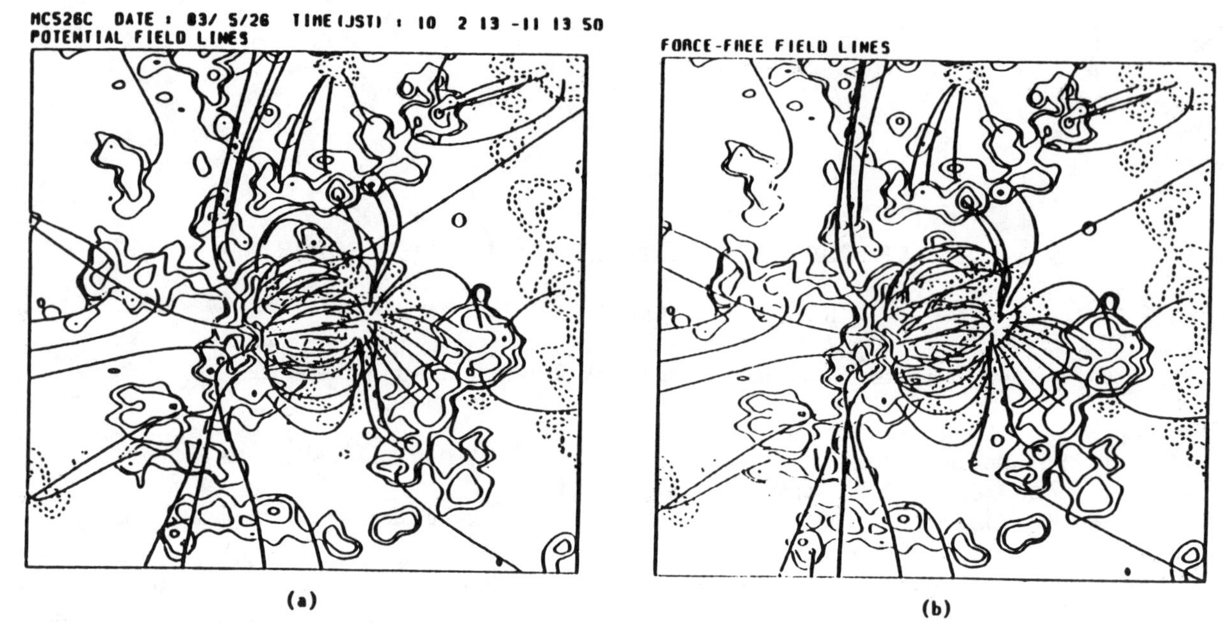

Fig. 1. Potential field lines (a) and force-free field lines (b). Potential field lines are calculated from the observed longitudinal magnetic field, while force-free field lines are computed using the distribution of electric currents. Field lines in these two figures have the same footpoints at one end of the field lines (from Sakurai and Makita, 1986).

the Marshall Space Flight Center (MSFC) (see, e.g., Hagyard and Rabin, 1986) showed convincingly that photospheric magnetic shear is a necessary, but not sufficient, precondition for flares. There is clearly some other condition, in addition to photospheric shear, that is necessary for flare occurrence; Hagyard and Rabin suggest that mass motions are probably involved.

Vector magnetograms provide photospheric vector magnetic field data for calculation of the potential and force-free coronal magnetic field configurations (Sakurai, 1979, 1981, 1982.) Figure 1, from Sakurai and Makita (1986), compares potential field lines (a) and force-free field lines (b) in an active region observed in 1983. For this example the magnetic free energy (associated with chromospheric and coronal currents) is calculated by Sakurai and Makita to be only 4×10^{29} erg, while the energy of the potential field (associated with currents below the photosphere) is 7.5×10^{31} ergs. Therefore the difference between the potential and force-free field morphologies in Figures 1a and b respectively, is small. Such calculations of the pre-flare coronal magnetic field can be compared with soft x-ray images obtained by means of the soft x-ray telescope instrument on Solar-A. This comparison may show that the observations cannot even be modeled in terms of a force-free field, in which case it will be very interesting to consider the locations of the implied current sheets.

It is interesting to speculate on how the role of flux emergence in the development of such current sheets could be explored using vector magnetic observations with good time resolution. If we follow the evolution of magnetic fields and assume $J \times B = 0$ initially, it should be possible to determine connectivity at a later time. Assuming no reconnection, and using our ability to track individual footpoints, we can infer the presence of current sheets, or at least thin high density currents. Knowing the connectivity and the topology, positions of current sheets can be inferred.

Two specific analytic examples of current sheet formation by continuous footpoint displacement have recently been derived by Low (1987). Low and Wolfson (1988) have shown that such current sheets can be finite in length, running along the separatrix lines of force that separate different flux systems (Figure 2). If such current sheets are locations of energetic electron acceleration and heating, their positions should be consistent with hard x-ray emission sites and the inferred connectivity, and they should be part of bright soft x-ray loops.

Craig and Sneyd (1986) and McClymont and Craig

(1987) have developed fully three-dimensional numerical methods for constructing force-free equilibria and for examining their stability properties. As a complement and extension to the work of Sakurai (1979, 1981, 1982), we expect to develop the capability for modeling the magnetic field of an active region with the goal of using this capability to compare the magnetic-field structure of selected active regions with the properties of flares that occur in those regions (see, e.g., Sakurai and Makita, 1986; Gary et al., 1987).

Many questions remain about the large-scale magnetic field morphology of flares and how it changes as flares take place. Comparison of three-dimensional coronal magnetic field calculations with the pre-flare and post-flare configurations revealed by soft x-ray images and Hα images will provide valuable insight into the restructuring of the magnetic field during a solar flares. Working with data from Big Bear Solar Observatory, LaBonte (1987) has recently provided interesting Hα evidence that this restructuring at a given location occurs only after the arrival of a flare ribbon, often long after the flare starts. The addition of vector magnetic field measurements in the photosphere and coronal x-ray images will clarify the meaning of this interesting result.

Magnetic Free Energy and Flares

Although it has never been shown directly, the energy released in a major solar flare ($\sim 10^{32}$ erg) is generally believed to be stored in electrical currents and nonpotential magnetic fields in the corona. The observational and theoretical understanding of this magnetic energy storage and conversion will be an important aspect of flare research at the next maximum. The basic theoretical ideas have been laid out by Low (1985), who has provided a necessary condition for an observed vector magnetic field to be force free, and shown how to measure its free magnetic energy content if it is force free. Low shows that the net Lorentz force in the coronal volume z > 0 above the boundary of the force-free region (z = 0) is zero if its three components vanish

$$\frac{1}{4\pi}\int\int_{z=0} B_x B_z dx dy = \frac{1}{4\pi}\int\int_{z=0} B_y B_z dx dy = \qquad (1)$$

$$\frac{1}{4\pi}\int\int_{z=0} (B_z^2 - B_x^2 - B_y^2) dx dy = 0$$

Of course, measurement limitations preclude satisfying equation (1) exactly, but the characteristic magnitude of the total Lorentz force exerted on the coronal volume if the atmosphere is not force free is given by

$$F_0 = \frac{1}{8\pi}\int_{z=0} (B_x^2 + B_y^2 + B_z^2) dx dy \qquad (2)$$

Hence a necessary condition for the observed field to be force free is that each of the integrals in equation (1) be much less than F_0.

Deep in the photosphere magnetic pressure does not dominate gas pressure, and it is therefore not likely that all photospheric magnetic fields are force free. On the other hand, in the chromosphere we expect active region fields to be force free. This motivates the development of the ability to measure chromospheric vector fields (see below).

Low (1985) has called attention to a force-free field relationship derived originally by Chandrasekhar (1961)

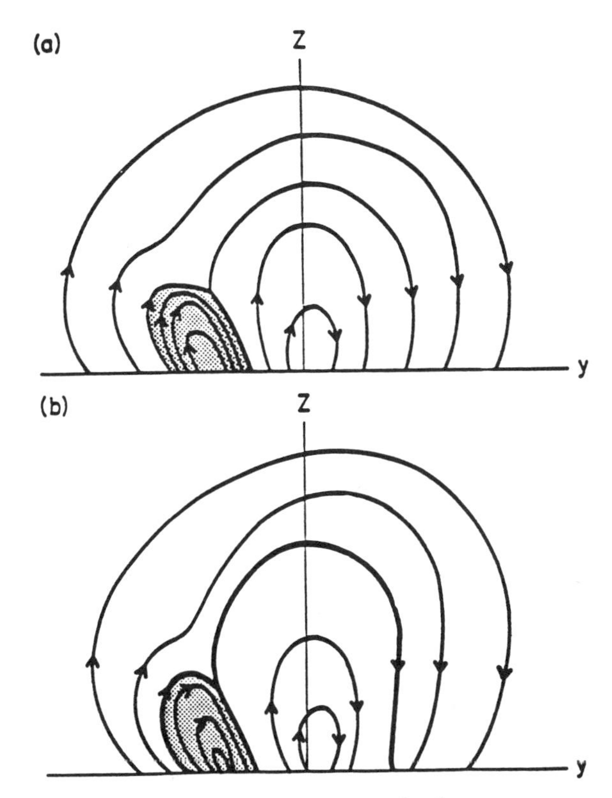

Fig. 2. Current sheet formation in the flux-emergence scenario. In Figure 2a, the shaded emergent flux and the background field are potential and a current sheet had formed where opposite fields come into contact. In Figure 2b, the emergent flux is highly twisted whereas the background field is potential. In addition to the current sheet formed at the line of contact between the two fields, there is also the current sheet formed along the separatrix line (from Low and Wolfson, 1988).

that gives the total magnetic energy in z > 0 in terms of the three components of the field at z = 0

$$\int\int\int_{z\geq o} \frac{B^2}{8\pi} dxdydz = \tag{3}$$

$$\frac{1}{4\pi}\int\int_{z=0} (xB_x + yB_y)B_z dxdy$$

The difference between the energy given by equation (3) and the energy contained in the potential field B_p, viz.

$$\int\int\int_{z\geq o} \frac{B_p^2}{8\pi} dxdydz = \tag{4}$$

$$\frac{1}{4\pi}\int\int_{z=0} (xB_{px} + yB_{py})B_z dxdy$$

is the free magnetic energy, which is believed to be the energy source for flares and active regions.

The Marshall and Tokyo groups have used these methods to measure the free magnetic energy in an active region that showed nonpotential magnetic features. Gary et al. (1987) found the total free energy in that active region to be of order 10^{32} erg, which was about 3σ above the measured noise level. As noted above, Sakurai and Makita (1986) measured the free energy in the active region in Figure 1 to be 4×10^{29} ergs, but did not estimate the uncertainty of the measurement. Moore, et al. (1987) showed that the uncertainty ΔE of a total energy measurement is related to the characteristic magnetic field strength B (of single measurement uncertainty ΔB) and characteristic active region dimension L by

$$\Delta E \approx \frac{1}{4\pi} B \, \Delta B \, \frac{L^3}{\sqrt{N}} \tag{5}$$

where N is the number of independent measurements contributing to the observational estimate of the integral. Equation (5) implies that if $B \approx 1000\ G$, $\Delta B \approx 150\ G$, $L \approx 4 \times 10^9$ cm, and $N \approx 400$, then $\Delta E \approx 10^{32}$ erg. For observational proof that magnetic free energy is liberated in a large flare this is not a sufficiently accurate measurement. Equation (5) shows that it can be improved by increasing the accuracy of an individual magnetic field measurement (through improved polarimetric instrumentation or analysis methods) or increasing the number of independent measurements (through enhanced spatial resolution).

From the Marshall result, it appears that an improvement of about an order of magnitude in the accuracy of magnetic free energy measurement will be necessary to enable the first convincing observational proof that the energy released in solar flares is derived from magnetic fields. Such improvements motivate both better ground-based vector magnetographs and their modification to make them suitable for space.

Photospheric Currents, X-Ray Structures, and Particle Acceleration

An exciting and valuable avenue of vector magnetic field research at the next solar maximum will be the measurement of photospheric currents in active regions. One can derive the electric current density **J** from

$$\mathbf{J} = \frac{c}{4\pi}\nabla \times \mathbf{B} \tag{6}$$

Such calculations have been made by both the Tokyo group (Sakurai and Makita, 1986) and Marshall group (e.g., Krall et al., 1982). The Marshall group has computed maps of longitudinal electrical current density in an interesting solar active region observed on April 6, 1980 (see, e.g., Hagyard et al., 1985; Hagyard, 1988). First, Hagyard et al. (1985) reported an approximate correspondence of peak current densities with bright Hα kernels of a flare observed several hours before the vector magnetogram was obtained. Subsequently Lin and Gaizauskas (1987) showed a closer spatial coincidence (to within the 2 arc sec limit of spatial resolution) between peak observed longitudinal current densities in the Marshall maps and the Hα kernels of a flare that occurred while the magnetogram was being obtained. Figure 3 shows both the map of current density (from Hagyard, 1988) and the off-band Hα image (Lin and Gaizauskas, 1987) obtained nearly simultaneously in April 1980. In the off-band Hα bright kernels coincide very closely with peaks in J_z. Recently Hagyard (1988) calculated the source field, i.e., the magnetic field produced in the photosphere by currents above the photosphere and their mirror currents below, and showed that the source field strongly resembled the field produced by two arcades of current-carrying loops crossing the magnetic inversion line.

Such interesting observations and analyses need to be carried out for a variety of flares during the coming solar maximum and compared to Solar-A soft x-ray and hard x-ray images. Comparisons such as the one shown in Figure 3 are ambiguous because the Hα data are ambiguous. The cause of the off-band Hα brightening is uncertain because the off-band response in a sunspot umbra must be different from a penumbra, which in turn is different from a plage. When Solar-A soft x-ray images are available along with J_z maps, they will show much more clearly where the corona is being

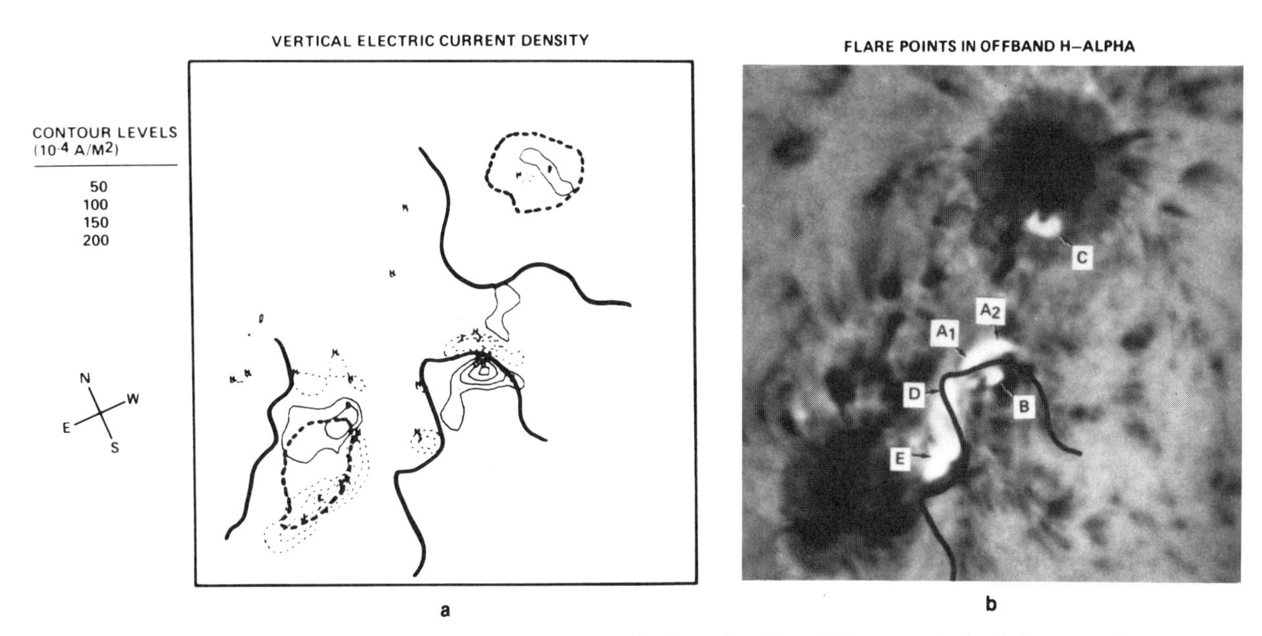

Fig. 3. Comparison of vertical currents (left) and off-band Hα images (right) in April 1980 (from Hagyard, 1988, and Lin and Gaizauskas, 1987).

heated than Hα does. When Solar-A hard x-ray images are available along with J_z maps, it will become much more clear where energetic electrons are precipitating than is the case when using off-band Hα.

The Imaging Vector Magnetograph

Magnetographs Versus Polarimeters

The instrument discussed here is an imaging vector magnetograph, not a Stokes polarimeter; its design tradeoffs favor imaging, not spectroscopy. In an imaging vector magnetograph, narrow-band images are obtained in each of the four Stokes parameters that characterize the polarized light from which vector magnetic fields are inferred. The images are obtained simultaneously over a field-of-view as large as an entire active region field system. Samples at all parts of the spectral line profile of a magnetically sensitive transition are not obtained simultaneously. Rather, they are obtained sequentially by tuning a narrow-band filter. The MSFC magnetograph (Hagyard et al., 1982) is the best example of an imaging vector magnetograph. In a Stokes polarimeter fully critically sampled spectral line profiles are obtained in each of the Stokes parameters. All spectral points on the line profile in each Stokes parameter are obtained simultaneously. Samples at all points of the solar active region are not obtained simultaneously, but sequentially by spatial rastering or scanning. Our own Haleakala Stokes Polarimeter (Mickey, 1985) is the best example of a Stokes polarimeter. What follows is a description of the expected performance characteristics of the imaging vector magnetograph, together with a sketch of the preliminary design.

Performance Characteristics

The imaging vector magnetograph, to be installed at Mees Solar Observatory (MSO) of the University of Hawaii, will have the following performance characteristics:

Spatial resolution: 1 arc sec. Detector pixel spacing of approximately 0.5 arc sec over a 4 x 5 arc min field-of-view. This high resolution will critically sample the high quality image typical at MSO early in the day.

Spectral resolution: 70 mÅ at 6000 Å. Tests using full line profiles from the Haleakala Stokes Polarimeter show that measurement at three positions within a spectral line with this passband minimizes distortion due to limited spectral sampling of the derived magnetic field.

Spectral range: 5000-7000 Å. This range includes both photospheric (e.g., Fe I λ6302) and chromospheric (e.g., Mg I λ5173) lines whose use for vector magnetic field measurement is well understood.

Temporal resolution: A complete magnetogram in a single line in 90 s. This resolution is determined primarily by the data recording speed; better resolution can be achieved over a smaller field-of-view.

Sensitivity: 10 Gauss longitudinal fields and 200 Gauss transverse fields in a few seconds. Simultaneous velocity measurements to 10 m s^{-1}. Temporal resolution can be traded for increased sensitivity.

Co-alignment: A simultaneous photospheric white-light image of the full field-of-view, for precise co-alignment with Solar-A high energy events and SXT images.

Telescope

The Imaging Vector Magnetograph will be built as a 20-cm refractor, very similar in size and configuration to the Haleakala Stokes Polarimeter, mounted on the MSO spar. The telescope and magnetograph portions of the instrument will both use on-axis design to eliminate instrumental polarization. Figure 4 shows the optical configuration of the magnetograph portion. The f/10 telescope objective forms the primary image at at the left side of the figure. At this point a rectangular field stop on the telescope axis selects the field to be observed and limits the light and heat entering the magnetograph portion of the instrument. The entire telescope is gimbal mounted, as the Haleakala Stokes Polarimeter is, so that it may be pointed at any region on the solar disk. A telecentric optical system produces an enlarged secondary image of the selected field, with an image of the entrance aperture near the rotating quarter-wave plate. Near the first lens of the magnifying optics there will be provision for insertion of calibration optics (not shown) and a lens which will form an image of the entrance aperture onto the detectors, as indicated by the dotted lines. This feature will enable flat-field calibration of the cameras. After the second lens of the magnifying optics the effective telescope aperture is reduced to f/125 in order to minimize broadening of the Fabry-Perot (FP) passband. Near the secondary image are located a broad-band interference filter, the rotating quarter-wave plate, and the FP filter. The interference filter protects the FP filter from excess heat. Since each part of the image has essentially the same view of the objective, the FP transmission profile does not change over the field. Such invariance is essential for minimizing the amount of processing needed for the conversion from Stokes-parameter images to magnetograms.

A reimaging lens pair then reduces the image scale to match the dimensions of the charge-coupled device (CCD) detectors, and also reduces the beam diameter to a size appropriate for the acousto-optic tunable filter (AOTF). The properties of AOTFs and the current state of AOTF technology are discussed in Katzka (1987). As shown in Figure 5, the AOTF divides the

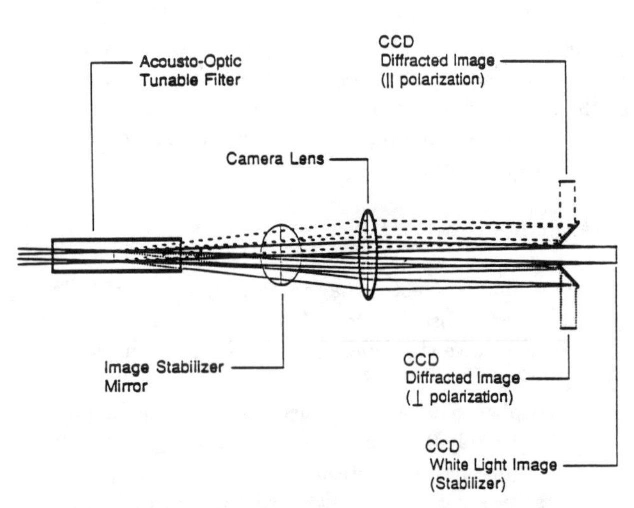

Fig. 4. The modulator section of the Imaging Vector Magnetograph. Vertical exaggeration is 10:1.

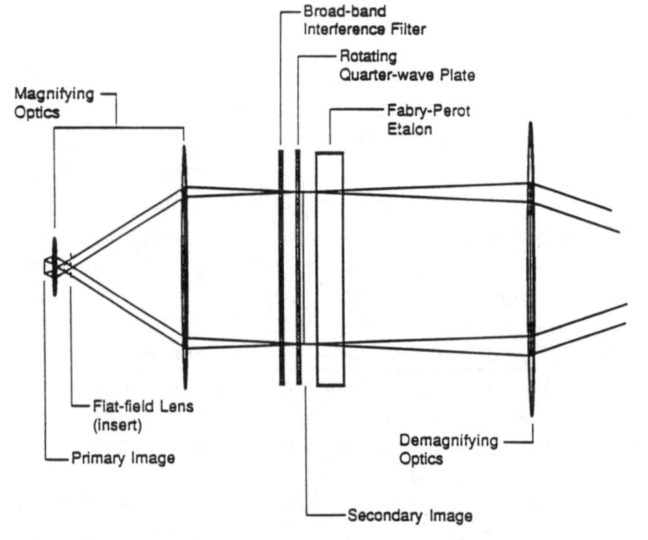

Fig. 5. The analyzer section of the Imaging Vector Magnetograph.

beam into three components: two orthogonally polarized beams containing wavelengths within the AOTF bandpass of 2 Å, which are each diffracted through an angle of about 5°, and an undiffracted beam, which contains all other wavelengths passed by the interference-filter/FP combination. All three beams are then reflected off an active mirror, which is used to provide image motion compensation, and are then imaged onto separate cameras. The undiffracted beam provides an image of the field being observed, which is used by the control computer to derive the control signals for the active mirror.

Monochromator

For accurate determination of solar vector magnetic fields and radial velocities it is necessary to observe with high spectral resolution at multiple points in each line profile. To provide rapid and flexible tunability and high spectral resolution we will use a servo-controlled FP etalon as the primary spectral element. Such etalons are produced commercially with realizable finesse of 50 or greater, stability of plate spacing to $\lambda/2000$, and millisecond tunability to any wavelength in a range of 3000 Å. The servo controller which maintains the plate spacing can be externally controlled to rapidly and accurately retune the etalon to a different wavelength.

An important question is the extent to which the transmission profile of the FP etalon is broadened by off-axis rays. Lites and Skumanich (1985) have shown that vector magnetic field measurements are sensitive to filter bandpass. The etalon bandpass will be approximately 50 mÅ for rays parallel to the optical axis. Tests we have done on Haleakala Stokes Polarimeter data using integral techniques described by Ronan et al. (1987) indicate that significant distortions of derived vector magnetic field parameters, particularly the azimuth angle, set in at bandpass values greater than about 70 mÅ. For this reason the FP etalon in this instrument is illuminated by an f/125 beam, which limits the bandpass broadening to about 20 mÅ, and enables a system spectral resolution of 70 mÅ.

To select the desired order of the etalon we will use an AOTF. This filter will have a passband of 2 Å, a contrast of 1000:1 and a large field-of-view. It will also be rapidly tunable over a wide wavelength range. AOTFs use a high-frequency acoustic wave to cause strain-induced index of refraction variations in a crystal, which diffract a light beam. In a birefringent crystal the diffraction exchanges the sense of linear polarization, and thus the direction of propagation; conservation of momentum requires that the sum of the acoustic and incident light wave vectors equal the wave vector of the orthogonally polarized diffracted wave. A single acoustic frequency diffracts a single light frequency. The ratio of frequencies is just the ratio of propagation speeds times the difference between the ordinary and extraordinary indices of refraction, giving acoustic frequencies of tens to hundreds of megahertz for optical filters. The acoustic waves are generated with piezo-electric transducers. The frequency of the AOTF can be switched in a few milliseconds, over a range of an octave.

The combination of an AOTF and a servo-controlled FP etalon has been proven in use at several different ground-based telescopes (including Hawaii) for solar, planetary, and cometary astronomy (Smith et al., 1987).

Polarization Analyzer

Our analyzer design is deliberately conservative. We will use a rotating quarter-wave retarder as the modulator in this system. We use this type of modulator in the Haleakala Stokes Polarimeter. All modulators have their problems, but our experience indicates that for this instrument the drawbacks of quarter-wave plates can be managed more easily than those of alternatives such as KD*P or piezo-optic modulators. .pp The AOTF will double as the analyzer: light of the proper frequency is diffracted about 5° to either side of the undiffracted beam, according to its state of linear polarization. Thus each of the diffracted beams will be intensity modulated (in the same way, but with opposite phase) according to the polarization state of the incoming light. Since this telescope is always used on-axis and contains no crystal optics other than the retarder and AOTF, our experience with the on-axis Haleakala Stokes Polarimeter indicates that instrumental polarization will be negligible. Spurious polarization signals would result from variations in timing of the retarder rotation or the camera read cycle; therefore such timing will be carefully controlled.

Detectors

A pair of high-resolution industrial CCD cameras will be used to record the modulated images. Each of the diffracted beams emerging from the AOTF will be imaged onto one of these cameras. No mechanical shutter is necessary; turning off the radio frequency signal to the AOTF turns off the diffracted beams imaged on the cameras. A third CCD camera will be used with the undiffracted broad-band image for purposes of image motion compensation. The 754 x 488 pixel detector arrays will permit observations of a 5.8 x 4.6 arc min field with pixel size of approximately 0.5 arc sec. The cameras will be clocked at approximately standard RS170 video rates, since the light level will be adequate to saturate the detectors in one 33 msec frame time. The video outputs from these cameras will be digitized to 8-bit precision, using variable-format digitizer boards in the host computer.

Data Acquisition

A 68020-based computer in a VME-bus chassis will be used to control filter wavelength settings, maintain synchronization between the retarder rotation and the cameras, provide overall sequencing and operator control functions, and store the data to tape. This software design builds on our experience with the Haleakala Stokes Polarimeter; the major difference is the use of dedicated image processing board sets to digitize and demodulate the video signals from the cameras and to provide real-time reduction to magnetic field, intensity, and velocity images.

A minimum modulation sequence consists of a half-rotation of the wave plate, during which the CCD cameras are read eight times. For each of these steps, the raw data frames will be combined into sum and difference frames, then added or subtracted into storage arrays according to the step in the modulation cycle. At this stage, if observing conditions do not justify use of the full instrumental resolution, binning can be done on the raw data to produce a larger effective pixel size. Four storage frames will be maintained, one for each Stokes parameter. Longer integrations may be obtained by accumulating for multiples of the basic modulation cycle. To derive magnetic field and velocity information, we will obtain Stokes frames at each of three positions in each spectral line to be observed. The Stokes parameter data can be combined in a straightforward way, similar to the reduction method we have used with Haleakala Polarimeter data (Ronan et al., 1987), to produce vector magnetic field (three images), intensity, and velocity maps with good accuracy in real time.

These five images will then be written to magnetic tape. Each image set will require 3.7 Mbyte of storage at the full instrumental resolution. We will use digital video tape as a storage medium. We will use the same Exabyte digital video recorders that we are presently integrating into other CCD systems for MSO. An image set can be written to tape in approximately 12 s, which is adequate for our planned observing cadence. Each cassette, which is about the size of a deck of cards, will hold 2 Gbyte of data, or about 500 image sets.

Our experience at Haleakala has shown us the value of establishing a simple basic mode of operation. We anticipate that during Solar-A and Max '91 this standard mode will be to observe two spectral lines in rapid sequence, in order to obtain information on the dependence of the vector magnetic field and electrical currents on height in the solar atmosphere. On the basis of the modeling that has been done by Lites, Skumanich, and collaborators, these should be the chromospheric Mg I 5173 and photospheric Fe I 6302 lines. These two lines can be observed in a span of 10 s, with the data held in memory until they can be written to tape (about 30 s).

During campaigns it may be valuable to add to the standard Solar-A and Max '91 observing mode certain special investigations. For example, it would be of considerable interest to observe a pair of related lines such as Mg I 5173/5184 or Fe I 6301/6302. Lites et al. (1988) have shown for the magnesium lines that a least-squares fit to both lines of the multiplet significantly reduces the error in the derived field. It seems likely that an integral reduction scheme such as we expect to use could also benefit from a two-line observation. Also, it would be valuable to make more detailed spectroscopic studies of slowly varying phenomena. The spectral resolution of this instrument is adequate to produce full Stokes profiles with reasonable accuracy, and its spatial resolution is superior to that of any existing Stokes polarimeter.

Analysis and Archiving

Off-line analysis of the magnetograph data will be done on an existing Sun workstation at our Honolulu facility. This machine has the computational capability needed for coordinate conversions, display, and analysis of these relatively large images. We will add additional peripherals in order to improve our ability to quickly and frequently access and display the data and exchange magnetograms with interested co-investigators and collaborators. It is important to be able to display the data as high quality hard copies, since some users will not want data in digital form, and high quality hard copies are necessary for publication. Experience shows that certain interesting datasets are accessed over and over, even during a single investigation. A digital optical disk drive will permit saving such datasets in a permanent, yet readily (randomly) accessible form. Studies of flux emergence and its relation to current sheets and high-energy particle acceleration require the ability to display successive magnetograms as movies. As shown by the work on photospheric velocity and magnetic fields done at Sacramento Peak by the Lockheed group, a video disk recorder is necessary to bring out the time dependences that are central to flux emergence and cancellation.

Given the digital nature of the raw data, and the facilities required for analysis, the medium of choice for data storage for up to a decade is the original 8 mm video cassette. Two years worth of daily vector magnetograms can be stored on a single cassette. For storage for more than a decade, the medium of choice is an optical disk. Optical disks are also likely to be a convenient medium of data exchange with interested co-investigators and collaborators, where large numbers of magnetograms are involved. Compact disks are already in regular use in distribution of large geophysical datasets by the World Data Center A in Boulder.

Acknowledgments. The work was supported by NASA grant NGL-12-001-011 to the University of Hawaii.

References

Canfield, R. C., E. R. Priest, and D. M. Rust, A model for the solar flare, in *Flare Related Magnetic Field Dynamics*, edited by D. M. Rust and Y. Nakagawa, NCAR, p. 361, 1974.

Chandrasekhar, S., *Hydrodynamic and Hydromagnetic Stability*, Oxford Univ. Press, 1961.

Craig, I. J. D., and A. D. Sneyd, A dynamic relaxation technique for determining the structure and stability of coronal magnetic fields, *Ap. J.*, *311*, 451-459, 1986.

Gary, G. A., R. L. Moore, M. J. Hagyard, and B. M. Haisch, Nonpotential features observed in the magnetic field of an active region, *Ap. J.*, *314*, 782-794, 1987.

Hagyard, M. J., Observed nonpotential magnetic fields and the inferred flow of electric currents at a location of repeated flaring, *Solar Phys.*, in press, 1988.

Hagyard, M. J., and D. M. Rabin, Measurement and interpretation of magnetic shear in solar active regions, *Adv. Space Res.*, *6*, (6), 7-16, 1986.

Hagyard, M. J., E. A. West, and J. B. Smith, Jr., Electric currents in active regions, in *Proceedings, Kunming Workshop on Solar Physics and Interplanetary Travelling Phenomena*, edited by Cornelis de Jager and Chen Biao, vol. 1, pp. 179-188, Science Press, Beijing, China, 1985.

Hagyard M. J., N. P. Cumings, E. A. West, and J. E. Smith, The MSFC Vector Magnetograph, *Solar Phys.*, *80*, 31-51, 1982.

Heyvaerts, J., E. R. Priest, and D. M. Rust, An emerging flux model for the solar phenomenon, *Ap. J.*, *216*, 123, 1977.

Katzka, P., Acousto-optic, electro-optic, and magneto-optic devices and applications, AOTF Overview: Past, Present, and Future, *Proc. SPIE*, *753*, 22, 1987.

Krall, K. R., J. B. Smith, M. J. Hagyard, E. A. West, and N. P. Cumings, Vector magnetic field evolution and associated photospheric velocity shear within a flare-productive active region, *Solar Phys.*, *79*, 59-75, 1982.

Kundu, M. R., and B. E. Woodgate, (editors), *Energetic Phenomena on the Sun, NASA Conf. Publ. 2439,* NASA, Washington, DC., 1986.

LaBonte, B. J., Flare-induced magnetic field changes in the chromosphere, *Solar Phys.*, *113*, 285-288, 1987.

Lin, Y, and V. Gaizauskas, Coincidence between Hα flare kernels and peaks of observed longitudinal electric current densities, *Solar Phys.*, *109*, 81, 1987.

Lites, B. W., and A. P. Skumanich, The inference of vector magnetic fields from polarization measurements with limited spectral resolution, in *Measurements of Solar Vector Magnetic Fields, NASA Conf. Publ. 2374*, edited by M. J. Hagyard, pp. 342-367, NASA, Washington, DC, 1985.

Lites, B. W., A. Skumanich, D. E. Rees, and G. A. Murphy, Stokes profile analysis and vector magnetic fields. IV. Synthesis and inversion of the chromospheric Mg I b lines, *Ap. J.*, *330*, 493-512, 1988.

Low, B. C., Modeling solar magnetic structures, in *Measurements of Solar Vector Magnetic Fields, NASA Conf. Publ. 2374*, edited by M. J. Hagyard, pp. 49-65, NASA, Washington, DC, 1985.

Low, B. C., Electric current sheet formation in a magnetic field induced by continuous magnetic footpoint displacements, *Ap. J.*, *323*, 358-367, 1987.

Low, B. C., and R. Wolfson, Spontaneous formation of electron current sheets and the origin of solar flares, *Ap. J.*, *324*, 574-581, 1988.

Martin, S. F., S. H. B. Livi, and J. Wang, The cancellation of magnetic flux. II. In a decaying active region, *Aust. J. Phys.*, *38*, 929, 1985.

McClymont, A. N., and I. J. D. Craig, The structure and stability of coronal magnetic fields, *Solar Phys.*, *113*, 131-136, 1987.

Mickey, D. L., The Haleakala Stokes Polarimeter, *Solar Phys.*, *97*, 223, 1985.

Moore, R. L., G. A. Gary, M. J. Hagyard, and J. M. Davis, Accuracy requirements for vector magnetic field measurements for solar flare prediction, *BAAS*, *18*, 1043, 1987.

Ronan, R. S., D. L. Mickey, and F. Q. Orrall, The derivation of vector magnetic fields from Stokes profiles: Integral versus least squares fitting techniques, *Solar Phys.*, *113*, 353-359, 1987.

Sakurai, T., A new approach to the force-free field and its application to the magnetic field of solar active regions, *Publ. Astron. Soc. Japan*, *31*, 209, 1979.

Sakurai, T., Calculation of force-free magnetic field with non-constant α, *Solar Phys.*, *69*, 343, 1981.

Sakurai, T., Green's function methods for potential magnetic fields, *Solar Phys.*, *76*, 301, 1982.

Sakurai, T., and M. Makita, Observation of magnetic field vectors in solar active regions, in *Hydrody-*

namic and Magnetohydrodynamic Problems in the Sun and Stars, edited by Y. Osaki, p. 53, University of Tokyo, Tokyo, 1986.

Smith, Wm. H., W. V. Schempp, C. P. Conner, and P. Katzka, Spectral imagery with an acousto-optic filter, *Publ. Astron. Soc. Pacific*, *99*, 1337-1343,1987.

Spicer D. S., An unstable arch model of a solar flare, *NRL Report 8036*, Washington, DC, 1976.

Sturrock, P. A. (editor), *Solar Flares - A Monograph from Skylab Solar Workshop II* Colorado Assoc. Univ. Press, Boulder, Colorado, 1980.

EMERGENCE OF ANCHORED FLUX TUBES THROUGH THE CONVECTION ZONE

George H. Fisher

Institute for Astronomy, University of Hawaii, Honolulu, HI 96822
and
Institute of Geophysics and Planetary Physics, LLNL, Livermore, CA 94550

Dean-Yi Chou

Physics Department, Tsing Hua University, Hsinchu 30043, Taiwan, R.O.C.

Alexander N. McClymont

Institute for Astronomy, University of Hawaii, Honolulu, HI 96822

Abstract. We model the evolution of buoyant magnetic flux tubes in the Sun's convection zone. A flux tube is assumed to lie initially near the top of the stably stratified radiative core below the convection zone, but a segment of it is perturbed into the convection zone by gradual heating and convective overshoot motions. The ends ("footpoints") of the segment remain anchored at the base of the convection zone, and if the segment is sufficiently long, it may be buoyantly unstable, rising through the convection zone in a short time. The length of the flux tube determines the ratio of buoyancy to magnetic tension: short loops of flux are arrested before reaching the top of the convection zone, while longer loops emerge to erupt through the photosphere. Using Spruit's convection zone model, we compute the minimum footpoint separation L_c required for erupting flux tubes. We explore the dependence of L_c on the initial thermal state of the perturbed flux tube segment and on its initial magnetic field strength. Following an investigation of thermal diffusion times and the dynamic rise times of unstable flux tube segments, we conclude that the most likely origin for magnetic flux which erupts to the surface is from short length scale perturbations ($L < L_c$) which are initially stable, but which are subsequently destabilized either by diffusion of heat into the tube or by stretching of the anchor points until L just exceeds L_c. In either case, the separation of the anchor points of the emergent flux tube should lie between the critical distance for a flux tube in mechanical equilibrium and one in thermal equilibrium. Finally, after comparing the dispersion of dynamic rise times with the much shorter observed active region formation time scales, we conclude that active regions form from the emergence of a single flux tube segment.

Introduction

In several recent theories of solar and stellar dynamos (Golub et al., 1981; Galloway and Weiss, 1981; DeLuca and Gilman, 1986) magnetic fields are generated not in the convection zone, but in a thin, convectively stable layer just below it. In order for this magnetic flux to emerge to the photosphere where it can be observed, it must first make its way through the convection zone. It is generally believed that magnetic buoyancy provides the force which pushes the magnetic flux toward the surface (Parker, 1979). Our goal in this paper is to use the thin flux tube approximation of Spruit (1981) to study the emergence of buoyant loops of magnetic flux whose ends are still anchored in the stable layer below the convection zone. Our approach is similar to that used by Moreno-Insertis (1986). Further details of our calculations may be found in Chou and Fisher (1988; henceforth Paper I) and Fisher et al. (1988; henceforth Paper II). Here we present only the essentials.

Description of the Model

In the thin flux tube approximation, the tube diameter is assumed to be smaller than any other relevant physical length scale, and the magnetic field is taken to be constant across the tube. Furthermore, since the magneto-acoustic transit time across the tube will be smaller than other time scales of interest, one may assume that there is always a balance between the gas pressure outside the tube and the sum of gas plus magnetic pressure inside the tube. One can think of a thin flux tube as a one-dimensional curve embedded in space. Our flux tube is tied down at both ends to the bottom of the convection zone, with the portion in between free to move in response to buoyancy, magnetic tension, and drag forces. We take the flux tube to lie in a vertical plane, defined by a horizontal x-axis at the base of the convection zone, and a vertical y-axis, and regard the height y of the flux tube as a function of x and time t. The ends of the flux tube segment are anchored at $x = 0$ and $x = L$. In all the calculations discussed in this paper, we assume that the flux tube initially ($t = 0$) lies flat at the bottom of the convection zone ($y = 0$). In the discussion below, subscript e refers to the external plasma of the convection zone, and subscript i refers to the plasma inside the flux tube.

The Basic Equations

The velocity of plasma in the flux tube consists of the speed of the tube itself in the direction normal to its length (denoted υ_n) and the speed of plasma moving along the flux tube (denoted υ_s). The equation of motion for the velocity component υ_n is given by

$$(\rho_e + \rho_i)\frac{d\upsilon_n}{dt} = \frac{B^2}{4\pi R} + \frac{(\rho_e - \rho_i)g}{\sqrt{1+(\partial y/\partial x)^2}} - \frac{C_D \rho_e |\upsilon_n| \upsilon_n}{\sqrt{\pi A}} \quad (1)$$

where the terms on the right-hand side of equation (1) correspond, respectively, to magnetic tension, buoyancy, and aerodynamic drag forces. The drag coefficient C_D is taken to be unity. Note that the inertial term contains the "added mass" of the displaced external plasma flowing around the flux tube (see, e.g., Landau and Lifshitz (1959) § 24). An insignificant centrifugal force due to parallel flows through bends in the flux tube has been neglected. Equation (1) must be supplemented by an equation relating the vertical motion of the flux tube to the velocity υ_n and the flux tube slope $\partial y/\partial x$. From geometrical considerations, we find at a fixed value of x

$$\frac{\partial y}{\partial t} = \upsilon_n \sqrt{1 + (\partial y/\partial x)^2} \quad (2)$$

The radius of curvature R, the magnetic field B, and the flux tube cross-sectional area A appearing in equation (1) are determined by

$$\frac{1}{R} = \frac{\partial^2 y/\partial x^2}{[1 + (\partial y/\partial x)^2]^{3/2}} \; ; \quad P_e - P_i = \frac{B^2}{8\pi} \; ; \quad B\,A = \Phi \quad (3)$$

where P_i and P_e are the internal and external gas pressures, respectively, and Φ is the total magnetic flux. The numerical techniques used for solving equations (1) and (2) are described in Paper I.

The parallel flows of the plasma inside the flux tube can be calculated from equations (1) and (2) and mass conservation arguments (see Paper I). These flows turn out to be sufficiently slow that the plasma can be taken to be in hydrostatic equilibrium. If we make the same assumption regarding the external plasma (i.e., ignoring effects of convection), then it is straightforward to determine all the thermodynamic and magnetic variables as functions of height alone. This procedure is outlined below. It is not at all obvious a priori that the neglect of convective motions is justified. The presence of these motions could affect flux tube evolution in two ways. First, convective motions could severely distort the shape of the flux tube. Second, convective motions introduce a turbulent pressure which perturbs the hydrostatic equilibrium and this could affect the calculated magnetic field strength. We consider both of these effects in Paper I and conclude that the first effect can be important for some of the thinnest flux tubes we study, whereas the second is not important for any of the cases we have studied.

The structure of the background external atmosphere is taken to be the convection zone model of Spruit (1974). This atmosphere is slightly superadiabatic due to convective heat transport. The information in Spruit's paper is sufficient to calculate the thermodynamic and magnetic quantities needed for our dynamic model. We assume that the plasma inside the flux tube is isolated from heat transfer and behaves adiabatically. There will of course be heat transport between the flux tube and its surroundings, but we have not included this explicitly in the model. Our approach in this paper is to approximate the effects of heat conduction by choosing a physically self-consistent value of the temperature difference between the flux tube and its environs (i.e., the parameter η introduced below) at the base of the convection zone. This will be described further in the section on time scales for emerging flux tubes.

It is straightforward to derive an equation for β [$\beta \equiv 8\pi P_e/B^2 = P_e/(P_e - P_i)$] in terms of the height variation of the temperature difference $\delta T \equiv T_e - T_i$ between the external and internal plasma; this avoids the numerical difficulty of subtracting two gas pressures which are nearly equal. Starting from the hydrostatic relations

$$\frac{dP_e}{dy} = -\rho_e\, g \; ; \quad \frac{dP_i}{dy} = -\rho_i\, g \quad (4)$$

one finds

$$\frac{1}{\beta(\beta-1)}\frac{d\beta}{dy} = -\frac{1}{\Lambda}\frac{\delta T/T_e}{1-\delta T/T_e} \quad (5)$$

where Λ is the external pressure scale height. This has the solution

$$\beta(y)^{-1} = \beta_0^{-1}\exp[-\gamma(y)] + \{1 - \exp[-\gamma(y)]\} \quad (6)$$

where $\beta_0 = \beta(0)$ and

$$\gamma(y) = \int_0^y \frac{dy'}{\Lambda(y')} \frac{\delta T/T_e(y')}{1 - \delta T/T_e(y')} \tag{7}$$

The temperature difference is simply

$$\delta T(y) = \delta T_0 - \int_0^y dy' \; (\nabla - \nabla_{ad}) \mu g / R_g \tag{8}$$

where $\delta T_0 = \delta T(0)$, $\nabla \equiv (\partial \ln T/\partial \ln P)$, $\nabla_{ad} \equiv (\partial \ln T/\partial \ln P)_S$, and μ and R_g are the mean mass per particle and gas constant. The temperature difference at the bottom of the convection zone δT_0 is given in terms of parameters η and B_0 by

$$\delta T_0 = \eta T_e(0)/\beta_0 \tag{9}$$

where $\beta_0 = 8\pi P_e(0)/B_0^2$. The quantity B_0 is the magnetic field strength at the base of the convection zone, and is one of the free parameters of the problem, while η is a dimensionless measure of the temperature difference at the base of the convection zone, and is constrained to lie between 0 and 1. When $\eta = 0$, the internal temperature is equal to that of the surrounding plasma, whereas for $\eta = 1$, the density inside the flux tube is equal to that outside. Together with L and Φ, specification of B_0 and η completely defines the flux emergence problem in our model.

Values of all these parameters are highly uncertain, so we have attempted to cover a wide range of parameter space. For most of our simulations, B_0 was chosen to be 10^4 G, 10^5 G, or 10^6 G, and the flux Φ was taken as 10^{18} Mx, 10^{20} Mx, or 10^{22} Mx. Our choice of these values is discussed further in Paper I.

Example of a Simulation

To demonstrate our model, we briefly discuss results of a single simulation. In this case $B_0 = 10^6$ G, $\Phi = 10^{18}$ Mx, $\eta = 0$, and $L = 2 \times 10^{10}$ cm. The position of the flux tube at numerous times is shown in Fig. 1. This flux tube is "unstable" and emerges from the top of the calculational domain without reaching an equilibrium configuration. Note that the upward motion of the flux tube slows significantly when its apex is roughly half-way through the convection zone. This indicates that the magnetic tension was almost strong enough to balance the buoyancy force. Had L in fact been only slightly smaller ($\lesssim 1.97 \times 10^{10}$ cm), the flux tube would have been "stable" and would have stopped rising when the height of the flux tube apex $y_a \approx 9 \times 10^9$ cm. We explore this interesting behavioral dichotomy in the following section.

The Critical Length Scale

From the discussion in the previous section, it is apparent that for given values of the other parameters, a

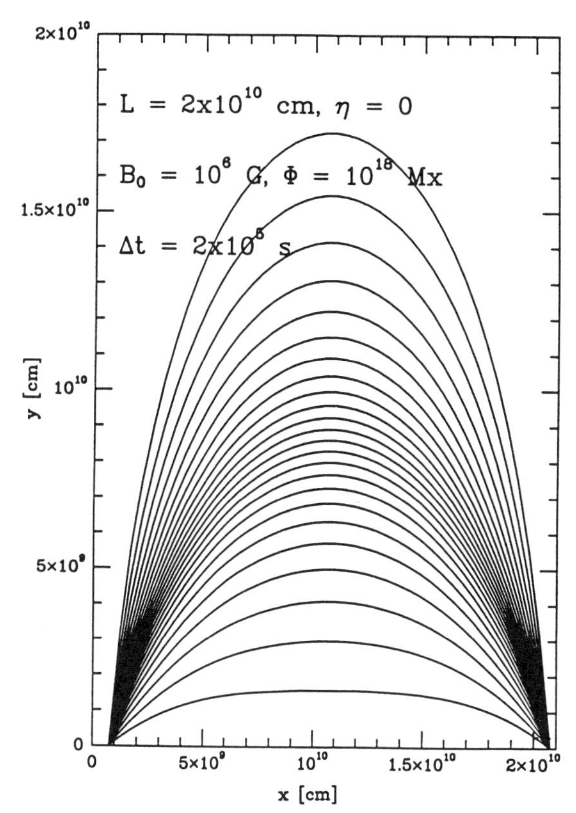

Fig. 1. Plot of flux tube shape at times separated by $2{\times}10^5$s.

critical length L_c exists. For $L < L_c$, the flux tube reaches a stable equilibrium, while for $L > L_c$ an unstable eruption to the photosphere occurs. Therefore, a knowledge of the dependence of L_c on the other parameters is very important for understanding the nature of flux emergence. The magnetic flux Φ affects the drag force and therefore the rate of rise, but it does not affect the relative balance between buoyancy and tension, so L_c is independent of Φ. To determine the dependence of L_c on η and B_0, we have used the numerical model described in the previous section. In Fig. 2, we plot L_c as a function of η for $B_0 = 10^4$ G and $B_0 = 10^6$ G.

There are several features of Figure 2 worthy of mention. First, note that for fixed B_0, L_c increases as η increases from 0 to 1. This is easily explained by the decrease in the relative strength of magnetic buoyancy compared to magnetic tension as the temperature inside the flux tube decreases. Thus the footpoint separation L over which magnetic tension can overcome buoyancy is increased (recall that for fixed apex height tension will scale roughly as L^{-2}). Second, as η approaches 1 for fixed B_0, L_c approaches a finite value, even though the buoyant force goes to zero at the base of the convection zone. Since for $\eta = 1$, the initial horizontal state is an equilibrium

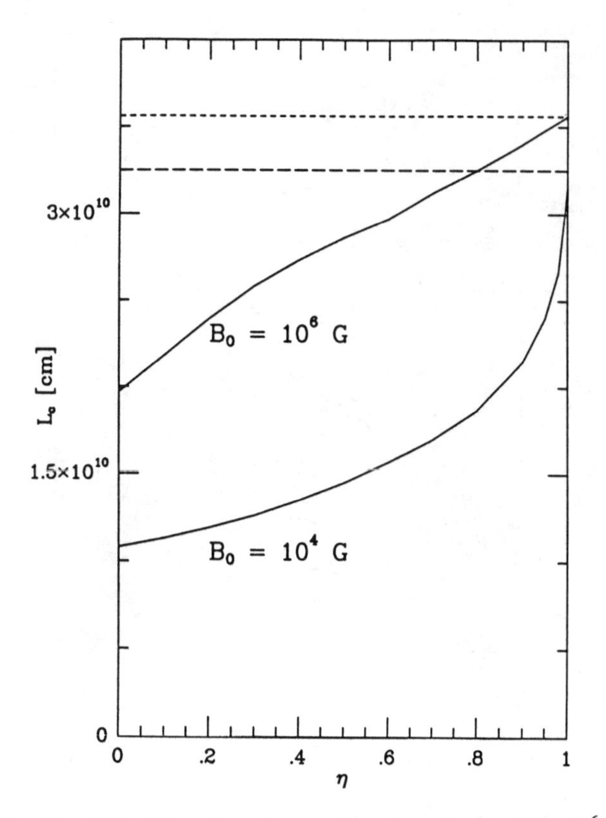

Fig. 2. Critical length scale as a function of η for $B_0 = 10^6$ and 10^4 G. The long-dashed horizontal line corresponds to the value from equation (10), and the short-dashed line corresponds to Parker's (1979) isothermal result of $2\pi\Lambda$.

configuration, one can perform a linear stability analysis of equations (1) and (2) (see Spruit and van Ballegooijen, 1982, and Paper II) to obtain L_c analytically

$$L_c = \sqrt{2/(1-\nabla)}\ \pi\ \Lambda \tag{10}$$

which is plotted as the horizontal long-dashed line in Figure 2. This analytic result corresponds very well to our numerical result for $\eta = 1$ and $B_0 = 10^4$ G, but is about 10% smaller than the critical length (for emergence of the flux tube through the photosphere) which we find numerically for $\eta = 1$ and $B_0 = 10^6$ G. The reason for this interesting discrepancy is discussed in Paper II.

Time Scales for Emerging Flux Tubes

If magnetic flux emerges from the convective overshoot region and evolves according to our model, only those flux tubes we have labeled "unstable" ($L > L_c$) can reach the photosphere. The time required for the flux tube to rise through the convection zone is therefore of great interest. We assume that a flux tube is formed in the stably stratified region beneath the convection zone, and begins its buoyant rise when perturbed upward into the convection zone by gradual heating and convective overshoot motions. Before perturbation, the flux tube must be in mechanical equilibrium ($\rho_i = \rho_e$), as otherwise it would adjust by rising or sinking adiabatically (Parker, 1979). The nature of a perturbation is described by the two parameters L and η. L is the length scale of the perturbation, and η measures its thermodynamic state. When $\eta = 1$, the perturbed flux tube is still in mechanical equilibrium ($\rho_i = \rho_e$), whereas if $\eta = 0$, the flux tube is in thermal equilibrium with its surroundings ($T_i = T_e$). The proper choice of η for the perturbation can be made by comparing the thermal diffusion time at the bottom of the convection zone τ_{th} to the dynamic rise time τ_d for unstable $\eta = 1$ flux tubes. If $\tau_{th} \gg \tau_d$, then $\eta = 1$ is the best description of the perturbation, whereas if $\tau_{th} \ll \tau_d$, then $\eta = 0$ is most appropriate. When the two are comparable, an intermediate value of η is called for. The thermal diffusion time scale is given by (see Paper II)

$$\tau_{th} \approx 0.1\ \frac{3}{16}\ \frac{\rho_e^2 C_p a^2 \kappa_R}{\sigma T_e^3} \tag{11}$$

where σ is the Stefan-Boltzmann constant, κ_R is the Rosseland mean opacity [cm²/g], and $a = [\Phi/(\pi B_0)]^{1/2}$ is the radius of the flux tube at the base of the convection zone. The dynamic rise time τ_d for $\eta = 1$ can be obtained from the numerical simulations. Note that the thinner (smaller Φ) flux tubes will have shorter thermal time scales and longer dynamic time scales. We find that τ_d does not depend strongly on L, provided L exceeds L_c by at least 5%. We therefore have used $L = 4 \times 10^{10}$ cm (cf. Figure 2) to compute our estimates of τ_d. Computed values of τ_d and τ_{th} as functions of Φ and B_0 are shown in Table 1, along with the resultant values for η. We also include in Table 1 the

TABLE 1. Flux Tube Time Scales

	B_0=10⁶ G	B_0=10⁵ G	B_0=10⁴ G
Uncorrected (η = 1) rise time τ_d (s)			
$\Phi = 10^{22}$ Mx	8.3×10^5	6.4×10^6	3.9×10^7
$\Phi = 10^{20}$ Mx	1.8×10^6	1.1×10^7	4.7×10^7
$\Phi = 10^{18}$ Mx	5.3×10^6	3.0×10^7	8.1×10^7
Thermal diffusion time τ_{th} (s)			
$\Phi = 10^{22}$ Mx	2.7×10^7	2.7×10^8	2.7×10^9
$\Phi = 10^{20}$ Mx	2.7×10^5	2.7×10^6	2.7×10^7
$\Phi = 10^{18}$ Mx	2.7×10^3	2.7×10^4	2.7×10^5
Self-consistent η			
$\Phi = 10^{22}$ Mx	1.0	1.0	1.0
$\Phi = 10^{20}$ Mx	0.0	0.0	0.5
$\Phi = 10^{18}$ Mx	0.0	0.0	0.0
Corrected rise time τ_r (s)			
$\Phi = 10^{22}$ Mx	8.3×10^5	6.4×10^6	3.9×10^7
$\Phi = 10^{20}$ Mx	6.7×10^5	3.1×10^6	8.6×10^6
$\Phi = 10^{18}$ Mx	2.1×10^6	9.8×10^6	1.7×10^7

"corrected" dynamical rise times (denoted τ_r) using these self-consistent values of η in the numerical simulations.

From the information in Table 1 and the critical length scales of Figure 2, it is now possible to construct a general picture of flux tube evolution. For given values of Φ and B_0, perturbation length scales divide naturally into three ranges: short, $L < L_c(\eta = 0)$; medium, $L_c(\eta = 0) < L < L_c(\eta = 1)$; and long, $L_c(\eta = 1) < L$. If $\tau_{th} \ll \tau_d$ then only short flux tubes are stable, while longer flux tubes rise to the photosphere on the dynamical time scale. But if $\tau_{th} \gg \tau_d$, both short and medium flux tubes are stable. Only the short tubes are "absolutely" stable, however, as the medium length tubes are able to rise quasi-statically on the thermal time scale as heat leaks into them. This is equivalent to gradually decreasing η from 1 to 0. However, before η reaches zero, a medium-L flux tube will suddenly find itself with $L > L_c$, and will then erupt to the photosphere on the much shorter dynamic rise time.

In the most straightforward (e.g., $\alpha\omega$ or $\alpha^2\omega$) kinematic dynamo models the oscillatory nature of the solar cycle is due to periodic conversion of poloidal to toroidal magnetic field and back via the mechanisms of stretching by differential rotation and the α-effect (which is usually attributed to net helicity in convective or convective overshoot motions). The dynamo period is therefore at least as great as the time scale necessary to reorient and stretch poloidal into toroidal field. For these dynamos to work, therefore, magnetic flux must remain in the dynamo region long enough for reorientation to occur, i.e., for a significant fraction of the solar cycle. The conclusions drawn below are based on this premise. An alternative picture is that the dynamo generates fields on a much shorter time scale, in which case our arguments below do not hold. In that case, however, the period of the solar cycle itself remains unexplained and some additional unknown mechanism must be invoked to account for it.

From a comparison of the time scales in Table 1 with the length of half a solar cycle (roughly 11 years or 3.4×10^8 s), some tentative conclusions about the nature of flux tube perturbations can be drawn. First, note that the rise time τ_r of unstable flux tubes is always much shorter than the duration of the solar cycle. If there were a continuous source of perturbations with length scales $L > L_c$, then magnetic flux could not remain stably submerged on time scales much longer than τ_r. One therefore concludes that such perturbations are either rare or nonexistent. Indeed, if convective overshoot motions are responsible for the flux tube perturbations, one would expect typical perturbation length scales to be of order the eddy size, which in mixing length theory is roughly the scale height near the bottom of the convection zone, i.e., 6×10^9 cm. This is significantly less than any of the L_c values in Figure 2. A perturbation with $L > L_c$ would require coherent action on the part of several adjacent eddies. Assuming this to be unlikely, the magnetic flux which does eventually emerge must originate from perturbations in the short or intermediate L range. We consider two possibilities.

First, if the separation L between anchor points is truly fixed in time, then only those perturbations in the intermediate L range can emerge at all, since short L perturbations result in completely stable structures. Furthermore, only those combinations of B_0 and Φ which give rise to $\eta = 1$ in Table 1 are viable, i.e., those flux tubes with $\Phi \approx 10^{22}$ Mx. In that case, we expect magnetic flux to emerge to the surface on thermal diffusion time scales and to have a footpoint separation between $L_c(\eta = 0)$ and $L_c(\eta = 1)$.

The second possibility is that there is stretching of the anchor point separation L past L_c after a stable short L perturbation has been made. There are two possible mechanisms for this. In the first instance, stretching could be accomplished by differential rotation, for example, if the anchor points differ in latitude. The second mechanism is thermal heating of the anchor points themselves. This may cause a gradual "unzipping" of the flux tube segment, until the footpoint separation exceeds L_c. The time scale for this to occur depends on a number of factors, including how deeply buried the remainder of the flux tube is in the overshoot region. This is discussed further in Paper II. Here again, for either mechanism, we expect the anchor point separation of emerging flux to lie in the range $L_c(\eta = 0) < L < L_c(\eta = 1)$.

As a final point, note that the dispersion in rise times for individual flux tubes with different values of Φ (a month or more-see Table 1) is much greater than the formation times for active regions (typically 2-3 days). We believe this indicates that active regions are formed from a single emerging flux tube. If many tubes of differing sizes were perturbed simultaneously, or if a single flux tube became highly fragmented near the bottom of the convection zone, one would expect the emergence of flux to be dispersed over a much greater time corresponding to the difference in rise times between small Φ (e.g., 10^{18} Mx) and large Φ (e.g., 10^{22} Mx) flux tubes. Any fragmentation therefore probably takes place in the topmost portion of the convection zone when the tube is emerging most rapidly.

Conclusions

We have developed a model, based on the thin flux tube approximation of Spruit (1981), for studying the emergence of magnetic flux through the convection zone when the footpoints of the flux loop are anchored a distance L apart in the stable layers below. Figure 1 shows the evolution of a flux tube whose footpoints are sufficiently far apart ($L > L_c$) to allow it to rise to the top of the convection zone. We have used our model to explore the critical length L_c separating flux tubes which form stable magnetic loops in the convection zone ($L < L_c$)

from those which erupt through the photosphere ($L > L_c$). Figure 2 shows the dependence of L_c on the temperature defect parameter η for two values of the field strength at the base of the convection zone B_0, 10^4 G, and 10^6 G. We find that for perturbations on length scales $L > L_c(B_0, \eta)$ the rise time is short compared to the duration of the solar cycle. Under the assumption that the solar cycle can be modeled as a kinematic (e.g., $\alpha\omega$ or $\alpha^2\omega$) dynamo operating below the convection zone, we conclude that such perturbations must therefore be rare or nonexistent. We speculate that the magnetic flux which does erupt through the photosphere forms initially from perturbations with $L < L_c$, resulting in stable structures. These are subsequently destabilized either by thermal diffusion or by stretching of the anchor points until L exceeds L_c. In either case, we expect that the anchor point separation L should fall roughly within the range $L_c(\eta = 0)$ to $L_c(\eta = 1)$. The rise time of flux tubes is the time scale for conduction of heat into the flux tube in the first case, and the stretching time in the second case. Finally, we argue that active regions are formed by the emergence of a single flux tube segment.

Acknowledgments. G. H. Fisher and A. N. McClymont were supported by NASA under grant NAGW86-4 and by NSF under grant ATM-86-19853. Dean-Yi Chou was supported by the NSC of ROC under grant NSC 77-0209-M007-01 and by NASA under grant NSG-7536 during his time at the University of Hawaii. G. H. Fisher was supported by the I.G.P.P. at Lawrence Livermore National Laboratory for several visits during which some of this work was completed.

References

Chou, D.-Y., and G. H. Fisher, Dynamics of anchored flux tubes in the convection zone, I. Details of the model, *Ap. J.*, in press, 1989.

DeLuca, E. E., and P. A. Gilman, Dynamo theory for the interface between the convection zone and the radiative interior of a star. Part I. Model equations and exact solutions, *Geophys. Astrophys. Fluid Dynamics*, *37*, 85, 1986.

Fisher, G. H., D.-Y. Chou, and A. N. McClymont, Dynamics of anchored flux tubes in the convection zone, II. A study of flux emergence, *Ap. J.*, in preparation, 1989.

Galloway, D. J., and N. O. Weiss, Convection and magnetic fields in stars, *Ap.J.*, *243*, 945, 1981.

Golub, L., R. Rosner, G. S. Vaiana, N. O. Weiss, Solar magnetic fields: The generation of emerging flux, *Ap. J.*, *243*, 309, 1981.

Landau, L. D., and E. M. Lifshitz, *Fluid Mechanics*, Pergamon Press, 1959.

Moreno-Insertis, F., Nonlinear time-evolution of kink-unstable magnetic flux tubes in the convective zone of the Sun, *Astron. Astrophys.*, *166*, 291, 1986.

Parker, E. N., *Cosmical Magnetic Fields*, Oxford Univ. Press, 1979.

Spruit, H. C., A model of the solar convection zone, *Solar Phy.*, *34*, 277, 1974.

Spruit, H. C., Motion of magnetic flux tubes in the solar convection zone and chromosphere, *Astron. Astrophys.*, *98*, 155, 1981.

Spruit, H. C., and A. A. van Ballegooijen, Stability of toroidal flux tubes in stars, *Astron. Astrophys.*, *106*, 58, 1982.

MAGNETOCONVECTION ON THE SOLAR SURFACE

G. W. Simon,[1] A. M. Title,[2] K. P. Topka,[2] T. D. Tarbell,[2] R. A. Shine,[2]
S. H. Ferguson,[2] H. Zirin,[3] and the SOUP Team[4]

Abstract. We describe and illustrate the first high-resolution observations of horizontal flows on the solar surface and their relation to magnetic field structure seen in the Sun's photosphere. The velocity data were deduced from white-light images obtained by the Solar Optical Universal Polarimeter (SOUP) instrument flown as part of NASA's Spacelab 2 mission (Space Shuttle flight 51-F, STS-19). Solar granules (with a typical size scale of 1 Mm and lifetime of 15 min) were used as tracers to measure larger-scale, longer-lived flows including mesogranules (6-12 Mm), supergranules (30 Mm), radial outflows from a sunspot, and streams (of length 50-100 Mm, width 5-10 Mm). These flows were compared to a 9-hour time series of the solar magnetic field obtained at the same time at the Big Bear Solar Observatory (BBSO). The flow field and the magnetic structure agree in remarkable detail. Indeed, the data suggest strongly that the flow field is a nearly perfect descriptor of the motion and evolution of the magnetic field (with the exception of the strongest fields within active regions which are able to inhibit the convection). If such measurements can be made synoptically from space, or under good seeing conditions from a ground-based observatory, it should be possible to pinpoint loci of magnetic mixing, twisting, and stress buildup, and thus predict the occurrence of solar flares, coronal heating, and mass ejections.

[1]Air Force Geophysics Laboratory, National Solar Observatory,* Sunspot, NM 88349
[2]Lockheed Palo Alto Research Laboratory (LPARL), Palo Alto, CA 94304
[3]California Institute of Technology, Pasadena, CA 91125
[4]L. Acton, D. Duncan, M. Finch, Z. Frank, G. Kelly, R. Lindgren, M. Morrill, N. Ogle (deceased), T. Pope, H. Ramsey, R. Reeves, R. Rehse, R. Wallace, LPARL; J. Harvey, J. Leibacher, W. Livingston, L. November, National Solar Observatory

*Operated by the Association of Universities for Research in Astronomy, Inc., under contract with the National Science Foundation. Partial support is provided by the USAF under a Memorandum of Understanding with the National Science Foundation.

Introduction

More than a quarter century ago, Leighton et al. (1962) discovered the solar supergranulation flow field. Shortly thereafter Simon and Leighton (1964) suggested that the supergranulation carries the solar magnetic field to the flow cell boundaries to form the well-known chromospheric network pattern long seen in calcium, hydrogen, and magnetic images.

Developments in magnetoconvection theory (Weiss, 1978; Meyer et al., 1979; Galloway and Weiss, 1981; Parker, 1982) showed how the motion of supergranules could redistribute and concentrate magnetic flux tubes in and below the solar surface. Recent works have extended the earlier studies to include motions in granules (Schmidt et al., 1985), and applied more elaborate two-dimensional and three-dimensional computational techniques (Proctor and Weiss, 1982; Galloway and Proctor, 1983; Nordlund, 1985a,b; Cattaneo, 1984; Hurlburt and Toomre, 1988). The theoretical analyses suggest that the observed magnetic and intensity structures in the surface and higher layers of the solar atmosphere depend on the nature of both large-scale (supergranular, mesogranular) and small-scale (granular) sub-surface flows.

Several observers (Simon, 1967; Muller and Mena, 1987; Title et al., 1987) have attempted to relate the motions of granules to magnetic structures, and many others (Vrabec, 1971; Smithson, 1973; Schroter and Wöhl, 1975; Mosher, 1977; Martin, 1988; Title et al., 1987) have measured the motions of individual magnetic field elements relative to the magnetic network.

Taken together, these observations imply that both the large-scale supergranular flow and the much smaller motion fields of individual granules help to determine the structure of magnetic field on the Sun's surface. Both the observations and theory suggest that the flow determines the evolution and distribution of the magnetic fields. However, in active regions where the field is strong, it is clear (Zwaan, 1978) that the magnetic field has a significant effect on the flow field.

The work described above provides the background and sets the stage for the remarkable observations from Spacelab 2 which we now describe in the following sections.

Data

The white-light data were obtained on 35 mm film by the Solar Optical Universal Polarimeter (SOUP) instrument flown on Spacelab 2 (NASA Space Shuttle mission 51-F). SOUP contained a 30-cm Cassegrain telescope and an active secondary mirror for image stabilization (Title et al., 1986). In this paper we used images from orbit 110 taken between 19:10:35 UT and 19:38:05 UT on August 5, 1985, in the vicinity of active region 4682. Frames were obtained every 2 s. The field-of-view covered an area 166 by 250 arc sec. The effective wavelength band of the observations is roughly 1000 Å centered on 6000 Å. The magnetic data were obtained by the Big Bear Solar Observatory (BBSO) on the same active region from August 5, 1985, 15:25:43 UT to August 6, 00:50:54 UT.

The high-quality granulation pictures taken by SOUP provided a unique opportunity to detect large-scale surface flows by direct displacement measurements of the local intensity pattern. We have applied correlation tracking methods to make the required measurements (November et al., 1987; November and Simon, 1988). The flow field is detectable because it advects the granulation pattern. That is, granulation serves as a tracer for the flow. Since granules typically last 10 to 15 min, measurements must be made in a time short compared to this lifetime. In addition, because the 5-min oscillation is also present in the movies (Title et al., 1986), measurements must be separated by considerably less than 2.5 min. We have used time differences between images of 10 to 60 s in this study. This seems adequate since velocities obtained from images up to 60 s apart do not differ significantly. Because the supergranulation flow ranges from 0.1 to 1 km s^{-1}, in a 30-s interval the local granulation pattern should move typically about 10 km or 15 milliarc sec. Due to atmospheric turbulence ("seeing"), the best ground-based imagery is rarely better than about 1 arc sec, and it contains distortions of magnitude comparable to the blurring. Clearly, with typical noise-to-signal ratios of about 100, measuring with confidence such small displacements had previously been very difficult. Details of the data reduction have been discussed by Simon et al. (1988).

Analysis

Shown in Figure 1 are (a) a SOUP image, (b) a BBSO magnetogram at nearly the same time, (c) the SOUP flow field (shown as vectors) overlayed on a gray-scale map of the divergence of the flow field, and (d) the SOUP flow field superposed on the magnetogram. Velocities generally lie in the range 100 to 800 m s^{-1} and represent the average value obtained by correlation tracking over the 28-min observation time of orbit 110. The divergence of the horizontal flow vector is approximately proportional to the average vertical velocity (November et al., 1987) and thus identifies cell interiors (sources or upflows) and boundaries (sinks or downdrafts). The numbers in Figures 1c and 1d indicate the centers of four strong cellular outflows; note that these centers are void of magnetic flux.

We see from Figure 1 the intimate relation between the flow field and the magnetic field. In relatively compact magnetic features, the flow field points toward the concentrations. In cell-like regions of the magnetogram the vectors of the flow field point radially outward from the cell centers toward the boundaries (network). We also see a third type of magnetic structure where the flow field converges, not to a sink point, but to a line. In the active region, other large-scale flows occur. The first and most striking discovery in the SOUP data was an annulus, about 5 arc sec wide extending from the edge of the sunspot penumbra into the surrounding photosphere, composed of radially out-streaming granules (Title et al., 1986). This confirmed a long-held opinion; namely, that because of its strong magnetic field, the spot inhibits normal convection to the surface, so one might expect that upflows would be diverted radially outward from the sunspot (Meyer et al., 1974). We observe that magnetic field motions across this annulus closely follow the flow vectors determined from the pattern of granular motions. This is the well known moat flow in which magnetic features flow radially out from the sunspot (Sheeley, 1969, 1971; Sheeley and Bhatnagar, 1971; Vrabec, 1971). We also observe that the pore region acts as a large sink, especially for flows streaming outward from the sunspot.

Another remarkable new result is the existence of streams (or currents), particularly in quiet Sun regions. We had thought that the Sun is covered by closely packed cellular structures of several scales (granules, mesogranules, supergranules) with old cells disappearing as new ones are formed. However, we see in Figure 1 that there are also several streams, some of which are 50-100 Mm in length and 5-10 Mm in cross section, where there exist no large scale cellular structures. The most striking of these begins at the left boundary of Figure 1 and extends halfway across the bottom part of the image.

We gain additional insight into the relationship between flows and magnetic fields by asking

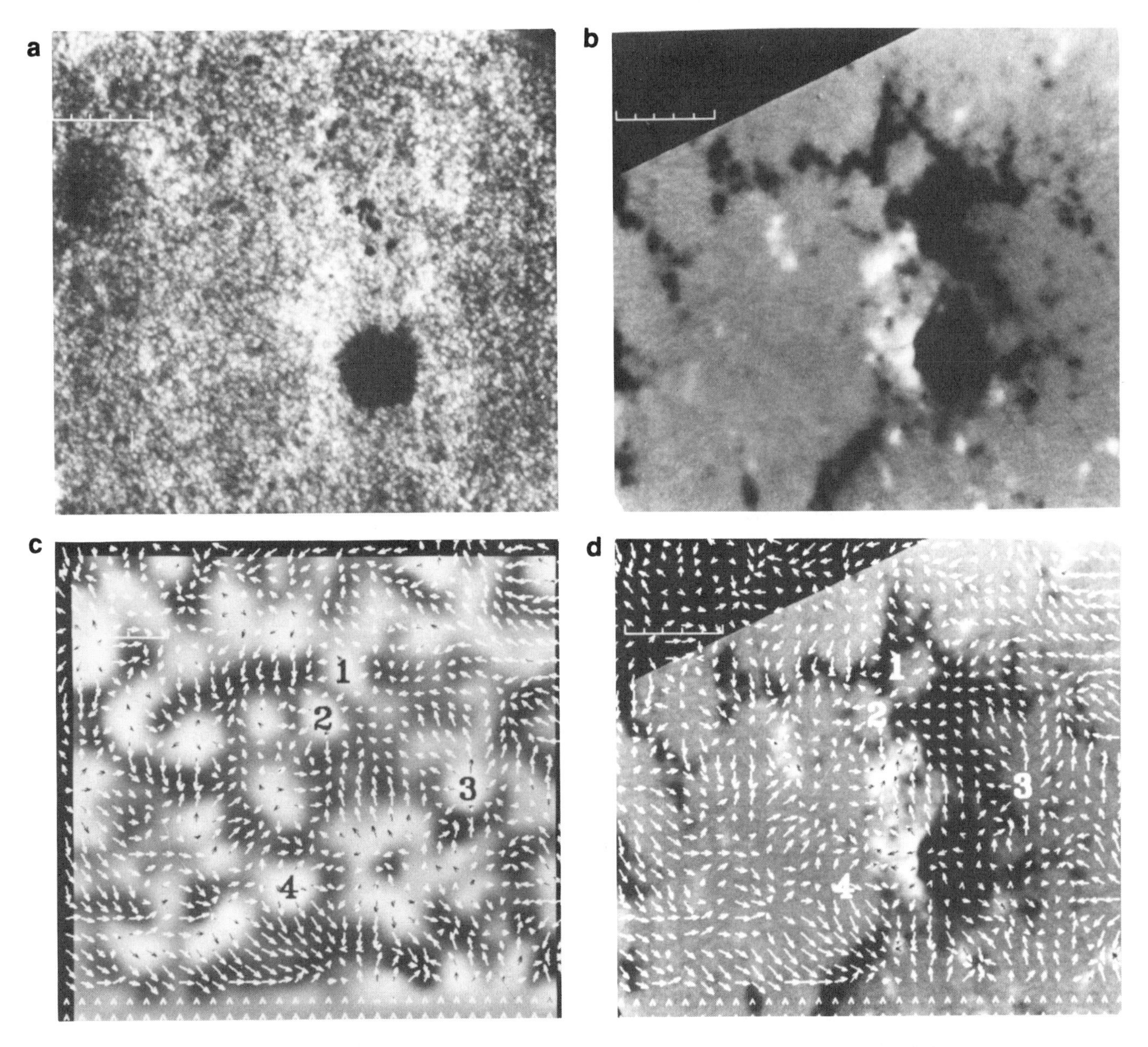

Fig 1. Comparison of SOUP and BBSO data obtained on August 5, 1985, at 19:28 UT. The field-of-view is 137 by 128 arc sec in the vicinity of active region 4682. A scale with 5 arc sec ticks is on each image. (a) SOUP white-light image, (b) BBSO magnetogram, (c) SOUP flow vectors superposed on a gray-scale image of the flow divergence (sources are bright, sinks dark), and (d) SOUP vectors superposed on the magnetogram. Four flow cells are identified in (c) and (d).

where the measured surface flow field would carry hypothetical free particles ("corks") originally distributed uniformly in the flow field. We calculated the cork flow by moving each cork according to the velocity of the local flow field at that cork's location. In Figure 2 we show the location of such corks overlaid on a magnetogram after 12 hours. The same four flow cells are marked, as in Figure 1. Most of the stable magnetic structures (i.e., those which show little or no motion during the nine hour movie) outside the sunspot are located at or near corks. And in places where we see magnetic flows, the movies show similar motions of corks and magnetic features. Initially the flow carries corks to the cell boundaries. Then the corks creep along the

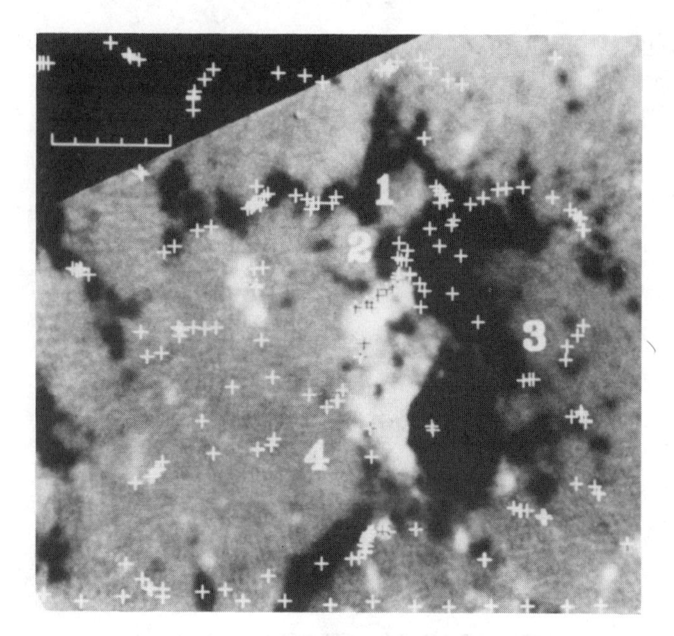

Fig 2. The same area as Figure 1, showing the positions of "corks" after 12 hours under the action of a constant flow field.

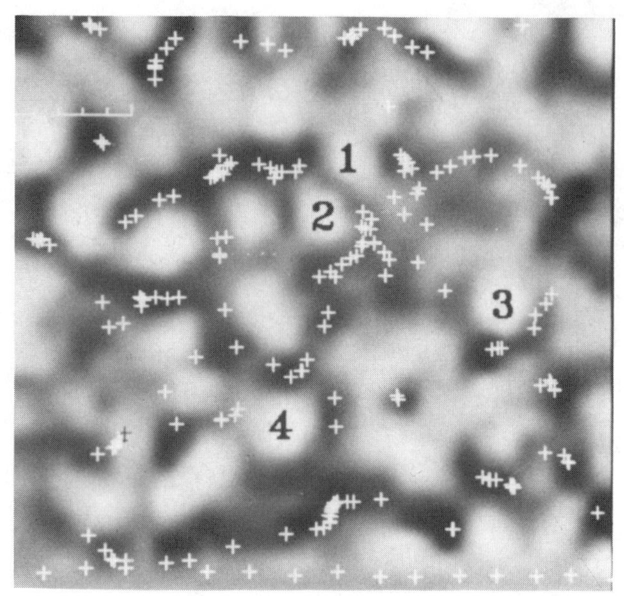

Fig 3. Divergence of the flow field of Figure 1 superposed on the corks after 12 hours. The cork pattern mainly shows a network of supergranular scale, in which mesogranules are embedded.

boundaries to sink regions which usually are vertices in the network pattern as seen in calcium, hydrogen, or magnetograms. We have made a magnetogram movie overlayed with the SOUP white-light flow field vectors, from which it is very clear that in the quiet Sun small magnetic field elements appear randomly near supergranule/mesogranule centers, rapidly move to cell boundaries, and then flow slowly along the boundaries, just as the cork model suggests. The movie also contains several examples in which a ring of magnetic network surrounding a cell increases in diamter; this suggests that the flow cell is similarly increasing in diameter.

It is interesting to note, however, that the magnetic (or cork) network sometimes shows a larger dominant cell size than do the flow maps. This is illustrated in Figure 3, where we have superposed a divergence map of the flow in Figure 1 and a snapshot of the corks after 12 hours. It shows that two or three smaller cells of mesogranular scale (November et al., 1981) are often contained within one traditional network cell. Cell sizes (center-to-center) in the flow divergence maps are often 6 to 12 Mm rather than the 30 Mm value usually associated with supergranulation.

Discussion

These simultaneous observations of white-light granulation and digital magnetograms have shown clearly the intimate interactions between surface motions and magnetic structures. We are able to demonstrate that the white-light granular flow field is a nearly perfect descriptor of the motion and evolution of the magnetic field. The flow field determined from one 28-min measurement is an excellent indicator of the motion of the corresponding magnetic field configuration and is valid for at least 4 hours before and after the SOUP observations. From the cork simulations we estimate that the magnetic pattern would require about 8 to 10 hours to develop, which suggests that the flow field and magnetic field have a lifetime longer than 10 hours as would be expected for large-scale supergranular structures.

Especially in quiet Sun regions, both the magnetograms and the cork simulation show an incomplete network: fully outlined cells are rare, and usually there are just enough markers in the boundaries to suggest a cellular pattern. The cork simulation shows that this is an intrinsic property of the flow patterns and not only a result of insufficient magnetic flux to complete the pattern.

These data suggest to us that flow along network boundaries may be an important feature in the evolution of the magnetic field pattern. This has implications for understanding phenomena such as coronal heating and the buildup of magnetic stresses in the network. First, flow along the network boundaries will tend to mix and twist the fields on very small scales. We have measured the vertical component of vorticity of the flow field and find that at some locations it reaches values of about 0.0001 s^{-1}. Thus substantial twist can be imparted to the magnetic field in only a few hours where the vorticity is large. The mixing and twisting will also be enhanced by local displacements of the field caused by randomly-directed motions and explosions of individual granules. Second, flow along the boundaries concentrates fields in vertices. These vertices are probably stable points in the flow field, so that new supergranules may form with a vertex at the previous boundary.

Such observations are valuable to theorists who have argued that dissipation and heating in magnetic regions depend critically on the spatial scale of the twisting of the flux tubes (van Ballegooijen, 1986; Parker, 1972, 1983). For example, Mikic et al. (1988) have used three-dimensional models to show how shearing photospheric flows might build up the energy of a magnetic arcade until it becomes unstable, forms current sheets, and then reconnects with rapid release of magnetic energy in the corona.

Just as the Doppler spectroheliogram observations of Leighton et al. (1962) made the phenomenon of supergranulation clearly recognizable near the limb, local correlation tracking can make horizontal flow patterns apparent at disk center. This latter geometry is much better suited to search for giant cell patterns (Simon and Weiss, 1968), banana cells (Hart et al., 1986a,b), and circumferential rolls (Ribes et al., 1985; Wilson, 1987). With this new technique we have already obtained the first clear maps of the mesogranulation pattern discovered by November et al. (1981), and characterized its effect in magnetic field evolution.

It is unfortunate that we are unable to make further measurements from space of these phenomena at the present time, since no suitable spacecraft is in orbit, nor is any planned for the next 5 years. In the interim, we are developing techniques and instrumentation to obtain similar data from ground-based observations. While atmospheric turbulence will degrade the image quality and make data reduction more difficult, we believe that under reasonably good seeing conditions (1 to 2 arc sec) this powerful technique will still be viable and will permit us to make significant progress in observing and explaining the buildup and dissipation of magnetic energy at the solar surface. Such observations have been underway for the past year, and are already beginning to yield promising results.

Acknowledgments. This work was supported in part by NASA contracts NAS8-32805 (SOUP) and NAS5-26813 (HRSO). Lockheed Independent Research funds provided support for the laser optical disk analysis system. Observations at Big Bear Solar Observatory are supported by NASA under grant NGL 05 002 034 and by the NSF Solar Terrestrial program under ATM-8513577.

References

Cattaneo, F., Ph.D. Thesis, University of Cambridge, 1984.

Galloway, D., and N. Weiss, Ap. J., **243**, 945, 1981.

Galloway, D., and M. Proctor, Geophys. Astrophys. Fl. Dyn., **24**, 109, 1983.

Hart, J., G. Glatzmaier, and J. Toomre, J. Fluid Mech., **173**, 519, 1986a.

Hart, J., J. Toomre, A. Deane, N. Hurlburt, G. Glatzmaier, G. Fichtl, F. Leslie, W. Fowlis, and P. Gilman, Science, **234**, 61, 1986b.

Hurlburt, N., and J. Toomre, Ap. J., **327**, 920, 1988.

Leighton, R., R. Noyes, and G. Simon, Ap. J., **135**, 474, 1962.

Martin, S., Solar Phys., in press, 1988.

Meyer, F., H. Schmidt, N. Weiss, and P. Wilson, M.N.R.A.S., **169**, 35, 1974.

Meyer, F., H. Schmidt, G. Simon, and N. Weiss, Astron. Astrophys., **76**, 35, 1979.

Mikic, A., D. Barnes, and D. Schnack, Ap. J., **328**, 830, 1988.

Mosher, J., Ph.D. Thesis, California Institute of Technology, 1977.

Muller, R., and B. Mena, Solar Phys., **112**, 295, 1987.

Nordlund, Å., in First Workshop on Theoretical Problems in High Resolution Solar Physics (München, September 1985), edited by H. Schmidt (München, Max Planck Institut für Astrophysik, MPA 212), p. 101, 1985a.

Nordlund, Å., in Workshop, Small Scale Magnetic Flux Concentrations in the Solar Photosphere, (Göttingen, October 1985), edited by W. Deinzer, M. Knölker, and H. Voigt (Göttingen, Vandenhöck and Ruprecht), p. 83, 1985b.

November, L., and G. Simon, Ap. J., 333, 1988.

November, L., J. Toomre, K. Gebbie, and G. Simon, Ap. J. (Letters), **245**, L123, 1981.

November, L., G. Simon, T. Tarbell, A. Title, and S. Ferguson, in Second Workshop on Theoretical Problems in High Resolution Solar Physics (Boulder, September 1986), edited by G. Athay (Washington, NASA Conference Publication 2483), p. 121, 1987.

Parker, E., Ap. J., **174**, 499, 1972.

Parker, E., Ap. J., **256**, 292, 1982.

Parker, E., Ap. J., **264**, 642, 1983.

Proctor, M., and N. Weiss, Rep. Prog. Phys., **45**, 1317, 1982.

Ribes, E., P. Mein, and A. Mangeney, Nature, **318**, 170, 1985.

Schmidt, H., G. Simon, and N. Weiss, Astron. Astrophys., **148**, 191, 1985.

Schröter, E. and H. Wöhl, Solar Physics, **42**, 3, 1985.

Sheeley, N., Solar Phys., **9**, 347, 1969.

Sheeley, N., in Proceedings of IAU Symposium 43, p. 310, 1971.

Sheeley, N., and A. Bhatnagar, Solar Phys., **19**, 338, 1971.

Simon, G., and R. Leighton, Ap. J., **140**, 1120, 1964.

Simon, G., Z. Ap. **65**, 345, 1967.

Simon, G., and N. Weiss, Z. Ap., **69**, 435, 1968.

Simon, G., A. Title, K. Topka, T. Tarbell, R. Shine, S. Ferguson, H. Zirin, and the SOUP Team, Ap. J., **327**, 964, 1988.

Smithson, R., Solar Phys., **29**, 365, 1973.

Title, A., T. Tarbell, G. Simon, and the SOUP Team, Adv. Space Res., **6**, 253, 1986.

Title, A., T. Tarbell, and K. Topka, Ap. J., **317**, 892, 1987.

Van Ballegooijen, A., Ap. J., **311**, 1001, 1986.

Vrabec, D., in Proceedings of IAU Symposium 43, p. 329, 1971.

Weiss, N., M.N.R.A.S., **183**, 63p, 1978.

Wilson, P., Solar Phys., **110**, 59, 1987.

Zwaan, C., Solar Phys., **60**, 213, 1978.

What new insight can be gained by understanding the three-dimensional structure and temporal variations of macroscale and microscale plasma and electromagnetic phenomena in the solar wind and in planetary magnetospheres?

SIMULATING THE MAGNETOSPHERE: THE STRUCTURE OF THE MAGNETOTAIL

Raymond J. Walker,[1] Tatsuki Ogino,[2] Maha Ashour-Abdalla[1,3]

Abstract. Global magnetohydrodynamic (MHD) simulations of the interaction between the solar wind, the magnetosphere, and the ionosphere are potentially useful as aids in interpreting spacecraft observations because they provide us with a tool to calculate the three-dimensional and time-dependent structure of the magnetosphere. To demonstrate the application of a global MHD model to the magnetosphere, we have simulated the structure of the magnetotail as a function of dipole tilt. We used a high resolution three-dimensional code with either 0.6 R_E or 1.0 R_E grid spacing. For small tilt angles the neutral sheet forms an arc across the tail in the Y-Z plane in GSM coordinates which is similar to that obtained from empirical models. The entire neutral sheet shifts in the north-south direction as the dipole tilt is changed. The shape of the nightside magnetosphere changes with tilt in order to preserve equal areas and hence equal magnetic flux in the northern and southern lobes.

Introduction

It is often difficult to interpret unambiguously spacecraft observations of the magnetosphere because the magnetosphere is a highly dynamic three-dimensional system and spacecraft data are limited to time series observations constrained to the spacecraft trajectory. Global magnetohydrodynamic (MHD) simulations of the interaction between the solar wind, the magnetosphere, and the ionosphere are potentially useful as aids in interpreting observations of this large, highly dynamic system because the resulting model is both three-dimensional and intrinsically time dependent.

[1] Institute of Geophysics and Planetary Physics, University of California, Los Angeles, CA 90024

[2] Research Institute of Atmospherics, Nagoya University, Toyokawa, Aiche 442, Japan

[3] Physics Department, University of California, Los Angeles, CA 90024

In a global MHD model we solve the MHD equations throughout the entire system as an initial value problem. The calculations are carried out on a mesh within the calculational domain or simulation box. An important feature of the models is that the boundary conditions are set at the edges of the simulation box and that the physical boundaries of the problem such as the bow shock and the magnetopause are formed naturally. In a typical simulation, at time t = 0, the solar wind is introduced at the inflow boundary (the left edge in Figure 1) and the subsequent time evolution of the system

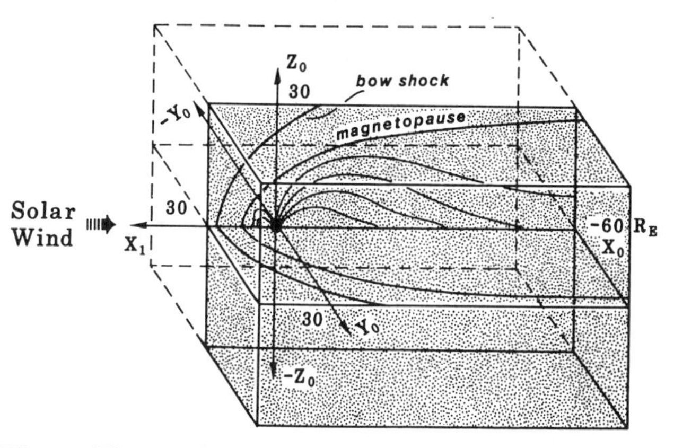

Fig. 1. The simulation box used in the tilted magnetosphere calculations. The solar wind enters the simulation box at $x = x_1$. The calculation was limited to the shaded area and mirror boundary conditions were applied at the noon-midnight meridian.

is calculated by self-consistently integrating the MHD equations. The resulting models offer the hope of integrating the diverse single point observations into a picture of the entire magnetospheric system.

The first global MHD model was developed 10 years ago (Leboeuf et al., 1978). This was a two-dimensional model. The same group developed the first three-dimensional code as well (Leboeuf et al., 1981). These simulations used a southward interplanetary magnetic field (IMF) and obtained a Dungey-type open magnetosphere. Another early model was developed by C. C. Wu and coworkers (Wu et al., 1981; Wu, 1983). Wu's later simulations used a computational grid with variable grid size so that higher resolution was obtained near magnetospheric boundaries. Wu modeled the magnetosphere for cases with no IMF and northward IMF. These earliest attempts at global MHD simulations reproduced many of the large-scale features of the magnetosphere.

Later models concentrated on more accurate representation and on including new physics in the simulations. Two-dimensional models were produced by Lyon and coworkers (Lyon et al., 1980, 1981) while three-dimensional models were produced by Brecht and coworkers (Brecht et al., 1981, 1982) and Ogino and coworkers (Ogino and Walker, 1984; Ogino et al., 1985, 1986b; Ogino, 1986). Brecht et al. (1981) studied the effects of IMF B_y on the configuration of the magnetotail. Ogino and coworkers have studied the generation of field-aligned currents in the Earth's magnetosphere and looked at cases with both northward and southward IMF. They also studied the effects of IMF B_y on the current systems. Recently Fedder and Lyon (1987) presented a new model. This model uses a grid numerically generated to provide high resolution in selected areas and has a simple model for ionospheric conductivity. All previous models used boundary conditions which treat the ionosphere as though it had infinite conductivity.

As we noted above, the global MHD models are intrinsically time dependent. The first three-dimensional, time-dependent magnetospheric calculation was by Brecht et al. (1982) who modeled tail reconnection. More recently Walker et al. (1987) used the Ogino code to simulate a magnetospheric substorm.

Global MHD simulation codes have been used to model magnetospheres other than the Earth's. Schmidt and Wegmann (1980, 1982; Wegmann et al., 1987), Fedder et al. (1984), Ogino et al. (1986a, 1988), and Sydora and Raeder (1988) investigated cometary magnetospheres. Brecht and Smith (1985) are developing a Venus simulation model. Walker and Ogino (1986) simulated the jovian magnetosphere. Most recently Linker (1987) developed a model of the interaction of jovian plasma with the moon Io.

Recently, we (Walker and Ogino, 1988) have tried to extend the simulation studies at the Earth by linking the model results to observations. We did this for two reasons. First the models have advanced to the point where we need to calibrate them. We need to learn the limitations of the calculations and to learn what new physics must be added to the models to properly model the magnetosphere. Second, we need to start using the models as a tool to interpret the spacecraft observations. In the first study, we compared plasma flow and field-aligned current observations from polar orbiting satellites with magnetospheric values from the simulations. In general we found good qualititive agreement between the model convection pattern and the flow along the spacecraft trajectory and good agreement between the observed locations of parallel currents and the current pattern predicted from the model. However, we noted some systematic differences as well. In the model we constrained the IMF to the Y_{GSM} - Z_{GSM} plane and assumed that the dipole tilt was zero. Predictably we found the best agreement between the calculation and observations when $B_x \approx 0$ and tilt $\approx 0°$ and that there were significant differences at large tilt angles and large values of B_x. Both the B_x component and the dipole tilt cause north-south asymmetries which were not included in the model.

In this paper we will demonstrate the use of global MHD simulations to study the magnetospheric configuration and report on our recent progress in including north-south asymmetries in the simulation by modeling the structure of the magnetotail as a function of dipole tilt.

Shape and Position of the Neutral Sheet and Plasma Sheet

That the shape and position of the neutral sheet and plasma sheet in the tail are influenced strongly by the tilt of the dipole axis was recognized from the earliest tail observations (Murayama, 1966; Speiser and Ness, 1967; Russell and Brody, 1967). When the dipole tilt (θ) is zero, the neutral sheet ($B_x = 0$) is close to the equator in geocentric solar magnetospheric (GSM) coordinates. The diurnal and annual variations in θ cause the neutral sheet to move northward and southward with respect to the equator. Near the Earth the neutral sheet lies in the magnetic equator while deeper in the tail the control of the tail configuration by the solar wind forces the neutral sheet to be parallel to the tail axis (see Figure 1a of Gosling et al., 1986). The place where this change occurs is frequently called the "hinging" distance and has been estimated to lie between 5.25 and 11 R_E (Murayama, 1966; Speiser and Ness, 1967; Russell and Brody, 1967; Fairfield and Ness, 1970; Bowling, 1974; Fairfield, 1980). Russell and Brody (1967) recognized that in the plane normal to the tail axis ($Y_{GSM} - Z_{GSM}$ plane), the surface was curved so that it was farthest from the GSM equator near the tail axis and returned to the axis near the magnetopause. Russell and Brody suggested that the neutral sheet was approximately a semicircle (see Figure 1b of Gosling et al., 1986).

More recently, Fairfield (1980) suggested that if the magnetopause diameter remains constant as the tilt changes, then the neutral sheet must cross the GSM equator near the magnetopause in order that the areas in the north and south lobes and hence the magnetic flux in the lobes be equal. Gosling et al. (1986) expanded on a suggestion by Bowling and Russell (1976) that the curvature of the neutral sheet decreases with increasing distance down the tail. In particular Gosling and coworkers developed a model based on ISEE observations in the near-Earth tail in which the neutral sheet crosses the equator much nearer the tail axis ($Y_{GSM} \approx \pm 12$ R_E) than it does in the model of Fairfield ($Y_{GSM} \approx \pm 18$ R_E) which is based on IMP data from deeper in the tail.

The general features of the hinged, curved neutral sheet have been confirmed theoretically by using a magnetohydrostatic equilibrium calculation by Voigt (1984) and by an earlier simulation by Wu (1984).

The Simulation Model

Our simulation model has been described in detail elsewhere (Ogino, 1986; Ogino et al., 1985, 1986b) so we will describe only the main features here. We have solved the MHD and Maxwell's equations as an initial value problem by using the modified two-step Lax-Wendroff scheme. The normalized resistive MHD equations that we solve are written as follows

$$\partial\rho/\partial t = -\nabla\cdot(\mathbf{v}\rho) + D\nabla^2\rho \quad (1a)$$

$$\partial\mathbf{v}/\partial t = -(\mathbf{v}\cdot\nabla)\mathbf{v} - (\nabla p)/\rho + (\mathbf{J}\times\mathbf{B})/\rho + \mathbf{g} + \mathbf{\Phi}/\rho \quad (1b)$$

$$\partial p/\partial t = -(\mathbf{v}\cdot\nabla)p - \gamma p\nabla\cdot\mathbf{v} + D_p\nabla^2 p \quad (1c)$$

$$\partial\mathbf{B}/\partial t = \nabla\times(\mathbf{v}\times\mathbf{B}) + \eta\nabla^2 B \quad (1d)$$

$$\mathbf{J} = \nabla\times(\mathbf{B} - \mathbf{B}_\mathrm{d}) \quad (1e)$$

where ρ is the plasma density, $\mathbf{v}$ is the flow velocity, p the plasma pressure, $\mathbf{B}$ the magnetic field, $\mathbf{J}$ the current density, $\mathbf{g}$ the gravity force, $\mathbf{\Phi} \equiv \mu\nabla^2\mathbf{v}$ the viscosity, $\gamma = 5/3$ the ratio of specific heats, $\eta = \eta_o(T/T_o)^{3/2}$ the resistivity, and T/T_o is the temperature normalized by its value in the ionosphere. The units for distance, velocity, and time are the Earth's radius, $R_E = 6.37\times10^6$ m, $v_A = 6.80\times10^6$ m s^{-1}, and the Alfvén transit time across one Earth radius , $\tau_A = R_E/v_A = 0.937$s. The numerical parameters are $\eta_o = 0.01$, and $\mu/\rho_{sw} = D = D_p = 0.003$, where ρ_{sw} is the solar wind density. The magnetic Reynolds number, which is the magnetic diffusion time divided by the Alfvén transit time, is $S = \tau_\eta/\tau_A = 100 - 1000$.

The viscosity and diffusion terms in (1a), (1b), and (1c) were added to reduce the MHD fluctuations which occur because there are unbalanced forces at time t = 0. These numerical fluctuations occur around the bow shock and the magnetopause and have scale sizes of one grid space. The values of D, D_p and μ where chosen to be large enough to suppress these fluctuations yet small enough that they do not significantly influence the global magnetospheric structure. The value of η was chosen to be larger than the numerical magnetic diffusion. For a more detailed discussion of the effects of the viscous and resistive terms see Ogino (1986).

In the simulation, a uniform solar wind with $n_{sw} = 5$ cm^{-3}, $v_{sw} = 300$ km s^{-1}, and $T_{sw} = 2\times10^5$ °K flows into a simulation box of dimensions $x_o \le x \le x_1$, $0 \le y \le y_o$ and $-z_o \le z \le z_o$ at $x = x_1$ where typically $x_o = -60$ R_E, $x_1 = y_o = z_o = 30$ R_E (Figure 1). Free boundary conditions, where the derivatives of all physical quantities are zero, are used at $x = x_o$, $y = y_o$, and $z = \pm z_o$. Mirror boundary conditions were used at $y = 0$. The ionospheric boundary condition imposed near the Earth was determined by requiring a static equilibrium (Ogino, 1986). All of the model parameters (ρ, v, p, B) were held constant for $\xi \equiv (x^2+y^2+z^2)^{1/2} < 3.5$ R_E and all perturbations were damped out by using a smoothing function near the ionosphere ($\xi \le 5.5$ R_E). The MHD equations were solved on either a $(N_x, N_y, N_z) = (90, 30, 60)$, (120, 40, 80) or (160, 50, 100) point grid. The boundary grid points are not included in these numbers. The mesh size was either 1 R_E or 0.6 R_E and the time step, Δt was taken to be $4\Delta x/v_A = 3.7$ s (for $\Delta x = 1$ R_E) or 2.25 s (for $\Delta x = 0.6$ R_E). This assured that the numerical stability criterion, ${v_g}^{max}\Delta t/\Delta x < 1$, where ${v_g}^{max}$ is the maximum group velocity in the calculation domain, was met.

The Magnetotail as a Function of Dipole Tilt

In Figure 2 we have plotted magnetic field lines calculated by using the high resolution version of the code for the case when $\theta = 0°$. For this and all of the remaining examples in the paper the IMF was set to zero.

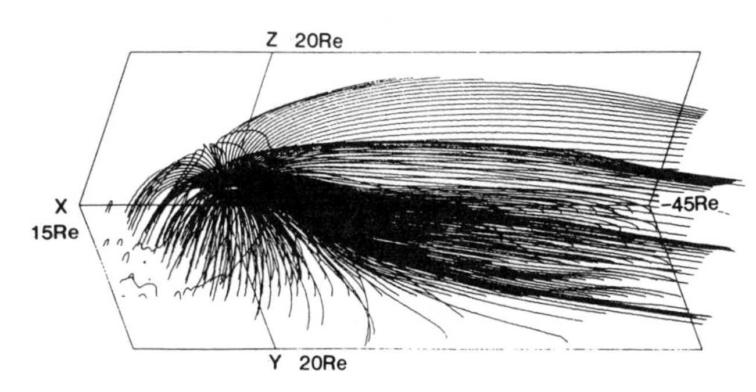

Fig. 2. Magnetic field lines for the case with zero dipole tilt and no IMF.

The magnetospheric configuration was plotted after 1536 time steps (~48 min). This is sufficient for the solar wind to flow approximately 1.8 times the length of the simulation box. We have "open" field lines in the lobes of the tail even though there is no IMF because of the free boundary conditions at the back plane of the simulation box. This configuration appears to be quasi-static, as it changes little during the last half of the run. Note that the closed field lines in the plasma sheet have a concave shape. This is characteristic of all of our runs and may be associated with maintaining a stable magnetotail configuration over the interval of the simulation.

We have plotted the field configuration for the case when the dipole tilt was ~ 20° in Figure 3. In this calculation we used the highest resolution version of the code with the largest number of grid points. The convex and rounded field lines map to the magnetopause. As we will see later these field lines are found in the region of the magnetosphere where the neutral sheet dips below the GSM equator. Nearer the axis of the magnetosphere the field configuration is very much as we would expect. Near the Earth the magnetic equator dominates and in the near-Earth plasma sheet the field lines start to bend away from the magnetic equator and become parallel to the solar wind.

The plasma and field configuration for the run in Figure 3 has been plotted in Figure 4. In the left column the magnetic field magnitude (B), the plasma pressure (P), and the density (ρ) have been plotted in the noon-midnight meridian. Cross sections in the $Y_{GSM} - Z_{GSM}$ plane at $x = -18.3$ R_E and $x = -36.3$ R_E are displayed in the middle and left panels. All of the quantities have been plotted as contour maps with the largest values shaded. The bent plasma sheet can be most clearly seen in the pressure panel on the left. The thickness of the central plasma sheet decreases with distance down the tail. However there is an extended plasma sheet characterized by lower pressures and densities which is north and south of the central plasma sheet and which increases in thickness with distance.

The neutral sheet and plasma sheet have the expected arched shape in the cross sectional plots. In these panels we can also see that the extended plasma sheet is thickest at the noon-midnight meridian. At $x = -36.3$ R_E the extended plasma sheet has a tongue of plasma extending mainly south from the central plasma sheet.

The entire magnetosphere develops north-south asymmetry for non-zero tilt. Note that the cross sections of the magnetosphere become "egg-shaped" in the middle and right panels. A careful examination of the areas in the north and south lobes shows that they are equal. Thus the entire magnetosphere changes its shape when the tilt changes in order to keep equal magnetic flux in the two tail lobes.

Figures 5a and 5b show the tail configuration in the noon-midnight meridian for tilt values between 15° and 90°. These plots represent the configuration 64 min after the simulation started. The grid spacing was 1 R_E in this run. The left panels show the magnetic field with small arrows whose length is proportional to the field magnitude. The region containing vec-

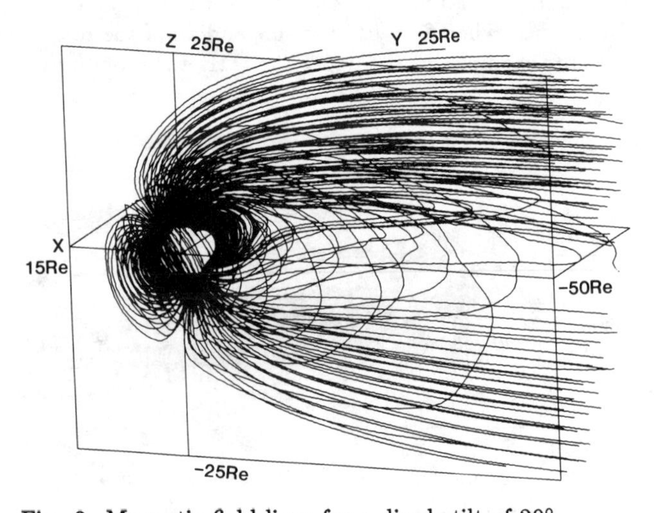

Fig. 3. Magnetic field lines for a dipole tilt of 20°.

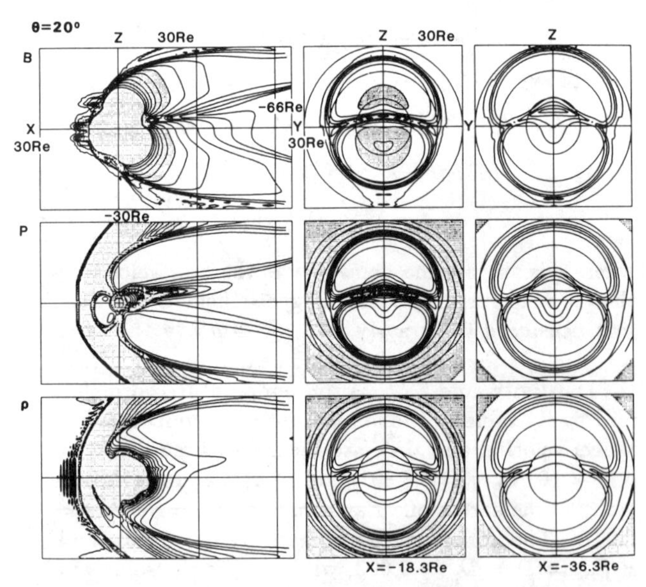

Fig. 4. Cross sectional plots of the magnetic field magnitude (B), the pressure (P) and the mass density (ρ). The left column shows values in the noon-midnight meridian while the center and right columns give the values at $x = -18.3$ R_E and $x = -36.3$ R_E, respectively. The contours with the largest values are shaded.

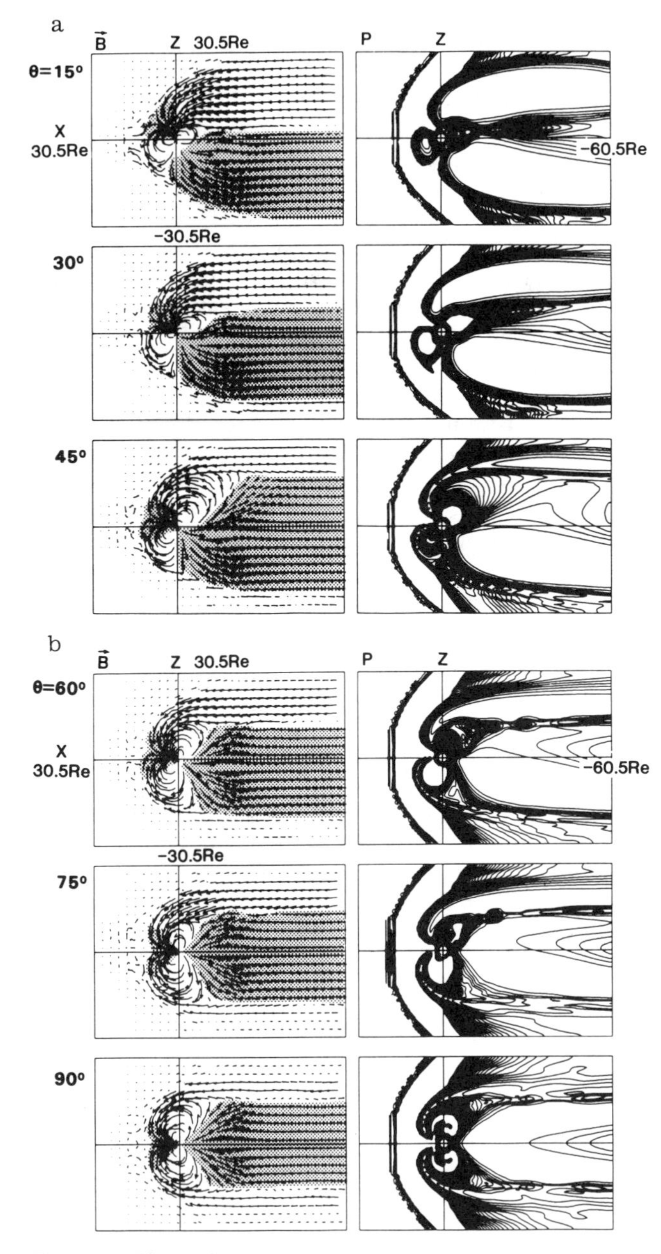

Fig. 5a. The tail configuration in the noon-midnight meridian for tilt angles of 15°, 30°, and 45°. The arrows in the left column give the direction and magnitude of the magnetic field. In the right column the plasma pressure values are contoured with the largest coutours shaded. b. Same as Figure 5a except for tilt angles of 60°, 75°, and 90°.

tors which point away from the Sun has been shaded. The right panels show a contour map of the plasma pressure. As the tilt increases from 15° to 45° the distance from the neutral sheet to the equator increases. When θ increases above 45° a second neutral sheet appears in the southern hemisphere. For $\theta = 90°$ the two neutral sheets are symmetric about the equator. The thickness of the extended plasma sheet increases with increasing tilt until $\theta = 45°$. At larger tilts the extended plasma sheet becomes narrower. Finally at larger tilts the plasma sheet pressure becomes structured and contains local pressure maxima and minima along its length. This feature is slowly propagating tailward. This suggests that this set of runs has not reached a quasi-steady state. However this should not markedly affect the determination of the location of the neutral sheet.

The configurations in the Y_{GSM} - Z_{GSM} planes at $x = -15.5$ R_E, $x = -30.5$ R_E and $x = -45.5$ R_E are displayed as a function of dipole tilt in Figures 6a - 6c. For each tilt the pressure is plotted in the left-hand column while the right-hand column shows the region of earthward ($B_x > 0$) and tailward ($B_x < 0$, shaded) magnetic field. The intersection of the earthward and tailward regions denotes the location of the neutral sheet. For $\theta = 0°$ the plasma sheet is in the GSM equator and near the Earth. As expected the thickness increases with distance away from the x axis. Deeper in the tail at $x = -45.5$ R_E there is a localized thickening of the plasma sheet near midnight. At larger θ the plasma sheet and neutral sheet form arcs with greater curvature for increasing tilt. The neutral

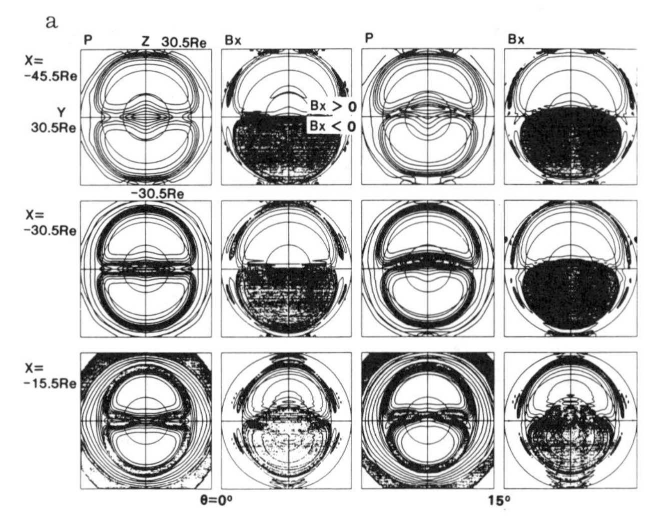

Fig. 6a. The tail configuration in the Y_{GSM}-Z_{GSM} plane at $x = -15.5$ R_E, $x = -30.5$ R_E, and $x = -45.5$ R_E. The left-most pair of columns was calculated for a tilt of 0° while the tilt was 15° for the right-most pair. In each pair the left column gives the plasma pressure. The right column of each pair contains contours of the B_x component of the magnetic field. $B_x < 0$ has been shaded.

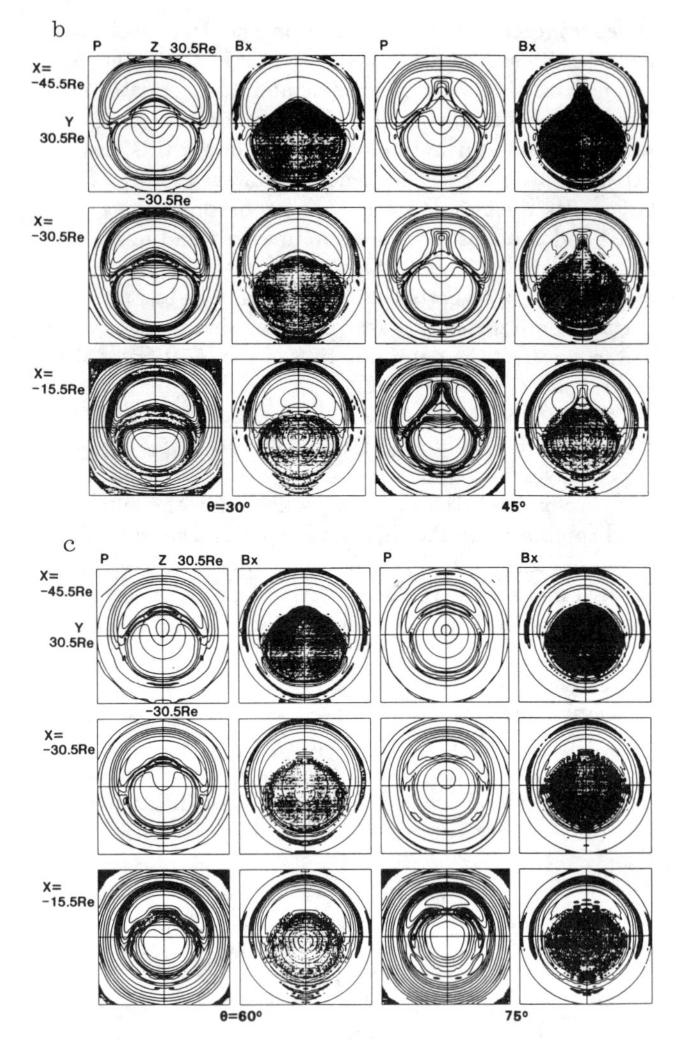

Fig. 6b. Same as Figure 6a except for dipole tilt angles of 30° and 45°. c. Same as Figure 6a except for dipole tilt angles of 60° and 75°.

sheet and plasma sheet cross the GSM equator at y coordinates away from midnight. The y value at which the neutral sheet crosses the equator increases with distance. This is most easily seen for $\theta > 15°$. For $\theta = 45°$ part of the plasma sheet extends northward toward the magnetopause. A narrow region of tailward pointing magnetic field also extends far above the equator. This feature is not sensitive to our choice of simulation region. When we repeated this calculation with the outer boundaries at $y_o = z_o = 40.5$ R_E instead of 30.5 R_E this plasma sheet extension did not change.

For $\theta > 45°$ the neutral sheet and plasma sheet have oval shapes. The two neutral sheets in the Figure 5 are the intersection of the noon-midnight meridian with the neutral sheet oval. At a tilt of 90° (not shown) the neutral sheet is a circle and the plasma sheet is a ring.

Summary and Conclusions

The configuration of the magnetotail produced from the simulation is consistent with that derived from satellites in the Earth's magnetosphere. As observed the central plasma sheet and neutral sheet move in the north-south direction as the dipole tilt changes. In addition the model plasma sheet and neutral sheet have a warped or arched cross section in the Y_{GSM} - Z_{GSM} plane as is observed. One feature not found in the empirical studies is that the shape of the entire magnetosphere changes as the tilt changes in order to keep the magnetic flux in the northern and southern lobes constant.

In Figure 7 we provide a direct comparison between the neutral sheet structure and location determined from the simulation and the empirical models of the neutral sheet position and shape. In the top row we have plotted the neutral sheet positions determined from the magnetic field plots in Figure 6 at $x = -30$ R_E as a solid line. The neutral sheet determined from the Fairfield (1980) model has been plotted as a dashed line. Similarly in the bottom row we have plotted the neutral sheet from the Gosling et al. (1986) study and the simulated neutral sheet at $x = -15$ R_E. We chose to compare the simulated values at $x = -30$ R_E with

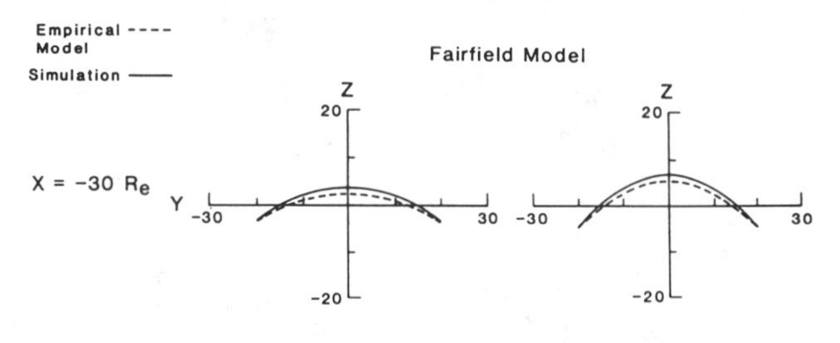

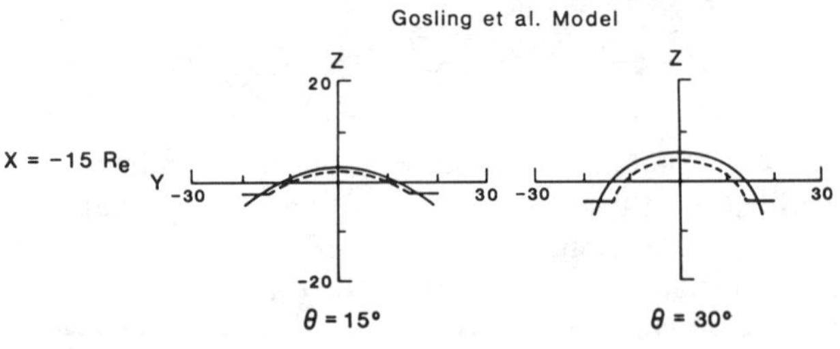

Fig. 7. A comparison between the simulated neutral sheet and that determined from empirical models. In the top row the neutral sheet from Fairfield's (1980) model has been plotted in the X_{GSM}-Y_{GSM} plane with a dashed line and the simulated neutral sheet at $x_{GSM} = -30$ R_E has been plotted with a solid line. In the bottom row the simulated neutral sheet at $x = -15$ R_E has been plotted along with the neutral sheet from the empirical model of Gosling et al. (1986).

the Fairfield model and the values at $x = -15$ R$_E$ with the Gosling et al. model because Fairfield used data from IMP spacecraft which are from deeper in the tail than the ISEE data used by Gosling et al.

All in all there is reasonable agreement between the empirical models and the simulation. In general the simulation curves lie above the curves from the empirical models. There are several possible reasons for this. First, there is considerable scatter in the observations. The scatter is typically 1 R$_E$ or more (cf. Gosling et al., 1986). Second, the grid spacing in the simulation creates a resolution problem for small features. Even though the grid spacing is as fine as 0.6 R$_E$, we probably cannot determine the location of the neutral sheet to better than two or three grid spaces. We currently are completing a study in which we use the simulation results to develop a new analytic model for the position of the neutral sheet and magnetopause.

Wu (1984) also compared his model with the empirical models. He chose to compare his calculation for $\theta = 30°$ at $x = -22$ R$_E$ with Fairfield's model. At that distance the neutral sheet is about 5 R$_E$ above the GSM equator in both the empirical model and the simulation. Using the curves in Figure 5, we estimate that the neutral sheet in our calculation is also about 5 R$_E$ above the equator at $x = -22$ R$_E$. This suggests that IMP observations may correspond to an average distance closer to the Earth than $x = -30$ R$_E$. In the simulation the neutral sheet is asympotically approaching a value of the distance from the neutral sheet which is larger than Fairfield's.

Voigt's (1984) magnetohydrostatic model produced a neutral sheet and a plasma sheet that moved with the tilt angle and had a curved cross section. As Wu (1984) has pointed out, the values for the distance between the neutral sheet and the equator from Voigt's model are generally smaller than the observed values. This is perhaps because the magnetopause cannot change its shape in Voigt's model. Voigt's calculations also indicated that the neutral sheet may be less curved in the deep tail than in the near-Earth tail. Gosling et al. (1986) reached a similar conclusion based on a comparison between the ISEE and IMP observations (see Figure 7). Simulation results in Figure 6 seem to confirm this in that near the Earth the neutral sheet crosses the GSM equator nearer the axis of the tail than it does in the deep tail.

Siscoe (1971) first noted that the neutral sheet would connect to the magnetopause for small pitch angles but would form a closed cylinder within the magnetosphere at $\theta = 90°$. Voigt et al. (1983) found a similar configuration in a pre-Voyager model of Uranus magnetosphere. Our calculation confirms this effect. The critical tilt angle above which the neutral sheet no longer reaches the magnetopause is about 45°. This critcial angle is probably never exceeded in the Earth's magnetosphere since $\theta < 35°$. However, there is at least the possibility of reaching the critical angle in Uranus' magnetosphere where the angle between the Z_{GSM} axis and the dipole becomes as large as 38°.

Future Code Development and Uses

As noted in the introduction this model which includes the dipole tilt is only one step in building a model without symmetry planes. Next we will investigate the effects of the IMF B_x component on the magnetospheric configuration. Then we will consider cases without symmetry planes and include the full IMF in the calculation. This hopefully will provide a tool that can be used to help organize spacecraft observations. It should be especially useful in multi-spacecraft missions such as the forthcoming International Solar Terrestrial Physics program.

In addition we plan to extend the tilted model to planetary magnetospheres, particularly Uranus. Models are possibly more important in studies of planetary magnetospheres because at the planets the observations are even more limited than at the Earth.

One area in which the model requires considerable improvement is at the ionospheric boundary. We need to add a much more realistic treatment of the ionospheric conductivity including a day-night asymmetry.

Acknowledgments.The work in the United States was supported by NASA Solar Terrestrial Theory Program grant NAGW-178 and NASA Uranus Data Analysis Program grant NAGW-1179. The work in Japan was supported by grants in aid from the Ministry of Education, Science and Culture.

References

Bowling, S. B., The influence of the direction of the geomagnetic dipole on the position of the neutral sheet, J. Geophys. Res., 79, 5155, 1974.

Bowling, S. B., and C. T. Russell, The position and shape of the neutral sheet at 30-R_E distance, J. Geophys. Res., 81, 270, 1976.

Brecht, S. H., and D. F. Smith, Three-dimensional simulations of the Venusian magnetosphere (abstract), EOS Trans. AGU, 66, 1037, 1985.

Brecht, S. H., J. G. Lyon, J. A. Fedder, and K. Hain, A simulation study of east-west IMF effects on the magnetosphere, Geophys. Res. Lett., 8, 397, 1981.

Brecht, S. H., J. G. Lyon, J. A. Fedder, and K. Hain, A time dependent three-dimensional simulation of the Earth's magnetosphere: Reconnection events, J. Geophys. Res. 87, 6098, 1982.

Fairfield, D. H., A statistical determination of the shape and position of the geomagnetic neutral sheet, J. Geophys. Res., 85, 775, 1980.

Fairfield, D. H., and N. F. Ness, Configuration of the

geomagnetic tail during substorms, J. Geophys. Res., 75, 7032, 1970.

Fedder, J. A., and J. G. Lyon, The solar wind-magnetosphere-ionosphere current voltage relationship, Geophys. Res. Lett., 14, 880, 1987.

Fedder, J. A., S. H. Brecht, and J. G. Lyon, MHD simulation of a comet, NRL Memo. Rpt., 5397, 1984.

Gosling, J. T., D. J. McComas, M. F. Thomsen and S. J. Bame, The warped neutral sheet and plasma sheet in the near-Earth geomagnetic tail, J. Geophys. Res., 91, 7093, 1986.

Leboeuf, J. N., T. Tajima, C. F. Kennel, and J. M. Dawson, Global simulations of the time-dependent magnetosphere, Geophys. Res. Lett., 5, 609, 1978.

Leboeuf, J. N., T. Tajima, C. F. Kennel, and J. M. Dawson, Global simulations of the three-dimensional magnetosphere, Geophys. Res. Lett., 8, 257, 1981.

Linker, J. A., The interaction of Io with the plasma torus, UCLA Ph.D. dissertation, 1987.

Lyon, J. G., S. H. Brecht, J. A. Fedder, and P. J. Palmadesso, The effect on the earth's magnetotail from shocks in the solar wind, Geophys. Res. Lett., 7, 712, 1980.

Lyon, J. G., S. H. Brecht, J. D. Huba, J. A. Fedder, and P. J. Palmadesso, Computer simulation of a geomagnetic substorm, Phys. Rev. Lett., 46, 1038, 1981.

Murayama, T., Spatial distribution of energetic electrons in the geomagnetic tail, J. Geophys. Res., 71, 5547, 1966.

Ogino, T., A three-dimensional MHD simulation of the interaction of the solar wind with the Earth's magnetosphere: The generation of field-aligned currents, J. Geophys. Res., 91, 6791, 1986.

Ogino, T., and R. J. Walker, A magnetohydrodynamic simulation of the bifurcation of the tail lobes during intervals with a northward interplanetary magnetic field, Geophys. Res. Lett., 11, 1018, 1984.

Ogino, T., R. J. Walker, M. Ashour-Abdalla, and J. M. Dawson, An MHD simulation of B_y-dependent magnetospheric convection and field-aligned currents during northward IMF, J.Geophys.Res., 90, 10,835, 1985.

Ogino, T., R. J. Walker, and M. Ashour-Abdalla, An MHD simulation of the interaction of the solar wind with the outflowing plasmas from a comet, Geophys. Res. Lett., 13, 929, 1986a.

Ogino, T., R. J. Walker, M. Ashour-Abdalla, and J. M. Dawson, An MHD simulation of the effects of the interplanetary magnetic field B_y component on the interaction of the solar wind with the Earth's magnetosphere during southward interplanetary magnetic field, J. Geophys. Res., 91, 10,029, 1986b.

Ogino, T., R. J. Walker, and M. Ashour-Abdalla, A three dimensional MHD simulation of the interaction of the solar wind with Comet Halley, UCLA IGPP Publ. 3047 , J. Geophys. Res., in press, 1988.

Russell, C. T. and K. I. Brody, Some remarks on the position and shape of the neutral sheet, J. Geophys. Res., 72, 6104, 1967.

Schmidt, H. U., and R. Wegmann, MHD-calculations for cometary plasmas, Comp. Phys. Comm., 19, 309, 1980.

Schmidt, H. U., and R. Wegmann, Plasma flow and magnetic fields in comets, in Comets, edited by L. L. Wilkering, p. 538, University of Arizona Press, Tucson, 1982.

Siscoe, G. L., Two magnetic tail models for Uranus, Planet. Space Sci., 19, 483, 1971.

Speiser, T. W., and N. F. Ness, The neutral sheet in the geomagnetic tail: Its motion, equivalent currents and field line reconnection through it, J. Geophys. Res., 72, 131, 1967.

Sydora, R. D., and J. Raeder, A particle MHD simulation approach with application to a global comet-solar wind interaction model, in Cometary and Solar Plasma Physics, edited by B. Buti, World Scientific Publ. Co., in press, 1988.

Voigt G.-H., The shape and position of the plasma sheet in the earth's magnetotail, J. Geophys. Res., 89, 2169, 1984.

Voigt, G.-H., T. W. Hill, and A. J. Dessler, The magnetosphere of Uranus: Plasma source, convection and field configuration, Ap. J., 266, 390, 1983.

Walker, R. J., and T. Ogino, A magnetohydrodynamic simulation of the interaction of the solar wind with the jovian magnetosphere (abstract), 2nd Neil Brice Memorial Symposium, 3, 1986.

Walker, R. J., and T. Ogino, Field-aligned currents and magnetospheric convection: A comparison between MHD simulations and observations, Modeling Magnetospheric Plasma, Geophys. Monogr. Ser., 44,edited by T. E. Moore and J. H. Waite, p.39, AGU, Washington, 1988.

Walker R. J., T. Ogino, and M. Ashour-Abdalla, A global magnetohydrodynamic model of magnetospheric substorms, SPI Conference Proceedings and Reprint Series, in press, 1988.

Wegmann, R., H. U. Schmidt, W. F. Huebner, and D. C. Boice, Cometary MHD and chemistry, Astron. Astrophys., 187, 339, 1987.

Wu, C. C., Shape of the magnetosphere, Geophys. Res. Lett. 10, 545, 1983.

Wu, C. C., The effects of dipole tilt on the structure of the magnetosphere, J. Geophys. Res., 89, 12, 11,048, 1984.

Wu, C. C., R. J. Walker, and J. M. Dawson, A three dimensional MHD model of the earth's magnetosphere, Geophys. Res. Lett., 8, 523, 1981.

INSTRUMENTATION FOR GLOBAL MAGNETOSPHERIC IMAGING VIA ENERGETIC NEUTRAL ATOMS

R. W. McEntire and D. G. Mitchell

Johns Hopkins University, Applied Physics Laboratory, Laurel, MD 20707

Abstract. The global imaging of magnetospheric hot plasma regions by the remote detection of their emission of energetic neutral atoms (ENAs) is a promising technique which awaits the development of appropriate instruments and flight opportunities. Measurements by the medium-energy particle instrument on ISEE 1 have demonstrated the feasibility of this approach, and an optimized instrument should provide a much richer scientific return. Many of the components and subsystems of such an instrument have a strong heritage or can be directly adapted from recent flight-proven instruments designed for more conventional measurements. Factors contributing to a particular choice of instrument configuration include scientific goals, orbit, expected signal and background levels, spacecraft stabilization, spacecraft resources, etc. The remote sensing of global magnetospheric dynamics is a fundamentally important scientific objective, and future planetary magnetospheric missions will probably include ENA imaging capability. Different mission parameters may make desirable fundamentally different sensor design approaches, but the technology for these sensors is now in hand.

Introduction

Energetic neutral atoms (ENAs) are atoms with energies of a few to hundreds of kiloelectron volts (keV) emitted from hot magnetized plasmas when energetic, singly-charged plasma ions undergo a charge exchange interaction with cold ambient neutral background gas atoms (Figure 1). The newly created ENAs, since they are no longer charged and therefore not affected by magnetic or electric fields, travel in ballistic trajectories (straight lines at these energies), so that all hot plasmas in a magnetospheric environment can be thought of as radiating ENAs continuously. The techniques we will discuss exploit this phenomenon by forming images as if the ENAs were photons. Furthermore, by analyzing the energy and species of the ENAs, these techniques can provide considerable information about the hot plasma source regions.

Until recently, trapped magnetospheric plasmas have been observable only in situ, so that one gets a single point or at best a few separate point measurements at any one time. Temporal and spatial variations are often difficult to distinguish, and a global perspective of magnetospheric plasma dynamics can only be inferred. With ENA imaging, global magnetospheric plasma dynamics can be directly and remotely observed. The spatial and temporal development of hot plasma composition and energy can be continuously monitored, providing for the first time the capability of monitoring the phenomenological relationship between the different plasma regions of the magnetosphere. Since ENA production is

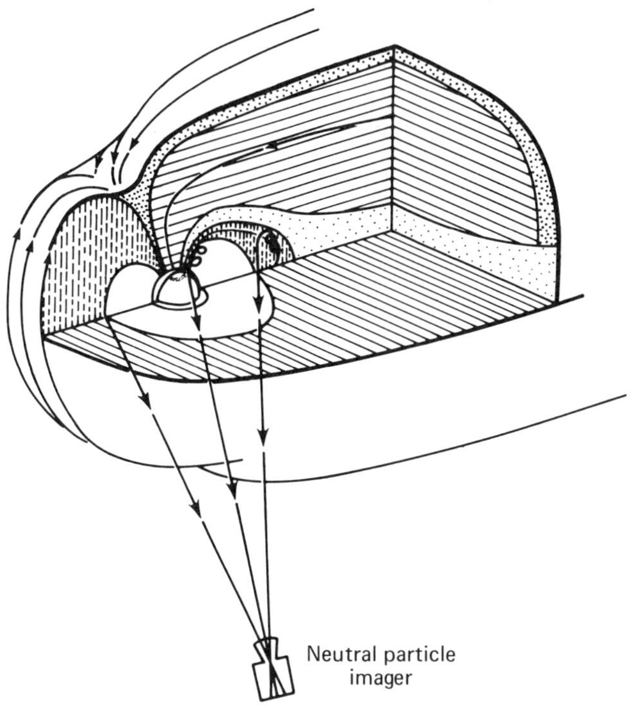

Neutral particle imager

Fig. 1. Schematic of the magnetosphere, with ENAs being emitted from the ring current and imaged remotely. Major plasma regions depicted are (1) ring current (vertical broken lines), (2) plasma mantle (dense stiple), (3) plasma sheet (sparse stiple), (4) equatorial plane (diagonal lines), and (5) plasmasphere (blank).

proportional to the geocoronal neutral gas density, which peaks at low altitudes, the best statistical measurements of ENA flux will come from the low-altitude regions of greatest importance in the study of magnetosphere/ionosphere/atmosphere coupling.

ENAs have been observed emanating from the magnetospheres of Jupiter and Saturn (Kirsch et al., 1981a,b; Cheng, 1986) as well as from Earth (reviewed by Tinsley, 1981). Analysis of IMP and ISEE observations at Earth (Roelof et al., 1985; Roelof, 1987) have strongly demonstrated the potential for ENA imaging. In all of the above cases ENAs were measured by instruments designed for energetic ion detection, during periods when the ambient energetic particle fluxes were very low and ENA fluxes could dominate the instrument response. In this paper, we discuss various generic types of instruments designed specifically for imaging and characterizing ENAs. Other papers in this volume discuss recent design and laboratory development efforts leading toward specific ENA instrumentation for future missions (Keath et al., 1989; Hsieh and Curtis, 1989; Curtis and Hsieh, 1989) and explore some of the contributions an ENA imager could make to a future mission to Saturn (Cheng and Krimigis, 1989).

For future missions to study the magnetospheres of either the Earth or the outer planets, ENA imager data with appropriate resolution, deconvolved using numerical modeling techniques (Roelof, 1987), should make it possible to monitor the global dynamics and distribution of the magnetospheric energetic plasma and neutral gas populations. Specific areas of interest would include substorm evolution in space and time; growth, evolution, and decay of energetic trapped particle populations; plasma sheet dynamics; magnetospheric convection; magnetosphere-ionosphere and magnetosphere-satellite coupling; and the dynamics of magnetospheric boundaries.

These goals will require an ENA imager whose characteristics (field-of-view, angular resolution, mass resolution, energy range, etc.) will, of necessity, represent a compromise chosen for each specific mission. We will now discuss the ENA and mission parameters which enter into these trade-offs.

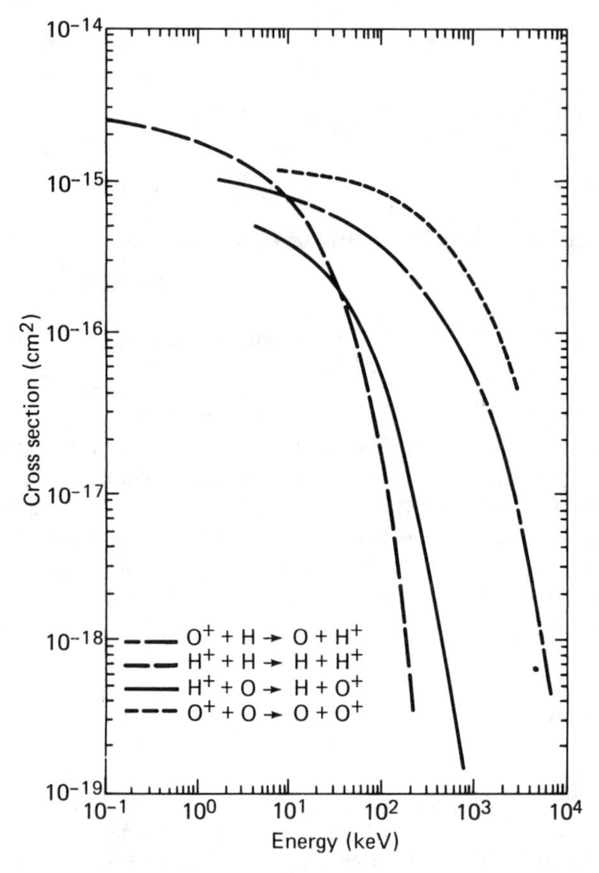

Fig. 2. Charge-exchange cross sections (σ_{10}) as a function of total ion energy for electron pickup by singly-charged energetic ions (H^+ and O^+) from cold (geocoronal and/or atmospheric) neutral hydrogen or oxygen. The four curves in the order shown in the legend are from: Phaneuf et al., 1978; McClure, 1966; Stier and Barnett, 1956; and Lo and Fite, 1970. Note that since the pickup cross section is a strong function of velocity, the cross sections for energetic oxygen (which is slow relative to the same energy hydrogen) extend to much greater total energy before dropping quickly.

ENA Parameters

ENAs are created by charge-exchange interactions between energetic magnetospheric ions and the neutral exosphere. Although other reactions occur, and may be important in other environments, the interactions shown in Figure 2 dominate ENA production in the Earth's magnetosphere. Other minor ENA species (e.g., helium, nitrogen, carbon, and heavier elements) are also of interest, from the perspective of tracing plasma sources (ionospheric or solar wind) and processes. Thus, composition resolution sufficient to identify *H*, *He*, *O*, and possibly *N*, *C* and heavier elements, is desirable in ENA imager instrumentation.

ENA production is a function of ion energy and species, and of neutral exospheric density and composition. The ENA unidirectional flux for species i is given by the line of sight integral

$$f_i(E)\ (\mathrm{cm}^2\ \mathrm{s\ str\ keV})^{-1} = \sum_k \sigma_{ik}(E) \int j_i\ (E)\ n_k(l)dl \tag{1}$$

where $n_k(l)$ is the density of the neutral exospheric gas constituents with which charge-exchange collisions are likely (primarily hydrogen, and at low altitudes, oxygen atoms), $j_i(E)$ is the directional singly-charged energetic ion flux along the direction of the line of sight at each point l along the line for species i within the source volume, and $\sigma_{ik}(E)$ is the energy-dependent charge exchange cross section σ_{10} for the species involved.

The energy range of an ENA imager will be limited by a variety of design considerations. Since ENAs carry the energy of their parent ions, the desired goals would be to image ENAs from less than 1 keV to many MeV in order to cover the full range of energetic particles in the magnetosphere. In practice an energy range from < 10 keV to more than 1 MeV should be achievable. Other techniques not discussed here may cover the very low energy range.

For all regions of the Earth's magnetosphere except low altitudes (≤2000 km) the only neutral species present in significant numbers is hydrogen. However, below 2000 km the relative importance of neutral atmospheric densities of H and O is a strong function of solar activity, as shown in Figure 3. Thus, modeling of low-altitude ENA fluxes can be complex and inherently time dependent. At high altitudes, the problem is more tractable, and there are both models and data from which we can compute neutral hydrogen densities and expected ion flux levels and spatial distributions. The fluxes of energetic ions are highly variable but extensively measured in the ring current most recently by the AMPTE/CCE spacecraft (Krimigis et al., 1985). The neutral hydrogen density conforms approximately to the Chamberlain model (Figure 4). As expected, it falls off quickly with radial distance,

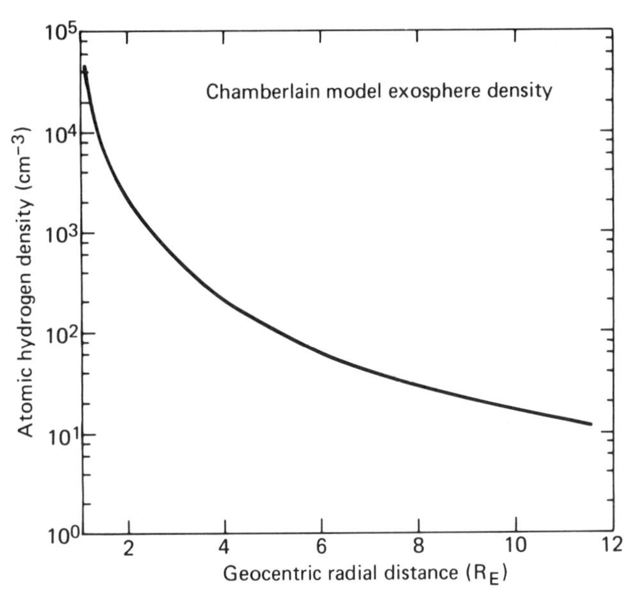

Fig. 4. Exospheric hydrogen density verses radial distance for the Chamberlain model at T = 1050, which provides the best fit to the Dynamics Explorer 1 geocoronal observations (from Rairden et al., 1986).

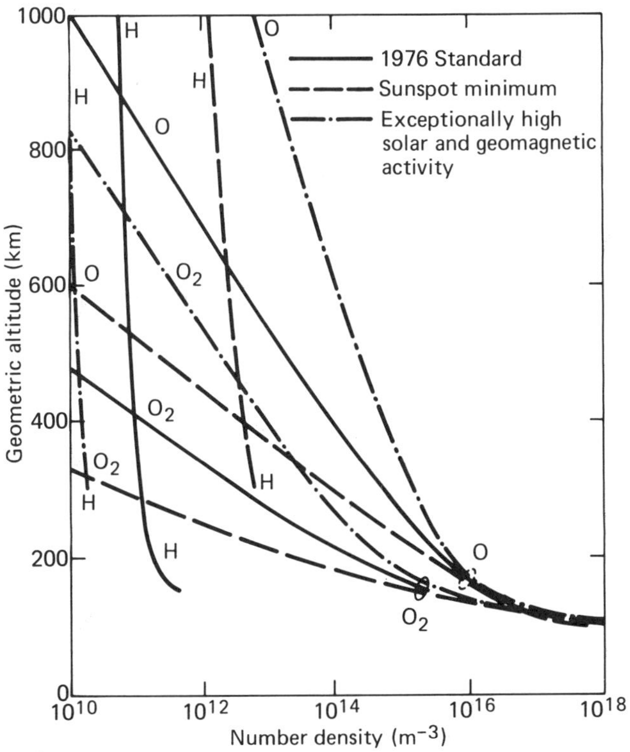

Fig. 3. Atmospheric density versus altitude for H, 0, and 0_2 for sunspot minimum, maximum, and average activity. Note that for different solar and geomagnetic conditions, either H or 0 can dominate the neutral densities from ~500 to ≳1000 km. However, even when 0 is dominant, the H scale height is much greater, so that above ~2000 km H always dominates (from U.S. Standard Atmosphere, 1976).

but with an increasing scale height with radius such that even at 10-12 R_E, n_H ~ 10 cm^{-3}. The high geocoronal density at low altitudes results in relatively enhanced ENA emission from these regions (equation (1)), but large source region scale lengths can result in significant ENA emission for most magnetospheric regions (Table 1).

The charge-exchange cross sections for the dominant species are given in Figure 2. The most important are those corresponding to H^+ and O^+ interacting with neutral exospheric hydrogen. The H^+ cross section begins to decrease strongly for proton energies above 10 keV, and drops precipitously above 50-100 keV. This is a most important constraint for ENA production, and it assures (equation (1)) that ENA hydrogen spectra will normally be concentrated below ~100-200 keV. The O^+ cross section drops off significantly only well above 1 MeV, and thus the ENA oxygen spectra will more directly reflect the source O^+ spectra up to much higher energies. However, the equilibrium flux of 0^+ itself drops rapidly at higher energies, or higher charge states become dominant (Spjeldvik and Fritz, 1978), so ENA observations will emphasize energies of a few megaelectron volts (MeV) and below.

The low-altitude region is of particular interest in monitoring the precipitation of energetic ions into the ionosphere/upper atmosphere. The relatively higher ENA flux from this region provides more favorable conditions for high time and spatial resolution imaging of this precipitation. Note that the presence of neutral oxygen at low altitudes (Figure 3) significantly increases the low-altitude, high energy ENA flux (see Figure 2). Also, since the exobase neutral density rises very

Table 1. Energetic Neutral Atom Fluxes

Approximate Distant (point source) Fluxes*				Extended source fluxes at the Earth**	
				Region	ENA flux ($cm^{-2}\ s^{-1}\ str^{-1}\ keV^{-1}$)
Earth	$\sim \frac{100}{R^2}$	Jupiter	$\sim \frac{400}{R^2}$	Low-altitude precipitation region	$\lesssim 1000$ ($\gtrsim 20$ keV)
Saturn	$\sim \frac{250}{R^2}$	Uranus	$< \frac{15}{R^2}$	Equatorial storm-time ring current (L = 3-5)	$\lesssim 15$ ($\gtrsim 30$ keV)
				Equatorial quiet-time ring current	~ 0.1-1 ($\gtrsim 30$ keV)
				Plasma sheet	~ 0.1-1 ($\gtrsim 20$ keV)

* Approximate integral fluxes ($cm^{-2}\ sec^{-1}\ keV^{-1}$) $\gtrsim 40$ keV. R is in planetary radii (Krimigis et al., 1988).
**Measured and modeled (Roelof, 1987), or estimated from ion fluxes and neutral densities.

quickly at low altitudes, the ENA emission from this region will have the more two-dimensional characteristic familiar from auroral emissions, providing the "phosphor screen" for auroral and sub-auroral energetic ion precipitation as the optical and UV aurora does for energetic electron precipitation.

In designing an imaging ENA instrument, one wishes to maximize angular and temporal resolution to follow the plasma dynamics at all time and spatial scales. Intrinsic time scales range from minutes (substorm onset) to days (ring current decay) and spatial scales from $>1\ R_E$ (outer magnetospheric structure) to >10 km in the auroral zone. However, as we will discuss in the instrumental section, low flux levels and strong background sources require compromises which limit practical resolution.

For an estimate of the ENA flux we rely on measurements by the ISEE 1 medium-energy particle instrument (MEPI) at Earth, by the Voyager LECP instrument at Jupiter and Saturn, and on modeling. At Earth, directional ENA fluxes above ~ 30 keV from the ring current range from <1 to $>10^3\ cm^{-2}\ s^{-1}\ str^{-1}\ keV^{-1}$, varying strongly with geomagnetic activity. There are indications that low altitude fluxes will be higher at times, perhaps by as much as a factor of 10^2, but such high fluxes would usually be of short duration since they would only occur when the parent ion pitch angle distribution loss cone was full, and charge exchange would then act as a strong loss process for that parent population. ENA fluxes at some finite level will be produced throughout the magnetosphere at all times, but practical instrumentation will have flux related limits below which statistical significance and/or signal-to-noise ratios become too low. For first-generation instrumentation, this practical limit is $\approx 1\ cm^{-2}\ s^{-1}\ str^{-1}\ keV^{-1}$. Rough estimates of ENA fluxes from the magnetospheres of the Earth, Jupiter, and Saturn are given in Table 1.

ENA Imaging - Mission Considerations •

For any mission on which an ENA imager will fly, the orbit and attitude stabilization of the spacecraft (and to a lesser extent, spacecraft resources) will have a major impact on the instrument design and achievable scientific goals.

Altitudes below ~ 400 km are not very useful, since the neutral density below that altitude is often high enough that multiple charge exchange (i.e., pick-up, stripping, and re-pick-up) will be common, and the original source region information will be lost; the particle velocity vector is effectively randomized while it is charged, and the imager will be bathed in a cloud of locally-generated ENAs.

As has been mentioned, however, relatively low altitudes are the source of the highest ENA fluxes. Low orbits in the range of 500 km to $\sim 2\ R_E$ would be best for obtaining high spatial resolution images of the low-altitude ENA emissions. For this purpose and for all orbits other than those with very high apogee, high orbit inclination is desirable. In order to maximize angular coverage from low altitudes, where the imager is surrounded by potential source regions, a spinning spacecraft would be best. One disadvantage is that a relatively short orbital period (typically ~ 100 min for 500-1000 km circular orbit altitudes) will hamper continuous monitoring of the development of substorm related events which evolve on timescales of 1 to 3 h. These altitudes are also useful for viewing the magnetospheric ring current region from beneath.

Orbits from ~ 2 to $\sim 10\ R_E$ are well suited to viewing the ring current, but spatial resolution of low-altitude precipitation regions as seen from far above will be quite limited. If the ring current and outer magnetosphere are of significant priority, a spinning spacecraft is again desirable for greatest field of view. For precipitation region studies, however, a three-axis oriented, Earth pointing spacecraft is best. These altitudes would be best for global viewing of the near-Earth manifestations of substorm and storm development.

Orbits above 10 R_E, out to perhaps 40 R_E, are best for a global perspective of the whole magnetosphere. For these orbits the continuous coverage of a three-axis oriented, Earth pointing instrument is highly desirable since ENA fluxes can be small, and temporal resolution will be determined by the observing time needed to accumulate reasonable statistics.

For most ENA instrument designs considered here, 2° is roughly the limit of the angular resolution attainable. At a distance of 1 R_E, an instrument with 2° angular resolution will resolve spatial structure down to ~200 km; at 10 R_E this figure is 2000 km and at 40 R_E it is 1.3 R_E. Beyond 40 R_E, the poor resolution of spatial structure would increasingly limit the usefulness of such an instrument.

Characteristics of Instrumentation to Image Energetic Neutral Atoms

Particle detectors designed to detect and image ENAs share many characteristics with, and use technology closely derived from, detectors now being flown to directly measure energetic charged particle populations in space. However, since ENAs are not deflected by magnetic or electric fields, imagers all rely upon straight path optical techniques and detectors sensitive to fast particles. In addition, the generally low flux of ENA (Table 1) requires instruments with large geometry factors and large detectors. These features introduce three potential sources of background which must be suppressed: photons (primarily EUV), ambient energetic charged particles, and penetrating cosmic radiation. Thus, almost every neutral particle imager design will carry out the functions of (1) charged particle rejection, (2) cosmic ray (penetrating particle) rejection, (3) photon (vacuum ultraviolet) suppression, and (4) ENA detection, with determination of the incident direction of each individual ENA event, and possibly the energy and mass (species) of each neutral atom. These generic ENA imager characteristics are discussed individually below, and then representative integrated instrument designs are developed.

Charged Particle Rejection

The remote source populations from which ENAs are emitted are, of course, energetic charged particles, and for most missions on which ENA imagers may be flown, there are likely to be significant local charged particle fluxes as well. In fact, in many cases the local energetic charged ion and electron fluxes in the imager environment can be orders of magnitude larger than the ENA flux to be measured. The actual detection of an incident energetic atom in any sensor ultimately depends on the energy lost during that incident particle's interaction with the matter of the sensor. This process depends on an equilibrium charge state which the incident atom almost immediately assumes once it interacts with matter, and essentially not at all on the incident particle's initial charge state before it impacted the sensor. Thus, to any detector (solid state, channel electron multiplier, etc.) an incident ENA and an incident energetic charged particle of the same energy and species are indistinguishable. If the local energetic charged particle environment represents a significant contamination to the ENA flux to be measured, then that charged particle background must be removed from the imager field-of-view, or greatly reduced, prior to any interaction with a detector in the imager.

Charged particles up to some energy can be deflected out of an ENA telescope field-of-view without affecting incident ENA using either magnetic or electric fields. In fact, particle telescopes designed to measure energetic ions in space have frequently included in their collimators permanent magnets to deflect incident electrons away from the telescope sensors. This works well with reasonably achievable magnetic fields up to electron energies of hundreds of keV. The amount of deflection is proportional to $qB(mE)^{-1/2}$, where q, m and E are the charge, mass, and energy of the particle involved, so that at equal energies electrons can be completely removed while the heavier-mass ions are only slightly deflected and continue on into the telescope. However, for ENA imagers it is necessary to remove incident ions as well, and magnetics capable of deflecting very energetic ions over a significant aperture could be unacceptably heavy for space flight.

Deflection of charged particles by an electric field ϵ is proportional to $\epsilon q/E$, and thus at equal energies is the same (in magnitude, if not direction) for electrons and singly-charged ions. The first segment of an ENA imager is typically a mechanical collimator which defines the overall telescope field-of-view. If different parts (usually closely-spaced plates) of the collimator are held at high relative potentials, charged particles up to some energy can be deflected into the collimator and stopped while ENAs pass through unaffected. Charged-particle rejection collimators with a number of different geometries have been proposed. One simple and effective configuration consists of a stack of parallel metal plates of length L and separation D, with L/D chosen to define a desired field-of-view. If alternate plates in this stack are held at high voltage to establish a potential V between each pair of plates, then all charged particles with energies below a rejection energy E_R will be deflected into the collimator walls; if $L/D \gg 1$ then

$$E_R \simeq qV\left[1 + \left(\frac{L}{4D}\right)^2\right] \qquad (2)$$

For example, if $L = 12$ cm and $D = 0.4$ cm then $E_R \approx 57\ V$ for singly-charged particles, and a potential difference of 10 kV across each pair of plates would prevent direct access of ions and electrons below ~570 keV. The actual situation is more complex since the deflected ions and in particular electrons will strike collimator plates and may scatter further into the instrument. Anti-scatter grooves or baffles are a key part of rejection collimator designs, and with appropriate care rejection factors of from 10^3 to $>10^5$ can be achieved

(Keath et al., 1989). Another advantage of electrostatic rejection is that an instrument using this technique can also measure the local charged particle flux if the rejection plate voltage is removed.

Penetrating Particle Rejection

ENA imagers will typically operate with foreground count rates of from <1 to a few hundred counts per second, and thus for many missions background rejection is a key design parameter. At locations near the Earth, geomagnetically trapped Van Allen belt electron fluxes greater than 3 to 5 MeV can range up to 10^2 to 10^5 cm^{-2} s^{-1}, and proton fluxes above 100 MeV can be greater than 10^4 cm^{-2} s^{-1}. An appropriate choice of orbit, and passive shielding of the instrument, can reduce the problem of energetic penetrators to a manageable level, but at a minimum a penetrating flux of cosmic rays of $\gtrsim 1$ cm^{-2} s^{-1} will be seen. For a simple, single-detector ENA imager, active anti-coincidence shielding around the detector can reduce false event rates to acceptable levels. More complex imagers may use two or more detectors to make a multiparameter measurement of each ENA event (e.g., time-of-flight and energy) and in such cases coincidence requirements and electronic processing will greatly reduce the number of accidental events from penetrators.

Photon Suppression

A number of ENA imager designs are based on either (1) the direct detection of incident energetic neutrals impacting on a position-sensing microchannel plate (MCP) detector assembly, or (2) the measurement of particle time-of-flight (TOF) between a very thin front foil and a rear sensor, usually a solid state detector that measures total energy (McEntire et al., 1985). In this latter case low-energy secondary electrons emitted when the incident particle traverses the front foil, and when it impacts the back detector, are accelerated and mapped onto MCPs to provide start and stop signals for TOF determination (and, using position-sensitive MCP assemblies, incident particle trajectory determination). Both designs can respond to incident energetic photons as well, either by direct detection at the MCP (case 1) or through photoelectrons emitted (with $\sim 1\%$ efficiency) when photons are transmitted through the front foil or are incident on the rear detector (Hsieh et al., 1980).

Ultraviolet photon fluxes in the terrestrial space environment can be large, with the dominant line being H Lyman α at 1216 Å (10.2 eV) with a direct flux at the Earth of $\sim 3 \times 10^{11}$ photons cm^{-2} s^{-1}. This radiation is resonantly scattered by the neutral hydrogen atoms of the Earth's exosphere to create the widespread Lyman α emission of the Earth's geocorona, with intensities ranging from ~ 500 Rayleighs to ~ 10 kiloRayleighs ($\sim 5 \times 10^7$ to 10^9 photons cm^{-2} s^{-1} str^{-1}) depending on look direction and orbit location (Rairden et al., 1983, 1986). This can create unacceptable background count rates in sensitive detectors, so ENA imager designs usually must attenuate incident photon fluxes to acceptable levels. This is normally done with a thin UV blocking foil. For TOF telescopes the front foil element serves both to stop Lyman α and to provide the secondary electron "start" pulse for penetrating ENAs. Foil thicknesses of 5-10 μg cm^{-2} or less are sufficient to generate secondary electrons (Clerc et al., 1973; Girard and Belore, 1977). For UV attenuation, the necessary foil thickness depends on the Lyman α foreground expected and the instrument susceptibility to single events in the microchannel plates. TOF telescopes require start-stop coincidences within ~ 100 ns for a valid event, and so are intrinsically resistant to uncorrelated photon-induced MCP pulses. Even so, attenuations of 10^{-3} or greater can be required, implying foil thicknesses >15 μg cm^{-2} (Hsieh et al., 1980; Keath et al., 1989). At these thicknesses scattering in the front foil can significantly reduce imager angular resolution, particularly for ENAs in the 10-200 keV range and for higher Z's (Hoğberg and Nordén, 1970), and the imager energy resolution and threshold are degraded. Thus, the photon background can be reduced as far as necessary, but for a TOF telescope this may involve a significant design tradeoff between background and angular resolution. For simpler imager designs in which the blocking foil can be placed just in front of the MCP (pinhole camera, etc.) angular resolution is not degraded, but without TOF coincidence requirements significantly thicker foils will be needed to reduce background.

A quite different technique for the suppression of EUV, the use of nuclear track filters, has been suggested by Gruntman and Leonas (1986a,b). These filters consist of thin (~ 1-2 μ) foils (much thicker, however, than the foils mentioned above, which have thicknesses in the 0.01 to 0.1 μ range) bombarded with energetic heavy ions and then etched to create a high density of straight cylindrical channels of known diameter. Overall filter open area can be $\sim 10\%$, and if the channel diameter is ≤ 1000 Å Lyman α would be almost completely removed, while ENAs could pass through the channels unaffected. Such filters would probably have acceptance angles of $\leq 10°$, however, and may severely scatter particles that interact with channel walls, so laboratory development to demonstrate the applicability of this technique is needed.

Imaging Instruments - Representative Designs and Techniques

In this section we will describe four different instruments which represent typical integrated configurations. The four are: (1) a collimated scanning telescope, (2) a pinhole camera, (3) a slit imager, and (4) a trajectory and composition analyzing instrument.

Collimated Scanning Telescope

This instrument images by collimating its field of view (FOV) to one pixel resolution, and sequentially scanning in angle, sampling each image pixel individually to build up a complete image. This approach was employed in the ISEE 1 MEPI which was designed not for ENAs but for three-dimensional sampling of energetic charged particles. It consists of a solid state detector and a collimator mounted on a scanning platform to scan in elevation angle while the spacecraft spin sweeps in azimuth. While simple in concept, and already in flight, this approach includes several drawbacks. A solid state detector is relatively insensitive to EUV, but the detection threshold (approximately 25 keV) is higher than desirable for ENAs. Energetic ions are not rejected, so ENAs can only be sensed over a limited portion of the orbit when the ambient charged particle flux is very low. While the angular resolution (approximately 13° × 22°) is good for energetic particle studies, it is poor for ENAs. The basic concept (the entire FOV of the instrument forming one pixel of the image) can be combined with charged particle rejection and a detector sensitive to lower energies, but the necessary scanning motor is an undesirable complication, and image temporal resolution will be poor since only one polar angular bin can be covered per spin.

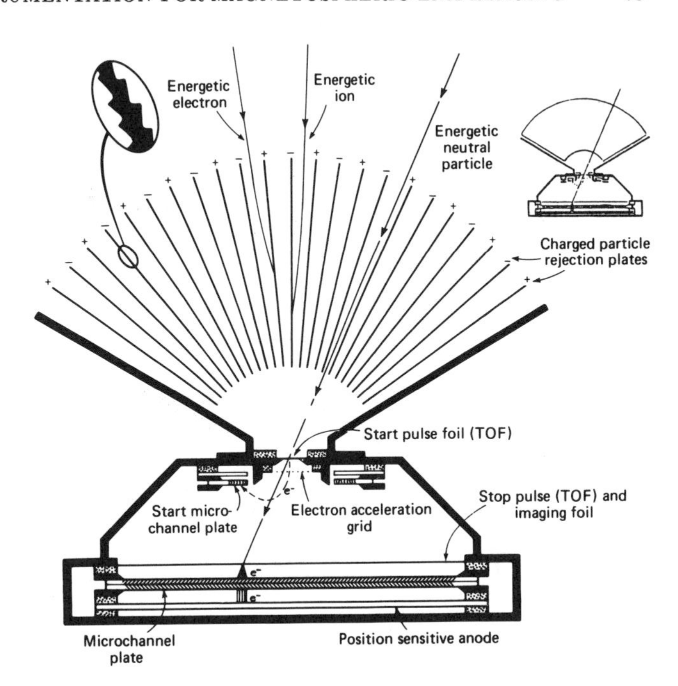

Fig. 5. ENA pinhole camera. Alternating potential on charged particle rejection plates deflects energetic ions and electrons from trajectories which would otherwise enter the pinhole. Plate surfaces are machined to inhibit forward scattering (detail). ENAs pass through the pinhole and strike a large foil close to a MCP. The foil both attenuates environmental EUV, which would otherwise saturate the MCP, and provides secondary electrons. These secondaries are amplified by the MCP and the position is recorded by a two-dimensional, position-sensitive anode. Optionally, a thin foil with a start MCP is placed in the pinhole opening, enabling the ENA velocity (TOF) to be measured. A cut in the other dimension shows the shape of the rejection plates (detail, upper right).

Pinhole Camera

In Figure 5 we show an imaging head based upon a pinhole camera concept. This approach is suited to a three-axis stabilized spacecraft where the imager stares continuously at the region of interest. Imaging is achieved by requiring the ENA to pass through a small aperture en route to the two-dimensional imaging detector. The detector consists of a secondary electron emitting thin foil, a microchannel plate to amplify the signal, and a position sensitive anode. Optionally, a similar foil can be placed in the pinhole aperture, providing secondary electrons for a timing pulse from a start-time microchannel plate so the ENAs time-of-flight through the instrument can be measured. This TOF system also provides background rejection for cosmic radiation. These same thin foils (e.g., carbon or parylene) attenuate EUV.

The fan of alternately charged aperture plates (viewed edge-on in the figure and extended over a similar angular range in the other dimension as shown in the detail) does not significantly collimate the ENA flux; their purpose is to prevent energetic charged particles from entering the aperture. Their surface is grooved (see detail) to inhibit forward scattering of the charged particles.

This instrument is capable of imaging ENAs (and, if a front foil and MCP are used, capable of providing TOF, and thus a velocity spectrum, over the entire image) while rejecting the three major sources of background (EUV, charged particles, and (with TOF, or an anticoincidence shield) cosmic rays).

Slit Imager

This instrument (Figure 6) is simply a variation on the pinhole camera just discussed, modified for application to a spinning spacecraft. The basic elements (thin foil, charged particle rejection plates, and imaging microchannel plate detector) provide the same functions as in the pinhole case.

In this instrument, the aperture is a long, narrow slit. Imaging is performed in one dimension by the detector, while the other dimension is collimated to a small angle corresponding to one pixel's width by the charged particle rejection plates, which are now parallel to each other and run perpendicular to the slit.

The instrument is mounted with the slit perpendicular to the spacecraft spin axis, so the one-dimensional imaging resolves structure in elevation angle, while the spacecraft spin sweeps the azimuth angle. Azimuthal resolution is then achieved by sectoring or time-tagging the instrument response; i.e., if the charged particle rejection plates collimate in azimuth to 2°, the response can be clocked into 180 sectors/spin to provide effective resolution of approximately 3° in azimuth,

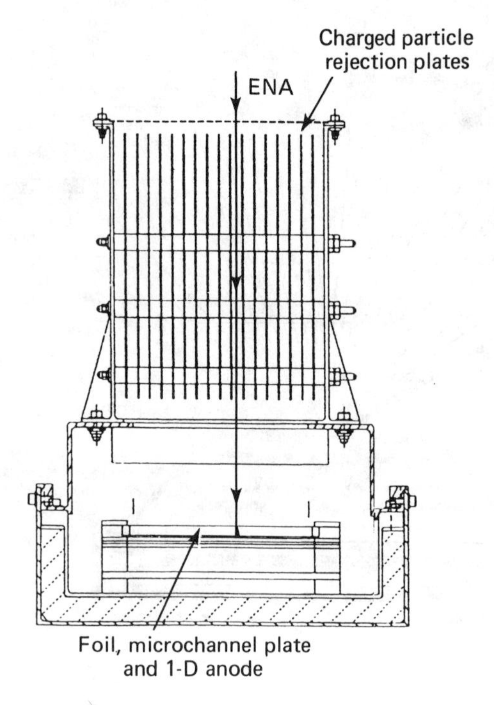

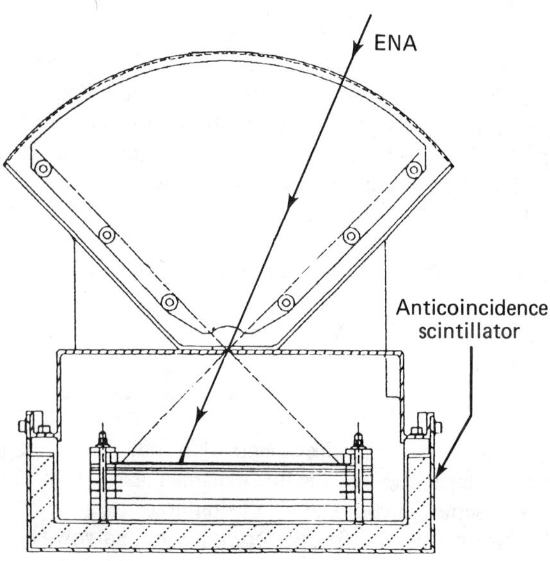

Fig. 6. Slit imager. A one-dimensional version of the pinhole, this imager works on the same principle (see text and Figure 5 caption for details). The primary differences shown here are: (1) the charged particle rejection plates are parallel, collimating the ENAs in one dimension and defining the instrument resolution in that dimension; and (2) inclusion of an anti-coincidence scintillator for cosmic ray background rejection. A grounded grid shields the high potential charged particle rejection plates from the ambient plasma, and in this version, no TOF system is shown.

with similar resolution in polar angle obtainable from the MCP position sensing anode.

Note that the slit imager, and the closely-related pinhole camera, can be operated as shown with UV blocking foils immediately in front of the MCPs, and without foils in the entrance pinhole or slit. Thus, scattering of low-energy ions in the foils is not a problem and ENA images can be obtained with resolution set only by the defining collimators and position-sensitive MCP assemblies. Realistic pixel resolutions of 2°-3° can be obtained down to integral energy thresholds of $\lesssim$10 keV (the penetration energy of typical foils used).

Trajectory and Composition Analyzing Imager

The previous instruments are rather simple and can provide good imaging of ENA fluxes. They do not, however, provide the energy and elemental resolution needed for comprehensive ENA imaging studies.

The instrument depicted in Figure 7 is based upon the AMPTE CCE medium-energy particle analyzer (MEPA) which was designed for composition and energy analysis (McEntire et al., 1985).

An incident ENA penetrates the thin grid-supported front foil and strikes the back solid state detector (SSD) where its energy is measured. Low-energy secondary electrons emitted from the surface of both the foil and the SSD when the particle penetrates are electrostatically accelerated and isochronally mapped, by an arrangement of high transparency grids, onto "start" and "stop" position-sensitive microchannel plate assemblies, whose fast output pulses allow measurement of the time-of-flight of the incident particle through the telescope. The measured particle time-of-flight (velocity) and energy determines the ENA mass, as demonstrated in Figure 8 using flight data from the MEPA instrument.

The direction from which the ENA originated, which corresponds to a particular pixel in the final image, is determined by recording the location of the secondary electron emission from both the front foil and the rear SSD. This trajectory determination can be made in either one or two dimensions depending, respectively, upon whether the charged particle rejection plates are arranged parallel (as shown) or in a fan (as in the pinhole camera). As in the other instruments, the one-dimensional approach is appropriate for a spinning spacecraft and the two-dimensional for a three-axis stabilized spacecraft. Again, EUV rejection by the front foil and the TOF measurement (and thus coincidence requirement) provides electronic rejection for most accidental events. This telescope design, using proven technology, is able to provide good images of incident ENAs as a function of energy and species over a fairly wide field-of-view. The technique does have disadvantages, however, including instrument complexity, poor angular resolution at low energy due to scattering in the front foil, and a relatively high ENA energy threshold ($\gtrsim$30 keV for protons, higher for heavier species) due to energy loss and scattering in the foil, energy thresholds on the rear SSD array, and pulse height defect in the SSD.

With the greatest flux of ENAs concentrated at low energies, poor angular, energy, and species resolution at these energies is a significant problem. One approach to providing species resolution at the lowest energies is to measure the particle velocity twice, before and after passing through a thin central foil, thus determining both the incident V and the ∇V due to energy loss in the foil (Schneider et al., 1975; Gruntman and Leonas, 1986b). This TOF × ∇TOF technique is discussed further by Hsieh and Curtis (1989). The pinhole and slit-type imagers, which do not need a front foil for trajectory determination, provide good angular resolution down to low ($\lesssim$10 keV) energies, but usually provide no energy or species resolution. These techniques can, of course, be combined with (precede) a TOF × E telescope to resolve this problem and provide good imaging $\lesssim$10 keV combined with energy and species resolution $\lesssim$30 keV.

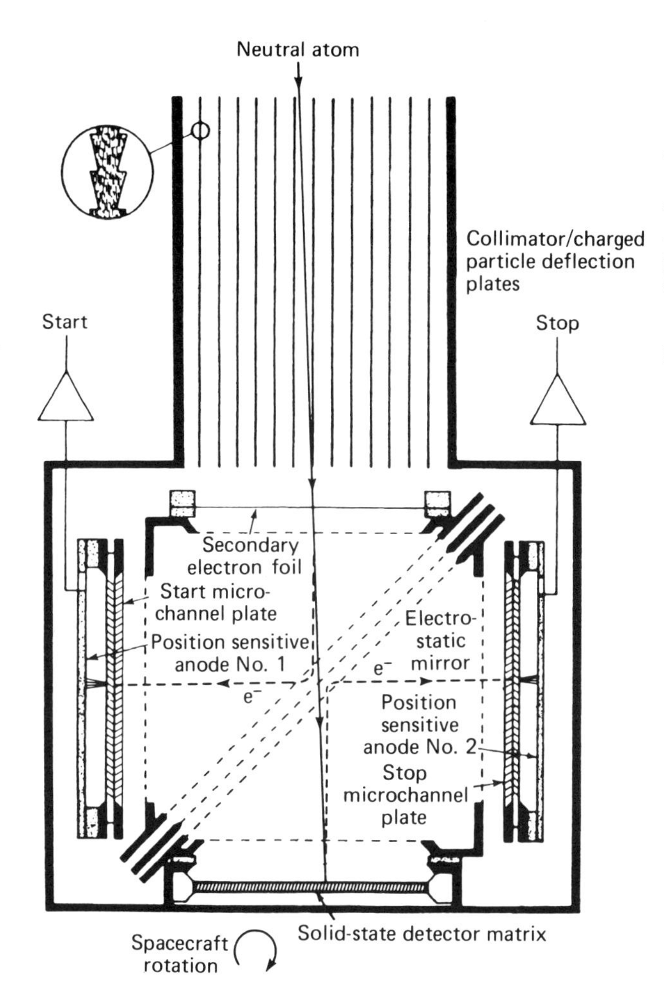

Fig. 7. Large front foil imager. As in previous designs (Figures 5 and 6), this imager includes energetic charged particle rejection plates. The ENA enter, collimated in one-dimension by the plates (or, in the two-dimensional version, from any angle within the collimator acceptance angle), and pass through a front foil. Secondaries are accelerated and deflected to a position-sensitive start MCP by potentials on the intermediate grid structure. Similarly, a stop position and time are recorded at the back of the instrument where the ENA energy is also measured with a solid state detector. This imager determines the ENA direction, energy, and TOF (composition) while increasing the geometry factor (sensitivity) by ~an order of magnitude over the pinhole and slit versions (Figures 5 and 6).

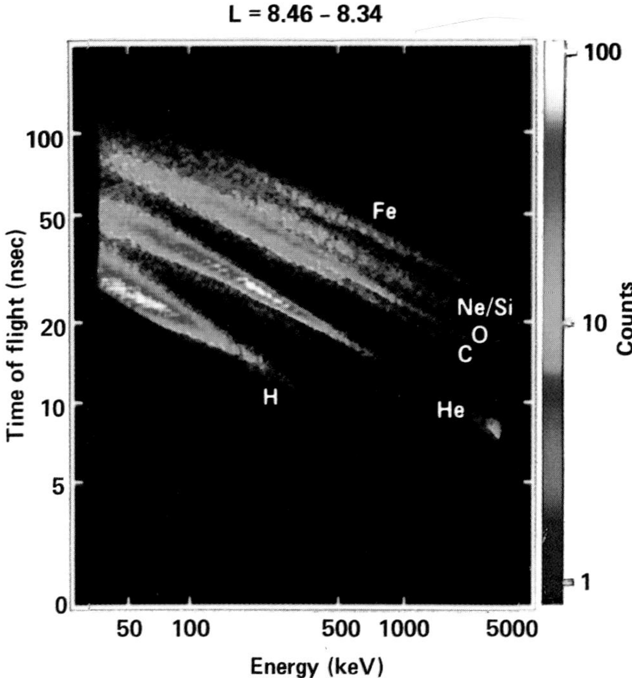

Fig. 8. Time-of-flight vs. energy matrix of magnetospheric ions obtained by the AMPTE CCE MEPA instrument on September 13, 1984 (L ~ 8.3-9). Tracks of protons, helium, carbon, oxygen, the neon-silicon group, and iron can be seen, as can the high-energy alpha source calibration peak. The TOF × E technique clearly separates the major species groups down to quite low energies.

A different kind of trajectory determination optics applicable to ENA imaging is the coded aperture technique (Gruntman and Leonas, 1986b; Caroli et al., 1987), which is discussed by Curtis and Hsieh (1989). Very briefly, the tech-

Table 2. Characteristics of ENA Imaging Techniques

Technique	Effective Geometry Factor/Pixel	Imaging Energy Threshold	Energy Measurement	Composition Measurement
Scanning Collimators	9×10^{-6}	$\gtrsim$25 keV	$\gtrsim$25 keV	No
Spinning one-dimensional Slit Imager	9×10^{-6}	$\gtrsim$10 keV	Integral	No
Spinning Front Foil TOF × E Imager	2×10^{-4}	$\gtrsim$20 keV*	$\gtrsim$30 keV	$\gtrsim$30 keV
Spinning one-dimensional Coded Aperture, TOF × E	1×10^{-4}**	$\gtrsim$10 keV	$\gtrsim$30 keV	$\gtrsim$30 keV
Pinhole, two-dimensional three-axis Stab. 90° × 90° FOV (Stepped through four positions to give 90° × 360° FOV)	9.7×10^{-6}	$\gtrsim$10 keV (no TOF) $\gtrsim$20 keV (TOF)	Integral	No
Two-dimensional, Large Front Foil TOF × E	4.6×10^{-3} (2.3×10^{-4})	$\gtrsim$30 keV*	$\gtrsim$30 keV	$\gtrsim$30 keV
Two-dimensional Coded Aperture, TOF × E	2.3×10^{-3} (1.1×10^{-4})**	$\gtrsim$10 keV	$\gtrsim$30 keV	$\gtrsim$30 keV

* Angular resolution increasingly degraded below ~100 keV, and for higher Z's, by scattering in front foil.

**Figure of merit for observation of a single point source. For extended sources coded aperture image reconstruction requires greater statistics, effectively degrading this figure of merit, although only for the low energies for which coded aperture angular reconstruction is needed for high resolution.

nique involves placing a carefully-chosen array of open and opaque cells in front of the instrument. This is the coded aperture. A point source will then cast the shadow of the closed cells on the image plane, and by comparing the shadow with the mask (via matrix algebra), the location of the point source can be determined. In the general case, an extended emission region can be thought of as the superposition of many point sources of varying intensity. By passing the image plane response through the same algebraic manipulation, the emission region structure is recovered. Advantages of this technique include good low-energy response (the ENA is not scattered in a front foil) and large geometry factor (the coded aperture can be 50% transparent) relative to a pinhole slit. Disadvantages include more image processing on the ground, ambiguities associated with "side-lobes," and the greater statistics necessary to deconvolve an extended source (i.e., for extended sources one may loose the advantage gained by the larger transmission area). This technique can also be combined with a TOF × E telescope to improve the low-energy imaging characteristics of the TOF × E system.

Instrument Sensitivity

The previous sections have outlined a variety of imaging techniques. Here we wish to briefly compare the relative sensitivity of those approaches. To accomplish that goal, we will assume a standard detector area of 60 cm^2, pixel resolution of 2° × 2°, and full angular coverage of 90° × 360°.

In Table 2 we list the effective geometry factors and resolution for each of the techniques discussed. The geometry factors include the detector area, angular FOV, any structure which reduces the transmission of ENAs, (e.g., grids, collimator plates, etc.), and duty cycle. By duty cycle we mean that portion of the total time to obtain an image spent viewing any particular pixel; for example, in the spinning/scanning telescope, collimated to $2° \times 2°$, the duty cycle $2°/90° \times 2°/360° = 1.23 \times 10^{-4}$. That, combined with a solid angle of $2° \times 2°$ ($= 1.22 \times 10^{-3}$ str) and an area of 60 cm^2 for the detector gives an effective geometry factor per pixel of $60 \times 1.22 \times 10^{-3} \times 1.23 \times 10^{-4} = 9.0 \times 10^{-6}$ cm^2 str, the Table 2 value.

As the table shows, there are essentially two classes of instruments in terms of sensitivity, those based on mechanically constraining the ENA trajectory (in the 10^{-5} cm^2 str/pixel range) and those which use the point of intersection of the ENA trajectory with a front foil (which gives a much larger front opening) to give the first of a two-point determination of the ENA arrival direction (in the 10^{-4} cm^2 str/pixel range).

Although some table entries show geometry factors in the 10^{-3} cm^2 str/pixel range, practical energetic charged particle background rejection collimators require subdivision of the aperture into many smaller apertures, reducing the per-pixel geometry factor by more than an order of magnitude to the values given in parentheses.

The instrument approaches discussed here are representative of first-generation ENA imager designs, but they are certainly not all inclusive. Other techniques, and permutations and modifications of the discussed techniques, are now under study by a number of groups. ENA imager instrumentation is an embryonic but rapidly developing field that should soon lead to a dramatically improved global view of magnetospheric plasma dynamics.

Acknowledgments. We benefited greatly in the preparationof this manuscript from discussions with E. C.Roelof and B. H. Mauk. The work was supported by NASA under grant NAGW-865.

References

Caroli, E., J. B. Stephens, G. DiCocco, L. Natalucci, and A. Spizzichino, Coded aperture imaging in X-and gamma-ray astronomy, *Space Sci. Res., 45*, 349-403, 1987.

Cheng, A. F., Energetic neutral particles from Jupiter and Saturn, *J. Geophys. Res., 91*, 4524-4530, 1986.

Cheng, A. F., and S. M. Krimigis, Energetic neutral particle imaging of Saturn's magnetosphere, these proceedings, 1989.

Clerc, H. G., H. J. Gehrhardt, L. Richter, and K. H. Schmidt, Heavy-ion induced secondary electron omission—A possible method for Z-identification, *Nucl. Instr. and Meth. 113*, 325-331, 1973.

Curtis, C. C., and K. C. Hsieh, Remote sensing of planetary magnetospheres: Imaging via energetic neutral atoms, these proceedings, 1989.

Girard, J., and M. Belore, Heavy ion timing with channel-plates, *Nucl. Instr. and Meth. 140*, 279-282, 1977.

Gruntman, M. A., and V. B. Leonas, Possibility of experimental study of energetic neutral atoms in interplanetary space, Preprint 1109, Space Research Institute, Academy of Sciences, Moscow, 1986a.

Gruntman, M. A., and V. B. Leonas, Experimental opportunity of planetary magnetosphere imaging in energetic neutral atoms, Preprint 1181, Space Research Institute, Academy of Sciences, Moscow, 1986b.

Hoğberg, G., and H. Nordén, Angular distributions of ions scattered in thin carbon foils, *Nucl. Instr. and Meth., 90*, 283-288, 1970.

Hsieh, K. C., E. Keppler, and G. Schmidtke, Extreme ultraviolet induced forward photoemission from thin carbon foils, *J. Appl. Phys., 51*, 2242-2246, 1980.

Hsieh, K. C., and C. C. Curtis, Remote sensing of planetary magnetospheres: Mass and energy analysis of energetic neutral atoms, these proceedings, 1989.

Keath, E. P., G. B. Andrews, A. F. Cheng, S. M. Krimigis, B. M. Mauk, D. G. Mitchell, and D. J. Williams, Instrumentation for energetic neutral atom imaging of magnetospheres, these proceedings, 1989.

Kirsch, E., S. M. Krimigis, J. W. Kohl, and E. P. Keath, Upper limits for X-ray and energetic neutral particle emission from Jupiter: Voyager 1 results, *Geophys. Res. Lett., 8*, 169-172, 1981a.

Kirsch, E., S. M. Krimigis, W. H. Ip, and G. Gloeckler, X-ray and energetic neutral particle emission from Saturn's magnetosphere, *Nature, 292*, 718-721, 1981b.

Krimigis, S. M., E. P. Keath, B. H. Mauk, A. F. Cheng, L. J. Lanzerotti, R. P. Lepping, and N. F. Ness, Observations of energetic ion enhancements and fast neutrals upstream and downstream of Uranus' bow shock by the Voyager 2 spacecraft, *Planet. Space Sci.*, in press, 1988.

Krimigis, S. M., G. Gloeckler, R. W. McEntire, T. A. Potemra, F. L. Scarf, and E. G. Shelley, Magnetic storm of September 4, 1984; A synthesis of ring current spectra and energy densities measured with AMPTE/CCE, *Geophys. Res. Lett., 12*, 320-332, 1985.

Lo, H. H., and W. L. Fite, electron-capture and loss cross sections for fast, heavy particles passing through gases, *Atomic Data, 1*, 305-328, 1970.

McClure, G. W., Electron transfer in proton-hydrogen-atom collisions: 2-17 keV, *Phys. Rev., 148*, 47-54, 1966.

McEntire, R. W., E. P. Keath, D. E. Fort, A.T.Y. Lui, and S. M. Krimigis, The medium energy particle analyzer (MEPA) on the AMPTE CCE Spacecraft, *IEEE Trans. Geosci. Remote Sensing, GE-23*, 230-233, 1985.

Phaneuf, R. A., F. W. Meyer, and R. H. McKnight, Single-electron capture by multiply charged ions of carbon, nitrogen, and oxygen in atomic and molecular hydrogen, *Phys. Rev. A, 17*, 534-545, 1978.

Rairden, R. L., L. A. Frank, and J. D. Craven, Geocoronal imaging with Dynamics Explorer: A first look, *Geophys. Res. Lett., 10*, 533-536, 1983.

Rairden, R. L., L. A. Frank, and J. D. Craven, Geocoronal imaging with Dynamics Explorer, *J. Geophys. Res., 91*, 13,613-13,630, 1986.

Roelof, E. C., Energetic neutral atom image of a storm-time ring current, *Geophys. Res. Lett., 14*, 652-655, 1987.

Roelof, E. C., D. G. Mitchell, and D. J. Williams, Energetic neutral atoms ($E \sim 50$ keV) from the ring current: IMP 7/8 and ISEE 1, *J. Geophys. Res., 90*, 10,991-11,008, 1985.

Schneider, W.F.W., B. Kohlmeyer, W. Pfeffer, F. Puhlhofer, and R. Bock, Properties of a time-of-flight telescope for heavy ions in the A = 100 range, *Nucl. Instr. and Meth., 123*, 93-98, 1975.

Spjeldvik, W. N., and T. A. Fritz, Theory for charge states of energetic oxygen ions in the Earth's radiation belts, *J. Geophys. Res., 83*, 1583-1595, 1978.

Stier, P. M. and C. F. Barnett, Charge exchange cross sections of hydrogen ions in gases, *Phys. Rev., 103*, 896-907, 1956.

U.S. Standard Atmosphere, 1976, NOAA document NOAA-S/T 76-1562, U.S. Government Printing Office, Washington, D.C., 1976.

Tinsley, B. A., Neutral atom precipitation—A review, *J. Atmos. Terr. Phys., 43*, 617-632. 1981.

SPECTROSCOPIC MEASUREMENTS OF SOLAR WIND PARAMETERS NEAR THE SUN

John L. Kohl, Heinz Weizer

Harvard-Smithsonian Center for Astrophysics, Cambridge, MA 02138

Stefano Livi

Max Planck Institut fuer Aeronomie, Katlenburg-Lindau, FRG

Abstract. Instrumentation and plasma diagnostic techniques are being developed to obtain a detailed empirical description of solar wind acceleration regions at heights between the coronal base and about 10 solar radii from sun center ($R_\odot$). The goal of this work is to determine a sufficient number of observational parameters to constrain, significantly, theories of solar wind acceleration, coronal heating, and solar wind composition. Although a substantial amount of data on the electron density structure of the corona already exists, there are only isolated measurements of other critical plasma parameters, except for observations of regions near the base of the corona. Ultraviolet spectroscopy provides a capability to expand greatly the number of plasma parameters that can be specified by means of remote sensing techniques. Ultraviolet measurements of spectral line profiles determine the random velocity distributions and effective temperature of protons, minor ions, and electrons. Ion densities and chemical abundances are derivable from the collisional component of the observed resonant line intensities. Outflow velocities can be determined from Doppler shifts and Doppler dimming of spectral lines. The instruments which are being developed for remote sensing of the extended corona consist of an occulted telescope system and a high resolution spectrometer. The basic design was proven on three sounding rocket flights. Initial data on proton temperatures and solar wind outflow velocities for heliospheric heights between 1.5 and 3.5 solar radii from sun center have been obtained. More powerful instruments are being developed for Spartan (a Shuttle-deployed subsatellite) and for the Solar Heliospheric Observatory mission.

Introduction

Although ultraviolet spectroscopic observations from space platforms during the past four decades (Baum et al., 1946; Goldberg et al., 1968) have dramatically increased our knowledge of the solar corona, ultraviolet spectroscopic diagnostics only have been applied, extensively, in studies of solar structures near the coronal base and in the deeper solar atmosphere. Instruments with increasingly greater spatial and spectral resolution continue to be developed (c.f. Brüeckner, 1979; Woodgate et al., 1980) for that region. There have been very few vacuum ultraviolet observations of the extended solar corona (i.e., the region beyond about 1.2 solar radii from sun center ($R_\odot$)). The first such observations of the extended solar corona occurred on March 7, 1970, when an objective grating spectrograph was flown on a sounding rocket into the path of totality of a natural solar eclipse. The initial results presented by Gabriel et al. (1971) emphasized the discovery of the unexpectedly bright HI Ly-α corona and the explanation of its production by resonant scattering of chromospheric Ly-α radiation. The attempt by G. Noci (personal communication, 1972) to derive outflow velocities from that data through use of a Doppler dimming analysis (Hyder and Lites, 1970) kindled an interest at the Center for Astrophysics in developing an instrument for observations of the extended corona in the absence of a natural eclipse.

An appropriate instrument, now called an ultraviolet coronagraph spectrometer, was designed and developed (Kohl et al., 1978) as a sounding rocket payload. It was flown together with a white-light coronagraph provided by the High Altitude Observatory on April 13, 1979, February 16, 1980, and July 20, 1982.

The first rocket flight provided measurements of the spectral line profile of HI Ly-α between 1.5 and 3.5 $R_\odot$ (Kohl et al., 1980). Data were obtained in a quiet coronal region and in a polar coronal hole. Proton temperatures in the quiet region derived from the line widths decreased with height from 2.6 x 10^6 K at 1.5 $R_\odot$ to 1.2 x 10^6 K at 4 $R_\odot$ (Withbroe et al., 1982a). These measurements combined with temperatures for lower heights, which were determined from earlier Skylab and eclipse data, suggested that there is a maximum in the quiet coronal proton temperature at about 1.5 $R_\odot$. Comparison of measured HI Ly-α intensities with those calculated using a representative model for the radial variation of the coronal electron density suggested that the solar wind was subsonic for $r < 4 R_\odot$ in the observed quiet coronal region. Comparison of the measured kinetic temperature to the predictions of a simple two-fluid model suggested that there is a small amount of proton heating and/or a nonthermal contribution to the motions of coronal protons between 1.5 and 4 $R_\odot$. The derived kinetic temperature for the coronal hole at 1.5 $R_\odot$ was 1.9 x 10^6 K compared to 2.5 x 10^6 K at the same height in the quiet region (Withbroe et al., 1986).

The mean temperature in the 1979 coronal hole was significantly higher, by ~69%, than that determined in a similar region which was observed in the 1980 rocket flight (Withbroe et al., 1985). Comparison of the measured kinetic temperatures for the 1980 hole with the predictions of a semiempirical two-fluid model suggested that there was also a small amount of proton heating or nonthermal contribution or both to the motions of coronal protons between 1.5 and 4 $R_\odot$ in the 1980 polar region. The apparent nonthermal component to the proton velocity had about the same radial dependance as might be expected for the transverse velocity component of Alfvén waves. The widths of the profiles place an upper limit of 110 ± 15 km s^{-1} on the rms magnitude of the line-of-sight component of velocities (thermal plus nonthermal) between r = 2.8 and 4 $R_\odot$. The electron densities in those two polar regions were similar and were a factor of $\sim$4 larger than in polar coronal holes observed at solar minimum. The flow velocities in both regions appeared to be subsonic for r < 4 $R_\odot$. The outward energy flux for the 1980 polar region, including the mechanical energy flux contributed by MHD waves, was estimated and used to predict flow velocities at 1 AU. If densities typically measured in situ in the solar wind near the ecliptic are applicable to polar flows, then the observed 1980 polar region is more likely to have been a source of low-speed than high-speed wind.

The polar region observed in 1982 appeared to have supersonic velocities at heights well within 3 $R_\odot$ (L. Strachan, personal communication, 1988). The results for all of the observed polar regions support the hypothesis that polar coronal holes observed at different times during the solar cycle can have much different temperatures, densities, and outflow velocities.

Following the 1982 flight, the rocket instruments were upgraded for use on Spartan, a Shuttle-deployed subsatellite. Although the original rocket observations were limited, essentially, to measurements of the spectral line profile and intensity of HI Ly-α, ultraviolet spectroscopy over a wide spectral range is practicable. For Spartan, spectroscopic diagnostics for electron temperature (based on measurements of electron scattered Ly-α) and for outflow velocities less than 250 km s^{-1} (based on measurements of the O VI resonance doublet) were added. Additional diagnostics are contemplated for more advanced instruments.

The ultraviolet spectroscopic diagnostics to be used for the extended corona have been described by Kohl and Withbroe (1982), Withbroe et al. (1982b) and Noci et al. (1987). The observational objective of the diagnostics is to obtain a detailed empirical description of the extended solar corona from the coronal base up to and possibly beyond 10 $R_\odot$. The diagnostics are based on measurements of the intensities and spectral line profiles of resonantly and Thompson scattered HI Ly-α, of collisionally excited and resonantly scattered lithium-like resonance lines such as O VI, Mg X and Si XII, and observations of primarily collisionally excited lines such as Fe XII.

The usual data analysis approach is to develop a model of the observed coronal plasma. The model is needed to account for temperature, density, and outflow velocity variations along the observed lines-of-sight. The individual plasma parameters in the model tend to be constrained by particular observables (e.g., the random velocity of a particle species is highly constrained by the spectral line profile). The models used for this purpose are relatively unconstrained by theory. For example, they include the physics of line formation and ionization balance, but do not include any treatment of heating or direct momentum deposition mechanisms.

In general, spectral line profiles provide the random velocity and temperature constraints. Spectral line shifts provide bulk velocities along the line-of-sight, and Doppler dimming of resonance lines specify the outflow velocity component that is perpendicular to the line of sight. Line profiles of species with low mass tend to be more sensitive to thermal velocities while heavier ions are strongly affected by nonthermal motions. Kohl and Withbroe (1982) have described how the intensity ratio of the members of a lithium-like resonance doublet can be used to determine the fractional amount of the emergent intensity that is due to collisional excitation as opposed to resonance scattering. The collisional component of the intensity tends to control the ion densities in the models while the resonantly scattered component controls the outflow velocity. (The resonantly scattered intensity depends on the wavelength overlap between the line profile of radiation entering the extended corona from deeper atmospheric layers and the profile for resonant scattering of that radiation by coronal ions. A bulk outflow velocity of coronal material introduces a Doppler shift between the two profiles, thus reducing the overlap and the resonantly scattered intensity.) With future instruments we expect to determine empirical models based on ultraviolet spectroscopic diagnostics that specify temperatures (thermal and nonthermal) for electrons, protons, HI, O VI, Mg X, Si XII, and Fe XII; densities for HI, O VI, Mg X, Si XII, and Fe XII; outflow velocities for the electron/proton plasma, O VI, Mg X, and Si XII; and chemical abundances for O, Si, Mg, and Fe.

The primary purpose of the present paper is to describe some of the observational problems that must be overcome by the instrument design and to describe an instrument that is capable of implementing the aforementioned observations.

The Observational Problem

An instrument for ultraviolet spectroscopy of the extended solar corona must be sensitive enough to detect intensities that are typically 1 x 10^{-6} of the solar disk intensities. It must also reduce the detected stray light level to about 8 orders of magnitude below the disk intensity at the wavelength of interest and to several additional orders of magnitude below the disk intensity at longer wavelengths. The spectral resolution should be high enough to resolve spectral line profiles for at least one ionization state of each of the most important elements observed in the solar wind. Time resolution for the more stable structures can be relatively long. Observations of coronal mass ejections (CMEs) require a capability to observe at least a few points along a CME's trajectory in a time that is compatible with the CME's passage through a spatial resolution element. Spatial resolution for visible light observations of the extended corona has traditionally been at about the 10 arc sec level. Spatial details down to this level have been observed in visible light (Poland, 1978). Lower spatial resolution can be useful for large-scale structures such as coronal holes and helmet streamers which typically have dimensions of several arc minutes.

Intensities as a function of distance r from sun center for a coronal hole and a quiet coronal region are shown in Figures 1 and 2, respectively. Streamers can be 3 to 6 times brighter than the quiet corona. Data points are from Smithsonian Astrophysical Observatory (SAO) rocket flights (Withbroe et al., 1982a; Kohl et al., 1984; L. Strachan, personal communication, 1988). Intensities of EUV lines were calculated with models based on the temperatures and densities measured in the rocket flights. The models used here to estimate the expected intensities are similar to those described

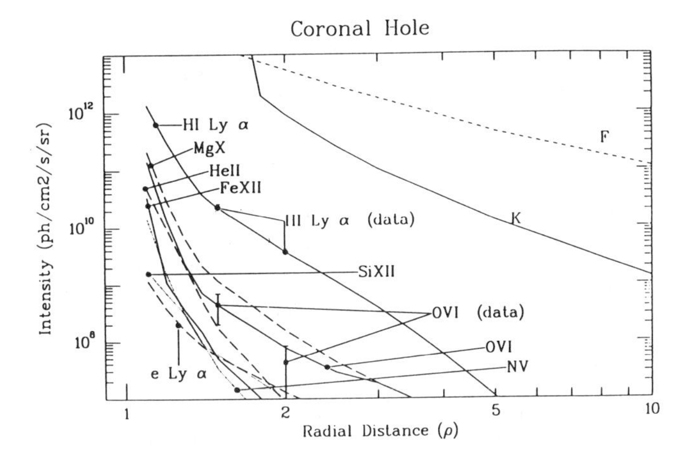

Fig. 1. An estimate of EUV intensities versus height in a coronal hole for the resonance lines of the indicated coronal ions. The estimates are based on measured Ly-α and OVI intensities (shown as data points) and Ly-α line profiles which were used to specify the parameters of the coronal model that yielded the intensities. The white light K and F coronae are also indicated.

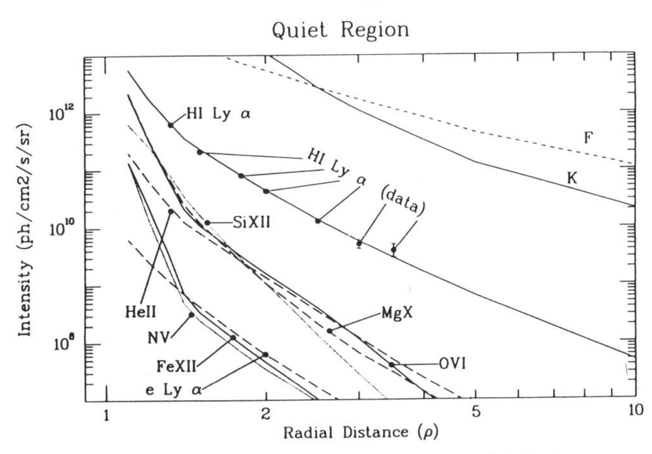

Fig. 2. An estimate of EUV intensities versus height in a typical quiet coronal region. The estimates are based on measured Ly-α intensities (shown as data points) which were used to specify the parameters of the coronal model that yielded the intensities. The visible K and F coronae are also indicated.

by Withbroe et al., 1982a. The coronal hole intensities represent a worse case where supersonic outflow occurs at 2 $R_{\odot}$. In that case Doppler dimming severely reduces the intensities for heavy ions. Higher intensities and count rates are expected in coronal holes where supersonic flow begins at larger heights. The intensities for the white light K and F coronae are from Allen (1973).

The required stray light rejection at the designated wavelength is illustrated in Figures 3 and 4 which show intensities in units of the corresponding disk intensity as a function of r for a typical coronal hole and a quiet coronal region. For EUV observations, it is best if the stray light level in the instrument is about a factor of 10 or more below the value in the figures. A capability for rejection of photospheric and chromospheric light at off-band wavelengths is also required. This is readily achievable with the present design which takes advantage of the wavelength selection properties of the diffraction grating and the long wavelength rejection of the solar blind detectors. In general, the EUV measurements require less stray light rejection than is needed for visible light. In the EUV the emission from the F Ly-α corona is also not expected to be a problem, even for the electron-scattered component of Ly-α (Withbroe et al., 1982b). As discussed earlier, the F coronal Ly-α emission can be easily removed from electron scattered Ly-α, because the F corona profile is much narrower than the electron scattered profile.

Electron, proton, and ion temperatures can be determined from measurements of the spectral line profiles

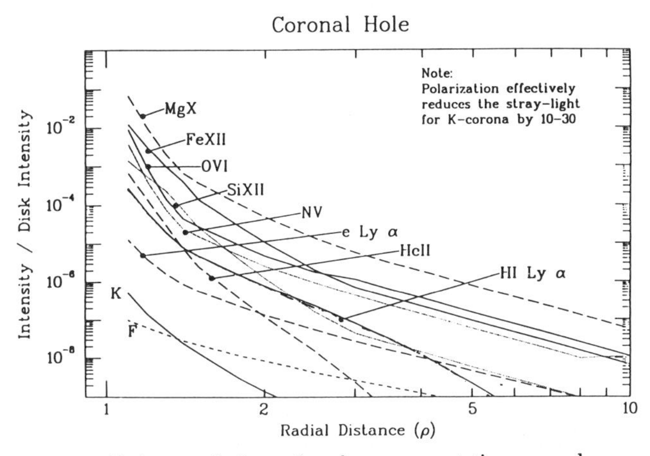

Fig. 3. Estimate of the ratio of a representative coronal hole intensity in the extended corona to the disk intensity for several spectral lines and the visible K and F coronae. The estimates are based on the intensities provided in Figure 1.

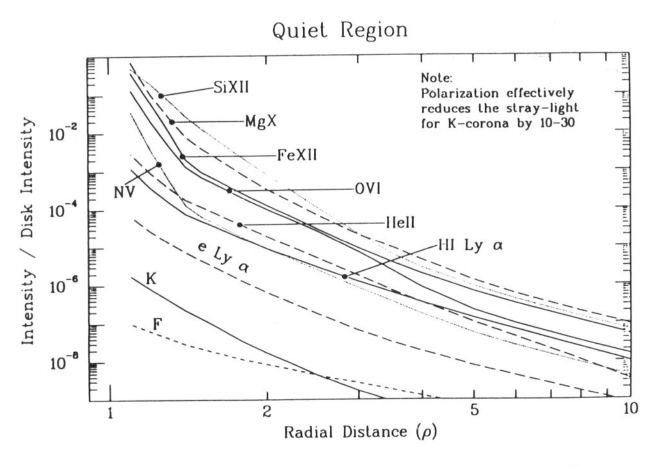

Fig. 4. Estimate of the ratio of a typical quiet coronal intensity to the disk intensity for several spectral lines and for the visible K and F coronae. The estimates are based on the intensities provided in Figure 2.

of HI Ly-α, Fe XII λ1242, O VI λ1032, Mg X λ610, and Si XII λ499. The shape of a spectral line depends on the velocity distribution of the particles emitting or scattering the measured photons. The velocities are produced by thermal motions, nonthermal motions (due, for example, to waves), and bulk outflow velocities in the line-of-sight. The coronal HI Ly-α profile is dominated by thermal motions (130 km s^{-1} at 10^6 K), while the more massive particles tend to be affected most strongly by nonthermal motions. Expected line widths are provided in Figure 5 along with measurements of the width of HI Ly-α made with the rocket UVCS in a polar coronal hole observed in 1980 (Withbroe et al., 1985). It can be seen from Figure 5 that relatively high spectral resolution is required to determine the line profiles of the ions of interest.

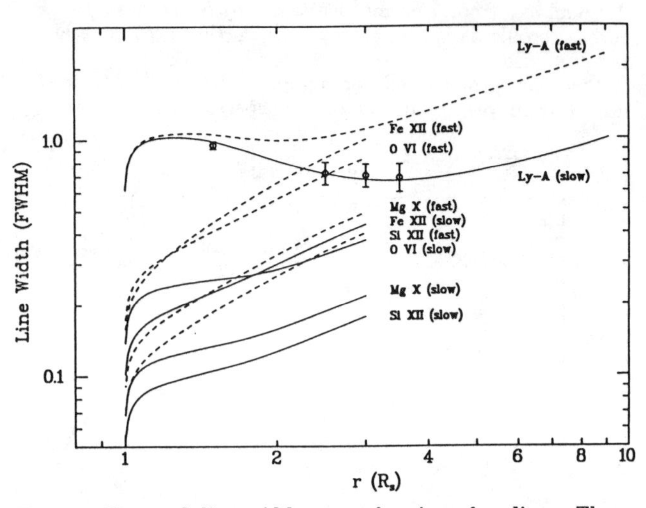

Fig. 5. Expected line widths as a function of radius. The solid curves are the calculated line widths for the electron, proton, and ion profiles for a polar coronal hole similar to that observed by the UVCS instrument in 1980. The short dashed curves are the calculated line widths for low (slow) speed wind for the O VI, Mg X, and Fe XII lines, and for the resonantly scattered component of HI Ly-α. The effect of MHD wave broadening is included in these curves. The points are measurements from the rocket instrument (Withbroe et al., 1985). The long dashed curves are the corresponding line widths calculated when sufficient MHD wave flux is included to generate high (fast) speed wind.

Instrument Description

The optical layout for a suitable ultraviolet coronagraph spectrometer (UVCS) for the aforementioned objectives is presented in Figure 6. The basic design, which is an enhanced version of SAO's rocket instrument, consists of an occulted telescope and a high resolution spectrometer assembly. Two off-axis paraboloidal telescope mirrors focus co-registered images of the extended corona onto the two entrance slits of the spectrometer assembly. The spectrometer system consists of two sections. One section is optimized for line profile measurements of HI Ly-α; its wavelength range is 1170-1260 Å and is extendable to 1130 and 1300 Å with grating motions. The other section is optimized for line intensity measurements of O VI λ1032, Mg X λ610, and He II λ304 in second order. The respective ranges are 981-1068 Å (extendable to 955 Å) and 604-635 Å. This arrangement provides optimal coatings for each range and permits selection of the appropriate slit widths for each type observation. For example, narrow slits for Ly-α and wider ones for the other ranges make the required observation times for Ly-α profiles and full intensities of weaker lines essentially equal.

The optical specifications of a suitable UVCS are listed in Table 1. The center line and roll axis of the instrument is usually pointed at sun center. One side of the rectangular entrance aperture acts as a linear external occulter and the other sides serve to limit the amount of light entering the instrument. The external occulter has a serrated edge. Light from the solar disk enters the aperture and is reflected by two mirrors into reflective wedge light traps. Coronal radiation from 1.2 to 10 $R_\odot$ passes through the aperture and illuminates the two telescope mirrors. For sun center pointing, the occulting edge and the telescope mirrors are such that radiation from 1.2 $R_\odot$ just reaches the edges of the mirrors and radiation from 10 $R_\odot$ fills them. A mechanism tilts the mirrors to scan

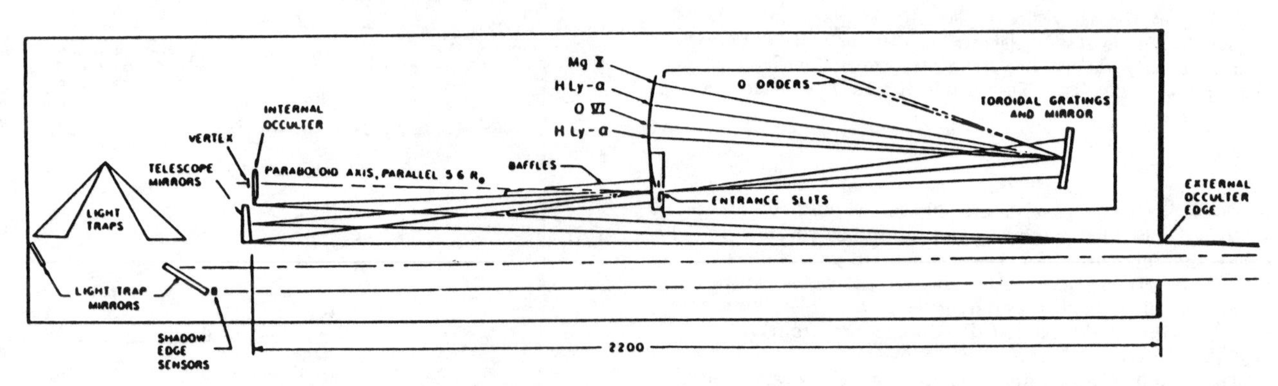

Fig. 6. The optical layout for an ultraviolet coronagraph spectrometer for spectroscopic measurements of the extended solar corona. The instrument consists of an occulted telescope and a high dispersion spectrometer assembly. A light trap captures the bright light from the solar disk that enters the instrument through the entrance aperture. A linear external occulter (which forms one edge of the entrance aperture) and an internal occulter, shields the telescope and spectrometer from the direct solar disk light.

TABLE 1. Optical Specifications

Telescopes		Spectrometer Sections	
Size	70 (disp) × 50 mm	Gratings Size	105 (disp) × 70 mm
Figure	Off-axis parabolic mirror	Figure	toroidal
Focal Length	750 mm	Nominal Radius	750 mm
Image Scale	0.218 mm/arc min	Radius Ratio	1.0259 (1216 Å)
Field-of-View	42′(tangential)×141′		1.0215 (1032 Å; 625 Å)
			1.0040 (visible)
		Ruling Frequency	2400 mm^{-1}
		Reciprocal Dispersion	5.63 Å/mm (1st order)
		Reciprocal Dispersion	2.82 Å/mm (2nd order)

the coronal images across the entrance slits. An internal occulter, just in front of the telescope mirrors, blocks light diffracted and scattered by the external occulter that would otherwise be imaged onto the slits and dominate the coronal radiation in many cases. This internal occulter is translated in front of the mirrors as they tilt to cover those parts of the mirrors that would reflect the unwanted light through the slits.

The spectrometer sections use toroidal concave diffraction gratings in Rowland circle mounts (Huber and Tondello, 1979) that produce stigmatic imaging at the subject wavelengths. The detectors for the EUV are multi-anode microchannel arrays (MAMAs), each with 360 by 1024 pixels on 0.025 mm centers (Timothy et al., 1981). The HI Ly-α section has two such detectors with MgF_2 windows (one for redundancy with a lens to correct for astigmatism). The other section has two windowless MAMAs, one at each of the two stigmatic wavelengths, O VI λ1032 and Mg X λ625.

By scanning the telescope mirror, a field of 141′ × 42′ is built up (see Figure 7). To complete the coverage, the instrument is rotated about its sun-center axis. The instrument design is capable of pointing adjustments for coalignment with the spacecraft sun sensor and for offset pointing adjustments for

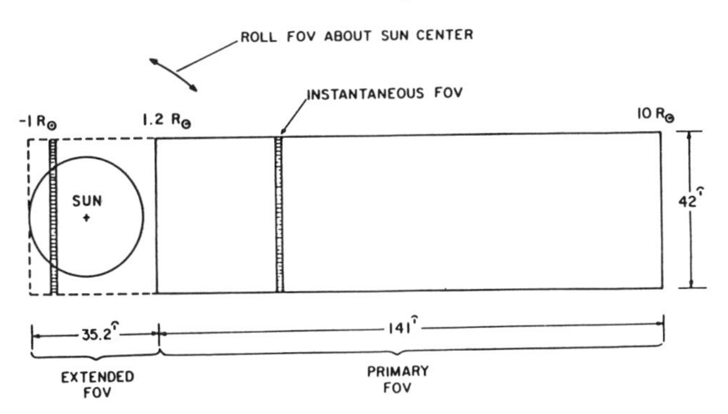

Fig. 7. The primary field-of-view of an ultraviolet coronagraph spectrometer typically extends from 1.2 to 10 solar radii from sun center. The instantaneous field-of-view is defined by the spectrometer entrance slit. Spatial resolution elements along the slit are defined by the pixels of array detectors located at the spectrometer focal surface. Rotation of the telescope mirror allows the instantaneous field-of-view to scan from 1.2 to 10 solar radii. Translation of the entire instrument permits it to observe the solar disk, and its rotation allows the field-of-view to be rolled about Sun center.

solar disk measurements or for observations with partial external occulting which increase the collecting area. Four photodiodes are located around the entrance to the solar disk light trap to detect decentering of the light patch and thus to act as a sun sensor that provides the data required for pointing adjustments.

Stray Light Suppression

The stray light suppression for the UVCS design has been measured in the laboratory and proven during three sounding rocket flights. The occulted telescope and spectrometer system of the UVCS provides the stray light suppression needed for EUV spectroscopy of the extended solar corona. For spectroscopy, it is advantageous to deviate from the circular occulting geometry of imaging coronagraphs in favor of a linear occulter system. The linear system provides a larger unvignetted collecting area at the expense of the field-of-view which, for spectroscopy, is already limited by the spectrometer entrance slit. The linear system also benefits from a simpler vignetting function (i.e., change in unvignetted collecting area with observed height). The purpose of this section is to describe the UVCS occulting system and its capability for suppressing stray light.

The basic optical arrangement is illustrated in Figure 8. There is a rectangular entrance aperture with knife edges on three sides and a serrated edge which acts as a linear external occulter (O_1). Light from the solar disk enters the instrument through the aperture, passes by a series of carefully designed light baffles, and is collected in a light trap. The telescope mirror is also rectangular and is aligned with two of its edges parallel to the O_1 edge. The geometrical shadow line, which is defined by the ray from 1 $R_\odot$ and by O_1, passes by the mirror at 1.59 mm beyond its edge. For the usual alignment, O_1 prevents coronal light from heliospheric heights <1.2 $R_\odot$ from striking the mirror. An observation of a spatial element at 2.5 $R_\odot$ is illustrated in Figure 8. The telescope forms an image of the extended corona on the spectrometer entrance slit. Light from the surfaces of the sunlight trap is prevented from entering the slit by a carefully designed entrance slit baffle (not shown) and the telescope mirror.

The telescope forms an image of the external occulter 388 mm inside the spectrometer. Diffracted light from the external occulter would be specularly reflected by the telescope and reach that image were it not for the spectrometer case which intercepts all but the part that passes through the entrance slit. Al-

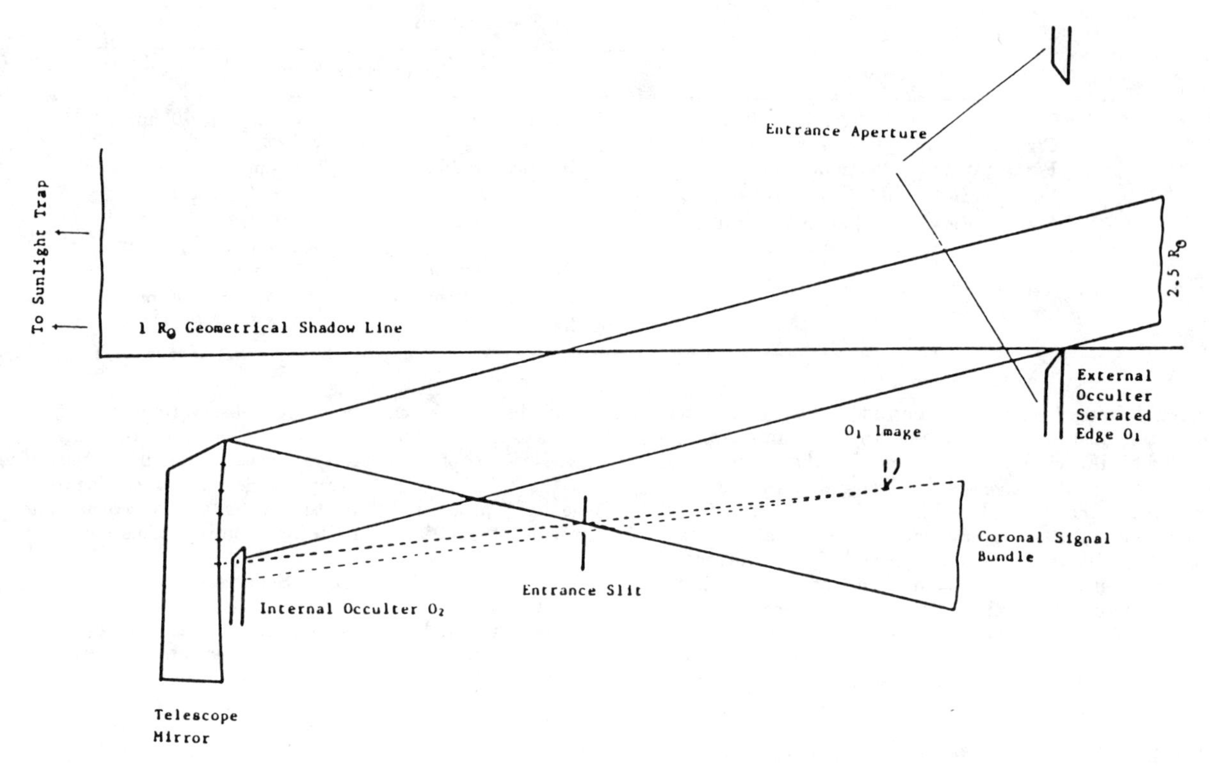

Fig. 8. The occulting geometery of the ultraviolet coronagraph spectrometer for an observation at 2.5 solar radii from sun center (not to scale).

though it is possible to block the light where it is imaged, it has been recognized that the light bundle that would enter the entrance slit can also be blocked with a carefully designed internal occulter (O_2) located near the surface of the telescope mirror. This arrangement reduces the stray light that enters the spectrometer, and it permits a single mechanism to provide coordinated mirror and internal occulter motions.

In order to scan the coronal image over the slit, the telescope mirror is rotated about an axis parallel to the external occulter. This also moves the image of the external occulter inside the spectrometer. It can be seen from the illustration that the bundle of diffracted light, which would pass through the slit (dashed lines), is always reflected from the innermost portion of the illuminated area of the telescope (i.e., away from the geometrical shadow line). The telescope mechanism moves the internal occulter across the face of the mirror so as to always intercept the specularly reflected bundle that would otherwise enter the slit.

Figure 9 illustrates the EUV (1216 Å) light flux across the surface of the telescope mirror due to diffracted solar disk light from the external occulter (Livi and Kohl, unpublished manuscript, 1988). The heavy solid curve is that expected from an ideal straight edge with the present geometry. Our laboratory measurements for the straight edge occulter from the rocket instrument are illustrated by the dashed curve. Our recent laboratory measurements for the linear serrated occulter for the Spartan UVCS are illustrated by the

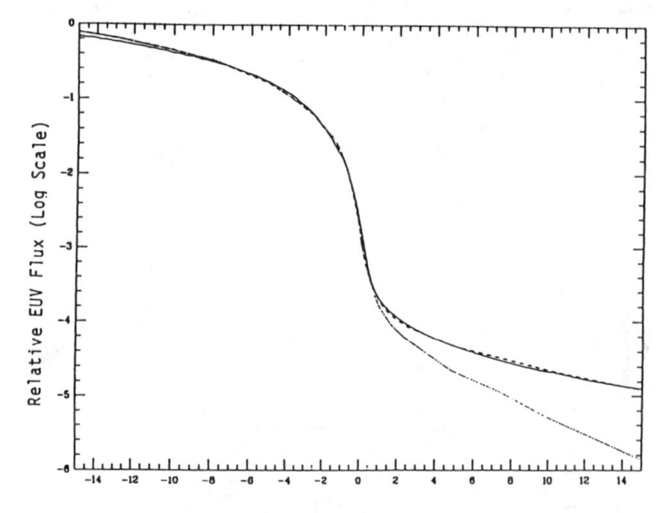

Fig. 9. EUV (1216 Å) light flux across the surface of the telescope mirror due to diffracted solar disk light from the external occulter. The heavy solid curve is that expected for an ideal straight edge with the present geometry, and the dashed curve is a smooth curve through measured points for a straight edge. The light solid curve is a smooth curve through measured points for a linear serrated occulter.

light solid curve. The ordinate is the fractional amount of the solar disk flux that impinges on the telescope mirror, and the abscissa is the distance (in mm) across the mirror. The geometrical shadow of O_1 is at the origin, and the edge of the mirror is normally at +1.59 mm.

Notice that the advantage of the serrated occulter improves with the distance from the geometrical shadow line. It is possible to improve the stray light suppression by pointing adjustments that move the geometrical shadow line away from the mirror (i.e., by over-occulting).

To explain the amount of light non-specularly reflected into the slit, it is necessary to know the properties of superpolished mirrors. Both the fraction of the incident flux and the angular distribution of the non-specularly reflected light are needed. The total integrated fraction is discussed by Elson et al. (1983). They provide values for the total integrated scattering (TIS) as a function of wavelength and surface correlation length σ. They discuss the general case where the wavelength (λ) is comparable to σ as well as the limiting cases where $\sigma \gg \lambda$ and $\sigma \ll \lambda$. Correlation lengths are typically as small as 1000 Å for supersmooth optical surfaces. The angular distribution and direction of the non-specularly reflected light is discussed in detail by Elmer (1976). The distribution is peaked in the direction of the specularly reflected light. The direction toward the UVCS spectrometer slit tends to be very near the peak of the distribution. The telescope mirror is designed to control scattering from its edges.

The entrance slit subtends a very small solid angle as seen from the telescope surface, and, consequently, only a small fraction of the non-specularly reflected light passes through the slit.

Table 2 contains the relevant data needed to compare the stray light level (termed "noise" in Table I) to the expected coronal signals. We illustrate the EUV suppression with HI Ly-α. For convenience, the photon rates are given in units of photons s^{-1} $[A_Z \cdot A_s/f^2]^{-1}$ where A_Z is the effective zonal telescope area (i.e., the area of the telescope from which most stray light emanates), A_s is the area of the observed spatial element at the entrance slit, and f is the telescope focal length. Coronal intensities are taken from Figures 1 and 2. It can be seen that the stray light suppression level is more than adequate for EUV wavelengths. The stray light is less than a few percent of the signal.

Stray light at off-band wavelengths inside the spectrometer case is further reduced by the off-band rejection of the spectrometer (typically 1 x 10^{-4} for wavelengths outside the observational range) and by the solar blind characteristics of the detector which must have a steeply decaying quantum efficiency longward of 2000 Å. This off-band rejection of bright UV and visible intensities is important for EUV observations.

Observational Opportunities

The basic optical design of the ultraviolet coronagraph spectrometer has been tested in the laboratory and proven on three sounding rocket flights. The first extensive application of ultraviolet spectroscopic diagnostics of the extended corona is planned for Spartan mission 201. This is expected to provide about 27 hours of coronal observations. Although the Spartan instrument is limited in its light collecting area, spectral coverage, instantaneous field-of-view (essentially one spatial element at a time), and spectral/spatial resolution, it is expected to increase significantly the present knowledge of the extended corona.

An exciting possibility for the Spartan instrument is to carry out spectroscopy and imaging during the polar passages of Ulysses. Coordinated measurements of solar wind acceleration parameters in polar regions of the solar corona with Spartan and in situ measurements of the polar wind from Ulysses would provide a unique opportunity to establish empirical constraints on polar wind models (Kohl et al., 1986).

The Solar Heliospheric Observatory (SOHO) is now expected to provide the first opportunity to carry out ultraviolet spectroscopy of the extended corona on an observing platform which contains a full complement of solar observing instruments. For the first time, the problems of solar wind generation and the dynamics of the solar transition region and corona can be attacked with a battery of instruments capable of detailed observations throughout the relevant regions of the solar atmosphere. Coordinated in situ measurements of solar wind composition will be virtually assured since such instruments are planned for the SOHO spacecraft.

The Pinhole Occulter Facility (Hudson et al., 1981) could accommodate an advanced ultraviolet coronagraph spectrometer that would have about an 80-fold increase in light collecting capability over a normal size instrument. The resulting increase in signal rates could be

TABLE 2. Stray Light Levels for HI Ly-α Profile

$R_\odot$	Stray Light Signal[1]	Coronal Signal[1]		Signal to Noise[2]	
		Quiet Corona	Coronal Hole	Quiet Corona	Coronal Hole
1.5	5.2×10^{7}	2.0×10^{11}	2.1×10^{10}	3800	400
2	7.8×10^{7}	1.4×10^{11}	9.0×10^{9}	1800	120
2.5	8.3×10^{7}	6.0×10^{10}	5.5×10^{9}	720	66
5	8.8×10^{7}	8.4×10^{9}	1.4×10^{8}	95	35[3]
10	9.0×10^{7}	1.4×10^{9}	-	16	-

[1] In photons s^{-1} $[A_z \bullet A_s/f^2]^{-1}$ (see text).
[2] Signal/noise for coronal streamers is about a factor of 3 better than for quiet regions.
[3] With over-occulting.

used to increase the spatial resolution to 1 arc sec or to provide very detailed and spatially extensive descriptions of rapidly time varying phenomena in the corona such as CMEs.

Acknowledgments. This program is supported by NASA under grant NAG 5-613 to the Smithsonian Astrophysical Observatory.

References

Allen, C. W., Astrophysical Quantities, 2nd ed., Athlone Press, University of London, 1973.

Baum, W. A., F. A. Johnson, J. J. Oberly, C. C. Rockwood, C. V. Strain, and R. Tousey, Solar ultraviolet spectrum to 88 kilometers, Phys. Rev., 70, 781, 1946.

Brüeckner, G. E., A high resolution view of the solar chromosphere and corona, in Highlights in Astronomy, 5, 557, 1979.

Elmer, H. J., Light scattering characteristics of optical surfaces, Ph.D. Thesis, University of Arizona, 1976.

Elson, J. M., J. P. Rahn, and J. M. Bennett, Relationship of the total integrated scattering from multilayer-coated optics to angle of incidence, polarization, correlation length, and roughness cross-correlation properties. Appl. Opt., 22, 3207, 1983.

Gabriel, A. H., W. R. S. Garton, L. Goldberg, T. J. L. Jones, C. Jordan, F. J. Morgan, R. W. Nicholls, W. H. Parkinson, H. J.B. Paxton, E. M. Reeves, C. B. Shenton, R. J. Speer, and R. Wilson, Rocket observations of the ultraviolet solar spectrum during the total eclipse of 1970 March 7, Ap. J., 169, 595, 1971.

Goldberg, L., R. W. Noyes, W. H. Parkinson, E. M. Reeves, and G. L. Withbroe, Ultraviolet solar images from space, Science, 162, 95, 1968.

Huber, M.C.E., and G. Tondello, Stigmatic performance of an EUV spectrograph with a single toroidal grating, Appl. Opt., 18, 3948, 1979.

Hudson, H.S., J. L. Kohl, R. P. Lin, R. M. MacQueen, E. Tandberg-Hanssen, and J. R. Dabbs, The Pinhole/Occulter Facility, NASA Technical Memorandum, NASA TM-82413, Marshall Space Flight Center, AL, 1981.

Hyder, C. L., and B. W. Lites, Hα Doppler brightening and Lyman-α Doppler dimming in moving Hα prominences, Solar Phys., 14, 147, 1970.

Kohl, J. L., and G. L. Withbroe, EUV spectroscopic plasma diagnostics for the solar wind acceleration region, Ap. J., 256, 263, 1982.

Kohl, J. L., E. M. Reeves, and B. Kirkham, The Lyman alpha coronagraph, in New Instrumentation for Space Astronomy, edited by K. van der Hucht and G. S. Vaiana, p. 91, Pergamon Press, Oxford, 1978

Kohl, J. L., H. Weiser, G. L. Withbroe, R. W. Noyes, W. H. Parkinson E. M. Reeves, R. M. MacQueen, and R. H. Munro, Measurements of coronal kinetic temperatures from 1.5 to 3 solar radii, Ap. J. Lett., 241, L117, 1980.

Kohl, J. L., H. Weiser, G. L. Withbroe, C. A. Zapata, and R. H. Munro, Evidence for supersonic solar wind velocities at 2.1 solar radii, Bull. Am. Astron. Soc., 16, 531, 1984.

Kohl, J. L., H. Weiser, G. L. Withbroe, G. Noci, and R. H. Munro, Coronal spectroscopy and imaging from Spartan during the polar passage of Ulysses, in The Sun and the Heliosphere in Three Dimensions, edited by R. G. Marsden, p. 39, Reidel Publ. Co., Dordrecht, 1986.

Noci, G., J. L. Kohl, and G. L. Withbroe, Solar wind diagnostics from Doppler enhanced scattering, Ap. J., 315, 706, 1987.

Poland, A. L., Motions and mass changes of a persistent coronal streamer, Solar Phys., 57, 141, 1978.

Timothy, J. G., G. H. Mount, and R. I. Bybee, Multianode microchannel arrays, IEEE Trans. Nucl. Sci., MS-28, 689, 1981.

Withbroe, G. L., J. L. Kohl, H. Weiser, G. Noci, and R. H. Munro, Analysis of coronal HI Lyman alpha measurements from a rocket flight on 13 April 1979, Ap. J., 254, 361, 1982a.

Withbroe, G. L., J. L. Kohl, H. Weiser, and R. H. Munro, Probing the solar wind acceleration region using spectroscopic techniques, Space Sci. Rev., 33, 17, 1982b.

Withbroe, G. L., J. L. Kohl, H. Weiser, and R. H. Munro, Coronal temperatures, heating and energy flow in a polar region of the sun at solar maximum, Ap. J., 297, 324, 1985.

Withbroe, G. L., J. L. Kohl, and H. Weiser, Analysis of coronal HI Lyman-alpha measurements in a polar region of the sun observed in 1979, Ap. J., 307, 381, 1986.

Woodgate, B. E., E. A. Tandberg-Hanssen, E. C. Bruner, J. M. Beckers, J. C. Brandt, W. Henze, C. L. Hyder, M. W. Kalet, P. J. Kenny, E. D. Knox, A. G. Michalitsianos, R. Rehse, R. A. Shine, and H. D. Tinsley, The ultraviolet spectrometer and polarimeter on the solar maximum mission, Solar Phys., 65, 73, 1980.

TEST PARTICLE MEASUREMENTS IN SPACE PLASMAS

Carl E. McIlwain

Center for Astrophysics and Space Sciences, University of California at San Diego, La Jolla, CA 92093

Abstract. Various ways of generating and sensing test particles for measurements of electric and magnetic fields in space plasmas are discussed. Some of these methods involve the test particles traversing great distances and are thus relatively immune from the local plasma disturbances invariably caused by the presence of a spacecraft. One technique is to send test particles on trajectories which bring them back to a detector on the same spacecraft after one or more gyro motions in the ambient magnetic field. It is shown that by measuring the angles of these trajectories and the times of flight as a function of the test particle energy, not only can the ambient electric field be determined over a wide dynamic range, but the magnetic field magnitude and the gradient of the magnitude can also be determined. The special and demanding requirements of a detector for gathering magnetically returned test particles have required the creation of a new and novel design.

Introduction

It is unecessary to review the importance of space plasma measurements in general and, specifically, the measurement of magnetic and electric fields in space plasmas, because the need for these measurements is universally recognized. The only question is: what techniques are best employed in making the measurements? Magnetic fields have, of course, been measured quite well for a long time. Electric fields have also been well measured in a few regimes, but, in particular, it has been found to be very difficult to obtain satisfactory measurements in the presence of the hot plasmas which are found in some of the most interesting regions of the ionosphere and magnetosphere.

It is the purpose of this paper to show how test particles can be used (1) to measure electric fields even in the presence of high fluxes of energetic particles, (2) to measure the magnitude of the magnetic field to an accuracy that has not been possible with standard instruments, and (3) to measure the gradient of the magnitude of the magnetic field and therefore obtain data more directly characterizing the current systems.

The design of a charged particle detector suitable for the sensing of test particles in the presence of the intense ambient particle fluxes will be described in some detail.

Ion Releases

One form of test particle measurement has been used for many years. This is the release of a cloud of ions and observing the fluorescent scattering of sunlight to follow the motion of the cloud in the ambient magnetic and electric fields. This technique, of course, has some limitations; for one thing it is necesssary to have an ion release and thus only a few spot measurements are feasible.

Direct Particle Beams Between Booms

Another use of test particles is to measure deflection of a beam in the electric and magnetic fields after the traverse of only a short distance. One possibility is simply to send beams from one part of the spacecraft to another, for example, between a triad of booms on a single spacecraft (see Figure 1). This actually has a number of advantages: it takes only a short time to travel between the booms and, therefore, fields of a quite high frequency can be measured. Since the distances are small, short wavelength waves can be measured. Furthermore, ions can be used. In particular, ions of different charges can be used, including ions with both positive and negative charges. This means that differential measurements can be made to determine the charge dependence as well as the energy dependence of the deflections.

The disadvantage is obvious: the particles never go very far from the spacecraft. This introduces some of the problems that bedevil the standard electric field measurements using double probes, namely, the problems associated with the fact that the presence of the spacecraft perturbs the ambient electric fields to first order. Perhaps by sending additional particle beams over shorter distances, the spacecraft-induced fields can be mapped. This would then provide the information needed for correcting the measurements made with long booms.

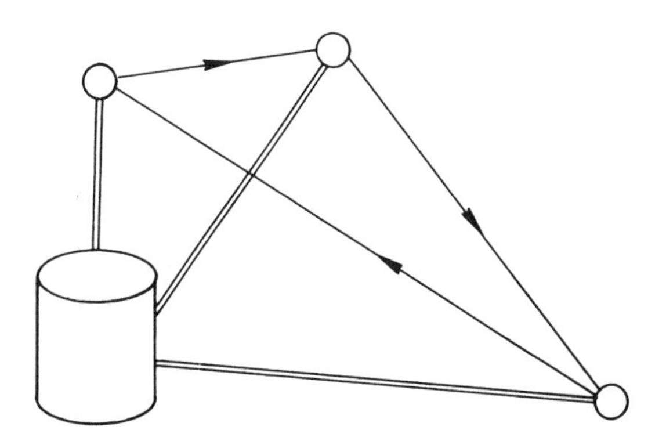

Fig. 1. Electric and magnetic fields can be sensed by projecting particles with various energies and charge states between booms on a single spacecraft.

Direct Particle Beams Between Spacecraft

One way of reducing the relative importance of the spacecraft-induced field is to send the particles between spacecraft (see Figure 2) and thus over much larger distances than possible with booms on a single spacecraft. This has the obvious disadvantage of requiring two or more spacecraft. It requires that the spacecraft orbits be slaved to each other, but has the advantage that if the spacecraft separations are intentionally varied, valuable wavelength information can be obtained.

Differential measurements using particles of various charges, masses, and energy can be used to reduce systematic errors and to separate the electric and magnetic components of the deflections.

Particle Beams Returned by the Magnetic Field

When test particles are projected in various directions perpendicular to the magnetic field, they will all execute a gyro motion and return to a common point one gyro period later. There is thus a one-dimensional focusing within the plane perpendicular to the magnetic field. If there is no gradient in the magnetic field and no electric field, the common point of return will be the starting point. In general, there is a drift which displaces the point of return from the projection point by a distance which is proportional to the perpendicular electric field and to the gradient of the magnetic field.

It can be shown that there are usually two directions in which particles can be projected so that they will be returned to a point on the same spacecraft by the actions of the magnetic and electric fields after executing one gyro turn. Further, it can be shown that the drift vector can be obtained either from the measurement of these two directions or from one of the directions plus the difference of the two times of flight. The drift due to the magnetic field gradient is proportional to energy. The energy dependence of the drift can therefore be used to separate the component of the drift due to the magnetic field gradient from the component due to the electric field.

A pioneering experiment of this kind was flown on the GEOS 2 spacecraft by a group at the Max Planck Institute for Extraterrestrial Physics led by Gerhard Haerendel. (Melzner et al., 1978; Baumjohann and Haerendel, 1985). In that case, the beams were directed outward from the spacecraft, with the spacecraft rotation changing the beam direction. With a time stationary electric field, two measurements per spin could be obtained (one each 3s). They found, however, like many others, that the electric field has components which are fast, often having large changes in 3s. In special circumstances, they were able to make the planned types of measurements. In fact, however, the primary way they were able to make measurements was by letting the electric field itself change the direction appropriate for return and observing the returns as the oscillating field swept the returning beam past the spacecraft.

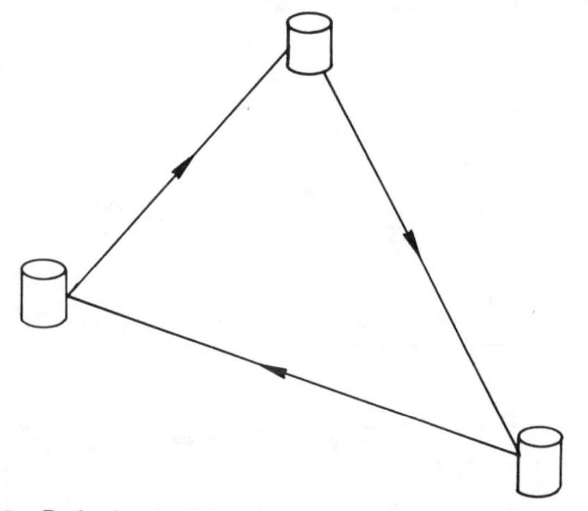

Fig. 2. Projecting particles with various energies and charge states between three or more spacecraft can provide additional information about plasma waves as well as providing greater freedom from perturbing spacecraft fields.

For the EQUATOR spacecraft in the OPEN program (now the Global Geospace Science program), it was decided to propose an experiment using a similar technique, but to keep the beams coming back to the detectors by continuously changing the direction of the beams. Unfortunately, the EQUATOR spacecraft was deleted from the program as a cost reduction measure.

More recently a test particle experiment was proposed for the Cluster program by a team of 17 investigators led by Goetz Paschmann at the Max Planck Institute for Extraterrestrial Physics. For this program, it is planned to use both the two angle technique and the one angle plus time of flight technique for measuring the drift. Two electron guns will be employed which are simultaneously servoed to send out time-coded beams so that the beams arrive at detectors on the opposite sides of the spacecraft after one or more gyro orbits.

The two angle technique of determining the drift vector is simply a triangulation using the vector between guns and detectors as the base line. When the drift step becomes large compared to the projected gun detector separation, the angle between the two beams becomes very small, and the triangulation for the size of the drift step becomes inaccurate (but note that the direction information remains accurate). Thus, paradoxically, the technique becomes insensitive when the drift step is large, but it becomes quite sensitive when the drift step is small (but not too small relative to the base line). Indeed, this technique can reliably measure much smaller electric fields in space plasmas than almost any other technique.

The region of the Earth's magnetosphere to be explored by the Cluster spacecraft encompasses a very wide range of parameters, and the triangulation technique alone will not be capable of obtaining the full range of desired measurements. Thus the time of flight technique will also be used. This technique has been developed by K. Tsuruda and his group in Japan. In the case of a very large drift step, the two beam directions are almost anti-parallel to each other and the difference in the time of flights will correspond to the time to go twice the drift step. This technique has now been successfully tested on a sounding rocket (Nakamura et al., 1987).

There are many aspects of this experiment that turn out to be quite difficult. One challenge is the design of the electron gun: it must be able to fire in at least a hemisphere of directions, to know what those directions are to within a fraction of a degree, and, of course, to be able to change that direction within milliseconds. It must also encode the beam such that the detector electronics can determine the time of firing to a fraction of a microsecond but at the same time produce no interference signals in the 0-0.5 MHz band of frequences.

Electron Beam Detector System

Even harder to design than a wide angle electron gun is a detector that can catch the electrons. It is one thing to direct a pencil beam in different directions; it is quite another to gather a whole bundle, i.e., a broad area of the beam, and coax that down to a sensor when the beam is coming from any arbitrary direction. Simultaneously there are many criteria which must be satisfied by the detector. It did not seem that the requirements for the test electron detection could be met by any reasonable modification of existing detector designs. An effort was therefore launched to conjure up a new configuration of electron optics.

The design goals included: (1) the capture of as many of the beam electrons as possible onto one or more sensor elements, while simultaneously minimizing the number of ambient electrons reaching those sensor elements; (2) the capability of adjusting the look directions within a few milliseconds to detect an electron beam coming from any direction within a hemisphere plus a substantial extension into the opposite hemisphere; (3) the measurement of relative intensities with an accuracy approaching that set by counting statistics with incoming fluxes varying in the range of 10^3 to 10^7 electrons cm^{-2} s^{-1}; (4) a timing resolution of 100 ns or better; (5) the production of as much angular information as possible to help determine the current trajectory between the gun and detector and thus help choose the optimum pattern of directions for the guns; (6) physical dimensions of less than 20 cm when the size is adjusted to capture at least 2 cm^2 of the beam from most beam directions; (7) employment of a relatively small sensor, such as a 4 cm diameter microchannel plate; (8) not being disturbed by solar UV and x-ray fluxes from any solar aspect angle; and (9) not generating any interfering magnetic or electric fields or electron fluxes within the frequency range of 0-0.5 MHz.

The last requirement entails that: (1) all electrodes with time-dependent potentials be shielded from the external plasma by an adequate arrangement of conductors and conducting grids; (2) conductors carrying time-dependent currents be paired or shielded to minimize direct interference or the generation of plasma waves; and (3) the time variable emission of photoelectrons and secondary electrons be strictly limited (in general this means that negative potentials cannot be used on any electrodes visible from the outside even if they are shielded by grids).

It was decided that to best satisfy goals (1) and (6), an electron optic design should be sought which achieves double angular focussing, i.e., performs the same function as a standard optical lens, namely, to serve as a phase-space transformer which maps the two angular coordinates of the ambient fluxes into two spatial coordinates at a focal surface. Now there are a number of electron optic designs being used in electron microscopes and electron cameras which provide this conversion, but none could be found which could also be arranged to simultaneously meet the other requirements. A double focusing design with an effective area of 2 cm^2 was employed (Mauk and McIlwain, 1975) to obtain high angular resolution within the source cone in geosynchronous orbit, but that design provided only limited angular coverage and thus cannot be modified to meet goal (2) above.

A design which seems to meet the design goals is illustrated in Figure 3. The thick lines represent the various conducting surfaces and the thin lines represent potential contours. The system is cylindrically symmetric, and thus can detect electrons coming from any azimuthal angle. After entering the system through a grid composed of a mesh of fine wires, the electrons are deflected in elevation angle by the electric field generated by a positive potential on the upper or lower deflection ring. As the electrons pass between the rings, they are focused in elevation angle by a positively biased aperture slit. The final design will probably be scaled to a somewhat larger size, but it will be shown below that adequate effective areas are obtained even with a maximum diameter of only 12 cm.

Figure 4 shows the trajectories of five 1-keV electrons which pass near the axis of the system for the case in which the

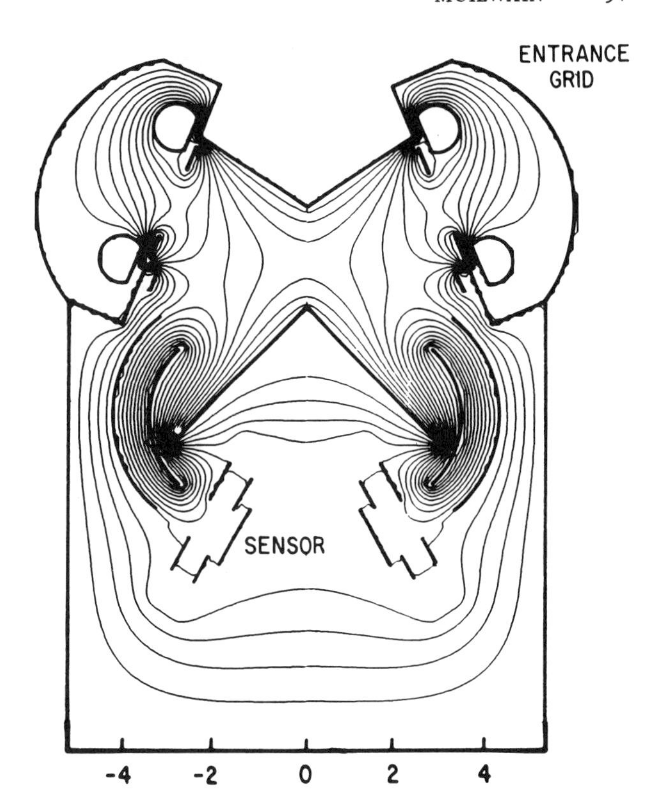

Fig. 3. Cross-sectional view of a preliminary detector design. The system is rotationally symmetric about the vertical axis. The thin lines correspond to potential contours at 200-V intervals with the upper deflection ring set to 2000 V.

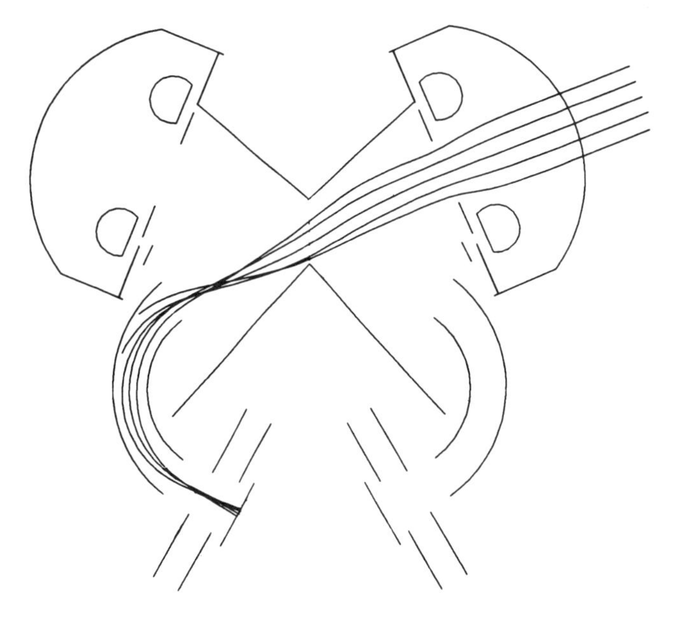

Fig. 4. Cross-sectional view of the detector with five electron trajectories which pass near the axis of symmetry for the case with zero potentials on the deflection rings. Note the initial convergence just ahead of the analyzer section.

potential on both rings is zero (and therefore cause no deflection from the initial 23° degree elevation angle). It can be seen that the entrance aperture serves as a lens which brings the electrons to an initial focus in the radial direction near the entrance of the toroidal energy analyzer section. The radii of curvature of the analyzer plates are chosen to focus the electrons a second time near the conical-shaped sensor.

Figure 5 is a top view of a set of 20 electron trajectories. It can be seen that the entrance slit has a relatively long positive focal length in the azimuthal direction, that the central region forms a negative azimuthal lens, and that the analyzer section brings the electrons to an approximate azimuthal focus near the sensor. Figure 6 is a perspective view of these trajectories looking nearly parallel to the direction of the initial beam.

In addition to maximizing the area of the beam which reaches the sensor, the shape of this area is of interest. Figure 7 shows the cross sections of the beam reaching the sensor with seven different potentials on the deflector rings. In each case, the figure corresponds to a cut perpendicular to the beam direction. It can be seen that the shape for the 86° elevation case is approaching the shape of an annular ring that occurs for a beam coming straight down the axis. It can be seen that a large effective area is maintained over a solid angle of about 10 str, which is about 80% of the total sky.

It is planned to employ a position-sensitive sensor. If the position resolution of the sensor is good, the sharpness of the focus in the two directions becomes of great interest because it will determine the signal-to-background ratio and the accuracy of the beam direction measurement. Figure 8 shows the mapping of the entering beam upon the sensor cone for the case with zero potentials on the deflector rings. It can be seen that the stippled region, which comprises an area of almost 2 cm^2, falls upon only 2 mm^2 at the sensor. Since the ambient electron fluxes perpendicular to the magnetic field are always broadly distributed in angle, the 1 keV ambient electrons are relatively unenhanced while at the same time, the beam fluxes are concentrated by almost a factor of 100.

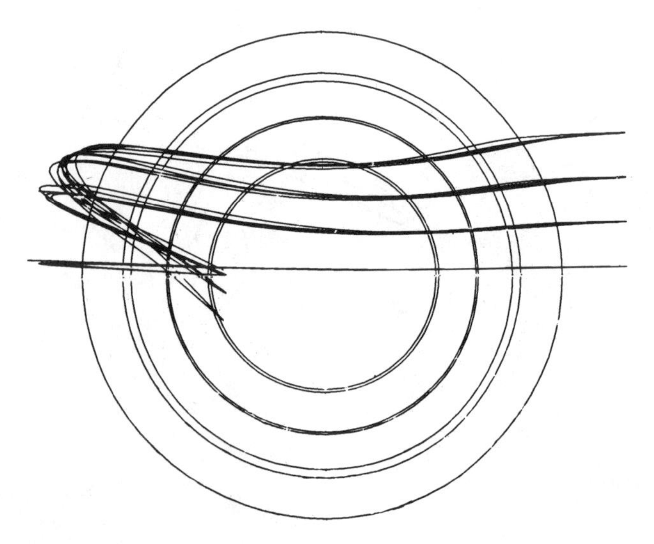

Fig. 5. Top view of 20 electron trajectories. Note that there is no intermediate convergence in the azimuthal displacements ahead of the analyzer section, only a single convergence near the sensor cone.

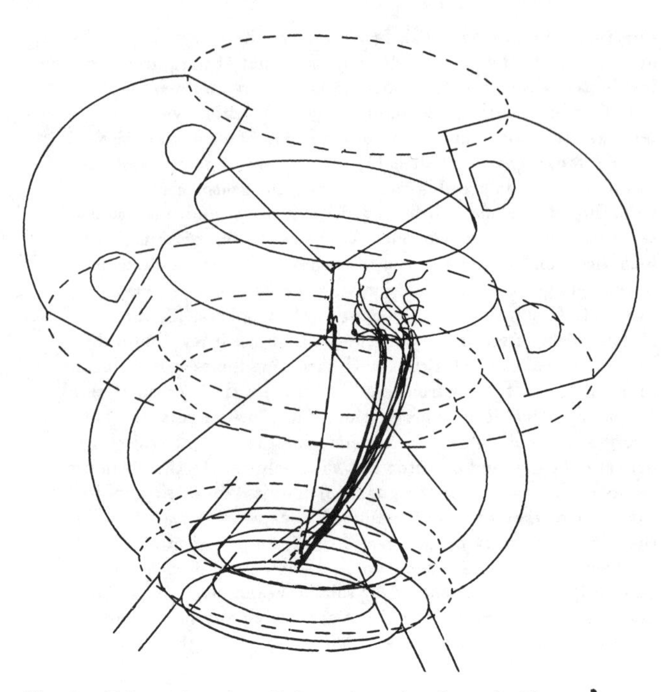

Fig. 6. Perspective view of the trajectories shown in Figure 5. The viewing angle is almost directly parallel to the beam direction.

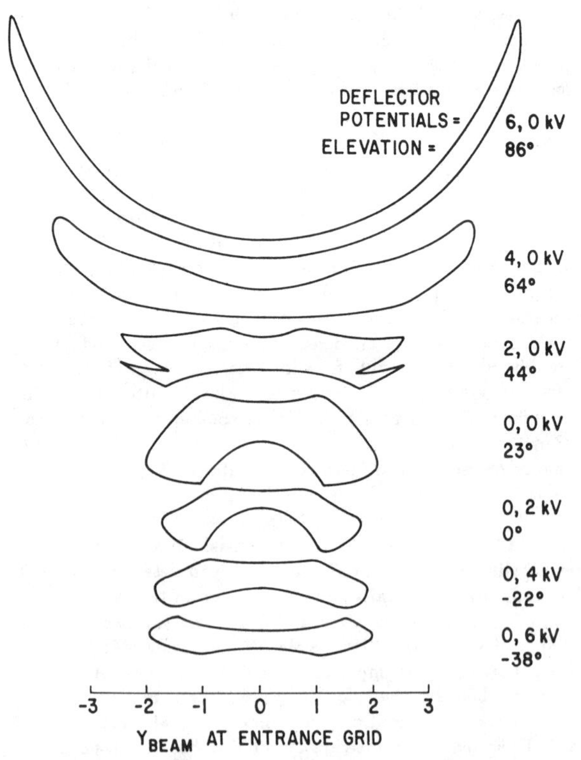

Fig. 7. Cross-sectional parts of beams reaching the sensor for seven sets of deflector ring potentials. In each case, the cross sections are taken perpendicular to the beams.

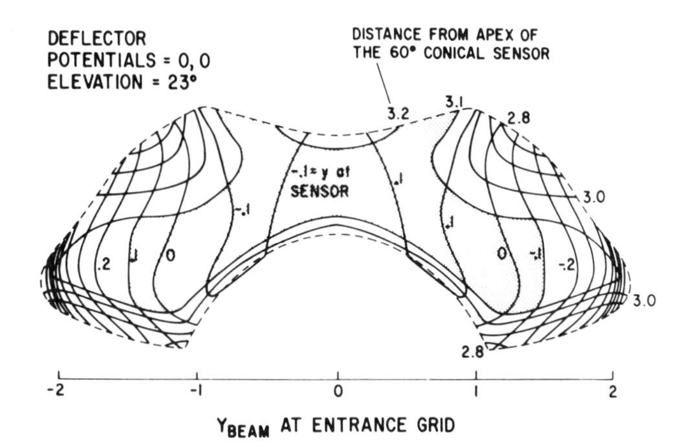

Fig. 8. Map showing where different parts of the incoming beam hit the sensor cone. The contours give the distance from the apex of the sensor cone and the displacement in the azimuthal direction. Mentally performing the transformation to obtain the mapping of the beam on the sensor, it can be seen that the image of the beam has a fairly broad halo, but that there is a sharp central peak.

The design shown here corresponds to opening up the apertures as far as possible while still satisfying goal (8) above. The desired insensitivity to solar photons is obtained by requiring them to make at least two bounces before reaching the sensor. The total flux of ambient electrons reaching the sensor is strongly determined by the width of these apertures because they set both the energy and angular acceptance ranges. While it is important to minimize the flux of ambient electrons at the sensor, it is also desirable to have a fairly wide angular acceptance in order to reduce the accuracy with which the deflectors must track fast variations in the beam elevation angle. Figure 9 shows the range of elevations accepted for a few sample cases.

To date, the detector design work has been confined to ray tracing in computer models. A prototype will be constructed in the near future to verify and refine the design. The prototype will, of course, also need an external mechanical support structure and electrical cables to the upper portion of the detector system. It should be possible to keep these supports and cables to thicknesses of only a few millimeters, in which case, the reduction in effective area will be negligible. Like an obstruction placed near an optical lens, the supports will cause only minor effects on the quality of the electron image, and, in particular, they will not cast shadows at the sensor.

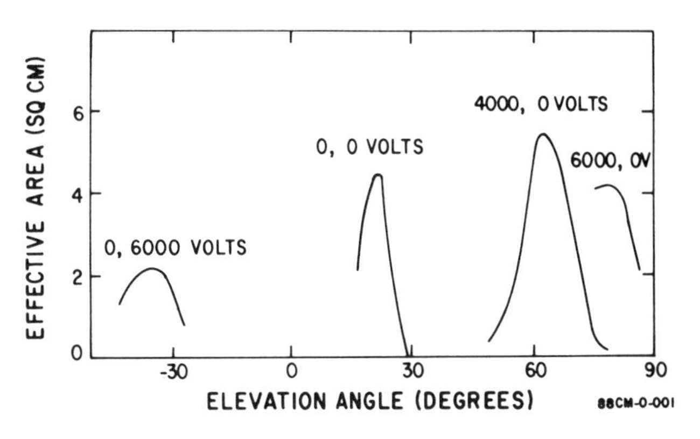

Fig. 9. In general, the deflector ring potentials will be controlled to maximize the area of the beam reaching the sensor. This figure shows that tracking errors of 4° or 5° will not seriously reduce the counting rates.

Conclusion

A number of test particle techniques are available to probe the electric and magnetic fields associated with the various plasma regimes found in the solar system. The technique of measuring the drift of test electrons gyrating at high altitudes in the earth's magnetosphere is now being refined for flight on the four Cluster spacecraft.

Acknowledgments. The assistance of the entire Cluster electron beam instrument team is acknowledged, with special thanks for the help provided by Jack Quinn and by Walker Fillius. Partial financial support was provided by NASA grant NGL 05-005-007, by NASA contract NAS 5-26854, and California Space Institute grant CS-59-87.

References

Baumjohann, W., and G. Haerendel, Magnetospheric convection observed between 0600 and 2100 LT: Solar wind and IMF dependence, *J. Geophys. Res., 90,* 6370-6378, 1985.

Mauk, Barry H., and Carl E. McIlwain, ATS-6 UCSD auroral particles experiment, *IEEE Trans. Aerosp. and Elec. Sys., AES 11,* 1125-1130, 1975.

Melzner, F., G. Metzner, and D. Antrack, The GEOS electron beam experiment S-329, *Space Sci. Instrum., 4,* 45-55, 1978.

Nakamura, M., H. Hayakawa, and K. Tsuruda, Electric field measurement by S-520-9 sounding rocket using the time of flight technique of lithium ion beam, *ISAS Research Note 367,* 1987.

COMPRESSIBLE DYNAMIC ALIGNMENT

R. B. Dahlburg and J. M. Picone

Laboratory for Computational Physics and Fluid Dynamics, Naval Research Laboratory, Washington, DC 20375

J. T. Karpen

E. O. Hulburt Center for Space Research, Naval Research Laboratory, Washington, DC 20375

Abstract. Dynamic alignment has been proposed to account for correlations between the magnetic and velocity fields of the solar wind. This dynamic alignment problem is part of a more general class of problems, related to self-organization in compressible magnetohydrodynamic (MHD) turbulence, which is not yet well understood. In previous work we demonstrated that dynamic alignment occurs in two-dimensional compressible turbulent magnetofluids (Dahlburg et al., 1988a). In this paper we discuss numerical simulations which further our understanding of dynamic alignment in compressible MHD. By varying the initial average Mach number, we determine the influence of compressibility on dynamic alignment in the "Orszag-Tang vortex," a frequently investigated two-dimensional MHD configuration. As the Mach number is raised we observe a time delay in growth of correlation between the magnetic field and the velocity field. As the Mach number approaches zero, the results for the compressible runs converge to the incompressible result. Our results indicate that the details of dynamic alignment depend on the solenoidality of the initial flow profile, as well as the degree to which the initial mechanical pressure resembles the appropriate incompressible mechanical pressure.

Introduction

Dynamic alignment has been proposed to explain the high degree of correlation often observed between the magnetic and velocity fields in the solar wind at distances ≤ 1 AU. Thus far, numerical simulation of dynamic alignment in turbulent magnetofluids has been done primarily for the incompressible case (Léorat et al., 1983; Matthaeus et al., 1983; Pouquet et al., 1986; Grappin, 1986). In the solar wind, however, the Alfvén speed and the sound speed are comparable. Hence, a full numerical treatment of the dynamic alignment phenomenon requires the inclusion of compressible effects.

Dynamic alignment is representative of a more general class of turbulent fluid behavior known as self-organization. Self-organization has been studied in detail for incompressible MHD (see the review by Hasegawa, 1985), but only recently for compressible magnetofluids. Horiuchi and Sato (1986) have investigated numerically the relaxation of a quiescent compressible magnetofluid with net initial magnetic helicity to a force-free state. Earlier Riyopoulos et al. (1982) studied a similar problem in two dimensions. Both of these numerical investigations relied on finite-difference methods. We have used a dealiased Fourier collocation method to show that dynamic alignment can occur in two-dimensional compressible magnetofluids (Dahlburg et al., 1988a). Similar investigations of turbulent compressible neutral fluids using spectral methods have recently been performed (Passot and Pouquet, 1987; Erlebacher et al., 1987; Canuto et al., 1987).

Our earlier simulations considered the effects of altering the initial degree of correlation. The runs were initialized with broadband random noise in the magnetic and velocity fields, and spatially uniform mechanical pressure and mass density distributions. For cases with a moderate degree of initial correlation we observed rapid growth of correlation and alignment followed by a phase of slower growth. In this paper we examine compressibility effects on dynamic alignment by varying the average Mach number. The initial conditions considered here are somewhat different from those of our previous study. To merge smoothly into the incompressible case, we here use a solenoidal initial velocity field, and the corresponding incompressible mechanical pressure distribution is used for the initial fluctuating mechanical pressure.

Formulation

We start with the nonlinear partial differential equations which govern the behavior of a two-dimensional, compressible, dissipative magnetofluid, written here in a dimensionless form:

$$\frac{\partial \rho}{\partial t} = -\nabla \cdot (\rho \mathbf{v}) \tag{1}$$

$$\frac{\partial (\rho \mathbf{v})}{\partial t} = -\nabla \cdot \Big[\rho \mathbf{v}\mathbf{v} - \mathbf{B}\mathbf{B} + \frac{1}{2}(p + |\mathbf{B}|^2)\mathbf{I} - \frac{\tau}{S_v}\Big] \tag{2}$$

$$\frac{\partial a}{\partial t} = \mathbf{v} \times \mathbf{B} + \frac{1}{S_r}\nabla^2 a \tag{3}$$

$$\frac{\partial E}{\partial t} = -\nabla \cdot \Big[(E+p)\mathbf{v} + (|\mathbf{B}|^2\,\mathbf{I} - 2\mathbf{B}\mathbf{B})\cdot \mathbf{v} - \frac{2}{S_v}\mathbf{v}\cdot\tau + \frac{2}{S_r}(\mathbf{B}\cdot\nabla\mathbf{B} - \nabla\mathbf{B}\cdot\mathbf{B}) - \frac{2}{(\gamma-1)\,S_v\,Pr\,A^2\,M^2}\nabla T\Big] \tag{4}$$

supplemented by an equation of state, $p = (\gamma - 1)U$, where: $\rho(\mathbf{x}, t) \equiv$ mass density, $\mathbf{v}(\mathbf{x}, t) = (u, v, 0) \equiv$ flow velocity,

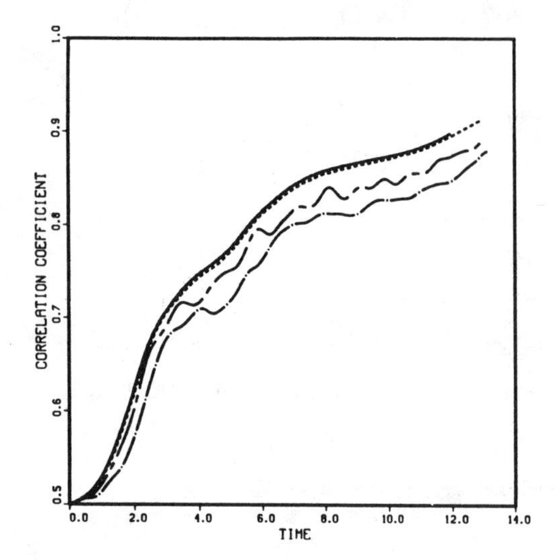

$p(\mathbf{x}, t) \equiv$ mechanical pressure, $\mathbf{I} \equiv$ unit dyad, $a(\mathbf{x}, t) \equiv z$ component of the magnetic vector potential, $\mathbf{B}(\mathbf{x}, t) = (B_x, B_y, 0) \equiv$ magnetic induction field, $U(\mathbf{x}, t) \equiv$ internal energy density, $E(\mathbf{x}, t) = \rho \mid \mathbf{v} \mid^2 + \mid \mathbf{B} \mid^2 + U \equiv$ total energy density, $T(\mathbf{x}, t) = \gamma A^2 M^2 p/2\rho \equiv$ temperature, $\tau(\mathbf{x}, t) = (\partial_j v_i + \partial_i v_j) - \frac{2}{3}\nabla \cdot \mathbf{v}\delta_{ij} \equiv$ viscous stress tensor, and $\gamma \equiv$ adiabatic ratio $= 5/3$. The thermal conductivity (κ), magnetic resistivity (η), and viscosity (μ) are constant and uniform, and Stokes relationship is assumed (cf. Peyret and Taylor, 1983). The important dimensionless numbers are: $S_v = \rho_o V_A L_o/\mu \equiv$ viscous Lundquist number (Dahlburg et al., 1983), $S_r = V_A L_o/\eta \equiv$ resistive Lundquist number, $M \equiv$ average Mach number, $Pr = C_p\mu/\kappa \equiv$ Prandtl number, and $A = (V_A^2/V_o^2)^{\frac{1}{2}} \equiv$ Alfvén number. In these definitions, ρ_o is a characteristic density, V_A is the Alfvén speed, L_o is a

Fig. 1. Correlation coefficient vs. time for all runs. The curves are parameterized by Mach number (M): Solid line – incompressible case ($M = 0$); Dashed line – $M = 0.2$; Long dash-short dash line – $M = 0.4$; and Dash-dot line – $M = 0.6$. As M increases a delay in correlation growth occurs. Note that as M approaches zero the compressible cases converge to the incompressible result.

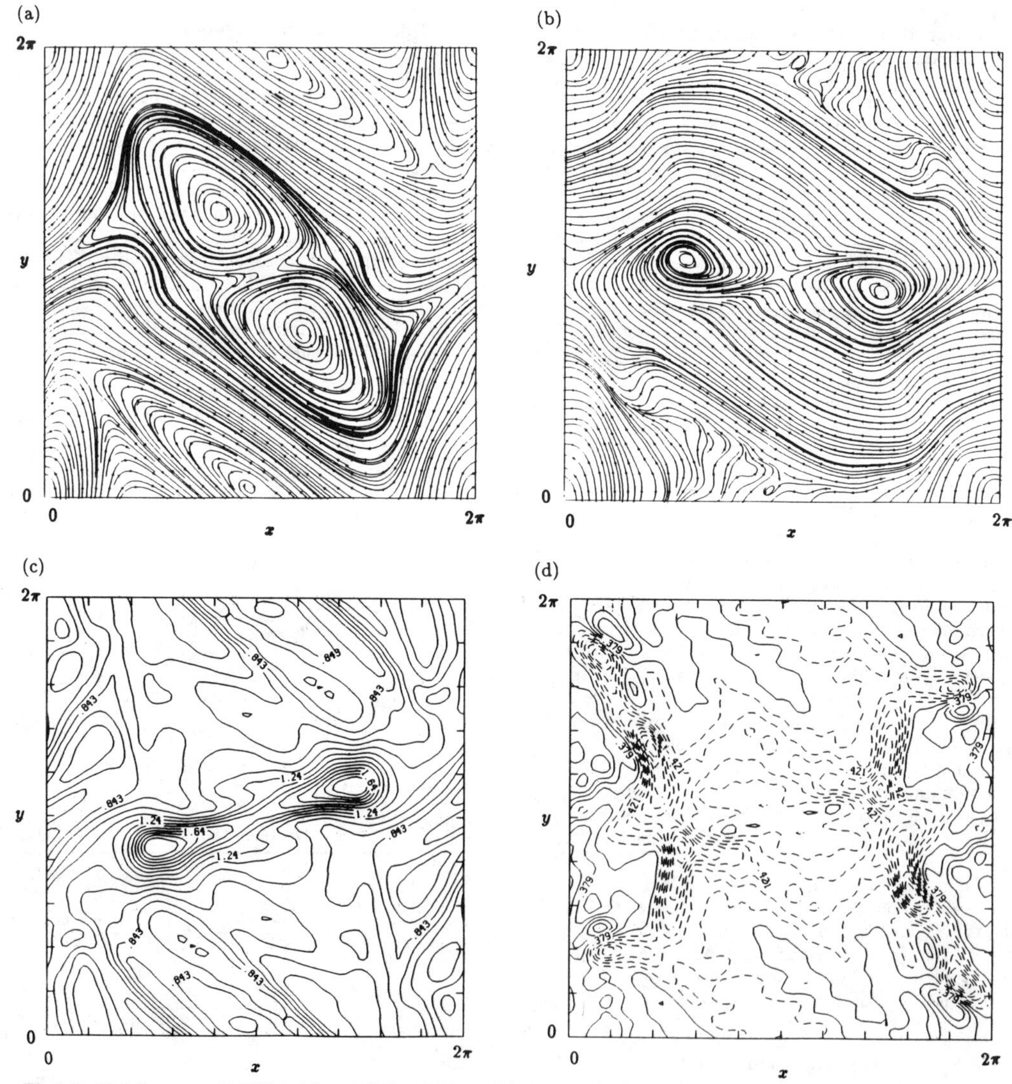

Fig. 2. Fields at $t = 2.17$ for $M = 0.6$ simulation. (a) Magnetic field. (b) Velocity field. (c) Mass density. (d) Local rate of dilatation.

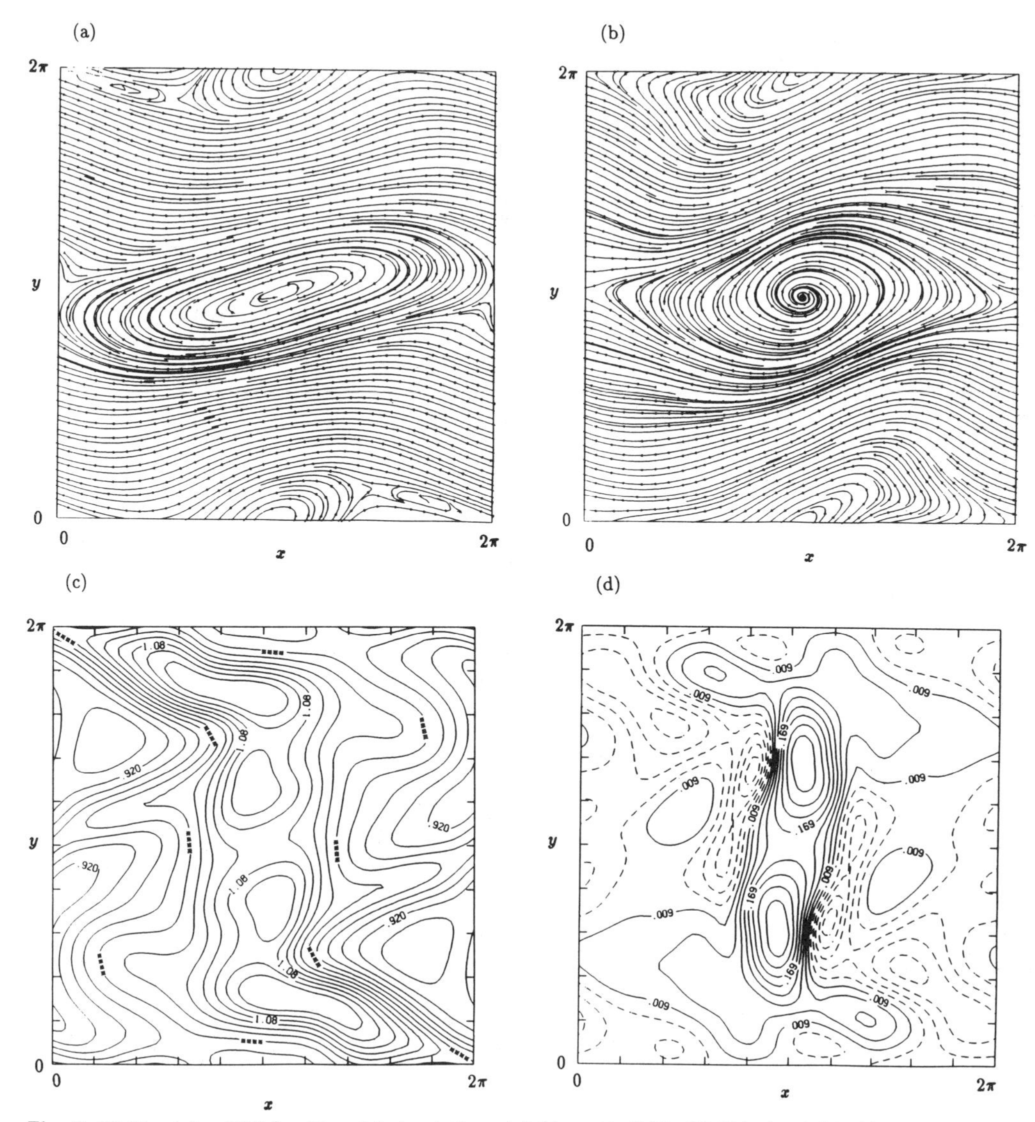

Fig. 3. Fields at $t = 13.2$ for $M = 0.6$ simulation. (a) Magnetic field. (b) Velocity field. (c) Mass density. (d) Local rate of dilatation. The growth of correlation between the magnetic field and the velocity field is evidenced by their resemblance to each other at this time.

characteristic length, C_p is the specific heat at constant pressure, and V_o is a characteristic flow speed. For the runs reported here, we set $S_v = S_r = 50$ and $Pr = A = 1$. The thermodynamic normalization sets $E_o = p_o = B_o^2/8\pi = \rho_o V_A^2/2$. Time ($t$) is measured in units of Alfvén transit times.

Periodic boundary conditions are enforced in both x and y. For initial conditions we use the "Orszag-Tang vortex" (Orszag and Tang, 1979):

$$\rho(x, y, t = 0) = 1 \tag{5}$$

$$\mathbf{v}(x, y, t = 0) = -\sin y\ \hat{\mathbf{e}}_x + \sin x\ \hat{\mathbf{e}}_y \tag{6}$$

and

$$\mathbf{B}(x, y, t = 0) = -\sin y\ \hat{\mathbf{e}}_x + \sin 2x\ \hat{\mathbf{e}}_y \tag{7}$$

where $\hat{\mathbf{e}}_x$ and $\hat{\mathbf{e}}_y$ are unit vectors in the x and y directions, respectively. The initial mechanical pressure, p, is decomposed into a mean part, p_m, and a fluctuating part, p_f. We set p_f equal to the appropriate incompressible pressure distribution, which is obtained by applying a divergence operator to the incompressible MHD equation of motion and then solving the resulting Poisson equation. We define M to be the ratio of the rms velocity to the rms sound speed. In our normalization:

$$M^2 = \frac{2\langle |\mathbf{v}_o|^2\rangle}{\gamma\langle(\frac{p_m + p_f}{\rho_o})\rangle} \tag{8}$$

where the subscript "o" here denotes the initial condition, and brackets denote an integral taken over the whole system. For our system (8) reduces to

$$M^2 = \frac{2}{\gamma p_m} \tag{9}$$

To solve the governing equations (1)-(4) we use a dealiased Fourier collocation method. Time is discretized by the modified Euler's method, a second-order Runge-Kutta scheme. The time step is determined by a compressible MHD CFL number ($N_{CFL} = sup\ \{((| \mathbf{v} | + C_S + V_A)\Delta t)/\Delta x\} \leq 0.3$, where $C_S \equiv$ sound speed). For our simulations, which use 64^2 Fourier modes, approximately 0.25 million words of core memory are required, and 0.1 s per time step on the NASA Ames Research Center CRAY-2. The incompressible results are computed by a two-dimensional version of the algorithm employed by Dahlburg et al. (1988b) for studying the turbulent decay of force-free magnetic fields.

Numerical Results

We report the results of numerical simulations of the Orszag-Tang vortex with different initial values of M, viz.: (1) incompressible ($M = 0$); (2) $M = 0.2$; (3) $M = 0.4$; and (4) $M = 0.6$. To determine the level of correlation in the magnetofluid we monitor the time evolution of a correlation coefficient, $\xi = 2\langle \mathbf{v} \cdot \mathbf{B}\rangle / (\langle | \mathbf{v} |^2\rangle + \langle | \mathbf{B} |^2\rangle)$. Initially $\xi = 0.5$ for the Orszag-Tang vortex. Figure 1 shows ξ as a function of t for the four runs. All of the runs exhibit the same behavior until about $t = 0.5$. At this time the compressible runs exhibit a phase of diminished growth of correlation, coincident with augmented cross-helicity decay. Subsequently the compressible runs show growth of correlation, but at any given time the level of correlation is less the greater the value of M. Note that as M approaches zero, the results for the compressible runs converge to the incompressible result.

Vector and contour plots of the fields assist in understanding the dynamic alignment process in compressible MHD. Figures 2 and 3 show several fields for the $M = 0.6$ case. For Figure 2, $t = 2.17$, at which time $\xi = 0.58$. Reconnection and convection of the initial magnetic field distribution are apparent. The formation of a central electric current sheet gives rise to Lorentz forces which create convection of additional magnetic flux into the reconnection zone. These features of the MHD evolution are similar to those of the incompressible case. However, mass is swept into the X-point and expelled in high-speed jets whose appearance is reminiscent of data on astrophysical jets. The jets produce local high-density regions, with normalized density jumps of about 2, which then are entrained into the exterior flow. Contour plots of the local rate of dilatation $(\nabla \cdot \mathbf{v})$ show the existence of regions of high compression in the magnetofluid. Compression regions have an extreme dilatation value of -1.8.

Figure 3 shows the same fields at $t = 13.2$, the conclusion of the $M = 0.6$ run. The magnetic field and velocity field are highly correlated at this time ($\xi = 0.879$) as can be seen by comparison of the magnetic field plot and the velocity field plot. These fields appear to be moving toward a one-dimensional state with little variation in x. At this time there is still substantial variation in the mass density, with the maximum fluctuations about 15% of the initial value. Plots of the local rate of dilatation show that compression and rarefaction are considerably diminished. The extreme dilatation in the compression regions is approximately -0.22 at this time.

Discussion

Our earlier series of numerical simulations were initialized with non-solenoidal flows and uniform mechanical pressure distributions. For moderate degrees of initial correlation, these simulations showed an initial burst of correlation growth. We conjectured that this burst was due to the initial tendency of the dissipative magnetofluid to flow parallel to the magnetic pressure gradient. The simulations reported in this paper are initialized with solenoidal flows and a fluctuating pressure distribution which corresponds to the incompressible pressure distribution. In this circumstance noticeable compressible effects are somewhat delayed. The $M = 0.2$ simulation behaves quasi-incompressibly. More noticeable differences in both the development and the details of the correlation occur in the $M = 0.4$ and $M = 0.6$ simulations. These results, combined with those of our earlier investigation (Dahlburg et al., 1988a), indicate the presence of several competing effects: (1) quasi-incompressible dynamic alignment, which occurs on the Alfvén time scale; (2) enhanced dynamic alignment occurring on the magnetoacoustic time scale due to the formation of non-solenoidal flows; and (3) Mach number dependent inhibition of alignment. Determination of the relative importance of the physical effects represented in our numerical simulations will require considerable further investigation. We note that the results reported in this paper might be altered by anisotropic and/or temperature-dependent transport coefficients.

Acknowledgments. We acknowledge major contributions by J. P. Dahlburg and J. H. Gardner. This work was sponsored by ONR and the NASA Solar-Terrestrial Theory Program.

References

Canuto, C., M. Y. Hussaini, A. Quarteroni, and T. A. Zang, *Spectral Methods in Fluid Dynamics*, Springer-Verlag, New York, 1987.

Dahlburg, R. B., T. A. Zang, D. Montgomery, and M. Y. Hussaini, Viscous, resistive magnetohydrodynamic stability computed by spectral techniques, *Proc. Nat. Acad. Sci. USA, 80*, 5798, 1983.

Dahlburg, R. B., J. M. Picone, and J. T. Karpen, Growth of correlation in compressible two - dimensional magnetofluid turbulence, *J. Geophys. Res., 93*, 2527, 1988a.

Dahlburg, R. B., J. P. Dahlburg, and J. T. Mariska, Helical magnetohydrodynamic turbulence and the coronal heating problem, *Astron. Astrophys., 198*, 300, 1988b.

Erlebacher, G., M. Y. Hussaini, C. G. Speziale, and T. A. Zang, Toward the large-eddy simulation of compressible turbulent flows, *ICASE Report, 87-20*, 1987.

Grappin, R., Onset and decay of two-dimensional magnetohydrodynamic turbulence with velocity-magnetic field correlation, *Phys. Fluids, 29*, 2433, 1986.

Hasegawa, A., Self-organization processes in continuous media, *Adv. Phys., 34*, 1, 1985.

Horiuchi, R., and Sato, T., Self-organization process in three-dimensional compressible magnetohydrodynamics, *Phys. Fluids, 29*, 4174, 1986.

Léorat, J., R. Grappin, A. Pouquet,and U. Frisch, Turbulence and magnetic fields, in *Stellar and Planetary Magnetism*, edited by A. M. Soward, p. 67, Gordon and Breach, New York, 1983.

Matthaeus, W. H., M. L. Goldstein, and D. C. Montgomery, The turbulent generation of outward traveling Alfvénic fluctuations in the solar wind, *Phys. Rev. Lett., 51*, 1484, 1983.

Orszag, S. A. and C. M. Tang, Small-scale structure of two-dimensional magnetohydrodynamic turbulence, *J. Fluid Mech., 90*, 129, 1979.

Passot, T. and Pouquet, A., Numerical simulation of compressible homogeneous flows in the turbulent regime, *J. Fluid Mech., 181*, 441, 1987.

Peyret, R. and T. D. Taylor, *Computational Methods for Fluid Flow*, Springer-Verlag, New York, 1983.

Pouquet, A., M. Meneguzzi, and U. Frisch, The growth of correlations in MHD turbulence, *Phys. Rev. A, 33*, 4266, 1986.

Riyopoulos, S., A. Bondeson, and D. Montgomery, Relaxation toward states of minimum energy in a compact torus, *Phys. Fluids, 25*, 107, 1982.

THERMAL INSTABILITY IN MAGNETIZED SOLAR PLASMAS

J. T. Karpen, S. K. Antiochos, J. M. Picone and R. B. Dahlburg

Naval Research Laboratory, Washington, DC 20375

Abstract. In astrophysical plasmas such as the solar corona or the interstellar medium, the radiation-driven thermal instability might explain the formation of cool, dense regions embedded in a hotter, more rarefied medium. In the present work, we extend our previous investigation of this phenomenon by simulating the response of a magnetized solar transition-region plasma to a spatially random magnetic-field perturbation, where the magnetic field is perpendicular to the computational plane. Our investigation has determined the effects of varying the plasma β and the heating mechanism. We find that the presence of the magnetic field, the value of the plasma β, and the heating rate significantly influence the size and number of the condensations, as well as the evolutionary time scale. The asymptotic final state is the same in all cases: a uniform, nearly static plasma in which the density and magnetic field strength are close to the initial values but the temperature has dropped to $\sim 10^4$ K. The time scale over which the plasma evolves to this final state increases greatly as β decreases and depends sensitively on the heating rate.

Introduction

Thermal instabilities have been invoked as possible explanations for a variety of astrophysical phenomena including interstellar clouds, condensations in planetary nebulae, and even galaxies (Field, 1965). This mechanism for forming condensed structures in a radiating fluid also might operate throughout the solar atmosphere, where cool and hot regions coexist over a wide range of spatial scales. Our primary goal is to investigate a possible physical mechanism for forming such inhomogeneities, by modeling the development of the radiation-driven thermal instability in a plasma whose characteristics are typical of the solar transition region.

In our earlier calculations (Dahlburg et al., 1987; Karpen et al., 1988a), random velocity or density perturbations were used to trigger the instability in a nonmagnetized plasma. As was found in earlier one-dimensional work (Oran et al., 1982), stable cold structures are formed which oscillate slightly around characteristic temperature and density values. The multi-dimensional calculations also indicate that thermal instability can produce filaments and associated turbulent flows, and that these structures can be stable over long time scales.

In the present work, we extend our investigation of thermal instability in solar plasmas into the magnetized regime. As in the earlier work, an equilibrium atmosphere of uniform temperature and density is set up in which the radiative losses are balanced by an initially uniform heating rate H_0. Here we impose a uniform magnetic field B_0 perpendicular to the plane of the model system. Thermal conduction in the computational plane can be assumed to be negligible. The equilibrium atmosphere is perturbed initially by small amplitude random noise in the magnetic field. We investigate the effects of the initial plasma β on the behavior of the instability, where β is the ratio of the thermal pressure to the magnetic pressure. For low- and high-β cases, we also determine the effects of different heating functions on the thermally unstable plasma. We describe this work in more detail in another paper (Karpen et al., 1989).

The Numerical Model

Computational Methods

The numerical model solves the two-dimensional, compressible, time-dependent conservation equations for mass, momentum, total energy, and magnetic induction (cf. Dahlburg et al., 1987 for a description of the basic computational methods). The plasma is assumed to be fully ionized with a mean ionic charge of 1.059. The optically thin radiative loss function $\Phi(T)$ is from J.C. Raymond (private communication, 1979) and is similar to that of Rosner et al. (1978).

We employ a format for perturbing the system about its equilibrium state similar to that used by Lilly (1969) in studies of the evolution of turbulence in Navier-Stokes fluids. The initial magnetic field is perturbed spatially with wave numbers distributed within a "thick square" in Fourier space (cf. Karpen et al., 1989) such that $3 \lesssim |\mathbf{k}| \lesssim 5\sqrt{2}$.

Three heating rates are used: local heating rate proportional to magnetic energy density $B^2/8\pi$, which represents heating due to magnetic energy dissipation (e.g., Sturrock and Uchida, 1981); local heating rate proportional to mass density ρ, which represents heating due to wave damping, particle-beam energy, or other collisional processes (e.g., Fisher et al., 1985); and constant volumetric heating, as used in our earlier simulations and in the linear analysis of Field (1965). The low-β, ρ-heating case was not simulated because it would evolve even more slowly than the high-β case, which itself was extremely slow compared with the other calculations. Therefore, we discuss only five calculations.

Initial Conditions

The computational plane is divided into 50 x 50 cells of fixed location, each of length $\Delta x = \Delta y = 1.837 \times 10^7$ cm. Periodic boundary conditions are imposed in x and y. Except for the addition of B_o, the initial, static equilibrium is the same as was used by Dahlburg et al. (1987) and Karpen et al. (1988): $\rho = 3.07 \times 10^{-14}$ g cm^{-3}, $T = 7.29 \times 10^5$ K, and $H_o = 2.72 \times 10^{-2}$ ergs cm^{-3} s^{-1}. Two magnetic-field strengths are used: $B_o = 1$ G ($\beta = 81$) and $B_o = 10$ G ($\beta = 0.81$). All calculations employ a magnetic field perturbation of maximum amplitude 0.05 B_o.

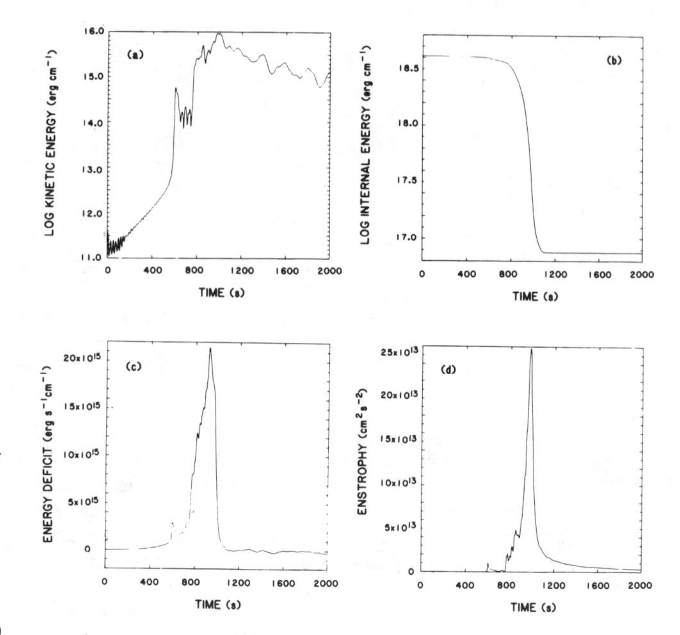

Fig. 1. Global quantities per unit length as a function of time for the high-β, B^2-heating calculation. (a) Kinetic energy. (b) Internal energy. (c) Instantaneous energy deficit (radiative losses minus heat energy input). (d) Enstrophy ($\int \omega^2 dA$, where ω is the vorticity and A is the system area).

Results

All calculations were similar in the gross features of the initial condensation phase and in the final state of the plasma. The intermediate portions differ substantially, and are discussed in detail in the following section. Here we outline the common behavioral patterns and physical characteristics of the calculations. To illustrate these general trends, we show results of the high-β, B^2-heating simulation. Figure 1 shows the temporal evolution of the global quantities, and Figures 2 and 3 contain contour plots of the temperature and magnetic field at selected times during the calculation. A nonlinear analytic model which further clarifies these results is presented by Karpen et al. (1989).

During the damping phase, which occupies the first few hundred seconds of the calculation, the ambient sound waves are damped as predicted by linear analysis (see Field, 1965; Oran et al., 1982). Thereafter, multiple condensations form within the system in all cases. Their spatial distribution retains the symmetry of the superposition of the lowest wave number modes present in the initial perturbation (i.e., the [0,3] and [3,0]) modes, as found in our earlier work. Hence, perturbing the magnetic field is similar to perturbing the density field, in that coupling to initially absent, lower wave number modes is negligible.

The condensation process is characterized by a rapid increase in the kinetic energy (5-7 orders of magnitude for the high-β calculations and 2-3 orders of magnitude for the low-β calculations), a 20-50% drop in the internal energy (independent of β), a sharp increase in the energy deficit (radiation minus heating) above the equilibrium level of zero, and the development of significant rotational flows in the system (i.e., increasing enstrophy).

A phase of linear instability growth is followed by the nonlinear transition phase, during which the condensations expand and coalesce while the surrounding plasma heats up to an extent dictated by the type of heating rate. The competing physical processes which characterize this complex phase are described in the next section.

The ultimate state of the plasma apparently is the same in all cases. During the relaxation phase, the cool regions expand until the entire plasma is at $T = 1.1$-1.4×10^4 K and the initial magnetic field strength and density have been recovered. In contrast to the unmagnetized results, there is a net loss of internal energy from the system as the final state is approached. As described in the next section, however, the

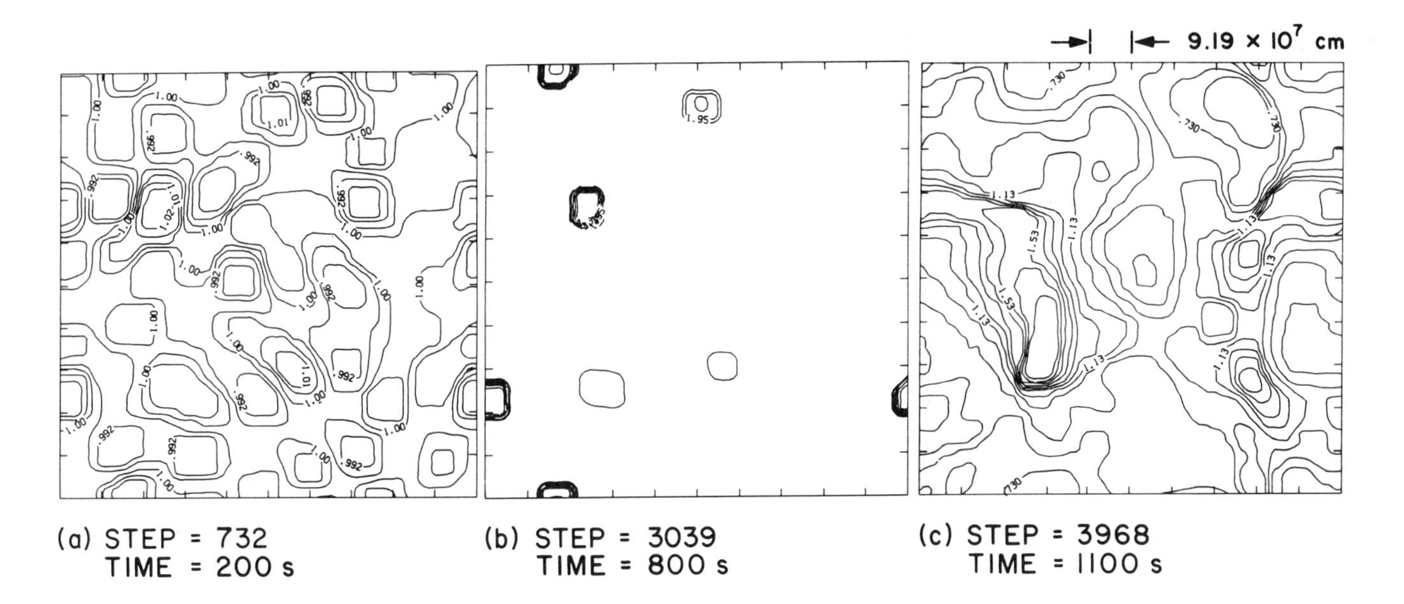

Fig. 2. Contour plots of temperature during the high-β, B^2-heating calculation. (a) The end of the damping phase (t = 200 s). The minimum and maximum values are 6.83×10^5 K and 7.64×10^5 K, respectively. (b) The end of the nonlinear transition phase (t = 800 s). The minimum and maximum values are 2.29×10^4 K and 7.74×10^5 K, respectively. (c) The end of the relaxation phase (t = 1100 s). The minimum and maximum values are 1.32×10^4 K and 2.02×10^5 K, respectively.

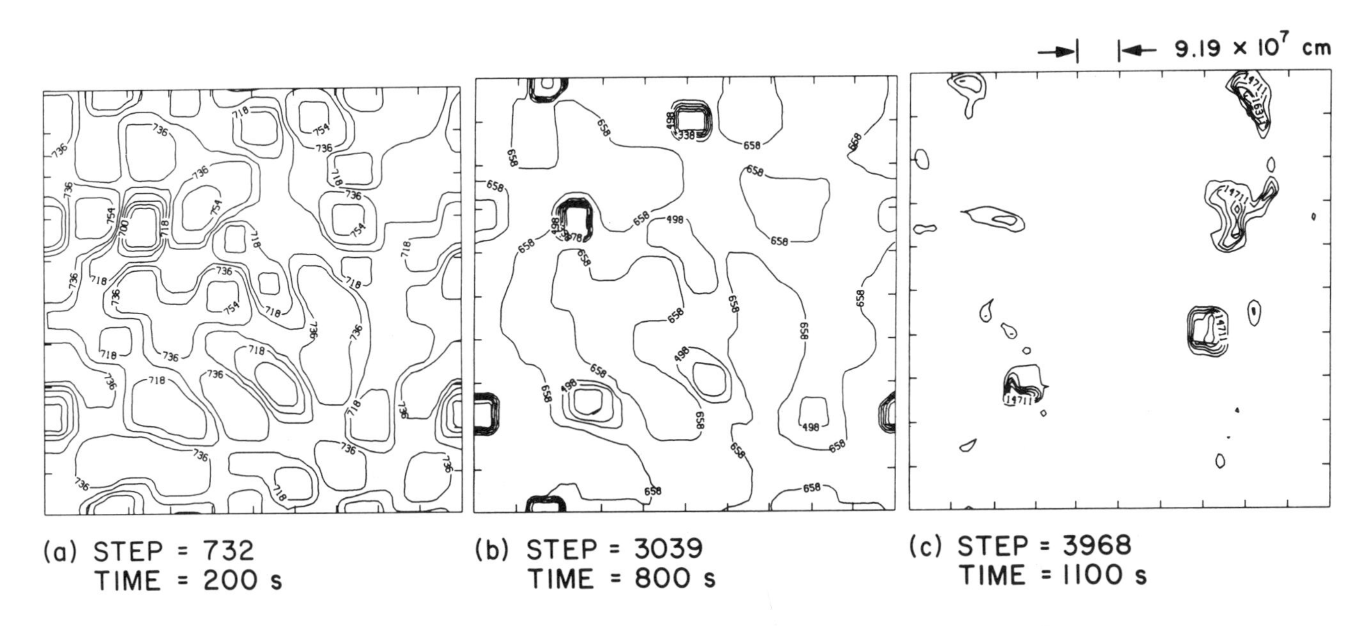

Fig. 3. Contour plots of magnetic field strength during the high-β, B^2-heating calculation. (a) The end of the damping phase (t = 200 s). The minimum and maximum values are 0.98 G and 1.03 G, respectively. (b) The end of the nonlinear transition phase (t = 800 s). The minimum and maximum values are 0.75 G and 6.77 G, respectively. (c) The end of the relaxation phase (t = 1100 s). The minimum and maximum values are 0.49 G and 1.90 G, respectively.

time scale over which this final state is attained and the physical conditions in the evolving plasma depend critically on the heating rate and the plasma β.

Discussion

The perpendicular magnetic field exerts a major influence on the system evolution through two processes: the suppression of thermal conduction in the computational plane, and the addition of the magnetic pressure to the thermal pressure field. We discuss each of these important processes in turn.

The suppression of thermal conduction in the computational plane is reflected in the morphology and long-term behavior of the plasma. The small-scale structure in the initial perturbation cannot be diffused in the present simulations, in contrast to the unmagnetized plasmas investigated previously. In fact, even the lowest fluctuation wave numbers introduced in the magnetized system were damped in the unmagnetized simulations before condensation on those scales could occur (cf. Karpen et al., 1988). In addition, energy cannot be redistributed through thermal gradients without thermal conduction. This is crucial in determining the form of the secondary equilibrium in the later stages of the calculation. For example, the unmagnetized plasma with constant heating produced stable, cool condensates surrounded by a much hotter, constant-temperature medium, while the present calculations only can reach equilibrium by cooling to a uniform temperature.

Despite the absence of cross-field thermal conduction in the model system, the gradients between the hot and cool regions cannot become discontinuous. Rather, these "transition zones" are broadened by the nonlinear radiative losses and by mass flows (for the high-β case). The finite width of this boundary region is responsible for the continual growth of the condensates, as we discuss later in this section.

The β value primarily affects the compressibility of the plasma. Although the magnetic field and the mass density are constrained to evolve similarly, the corresponding pressures do not: the magnetic pressure is proportional to B^2 while the thermal pressure is proportional to ρ. As mass and magnetic flux are convected into the condensing region, the magnetic pressure inhibits compression by building up faster than the thermal pressure. As a result, cool regions are formed in the low-β systems without the large density increase typical of the high-β condensates. In addition, the flows are much slower and less rotational in the low-β calculations.

The heating process is critical in determining the evolutionary time scale of the magnetized plasma. This is consistent with 1D unmagnetized calculations performed by Dahlburg and Mariska (1988). Regardless of the plasma β, all cases eventually find the same equilibrium state. The physical characteristics of the evolving plasma differ, however, before this final state is achieved. We discuss the effects of the heating rate in order of increasing time scale.

B^2-heating case. In these simulations, the plasma surrounding the condensates never gets much hotter than the initial temperature, so $\Phi(T)$ drops slightly at most. We see a greater decrease in the density and magnetic field strength of the surrounding environment as the condensates form, thus quadratically decreasing both the radiative losses and the heating rate. In the "transition zone" next to the condensates, however, radiative losses always exceed the heat input. The resultant thermal pressure deficit drives flows of hot material toward the cool regions. The net cooling induced by the flows is so strong, in this case, that the entire hot region cools nearly simultaneously until the temperature is below the initial hot-equilibrium value. Consequently, the whole system cools rapidly via thermal instability until radiation and heating are balanced again (at $T \sim 1.1\text{-}1.4 \times 10^4$ K) and the initial, uniform density and magnetic field strength are restored.

ρ-heating case. These calculations exhibit behavior intermediate to the other heating cases. As the plasma surrounding the original condensates heats up at a moderate pace, the cool regions grow through the following processes. As described above, flows of hot plasma are driven toward the condensates by a thermal pressure deficit. This hot plasma expands and, hence, cools sufficiently to induce local thermal instability. In essence, a "thermal instability front" propagates outward from each condensate at a rate determined by the pressure deficit next to the cool region and the local compressibility. The linear density dependence of the heating rate ensures that heating dominates over the combined cooling processes (i.e., radiation and expansion) longer than in the B^2-heating calculations. In addition, the rarefied regions become hotter and smaller in the ρ-heating case while they are being eliminated by the expanding instability fronts. Expansion cooling plays two roles here: near the condensates, it feeds the growing instability fronts by cooling the local plasma below the hot equilibrium temperature; elsewhere in the hot region, it gradually counteracts the effects of heating and, again, lowers the peak temperature sufficiently to enter the radiation-dominated regime (i.e., to induce thermal instability away from the condensations).

Constant heating. The cool regions grow slowly while the surrounding medium heats up rapidly. Unlike the other cases, the heating rate is not affected by the local physical properties. As a consequence, heating dominates in the rarefied regions for an even longer interval before expansion cooling becomes important, and the rarefied plasma becomes even hotter than in the ρ-heating calculation. Expansion cooling operates in two ways here, as in the ρ-heating case.

Acknowledgments. This research is supported by the Office of Naval Research and the NASA Solar-Terrestrial Theory Program. The calculations were performed on the NRL Cray X-MP with a generous grant of computer time under the 6.1 Production Run program. We thank C. R. DeVore and J. H. Gardner for helpful discussions.

References

Dahlburg, R. B., C. R. DeVore, J. M. Picone, J. T. Mariska, and J. T. Karpen, Nonlinear evolution of radiation-driven thermally unstable fluids, Ap. J., 315, 385, 1987.

Dahlburg, R. B., and J. T. Mariska, Influence of the heating rate on the condensational instability, Solar Phys., 117, 51, 1988.

Field, G. B., Thermal instability, Ap. J., 142, 531, 1965.

Fisher, G. H., R. C. Canfield, and A. N. McClymont, Flare loop radiative hydrodynamics. VII. Dynamics of the thick-target heated chromosphere, Ap. J., 289, 434, 1985.

Karpen, J. T., J. M. Picone, and R. B. Dahlburg, Nonlinear thermal instability in the solar transition region, Ap. J., 324, 590, 1988.

Karpen, J. T., S. A. Antiochos, J. M. Picone, and R. B. Dahlburg, Nonlinear thermal instability in magnetized solar plasmas, Ap. J., 338, 493, 1989.

Lilly, C., Numerical simulation of two-dimensional turbulence, Phys. Fluids Suppl. II, 240, 1969.

Oran, E. S., J. T. Mariska, and J. P. Boris, The condensational instability in the solar transition region and corona, Ap. J., 254, 349, 1982.

Rosner, R., W. H. Tucker, and G. S. Vaiana, Dynamics of the quiescent solar corona, Ap. J., 220, 643, 1978.

Sturrock, P. A., and Y. Uchida, Coronal heating by stochastic magnetic pumping, Ap. J., 246, 331, 1981.

THE PLASMA ENVIRONMENT AT SATURN: PROGRESS AND PROBLEMS

John D. Richardson

Center for Space Research, Massachusetts Institute of Technology, Cambridge, MA 02139

Abstract. Although much progress has been made toward understanding the plasma environment near Saturn, many questions remain. This paper reviews the current state of observations and theory in Saturn's magnetosphere and highlights remaining problems. Observations and modeling give a consistent picture of the inner magnetosphere in which an anisotropic plasma is tightly confined near the equatorial plane. The primary heavy ion source is sputtered neutrals from the water ice surfaces of the Saturnian satellites, which yield a plasma with approximately equal densities of atomic and molecular ions. The plasma density is limited by recombination and charge exchange loss inside L = 8, whereas transport loss dominates outside this. A model of the plasma distribution in the region inside 12 Saturn radii (R_s) is used to produce contour plots of plasma density which are consistent with plasma observations from both the Voyager 1 and 2 spacecraft. The magnetosphere is more dynamic outside 12 R_s, with blobs of hot plasma interspersed with outward moving cold plasma. The dynamics, composition, and source of this hot plasma are not well understood. The Cassini mission, tentatively scheduled for launch in the mid-1990's, should resolve many of the outstanding issues.

Introduction

Three spacecraft traversed the magnetosphere of Saturn, Pioneer 11 and Voyagers 1 and 2 (Frank et al., 1980; Bridge et al., 1981, 1982). All three spacecraft sampled plasma conditions: Pioneer 11 measured ions in the energy/charge range 100 eV to 8 keV, and the two Voyagers measured both ions and electrons with energy/charge 10-5950 eV. Neither plasma instrument could directly determine composition, and spatial coverage obtained by each was of course limited by the spacecraft trajectory and orientation.

The basic information needed to understand the magnetospheric plasma environment are the ion and electron distribution functions (which give densities, temperatures, and velocity), the spatial distribution of plasma, the role of time-dependent processes, the ion composition, and sources and sinks for both energy and particles. In this paper I review the extent to which this information is known through either direct measurement or theoretical modeling and highlight the outstanding problems which remain.

Ion Distributions and Fluid Parameters

The Voyager plasma instrument (PLS) measures ion and electron current as a function of energy/charge using four modulated-grid Faraday cups (see Bridge et al., 1977). Since electrons are subsonic, a reasonable measure of the reduced distribution function between 10 and 5950 eV is acquired if one can neglect or compensate for effects such as spacecraft charging and ion feedthrough. Figure 1 shows a plot of electron densities and temperatures from Voyager 2 adapted from Sittler et al. (1983). The electron spectra are modeled as a cold Maxwellian population plus a hot non-Maxwellian tail. The cold temperature and density in Figure 1 are derived from fits to a Maxwellian; a moment calculation was then performed on the remaining signal to find the density and average energy of the hot component. The effects of spacecraft charging and ion feedthrough were neglected in these calculations; comparison with ion densities indicates the spacecraft potential is approximately 10-20 volts negative (Richardson and Sittler, 1988), resulting in an underestimate of the electron density by a factor of about 2. Inside L = 5 the bulk of the electrons fall below the energy threshold of the detector, so densities are not reliable in this region and the inferred temperature is 1 eV or less. Elsewhere the relative densities are accurate, however, and give a good overview of the magnetosphere.

The magnetosphere can be divided into two major regions (seen most readily in Figure 1), the plasma sheet in which the density decreases fairly smoothly with radius and the plasma mantle where plasma densities and temperatures can change by a factor of 10 in less than the

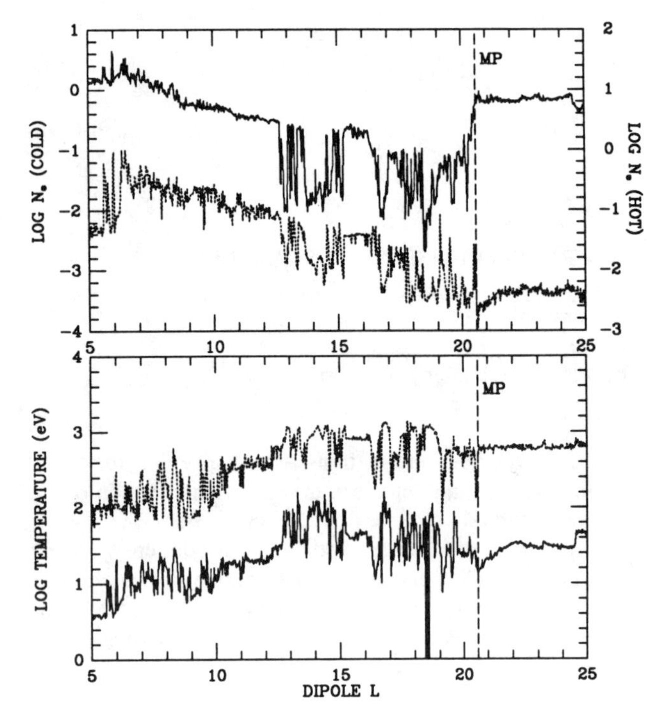

Fig. 1. Densities and temperatures for the thermal (solid line) and hot (broken line) electron populations observed by the PLS instrument on Voyager 2. The magnetopause (MP) location is indicated by the dashed line.

instrument sampling time (96 s). The transition occurs at L = 16 in the Voyager 1 data and at L = 12 in the Voyager 2 data. Electron temperatures increase with L, from less than 1 eV in the inner magnetosphere to 500-800 eV in the outer magnetosphere. The fraction of hot electrons increases with L, and the hot electrons dominate the plasma pressure outside of L = 7 (Sittler et al., 1983). Plasma density and electron temperature are strongly anti-correlated in the mantle region.

The ions are generally transonic or supersonic. This is both a problem and an opportunity. It means that ion distributions are not measured unless the detectors are looking into the plasma flow, but if the detectors are properly oriented the simultaneous measurement of ion currents in all four detectors allows the bulk flow velocity and in some cases the pitch angle anisotropy to be determined, as well as densities and temperatures. Furthermore, if the energy separation of ions due to their different masses (ions are assumed to be co-moving) is greater than the ion thermal energies, then ion composition can also be determined. Two sample spectra are shown in Figure 2 from L = 14.3 and L = 15.3. One has two cold, well-resolved current peaks; the other is a featureless hot spectrum. Current peaks are generally resolved in the inner magnetosphere (L < 12). In this region plasma parameters vary slowly and smoothly. Outside of L = 12

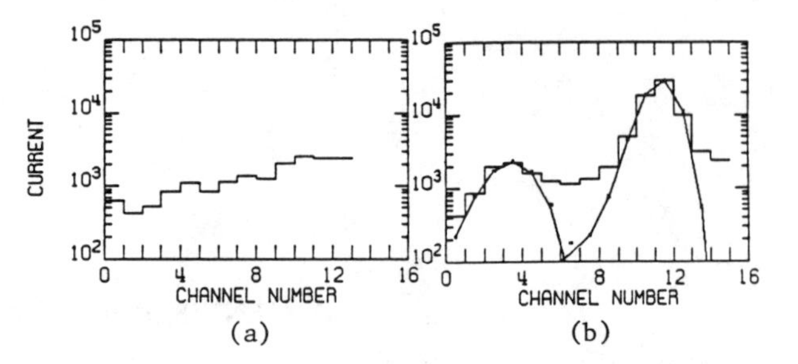

Fig. 2. Two sample ion spectra from Voyager 1. The current is plotted in femtoamps (10^{-15}) and the voltage runs logarithmicly from 10-5950 V. Superimposed on Figure 2b is the best fit to the data obtained assuming Maxwellian distributions of H^+ and O^+ ions.

spectra, and therefore plasma conditions, are variable, and the spacecraft moves from regions of hot (Figure 2a) to cold (Figure 2b) plasma. When cold peaks are present, the lower energy peak is presumed to be protons, and the higher energy peak is then due to heavy ions with mass near 16, either water group (O^+, OH^+, H_2O^+) or nitrogen containing ions (N^+, NH_3^+).

Richardson (1986) analyzed a representative sampling of spectra containing at least one cold peak, using a nonlinear least squares fitting routine to find the simulated Maxwellian distribution which best matched the observed currents. For purposes of this analysis the heavy ion was assumed to have a mass of 16 (see discussion of composition below); use of masses 14-18 has a small effect on the numbers obtained but does not effect the qualitative results. A sample fit resulting from this analysis is superimposed on the data in Figure 2b. Figure 3 shows the azimuthal velocity, heavy ion density, and temperature derived from the fitting procedure plotted versus dipole L. It should be remembered that these values are only representative of the average plasma conditions inside L = 12, since the hot spectra outside L = 12 are not fit. Non-azimuthal velocities are small except in the plasma mantle where outward flow is observed. The ion temperatures increase monotonically with L out to L = 12, then on average decrease slowly. The density is strongly dependent on both latitude and radial distance, with densities increasing toward Saturn and strongly peaked at the equator. The peak plasma density observed was 100 cm^{-3} at the Voyager 2 ring plane crossing at L = 2.7 (Lazarus and McNutt, 1983). The density and temperature of both ions and electrons observed by Voyager 1 inbound are greater than those measured by Voyager 1 outbound at the same L shells. The velocity is below that expected for corotation even in the inner magnetosphere. The implication of these observations for plasma and energy sources and sinks and the plasma distribution will be discussed below.

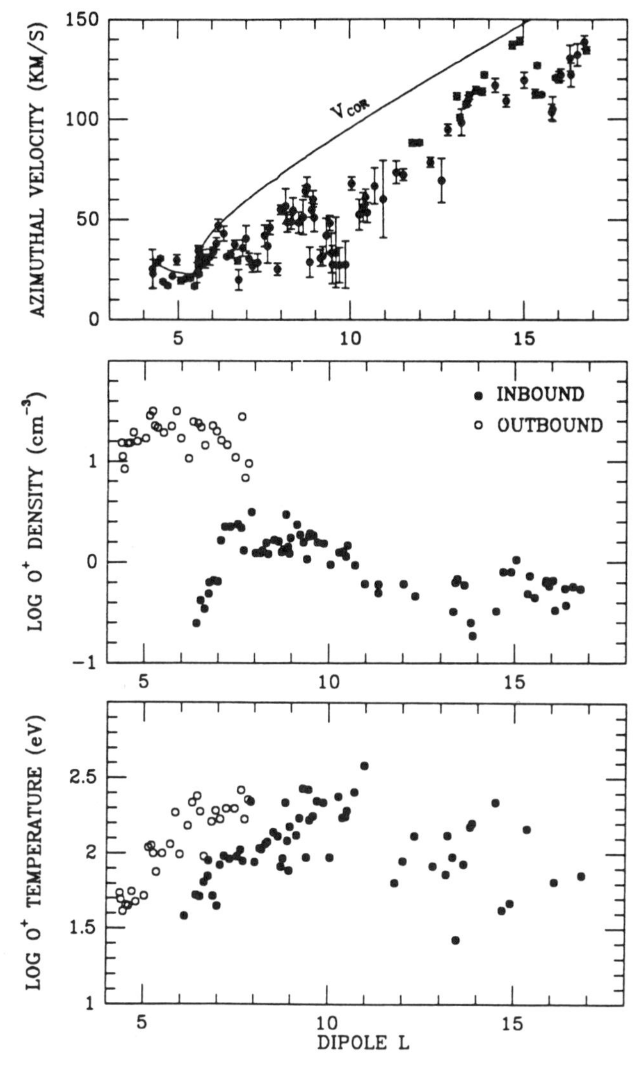

Fig. 3. Azimuthal velocity and heavy ion (mass = 16) densities and temperatures derived from the Voyager 1 PLS measurements.

Composition

As indicated above, Faraday cups can only determine ion composition if the plasma is supersonic and corroborating evidence indicating the likely composition is available. The possible sources of plasma at Saturn are sputtering and micrometeorite erosion of the inner moons and rings, which have water ice surfaces, and escape of atmospheric and ionospheric constituents from Saturn and Titan. Saturn could be a source of H, H^+, H_2^+, and H_3^+. Titan has a predominately nitrogen atmosphere and could be a source of H, H_2, N, H^+, N^+, N_2^+, and/or H_2CN^+.

Voyager flew directly through Titan's wake about 4400 km downstream from Titan. Pickup H^+ was observed prior to entering the wake region, and in the wake there is a direct outflow of cold heavy ions (Hartle et al., 1982). Although their identity was not uniquely determined, these ions probably have mass 28. Assuming this mass designation is correct the ions could be either N_2^+ or H_2CN^+.

The hot spectra in the plasma mantle and outer plasma sheet are presumably hot pickup ions, but again the composition is unknown. In the inner magnetosphere the situation is more fortuitous. The likely plasma source is neutrals sputtered off the surfaces of the inner moons and rings, giving neutral H, H_2, O, OH, H_2O, and O_2. This is consistent with the the identification of the light and heavy ion peaks in the spectra as protons and ions with mass 16. The ions in several spectra at the outer edge of the plasma sheet are sufficiently cold to determine that they have mass 16-18 rather than mass 14 (Richardson, 1986), indicating an inner moon instead of a Titan source. The atomic and molecular chemistry expected to occur in the inner magnetosphere has been modeled by Richardson et al. (1986). They find that the light ions should be about 75% H^+ and 25% H_2^+, and the heavy ions should consist of 40% O^+, 40% H_2O^+, 18% OH^+, and 2% O_2^+. In situ plasma measurements with good mass resolution are required to test these results.

Plasma and Energy Sources and Sinks

Observational and theoretical evidence both suggest that ions are being added to the magnetosphere throughout the region inside L = 12. Calculations of neutral density profiles indicate that the neutral cloud spreads out from the individual satellites and forms a thin disk of neutrals in the equatorial plane (Barton, 1983). These neutrals are ionized and gain perpendicular energy approximately equal to the local rotational energy. Ion data in Figure 3 support this interpretation; the deviation of the azimuthal flow velocities from strict corotation throughout the inner magnetosphere suggests plasma is being injected into the magnetosphere throughout this region. The increase of ion temperature with L out to L = 12 also indicates local plasma production throughout this region, since plasma distributed in L via transport with no sources would have a $T \propto L^{-3}$ dependence due to conservation of the first adiabatic invariant. Outside L = 12 the temperature of the ions decreases, and plasma sources (particularly for heavy ions) are probably small in this region.

The primary loss mechanism for plasma inside L = 8 is dissociative recombination and charge exchange (Richardson et al., 1986). Molecular ions are efficiently removed via dissociative recombination, and atomic ions are lost via charge exchange with atomic and molecular neutrals. Modeling the sputtering and the chemistry of the Enceladus, Tethys, and Dione tori shows that equilibrium densities resulting from sputtering sources and losses due to recombination and charge exchange match the observations well. Radiative cooling may be an important energy sink inward of Tethys, resulting in the very low electron temperatures close to Saturn (Eviatar, 1984).

The increase in electron temperature outside L = 8 reduces the effectiveness of dissociative recombination and then transport plays the major role in removing plasma. The transport process is probably sporadic. The observation of water group ions in the outer plasma sheet requires fast transport outward from Rhea to avoid charge exchange loss with the Titan H torus, and continuous rapid transport would quickly deplete the plasma density. Periodic buildup of plasma in the inner magnetosphere followed by direct outflow of flux tubes may occur. Some type of time-dependent process is indicated by the rapid transitions between hot to cold plasma observed from L = 12 to the outer edge of the plasma sheet. Outside the plasma sheet, the density peaks observed throughout the plasma mantle are blobs of plasma which become detached from the plasma sheet due to a centrifugally-driven flute instability (Goertz, 1983). An alternative explanation for the plasma peaks observed by Voyager 1 near Titan is that plumes of material escape from Titan and wrap completely around Saturn, with radial displacements from Titan's orbital distance caused by fluctuations in the solar wind pressure (Eviatar et al., 1982).

Broadfoot et al. (1981) observed a cloud of neutral H surrounding Titan, and theoretical models also predict the escape of H_2 and N. Ionization of these neutrals forms plasma, which is heated by tapping Saturn's rotational energy. This plasma would be quickly removed from the mantle via transport; it has been suggested, however, that N^+ from Titan could play an important role in the Saturnian aurora (Barbosa and Eviatar, 1986).

A recent alternative interpretation of the UVS data is that a large cloud of H escapes from Saturn's atmosphere, forming a dense cloud extending past Titan (Shemansky and Smith, 1982; Shemansky et al., 1985). This would imply large densities of H (> 25 cm^{-3}) in the inner magnetosphere (see Hilton and Hunten, 1988); the presence of this much H seems inconsistent with the proton to heavy ion ratio observed by PLS in the inner magnetosphere (Richardson and Eviatar, 1987).

Spatial Distribution

With only a few spacecraft flybys it is very difficult to separate longitudinal, local time, time-dependent, and radial variations in plasma parameters. Since the ionization lifetime for neutrals is much longer than the planetary rotation time, azimuthal symmetry is assumed. If time independence is also assumed, which may be reasonable inside L = 12, then the radial and latitudinal structure can be modeled. The distribution of a multi-component plasma along the magnetic field results from a balance between the centrifugal force, the pressure gradient force, the mirror force, and the ambipolar electric field (see Vasyliunas, 1983). Calculating this distribution requires as input the anisotropy, temperature and density of each plasma component, and the azimuthal velocity. Anisotropies have been given at the Voyager 2 ring plane crossing by Lazarus and McNutt (1983), who compared the measured perpendicular ion temperature with the parallel temperature derived from the observed scale height. Richardson and Eviatar (1988) also determined anisotropies by a combination of directly fitting observations and by comparison of inbound and outbound observations from Voyager 1. Richardson and Sittler (1988) use these input parameters to construct the density contours for protons, heavy ions, and electrons shown in Figure 4. Superimposed on the contours are the portions of the Voyager trajectories for which ion data are available. The heavy ion density increases toward Saturn and is tightly confined to the equatorial plane inside L = 8. Outside L = 8 increasing ion

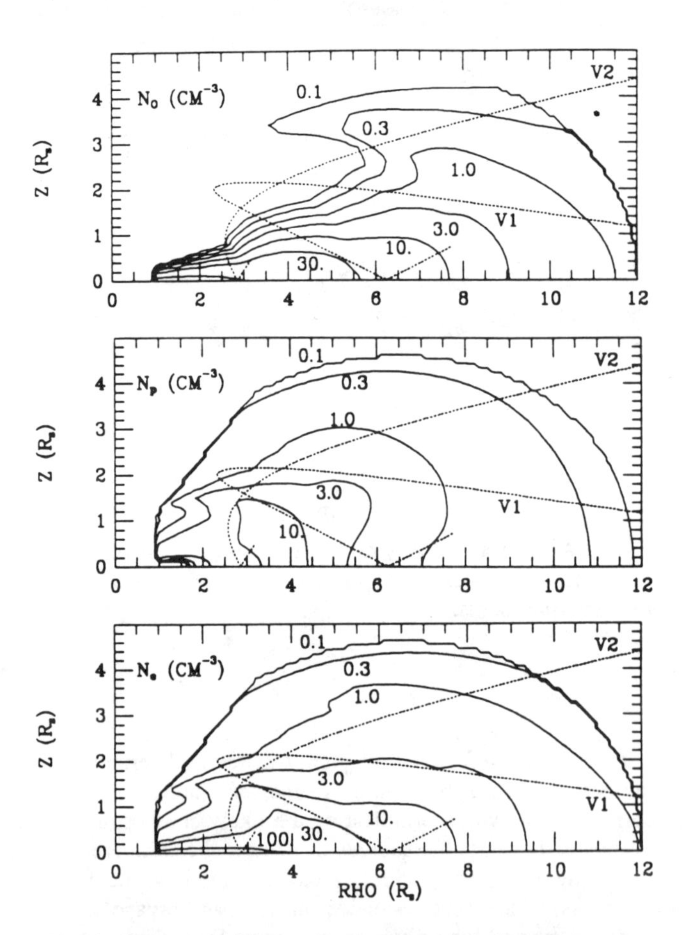

Fig. 4. Contours of heavy ion, proton, and electron density in Saturn's inner magnetosphere. Rho is distance from the spin axis; Z distance from the equator. The outer boundary line indicates the boundaries of the calculation at L = 12, latitude = 50°, and rho = 1.

temperatures and decreasing anisotropies allow the heavy ions to spread in latitude. The protons are less tightly bunched at the equator, and the peak density is at L = 4. Note that between L = 5 and L = 8 the ambipolar effect causes proton densities to increase away from the equator. These results show that all the Voyager ion data can be fit with a steady state, azimuthally symmetric model.

Outstanding Issues

Although we have arrived at a reasonably consistent picture of the plasma environment at Saturn, especially for the inner magnetosphere, many questions remain. Even in the inner magnetosphere only coarse determinations of composition are available from observations, with the details filled in by theoretical models. The presence of neutrals in the inner magnetosphere is inferred from plasma observations and sputtering theory; direct measurements would provide a check on the sputtering results and chemistry and directly identify the plasma source. The composition in the hot outer regions of the magnetosphere is essentially unknown; this composition is crucial for understanding Titan's role in the magnetosphere. Identification of the constituents in Titan's plume will give valuable information on Titan's upper atmosphere and ionosphere.

The latitudinal structure shown in the contour plots is also more inference than direct measurement. Direct measurement of anisotropies is necessary, as well as sampling at diverse latitudes, longitudes, and local times. The data indicate that time-dependent processes are important outside L = 12. Long-term monitoring is necessary to study variations in the radial extent of the plasma sheet, the formation and transport of the plasma blobs outside L=12, the effect of variations in solar wind pressure on the system, and possible time variations of plasma and neutral density in the inner magnetosphere.

All of these observations call for an orbiter. Cassini has been proposed as a combined orbiter-Titan atmosphere probe mission to be launched in the mid-1990's. It would make many passages through the magnetosphere and close approaches to all the important satellites, and should answer many of our remaining questions about the planet. From a magnetospheric point of few it does have one disadvantage: a large apogee results in only one traversal of the magnetosphere every 30 days. Therefore it will be important to have remote sensing instruments such as an energetic neutral atom imager and an ultraviolet spectrometer to monitor magnetospheric events between in situ measurements.

Acknowledgments. Thanks to E. C. Sittler, Jr., for the electron data shown in Figure 1. The work at MIT was supported by NASA and performed under contract 957781 from JPL to MIT.

References

Barbosa, D. D., and A. Eviatar, Planetary fast neutral emission and effects on the solar wind: A cometary exosphere analog, *Astrophys. J.*, *310*, 927, 1986.

Barton, L. A., Magnetospheric ion erosion of the icy satellites of Saturn, M.S. thesis, Univ. of Virginia, Charlottesville, 1983.

Bridge, H. S., J. W. Belcher, R. J. Butler, A. J. Lazarus, A. M. Mavretic, J. D. Sullivan, G. L. Siscoe, and V. M. Vasyliunas, The plasma experiment on the 1977 Voyager mission, *Space Sci. Rev.*, *21*, 259-287, 1977.

Bridge, H. S. et al., Plasma observations near Saturn: Initial results from Voyager 1, *Science*, *212*, 217-224, 1981.

Bridge, H. S. et al., Plasma observations near Saturn: Initial results from Voyager 2, *Science*, *215*, 563-570, 1982.

Broadfoot, A.L. et al., Extreme ultraviolet observations from Voyager 1 encounter with Saturn, *Science*, *212*, 206-211, 1981.

Eviatar, A., Plasma in Saturn's magnetosphere, *J. Geophys. Res.*, *89*, 3821-3828, 1984.

Eviatar, A., G. L. Siscoe, J. D. Scudder, and J. D. Sullivan, The plumes of Titan, *J. Geophys. Res.*, *87*, 8091-8103, 1982.

Frank, L. A., B. G. Burek, K. L. Ackerson, J. H. Wolfe, and J. D. Mihalov, Plasma in Saturn's magnetosphere, *J. Geophys. Res.*, *85*, 5695-5708, 1980.

Goertz, C. K., Detached plasma in Saturn's turbulence layer, *Geophys. Res. Lett.*, *10*, 455-458, 1983.

Hartle, R. E., E. C. Sittler, Jr., K. W. Ogilvie, J. D. Scudder, A. J. Lazarus, and S. K. Atreya, Titan's ion exosphere observed from Voyager 1, *J. Geophys. Res.*, *87*, 1383-1394, 1982.

Hilton, D. A., and D. M. Hunten, A partially collisional model of the Titan hydrogen torus, *Icarus*, *73*, 248-268, 1988.

Lazarus, A. J., and R. L. McNutt, Jr., Low-energy plasma ion observations in Saturn's magnetosphere, *J. Geophys. Res.*, *88*, 8831-8846, 1983.

Richardson, J. D., Thermal ions at Saturn: Plasma parameters and implications, *J. Geophys. Res.*, *91*, 1381-1389, 1986.

Richardson, J. D., and A. Eviatar, Limits on the extent of Saturn's hydrogen cloud, *Geophys. Res. Lett.*, *14*, 999-1002, 1987.

Richardson, J. D., and A. Eviatar, Observational and theoretical evidence for anisotropies in Saturn's magnetosphere, *J. Geophys. Res.*, in press, 1988.

Richardson, J. D., and E. C. Sittler, Jr., The PLS Saturn plasma model, manuscript in preparation, 1988.

Richardson, J. D., A. Eviatar, and G. L. Siscoe, Satellite tori at Saturn, *J. Geophys. Res.*, *91*, 8749-8755, 1986.

Shemansky, D. E., and G. R. Smith, Whence comes the

"Titan" hydrogen torus (abstract), *Eos Trans. AGU, 63*, 1019, 1982.

Shemansky, D. E., G. R. Smith, and D. T. Hall, Extended atomic hydrogen cloud in the Saturn system and a possible analogous distribution at Uranus (abstract), *Eos Trans. AGU, 66*, 1108, 1985.

Sittler, E. C., Jr., K. W. Ogilvie, and J. D. Scudder, Survey of low-energy plasma electrons in Saturn's magnetosphere, *J. Geophys. Res., 88*, 8848-8870, 1983.

Vasyliunas, V. M., Plasma distribution and flow, Chapter 11 in *Physics of the Jovian Magnetosphere*, edited by A. J. Dessler, Cambridge Univ. Press, New York, 1983.

What is the role of turbulence in the transfer of energy in solar wind and magnetospheric plasmas?

NUMERICAL SIMULATION OF INTERPLANETARY AND MAGNETOSPHERIC PHENOMENA: THE KELVIN-HELMHOLTZ INSTABILITY

Melvyn L. Goldstein, D. Aaron Roberts

Code 692, Laboratory for Extraterrestrial Physics, NASA Goddard Space Flight Center, Greenbelt, MD 20771

William H. Matthaeus

Bartol Research Institute, University of Delaware, Newark, DE 19716

Abstract. A series of numerical simulations are described which comprise part of a study of processes important in both interplanetary and magnetospheric physics. One goal is to understand the evolution and small-scale structure of the Kelvin-Helmholtz instability in the presence of sheared magnetic fields, including magnetic neutral sheets which can be the site of turbulent magnetic reconnection. These simulations use a two-dimensional spectral method code to solve the equations of incompressible magnetohydrodynamics with magnetic and kinetic Reynolds numbers of 1000. We find that the presence of magnetic fields in the vicinity of the velocity shears greatly influences the development of the nonlinear Kelvin-Helmholtz instability. Configurations that model the boundary between fast and slow speed streams in the solar wind are discussed, and we present evidence that stream shear may produce magnetohydrodynamic (MHD) turbulence in the solar wind. In simulations designed to study the magnetosheath-magnetopause boundary in Earth's tail, we find that large vortical structures are formed that resemble observed plasma flows.

Introduction

The solar wind and the magnetosphere provide a natural and accessible laboratory in which to study many nonlinear plasma phenomena of relevance in astrophysics. Two such phenomena are magnetic reconnection and the Kelvin-Helmholtz instability, both of which have be studied using in situ satellite experiments, analytic theory, modeling, and numerical simulations. Although the Kelvin-Helmholtz instability is driven by velocity shear, in fluids containing magnetic fields the stability of the shear interface and the rate at which the instability can evolve is modulated by the magnitude and direction of the magnetic field. If there is a magnetic neutral sheet in the vicinity of the velocity shear, then a complex interaction between the Kelvin-Helmholtz instability and magnetic reconnection ensues.

Magnetic reconnection by itself appears to play a crucial role in determining the interaction between the solar wind and the magnetosphere, perhaps most importantly via "flux transfer events" (Russell and Elphic, 1978; Cowley, 1976; Russell 1984) upstream of the Earth's bow shock. Flux transfer events provide a means by which solar wind magnetic fields can become topologically connected to the Earth's magnetosphere, thereby allowing for an exchange of plasma and energy and momentum between the interplanetary medium and the magnetosphere (see, for example, Dungey, 1961; Sonnerup, 1984). As we discuss below, there are indications from particle observations in the dayside magnetosheath and from numerical models that the Kelvin-Helmholtz instability may enhance the propensity of the boundary to reconnect. In the magnetotail reconnection may be the central mechanism by which magnetic substorms are energized although this is still a somewhat controversial model (see, for example, Nishida, 1984). Time-dependent models of reconnection and their application to substorms have been discussed by Owen and Cowley (1987). A recent survey of theoretical work on the linear and nonlinear tearing mode theory as it pertains to reconnection can be found in Hones (1984) and Butler and Papadopoulos (1984). The later reference also contains a discussion of turbulent reconnection.

The Kelvin-Helmholtz instability, which is the major focus of this paper, is excited at the boundary between two parallel flows (Chandrasekhar, 1961). In magnetospheric physics the Kelvin-Helmholtz instability has been conjectured to produce a viscous boundary layer (for example, Miura, 1984, 1987) that could be a major mechanism responsible for the exchange of momentum and energy between the solar wind and the magnetosphere (Axford and Hines, 1961). Whereas magnetic reconnection between the solar wind and the magnetosphere preferentially takes place at the front of the magnetosphere, the Kelvin-Helmholtz instability is likely to be most important at the magnetopause in the tail (see, for example, Southwood, 1979; Sonnerup, 1980; Sckopke, 1981; Miura, 1984, 1987; Wu, 1986). Because the magnetic field generally is not parallel to the velocity shear at the inner boundary of the magnetopause, it is the more likely site of the excitation of the Kelvin-Helmholtz instability (Lee et al., 1981).

More evidence for the importance of the Kelvin-Helmholtz instability in the controlling magnetospheric dynamics came during the ISEE 1 and 2 missions. Hones et al. (1978, 1981) reported the presence of large scale plasma vortices in the early morning sector of the plasma sheet. On at least one occasion this flow, which had a scale of several R_E, lasted for 9 hours. Hones et al. (1981) described the existence of a vortex "roll" or "street" that had a wavelength of 24-20 R_E and a speed of several hundred km/s. Ogilvie and Fitzenreiter (1988), using the vector electron spectrometer data from ISEE 1, have studied the stability of the magnetopause and boundary layer. They found evidence that the layer contains a surface wave perhaps generated by the Kelvin-Helmholtz instability. Another observation of a large-scale vortex, this time in the deep tail (at X_{GSM} = -217 R_E) was observed by ISEE 3 and has been described by Sanderson et al. (1984, 1986). Plasmoids, which closely resemble the O-regions produced by magnetic reconnection, are also found in the magnetotail. It has been suggested (see Nishida, 1984) that the plasmoids are produced close to the near-Earth neutral sheet and afterward move tailward at velocities of several hundred km/s. In this situation, an interaction between vortices and plasmoids in the neutral sheet is likely.

Westward traveling surges and other auroral forms may also be manifestations of a Kelvin-Helmholtz instability generated in the velocity shear zone at the interface between the low-latitude boundary layer and the central plasma sheet (Rostoker, 1987). The occurrence of these auroral forms in Rostoker's model is then associated with times during which magnetic reconnection produces an enhanced momentum density in the central plasma sheet. Investigating the close relationship between the Kelvin-Helmholtz instability and reconnection is also the focus of simulations reported by La Belle-Hamer et al. (1988) which relate patchy reconnection at the dayside magnetopause to enhanced reconnection rates produced when vortices, excited by the Kelvin-Helmholtz instability, compress the current sheet. In the front of the magnetosphere, periodicities in particle data with frequencies in the range 2×10^{-3} to 8.3×10^{-2} Hz have led Frahm (1984) to conjecture that large-amplitude waves produced by the Kelvin-Helmholtz instability are injecting ions into the magnetosphere. The possible interaction between the vortex rolls and plasmoids suggests that boundary layer instability, vortex roll formation, and plasmoids are all closely related.

The Kelvin-Helmholtz instability appears to play an important role in determining the dynamical evolution of the solar wind as a turbulent medium. Coleman (1968) first suggested that velocity shear at the interface between high-speed and low-speed interplanetary streams could be the free energy source that produced turbulent cascades of magnetic energy resulting in Kolmogoroff-like $k^{-5/3}$ energy spectra (Kolmogoroff, 1941; Batchelor, 1970). This view contrasted with the one suggested by Belcher and Davis (1971) who noted that the sign of the high degree of correlation often seen between the fluctuating interplanetary magnetic field **b** and velocity field **v** implied that Alfvénic fluctuations were propagating away from the sun. Belcher and Davis concluded that the observed Alfvénic fluctuations had to have been generated in the solar atmosphere, inside the transonic point because they expected turbulent generation of Alfvénic fluctuations to produce both inward and outward senses of propagation. That this supposition is not always the case was pointed out by Matthaeus et al. (1983).

Early consideration of the importance of the Kelvin-Helmholtz instability in generating interplanetary turbulence were based on linear theory. For example, Parker (1964) showed that for flow and wave vectors parallel to the magnetic field, the instability criterion for excitation of the Kelvin-Helmholtz instability with wave vectors parallel to the velocity shear required that the sound speed exceed the Alfvén speed; a situation that does not commonly arise. In addition, because the linear Kelvin-Helmholtz instability is stable to the generation of fluctuations with wavelengths smaller than the thickness of the velocity shear layer, it should not be excited in the vicinity of 1 AU, where the stream boundaries are quite thick (see, for example, Bavassano et al., 1978).

There are, however, other studies that suggest that the Kelvin-Helmholtz instability might be important in interplanetary processes (Burlaga, 1969; Neugebauer et al., 1986; Korzhov et al., 1984). For example, Korzhov et al. noted that while some calculations of stream evolution predict that the stream interfaces should become steeper between 0.3 and 1 AU, in situ measurements (Rosenbauer et al., 1977; Burlaga, 1979) showed that the interfaces are actually steeper in the inner heliosphere than at 1 AU. Korzhov et al. suggested that the Kelvin-Helmholtz instability was a likely mechanism for smoothing the velocity profiles as the solar wind flowed outward. They also pointed out that Parker's analysis was too restrictive, and that oblique waves with shorter wavelengths than considered by Parker could become unstable. At present, the excitation of the Kelvin-Helmholtz instability in MHD and its relationship to the onset of turbulence has not yet been studied as intensively as has the corresponding phenomenon in magnetic reconnection (see, for example, Matthaeus and Lamkin, 1985, 1986). This paper contains preliminary results of such a study using numerical simulations (Goldstein et al., 1987b).

Recently, Roberts et al. (1987a,b) undertook an extensive analysis of solar wind data obtained throughout the heliosphere from 0.3–20 AU from the Helios and Voyager spacecraft. They found that inside 1 AU, Alfvénic fluctuations are predominantly outward propagating as reported by Belcher and Davis (1971), but the magnitude of the correlation between **b** and **v** rapidly decreased with increasing distance from the sun. Farther out, the predominance of outward propagation changed to a mixture of inward and outward propagating fluctuations. This evolution of the interplanetary plasma, attributed by Roberts et al. to the development of turbulence, was roughly correlated with the presence of velocity gradients in the flow. They concluded that turbulent fluctuations were indeed generated in the solar atmosphere inside the sonic point, as suggested by Belcher and Davis, but, as the solar wind flows into the heliosphere, velocity shears apparently generate new turbulence, as envisioned by Coleman. The in situ generation of turbulence is characterized by reduced correlations between **b** and **v**, and by an increased mixture of both inward and outward propagation.

In the present paper, we begin the task of trying to understand the Kelvin-Helmholtz instability and its relationship to both solar wind turbulence and magnetospheric phenomenon. We first concentrate on the role of the Kelvin-Helmholtz instability in generating turbulence without the complications of magnetic neutral sheets. For example, as pointed out above, in linear theory, shear layers are stable to the excitation of waves with wavelengths smaller than the scale of the velocity gradient. In contrast, turbulence implies the existence of a direct cascade of energy from large to small scales. Furthermore, the presence of a magnetic field parallel to the velocity shear is known to suppress the instability. The effect this might have on the generation of turbulence has not been studied until now. We focus on these are two issues before discussing results from the simulations containing magnetic neutral sheets. The relevance of these simulations to solar wind observations is discussed in detail below; however, the outstanding problem that these numerical experiments are designed to elucidate is the extent to which the solar wind is a dynamically evolving turbulent magnetofluid.

We first construct simulations in which the magnetic shear is relatively weak but parallel to the velocity shear. This situation is roughly analogous to stream shears in the inner heliosphere. As we shall see below, while the magnetic field configuration in these simulations inhibits the development of the large-scale vortex rolls, nevertheless a turbulent cascade of energy from small to large wave numbers develops rapidly. Furthermore, the evolution of the cross helicity in the second simulation described below resembles Roberts et al. (1987a,b) description of interplanetary observations.

In a second set of simulations, we turn to more complex configurations and include a magnetic neutral sheet close to the velocity shear layer to study the interplay between the Kelvin-Helmholtz instability and turbulent reconnection; this situation appears relevant for understanding boundary layer phenomena in the Earth's magnetosphere. The initial conditions in these experiments are constructed be analogous to two cuts through the Earth's magnetotail; one in the north-south plane (if a magnetic neutral sheet is included in the center of the simulation domain), and the other in the east-west plane. While these simulations are preliminary, ultimately we hope to answer questions such as the following: (1) Are magnetotail observations related to vortices generated by the Kelvin-Helmholtz instability? (2) How do boundary layers affect the generation of small-scale fluctuations? and(3) What is the role of small-scale fluctuations in the transport of momentum and energy from the solar wind into the magnetosphere?

Method

We solve the equations of incompressible MHD in two dimensions using an algorithm designed to accurately evaluate small-scale fluctuations so that turbulent cascades of energy can be studied. In appropriate dimensionless units, the equations of MHD are

$$\frac{\partial \mathbf{v}}{\partial t} + \mathbf{v}\cdot\nabla\mathbf{v} = \mathbf{B}\cdot\nabla\mathbf{B} - \nabla P + \nu\nabla^2\mathbf{v} \qquad (1)$$

$$\frac{\partial \mathbf{B}}{\partial t} + \mathbf{v}\cdot\nabla\mathbf{B} = \mathbf{B}\cdot\nabla\mathbf{v} + \eta\nabla^2\mathbf{B} \tag{2}$$

$$\nabla\cdot\mathbf{v} = 0 = \nabla\cdot\mathbf{B} \tag{3}$$

where $\mathbf{v}$ is the velocity field, $\mathbf{B} = \mathbf{B_o} + \mathbf{b}$ is the total magnetic field— mean plus fluctuating, P is the total pressure, ν is the reciprocal of the mechanical Reynolds number R, η is the reciprocal of the magnetic Reynolds number R_m, and the mass density is assumed uniform and constant. The unit of time is the "eddy turnover time" for a unit velocity field of unit length scale.

The algorithm accurately conserves the known global (or "rugged") invariants of ideal incompressible MHD in two dimensions: total energy (per unit mass)

$$E = \frac{1}{2}\int d^2x\ (\mathbf{v}^2 + \mathbf{b}^2) \tag{4}$$

the cross helicity

$$H_c = \frac{1}{2}\int d^2x\ (\mathbf{v}\cdot\mathbf{b}) \tag{5}$$

and the mean-squared vector potential

$$A = \frac{1}{2}\int d^2x\ a^2 \tag{6}$$

(Fyfe and Montgomery, 1976; Kraichnan and Montgomery, 1979).

In two dimensions we let $\mathbf{B} = (B_x,B_y,0)$, $\mathbf{v} = (v_x,v_y,0)$, and $\partial/\partial z = 0$ for all variables so that equations (1)-(3) can be simplified by writing $\mathbf{v}$ and $\mathbf{B}$ in terms of a stream function $\psi(x,y,t)$ and a vector potential $\mathbf{a} = a(x,y,t)\hat{e}_z$ so that $\mathbf{v} = \nabla\times\hat{e}_z\psi$ and $\mathbf{B} = \nabla\times\hat{e}_z a$. With these definitions the dimensionless vorticity and current are $\nabla\times\mathbf{v} = \omega\hat{e}_z$ and $\nabla\times\mathbf{B} = j\hat{e}_z$ with $j = -\nabla^2 a$ and $\omega = -\nabla^2\psi$. Equations (1) and (2) become

$$\frac{\partial\omega}{\partial t} + \mathbf{v}\cdot\nabla\omega = \mathbf{B}\cdot\nabla j + \nu\nabla^2\omega \tag{7}$$

$$\frac{\partial a}{\partial t} + \mathbf{v}\cdot\nabla a = \eta\nabla^2 a \tag{8}$$

and equation (3) is satisfied automatically. In these variables, the three ideal "rugged" invariants (equations (4-6)) can be rewritten in terms of a and ψ as

$$E \equiv \frac{1}{2}\int[(\nabla a)^2 + (\nabla\psi)^2]\ d^2x \tag{9}$$

$$H_c \equiv \frac{1}{2}\int\nabla a\cdot\nabla\psi\ d^2x \tag{10}$$

and

$$A \equiv \frac{1}{2}\int a^2\ d^2x \tag{11}$$

All of simulations described here used 256×256 Fourier modes and were run on the Cyber 205 at the Goddard Space Flight Center.

Equations (7) and (8) are solved as an initial value problem in the xy plane on a periodic square with sides 2π. In the solutions the fields are represented as truncated Fourier series so that, for example, $j(\mathbf{k},t) = k^2a(\mathbf{k},t)$, where $a = \sum_k a(\mathbf{k},t)\exp(i\mathbf{k}\cdot\mathbf{x})$. With a magnetic and kinetic Reynolds numbers as large as 1000, the dissipation wave number K_d is less than k_{max}, where $k_{max} = 120$ is the largest non-zero wave number after the equations are dealiased using the isotropic truncation algorithm (Patterson and Orszag, 1971). The dissipation wave number is given in terms of the mean square vorticity $\Omega = \langle\omega^2\rangle$ and mean square current $J = \langle j^2\rangle$ as (Matthaeus and Montgomery, 1981; Ting et al., 1986)

$$K_d = [2(\Omega + J)R^2]^{1/4} \tag{12}$$

J and Ω can be computed from the simulation. In these simulations K_d ranges from 50–75 (cf. Goldstein et al., 1987a).

Four simulations are described below. Two are designed to investigate the development of turbulence at stream shears in the presence of dc magnetic fields. One of these cases is designed to follow the evolution of an initially highly aligned flow at large wave numbers. These experiments contain only weak magnetic gradients, stronger velocity gradients, and no neutral sheets. The second two simulations contain relatively strong gradients in both **b** and **v**. These latter two experiments represent idealized cuts through the tail of the Earth's magnetosphere in the noon-midnight plane and in the plane of the plasma sheet.

Results

Solar Wind

In the inner heliosphere the boundary between solar wind streams is fairly sharp (Rosenbauer et al., 1977; Burlaga, 1979) and magnetic energy is small compared to the kinetic energy. These properties are modeled in the first two simulations whose initial configurations are shown schematically in Figure 1. The square, 2π on a side, contains two velocity layers represented by the large gray arrows (a minimum of two shear layers is necessitated by the periodic boundary conditions) at $y = \pi/2$ and $3\pi/2$. Note that the center of momentum frame of reference is used.

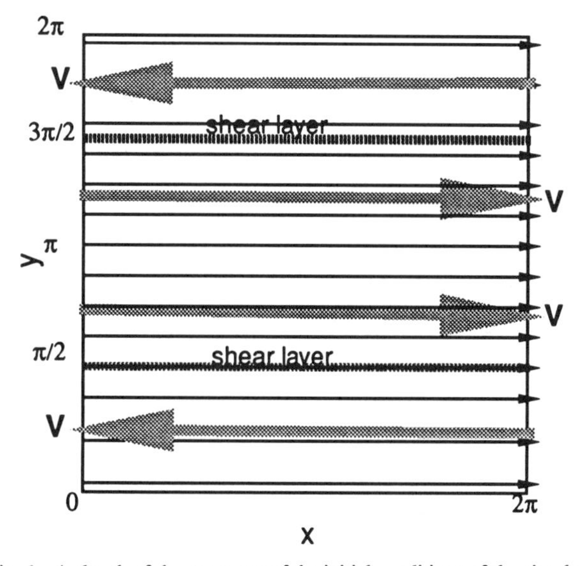

Fig. 1. A sketch of the geometry of the initial conditions of the simulation described in the text. The plane of the simulation includes two moderate velocity shear layers and a strong dc magnetic field . The magnetic field has a weak current sheet close to the velocity shear. Two different initial conditions based on this model were used in the simulations.

The fluctuating magnetic field in this case was relatively weak throughout the box and the gradient of the magnetic field was defined using 8 Fourier modes while the velocity gradient, which is somewhat sharper, was defined using 16 Fourier modes. In the first experiment, the kinetic energy was 47 times larger than the magnetic energy. As with all the simulations discussed in this paper, the fluctuating energy was initially normalized to 1. A mean magnetic field $\mathbf{B_o} = B_o\hat{e}_x = 1.244$ was included. To initiate the evolution of the system, randomly phased isotropic noise was added (out to $|\mathbf{k}| = 16$) which contained 1% of the energy in the fluctuating fields. In this simulation the magnetic and velocity fields were initially fairly uncorrelated ($\sigma_c \equiv 2H_c/E = 0.12$).

Nonlinear couplings transfer energy out to large wave numbers almost immediately. We refer to the energy spectrum as a function of $|\mathbf{k}|$, averaged over annular rings in k-space, as the modal energy spectrum. This spectrum at T = 1 demonstrates the rapid nonlinear energy transfer (Figure 2, top panel). By the end of this run, however, dissipation has damped the energy at the highest wave numbers (K_d is ≈ 60). The spectrum at T = 22 is shown in the bottom panel of Figure 2. Because of the strong dissipation, only a fairly short portion of what might be a power law spectrum can be resolved. This behavior of the spectrum is in contrast to what is expected in linear theory (Sen, 1963). This rapid excitation of small scales is also absent in simulations of the Kelvin-Helmholtz instability which use low order finite difference algorithms (Miura, 1984, 1987; La Belle-Hamer et al., 1988) because they are not optimized for turbulence studies.

The dynamical activity in this simulation is reminiscent, in the nomenclature of Ting et al. (1986), of their in Region II, where the magnetic field is dynamically unimportant and the magnetofluid tends to behave somewhat like a two-dimensional Navier-Stokes fluid. In that context, it is interesting to note that the index of the energy spectrum, to the extent that it can be defined, appears to more closely approximate the Kolmogoroff value of -3 expected for Navier-Stokes turbulence than the value of -5/3 predicted for an MHD fluid (see, Kraichnan and Montgomery, 1979). In three-dimensional MHD turbulence the Kolmogoroff inertial range index for a Navier-Stokes fluid is still –5/3, but Kraichnan's argument as it applies to three dimensions predicts -3/2 for a magnetofluid. The observed solar wind inertial range spectrum tends to have a power law closer to -5/3 than -3/2 (Matthaeus and Goldstein, 1982) and the characteristic parameters of the solar wind fluid also place it in Region II. Note, however, that the application of the classification scheme of Ting et al. (1986) to both these simulations and solar wind observations is suspect because their analysis was carried out with no dc magnetic field, and, as shown below, the presence of a dc magnetic field has significant dynamical effects.

Figure 3 depicts the evolution of the velocity profile at T = 1, 7.5, and 22.5 (the end of this run). The figure shows a vertical cut of $V_x(x_o,y)$ through the simulation box 50 points from the origin in x (cf. Figure 1) at $x_o = 1.23$. Note that the velocity gradient remains relatively steep until late in the run.

Another important feature of these simulations is the evolution of the spectrum toward a highly anisotropic state. The anisotropy may arise because the cascade of magnetic energy in the direction parallel to the mean magnetic field is inhibited, an effect first studied by Shebalin et al. (1983). Shebalin et al. used an MHD spectral method code to show that for strong anisotropies to occur, only a rather modest dc magnetic field need be present. Similar anisotropic spectra have been seen in neutral sheet simulations by Matthaeus and Lamkin (1985) (without a dc magnetic field) and in various combinations of neutral sheets and velocity shear by Goldstein et al. (1987b) (in the presence of a dc magnetic field). In extreme cases of very strong dc magnetic fields, the MHD equations can be rewritten in a form sometimes useful in theoretical treatments of laboratory plasma phenomena (Strauss, 1976; Montgomery, 1982). When velocity shears parallel to the dc magnetic field are present, the shear itself tends to produce anisotropic spectra. In the present situation it is not clear which process is dominant.

The initial noise spectrum is isotropic in all runs reported here. By T = 2 (upper panel of Figure 4) we see that the two-dimensional spectrum of

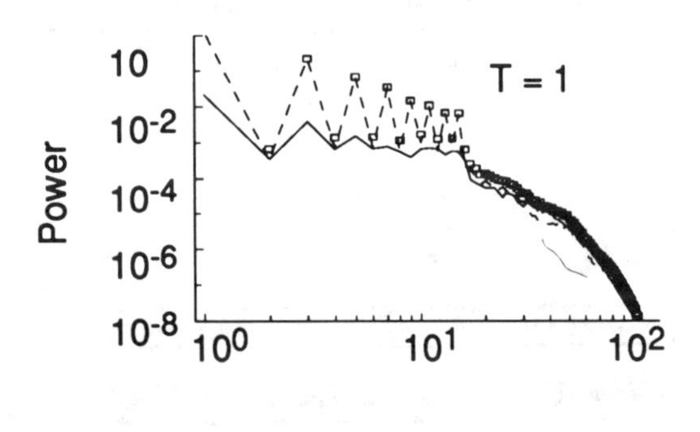

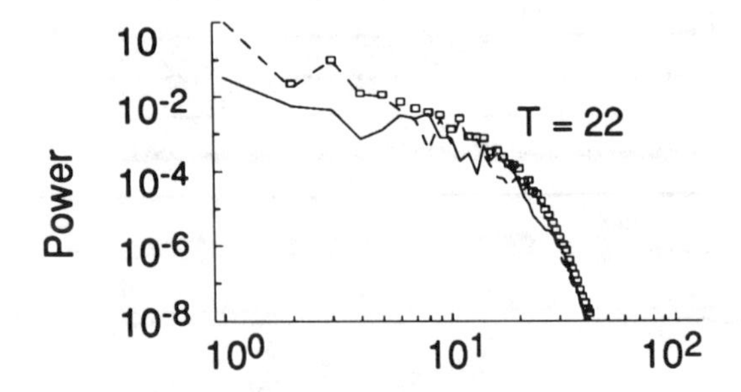

Fig. 2. The modal energy spectrum at T = 1.0 (top) and T = 22 (bottom) from the first of the initial conditions described in the text that are based on the sketch in Figure 1. At t = 0 only modes with $|\mathbf{k}| \leq 16$ were nonzero. Solid lines represent energy in magnetic fluctuations, dashed lines energy in velocity fluctuations, and the boxes, total energy.

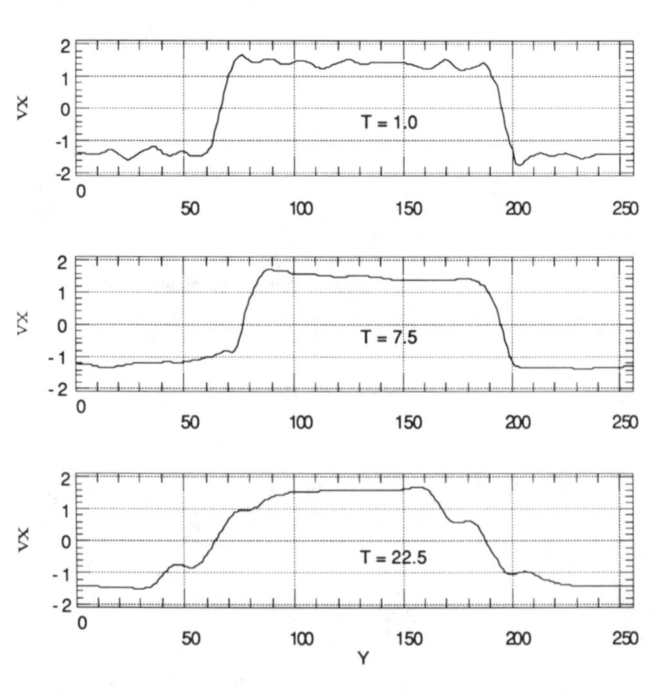

Fig. 3. A cut through the simulation box at T = 1, 7.5 and 22.5 showing $V_x(x_o,y)$ at $x_o = 1.23$.

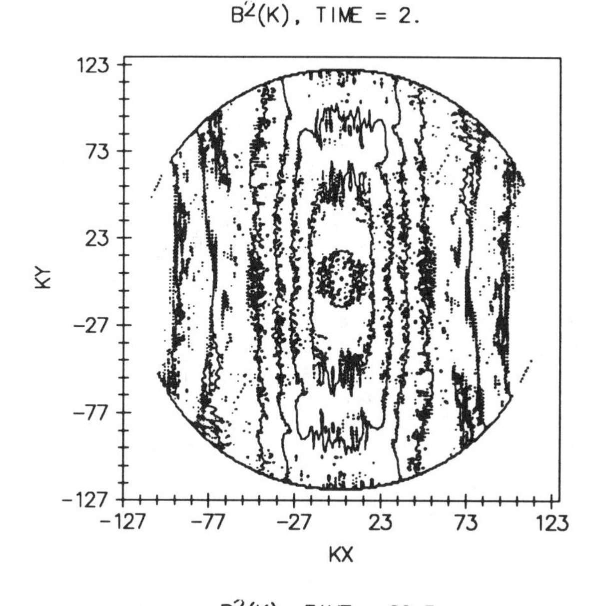

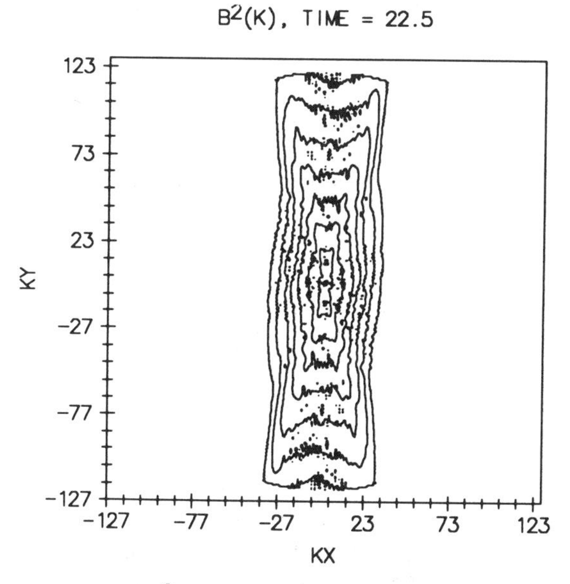

Fig. 4. Contours of $B^2(\mathbf{k})$ at T = 2 (top) and T = 22.5 (bottom). Initially the velocity shear was defined only for modes such that $k_y \leq 16$.

$|B(\mathbf{k})|^2$ is highly anisotropic except in the vicinity of k_x and $k_y \approx 0$. Energy has been transferred more rapidly in the k_y direction, orthogonal to the dc magnetic field and the vorticity layers, than in the k_x direction, parallel to $\mathbf{B_0}$. By T = 22.5 (lower panel of Figure 4) this process has reached an extremely anisotropic state. The possible implications for solar wind phenomenology are intriguing and have been discussed in detail by Matthaeus et al. (1988) who present evidence that such an anisotropy exists in the solar wind.

In the second simulation we examine the evolution of cross helicity in the vicinity of velocity shear and the extent to which the Kelvin-Helmholtz instability mediates the transformation of the solar wind from a highly aligned state (in which $\mathbf{v} \approx \mathbf{b}$) in the inner heliosphere to a more mixed, significantly less aligned state between 0.3 and 1 AU. In this run the ratio of fluctuating kinetic energy to the fluctuating magnetic energy was ≈ 50; the total fluctuating energy was normalized to unity, and the background dc magnetic field was $\mathbf{B_0} = 1$. The shear layers contained 98% of the fluctuating energy, but in this case the gradients were very weak since they were defined using only four Fourier modes. The placement of the vortex sheets was offset slightly from the current sheets, centered at $\pi/4$ and $3\pi/4$, to keep the overall cross helicity small ($\sigma_c = 0.17$). Above k = 4, the simulation initially contained highly correlated noise ($\sigma_c = 1$) with a k^{-4} spectrum. Below k = 4 the noise spectrum was k^{-1}. In this run $\Omega = 2.2$, J = 0.41, $R = R_m = 1000$, and $K_d \cong 50$.

The low k coherent velocity modes at k = 1 and k = 3 drive a direct cascade of energy out to high wave numbers where they are dissipated. This is illustrated in the top and bottom panels (T = 0 and 7, respectively) in Figure 5. The importance of the high k modes at late times can be

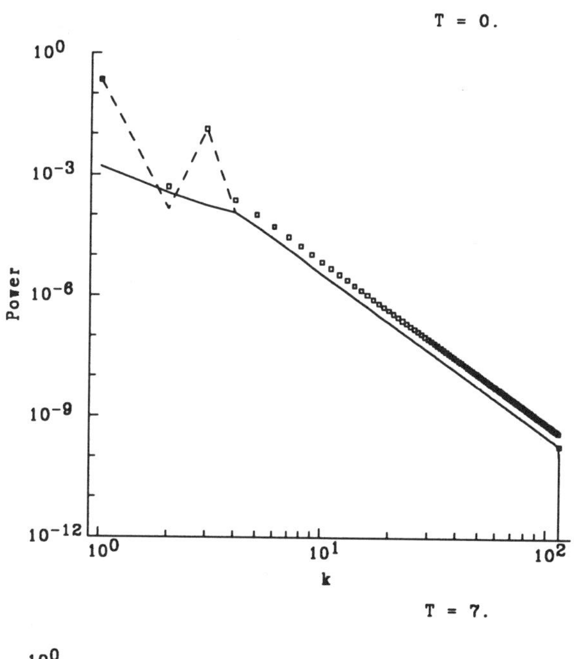

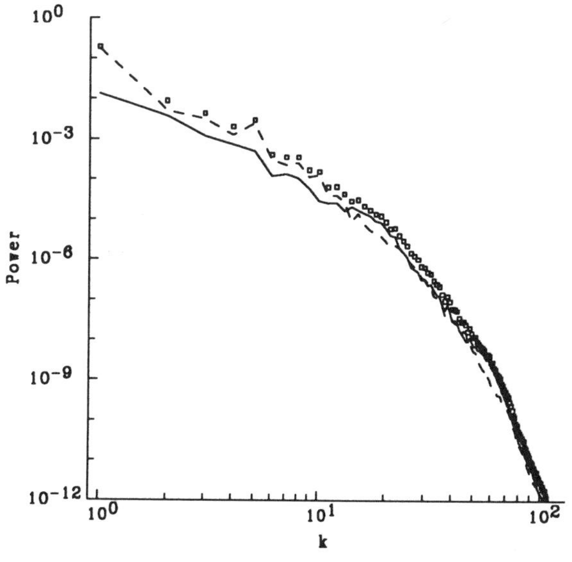

Fig. 5. The modal energy spectrum at T = 1.0 (top) and T = 7 (bottom) from the second simulation described in the text.

seen in the shape of the velocity shear layer. In Figure 6 velocity profiles $V_x(x_o,y)$ are plotted (again at $x_o = 1.23$) at T = 1 and T = 7. By T = 7 at this value of x_o, the gradient is actually steeper than it was initially (at least in this region of the simulation box).

In Figure 7 we have plotted the normalized helicity σ_c as a function of wave number at T = 0 and T = 7. As stated above, initially all modes with k > 4 were completely aligned (i.e., $\mathbf{b}(\mathbf{k}) = \mathbf{v}(\mathbf{k})$ and $\sigma_c = +1$). By T = 7, however, only the modes with the highest wave numbers remain aligned. The alignment of the lower part of the spectrum is significantly decreased, and for k < 20 is essentially zero. Figure 8 gives a clearer picture of the role played by the velocity shear in destroying correlations. The figure is a plot of σ_c at times T = 2, 3, and 7. In constructing the normalized helicity for this plot, the magnetic and velocity fields were spatially filtered by choosing only those values of $\mathbf{b}(\mathbf{k})$ and $\mathbf{v}(\mathbf{k})$ for which $|\mathbf{k}|$ fell within the band $10 < |\mathbf{k}| < 30$. The vectors $\mathbf{b}(\mathbf{k})$ and $\mathbf{v}(\mathbf{k})$ were then inverse Fourier transformed to arrive at the filtered values of $\mathbf{b}(\mathbf{x})$ and $\mathbf{v}(\mathbf{x})$ used to form σ_c. The normalized cross helicity was constructed by computing an average of $\mathbf{b}(\mathbf{x})\cdot\mathbf{v}(\mathbf{x})$ over 20 points and dividing by the average energy in this wave number band over the same 20 points. One can easily see from the figure the location of the two vorticity sheets. Initially nearly all values of $\mathbf{b}(\mathbf{x})$ and $\mathbf{v}(\mathbf{x})$ in this band of spatial values were completely aligned. By T = 2 the values of σ_c in close to the velocity shear layers are significantly reduced. The process continues (T = 3) and by T = 7 (the end of this run) the alignment has been systematically destroyed.

The importance of this plot as a diagnostic of the evolution of the cross helicity lies in its similarities to the analysis of σ_c presented by Roberts et al. (1987a,b) who showed that running values of σ_c obtained from Helios 1 data in 1978 systematically decrease as the spacecraft moved from 0.3 to nearly 1 AU (see Figure 2 of Roberts et al., 1987b). The similarity between the behavior of solar wind data and this simulation suggests that in the solar wind the high values of σ_c reflect a highly aligned initial state in the solar atmosphere that is rapidly eroded between the corona and 1 AU by, perhaps, the excitation of a Kelvin-Helmholtz instability at the location of stream shears.

Roberts et al. also illustrated the evolution of σ_c from 0.3-20 AU in the form of percentage distributions from both Helios and Voyager spacecraft. In Figure 9 (top), we have reproduced a panel from their paper. The distributions were computed using 3-hour estimates of σ_c (cf. their Figure 9). The curves in the top panel of Figure 9 are percentage distributions of

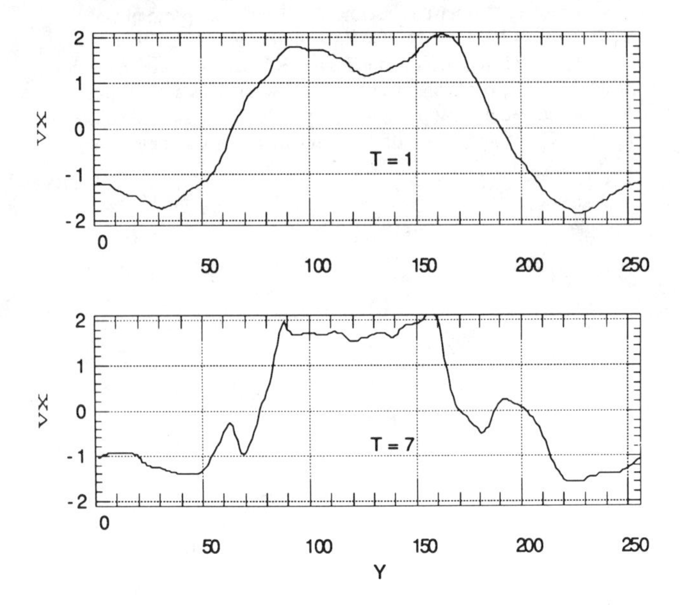

Fig. 6. A cut through the simulation box at T = 1 and 7 showing $V_x(x_o,y)$ at $x_o = 1.23$.

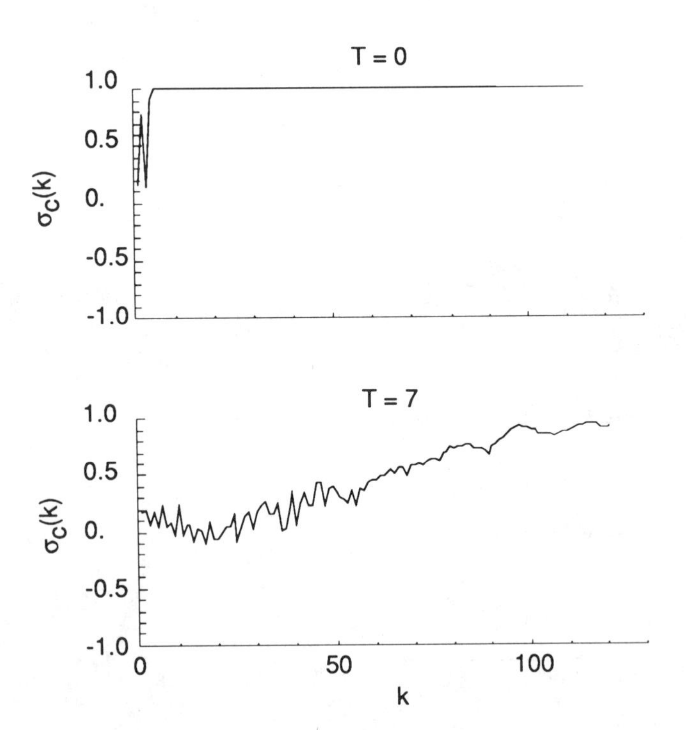

Fig. 7. Spectrum of σ_c. At T = 0 (top panel) the magnetic and velocity fields were aligned at high wave numbers. The final state of the run (T = 7) is shown in the bottom panel.

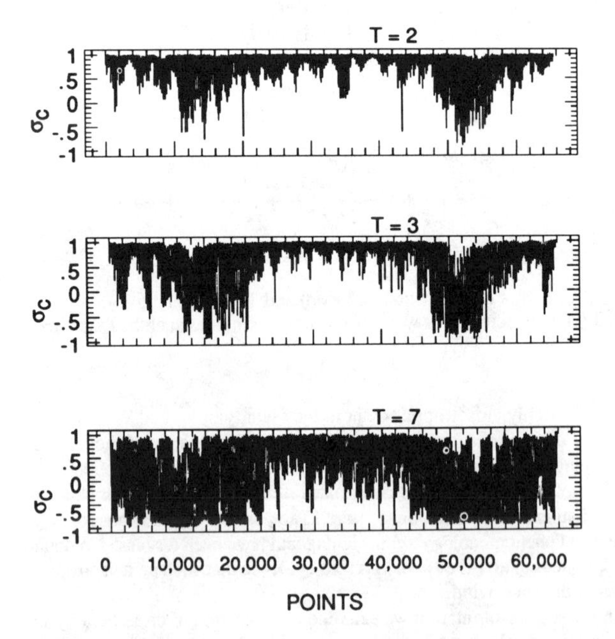

Fig. 8. Plot of localized values of σ_c across the simulation domain at T = 2, 3, and 7. The cross helicity is evaluated for the k = 10 to k = 30 band of wave numbers. Each horizontal cut through the simulation box is sampled in turn, so the regions around points 15,000 and 55,000 show many cuts through the region of maximum shear.

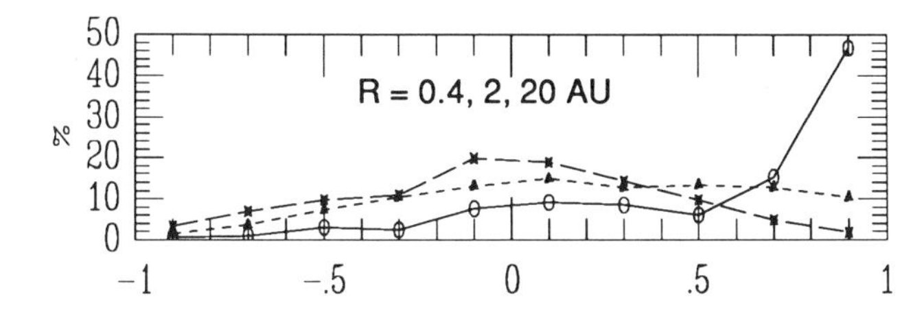

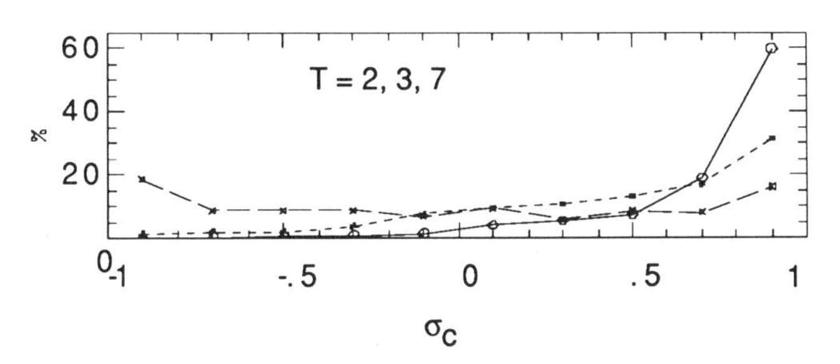

Fig. 9. Histograms of occurrence probably of σ_c constructed in the top panel from Roberts et al. (1987b), and in the bottom panel from the data shown in Figure 8 for points 8,000 to 20,000.

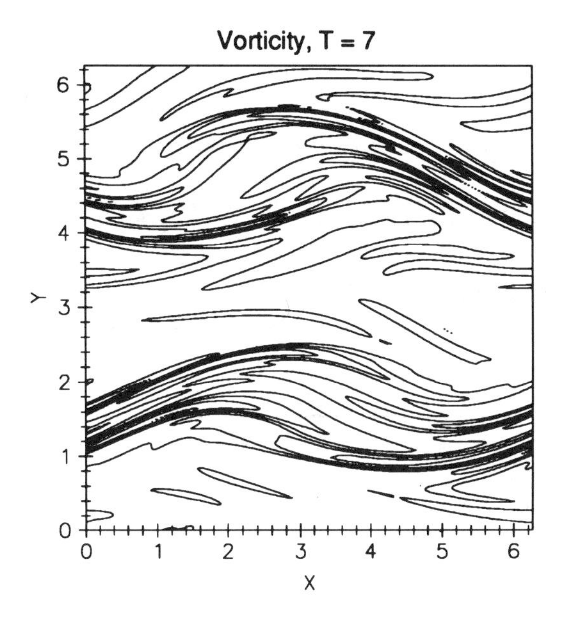

Fig. 10. Contours of vorticity at T = 7 from the second solar wind simulation.

σ_c in bins 0.2 wide obtained from data sets acquired at 0.4, 2, and 20 AU. Note the rapid decrease in the occurrence of high values of the cross helicity which is evident even by 2 AU. By 20 AU, the most probable value for σ_c is 0. If we equate distance in the heliosphere with time in the simulations, then we see a similar evolution of σ_c in the bottom panel of Figure 9. The plot is also a percentage distribution of σ_c values this time compiled from a region of the simulation box near one of the shear layers (corresponding to points 8000–20,000 of the data shown in Figure 8). The lines connected by circles, triangles, and squares are from T = 2, 3, and 7, respectively. The solar wind and simulation data both show the same rapid decline in values of σ_c near 1.

The nonlinear effects seen in these two simulations take place in a parameter regime where the background dc magnetic field acts to suppress the instability. Although turbulence effects are clearly important, the field does effectively prevent the formation of vortex rolls. In Figure 10 we have plotted contours of vorticity at T = 7. While the shear layer, which at T = 0 was parallel to the x-axis, has been deformed, there are no well defined vortex rolls present. This is in marked contrast to what occurs in simulations in an ordinary fluid without any magnetic field. In Figure 11 we show a contour plot of vorticity at T = 2 from a run with initial conditions similar to those of the first "solar wind" simulation discussed above except that in this case all magnetic fields were set to zero. Vortex roll-up and subsequent pairing is a very rapid process in these circumstances.

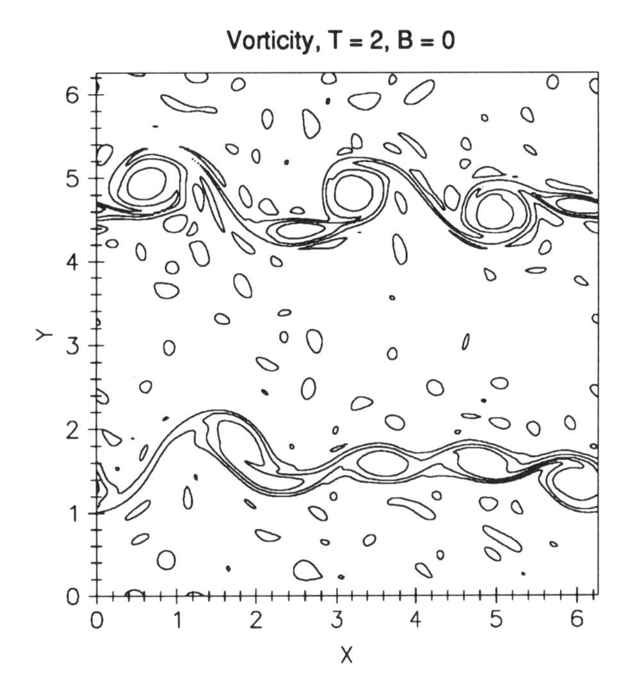

Fig. 11. Contours of vorticity at T = 2 from a simulation in which both the fluctuating and dc magnetic fields were set to zero.

Magnetosphere

Figure 12 shows the configuration of the simulations designed to study the interaction between velocity shear layers and strong magnetic current sheets, including neutral sheets. This simulation represents an idealized cut through the Earth's magnetotail in the noon-midnight meridian. Two strong velocity and magnetic shear layers are present in the upper and lower regions of the box. In addition a magnetic neutral sheet is in the center of the box. The dc magnetic field B_o was -0.15, the ratio of kinetic to magnetic energy was 1.94, and $H_c = 0.09$.

Figure 13 contains plots of modal energy spectra at T = 1 (top) and at the end of the run (T = 13.5), bottom. Recall that initially the spectrum is nonzero only for $|\mathbf{k}| \leq 16$. A substantial cascade of energy to small scales has already begun within one eddy turnover time. By T = 13.5, the tendency for the energy to populate high wave numbers is very clear. Although there is the suggestion of the formation of a power law spectrum, there was probably too much dissipation in this simulation for the Kolmogoroff limit to become well defined.

Another interesting aspect of these spectra is the fact that at high wave numbers the magnetic energy exceeds the kinetic energy at late times

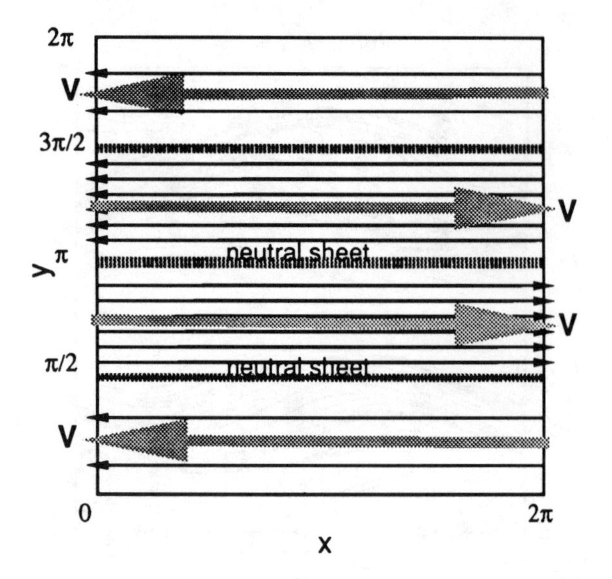

Fig. 12. Similar to Figure 1, but for the first of two "magnetosphere" simulations. The plane of the simulation includes two moderately strong velocity shear layers and a weak magnetic field in the central region with a current sheet close to the velocity shear. A second magnetic neutral sheet lies in the center of the box.

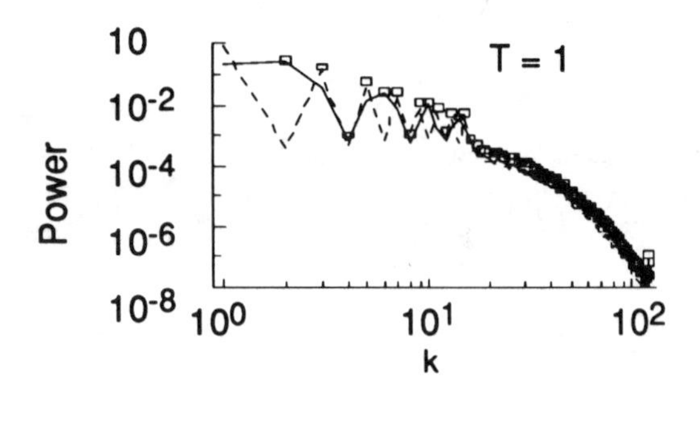

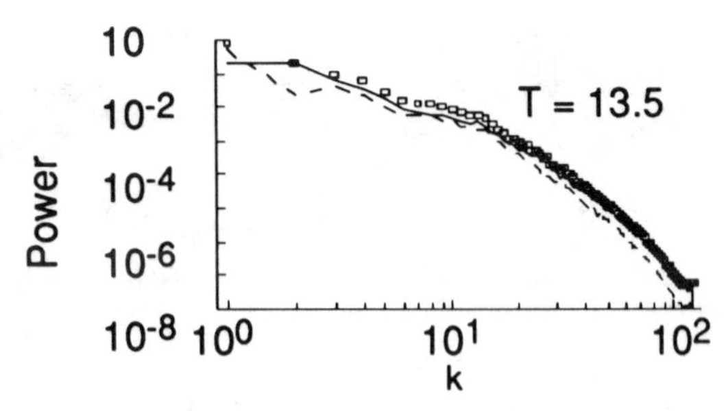

Fig. 13. Modal energy spectra at T = 1 and T = 13.5. At T = 0 only modes with |k| ≤ were nonzero.

although the initial conditions were equipartitioned. Consequently, the Alfvén ratio defined as $r_A \equiv E_V/E_B$ is less than one. Kraichnan (1965) has argued that in the inertial range of fully developed three-dimensional isotropic and homogeneous turbulence, magnetic and kinetic energy should become equipartitioned. In studies of turbulence in the solar wind, Matthaeus and Goldstein (1982) and Roberts et al. (1987a) found that r_A was systematically less than unity. There have been no analogous studies made in the magnetosphere.

By T = 4.5 magnetic reconnection is well established at both the central and lower neutral sheets. In Figure 14, magnetic field lines (contours of constant a) are shown at times T = 4.5 (top panel) and T = 13.5 (bottom panel). At T = 4.5 in addition to the large O-regions (reminiscent of plasmoids) evident at the two neutral sheets, there are also smaller O-regions, or bubbles. These have been seen in previous high Reynolds number, high spatial resolution simulations and the phenomenon has been discussed in detail by Matthaeus and Lamkin (1985, 1986). By T = 13.5 a large vortical pattern has developed in the lower third of the simulation domain and the magnetic fields in the "tail lobes" have reconnected with the "magnetosheath" fields. The structure of this pattern will become more evident below.

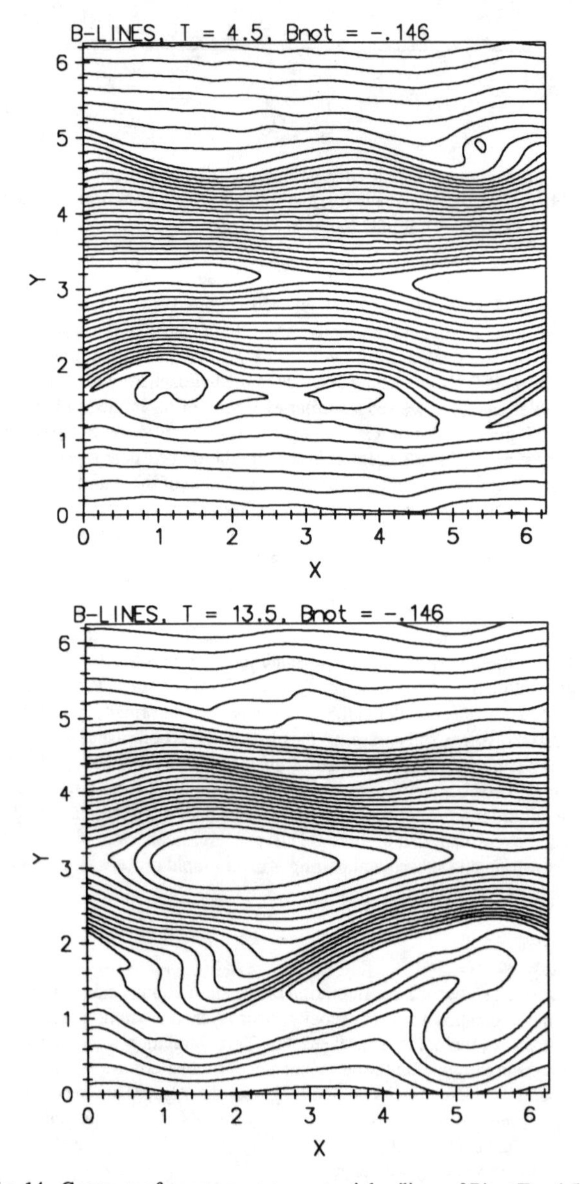

Fig. 14. Contours of constant vector potential a (lines of B) at T = 4.5 (top) and at T = 13.5 (bottom).

The rapid excitation of small-scale structures is seen clearly in the contour plots of the current j at T = 4.5 and T = 9 plotted in Figure 15. The intense currents at the central X-point is evident at T = 4.5. The small-scale concentrations of current at the lower neutral sheet are associated both with reconnection and with strong concentrations of vorticity produced by the rapidly developing Kelvin-Helmholtz instability. Note also that even in the upper current sheet at T = 4.5, where the parallel oriented magnetic fields might be expected to suppress the Kelvin-Helmholtz instability, strong concentrations of j appear as well as does an indication of vortex roll-up. By T = 9 the strong vortical flows in the lower current sheet have distorted the central X- and O-regions and have sheared the upper current sheet.

The vorticity contours follow the pattern described for the current. Contours of vorticity are plotted in Figure 16 at T = 4.5 and 10. There is

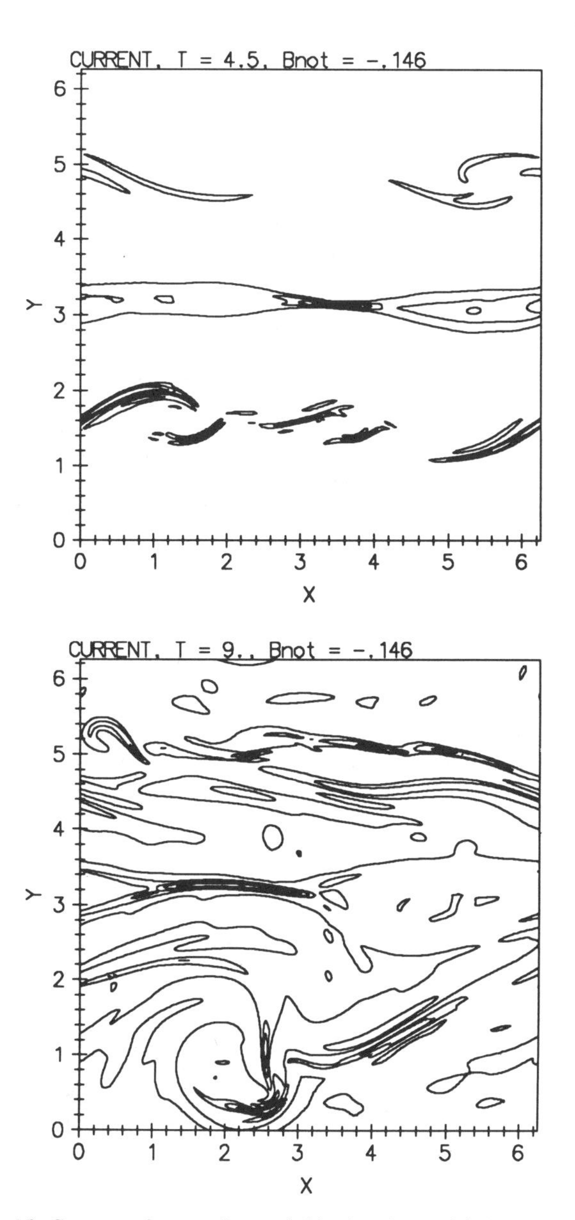

Fig. 15. Contours of current j at t = 4.5 (top) and at t = 9.0 (bottom).

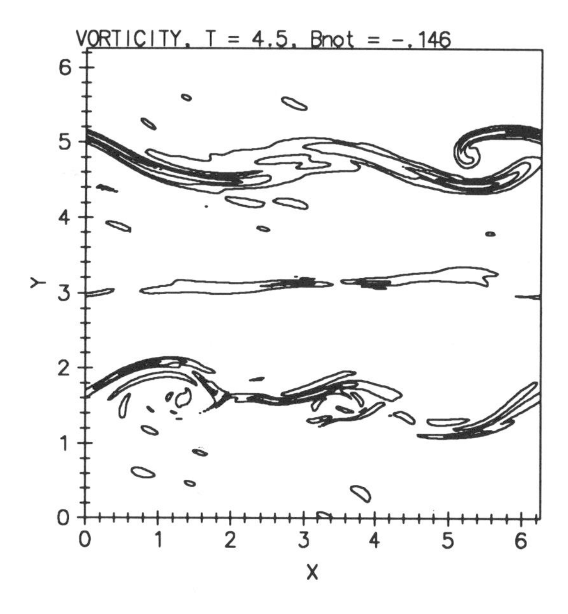

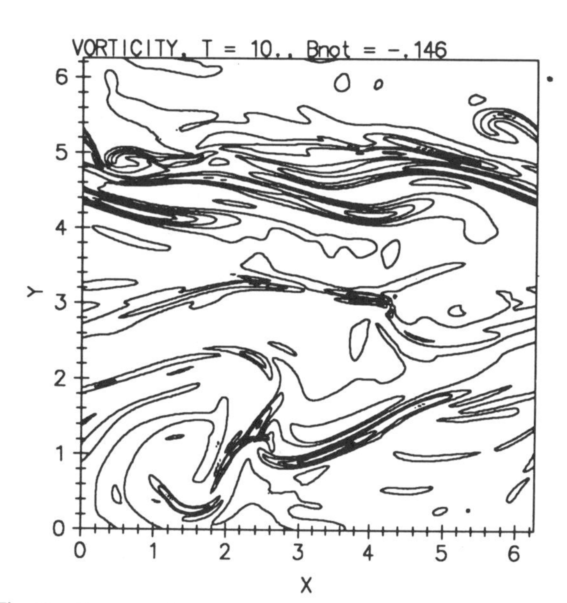

Fig. 16. Contours of vorticity ω at t = 4.5 (top) and at t = 10.0 (bottom).

a well defined quadrupole associated with the X-point in the central neutral sheet at T = 4.5. This pattern has been described previously by Matthaeus (1982) and Matthaeus and Lamkin (1986). The large vortices tend to lie in regions of the simulation square which initially contained weaker magnetic fields, suggesting that in the magnetotail, so long as one is far from the plane of the plasma sheet, it is the magnetosheath rather than the tail lobes that will be more turbulent.

The second case studied was modeled after the observations reported by Hones et al. (1978, 1981). We set the magnetic field very close to zero in the central portion of the simulation square, in analogy to the plane of the tail near the neutral sheet. (The addition of a small component of **B** in the $\hat{e}_z$ direction, perpendicular to the plane of these simulations would have no effect on the physics of these simulations.) The upper and lower quarters of the square contain stronger magnetic fields to represent the magnetosheath. The velocity fields had the same orientation as in the first case (cf. Figure 12) but the ratio of fluctuating kinetic energy to fluctuating magnetic energy was 46. The large value arises because **B** is nearly zero in half the simulation domain. The field is also weak in the remainder of the domain when compared to the intensity used in the central region (the "tail lobes") in the first run described above. The cross helicity $H_c = 0.12$ and B_o was -0.178. The x-component of **v**, initially a function only of y apart from 1%added noise, was defined with 16 Fourier modes.

Two stages in the dynamical development of the fluid are shown in Figure 17. The upper plot shows magnetic field lines at T = 2. The meandering contours in the center reflect the small noise level that was superimposed on the very weak magnetic field there. Already the boundary layers between the weak and strong field regions have become unstable. The upper panel of Figure 18 shows the behavior of the vorticity at the same time. The surface of the boundary layer is now seen to consist of fairly well developed vortex rolls with small-scale concentrations of vorticity very similar to those in the simulation with no magnetic field shown in Figure 11. The energy spectra of the fields show a rapid cascade of magnetic and kinetic energy to high wave numbers.

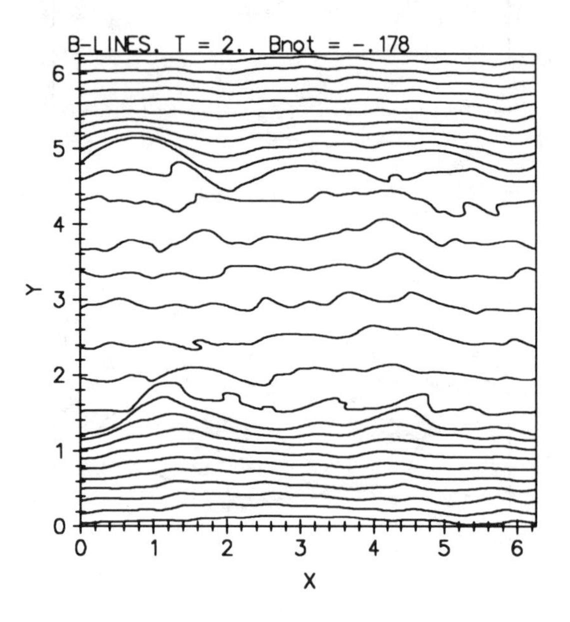

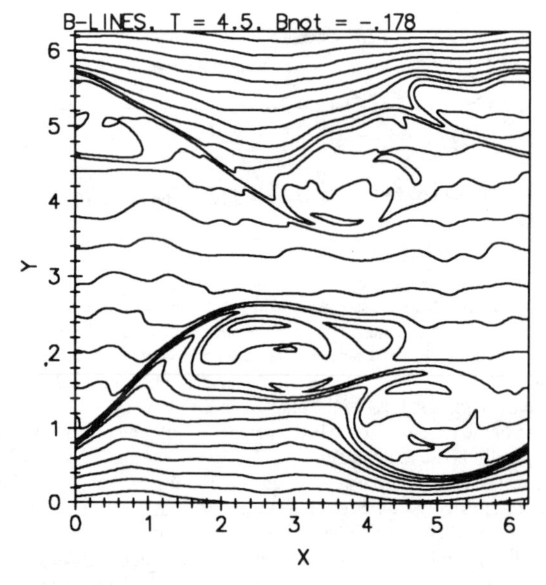

Fig. 17. Contours of a from the second simulation at T = 2 (upper panel) and at T = 4.5 (bottom panel).

By T = 4.5 (the end of this run) the dynamical evolution has greatly distorted the large-scale structure initially present. The distortions of the boundary now reach far into the central region, as can been seen the contours of a plotted in the lower panel of Figure 17. The vortices in the upper and lower regions now occupy much of the central region. The structure of the large-scale vorticity is shown in the contours of ω plotted in the lower panel of Figure 18. Strong enhancements in the vorticity at small scales are also apparent in the upper boundary layer and there is some evidence that magnetic reconnection is occurring there and elsewhere in the simulation domain. Perhaps the most visible feature in these figures is the fact that both the upper and lower boundary layers show a evidence of vortex pairing.

Another view of the flow system is shown in Figure 19 which is a plot of stream lines at T = 2 and 4.5. The early development of a string of vortices is visible in the upper panel while a later stage of the flow is shown in the lower panel. Note that the vortices tend to lie toward the weak field side of the boundary, reminiscent of Hones et al. (1978, 1981) observations. The vortex pairing is especially evident at the lower boundary.

Discussion and Conclusions

In this preliminary report, we have tried to emphasize those aspects of these numerical experiments that potentially have implications for magnetospheric and interplanetary research. A common feature of these simulations is the rapid cascade of energy from the long wavelength, low wave number, energy containing scales, toward short scales and out to the dissipation regime. The inertial range of this energy spectrum takes the approximate form of a power law, although, with Reynolds numbers of 1000 and a dissipation number K_d of approximately 50, the power law portion of the spectrum is not as well defined as it is in solar wind spectra. In those situations in which the initial configuration contains substantially more kinetic than magnetic energy, it appears that the index of the power law in the inertial range, to the extent that it can be determined, is close to that expected for a two-dimensional Navier-Stokes fluid in the cascade controlled by enstrophy (see, Kraichnan and Montgomery, 1979), viz. k^{-3}. This is in contrast to the $k^{-3/2}$ or $k^{-5/3}$ cascade controlled by energy expected for two-dimensional MHD in other circumstances (see, Kraichnan, 1965) This may be related to the fact that careful determination of the power spectral index of three-dimensional interplanetary turbulence (Matthaeus and Goldstein, 1982) shows that the spectral shape is generally closer to $k^{-5/3}$, the predicted value for a Navier-Stokes fluid, than to the prediction made by Kraichnan (1965) for an MHD fluid, which is $k^{-3/2}$.

A second general feature of all of these simulations is the appearance of highly anisotropic spectra. This may be due to either an intrinsic anisotropy in the spectral transfer produced by the Kelvin-Helmholtz configuration, or, perhaps, by the presence of a dc magnetic field which is known to produce highly anisotropic spectra (Shebalin, 1983). Evidence of a similar anisotropy has been found in a two-dimensional correlation function constructed from ISEE 3 data by Matthaeus et al. (1988). Such an anisotropy, in which magnetic energy preferentially couples to wave vectors orthogonal to the direction of the dc magnetic field, has important implications for charged particle propagation and for the evolution of the direction of minimum variance of interplanetary fluctuations in the heliosphere (see the discussion in Matthaeus et al., 1988). With the forthcoming Cluster mission of four spacecraft, it should be possible to

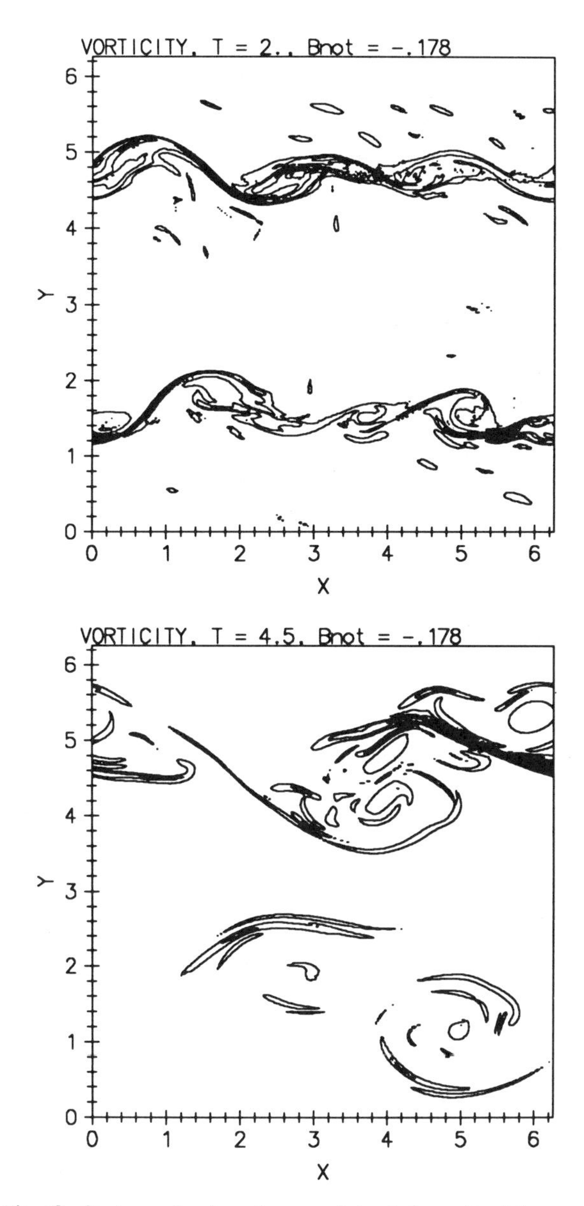

Fig. 18. Contours of ω from the second simulation at T = 2 (upper panel) and at T = 4.5 (bottom panel).

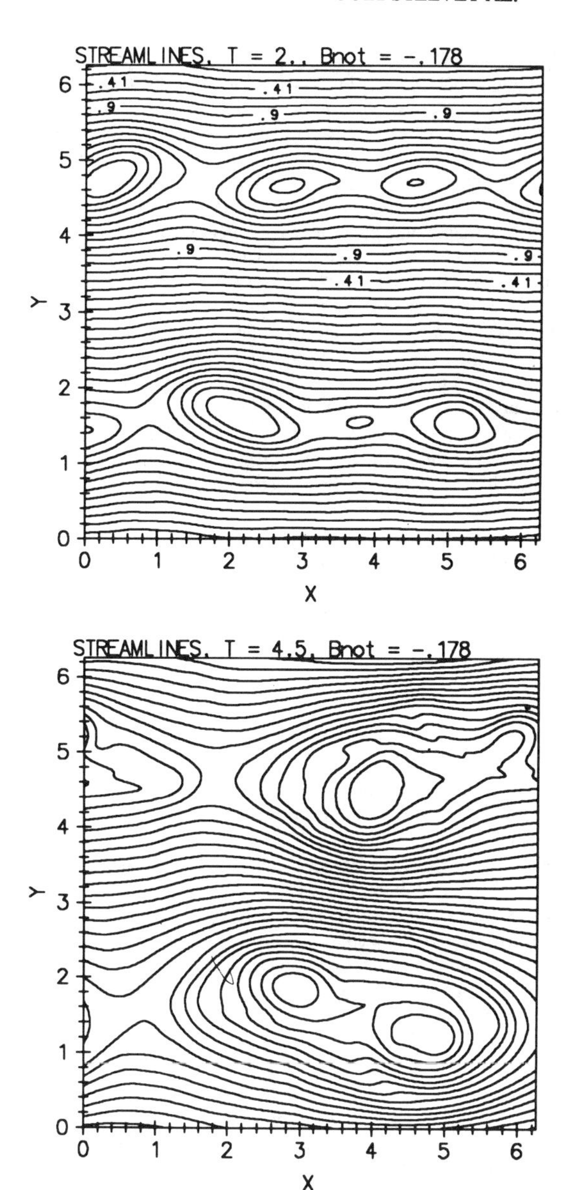

Fig. 19. Plot of ψ at t = 2 (top panel) and t = 4.5 (bottom panel) for the second "magnetosphere" simulation described in the text.

determine the three-dimensional structure of velocity and magnetic field correlation functions in the vicinity of velocity shear layers.

Nonlinear interactions play an essential role in accounting for the observed evolution of these simulations over periods of several to several tens of characteristic times. Instability descriptions such as the linear Kelvin-Helmholtz or tearing instabilities have as common features (Chandrasekhar, 1961; Matthaeus and Lamkin, 1986) that the wave number measured along the initial vortex or current sheet labels eigenmodes. Consequently, no single eigenmode of the instability can describe phenomena such as current filamentation or vortex pairing, features we have described as appearing in one or two characteristic times and that involve the excitation of transverse modes. It will be interesting, upon the launch of the Geotail and Cluster satellites, to see if evidence can be found for either current filamentation or vortex pairing.

The smallest wavelengths excited are significantly smaller than the scale size of the velocity gradients. This is an intrinsically nonlinear phenomenon that appears in these simulations because the algorithm is designed to compute accurately solutions of the MHD equations at high wave numbers. This is probably the reason that the vortex rolls seen in these simulations do not have the smooth, nearly perfectly round shapes seen in simulations that employ second-order finite difference algorithms (e.g., Miura, 1984, 1987; La Belle-Hamer et al., 1988). The nonlinear coupling to small scales also enhances the development of turbulent spectra in situations where the development of the Kelvin-Helmholtz instability should be inhibited, as is the case when a strong magnetic field is parallel to the flow velocity.

The issue of the relevance of linear theory relative to nonlinear (turbulence) theory is significant in drawing conclusions as to whether sheared velocity and magnetic configurations such as those described here might be important sources of dynamical activity in the magnetosphere and in the solar wind. In spite of the obvious limitations of the numerical models and the simulation geometry we have used, perhaps the most significant conclusion that can be drawn from these preliminary studies is that nonlinear effects are indeed expected to be rapid and powerful dynamical drivers in such geometries. Our restrictions to incompressibility and two-dimensional periodic geometry are compromises in the physics that are accepted for obvious reasons of computational economy. As new algorithms (e.g., Ghosh and Matthaeus, 1988) and more capable computational resources becomes available, these restrictions will need to be lifted to simulate more realistically turbulent processes of relevance to the solar wind and the magnetosphere. Nevertheless our intuition from the perspective of turbulence theory is that three-dimensional and compressible magnetofluids will be even more prone to turbulence effects than two-dimensional and incompressible ones, since each extension of the model incorporates more interacting degrees of freedom. Consequently, we believe that the present results give strong indications that shear-driven dynamical processes can rapidly and significantly affect the state of the solar wind and magnetosphere. A more complete account of these simulations and related ones is planned.

Acknowledgments: This work was supported, in part, by the Solar Terrestrial Theory Program at the Goddard Space Flight Center and by National Science Foundation grant ATM-8609740 to the Bartol Research Institute. D. A. Roberts is an NAS/NRC Research Associate.

References

Axford, W. I., and C. O. Hines, A unifying theory of high-latitude geophysical phenomena and geomagnetic storms, Can. J. Phys., 39, 1433, 1961.

Batchelor, G. K., Theory of Homogeneous Turbulence, Cambridge University Press, New York, 1970.

Bavassano, B., M. Dobrowolny, and G. Moreno, Local instabilities of Alfvén waves in high speed streams, Solar Phys., 57, 445, 1978.

Belcher, J. W., and L. Davis, Large-amplitude Alfvén waves in the interplanetary medium, 2, J. Geophys. Res., 76, 3534, 1971.

Burlaga, L. F., Large velocity discontinuities in the solar wind, Solar Phys. 7, 72, 1969.

Burlaga, L. F., Magnetic fields, plasmas, and coronal holes:The inner solar system, Space Sci. Rev., 23, 201, 1979.

Butler, D. M., and K. Papadopoulos, Solar Terrestrial Physics: Present and Future, NASA Reference Publication 1120, 1984.

Chandrasekhar, S., Hydrodynamic and Hydromagnetic Stability, Clarendon, Oxford, 1961.

Coleman, P. J., Turbulence, viscosity, and dissipation in the solar wind plasma, Ap. J., 153, 371, 1968.

Cowley, S. W. H., Comments on the merging of non-antiparallel fields, J. Geophys. Res., 81, 3455, 1976.

Dungey, J. W., Interplanetary magnetic field and the auroral zones, Phys. Rev. Lett., 6, 47, 1961.

Frahm, R. H. Cusp particle detection and ion injection source oscillations, M. S. Thesis, Rice University, Houston, TX,1984.

Fyfe, D., and D. Montgomery, High beta turbulence in two dimensional magnetohydrodynamics, J. Plasma Phys., 16, 181, 1976.

Goldstein, M. L., D. A. Roberts, and W. H. Matthaeus, Numerical simulation of the generation of turbulence from cometary ion pickup, Geophys. Res. Lett., 14, 860, 1987a.

Goldstein, M. L., D. A. Roberts, S. Ghosh, and W. H. Matthaeus, Numerical simulations of solar wind and magnetospheric phenomena, in Proc. 21st ESLAB Symposium, Bokesjø, Norway, ESA SP-275, edited by B. Battrick and E. J. Rolfe, p. 115, , 1987b.

Ghosh, S., and W. H. Matthaeus, Algorithms for simulations of compressible magnetohydrodynamic turbulence, Proc. Third International Conference on Supercomputing, Boston, Mass., May 15-20, p. 343, 1988.

Hones, E. W., Jr. (editor), Magnetic Reconnection in Space and Laboratory Plasmas, Geophys. Monogr. Ser., vol. 30, AGU, Washington, D.C., 1984.

Hones, E. W., Jr., G. Paschmann, S. J. Bame, J. R. Asbridge, N. Sckopke, and K. Schindler, Vortices in magnetospheric plasma flow, Geophys. Res. Lett., 5, 1059, 1978.

Hones, E. W., Jr., J. Birn, S. J. Bame, J. R. Asbridge, G. Paschmann, N. Sckopke, and G. Haerendel, Further determination of the characteristics of magnetospheric plasma vortices with ISEE 1 and 2, J. Geophys. Res., 86, 814, 1981.

Kolmogoroff, A. N., The local structure of turbulence in incompressible viscous fluid for very large Reynolds numbers, C. R. Acad. Sci. URSS, 30, 301, 1941.

Korzhov, N. P., V. V. Mishin, V. M. Tomozov, On the role of plasma parameters and the Kelvin-Helmholtz instability in a viscous interaction of solar wind streams, Planet. Space Sci. 32, 1169, 1984.

Kraichnan, R., The inertial range spectrum of hydromagnetic turbulence, Phys. Fluids, 8, 1385 1965.

Kraichnan, R. H., and D. Montgomery, Two-dimensional turbulence, Rep. Prog. Phys., 43, 547, 1979.

La Belle-Hamer, A. L., Z. F. Fu, and L. C. Lee, A mechanism for patchy reconnection at the dayside magnetopause, Geophys. Res. Lett., 15, 152, 1988.

Lee, L. C., K. Albano, and J. R. Kan, Kelvin-Helmholtz instability in the magnetopause-boundary layer region, J. Geophys. Res., 86, 814, 1981.

Matthaeus, W. H., Magnetic reconnection in two dimensions, Geophys. Res. Lett., 9, 660, 1982.

Matthaeus, W. H., and M. L. Goldstein, Measurement of the rugged invariants of magnetohydrodynamic turbulence in the solar wind, J. Geophys. Res., 87, 6011, 1982.

Matthaeus, W. H., and S. L. Lamkin, Rapid reconnection caused by finite amplitude fluctuations, Phys. Fluids, 28, 303, 1985.

Matthaeus, W. H., and S. L. Lamkin, Turbulent magnetic reconnection, Phys. Fluids, 29, 2513, 1986.

Matthaeus, W. H., and D. C. Montgomery, Nonlinear evolution of the sheet pinch, J. Plasma Phys., 25, 11, 1981.

Matthaeus, W. H., M. L. Goldstein, and D. C. Montgomery, Turbulent generation of outward-traveling interplanetary Alfvén fluctuations, Phys. Rev. Lett., 51, 1484, 1983.

Matthaeus, W. H., M. L. Goldstein, and D. A. Roberts, Evidence for the presence of quasi-two-dimensional nearly-incompressible fluctuations in the solar wind, J. Geophys. Res., submitted, 1988.

Miura, A., Anomalous transport by magnetohydrodynamic Kelvin-Helmholtz instabilities in the solar wind-magnetosphere interaction, J. Geophys. Res., 89, 801, 1984.

Miura, A., Simulation of Kelvin-Helmholtz instability at the magnetospheric boundary, J. Geophys. Res., 92, 3195, 1987.

Montgomery, D., Major disruptions, inverse cascades, and the Strauss equations, Physica Scripta, T2/1, 83, 1982.

Neugebauer, M., C. J. Alexander, R. Schwenn, A. K. Richter, Tangential discontinuities in the solar wind: Correlated field and velocity changes and the Kelvin-Helmholtz instability, J. Geophys. Res., 91, 13694, 1986.

Nishida, A., Reconnection in Earth's magnetotail: an overview, in Magnetic Reconnection in Space and Laboratory Plasmas, Geophys. Mongr. Ser., vol. 30, edited by E. W. Hones, Jr., p. 159, AGU , Washington, D. C., 1984.

Ogilvie, K. W., and R. J. Fitzenreiter, The Kelvin-Helmholtz instability at the magnetopause, in preparation, 1988.

Owen, C. J., and S. W. H. Cowley, Simple models of time-dependent reconnection in a collision-free plasma with an application to substorms in the geomagnetic tail, Planet. Space. Sci., 35, 451, 1987.

Parker, E. N., Dynamical properties of solar and stellar winds. III. The dynamics of coronal streamers, Ap. J., 139, 690, 1964.

Patterson, G. S., and S. A. Orszag, Spectral calculations of isotropic turbulence: Efficient removal of aliasing interactions, Phys. Fluids, 14, 2358, 1971.

Roberts, D. A., L. W. Klein, M. L. Goldstein, and W. H. Matthaeus, The nature and evolution of magnetohydrodynamic fluctuations in the solar wind: Voyager observations, J. Geophys. Res., 92, 11,021, 1987a.

Roberts, D. A., M. L. Goldstein, L. W. Klein, and W. H. Matthaeus, Origin and evolution of fluctuations in the solar wind: Helios observations and Helios-Voyager comparisons, J. Geophys. Res., 92, 12,023, 1987b.

Rosenbauer, H., R. Schwenn, E. Marsch, B. Meyer, H. Miggenrieder, M. D. Montgomery, K. H. Mulhausen, W. Pilipp, W. Voges, and S. M. Zink, A survey of initial results of the Helios plasma experiment, J. Geophys. Res., 42, 561, 1977.

Rostoker, G., The Kelvin-Helmholtz instability and its role in the generation of the electric currents associated with Ps 6 and westward traveling surges, University of Alberta preprint, 1987.

Russell, C. T., Reconnection at the Earth's magnetopause: Magnetic field observations and flux transfer events, in Magnetic Reconnection in Space and Laboratory Plasmas, Geophys. Mongr. Ser., vol. 30, edited by E. W. Hones, Jr., p. 124, AGU, Washington, D. C., 1984.

Russell, C. T., and R. C. Elphic, Initial ISEE magnetometer results: Magnetopause observations, Space Sci. Rev., 22, 681, 1978.

Sanderson, T. R., P. W. Daly, K. -P. Wenzel, E. W. Hones, Jr., and S. J. Bame, Energetic ion observations of a large scale vortex in the distant geotail, Geophys. Res. Lett., 11, 1094, 1984.

Sanderson, T. R., P. W. Daly, K. -P. Wenzel, E. W. Hones, Jr., and E. J. Smith, Observations of a large-scale vortex-like structure in the deep-tail plasma sheet boundary layer, in Solar Wind-Magnetosphere Coupling, edited by, Y. Kamide and J. A. Slavin, p. 739, Terra Scientific Publ. Co., Tokyo, 1986.

Sckopke, N., G. Paschmann, G. Haerendel, B. U. Ö. Sonnerup, S. J. Bame, T. G. Forbes, E. W. Hones, Jr., and C. T. Russell, Structure of the low-latitude boundary layer, J. Geophys. Res., 86, 2099, 1981.

Sen, A. K., Stability of hydromagnetic Kelvin-Helmholtz discontinuity, Phys. Fluids, 5, 1154, 1963.

Shebalin, J. V., W. H. Matthaeus, and D. Montgomery, Anisotropy in MHD turbulence due to a mean magnetic field, J. Plasma Phys., 29, 525, 1983.

Sonnerup, B. U. Ö., Theory of the low latitude boundary layer, J. Geophys. Res., 85, 2017, 1980.

Sonnerup, B. U. Ö., Magnetic field reconnection at the magnetopause: an overview, in Magnetic Reconnection in Space and Laboratory Plasmas, Geophys. Mongr. Ser., vol. 30, edited by E. W. Hones, Jr., p. 92, AGU, Washington, D. C., 1984.

Southwood, D. J., Magnetopause Kelvin-Helmholtz instability, in Proc. Magnetospheric Boundary Layers Conf., Alpach, 11-15 June 1979, ESA SP-48, 1979.

Strauss, H. R., Nonlinear, three-dimensional magnetohydrodynamics of noncircular tokamaks, Phys. Fluids, 19, 134, 1976.

Ting, A. C., W. H. Matthaeus, and D. Montgomery, Turbulent relaxation processes in magnetohydrodynamics, Phys. Fluids, 29, 3261 1986.

Wu, C. C., Kelvin-Helmholtz instability at the magnetopause boundary, J. Geophys. Res., 91, 3042, 1986.

MHD INTERMEDIATE SHOCKS AND THE MAGNETOPAUSE

C. C. Wu

Department of Physics, University of California at Los Angeles, Los Angeles, CA 90024

Abstract. Contrary to the usual belief that MHD intermediate shocks are extraneous, we have recently shown by numerical solutions of dissipative MHD equations that intermediate shocks are admissible and can be formed through nonlinear steepening from continuous waves. Some aspects of intermediate shocks are presented. Since a transverse magnetic field can only rotate across "intermediat shocks" in dissipative MHD, one expects an intermediate shock, instead of a rotational discontinuity, to exist at the magnetopause.

Introduction

Magnetohydrodynamics (MHD) intermediate shocks are shocks with shock frame fluid velocities greater than the Alfvén speed ahead and less than the Alfvén speed behind. Or equivalently, across intermediate shocks the sign of the tangential component of the magnetic field changes. These shocks had been considered extraneous, or nonevolutionary, or unstable, and they had been thought not to correspond to physical reality (Kantrowitz and Petschek, 1966; Jeffrey and Taniuti, 1964).

However, we have recently shown by numerical solutions of dissipative MHD equations that MHD intermediate shocks can be formed from continuous waves (Wu, 1987). Thus according to the formation argument which requires that physical shocks be formed from continuous waves through a more realistic description, the MHD intermediate shocks should be considered physical.

In this paper, we present some results from our study of the intermediate shocks. First, we show that intermediate shocks can be formed by wave steepening (Wu, 1987, 1988c). Second, we show that a rotational discontinuity is not stable in dissipative MHD, and it evolves to an intermediate shock of the kind whose shock frame fluid velocities are subfast, super-Alfvénic, superslow ahead, and subfast, sub-Alfvénic, superslow behind the shock (Wu, 1988b). Thus in regard to the the magnetospheric physics, our study suggests that instead of a rotational discontinuity, an intermediate shock exists at the magnetopause.

Basic Equations and the Numerical Method

The dissipative MHD equations with resistivity and viscosity, but ignoring thermal conductivity, are (Landau and Lifshitz, 1960)

$$\frac{\partial \rho}{\partial t} = -\nabla \cdot (\rho \mathbf{v}) \tag{1}$$

$$\frac{\partial(\rho \mathbf{v})}{\partial t} = -\nabla \cdot (\rho \mathbf{v}\mathbf{v} + \mathbf{I}(p + \frac{B^2}{2}) - \mathbf{B}\mathbf{B}) + \nu \nabla^2 \mathbf{v} + (\mu + \frac{\nu}{3})\nabla(\nabla \cdot \mathbf{v}) \tag{2}$$

$$\frac{\partial \mathbf{B}}{\partial t} = -\nabla \times \mathbf{E} \tag{3}$$

$$\frac{\partial \epsilon}{\partial t} = -\nabla \cdot ((\frac{\rho v^2}{2} + \frac{p}{\gamma - 1} + p)\mathbf{v} + \mathbf{E} \times \mathbf{B}) + \nabla \cdot \sigma \cdot \mathbf{v} \tag{4}$$

Here, σ is the viscous tensor, $\sigma_{ik} = \nu(\partial v_i/\partial x_k + \partial v_k/\partial x_i - 2/3\delta_{ik} \partial v_l/\partial x_l) + \mu\delta_{ik}\partial v_l/\partial x_l$. The mass density, pressure, velocity, and magnetic field are denoted by $\rho, p, \mathbf{v}$, and $\mathbf{B}$, respectively; γ is the ratio of the specific heats, and the energy density (ϵ) is given by $\epsilon = \rho v^2/2 + B^2/2 + p/(\gamma - 1)$. The electric field $\mathbf{E}$ is given by Ohm's law, $\eta\mathbf{J} = \mathbf{E} + \mathbf{v} \times \mathbf{B}$, with the current density $\mathbf{J} = \nabla \times \mathbf{B}$ and η constant resistivity. ν and μ are the two coefficients of viscosity. In addition to these equations, we require $\nabla \cdot \mathbf{B} = 0$. In these equations, we have used units so that factors such as 4π and c do not appear. In this paper, we only consider one-dimensional flow along the x direction, then $\partial/\partial y = \partial/\partial z = 0$, and B_x is a constant. When $\eta = \nu = \mu = 0$, these equations are ideal MHD equations.

To solve these MHD equations, we used an Eulerian uniform grid and an explicit conservative two-step Lax-Wendroff difference scheme (Richtmyer and Morton, 1967). The treatment of the dissipative terms is also explicit. Therefore, the time step size is limited by the Courant-Friedrichs-Lewy (CFL) condition for the flow and by the diffusion time scale. In the first case, periodic boundary conditions were employed; while in the second case, the numerical boundaries were placed far away from the origin that no waves would reach the boundaries within the calculation time and that the boundary values remained constant in time.

Note that the ideal set of MHD equations is not well defined when discontinuities are formed. On the other hand, the initial value problems for the dissipative MHD equations may be well-posed. The solutions of the dissipative MHD equations are smooth all the time. In particular, a shock is not a discontinuity, but a thin region in our level of description. In other words, the shock structure is resolved in our calculations. This is a contrast to a shock capturing scheme which solves the ideal set of equations to obtain physical jump conditions across shocks but not the details of the shock transition layer. However, nonphysical shock could be captured in some schemes. To use a shock-capturing scheme, we already assume what the physical jump relations are. Furthermore, as was shown in Wu (1988a,b), the discontinuity theory is not a good approximation to dissipative MHD.

Results

In the first example, we present the evolution of a slow-mode simple wave solution in Figures 1 and 2. The initial condition is given by $B_y = 0.5\sin(2\pi x)$, $0 \leq x \leq 1$, with periodic boundary conditions. The other quantities were then numerically obtained from simple wave relations (Wu, 1987). We have chosen such a simple initial distribution in which only one wave is involved initially so that one can easily identify the characteristics of the wave and the properties of the shocks formed. However, the result which shows that intermediate shocks can be formed through nonlinear

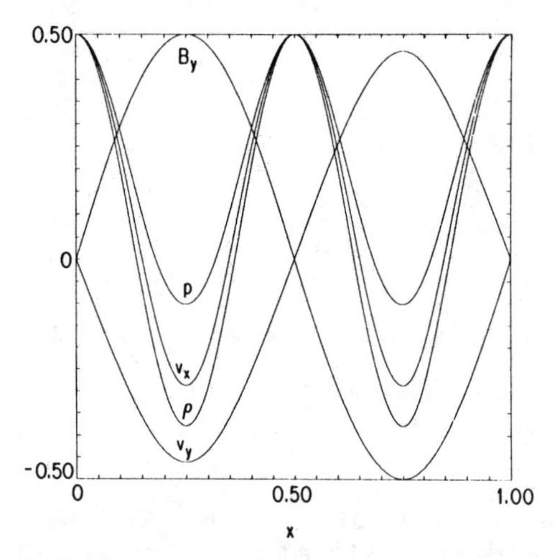

Fig. 1. Initial distributions. The scale along the vertical axis is for B_y. For ρ, the scale is from 0.78 to 1; for v_x, -0.26 to 0; for v_y, -0.6 to 0.6; for p, 0.5 to 1.

steepening processes is quite general (Wu, 1988c). All quantities were normalized with respect to the following prescribed values at $B_y = 0$: $B_x = 1, \rho = 1, p = 1, \mathbf{v} = 0, B_z = 0$. Note that $v_z = 0$, $B_z = 0$; the problem is coplanar. We used $\gamma = 5/3$, thus the sound speed at $B_y = 0$ is $\sqrt{5/3}$. Since the sound speed is greater than the Alfvén speed at $B_y = 0$, the slow simple wave can cross $B_y = 0$. The initial distributions are plotted in Figure 1. The solution at $t = 2$, shown in Figure 2, is obtained with $\eta = 0.0004$, and $\mu = \nu = 0$. The calculation was carried out with 6400 grid points ($\Delta x = 1/6400$) and $\Delta t = 1/40000$; Δx is the grid spacing and Δt is the time step size. The convergence of the solution has been checked. In Figure 2 shocks can easily be identified. The shocks are slow intermediate shocks! The first thing that one notices is that B_y changes sign across the shocks. As was done in Wu (1987), the shocks were found to fit well with the Rankine-Hugoniot jump relations, and in the shock frame the fluid velocities are superslow, super-Alfvénic, and subfast ahead, and subslow, sub-Alfvénic, and subfast behind.

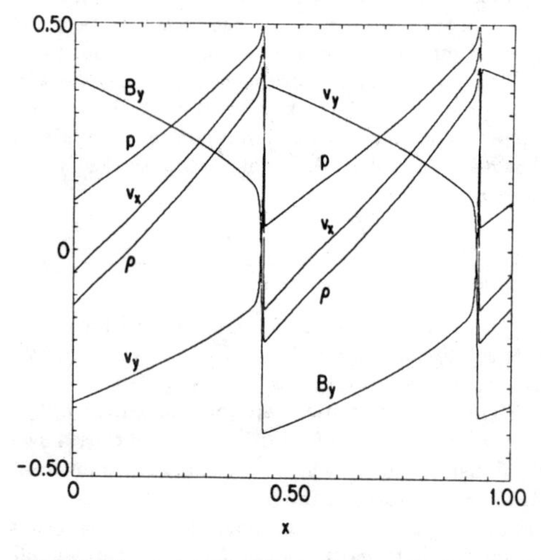

Fig. 2. Distributions at t=2. The scales are the same as in Fig 1.

The above case concerns the coplanar case in which both $v_z = 0$ and $B_z = 0$. According to Kantrowitz and Petschek (1966), the coplanar case is a singular one. They argued that any small value of v_z or B_z would split the intermediate shocks. However, by introducing B_z (v_z) perturbations in the calculation, we find that the solution is a smooth function of B_z (v_z) (Wu, 1987). This contradicts their arguments. The shock-like solution obtained in the full MHD case is not coplanar and thus does not satisfy the Rankine-Hugoniot relations. In other words, the result shows that there is a larger class of shock structures in the time-dependent dissipative MHD equations than are given by the Rankine-Hugoniot relations. Due to these solutions, intermediate shock solutions have a neighboring solution when the system is not coplanar. In contrast, the conventional view only allows the shock solutions that satisfy the Rankine-Hugoniot relations, and therefore considers the intermediate shock solution singular.

In the conventional view, when the normal component of magnetic field is nonzero, the transverse component of magnetic field can only rotate through rotational discontinuities (intermediate waves). Therefore the classical reconnection model of Levy et al. (1964) requires rotational discontinuity surfaces to stand at the magnetopause. However, the next example will show that a rotational discontinuity is not stable in dissipative MHD, and the discontinuity will evolve to a time-dependent "intermediate shock."

In the second example, we start with a rotational discontinuity, or, in fact, a finite-width intermediate wave. For the rotational discontinuity with characteristic speeds $v_x \pm B_x/\sqrt{\rho}$, one has the following jump relations (Jeffrey and Taniuti, 1964):

$$[\mathbf{v}_t] = \mp sgn(B_x)[\mathbf{B}_t/\sqrt{\rho}] \tag{5}$$

$$[p] = [\rho] = [entropy] = [B^2] = [v_x] = 0 \tag{6}$$

Here, angle brackets denote a jump. Specifically, we have used the following representative distributions for **B** and **v**

$$B_y(x) = B_t \cos(\theta(x)) \tag{7}$$

$$B_z(x) = B_t \sin(\theta(x)) \tag{8}$$

with $\theta(x) = (\theta_l + \theta_r)/2 - (\theta_l - \theta_r)\tanh(x)/2$ and B_t a constant. According to the wave relations, equation (5), and $v_y(x) = B_y(x)/\sqrt{\rho_0}$, $v_z(x) = B_z(x)/\sqrt{\rho_0}$. As was given in (6), ρ, p, v_x, and B_x have constant initial distributions ρ_0, p_0, v_{x0}, and B_{x0}, respectively. B_{x0} is assumed positive. These distributions represent an intermediate wave moving with velocity $v_{x0} - B_x/\sqrt{\rho_0}$, which rotates the transverse magnetic field from θ_l to θ_r within a transition region of width 2. The smooth transition is required to assure the validity of the dissipation terms, but the solution depends on the structure of the region (Wu, 1988b). Figure 3 shows the result of such a calculation with the initial conditions given by (7) and (8) with $\theta_l = 0$, $\theta_r = -120°$, $p_0 = 1$, $\rho_0 = 1$, $B_t = 2$, $B_{x0} = 1$, $v_{x0} = 0$, and $\gamma = 5/3$. The conditions represent an intermediate wave moving to the left with speed $v_{x0} - B_{x0}/\sqrt{\rho_0}$, which is -1. We have used constant dissipation coefficients $\eta = \nu = 0.05$ and $\mu = 0$. In terms of the constant state Alfvén speed and the half width of the discontinuity, the magnetic Reynolds' number and ordinary Reynolds' number are 20. The calculations were carried out with 14400 grid points covering the region $-360 \le x \le 360$ (thus the spatial spacing $\Delta x = 1/20$) and with the time step size $\Delta t = 60/4000$.

If there is no dissipation, the intermediate wave will just move to the left with constant velocity -1. By including resistivity, for example, B_y and B_z in general do not dissipate at the same rate, for the dissipation process depends on the gradients. Therefore $|\mathbf{B}|$ is no longer constant. This then affects other wave modes through

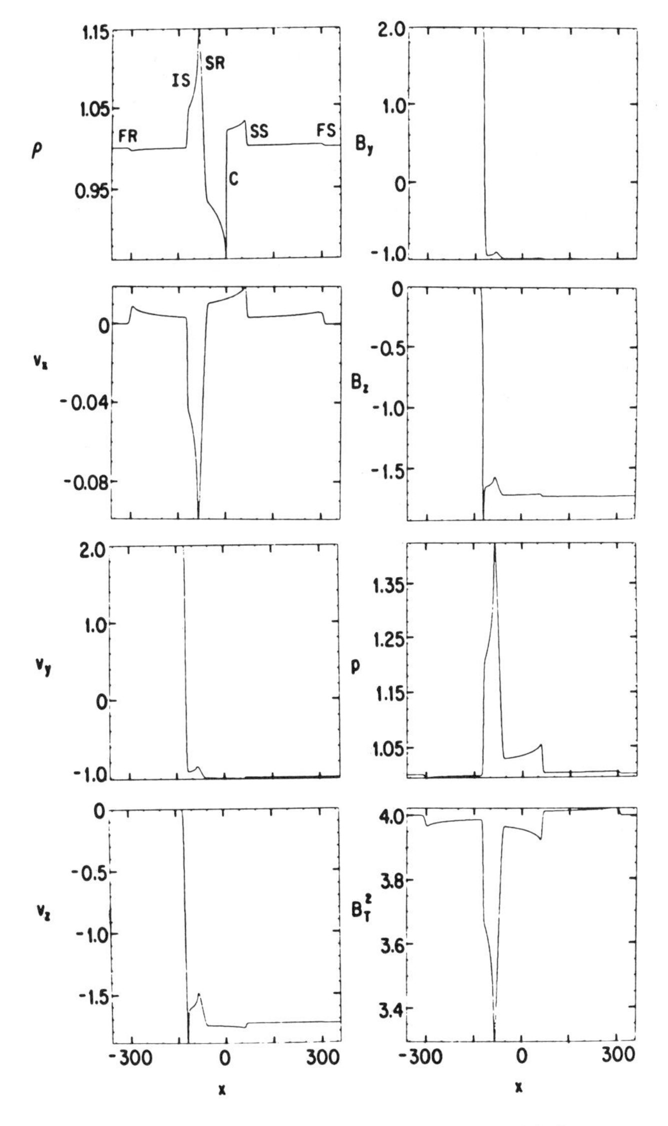

Fig. 3. Solution at t=120 for the noncoplanar case with θ_r = -120° and with $\eta(\nu)$ = 0.05. The symbols FR, IS, SR, C, SS, and FS denote a fast rarefaction, an intermediate shock, a slow rarefaction, a contact discontinuity, a slow shock, and a fast shock, respectively. B_T^2 is defined as $B_y^2 + B_z^2$.

the magnetic pressure term in the momentum equations and the magnetic energy term in the energy equation. As is shown in Figure 3, this initial perturbation does not relax and the solution does not go back to the original discontinuity as would be the case for stable shocks. This represents one essential difference between rotational discontinuities and stable shocks.

The solution at t = 120 is shown in Figure 3. One notices immediately that the solution is not a pure rotational discontinuity, which would just move to the left with constant velocity -1. We have computed cases with $\theta_r = -180^\circ$, $\theta_r = -160^\circ$, -140°, and -120° and found that the solutions are a continuous function of θ_r. The waves in the solution can then be identified by relating them to the waves in the limit $\theta_r = -180^\circ$, or by considering the characteristic speeds of the fast, intermediate, and slow waves in relation to the speeds of the discontinuities. As in the case $\theta_r = -180^\circ$, the waves in Figure 3 are identified as a "fast rarefaction wave," an "intermediate shock," etc. However, unlike the $\theta_r = -180^\circ$ case, the solution does not exactly include waves separated by constant states. The states between the waves are not constant in this case. This indicates that the dimensional parameters introduced by the dissipation terms play a role. Since the intermediate shock involves a rotation of the magnetic field, it does not satisfy the Rankine-Hugoniot jump relations, and represents a time-dependent shock-like solution of the dissipative MHD equations. The intermediate shock obtained is of the kind whose shock frame fluid velocities are subfast, super-Alfvénic, superslow ahead, and subfast, sub-Alfvénic, superslow behind the shock. For this case where the sound speed is greater than the Alfv́en speed, the magnitude of the transverse magnetic field decreases across the shock. (In the case where the sound speed is less than the Alfvén speed, the magnitude of the transverse magnetic field can increase across the intermediate shock of the same kind.)

We have carried out calculations with different initial rotation angles. When the rotation angle is greater than 180°, the time development becomes much more complicated than the cases above. Initially, an intermediate shock with the initial magnetic field rotation angle inside the shock layer is formed, but this structure is not stable. Since intermediate waves converge on the intermediate shock, the intermediate wave behind the shock will bring the z component of magnetic flux to the intermediate shock. Thus eventually the z component of the magnetic field inside the intermediate shock will have the same sign as the state behind the shock, and the system relaxes to a rotation of less than 180°.

Summary and Discussion

In the paper we have shown by numerical examples some aspects of the intermediate shocks. In other papers, we have considered the stability and structure of these shocks (Wu, 1988a,c). One implication of the study is that one expects an intermediate shock, instead of a rotational discontinuity, to exist at the magnetopause since a transverse magnetic field can only rotate across "intermediate shocks" in the dissipative MHD. Furthermore, since the angle of the magnetic field rotation is generally not 180°, the dayside reconnection process may take place not in a steady state fashion as proposed in the classical model of Levy et al. (1964), but in a time-dependent way involving intermediate shocks. Although more realistic modeling of the dayside reconnection process is needed, the suggestion has the following consequences. One is that one has a different signature from that of a rotational discontinuity to compare with space observations, since the nature of an intermediate shock may be quite different from a rotational discontinuity. In general, an intermediate shock is compressive and involves increases in pressure, density, and entropy across the shock. In addition, across the shock, the magnitude of the magnetic field decreases in high β plasmas. However, since a rotational discontinuity is a limiting case of the intermediate shocks, when an intermediate shock is close to a rotational discontinuity, their properties will be close. For checking rotational discontinuities, the tangential component test associated with the jump relation (5) has been used (Sonnerup et al., 1981). Although the relation may not be strictly valid for noncoplanar intermediate shocks, it is approximately satisfied for the intermediate shock shown in Figure 3. Another consequence is that the sense of the magnetic field rotation across the magnetopause should depend on the relative orientation of the magnetosheath and magnetospheric field and the rotation should be less than 180°. This result is consistent with the findings of Berchem and Russell (1982) based on ISEE 1 and 2 observations.

Acknowledgments. This work was supported by the National Science Foundation grant ATM 85-21125. Computing support was partly provided by the San Diego Supercomputer Center, which is sponsored by the National Science Foundation.

References

Berchem, J., and C. T. Russell, Magnetic field rotation through the magnetopause: ISEE 1 and 2 observations, J. Geophys. Res., 87, 8139, 1982.

Jeffrey, A., and T. Taniuti, Non-Linear Wave Propagation, Academic, Orlando, Florida., 1964.

Kantrowitz, A. R., and H. E. Petschek, MHD characteristics and shock waves, in Plasma Physics in Theory and Application, edited by W.B. Kunkel, pp. 148-206, MaGraw-Hill, New York, 1966.

Landau, L. D., and E. M. Lifshitz, Electrodynamics of Continuous Media, Addison-Wesley, Reading, Massachusetts, 1960.

Levy, R. H., H. E. Petschek, and G. L. Siscoe, Aerodynamic aspects of the magnetospheric flow, AIAA J., 2, 2065, 1964.

Richtmyer, R. D., and K. W. Morton, Difference Methods for Initial-Value Problems, 2nd ed., Wiley-Interscience, New York, 1967.

Sonnerup, B. U. Ö., G. Paschmann, I. Papamastorakis, N. Sckopke, G. Haerendel, S. J. Bame, J. R. Asbridge, J. T. Gosling, and C. T. Russell, Evidence for magnetic field reconnection at the Earth's magnetopause, J. Geophys. Res., 86, 10049, 1981.

Wu, C. C., On MHD intermediate shocks, Geophys. Res. Lett., 14, 668, 1987.

Wu, C. C., The MHD intermediate shock interaction with an intermediate wave: Are intermediate shocks physical?, J. Geophys. Res., 93, 987, 1988a.

Wu, C. C., Effects of dissipation on rotational discontinuities, J. Geophys. Res., 93, 3969, 1988b.

Wu, C. C., Formation, stability and structure of intermediate shocks, in preparation, 1988c.

What determines the composition and charge state of solar-wind and magnetospheric ions?

COMPOSITION OF THE SOLAR WIND

P. Bochsler and J. Geiss

Physikalisches Institut, University of Bern, CH-3012 Bern, Switzerland

Abstract. Under normal circumstances the composition of the solar wind is mainly determined by processes occurring in the chromosphere and the transition region. From the apparent systematics of solar wind abundances with respect to the first ionization potentials of the elements, it is concluded that these processes involve an efficient ion-neutral separation. Detailed models of these processes have been partially successful in reproducing solar wind abundances. However, some features, e.g., the relatively low carbon abundance of solar energetic particles (SEPs) and solar wind ions, remain difficult to explain. The isotopic composition of solar wind neon differs significantly from SEP-neon. It is not known what causes the discrepancy, and this raises the question of how closely the solar wind reflects solar surface isotopic abundances. The question can probably be solved only with precise measurements of other isotopic ratios in the solar wind by new techniques. The charge states of solar wind ions are established during the transfer of matter through the corona. The charge state of elements depends sensitively on the energy distribution of electrons in the lower corona. It is expected that future observations can yield more detailed information about charge state distributions and that charge states will therefore serve as an efficient diagnostic tool for varying conditions in the inner corona.

Introduction

Does the composition of the solar wind reflect solar surface abundances? In discussing this question one has to take into account not only the uncertainties of solar wind measurements but also the quality of solar surface abundance data.

With respect to the isotopic composition we find that because of the lack of direct and reliable measurements in the outer convective zone and the solar atmosphere, the isotopic composition of the solar wind is the best available source of information.

With respect to the elemental composition of the solar wind we have learned from refined measurement techniques during the past decade that the He/H ratio is not the only case for which the long-time average of the solar wind differs from the solar surface value by a factor of 2 or more. As will be discussed in the following sections, similar deviations are observed for the elements with low first ionization potential (FIP). On the other hand, it has been found that on the average solar energetic particles (SEPs) show abundances remarkably similar to solar wind ions.

There exists no consistent model of the processes which produce the temporal variation of solar wind and SEP abundances and which could furthermore explain the differences between the long-time average fluxes and the solar surface composition as determined by optical methods.

The observational data demonstrate that there must be several physical mechanisms involved which interact in a complicated way and affect chemical abundances in the corona and in the solar wind. These processes must be located in the upper chromosphere, the transition region, and in the low corona. The Alfvén point (at 10 to 40 solar radii where the flow speed reaches the Alfvén speed) can be considered as a point-of-no-return so that no substantial changes of composition occur further out. However, streams with different composition can interact and thereby modify the spatial variations of the composition of the interplanetary medium. The average composition of the bulk outflow is not affected until mixture with the local interstellar medium becomes important.

Presently, one distinguishes between two basic processes which can fractionate the solar wind composition with respect to solar surface material:

1. Fractionation due to atomic properties of elements, e.g., ionization properties, molecule formation, ion-atom interaction exchanging momentum and charge. These processes, if they occur, must take place in regions of low temperature and relatively high density, e.g., the transition zone, and the upper chromosphere.

2. Fractionation among ions due to different masses or charges or mass per charge ratios. Action of electric and magnetic fields (or waves) on ions or by ion-ion interaction. Such processes might take place anywhere form the solar surface

TABLE 1. Normalized Elemental Abundances in the Solar System, in the Corona (as Derived from SEPs), and in the Solar Wind

Element	First Ionization Potential [V]	Solar System	SEP-derived Corona	Solar Wind	Ref.
H	13.595	1350	-	1900 ± 400	a
He	24.481	108	72 ± 3	75 ± 20	b
C	11.256	0.60	$0.414^{+0.044}_{-0.040}$	0.43 ± .02	c
N	14.53	0.123	0.123 ± 0.009	0.15 ± .06	c
O	13.614	≡ 1.00	≡ 1.00	≡ 1.00	
F	17.418	4.2(-5)	$4.9(-5)^{+5.1(-5)}_{-4.9(-5)}$	-	
Ne	21.559	0.142	$0.138^{+0.015}_{-0.014}$	0.17 ± 0.02	b,d
Na	5.138	2.8(-3)	$0.0117^{+0.0012}_{-0.0011}$	-	
Mg	7.644	0.053	0.192 ± 0.011	-	
Al	5.984	4.2(-3)	0.0147 ± 0.0007	-	
Si	8.149	0.050	$0.176^{+0.010}_{-0.011}$	0.22 ± .07	e
P	10.484	5.2(-4)	$8.61(-4)^{+1.1(-4)}_{-1.3(-4)}$	-	
S	10.357	0.026	$0.0426^{+0.0018}_{-0.0016}$	-	
Cl	13.01	2.6(-4)	$4.19(-4)^{+1.5(-4)}_{-1.4(-4)}$	-	
Ar	15.755	4.6(-3)	$4.24(-3)^{+7.4(-4)}_{-6.3(-4)}$	4.0(-3) ± 1.0(-3)	d
K	4.339	1.88(-4)	$6.87(-4)^{+3.7(-4)}_{-2.8(-4)}$	-	
Ca	6.111	3.00(-3)	$0.0144^{+0.0025}_{-0.0021}$	-	
Sc	6.54	1.7(-6)	$5.5(-5)^{+9.7(-5)}_{-5.5(-5)}$	-	
Ti	6.82	1.2(-4)	$8.6(-4)^{+2.8(-4)}_{-2.2(-4)}$	-	
V	6.74	1.5(-5)	$8.5(-5)^{+1.2(-4)}_{-8.5(-5)}$		
Cr	6.764	6.7(-4)	$3.2(-3)^{+6.8(-4)}_{-5.8(-4)}$	-	
Mn	7.432	4.7(-4)	$1.20(-3)^{+6.9(-4)}_{-4.7(-4)}$	-	
Fe	7.87	0.045	$0.223^{+0.030}_{-0.026}$	$0.19^{+.09}_{-.06}$	f
Co	7.86	1.11(-4)	<3.2(-3)	-	
Ni	7.633	2.45(-3)	$8.2(-3)^{+1.4(-3)}_{-1.3(-3)}$	-	
Cu	7.724	2.56(-5)	$1.0(-4)^{+1.5(-4)}_{-1.0(-4)}$	-	
Zn	9.391	6.30(-5)	$2.8(-4)^{+1.5(-4)}_{-1.3(-4)}$	-	

The solar system abundances are taken from the compilation of Anders and Ebihara (1982) except for Ne and Ar which are "local galactic" values, given by Meyer (1987). The SEP-derived coronal abundances are from Breneman and Stone (1985), and the SEP-helium value is from Cook et al. (1984). Solar wind: (a) Bame et al. (1975), (b) Bochsler et al. (1986), (c) Gloeckler et al. (1986), (d) Geiss et al. (1972), (e) Bochsler (1987), (f) Schmid et al. (1988).

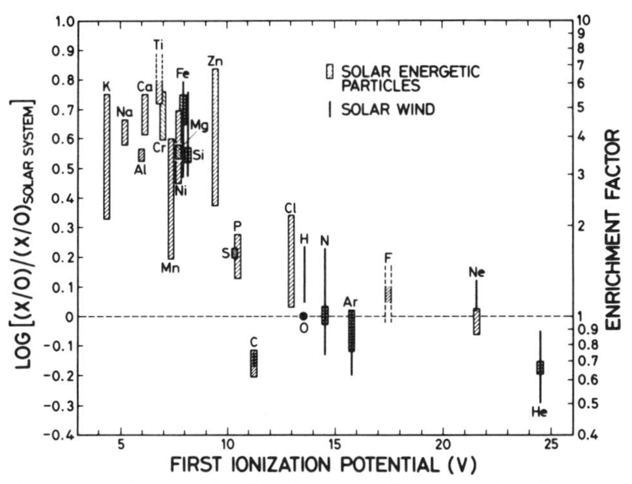

Fig. 1. Elemental abundances of SEP and solar wind relative to photosphere vs. first ionization potential (for references see Table 1).

to the interplanetary medium but they are most important in the corona.

Obviously, the charge state of an ion largely determines its fate in the inner corona. On the other hand, seen in a global context, the charge state distribution of the ions determines the flow of the solar wind in the corona. For instance, if there were no helium with its relatively low Coulomb drag factor, the bulk properties of the solar wind would under most conditions be quite different from what is actually observed. Current models involving the solution of the continuity and momentum equation individually for each species and for different flow geometries can account for many details of the observed charge state distribution (Bürgi and Geiss, 1986).

Observations

Table 1 is a summary of elemental abundances in the solar system, the corona as derived from SEP observations, and abundances in the solar wind. We have used the solar system values of Anders and Ebihara (1982) for reference, assuming that the compilation of these authors reflects solar surface abundances to the best available precision. In their tables, wherever applicable, meteorite data are given priority over solar data. This seems justified in view of the better precision. Anders and Ebihara demonstrate that except for Li and the more volatile elements, the solar values and the abundances of C1 carbonaceous chondrites are consistent.

Figure 1 shows a comparison of SEP and solar wind elemental abundances with solar system abundances. In Table 2 we compare the isotopic composition of some light elements in SEPs and solar wind. For the elements in this comparison there is no independent "solar system" value except measurement in terrestrial and meteoritic materials. The values are compatible with solar wind except for the noble gases. In the case of helium the discrepancy is understood; in the case of neon, terrestrial or meteoritical values cannot be considered as representative for the solar system since neon is depleted in these samples by many orders of magnitude relative to refractory

TABLE 2. Comparison of isotopic ratios of some light elements for solar energetic particles and solar wind

	SEP	Ref.	Solar Wind	Ref.
$^{3}He/^{4}He$	$<2.6 \cdot 10^{-3}$	a	$(4.9 \pm 0.5) \cdot 10^{-4}$	b, c, d
$^{13}C/^{12}C$	$(9.5 \pm {}^{4.2}_{2.9}) \cdot 10^{-3}$	a	$(1.11 \pm .02) \cdot 10^{-3}$	e
$^{15}N/^{14}N$	$(8 \pm {}^{10}_{5}) \cdot 10^{-3}$	a	$(4.0 \pm 0.3) \cdot 10^{-3}$	f
$^{21}Ne/^{20}Ne$	<0.014 $(2.62 \pm 0.08) \cdot 10^{-3}$	a	$(2.4 \pm 0.3) \cdot 10^{-3}$	b, d
$^{22}Ne/^{20}Ne$	$(0.109 \pm {}^{0.026}_{0.019})$	a		
			$(7.30 \pm 0.16) \cdot 10^{-2}$	b, d
	(0.088 ± 0.002)	d		
$^{25}Mg/^{24}Mg$	$(0.148 \pm {}^{0.046}_{-0.026})$	a	-	
$^{26}Mg/^{24}Mg$	$(0.148 \pm {}^{0.043}_{-0.025})$	a		

(a) Mewaldt et al. (1984), (b) Geiss et al. (1972), (c) Coplan et al. (1984), (d) Benkert et al. (1988), (e) Becker (1980), (f) Geiss and Bochsler (1982).

elements. By no means is it possible to exclude a severe isotopic fractionation of neon in these samples.

The relation between average elemental abundances in SEPs and the first ionization potential (FIP) of the elements has become evident during the last 15 years (Crawford et al., 1972; Hovestadt, 1974; Meyer, 1981; Cook et al., 1984; Breneman and Stone, 1985; Meyer, 1985). The similarity of average solar wind elemental abundances and charge states with SEPs has been taken as evidence that both, SEPs and solar wind particles, are essentially coronal material which has been accelerated to different energies by differing mechanisms (Breneman and Stone, 1985; Meyer, 1985). The sorting according to the FIP and the magnitude of the fractionation suggests that this effect has to do with incomplete ionization of elements. The fractionation might occur in the upper chromosphere or the transition region during the initial acceleration, possibly coupled with a separating transport mechanism.

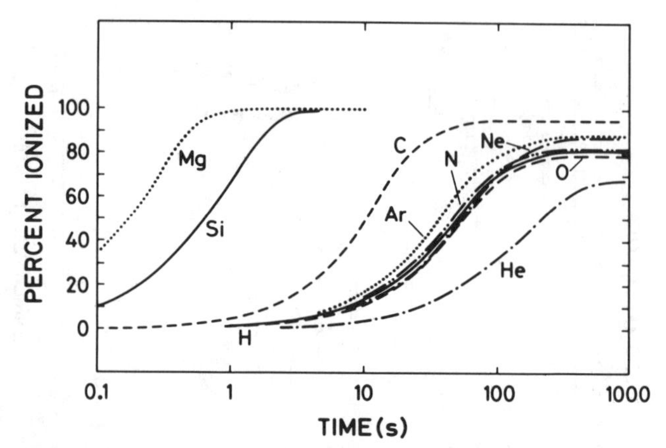

Fig. 3. Ionization of various elements as a function of time in an optically thin layer at the solar surface. A constant temperature of 10^4 K and a constant gas density $n(H) + n(H^+) = 10^{10}$ cm^{-3} were assumed (from Geiss and Bochsler, 1985).

Processes Occurring in the Chromosphere and in the Transition Region

Of course, sorting of abundances according to FIPs is not yet an explanation of a physical mechanism but the finding provides an important hint for further investigation.

Along these lines Geiss (1982) has investigated ion-neutral separation processes in the chromosphere-transition region. Vauclair and Meyer (1985) and Geiss and Bochsler (1985) have evaluated mechanisms to explain the influence of incomplete ionization in more detail. In the percolator model of Vauclair and Meyer (1985) ions are boiled up to the corona through spicules. They and the neutrals are prevented from gravitational settling to the photosphere by large-scale horizontal magnetic fields which serve as filters. An illustration of the scheme of Vauclair and Meyer is given in Figure 2. The proposed mechanism succeeds in separating ions from neutrals provided that it works for sufficiently long times, say on a scale of the order of several days, which seems somewhat unrealistic. Diffusion times could be shorter if the process operates in a less dense medium. In that case, however, the radiation field at the solar surface can affect the ionization balance in the medium. Geiss and Bochsler (1985) have therefore approached the problem from a different viewpoint and investigated ionization times for different ions in a gas under the influence of the unimpeded solar surface radiation field. These calculations were done for nine elements taking into account the most important atomic levels which contribute to ionization under typical conditions in the chromosphere (n_H: 10^{10} cm^{-3}, $T = 10^4$ K). Figure 3 is an illustration of their results. There is, of course, a relation, but no one-to-one correspondance between the FIPs and the ionization times. For instance, Ne ionizes faster than O although its ionization potential is higher by 8 eV. Geiss and Bochsler (1986), using these ionization times, have estimated the required diffusion lengths across magnetic flux tubes in order to separate neutrals from ions and thus to obtain the observed fractionation. Scales of structures must typically not be larger than 100 km at the chosen density of 10^9 cm^{-3} in order to separate the group C, N, O, Ne, Ar from the low FIP elements Mg, Si, Fe. Detailed models involving incomplete ionization and separation of ions from atoms by transport of atoms out of magnetic structures are in preparation (Von Steiger and Geiss, 1988); here we just give some sketches of possible separation mechanisms (Figure 4). Diffusion in a static medium can be speeded up in several ways: Motion and/or acceleration of a magnetic structure across the partly ionized medium could enhance the ion/neutral ratio in the structure. Similarly, turbulent motion or wave action in the magnetic structure could efficiently lower diffusion times of neutrals without releasing ions and thus accelerate the separation process.

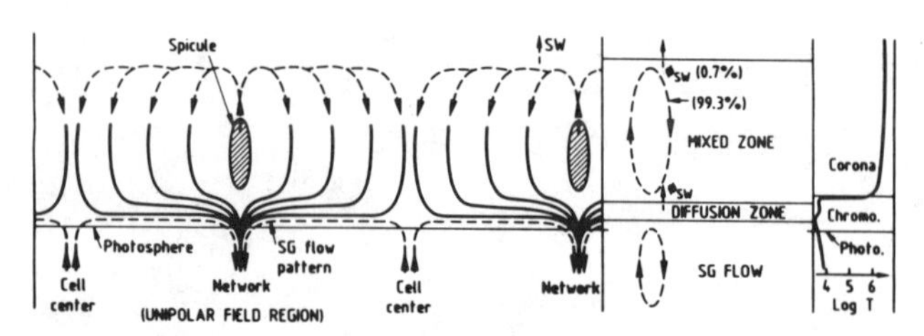

Fig. 2. Mechanism for separation of ions from neutrals by gravitational settling across magnetic field structures (from Vauclair and Meyer, 1985).

Whereas it is possible to reproduce the pattern of observed solar wind abundances in a very

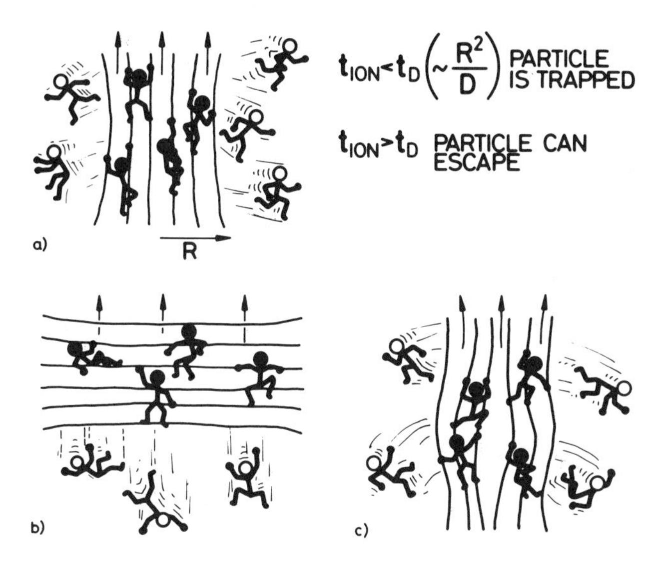

Fig. 4. Possible schemes to separate ions from neutral in the solar atmosphere. (a) Diffusion of neutrals across magnetic structures. (b) Motion or acceleration of magnetic structures containing ions and neutrals. (c) Turbulent motion in magnetic structures enhances diffusion of neutrals.

general manner with these models, so far it has been difficult to account for the observed details. For instance Gloeckler et al. (1986) have determined a precise average C/O ratio with AMPTE/CCE (Active Magnetospheric Particle Tracing Experiment) for a set of short periods in which AMPTE/CCE entered the magnetosheath of the Earth. The ion population of the magnetosheath is considered to reflect the solar wind and their value of 0.45 is consistent with the value of Breneman and Stone (1985) for SEPs. It is a bit lower but certainly not higher than the photospheric value of 0.59 derived by optical means (Grevesse, 1984; Harris et al., 1987). From Figure 3 it is evident that it is very difficult to explain the underabundance of C with respect to O for a model which only involves ionization times as a criterion to separate otherwise roughly similar species from each other.

In order to derive photospheric abundances from SEP abundances Breneman and Stone (1985) used a Q/M dependent fractionation in addition to a correction for the different ionization potentials of the elements (see also Meyer, 1985). With this parameterization they succeeded in reproducing photospheric abundances from SEP abundances with the notable exception of C which still seems to be underabundant by a factor of 2.

Among the more abundant elements, C is one of the first to form stable compounds even at relatively high temperatures. CO and CH have considerably higher FIPs than C, H, and O and thus one could be led to suspect that by forming such compounds before ionization, carbon could remain in the group of high FIP elements. However, even at the minimum temperature in the solar atmosphere these compounds are unstable; furthermore, forming any stable compound among minor species in the relatively thin medium would take days.

Quite a different approach to explain SEP abundances has been taken by Mullan and Levine (1981). These authors investigate an injection process whereby collapsing magnetic neutral sheets can accelerate particles. The fractionation mechanism operates by the competition between efficient acceleration and heating and subsequent losses of ions due to diffusion. Since the collision cross section per mass of an ion goes as Z^2/A, heavily charged ions will be more readily lost than weakly charged ions. This model is able qualitatively to reproduce the abundance systematics of SEPs. Meyer (1985) has criticized the investigation of Mullan and Levine (1981) for the assumption of an equilibrium ionization state, whereas in a realistic picture of expanding loops, the site of the proposed injection process, a dramatic non-equilibrium situation might prevail. The fractionation between different ions achieved by this mechanism is often of several orders of magnitude, whereas the real effects are of the order of a factor 10. This in itself is not a problem because overfractionation can be reduced by making the process less efficient or by mixing with unfractionated material. However, if such an explanation is invoked, serious discrepancies with observation become conspicuous. For instance, the model of Mullan and Levine (1981) produces relatively similar enrichments for Na and Ne, whereas in the SEP composition Na is found to be highly enriched and Ne not at all (cf. Table 1 and Figure 1). Also, the process does not account for the similarity of SEP and solar wind abundances. In any case, Mullan and Levine (1981) have demonstrated that using atomic properties but not involving the FIP can lead to fractionation for seed particles which in a global way resembles the SEP observation.

SEP ions differ from solar wind ions with respect to the isotopic composition of He and Ne. In rare cases the $^3He/^4He$ ratio in SEPs reaches 1, i.e., ^3He is enriched by 3 orders of magnitude over the solar surface value. An explanation involving resonant ion-cyclotron heating of ^3He in flares has been proposed by Fisk (1978). Although the discrepancy between the $^{20}Ne/^{22}Ne$ ratio in SEPs and the solar wind is only of the order 30%, it is difficult to understand this as a consequence of a similar mechanism operating on ^{22}Ne in solar flares. Using the expressions of Mason et al. (1980) and Luhn (1986) it can be seen that Ne isotopes could only be fractionated in a very limited temperature range without simultaneously affecting Mg isotopes. The Mg isotopic ratios in solar flares are, however, consistent with the solar system values (Mewaldt et al., 1984).

Mullan (1983) uses the model of Mullan and Levine (1981) of collapsing neutral sheets to preferentially inject ^{22}Ne into the energetic particle population. ^{22}Ne has a smaller Coulomb cross section per mass than ^{20}Ne and is thus better protected against diffusion losses. The simultaneous fractionation of other isotopic ratios for which there is no evidence of fractionation is circumvented by appropriate selection of the ionization temperature.

The FIP of an element is directly related to its chemical properties, e.g., to its volatility. Occasionally, it has been speculated that interplanetary grains entering the solar corona could thereby enrich the corona with low FIP elements. Meyer (1985) and Geiss and Bochsler (1986) have recently discussed such possibilities. At present, it is not possible to completely exclude a contamination of the corona by interplanetary grains. The problem could be tackled in several ways:

1. Estimation of the total amount of inflowing grains
2. Investigation of the abundance pattern of SEPs and solar wind ions in detail
3. Investigation of charge states of SEPs and solar wind ions
4. Observation of latitudinal effects in the corona
5. Comparison of SEPs vs. solar wind for depth effects in the corona.

It is difficult to make an assessment with respect to (1); at present, (2) and (3) look more promising, and (4) and (5) can be investigated in detail with the out-of-ecliptic mission ULYSSES. In our opinion, the present evidence indicates that the contamination of the corona by interplanetary grains could not produce the large abundance differences between the photosphere on one hand and SEPs and the solar wind on the other.

Processes Occurring in the Corona

Geiss et al. (1970) have investigated the momentum equation for minor ions in the solar wind, including the effects of the separation E field and Coulomb drag by protons. They could demonstrate that for heavy ions, friction with the outstreaming protons is an important contribution in the momentum balance of these particles. Furthermore, it became evident that $^4He^{++}$ with a very unfavourable Coulomb drag factor has to be accelerated by additional forces in order to reach the observed velocity excess of $^4He^{++}$ over H^+ at 1 AU. The most obvious candidate is wave particle interaction in the corona (e.g., Hollweg, 1981). Borrini et al. (1983) have reviewed the observational facts and theoretical models on helium abundance variation. Strong variations of the helium abundance are unambiguously related to complicated magnetic structures at the solar surface, whereas homogeneous magnetic patterns as observed in coronal holes, produce stable helium abundances. We support the view presented by Axford (1986) who points out that the slow stream solar wind is a mixture of plasma emanating from different surface structures which are connected to the solar wind flow by rapidly varying footpoints of the flow tubes. Similar conclusions which require the treatment of the slow speed solar wind as a stochastic phenomenon have resulted from the detection of weakly ionized species (Schwenn et al., 1980; Zwickl et al., 1982; Bochsler, 1983) and the observation of jets near the solar surface (Brueckner and Bartoe, 1983).

Noci and Porri (1983) have pointed out that helium in the corona plays an important role in the momentum balance of the solar wind. Bürgi and Geiss (1986) have investigated the momentum and ionization/recombination equation for major and minor species. They found that observational constraints on electron densities in the inner corona and the charge state of heavy elements at 1 AU demand solutions of the momentum equation which show a strong helium enrichment in the low corona (typically by an order of magnitude). This has the effect that the Coulomb drag exerted on minor ions by He^{++} and H^+ become comparable. As a consequence $^4He^{++}$ tends to collect minor ions in the flow and, hence, the gravitational settling and enrichment of the innermost corona will comprise He and minor ions (Bürgi and Geiss, 1986). Explosive events at the solar surface (Hirshberg et al., 1972) and/or wave action (Hollweg, 1981) might eventually drive parts of this layer into the outer corona and enrich the ambient solar wind flow with ^4He and heavier elements.

If the varying relative strength of wave-particle and particle-particle interaction is responsible for a large part of the fluctuations in the He/H ratio of the solar wind then it could certainly influence, although to a lesser extent, other elemental and also isotopic ratios in the solar wind. Up to now it has been difficult to observe such effects. The correlation of fluxes of different species in the solar wind is quite strong. For instance, in the case of He- and O fluxes, Bochsler et al. (1986) found a correlation coefficient $r = 0.755 \pm 0.005$ (Figure 5). The picture illustrates that whereas fluxes can vary by typically a factor of ten, the He/O ratio varies by a factor of 1.5 only, an order of magnitude in precision which is not easy to establish with certitude for individual measurements with presently used techniques. On the other hand, it seems that the temporal variance of measured helium fluxes and thus of the He/H ratio is larger than the variance of the O/H ratio, despite the possible influence of measurement uncertainties. The strong correlation of He and O fluxes in the solar wind indicates that wave action generally does not discriminate much between either species.

Charge State Distribution of Elements

The analysis of charge state distribution of elements in the solar wind at 1 AU yields extremely important insight into the dynamics of the inner corona. Hundhausen et al. (1968) have shown that it is this region where the charge state of ions is determined. In the following, Bame et al.

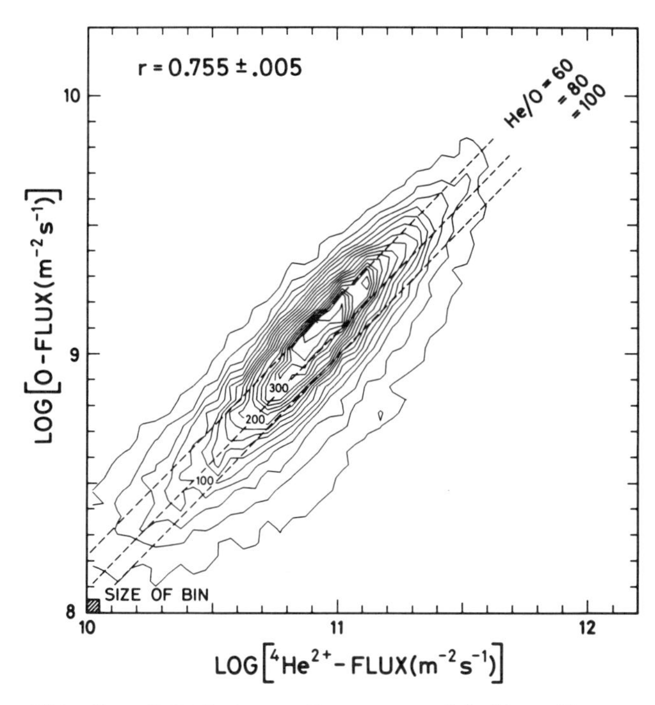

Fig. 5. Correlogram of oxygen and helium fluxes. The dashed lines are for equal He/O ratios. The ridge defined by the most frequent occurrences runs along a line with He/O = 70 (from Bochsler et al., 1986).

(1974) have derived temperatures and temperature gradients in the corona from a comparison of charge states of oxygen, silicon, and iron ions. For these early models and calculations it was assumed that the velocity distribution of the electron gas which essentially rules the ionization and recombination processes is Maxwellian. More recently, the influence of non-Maxwellian exospheric tails in the electron distribution on charge states has been investigated thoroughly (cf. Owocki et al., 1983; Bürgi, 1987). Bürgi (1987) has shown that the presence of a non-Maxwellian tail in the electron distribution can possibly only be detected by deviations of the charge state distribution of silicon (or heavier elements). Observations of the charge states of C, N, O, Ne, and Mg can be interpreted either way.

Outlook

Although in the last decade improved instrumentation has brought a wealth of new data and theoretical insight into the mechanisms which drive the solar wind, our knowledge no longer corresponds to the technically possible instrumentation of today. From AMPTE (Gloeckler et al. 1986) we have experienced the gain of high quality information achieved with the newly adopted time-of-flight technique. The next epoch of solar wind composition measurements, beginning with the Russian mission to Mars, PHOBOS, and ULYSSES, the joint NASA/ESA out-of-ecliptic mission, will initiate another step forward.

As a long-range perspective we consider the establishment of points of reference for the solar isotopic composition of a large number of elements. On a wide variety of samples from the solar system the isotopic composition of elements has been compared at a level of precision up to 10^{-5}. Isotopic measurements on the largest single reservoir in the solar system, the sun, are still in their infancy. The potentially best determinations could be obtained from the solar wind. This has been demonstrated with the measurement of the isotopic composition of light noble gases with the Apollo Solar Wind Composition Experiment (Geiss et al., 1972) and the detailed investigation of surface correlated elements in lunar soil (cf. Eberhardt, 1974). The first direct measurement of $^3He/^4He$ in the solar wind had implications reaching beyond the history of the solar system. The investigation of nitrogen isotopes in lunar soils produced surprising evidence for a secular trend which is still a matter of controversy (Kerridge, 1975; Becker and Clayton, 1975; Geiss and Bochsler, 1982). The case of neon isotopes has been briefly addressed in the previous sections. Also here, the final word has not been said. Once the uncertainties of the solar isotopic composition have been narrowed down to a value which is comparable to or smaller than the natural variation in the solar system, it will be possible to use it as a new point of reference. We might say that in the field of isotopic studies the transition to the heliocentric system has not yet occurred. It is clear that this transition will have a strong impact on our understanding of the history of the solar system. A first step toward this long-range goal is a better understanding of the fractionation processes for SEP and solar wind isotopic abundances.

The tasks to be accomplished in the future include therefore:

- measurement of solar wind elemental abundances in the FIP range from Ly-α to the ionization energy of H, e.g., hereto unmeasured species such as P, S, Cl and better measurements of C and O.
- measurement of some isotopic ratios with high time resolution and sufficient precision to detect the probably small but real variations in the solar wind. As a guideline for time resolution we may use the currently achieved resolution with optical instruments. One arc sec corresponds to 7 min. in solar rotation.
- detailed analysis of charge state distribution will provide further insight for processes occurring in the inner corona, especially if combined with optical measurements, e.g., the Doppler dimming technique (Withbroe et al., 1982).

Acknowledgments. The authors are grateful for discussions with A. Bürgi and R. Von Steiger. I. Peter prepared the camera-ready manuscript. This work was in part supported by the Swiss National Science Foundation.

References

Anders, E., and M. Ebihara, Solar-system abundances of the elements, Geochim. Cosmochim. Acta, 46, 2363, 1982.

Axford, I. W., Lecture presented at the University of Bern, September 4, 1986.

Bame, S., J. R. Asbridge, W. C. Feldman, and P. D. Kearney, The quiet corona: Temperature and temperature gradient, Solar Phys., 35, 137, 1974.

Bame, S. J., J. R. Asbridge, W. C. Feldman, M. D. Montgomery, and P. D. Kearney, Solar wind heavy ion abundances, Solar Phys., 43, 463, 1975.

Becker, R. H., and R. N. Clayton, Nitrogen abundances and isotopic compositions in lunar samples, Proc. 6th Lunar Sci. Conf., 2131, 1975.

Becker, R. H., Evidence for a secular variation in the $^{13}C/^{12}C$ ratio of carbon implanted in lunar soils, Earth Planet. Sci. Lett., 50, 189, 1980.

Benkert, J. P., H. Baur, A. Pedroni, R. Wieler, and P. Signer, Solar He, Ne and Ar in regolith minerals: all are mixtures of two components, 19th Lun. Planet. Sci. Conf., preprint, 1988.

Bochsler, P., Mixed solar wind originating from coronal regions of different temperatures, in Solar Wind Five, NASA CP-2280, edited by M. Neugebauer, p. 613, 1983.

Bochsler, P., Silicon abundance in the solar wind (abstract), International Union of Geodesy and Geophysics (IUGG), Vancouver, 658, 1987.

Bochsler, P., J. Geiss, and S. Kunz, Abundances of carbon, oxygen, and neon in the solar wind during the period from August 1978 to June 1982, Solar Phys., 103, 177, 1986.

Borrini, G., J. T. Gosling, S. J. Bame, and W. C. Feldman, Helium abundance variations in the solar wind, Solar Phys., 83, 367, 1983.

Breneman, H. H., and E. C. Stone, Solar coronal and photospheric abundances from solar energetic particle measurements, Astrophys. J., 299, L57, 1985.

Brueckner, G. E., and J.-D.-F. Bartoe, Observations of high energy jets in the corona above the quiet sun, the heating of the corona and the acceleration of the solar wind, Astrophys. J., 272, 329, 1983.

Bürgi, A., Effects of non-Maxwellian electron velocity distribution functions and non-spherical geometry on minor ions in the solar wind, J. Geophys. Res., 92, 1057, 1987.

Bürgi, A., and J. Geiss, Helium and minor ions in the corona and solar wind: Dynamics and charge states, Solar Phys., 103, 347, 1986.

Cook, W. R., E. C. Stone, and R. E. Vogt, Elemental composition of solar energetic particles, Astrophys. J., 279, 827, 1984.

Coplan, M. A., K. W. Ogilvie, P. Bochsler, and J. Geiss, Interpretation of ^{3}He abundance variations in the solar wind, Solar Phys., 93, 415, 1984.

Crawford, H. J., P. B. Price, and J. D Sullivan, Composition and energy spectra of heavy nuclei with 0.5 < E < 40 MeV per nucleon in the 1971 January 24 and September 1 solar flares, Astrophys. J., 175, L149, 1972.

Eberhardt, P., The solar wind as deduced from lunar samples, Solar Wind Three, edited by C. T. Russell, p. 58, 1974.

Fisk, L. A., ^{3}He-rich flares: A possible explanation, Astrophys. J., 224, 1048, 1978.

Geiss, J., Processes affecting abundances in the solar wind, Space Sci. Rev., 33, 201, 1982.

Geiss, J., and P. Bochsler, Nitrogen isotopes in the solar system, Geochim. Cosmocim. Acta, 46, 529, 1982.

Geiss, J., and P. Bochsler, Ion composition in the solar wind in relation to solar abundances, in Rapports isotopiques dans le système solaire, Cepadues-editions, p. 213, 1985.

Geiss, J., and P. Bochsler, Solar wind composition and what we expect to learn from out-of-ecliptic measurements, in The Sun and the Heliosphere in Three Dimensions, edited by R. G. Marsden, p. 173, D. Reidel Publishing Company, 1986.

Geiss, J., F. Bühler, H. Cerutti, P. Eberhardt, and Ch. Filleux, Solar wind composition experiment, Section 14, in Apollo 16 Preliminary Science Report, NASA SP-315, 1972.

Geiss, J., P. Hirt, and H. Leutwyler, On acceleration and motion of ions in corona and solar wind, Solar Phys., 12, 458, 1970.

Gloeckler, G., F. M. Ipavich, D. C. Hamilton, B. Wilken, W. Stüdemann, G. Kremser, and D. Hovestadt, Solar wind carbon, nitrogen and oxygen abundances measured in the earth's magnetosheath with AMPTE/CCE, Geophys. Res. Lett., 13, 793, 1986.

Grevesse, N., Accurate atomic data and solar photospheric spectroscopy, Physica Scripta, T8, 49, 1984.

Harris, M. J., D. L. Lambert, and A. Goldman, The $^{12}C/^{13}C$ and $^{16}O/^{18}O$ ratios in the solar photosphere, M.N.R.A.S., 224, 237, 1987.

Hirshberg, J., S. J. Bame, and D. E. Robbins, Solar flares and solar wind helium enrichments: July 1965 - July 1967, Solar Phys., 23, 467, 1972.

Hollweg, J. V., Helium and heavy ions, in Solar Wind Four, edited by H. Rosenbauer, p. 414, 1981.

Hovestadt, D., Nuclear composition of solar cosmic rays, in Proc. Solar Wind Three, edited by C. T. Russell, p. 2, 1974.

Hundhausen, A. J., H. E. Gilbert, and S. J. Bame, Ionization state of the interplanetary plasma, J. Geophys. Res., 73, 5485, 1968.

Kerridge, J. F., Solar nitrogen: Evidence for a secular increase in the ratio of nitrogen 15 to nitrogen 14, Science, 188, 162, 1975.

Luhn, A. M., Die Ladungszustände solarer energetischer Teilchen, MPE Report, 195, Max-Planck-Institut für Physik und Astrophysik, Garching, W. Germany, 1986.

Mason, G. M., L. A. Fisk, D. Hovestadt, and G. Gloeckler, A survey of ∿ 1 MeV nucleon^{-1} solar flare particle abundances, $1 \leq Z \leq 26$, during the 1973-1977 solar minimum period, Astrophys. J., 239, 1070, 1980.

Mewaldt, R. A., J. D. Spalding, and E. C. Stone, A high-resolution study of the isotopes of solar flare nuclei, Astrophys. J., 280, 892, 1984.

Meyer, J. P., A tentative ordering of all available solar energetic particles abundance observations: II. Discussion and comparison with coronal abundances, Proc. 17th Int. Cosmic Ray Conf., 3, 149, 1981.

Meyer, J. P., The baseline composition of solar energetic particles, Astrophys. J. Suppl., 57, 151, 1985.

Meyer, J. P., Everything you always wanted to ask about local galactic abundances, but were afraid to know, to appear in Symposium on the origin and distribution of the elements (New Orleans), 1988.

Mullan, D. J., Isotopic anomalies among solar energetic particles: Contribution of preacceleration in collapsing magnetic neutral sheets, Astrophys. J., 268, 385, 1983.

Mullan, D. J., and R. H. Levine, Preacceleration in collapsing neutral sheets and anomalous abundances of solar flare particles, Astrophys. J. Suppl., 47, 87, 1981.

Noci, G., and A. Porri, Models of the solar wind acceleration region, Paper 4L.04 presented at the 18th IAGA General Assembly Meeting, Hamburg, 1983.

Owocki, S. P., T. E. Holzer, and A. J. Hundhausen, The solar wind ionization state as a coronal temperature diagnostic, Astrophys. J., 275, 354, 1983.

Schmid, J., P. Bochsler, and J. Geiss, Abundance of iron ions in the solar wind, Astrophys. J., in press, 1988.

Schwenn, R., H. Rosenbauer, and K. H. Mühlhäuser, Singly-ionized helium in the driver gas of an interplanetary shock wave, Geophys. Res. Lett., 7, 201, 1980.

Vauclair, S., and J. P. Meyer, Diffusion in the chromosphere and the composition of the solar corona and energetic particles, Proc. 19th Int. Cosmic Ray Conf., 4, 233, 1985.

Von Steiger, R., and J. Geiss, Ion-atom separation in the solar chromosphere: Transport across a magnetic field, in preparation, 1988.

Withbroe, G. L., J. L. Kohl, and H. Weiser, Probing the solar wind acceleration region using spectroscopic techniques, Space Sci. Rev., 33, 17, 1982.

Zwickl, R. D., J. R. Asbridge, S. J. Bame, W. C. Feldman, and J. T. Gosling, He^{+} and other unusual ions in the solar wind: A systematic search covering 1972-1980, J. Geophys. Res., 87, 7379, 1982.

SPACE PLASMA MASS SPECTROSCOPY BELOW 60 keV

D. T. Young

Southwest Research Institute, San Antonio, TX 78284

Abstract. The next generation of missions in solar system plasma physics requires much more detailed and quantitative measurements than have been possible in the past. It has been found that magnetospheric plasmas, long considered to be primarily hydrogenic, contain large admixtures of "heavy" ions that are critical to many magnetospheric processes and hence must be included in plans for future measurements. Complexity of composition is only one attribute of space plasmas: they also exhibit diverse velocity distributions requiring 4π sr coverage, as well as strong spatial and temporal variations requiring high time resolution. Finally, the upper energy limit of these plasmas is now considered to be in the range of 60 keV (a typical limit for electrostatic analysis) to 100 keV (the limit set by ESA/Cluster mission requirement), which demands novel approaches to mass discrimination techniques. All of these factors suggest that missions currently under consideration by the space plasma community will require instruments far in advance of those used previously. This paper reviews the state of plasma mass spectroscopy, and offers one way to quantify the necessary advancements by introducing a "quality-factor" for plasma instrumentation. We demonstrate that the science objectives of the Cluster and Polar spacecraft of the International Solar Terrestrial Program require orders of magnitude improvement in performance compared to earlier generations of ion mass spectrometers. The needed improvements require new instrument designs that will place added demands on spacecraft resources. Upon examining the resources that are available to such instruments, we find little or no improvement over the 20-year period covered by this review.

The Basis of Instrument Design Requirements

Most spacecraft orbits are chosen so as to intersect as many different plasma regimes within the terrestrial magnetosphere as are consistent with mission goals. For example, ESA/Geos 1 and 2, ISEE 1 and 2, DE 1, AMPTE/CCE, Cluster, and Polar are all multi-regime missions. Thus a fundamental problem of measurement is how to accommodate the resulting wide range of plasma parameters exemplified by the contents of Table 1 (freely adapted from ESA, 1985). Compounding the demand for dynamic range is the time variation inherent in each population, as well as, very often, motion of the plasma and its boundaries at speeds of tens to hundreds of km s^{-1} relative to the spacecraft. The Cluster mission will have to contend with these problems. It represents, in principle, a significant step forward as it seeks to address "the true temporal and three-dimensional spatial structures which occur on spatial scales of several 100 to several 1000 km" (ESA, 1985).

The purpose of this review is to examine briefly the subject of mass spectroscopy as applied to space instrumentation, and to assess the impact of mission requirements, such as those of Cluster, on plasma composition measurements. We wish to inquire into the nature of instrument design and optimization under constraints of both ion optics and finite spacecraft resources. Such an inquiry raises the issue of availability and allocation of spacecraft resources, something which is rarely examined, much less challenged, by space scientists. It also leads us to try to formulate quantitative estimates of instrument performance which might ultimately prove useful in deciding what are the actual limits on measurement capabilities. We shall see that instrumentation proposed for the Cluster and Polar spacecraft represents significant improvements which are needed in order to meet ambitious science requirements. Curiously, however, spacecraft resource allocations for instrument power, weight and telemetry have remained largely unchanged over the 20-year period 1975–1995.

Perhaps the most pressing design requirement is for three-dimensional distribution function measurements of ion composition. In one approach to obtaining full angular coverage, plasma analyzers have been designed with 360° fan-shaped fields-of-view. The 360° fan rotates with the spacecraft (Figure 1b) thereby sweeping out a 4π sr field-of-view (FOV). Although this method restricts a single instrument to obtaining a three-dimensional distribution once per half spin, the resulting time resolution may be adequate if the spin rate is sufficiently high. There also exist analyzer designs, intended primarily for three-axis stabilized spacecraft, that electrostatically scan the field-of-view over a little less than 2π sr (Bame *et al.*, 1989) or

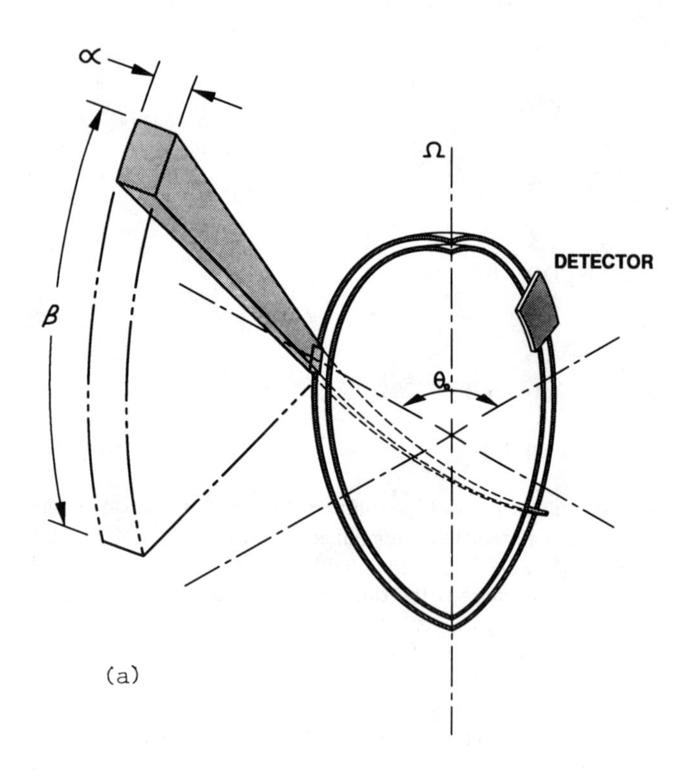

Fig. 1a. Schematic of a flat-aperture spherical section analyzer with a section angle, θ_o, of 90° which corresponds approximately to the particle bending angle in the electric field. Analyzers are usually oriented on spacecraft so that the Ω-axis is parallel to the spacecraft spin axis.

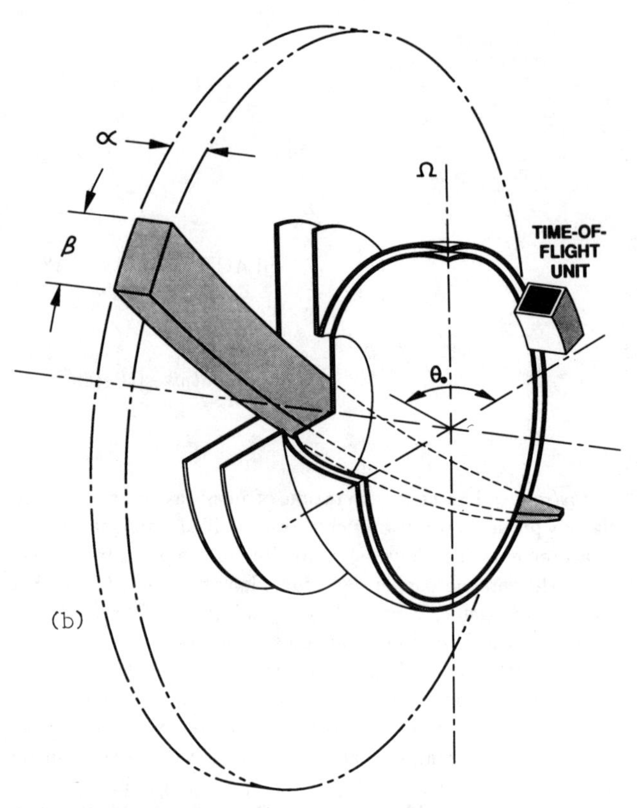

Fig. 1b. Schematic of a symmetric "top-hat" analyzer having a 360° FOV and a toroidal plate geometry similar to the HPCA discussed in (Young *et al.*, 1988a,b) and to designs of Carlson *et al.* (1983), Carlson (1988), and Sablik *et al.* (1988). Note that β acceptance is 360° in comparison to Fig. 1a. Analyzer orientation is usually with Ω-axis parallel to spin axis.

use multiple-nested sensors for this purpose (McFadden *et al.*, 1988). It is therefore possible in principle to escape the field-of-view restrictions imposed by 3-axis oriented spacecraft as well as the temporal resolution limits dictated by spacecraft spin. However, we confine this discussion to fixed, fan-shaped FOVs. There is no reason in principle why the arguments presented here on quality factors cannot be extended to scanned FOVs.

The term mass spectrometry applies to measurements in which a mass spectrum is obtained by a relatively slow scan of voltage such that a only single mass peak is observed at any instant. Spectroscopy originally referred to the image of a full mass spectrum displayed on a photographic plate or, nowadays, on an imaging detector such as a microchannel plate (MCP). Spectroscopy is a more general term which includes the above as well as rf and time-of-flight techniques.

Finally we note that the evolution toward full three-dimensional fields-of-view and extra-large geometric factors comes at a cost to other spectroscopy goals, particularly the high angular and mass resolution needed, for example, for detailed studies of auroral ion distributions and acceleration processes. And yet, one must almost certainly be sacrificed to obtain the other: large areal or angular acceptance is never synonymous with high mass resolution for instruments of a given size.

Design of Plasma Composition Analyzers

Up to now there have been two categories of plasma mass spectrometers that meet the requirements mentioned in the previous section:

- Energy/charge + magnetic analysis (E/Q-M/Q)
- Energy/charge + time-of-flight analysis (E/Q-TOF)

A third category, namely instruments using an energy/charge + TOF + total energy (E/Q-TOF-E) technique, are generally better suited to higher ion energies and are not dealt with here (see, e. g., Gloeckler, 1988). Still other designs have been developed to meet the conditions of solar wind interactions with non-magnetized bodies in which a wide range of distribution functions are encountered on very short time scales such as during cometary or planetary encounters (Neugebauer, 1989). In the following sections we review briefly the principles of, and

DOUBLE-FOCUSING MASS SPECTROMETERS

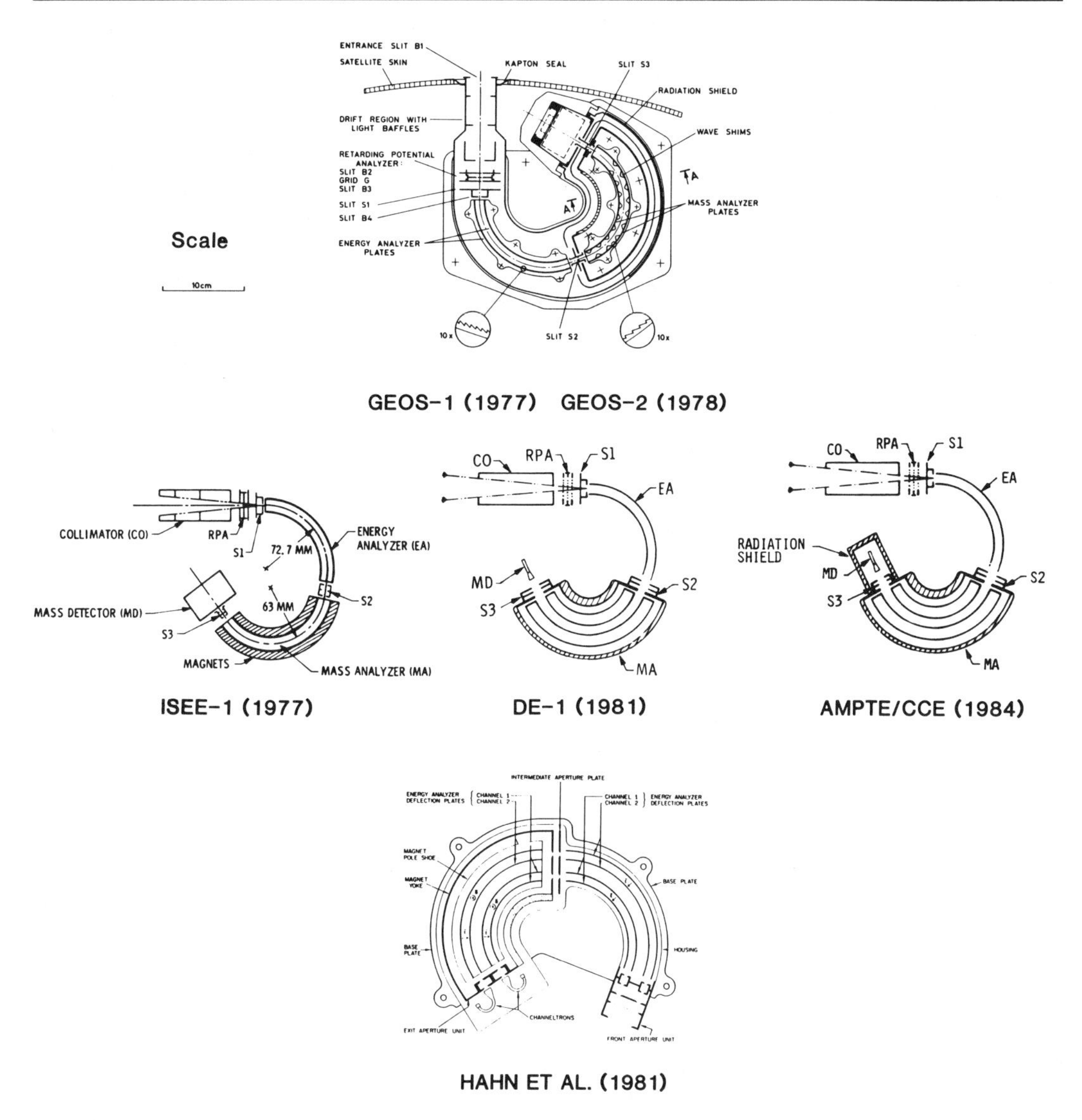

Fig. 2a. Schematic diagrams of the main family of double-focusing mass spectrometers and spectrographs used for satellite-borne plasma measurements. All cross sections are taken in the dispersive plane of the instruments. Geos 1 and 2 are described in Balsiger *et al.* (1976); the ISEE 1, DE 1, and AMPTE/CCE instruments in Shelley *et al.* (1978, 1981, 1985), respectively; and FIMS in Hahn *et al.* (1981).

latest developments in, the two approaches most often taken for magnetospheric plasma investigations.

Advances in Magnetic Analyzers

Within the range of parameters of interest here, E/Q-M/Q spectrometers of the so-called Geos-type have played a significant role in magnetospheric research and they continue to be used and to evolve (Balsiger, 1983, Wilken *et al.*, 1982). In Figure 2a we show a selection of Geos-type spectrometers. As originally conceived, this design provides double focusing (i.e., ion rays which diverge slightly in angle and energy at the entrance to the analyzer are focused at the detector) along a line in energy-mass space. A pre-acceleration voltage of several kV is used to give ions enough momentum to pass through a magnetic field of typically $\sim$ 700 to 2000 Gauss. Pre-acceleration reduces the range of particle momenta which must be handled by the magnet and keeps the entire analyzer fairly close to focusing conditions over a larger energy range. This device is not seriously out of focus even at higher energies, where the dispersion caused by magnetic forces decreases significantly, because of the action of the $\sim 130°$ sector electric field within the magnetic field. In short, this is an excellent example of trade-offs which can be made in optimizing an instrument with respect to several key performance parameters. The basic Geos design has been flown on ESA/Geos 1 and 2, ISEE 1, DE 1 and AMPTE/CCE (Balsiger *et al.*, 1976; Shelley *et al.*, 1978, 1981, 1985). Hahn *et al.* (1981) developed a version in which two mass peaks could be detected simultaneously, thus making the first move toward a spectrographic capability.

More advanced designs involve the use of spectrographic principles to image a range of M/Q values simultaneously. Moore (1977) and Coplan *et al.* (1984) have both put forward instruments based on the Mattauch-Herzog principle (Figure 2b). Recently a toroidal ion mass spectrograph (TIMS) has been evolved from the cylindrical double focusing Geos-type geometry. The TIMS is double focusing at all masses; it images a wide field-of-view in the polar direction (Figure 2b), and improves time resolution by providing a spectrographic mass and angle imaging capability. In order to obtain double focusing over a large M/Q range, it is necessary to use two electrostatic analyzers before the magnet section. The first forms an intermediate image in azimuth, which is used to define the energy acceptance, with a final, astigmatic image located at the detector (Young *et al.*, 1987a,b; Ghielmetti and Young, 1987). This type of device is capable of very good mass resolution ($M/\Delta M > 10$) and has the inherent feature of most magnetic designs that "ghost" mass peaks are suppressed.

Although the toroidal analyzer described by Ghielmetti and Young has a polar acceptance of $\pm 30°$, or roughly 6 times that of its cylindrical predecessors, this still was not sufficient for the demands of the Cluster and Polar missions. A magnetic spectrograph, based on non-standard toroidal optics (e.g., not on the usual paraxial ray Mattauch-Herzog principle), with a full 360° FOV, has been designed recently by the Lockheed group (Shelley *et al.*, 1988). Although it is conceptually similar to the TIMS, in reality a more sophisticated modeling approach has been required to achieve double focusing. A prototype has been built, tested, and ray traced and the completed instrument is scheduled to fly on the Polar spacecraft. Cylindrically symmetric spectrographic geometries have also been described by Coplan *et al.* (1984) (Fig. 2b) and Lundin *et al.* (1989).

A 360° FOV magnetic analyzer might seem at first to be impractical due to requirements for large magnets or to the existence of large stray magnetic fields. However, the closed configuration of the magnetic field (Figure 2b) creates a self-contained loop of magnetic flux nearly all of which is available for dispersing ion trajectories. In designs of the Geos- and TIMS-type, the magnetic field was required to be uniform over a height of ~ 3 cm (Figure 2c) and was terminated by high permeability pole pieces and a relatively large iron yoke. The yoke created a circuit for magnetic flux that was relatively heavy (1 kg) and inefficient in terms of particle analysis. In contrast, the 360° field does not conduct magnetic flux through an iron yoke, but rather maintains it in a loop. This increases the area of flux available for particle analysis by roughly an order of magnitude with no increase in magnet weight. Together with the use of sophisticated rare Earth magnetic materials, the magnet and related assembly weigh under 1 kg.

Advances in Time-of-Flight Analyzers

Time-of-flight (TOF) analyzers were initially developed for a wide variety of laboratory applications and have been adapted for space flight by several groups (Gloeckler and Hsieh, 1979; Gloeckler *et al.*, 1985; Hovestadt *et al.*, 1978; Fritz *et al.*, 1985; McEntire *et al.*, 1985; Wilken *et al.*, 1987). As is the case for magnetic analyzers, TOF can be made to operate in a number of configurations, often in combination with E/Q and total energy analysis, the latter making use of solid state detectors (SSD) in an E/Q-TOF-E scheme (Gloeckler, 1988). The conceptually simpler E/Q-TOF alternative, first flown on the Giotto mission (Wilken *et al.*, 1987), is more suitable for less energetic plasmas (because SSD energy thresholds do not have to be overcome) and situations in which charge state measurements and high M/Q resolution are not required but high time resolution is. The Giotto variant relies only on microchannel plates (MCPs) for particle detection rather than solid state detectors. The E/Q-TOF analyzer, although able to separate only a few solar wind mass per charge species, is perfectly adequate for many magnetospheric missions and has the advantages of requiring lower acceleration voltage (since no SSD detection threshold needs to be overcome) and hence lower instrument volume, as well as somewhat less sophisticated electronics for coincidence and decoding circuitry.

The E/Q-TOF analyzer developed for the LOMICS (low-energy mass composition spectrometer) instrument on CRESS (Fritz *et al.*, 1985) began as a joint development effort between Los Alamos National Laboratory and Max Planck Institute for Aeronomy for the Implanted Ion Sensor (IIS) on Giotto and bears some resemblance to the latter as shown in Figure 3. We refer the reader to Wilken *et al.* (1987) for a description of the IIS and discuss here only the LOMICS design and its evolution into the hot plasma composition analyzer (HPCA), a 360° FOV instrument designed for Cluster-type missions and described elsewhere in this volume (Young *et al.*, 1989). A conceptually similar design has also been developed by E. Mobius and colleagues for Cluster (Carlson, 1988).

Figures 1b and 4 show the principle of the HPCA design. Ions are first analyzed by E/Q and focused by the toroidal analyzer (Young *et al.*, 1988). They are accelerated by -15 kV into a thin ($1\ \mu g\, cm^{-2}$) carbon foil mounted on a 90% transmission grid. Ions straggle in the carbon foil, losing an average of ~ 0.5 to 1.5 keV depending on species and energy, and scatter through an rms angle of roughly Z degrees where Z is the nuclear charge of the incident ion. Laboratory tests have shown that although straggling degrades TOF resolution somewhat, mass resolution ($M/\Delta M$) of four is easily achieveable with a compact (5.0 cm long) system. As ions exit the foil they have a high probability of emitting one or more secondary electrons which are collected onto a start MCP (Figure 4). This gives a pulse whose leading edge triggers the time measurement cir-

DOUBLE-FOCUSING MASS SPECTROGRAPHS

MATTAUCH-HERZOG OPTICS

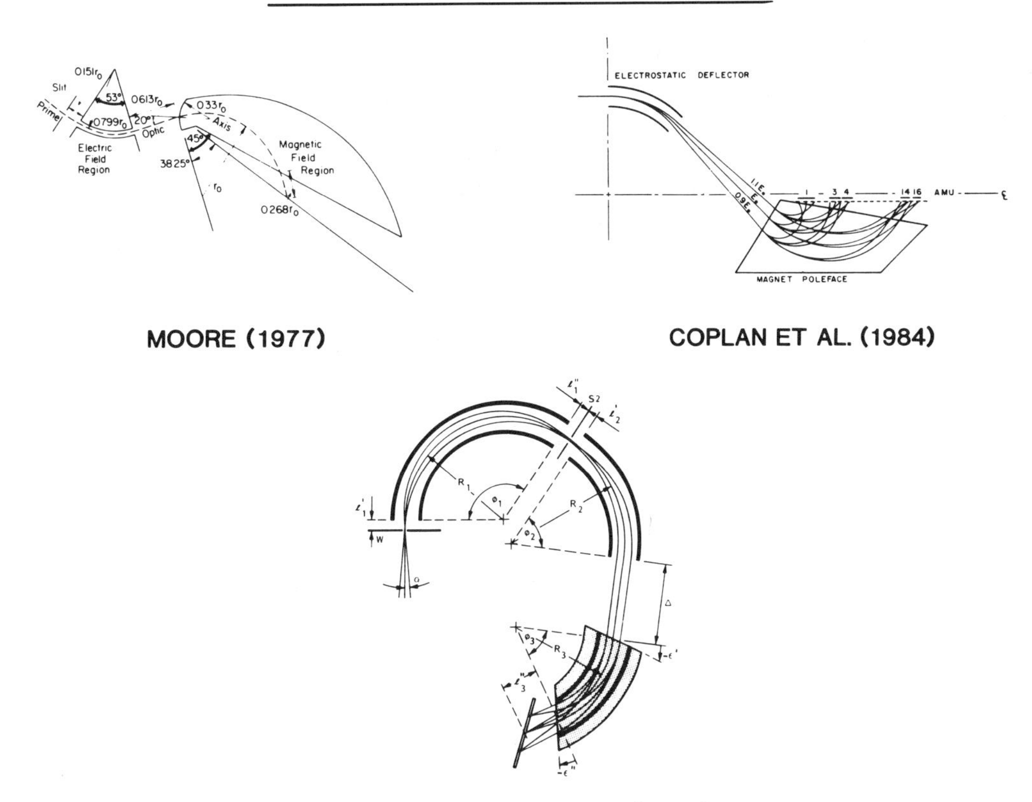

NON-STANDARD TOROIDAL OPTICS

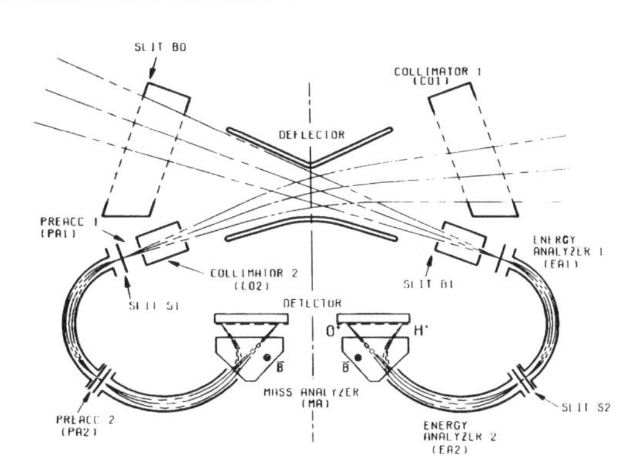

GHIELMETTI AND SHELLEY (1988)

Fig. 2b. Schematic diagrams of double-focusing Mattauch-Herzog type spectrographs described in Moore (1977), Coplan *et al.* (1984), and Ghielmetti and Young (1987). In the plane out of the page, Moore's geometry is flat, whereas that of Coplan *et al.* is cylindrical and that of Ghielmetti and Young is toroidal. The TIMAS/Polar design uses a non-standard toroidal focusing configuration.

DOUBLE-FOCUSING/ELEVATION VIEWS

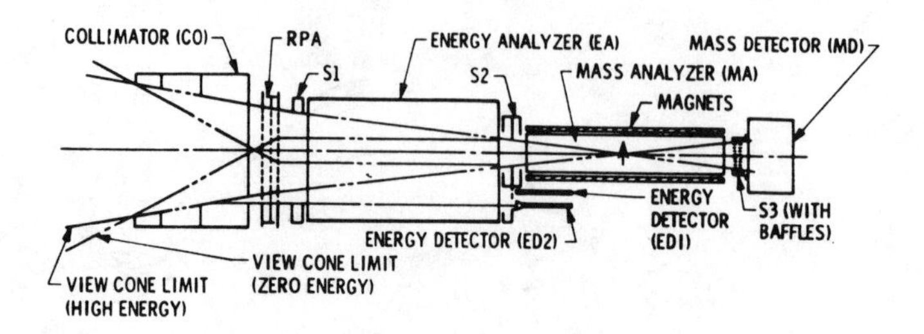

GEOS–CYLINDRICAL SPECTROMETER

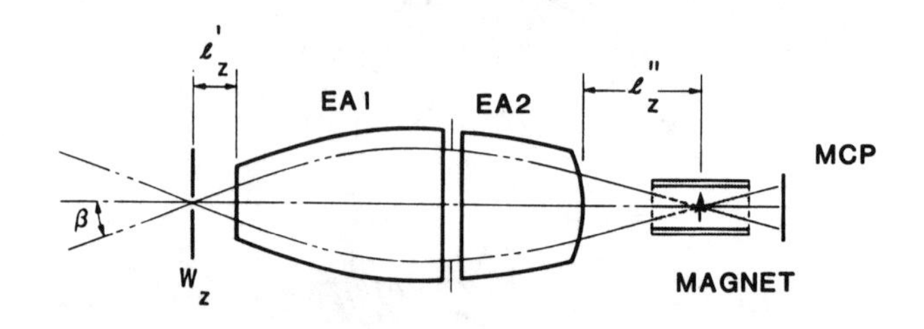

TIMS–TOROIDAL SPECTROGRAPH

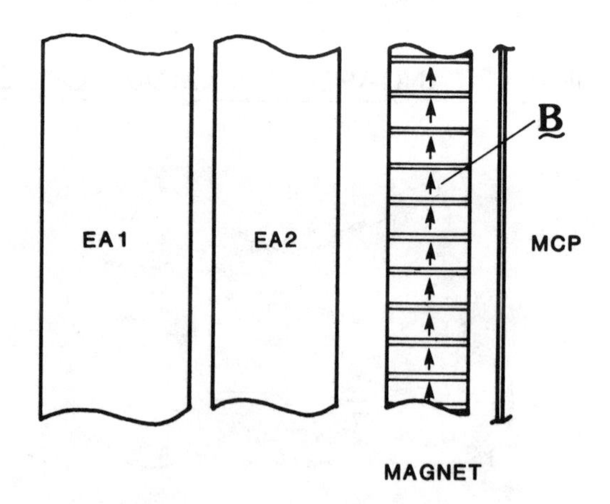

MAGNET

TIMAS–360 Deg. TOROIDAL SPECTROGRAPH

Fig. 2c. "Rolled out" elevation views of the principal instrument types shown in Figures 2a and 2b. The Geos, ISEE 1, and FIMS instruments are primarily of cylindrical geometry; TIMS is toroidal with β direction acceptance of $\sim \pm 30°$; TIMAS is fully 360° in "β" viewing, achieved in part by making the magnetic field structure periodic in the β-direction.

TIME-OF-FLIGHT SPECTROMETERS

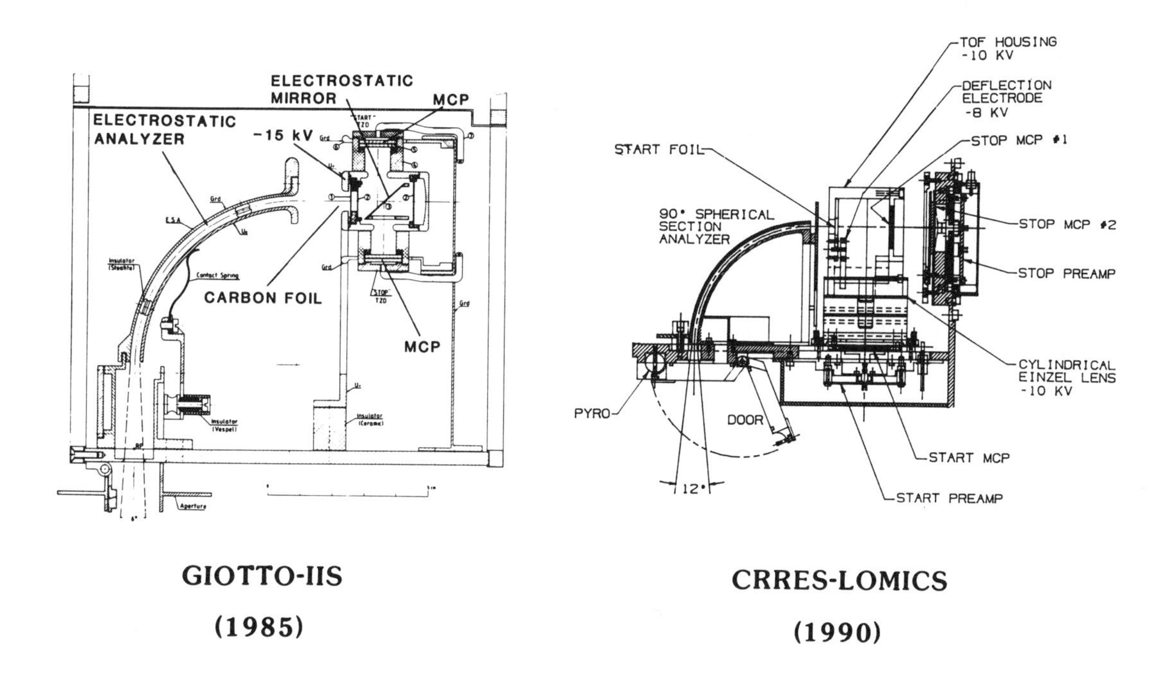

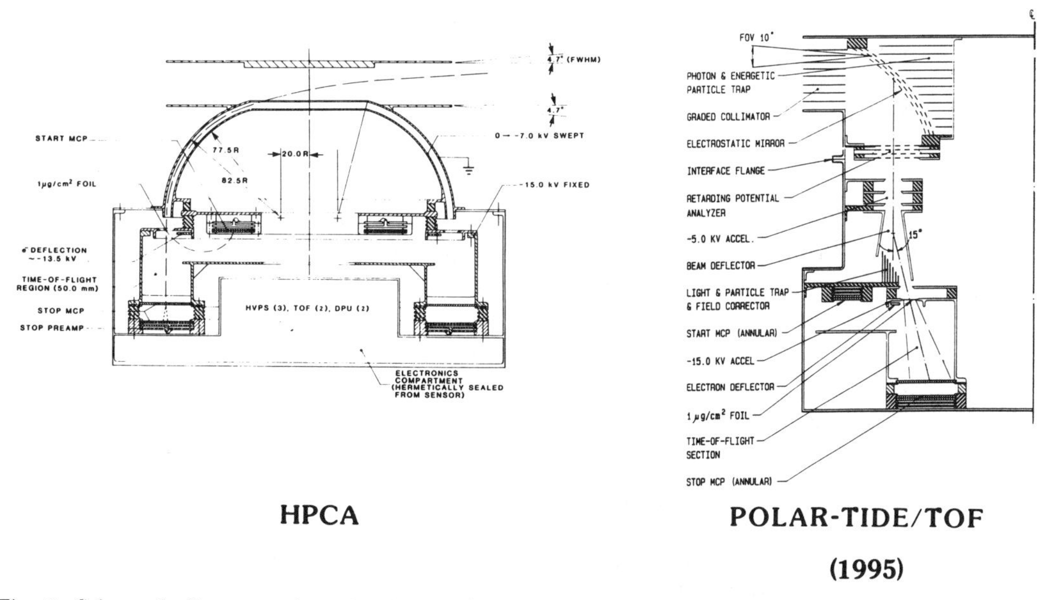

Fig. 3. Schematic diagrams of one family of E/Q-TOF analyzers. As in Figure 2a, cross sections are in the energy-dispersive plane of the instruments. The Giotto/IIS is described in Wilken *et al.* (1987), the CRRES-LOMICS in Fritz *et al.* (1985), and the HPCA in Young *et al.* (1989). The Polar-TIDE/TOF is described in Chappell *et al.* (1988).

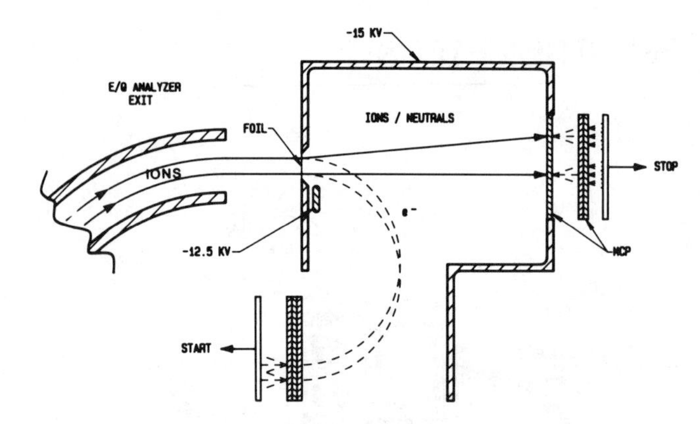

Fig. 4. Principle of the HPCA TOF analyzer showing location of the carbon foil, TOF region at -15 kV potential, and start and stop MCPs.

cuitry. Ions (or neutrals if the incident ion exits in this state) proceed through the TOF region and strike the first section of the stop-MCP. Electrons from this section are accelerated across the -15 kV potential and strike the remainder of the stop-MCP stack. The stop-pulse terminates the TOF measurement and initiates processing of the TOF signal.

Without going into further detail, it can be seen that the TOF method provides a nearly instantaneous measurement of ion E/Q and TOF which can be converted within an instrument into M/Q information. There is no M/Q scanning, as in a spectrometer design. A second advantage is that the foil takes the place of the object slit used in magnetic instruments. Since the foil width is generally $\gtrsim 5$ mm whereas a typical object slit is $\lesssim 1$ mm, there is a clear gain in throughput with the TOF system. Finally, because the system requires a coincidence between start and stop pulses within the longest TOF time (about 100 ns), it provides good noise rejection, e.g., against penetrating background radiation. (There is also circuitry protecting against certain false events as well as start pulses without valid stops, etc.) The TOF system shown in Figure 4 readily lends itself to a rotationally symmetric design capable of viewing the full exit aperture of a 360° FOV E/Q analyzer as shown in Fig. 1b (Young *et al.*, 1989; Bame *et al.*, 1989).

Towards a Quantitative Assessment of Instrument Quality

We now want to build on this brief survey of satellite-borne mass spectroscopy and try to specify an instrument quality factor, Q. The idea of a Q factor is not current in the field of space particle measurements but is used fairly extensively in laboratory particle optics (Wollnik, 1971; Wang, 1972; Heddle, 1971). Although Q factors were not introduced explicitly, Decreau *et al.* (1975), Balsiger *et al.* (1976), and Ghielmetti and Young (1987) have all discussed the optimization of ion optical designs for satellite use.

Perhaps the most commonly used measure of a space plasma analyzer is its "geometric" factor: the bigger, the better. As generally conceived, geometric factor is the product of instrument sensitive area and acceptance solid angle. The "energy-geometric" factor also includes energy resolution, $\Delta E/E$, in this product. Although geometric factor often serves as an indicator of potential instrument performance in tenuous hot plasmas (e.g., the central plasma sheet and ring current) it may fail elsewhere. For example, in solar wind and beam-like distributions instrument attributes of energy and angular resolution are often more important (Neugebauer, 1989). The common measure of a mass spectrometer is its mass resolution, $M/\Delta M$. For a given instrument size, geometric factor and mass resolution are inversely related. Herein lies the core of the problem in space plasma spectroscopy. How can these factors be brought into harmony to produce a design that has "enough" geometric factor with "just the right" resolution, and so on? That is the subject of this section.

In Table 1 we list measurement parameters that characterize most instrument designs. We have introduced definitions of energy, mass and angular resolution in the form of $E/\Delta E$, $M/\Delta M$, and $1/\Delta a$, respectively, rather than their reciprocals, which are more widely used. This definition of mass resolution is common (Duckworth and Ghoshal, 1963) and we introduce $E/\Delta E$ and $1/\Delta a$ to be consistent. Thus the concept of a "high resolution" device, in the sense of a measurement which shows finer detail, is consistent with this definition.

Table 2 gives the usual list of design parameters and then attempts to list, on one hand, instrument characteristics and, on the other, plasma characteristics that are most important in defining that particular parameter. Table 3 lists instrument resource requirements and the major elements of instrument design which affect them most strongly. Essentially all of the items listed in Tables 2 and 3 must be considered by the experiment designer in an interactive fashion which can be touched on here only briefly.

One cannot expect to optimize an instrument precisely with respect to so many parameters. We will attempt to show instead how the major components of performance can be arranged to produce a Q factor for two classes of instruments: (1) the spherical section E/Q analyzer and (2) an ion composition analyzer.

Q-Factor for the E/Q *Analyzer*

Using expressions developed by Gosling *et al.* (1978) we can write for the energy-geometric factor, G_E, of a spherical section analyzer

$$G_E = A_e \langle \Delta a \frac{\Delta E}{E} \rangle \int_{\Delta\beta} cos\beta d\beta \qquad (1)$$

where A_e is the analyzer entrance aperture area, Δa and $\Delta\beta$ are acceptance in a and β (Figure 1a) and the average $\langle \Delta a \, \Delta E/E \rangle$ is taken over the full acceptance response of the analyzer. An approximation for the latter is given by Gosling *et al.*

TABLE 1. Desired Parameter Coverage and Resolution for Magnetospheric Plasma Composition Analyzers

Parameter	Desired Range		Units	Desired Resolution[a]		Units
	Lower	Upper		Highest	Lowest	
Energy/charge[b]	$\lesssim 1$	$\sim 60,000$	eV/e	33	3	eV/eV
Mass/charge[c]	1	32	amu/e	16	1.5	amu/amu
Angular	–	4π	sr	$60/\pi$	$3/\pi$	rad^{-1}
Temporal	–	–		1	0.03	Hz
Flux	< 1	$> 10^9$	$(cm^2\, sr\, s\, eV)^{-1}$	–	–	
Density	$< 10^{-3}$	$> 10^3$	cm^{-3}	–	–	

[a] Energy resolution is $E/\Delta E$; mass resolution is $M/\Delta M$ defined at 1% of peak height, angular resolution is $1/\Delta a$; temporal resolution refers to full phase-space coverage.

[b] Lower limit defined by spacecraft potential. Upper limit is given as 100 keV in ESA (1985) but is generally around 60 keV for instruments discussed here.

[c] Upper limit is set by molecular species ($O_2{}^+$) seen in terrestrial magnetosphere. Heavier species such as CO^+, SO^+, or even Ba^+ can be accommodated by some designs.

$$\langle \Delta a \frac{\Delta E}{E} \rangle = \frac{1}{4K^2} csc^3 \frac{\theta_o}{2} \left[\frac{7}{8} + cos \frac{\theta_o}{2} \right] \quad (2)$$

where θ_o is the "local" bending angle of the analyzer and $K = R_o/\Delta R$ is the analyzer constant for a spherical section of mean radius R_o and plate spacing ΔR. Strictly speaking, the above expressions are valid only for a spherical section analyzer with a flat entrance aperture subtending an arc $\psi \lesssim 10°$ (Figure 1a). However, Carlson *et al.* (1983) argue that this formula

TABLE 2. Determinants of Instrument Parameters

Design Parameter		Primary Instrument Determinants	Primary Plasma Drivers
Geometric factor		Slit or TOF foil width, analyzer arc length and relative plate spacing, plate radius	Available flux, abundance of minor species
Energy	range	Maximum voltage, absolute plate spacing	Temperature, bulk velocity
	resolution	Plate arc length and relative spacing, slit widths	Temperature, Mach number
Angular	range	Analyzer geometry and orientation	Anisotropy
	resolution	Plate arc length and relative spacing, detector spatial resolution	Anisotropy
Mass	range	Magnet strength, voltage range, detector size	Species present
	resolution	Magnet strength, focusing, energy and angle resolution, TOF foil-thickness, TOF pre-acceleration voltage	Species present
Temporal resolution		Geometric factor, voltage sweep rate, signal/noise	Total flux, temporal and spatial fluctuations, relative plasma-spacecraft velocity

TABLE 3. Drivers of Instrument Resources

Instrument Resource	Primary Resource Drivers
Mass	Geometric factor, high voltage stand-off distances, FOV, mass resolution capability, onboard data processing, radiation shielding, special mechanisms (dust covers, etc.)
Power	High voltage levels, voltage sweep rate, onboard computation speed, TOF circuitry, MCP detector area and strip current, special mechanisms (e.g., motors)
Telemetry	Number of pixels in phase-space volume, onboard processing and compression, burst memory
Reliability	Mechanisms, electronic complexity, structural robustness, high voltage design, test history
Cost	State-of-the-art technologies, quality assurance program, complexity of design

also provides a reasonable approximation for spherical axisymmetric ("top-hat") analyzers and Young *et al.* (1988) present similar arguments for its application to mildly toroidal configurations.

Using equations (1) and (2) one can easily show that the geometric factor of this type of analyzer is proportional to

$$G \sim \frac{R_o^2}{K^3} \psi_o f(\theta_o) \overline{\Delta\beta} \tag{3}$$

where $f(\theta_o)$ is given by equation (2), $\overline{\Delta\beta}$ is the integral in equation (1), and the entrance area A_e is given by

$$A_e = \Delta R R_o \psi_o = \frac{R_o^2}{K} \psi_o \tag{4}$$

The work of Gosling *et al.* (1978), Paolini and Theodoridis (1967) and others shows that the angular and energy resolutions, $1/\Delta a$ and $E/\Delta E$, are related and, for an analyzer with a large bending angle and analyzer constant, i.e., with $\theta_o \gtrsim 90°$ and $\Delta R \ll R_o$, we have approximately

$$\frac{1}{\Delta a} \sim \frac{1}{2}\frac{\Delta E}{E} \sim \frac{R_o}{\Delta R} \tag{5}$$

Since angular and energy resolution are inextricably linked we refer to the product $\langle \Delta a \Delta E/E \rangle^{-1}$ as the "resolution" of a spherical (or mildly toroidal) analyzer. Again note that resolution is used here in a sense reciprocal to its common usage.

We now require an expression for a quality factor, Q, which embodies optimum performance features. In general, for a spectrometer we require the highest possible signal (count rate) for a given resolution, where the latter is chosen by consideration of the particle populations under study. We therefore tentatively put forward a Q factor

$$\text{Q} = \text{geometric factor} \times \text{resolution} \tag{6}$$

and require that it be a maximum for a given analyzer. Equation (6) is quite similar to expressions developed by Wollnik (1971), Steckelmacher (1973), and Heddle (1971) for laboratory ion and electron optical devices. One element is still missing from this expression however: to be useful for our purpose, a Q factor must express something of the unique optimization requirements faced by spaceflight instruments. This suggests including instrument mass as a parameter. If we consider that the mass of analyzer plates scales as their surface area, then mass $\sim R_o^2$. Typically as analyzer volume increases so does the area of detectors required to cover the exit aperture, provided the analyzer is scaled so that resolution is constant. It is therefore arguable that total sensor mass also varies roughly as R_o^2. Then for Q factor we have instead of equation (6) for an electrostatic analyzer (ESA)

$$\text{Q}_{\text{ESA}} = \frac{\text{geometric factor} \times \text{resolution}}{\text{analyzer mass}}$$

and from equations (2) and (3) we obtain

$$\text{Q}_{\text{ESA}} = \frac{A_e}{A_s} \cdot \overline{\Delta\beta} \tag{7}$$

Note that the expression $\langle \Delta a \Delta E/E \rangle$, which appears in both the geometric factor and resolution, has cancelled out. Because instruments are flown with a variety of polar angle acceptances we rewrite the above equation as Q factor per unit of polar angle

$$\frac{\text{Q}_{\text{ESA}}}{\overline{\Delta\beta}} = \frac{A_e}{A_s} \tag{8}$$

From this perspective the optimum analyzer for any given resolution is the one that maximizes entrance area per unit of analyzer plate surface area. Thus it is easily seen that the symmetric analyzers described by Carlson *et al.* (1983), Sablik *et al.* (1988) and Young *et al.* (1988) offer a sizable gain in Q over earlier flat-aperture designs. Consider the following comparison between a 90°-bending angle device of the latter type, with a symmetric analyzer of $\sim$ 75° effective bending angle (see Young *et al.*, 1987 for a discussion). The two sets of analyzers have qualitatively similar fields-of-view (neglecting the $cos\theta$ loss of transmission at the edges of the flat-aperture field-of-view) and resolution (for equal values of K) but A_e is $\sim 2\pi R_o \Delta R cos\theta_o$ for the symmetric analyzer and $2R_o \Delta R \psi_o$ for two of the standard analyzers (we take two of the standard analyzers in order to give roughly comparable FOVs and hardware size). Note that $A_e \to 0$ as $\theta_o \to 90°$ for the symmetric analyzer because the "aperture" closes up as $\theta_o \to 90°$. Thus if we include the polar angle transmission as written in equation (7)

$$\frac{Q_{\mathrm{sym}}}{Q_{\mathrm{flat}}} \sim \frac{\pi cos\theta_o \Delta\beta_{\mathrm{sym}}}{\psi_o \Delta\beta_{\mathrm{flat}}} \tag{9}$$

Now Gosling *et al.* have shown that $\psi_o > 10°$ in total aperture width produces appreciable skewing of the energy-angle transmission passband in flat-aperture analyzers. Taking this as a typical limiting value and using for symmetric analyzers $\theta_o \sim 75°$, then

$$\frac{Q_{\mathrm{sym}}}{Q_{\mathrm{flat}}} \sim \frac{\pi^2 cos75°}{0.3} = 8.5 \tag{10}$$

Thus the relatively simple expedient of using a symmetric analyzer in place of the more conventional spherical section improves measurement capabilities by nearly an order of magnitude for the same quality measurement in terms of angular and energy resolution and for the same weight of instrumentation. An added bonus is freedom from the energy-angle skewing in transmission that occurs with conventional spherical analyzers (Carlson *et al.*, 1983; Sablik *et al.*, 1988).

Q Factor for Mass Spectroscopy

If we now apply a similar analysis to space mass spectroscopy, several other considerations enter, and, because the complexity of the instrument necessarily increases, it becomes more difficult to express a Q factor in closed form. Obviously mass resolution $M/\Delta M$ must be included, but also the concept of signal-to-noise (S/N) ratio and the issues of spectrometry vs. spectrography and overall FOV become important. As mentioned earlier there are currently two distinct types of satellite-borne mass spectrographs that must be considered as well: devices based on magnets and those relying on time-of-flight.

Resolution in mass spectroscopy is generally defined as the ratio of image dispersion to image width. In time-of-flight spectroscopy dispersion is temporal rather than spatial; however, the principle is basically the same. Analysis of magnetic spectrometer optics (Duckworth and Ghoshal, 1963) shows that, to first order, mass resolution for a focusing instrument can be written

$$\frac{M}{\Delta M} = \frac{R_m}{s_1 + s_2 + R_m(\Delta E/E) + I.E.} \tag{11}$$

where s_1 and s_2 are object and image slit widths, R_m is the radius of curvature of the central ray in the magnetic field, and $I.E.$ represents image errors due to aberrations. In spectrographs the image "slit" width becomes the pixel size of the imaging device. Clearly $M/\Delta M$ improves with increasing R_m (which implies a larger, heavier magnet) and decreasing slit size (which causes geometric factor to decrease proportionately). Certain aberrations can be reduced by making homogeneous field regions larger and more uniform with respect to beam cross section and by making fringe fields correspondingly smaller. On the other hand, some aberrations increase with increasing R_m. In any case increasing R_m results in increased instrument size and weight.

Analytical expressions for resolution exist for rf and gated time-of-flight spectrographs (Wollnik and Matsuo, 1981). Because of scattering and straggling, the use of thin foils introduces a stochastic element to the calculation of resolution (Gloeckler and Hsieh, 1979). Generally,

$$\frac{M}{\Delta M} = \frac{1}{2}\frac{T_o}{\Delta T} = \frac{1}{2}\left[\sum \frac{1}{\delta T_i^{\,2}}\right]^{\frac{1}{2}} \tag{12}$$

where δT_i are individual terms of the form $\Delta T/T_o$ (Young *et al.*, 1989). Here T_o is the mean time-of-flight through the analyzer and ΔT is the spread in flight times introduced by any of several effects: energy dispersion caused by the finite passband of the E/Q analyzer and by energy straggling in the foil, angular dispersion caused by the same two effects, statistical noise introduced by timing electronics and, finally, scatter in the trajectories of electrons collected from the surfaces generating start and stop signals. These terms are assumed to be statistically independent so that the δT_i's add as the root sum of squares.

It is worth pointing out that there is a class of TOF analyzers in which the transmitted beam is focused in energy-angle and time (so-called triply focused or isochronous TOF systems, cf. Poschenrieder, 1972; Wollnik and Matsuo, 1981; Sakurai *et al.*, 1985a,b) and certain δT_i's in (12) are thereby reduced. This effect is analogous to double focusing in magnetic mass spectroscopy except that conditions are sought in which temporal variations among the transmitted ion trajectories are minimized. The TOF designs discussed here have an inherent peak width due to scattering in the foil that is used to generate the start of the TOF measurement; however, there is no reason why the principles of triple focusing cannot be applied to foil systems as well (Young *et al.*, 1987c). Other approaches to improving resolution include increasing the TOF path length, which is the basic source of particle dispersion, but this can be done only at the expense of instrument size and weight. Carbon foils can be made thinner (e.g. $< 1.0\,\mu\mathrm{g\,cm^{-2}}$) and acceleration voltages higher (up to $\sim$ 50 kV) but again at the expense of reliability, power, weight, and so on.

If we turn to S/N ratios we encounter a concept that is much more critical to M/Q than it is to E/Q analysis. There are two aspects to S/N: overall background noise and spurious or "ghost" mass peaks. In the latter we include not only separate pseudo-mass peaks, but also "shoulders," "tails," and similar imperfections in peak shape.

Background in both types of satellite-borne composition instruments originates in a number of sources:

- internal scattering of radiation incident on the instrument aperture: this includes UV, ions outside the analyzer E/Q passband, and electrons;
- field emission of electrons from internal analyzer surfaces and detectors;
- intrinsic detector background (dark current);
- penetrating fluxes of very high-energy radiation: electrons ($\gtrsim$ 1 MeV), ions ($\gtrsim$ 100 MeV), and bremsstrahlung;

- scattering of ions within the E/Q passband of the analyzer but outside the M/Q region of interest;
- electronics noise including random TOF coincidences as well as false mass peak image registration in magnetic spectrographs.

The sources of background and ghost peaks in the two types of analyzers are not entirely dissimilar: background refers to randomly occurring spurious signals spread across the entire M/Q spectrum, whereas ghosts are due to correlated signals showing up at a given M/Q value to create a spurious peak, often due to scattering in angle and energy within the instrument. In very carefully designed magnetic analyzers it is possible to reduce ghost peaks to $< 5 \times 10^{-5}$ (Balsiger *et al.*, 1976) of the main peak amplitude. On the other hand, background due to penetrating radiation within the Earth's radiation belts at L $\simeq$ 2.5–4.5 may be so severe as to preclude measurements by magnetic analyzers that do not have anti-coincidence schemes. TOF analyzers using triple coincidence systems have excellent background rejection but the price paid for this is relatively low efficiency. Wilken and Studemann (1984) give a good discussion of the origin of background and ghosting in TOF systems.

Measurement cycle time is another element in our discussion of space plasma mass spectroscopy. Here we may distinguish between true spectrometers, in which the M/Q spectrum is scanned with a certain duty cycle, and spectrographs, in which all ion species are measured simultaneously. Magnetic spectrographs with photographic plates as recording media are among the oldest instruments in laboratory mass spectroscopy. Imaging MCP technology has made spectrograph designs feasible in space. TOF instruments should properly be classed as spectrographs as well: although ion species are not imaged simultaneously they are sampled at random and each event is processed in a period which is usually fairly short, i.e., $\lesssim 50\,\mu s$ being typical and $\sim 2\mu s$ being feasible with recent "flash" analog-to-digital converters (J. Cessna, private communication, 1988).

Instrument cycle speed is determined by several considerations. By a "cycle" we understand a full sample of four-dimensional phase space consisting of two angles, energy and mass. The class of spectrographs under discussion here generally image in one angle and mass and rely on "scanning" to collect information in the second angle and energy. Mass spectrometers such as the Geos-type must scan the mass dimension as well. As mentioned earlier, angular scanning is achieved through spacecraft rotation or other methods and thus has a certain period associated with it; most commonly on magnetospheric missions this is either one half or one spacecraft spin period. Energy scanning is accomplished by rapidly varying analyzer plate voltages at rates up to $\sim 3 \times 10^5$ V/s^{-1}. One usually attempts to obtain a sufficient number of energy sweeps (e.g., 8, 16, or 32) in a single satellite spin to avoid having gaps in instrument coverage caused by the finite FOV in the plane orthogonal to the spin axis (Figure 1). Once a strategy is devised for sampling energy and angle there are still two items to be considered: (1) obtaining a statistically significant sample of the particle distribution (e.g., $\sim 10^3$ to 10^4 counts per distribution sample, corresponding to statistical errors of $\sim$ 3% to 1% respectively); and (2) finding sufficient electrical power to drive high voltage supplies and on-board data processing and control logic, at the chosen rate. Of course the requirement for a statistically significant counting sample is equivalent to the requirement on geometric factor, whereas item (2) impacts both available power and electronics technology. For example, the power required for commonly used CMOS logic circuitry tends to be proportional to the speed at which it is run. Although the actual instrument sampling rate may be fairly slow (e.g., $\sim$ 100 Hz), the computational (and electrical) power required to translate four-dimensional phase-space measurements into more transmittable data such as moments or compressed spectral samples may be formidable. Data compression ratios of 100:1 are not unheard of.

The overall cycle time of a spectrometer-type instrument may be written

$$T_{C1} = T_A \cdot T_E \cdot T_M$$

where T_A is the angle scan period, and T_E and T_M the energy and mass scan periods, respectively. A spectrograph that images or otherwise samples all masses simultaneously gives simply

$$T_{C2} = T_A \cdot T_E$$

So there is a gain by a factor T_M over spectrometers.

A final qualitative factor in mass spectroscopy is the instantaneous FOV. Recall that this factor entered our discussion of the E/Q analyzer in the form of the imaged (polar) angle β. Mass analyzers represent an instrument type in which historically the FOV has been small (a) because rays admitted at large β angles are difficult to handle in systems that are non-imaging in that direction (e.g., due to ion scattering) and (b) because of aberrations introduced by large angles. It is significant, therefore, that the generation of mass resolving instruments represented at this conference (Lundin *et al.*, 1989; Young *et al.*, 1989) and planned for new missions (Shelley *et al.*, 1988) has 360° fields-of-view. We therefore argue that if one demands a 360° FOV device, then FOV effectively enters the Q factor twice: once in the geometric factor and a second time because, for Cluster-type instruments, an instantaneous 2π-radian measurement is of significantly greater value than a unidirectional measurement made with an equally large geometric factor. In fact an instrument with a restricted FOV, no matter how large its geometric factor, would not meet today's requirements for fully three-dimensional measurements. However, as noted above, this consideration should be recognized as a trade-off that works against other, possibly desirable, attributes such as small image errors and high angular resolution.

If we now put all of these effects together, we have for a Q factor

$$Q_{MS} = \frac{\text{geometric factor} \times \text{energy-angle-resolution} \times \text{mass resolution} \times \text{S/N} \times \text{FOV}}{\text{instrument mass} \times \text{cycle time}}$$

TABLE 4. Comparison of Mass Spectrometer and Mass Spectrograph Q Values

Parameter	Mass Spectrometer (Goes-type)	TOF Spectrograph (HPCA-type)	Units
Geometric factor	3×10^{-4}	5×10^{-2}	cm^2 sr eV/eV
Efficiency	0.3	0.2	counts/ion
$M/\Delta M$	4	4	amu/amu
$E/\Delta E$	16	5	eV/eV
$1/\Delta a$	9	7.5	rad^{-1}
S/N	?	?	–
FOV (β)	0.17	6.28	rad
Mass	5.0	8.0	kg
Cycle time	360	2	s
Q–factor	4.8×10^{-6}	0.6	cm^2 sr(kg s)$^{-1}$

It is of interest to use this Q factor to compare double focusing spectrometers flown in the 1970's on Geos, ISEE 1 and other spacecraft with 360° FOV time-of-flight spectrographs currently under development. Table 4 summarizes characteristics of each at an energy ~ 5 keV/e which is above typical magnetic analyzer pre-acceleration voltages. As might be expected, the 360° FOV time-of-flight analyzer comes out well ahead. However, despite all of the above arguments, Table 4 should not be taken too literally: one design is not necessarily five orders of magnitude better than the other and would not lead to that order of magnitude improvement in scientific knowledge. In reality they are different classes of instruments designed in different eras for different mission requirements. We do argue, however, that in some sense a significant, qualitative improvement in mass spectroscopy has taken place and is available for future missions. The reader should also bear in mind the down side of tradeoffs made to achieve these designs. In the end, the real usefulness of Q factor analysis may lie in comparisons of instruments which are fairly similar and tradeoff decisions are perhaps less clear-cut.

Toward a Q Value for Spacecraft

In a field where instruments are so strictly constrained by spacecraft resources that designs are pared to the nearest gram and milliwatt, it is worth inquiring, if only briefly, into the nature of these constraints. The outcome of this inquiry will be that, for whatever reasons, the design of scientific spacecraft, as measured by resources allocated to the scientific payload, has not advanced at anything approaching the rate of the instruments discussed above.

What are the performance characteristics by which we might judge spacecraft design? To first order they are the weight, power, and telemetry rate made available to the instrument payload and the cost (in dollars) to the community at large. Because of the difficulty of obtaining costs, we neglect this factor and, in Table 5, list other factors for a number of repre-

TABLE 5. Spacecraft Payload Resources

Spacecraft	No.	Power (W)			Weight (kg)			TLM*	
(year)	Exps.	S/C	Exp	Exp/S/C	S/C	Exp	Exp/S/C	(kbits/s)	Q
ESA/Geos 1,2 (1977,1978)	9	110	40	0.36	303	40	0.13	12.0	0.042
ISEE 1 (1977)	13	?	76	$\lesssim 0.5$	340	72	0.26	16.4	< 0.18
Giotto (1985)	10	200	51	0.26	574	59	0.10	20.0	0.023
CRRES (1990)	20	300	198	0.66	1500	254	0.17	14.5	0.10
Cluster (1995)	8	143	47	0.33	426	47	0.11	20.0	0.033

* Telemetry rates exclude wideband wave and optical data for GEOS and Giotto respectively. The ratio $TLM_{Exp}/TLM_{s/c} = 0.9$ has been adopted for all examples in this table.

sentative spacecraft flown in the past ten years or planned for the next ten. The table gives payload mass, power, telemetry (TLM), and a Q value for spacecraft using the following criteria which, we freely admit, could be the subject of some debate

$$Q_{s/c} \sim \frac{P_{exp} \times M_{exp} \times TLM_{exp}}{P_{s/c} \times M_{s/c} \times TLM_{s/c}}$$

Here the values with subscript "exp" refer to requirements for the entire experiment payload. Thus $Q_{s/c}$ is simply the ratio of resources available to experiments vs. that of the entire spacecraft. The best possible spacecraft would be one in which $Q_{s/c} = 1$, with all resources devoted to the payload.

It is clear from Table 5 (as it is intuitively clear to many space experimentalists) that the essential quality of resources provided by spacecraft has increased very little or not at all during the 20-year period covered by our discussion (1975–1995). This situation might be acceptable if the product had become cheaper or more readily available, but this is simply not the case. The fact is that a two-spacecraft mission of $400 million planned for the 1990's offers no better performance to the experimenter than a similar mission (ISEE 1, 2) costing $100 million during the 1970's. Even if the spacecraft performance figures used here are incorrect by factors of 2 to 3, one is still left with an enormous gap between the increased performance of plasma instruments and the lagging performance of modern spacecraft. Thus the burden of improved performance needed to meet present and future science objectives is more and more pushed onto the experimenter. Power and telemetry provide an excellent case in point: A mission like Cluster demands 2-s time resolution for full spectrum measurements. Therefore, larger geometric factors are found, the sampling process is speeded up, techniques like TOF are introduced (with heavy demands for high speed analog circuitry), and so on. And what does one do with these data? Telemetry rates are about the same (Table 5) so the number-crunching requirement within experiments increases drastically, to the point where computation rates are very high. This approach in turn requires more experiment power and weight (including radiation shielding for the relatively "soft" high speed components), which are simply not available from present-day spacecraft.

Conclusions

It is my contention that plasma composition measurement capability has improved significantly with the introduction of both TOF and magnetic 360° FOV mass spectrographs, even though these are not necessarily the only tradeoffs that can be made in the quest for "better" mass spectroscopy. Planned instruments seem to be fully capable of answering the stated measurement needs of the next-generation scientific spacecraft. Furthermore, there appear to be methods to quantify the quality of these, or any other, plasma particle instruments. I suggest that a study of Q factors should be undertaken to provide an objective and systematic basis for judging instrument designs and tradeoffs. Finally, it appears that scientific spacecraft, or at least those made available for magnetospheric research, have evolved very little over the time period under discussion (1975–1995). This situation places an increasing burden on instrument designers and experiment teams and could eventually stifle the advancement of space plasma physics research.

Acknowledgments. This review has grown out of discussions with many colleagues over many years, most notably A. Ghielmetti, J. Burch, S. Bame, E. Shelley, J. Cessna and N. Eaker to whom I am grateful. In particular I wish to thank A. Ghielmetti and J. Burch for critical readings of the manuscript. I also thank L. Spink and R. Spinks for typing and acknowledge the support of Southwest Research Institute and the U.S. Department of Energy.

References

Balsiger, H., P. Eberhardt, J. Geiss, A. Ghielmetti, H. P. Walker, D. T. Young, H. Loidl, and H. Rosenbauer, A satellite-borne ion mass spectrometer for the enrgy range 0 to 16 keV, *Space Sci. Instrum., 2*, 499, 1976.

Balsiger, H., Recent developments in ion mass spectrometers in the energy range below 100 keV, *Adv. Space Res., 2*, 3, 1983.

Bame, S. J., R. H. Martin, D. J. McComas, J. L. Burch, J. A. Marshall, and D. T. Young, Three-dimensional plasma measurements from three-axis stabilized spacecraft, this volume, 1989.

Carlson, C. W., D. W. Curtis, G. Paschmann, and W. Michael, An instrument for rapidly measuring plasma distribution functions with high resolution, *Adv. Space Res., 2*, 67, 1983.

Carlson, C. W., Three-dimensional plasma measurement techniques, invited paper presented at 1988 Yosemite Conference on Outstanding Problems in Solar System Plasma Physics: Theory and Instrumentation, Yosemite National Park, California, February 2-5, 1988.

Chappell, C. R., et al., The Thermal Ion Dynamics Experiment (TIDE), Marshall Space Flight Center proposal, Huntsville, Alabama, January 1988.

Coplan, M. A., J. H. Moore, and R. A. Hoffman, Double focusing ion mass spectrometer of cylindircal symmetry, *Rev. Sci. Instrum., 55*, 537, 1984.

Decreau, P., R. Prange, and J. J. Berthelier, Optimization of toroidal electrostatic analyzers for measurement of low energy particles in space, *Rev. Sci. Instrum., 46*, 995, 1975.

Duckworth, H. E., and S. N. Ghoshal, High resolution mass spectroscopes, in *Mass Spectrometry*, edited by C. A. McDowell, p. 201, McGraw-Hill, New York, 1963.

European Space Agency, Future Scientific Program Study Office, "Cluster-Study in Three Dimensions of Plasma Turbulence and Small-Scale Structure", ESA SCI(85)8, European Space Agency, Noordwijk, Holland, 1985.

Fritz, T. A., D. T. Young, et al., The mass composition instruments (AFGL-701-11) on CRRES, in *CRRES/SPACERAD Experiment Descriptions*, edited by M. S. Gussenhoven, E. G. Mullen, and R. C. Sagalyn, p. 127, AFGL-TR-85-0017, Air Force Geophysics Laboratory, Hanscom AFB, Massachusetts, 1985.

Ghielmetti, A. G., and D. T. Young, A double focusing toroidal mass spectrograph for energetic plasmas: I. First order theory, *Nucl. Instrum. Meth., A258*, 297, 1987.

Gloeckler, G., and K. C. Hsieh, Time-of-flight technique identification at energies from 2 to 400 keV/nucleon, *Nucl. Instrum. Meth.*, *165*, 537, 1979.

Gloeckler, G., *et al.*, The Charge-Energy-Mass (CHEM) spectrometer for 0.3 to 300 keV/e ions on AMPTE-CCE, *IEEE Trans. Geosci. Remote Sensing, GE-23*, 234, 1985.

Gloeckler, G., Measurements of the charge state and mass composition of hot plasmas, suprathermal ions and energetic particles using the time-of-flight vs. energy technique, invited paper presented at 1988 Yosemite Conference on Outstanding Problems in Solar System Plasma Physics: Theory and Instrumentation, Yosemite National Park, California, February 2-5, 1988.

Gosling, J. T., J. R. Asbridge, S. J. Bame, and W. C. Feldman, Effects of a long entrance aperture upon the azimuthal response of spherical section electrostatic analyzers, *Rev. Sci. Instrum.*, *49*, 1260, 1978.

Hahn, S. F., J. L. Burch, and W. C. Feldman, Development of a fast ion mass spectrometer for space research, *Rev. Sci. Instrum.*, *52*, 247, 1981.

Heddle, D. W. O., A comparison of the étendue of electron spectrometers, *J. Phys. E: Sci. Instrum.*, *4*, 589, 1971.

Hovestadt, D., *et al.*, The nuclear and ionic charge distribution particle experiments on the ISEE-1 and ISEE-C spacecraft, *IEEE Trans. Geosci. Electron., GE-16*, 166, 1978.

Lundin, R., S. Olsen, A. Zakharov, E. Dubinin, N. Pissarenkko, and R. Pellinen, General description of the ASPERA experiment on the Soviet Phobos spacecraft, this volume, 1989.

McEntire, R. W., E. P. Keath, D. E. Fort, A. T. Y. Lui, and S. M. Krimigis, The medium-energy particle analyzer (MEPA) on the AMPTE CCE spacecraft, *IEEE Trans. Geosci. Remote Sensing, GE-23*, 230, 1985.

McFaddin, J. P., R. P. Lin, and C. W. Carlson, A 4π steradian angle-imaging particle detector, invited paper presented at 1988 Yosemite Conference on Outstanding Problems in Solar System Plasma Physics: Theory and Instrumentation, Yosemite National Park, California, February 2-5, 1988.

Neugebauer, M., Ion spectrometers for studying the interaction of the solar wind with nonmagnetic bodies, this volume, 1989.

Paolini, F. R., and G. C. Theodoridis, Charged particle transmission through spherical plate electrostatic analyzers, *Rev. Sci. Instrum.*, *38*, 579, 1967.

Poschenrieder, W. P., Multiple-focusing time-of-flight mass spectrometers Part II. TOFMS with equal energy acceleration, *Int. J. Mass Spectr. Ion Phys.*, *9*, 357, 1972.

Sablik, M. J., D. Golimowski, J. R. Sharber, and J. D. Winningham, Computer simulation of a 360° field-of-view "top-hat" electrostatic analyzer, *Rev. Sci. Instrum.*, *59*, 146, 1988.

Sakurai, T., T. Matsuo, and H. Matsuda, Ion optics for time-of-flight mass spectrometers with multiple symmetry, *Int. J. Mass Spectr. Ion Proc.*, *63*, 273, 1985a.

Sakurai, T., Y. Fujita, T. Matsuo, H. Matsuda, I. Katakuse, and K. Miseki, A new time-of-flight mass spectrometer, *Int. J. Mass Spectr. Ion Proc.*, *66*, 283, 1985b.

Shelley, E. G., R. D. Sharp, R. G. Johnson, J. Geiss, P. Eberhardt, H. Balsiger, G. Haerendel, and H. Rosenbauer, Plasma Composition Experiment on ISEE-A, *IEEE, Trans. Geosci. Electron., GE-16*, 266, 1978.

Shelley, E. G., D. A. Simpson, T. C. Sanders, H. Balsiger, and A. Ghielmetti, The energetic ion composition spectrometer (EICS) for the Dynamics Explorer-A, *Space Sci. Instrum.*, *5*, 443, 1981.

Shelley, E. G., A. Ghielmetti, E. Hertzburg, S. J. Battel, K. Altwegg-von Burg, and H. Balsiger, The AMPTE/CCE Hot-Plasma Composition Experiment (HPCE), *IEEE Trans. Geosci. Remote Sensing, GE-23*, 241, 1985.

Shelley, E. G., H. Balsiger, J. L. Burch, C. W. Carlson, J. Geiss, A. Ghielmetti, A. Johnstone, O. W. Lennartsson, G. Paschmann, W. K. Peterson, H. Rosenbauer, B. A. Whalen, and D. T. Young, Toroidal Imaging Mass-Angle Spectrograph Experiment for the Polar Spacecraft of the Global Geospace Science Mission, LMSC-D089358, Lockheed Missiles and Space Co., Inc., Palo Alto, California, February 1988.

Steckelmacher, W., Energy analyzers for charged particle beams, *J. Phys. E: Sci. Instrum.*, *6*, 1061, 1973.

Wang, K. L., Optimum transport in charged particle devices, *J. Phys. E: Sci. Instrum.*, *5*, 1196, 1972.

Wilken, B., T. A. Fritz, and W. Studemann, Experimental techniques for ion composition measurements in space, *Nucl. Instrum. Meth.*, *196*, 161, 1982.

Wilken, B., and W. Studemann, A compact time-of-flight mass-spectrometer with electrostatic mirrors, *Nucl. Instrum. Meth.*, *222*, 587, 1984.

Wilken, B., W. Weiss, W. Studemann, and N. Hasebe, The Giotto Implanted Ion spectrometer (IIS): Physics and technique of detection, *J. Phys. E: Sci. Instrum*, *20*, 778, 1987.

Wollnik, H., A "Q-value" for particle spectrometers, *Nucl. Instrum. Meth.*, *95*, 453, 1971.

Wollnik, H., and T. Matsuo, A Q-value for energy-focused, time-of-flight spectrometers, *Int. J. Mass Spectr. Ion Phys.*, *37*, 209, 1981.

Young, D. T., J. A. Marshall, J. L. Burch, T. L. Booker, A. G. Ghielmetti and E. G. Shelley, A double focusing toroidal mass spectrograph for energetic plasmas: II. Experimental results, *Nucl. Instrum. Meth.*, *A258*, 304, 1987a.

Young, D. T., A. G. Ghielmetti, E. G. Shelley, J. A. Marshall, J. L. Burch, and T. L. Booker, Experimental tests of a toroidal electrostatic analyzer,*Rev. Sci. Instrum.*, *58*, 501, 1987b.

Young, D. T., S. J. Bame, B. L. Barraclough, J. T. Gosling, D. J. McComas, M. F. Thomsen, J. L. Burch, J. A. Marshall, and J. H. Waite, Instruments to measure 3-Dimensional Distributions of Electrons and Mass-Resolved Ions on 3-Axis Stabilized Spacecraft Without the Use of Scan Platforms, Proposal No. ET-87-34, Los Alamos National Laboratory, November, 1987c.

Young, D. T., S. J. Bame, M. F. Thomsen, R. H. Martin, J. L. Burch, J. A. Marshall, and B. Reinhard, A 2π-radian field-of-view toroidal electrostatic analyzer, *Rev. Sci. Instrum.*, in press, 1988.

Young, D. T., J. A. Marshall, J. L. Burch, S. J. Bame, and R. H. Martin, A 360° field-of-view toroidal ion composition analyzer using time-of-flight, this volume, 1989.

REMOTE SENSING OF PLANETARY MAGNETOSPHERES: MASS AND ENERGY ANALYSIS OF ENERGETIC NEUTRAL ATOMS

K. C. Hsieh and C. C. Curtis

Department of Physics, University of Arizona, Tucson, AZ 85721

Abstract. Remote sampling of magnetospheric ions by the detection of energetic neutral atoms (ENAs) can provide a global view of the composition and dynamics of the ions. We describe a tandem time-of-flight analyzer in terms of its application to Earth and Saturn.

Introduction

The composition and transport of the energetic ion populations within planetary magnetospheres have been investigated for three decades. Phenomena at specific locations and times have been observed and frequently understood, but measurements involving in situ observations usually suffer the difficulty of separating temporal from spatial variations. Even with multiple-spacecraft missions, this dilemma persists. One solution is to sample the ion populations in different regions of the magnetosphere simultaneously. This requires observation from an exterior vantage point, i.e., remote sensing.

The carriers of information from the magnetically confined ions must be electrically neutral. The overlap regions of the planet's extended neutral atmosphere and its magnetosphere give birth to the necessary messengers. They are the energetic neutral atoms (ENAs), produced when the singly charged magnetospheric ions exchange charge with the ambient neutral atoms. In the case of Earth, it has long been recognized that charge exchange between the energetic ions and the exospheric and geocoronal H atoms is an effective means to dissipate the energies in the ring current during a magnetic storm (Dessler and Parker, 1959; Dessler et al., 1961). Early direct detection of keV H can be traced back to ground-based aurora observations by Meinel (1951) and rocket measurements by Wax et al. (1970), but these endeavors were not continued. Later observations in ion fluxes supported the role of charge exchange (Moritz, 1972; Hovestadt et al., 1972; Shelley et al., 1972; Smith et al., 1976). More recently, convincing observations of ENAs from Jupiter (Kirsch et al., 1981a), Saturn (Kirsch et al., 1981b), and Earth (Roelof et al., 1985; Roelof, 1987) have firmly established the importance of charge exchange in planetary magnetospheres and given credence to the possibility of remotely sensing planetary magnetospheres via ENAs.

Remote sensing via ENAs consists of determining the arrival direction, as well as the mass and energy of the escaping particles. The former is discussed by Curtis and Hsieh (1989). This report concentrates on the mass and energy analysis of ENAs.

Estimated ENA Flux

We first show what one can expect as the "typical" ENA flux, one from Earth and one from Saturn.

The ENA flux from Earth (Figure 1a) was estimated by taking the ring current ion fluxes measured by AMPTE (Gloeckler and Hamilton, 1988); the charge exchange cross sections measured by Freeman and Jones (1974), Phaneuf et al. (1978), and Olsen et al. (1977); the geocoronal model of Rairden et al. (1986); and by assuming a uniform distribution of the ion flux within the L = 3 and L = 7 shells. This simple model gives ~16 H atoms cm^{-2} s^{-1} between 10 and 100 keV, ~3 O atoms cm^{-2} s^{-1} between 60 and 200 keV, and ~9 x 10^{-2} He atoms cm^{-2} s^{-1} between 13 and 100 keV at L1 point, 1.5 x 10^6 km from Earth.

The estimated ENA flux from Saturn (Figure 1b) was taken from a model (Hsieh and Curtis, 1988) which used the Voyager ion measurements (Krimigis et al., 1983), the cross sections mentioned above, neutral H distribution (Hilton and Hunten, 1988), and an L-B dependence of the ions that qualitatively agreed with the observations of Voyager 2 (Armstrong et al., 1983). Since Voyager did not provide mass resolution, only the (H^+, H) interaction and the production of energetic H is considered. The estimates are compared with that of Cheng (1986) and the measurement of Kirsch et al. (1981b). The differences between the model and the measurement could be due to several factors, one being the inclusion of energetic O in the measurement, which would suggest an O spectrum similar to that in Figure 1a.

Instrument Description

Figure 2 shows the principal parts of the imaging spectrometer for ENA (ISENA) that could provide the necessary low threshold, high sensitivity, and adequate mass and energy resolutions. The imaging capability of the ISENA is discussed by Curtis and Hsieh (1989).

Arriving ENAs pass through the collimator, which serves as an ion trap when appropriate potentials are applied to adjacent stacks. (This reduces possible

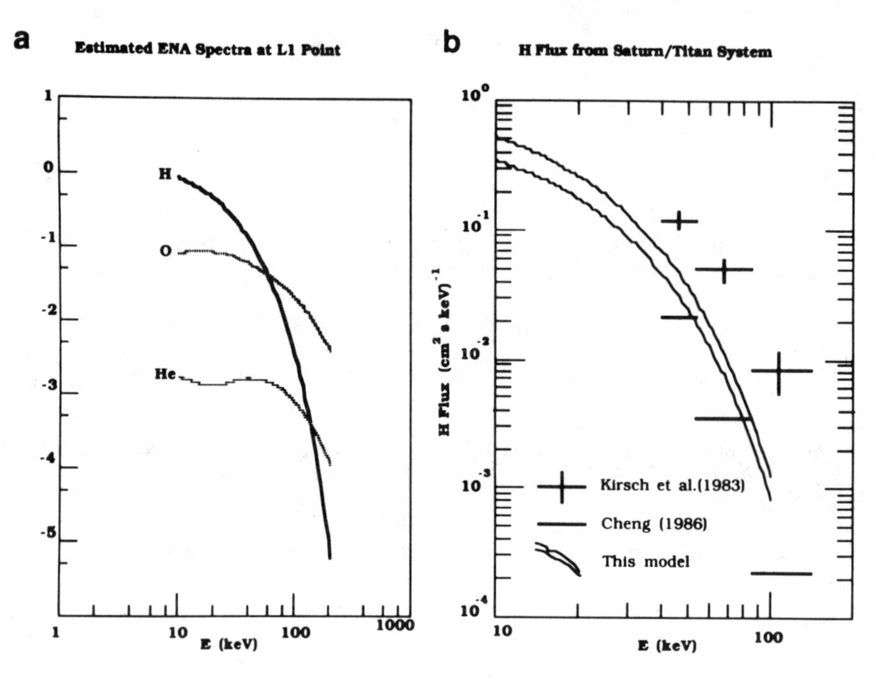

Fig. 1. Estimated ENA spectra for (a) Earth at L1 point and (b) Saturn at 60 R_S.

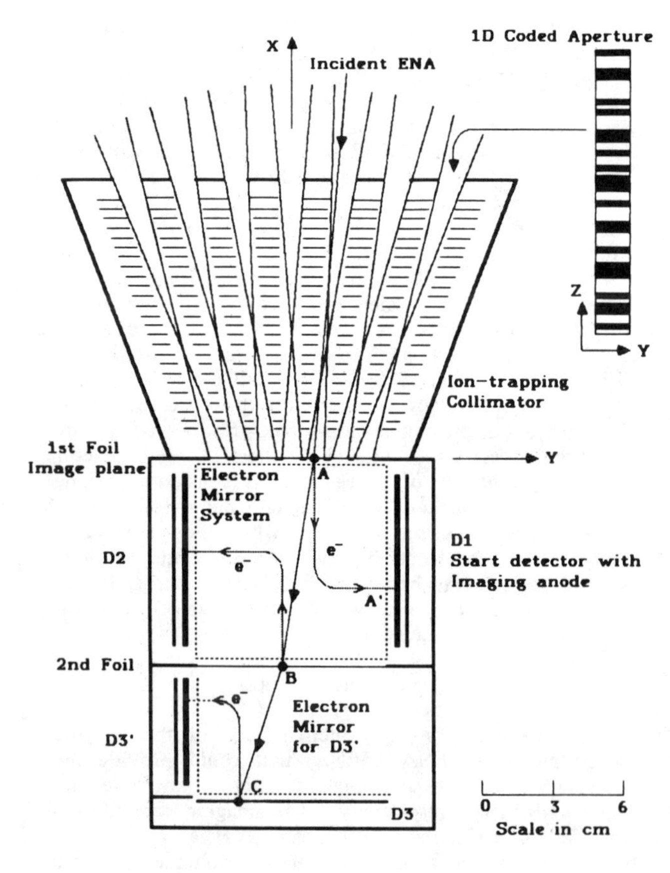

Fig. 2. Schematic drawing of the ISENA.

contamination by ions of the same mass and energy.) Undeflected by the electric fields, an incident ENA of sufficient energy penetrates the first foil and releases secondary electrons. A grid maintained at a positive potential (~200 eV) accelerates the electrons into an electron mirror, e.g., a 45° electrostatic mirror (Wilken and Stüdemann, 1984), which redirects them onto a microchannel plate (MCP) detector D1.

Meanwhile, the ENA emerges from the first foil in a scattered direction with less energy and perhaps an altered charge state. If it emerges with a positive charge, it must overcome the electron mirror barrier potential to reach the second foil. The secondary electrons generated there are accelerated and guided onto D2 by the same mirror system. The time elapsed between the D1 and D2 signals constitutes the first time-of-flight (TOF1) measurement of the incident ENA.

The D2 signal also serves as the start signal for the second time-of-flight (TOF2) measurement, if the incident particle has sufficient energy after penetrating the second foil to reach D3 within a pre-set time. D3 is a solid-state detector having thresholds ~30 keV for H and ~80 keV for O, after taking the foils into account. Particles of residual energies below the thresholds can still have their TOF2 registered by the secondary electrons emitted from the front surface of D3 and guided into D3', an MCP, in a manner similar to D1 and D2. Particles of higher residual energies generate signals in D3, in addition to their TOF1 and TOF2 measurements. The section that begins with D2 is similar to many of the existing TOF ion analyzers (e.g., Gloeckler and Hsieh, 1979; Wilken and Stüdemann, 1984; McEntire et al., 1985; Möbius et al., 1985). The advantage of 2TOF is the lowering of the thresholds for mass analysis.

The energy thresholds and intervals for analysis are set by the properties of the foils and the flight distances. The sensitivities, on the other hand, depend on the size of the detectors and the design of the electron mirror system, in addition to the choice of the first foil. One 2TOF design uses a first foil having ~80 Å Au on top of ~820 Å C, the second foil ~280 Å C, the first TOF distance 8.7 cm, the second 5.25 cm, and all three detectors 7 x 9 cm^2.

Using TRIM87, an evolving code being developed by Ziegler et al. (1985) for the penetration of particles in matter, we obtain an idealized response of 2TOF (Figure 3) to the two most likely ENA species for both Earth and Saturn. H between ~8 and ~11 keV and O between ~65 and ~70 keV are too slow to complete TOF2 within 256 ns. Their TOF2s are set to 256 ns, and they appear in the upper right corner of Figure 3. At higher energies, the particles will have both TOF1 and TOF2 <256 ns, thus leaving tracks in the TOF2-TOF1 plane. The "thickness" of the tracks shown in Figure 3 simulates scattering in the foils, which causes the exit direction to deviate as much as 15° from the incident direction. Such a spread is much larger than the uncertainties in the TOF electronics, typically ~1 ns, but much less than the effect of straggling. Figure 4 shows the relative spreads in TOF1 for the two species having similar TOF2s (15 keV H with 100 keV O, and 20 keV H with 200 keV O) when both scattering and straggling are included in the TRIM87 calculation. The track width (spread in TOF1) of H at 20 keV is markedly wider than that of O at 200 keV. Although they overlap, the presence of 20 keV H cannot escape detection, especially when its expected flux at 20 keV is orders-of-magnitude higher than the O flux at 200 keV (see Figure 1). At 15 keV, the H track is better separated from the Os of 100 keV. By having two TOFs, the measurable H spectrum is extended to ~10 keV and well-separated from O, where the Os (>80 keV) are clearly identified by their signals in D3. Because of the shape of the H spectrum (Figure 1), lowering the threshold also means raising the sensitivity of 2TOF.

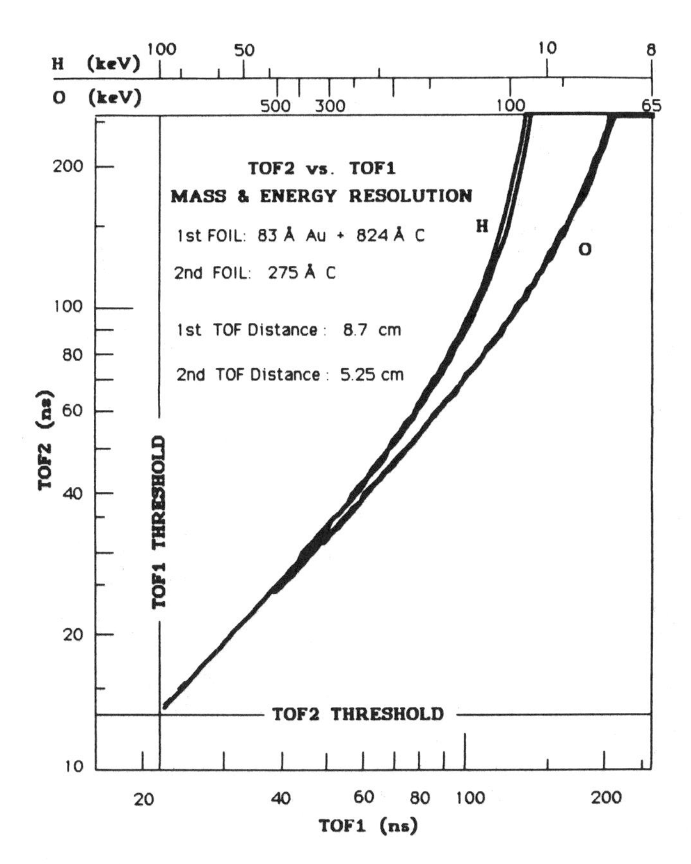

Fig. 3. The mass and energy resolution of the ISENA's 2TOF analyzer. The most likely ENA species H and O are identified by the loci in the TOF2-TOF1 plane.

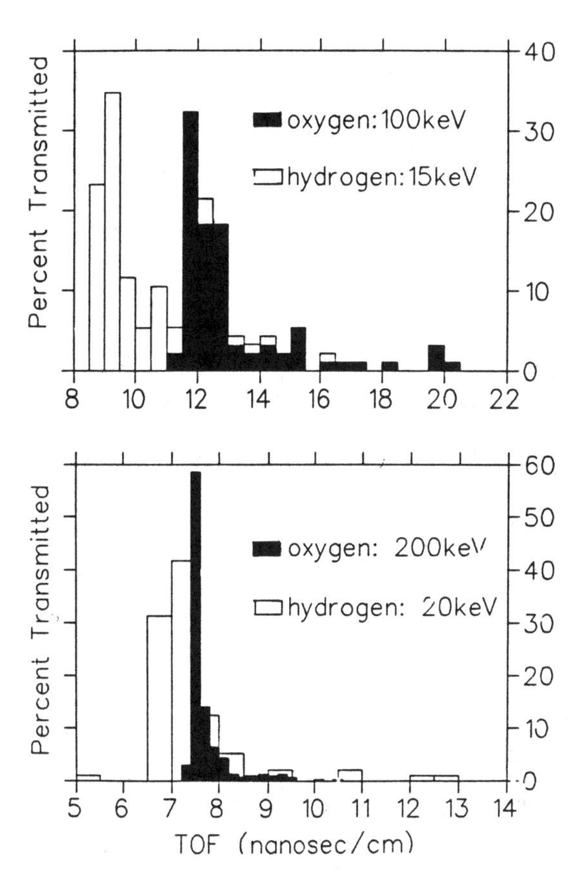

Fig. 4. Simulated separation of H from O in TOF1 when both scattering and straggling are accounted for.

The sensitivities of all TOF analyzers depend on the secondary electron yields from the foils and the transmission of the grids used for the electron mirrors. The estimated sensitivities for 2TOF are: 7% for 10 keV H, increasing to 19% for 100 keV H, and 27% for 70 keV O, rising to 50% for 400 keV. The higher sensitivities for O are due to the more efficient secondary electron yield by heavier particles. With these sensitivities, the dimension of the exposed areas of the first foil, hence, D1 (7 x (0.8 x 7.0) cm^2), and the spectra (Figure 1), we expect to detect in one second ~50 H between 10 and 100 keV and ~30 O between 65 and 200 keV in the case of Earth and ≳15 H between 10 and 100 keV in the case of Saturn.

These event rates are quite acceptable, especially when the noise/signal ratio is low. The ISENA takes appropriate measures against all three noise sources: (a) ions of mass and energy identical to ENAs, (b) random events due to ions of higher energies, cosmic rays, and detector noise, and (c) EUV from the planetary atmosphere and plasmasphere.

(a) Two mechanisms can reject the ions that mimic ENAs; one is natural and the other is part of the design. Remote sensing is done outside of the magnetosphere; the magnetic field greatly reduces the flux of those ions which could otherwise reach the ISENA. This enabled Kirsch et al. (1981a,b) to claim that what the Voyager ion detector observed before entering the magnetospheres of Jupiter and Saturn were ENAs. However, when the spacecraft is magnetically connected to the magnetosphere, an additional rejection mechanism is required. This is done by the collimator when adjacent stacks of baffles are biased at opposite potentials. For example, at +8 kV and -8 kV, the electric field in the narrow and long channels between the stacks of baffles can sweep away ions <160 keV/e, thus freeing the ENAs <160 keV/e from contamination.

(b) Unlike the ions of mass and energy identical to the ENAs, random noise events do not concentrate on the ENA tracks in the TOF2-TOF1 plane. The tandem TOF entails the most effective means to reject random noise, i.e., coincidence requirement for the true events. To have a TOF1 event requires a D1 signal followed by a D2 signal within a preset interval τ_1. Likewise, a TOF2 event requires a D2 signal followed by a D3' signal within a preset interval τ_2. These random events may appear in the TOF2-TOF1 plane, if they share the same D2 signal, i.e., a triple coincidence. As seen in Figure 3, τ_1 = 256-22 = 234 ns and τ_2 = 256-13 = 243 ns. The chance coincidence per second is then

$$C_3 \sim \tau_1 \tau_2 N_1 N_2 N_3 \sim 6 \times 10^{-14} N_1 N_2 N_3$$

where N_1, N_2, and N_3 are the random rates in the respective detectors. If all the random noise sources contribute even as much as 10^4 counts per second in each detector, the chance coincidence is still merely 0.06 s^{-1}, ~4 x 10^{-3} of the estimated ENA rate of 15 H s^{-1} in the case of Saturn and ~8 x 10^{-4} of the total estimated ENA rate in the case of Earth. In the region suitable for remote sensing, outside of the magnetosphere, there are not enough energetic ions and cosmic rays to produce even 100 counts s^{-1} in any of the detectors, except during solar flare events. Therefore, random noise poses no threat to our detection of ENAs.

(c) The real threat is the EUV photons. At 60 R_S from Saturn (Broadfoot et al., 1981; Sandel et al., 1982), we expect 8 x 10^7 photons s^{-1} to fall on the 34 cm^2 of exposed first foil. At L1 point of Earth (Rairden et al., 1986), we expect 4 x 10^6 photons s^{-1}. These photons can contribute to noise in two ways: (i) by production of secondary electrons in the thin foils, thus triggering the respective MCPs in the same manner as ENAs and (ii) by reflection from the interior structures and striking the MCPs directly. The latter is insignificant, because the most probable reflectors that can direct photons to the MCPs are the grid wires of the electron mirrors, which are extremely small in surface area and low in reflectance (<<10% at 1216 Å), and the MCPs' response to UV photons is much smaller than to the post-accelerated electrons. We consider then only the former. The obvious solution is attenuation by the proper choice of the first foil. For the sample given here, i.e., 83 Å Au on 824 Å C as the first foil, the transmission for 1216 Å photons is ~2 x 10^{-4} (Hunter, 1985; Lynch and Hunter, 1985; Hsieh et al., 1980). The photoemission yield from carbon and other thin films by 10 eV photons is ~10^{-2} electrons/photon in the backward direction (Feuerbacher and Fitton, 1972), and from carbon foils ~10^{-2} electrons/transmitted photon in the forward direction (Hsieh et al., 1980). The first foil then converts the 8 x 10^7 photons s^{-1} to ~2 x 10^2 secondary electrons s^{-1} into the electron mirror. Even if we ignore all other <1 efficiencies and the additional attenuation of the second foil for D3, the absolute maximum noise these electrons could cause in any of the detectors is 2 x 10^2 s^{-1}, which would only give a chance coincidence rate of 2 x 10^{-7} s^{-1}! Therefore, we not only can block the EUV flux, but can expect to use an even thinner first foil to lower the threshold without significantly raising the noise level.

Discussion

The threshold of any ENA analyzer should be low enough to detect H < 30 keV and O < 100 keV since they are more sensitive to the changes in the conditions of the magnetosphere and the flux is higher in favor of detection. However, the ISENA cannot push the detection of ENAs down to the region of corotating fast ions, <1 keV for H and <20 keV for O, so that the inner magnetosphere could be studied (ESA/NASA Assessment Study on Cassini, EAS Ref. SCI(85) 1, 1985). The thickness of the first foil, the EUV filter, stands in the way. Even assuming that the first foil could be thin enough to admit the low-energy particles, post-acceleration would be needed. This requires ionizing the incident neutrals, which is most effectively done by stripping in the first foil. For H, the efficiency might drop by an order of magnitude, depending on energy (Gloeckler and Hsieh, 1979). For O the problem is worse; the emerging particle could be in any of nine charge states with a probability distribution depending on energy. Post-acceleration would then seriously scramble the original spectrum.

The ISENA and other ENA energy analyzers must accept the limitations on the threshold and relinquish that portion of the spectrum and that region of space to other modes of observation. EUV imaging is an obvious candidate, since He^+ and O^+ emit in 304 Å and 834 Å, respectively. And 1216 Å emission provides information on the spatial distribution of H density, which is needed for ENA data interpretation.

Conclusion

Remote sensing of planetary magnetospheres by ENAs is a good way to investigate the composition and transport of the ions in the magnetosphere, complementing in situ observations. The mass and energy analysis capabilities of the ISENA, an imaging spectrometer for ENAs, were described. We showed that a tandem TOF analyzer 2TOF can (a) have sufficient sensitivity to detect the expected ENA fluxes, specifically, in the case of Earth and Saturn; (b) resolve the most likely candidates, H and O, with energy indentifications in the interval ~10 to 100 keV for H and ~60 to 400 keV for O; (c) effectively reject all expected sources of noise; and (d) even lower the energy threshold, below 8 keV for H and below 60 keV for O, without threatening the ENA analysis with the noise due to planetary EUV photons.

Acknowledgments. This paper is dedicated to Wolfgang Stüdemann, who first suggested the concept of a tandem TOF in 1984. We thank C. Y. Fan for helpful discussions. This work is in part supported by NASA grant NAG5-625.

References

Armstrong, T. P., M. T. Paonessa, E. V. Bell, II, and S. M. Krimigis, Voyager observations of Saturnian ion and electron phase space densities, J. Geophys. Res., 88, 8893-8904, 1983.

Broadfoot, A. L., et al., Extreme ultraviolet observations from Voyager 1 encounter with Saturn, Science, 212, 206, 1981.

Cheng, A. F., Energetic neutral particles from Jupiter and Saturn, J. Geophys. Res., 9, 4524-4530, 1986.

Curtis, C. C., and K. C. Hsieh, Remote sensing of planetary magnetospheres: imaging via energetic neutral atoms, these proceedings, 1989.

Dessler, A. J., and E. N. Parker, Hydromagnetic theory of geomagnetic storms, J. Geophys. Res., 64, 2239, 1959.

Dessler, A. J., W. W. Hanson, and E. N. Parker, Formation of the geomagnetic storm main-phase ring current, J. Geophys. Res., 66, 3631, 1961.

Feuerbacher, H. E., and B. Fitton, Experimental investigation of photoemission from satellite surface materials, J. Appl. Phys., 43, 1563, 1972.

Freeman, R. L., and E. M. Jones, Atomic collision processes in plasma physics experiments: Analytic expressions for selected cross-sections and Maxmillian rate coefficients, UKAEA Culham Laboratory CLM-R 137, 1974.

Gloeckler, G., and K. C. Hsieh, Time-of-flight techniques for particle identification at energies from 2 to 400 keV/nucleon, Nucl. Instrum. Methods, 165, 537, 1979.

Gloeckler, G., and D. C. Hamilton, AMPTE ion composition results, Physica Scripta, T18, 73-84, 1987.

Hilton, D., and D. M. Hunten, A partially collisional model of the Titan hydrogen torus, Icarus, 73, 248-268, 1988.

Hovestadt, D., B. Hausler, and M. Scholer, Observation of energetic particles at very low altitudes near the geomagnetic equator, Phys. Rev. Lett., 28, 1340, 1972.

Hsieh, K. C., and C. C. Curtis, A model for the spatial and energy distributions of energetic neutral atoms produced within the Saturn/Titan plasma system, Geophys. Res. Lett., in press, 1988.

Hsieh, K. C., E. Keppler, and G. Schmidtke, Extreme ultraviolet induced forward photoemission from thin carbon foils, J. Appl. Phys., 5, 2242-2246, 1980.

Hunter, W. R., Measurement of optical constants in the vacuum ultraviolet spectral region, in Handbook of Optical Constants of Solids, edited by E. D. Palik, pp. 69-88, Academic Press, 1985.

Kirsch, E., S. M. Krimigis, J. W. Kohl, and E. P. Keath, X-ray and energetic neutral particle emission from Jupiter: Voyager 1 results, Geophys. Res. Lett., 8, 169, 1981a.

Kirsch, E., S. M. Krimigis, W. H. Ip, and G. Gloeckler, X-ray and energetic neutral particle emission from Saturn's magnetosphere, Nature, 292, 718, 1981b.

Krimigis, S. M., J. F. Carbary, E. P. Keath, T. P. Armstrong, L. J. Lanzerotti, and G. Gloeckler, General characteristics of hot plasma and energetic particles in the Saturnian magnetosphere: results from the Voyager spacecraft, J. Geophys. Res., 88, 8871-8892, 1983.

Lynch, D. W., and W. R. Hunter, Comments on the optical constants of metals and an introduction to the data for several metals, in Handbook of Optical Constants of Solids, edited by E. D. Palik, pp. 275-367, Academic Press, 1985.

McEntire, R. W., E. P. Keath, D. E. Fort, A. T. Y. Liu, and S. M. Krimigis, The medium-energy particle analyzer (MEPA) on the AMPTE spacecraft, IEEE Trans. Geosci. & Remote Sensing, GE-23(3), 230, 1985.

Meinel, A. B., Doppler-shifted auroral hydrogen emission, Ap. J. 113, 50-54, 1951.

Möbius, E., et al., The time-of-flight spectrometer SULEICA for ions of the energy range 5-270 keV/charge on the AMPTE IRM, IEEE Trans. Geosci. & Remote Sensing, GE-23(3), 274, 1985.

Moritz, J., Energetic protons at low equitorial altitudes: A newly discovered radiation belt phenomenon and its explanation, Z. Geophys., 38, 701, 1972.

Olson, R. E., A. Salap, R. A. Phaneuf, and F. W. Meyer, Electron loss by atomic and molecular hydrogen in collisions with $^3He^{++}$ and $^4He^+$, Phys. Rev. A, 16, 1867-1872, 1977.

Phaneuf, R. A., F. W. Meyer, and R. H. McKnight, Single-electron capture by multiply charged ions of carbon, nitrogen, and oxygen in atomic and molecular hydrogen, Phys. Rev. A, 17, 534-545, 1978.

Rairden, R. L., L. A. Frank, and J. D. Craven, Geocoronal imaging with Dynamics Explorer, J. Geophys. Res., 91, 13,613, 1986.

Roelof, E. C., Energetic neutral atom image of a storm-time ring current, Geophys. Res. Lett., 14, 652, 1987.

Roelof, E. C., D. G. Mitchell, and D. J. Williams, Energetic neutral atoms (E~50 keV) from the ring current, IMP 7/8 and ISSE-1, J. Geophys. Res., 90, 10,91, 1985.

Sandel, B. R., et al., Extreme ultraviolet observations from Voyager 2 encounter with Saturn, Science, 215, 548-553, 1982.

Shelley, E. G., R. G. Johnson, and R. D. Sharp, Sattelite observation of energetic heavy ions during a geomagnetic storm, J. Geophys. Res., 77, 6104, 1972.

Smith, P. H., R. A.Hoffman, and T. A. Fritz, Ring current proton decay by charge exchange, J. Geophys. Res., 81, 2701, 1976.

Wax, R. L., W. R. Simpson, and W. Bernstein, Large fluxes of 1-keV atomic hydrogen at 800 km, J. Geophys. Res., 75, 6390, 1970.

Wilken, B., and W. Stüdemann, A compact time-of-flight mass spectrometer with electrostatic mirrors, Nucl. Instrum. Methods, 222, 587-600, 1984.

Ziegler, J. F., J. P. Biersack, and U. Littmark, The stopping range of ions in solids, in The Stopping and Ranges of Ions in Matter, Vol. 1, edited by J. F. Ziegler, Pergamon Press, 1985.

INSTRUMENTATION FOR ENERGETIC NEUTRAL ATOM IMAGING OF MAGNETOSPHERES

E. P. Keath, G. B. Andrews, A. F. Cheng, S. M. Krimigis, B. H. Mauk, D. G. Mitchell, and D. J. Williams

The Johns Hopkins University Applied Physics Laboratory, Johns Hopkins Road, Laurel, MD 20707

Abstract. The imaging of magnetospheres using energetic neutral atoms (ENAs), produced by charge exchange between energetic ions and neutral atoms, is a promising new experimental technique for investigating the global dynamics of hot magnetospheric plasmas. We will give here an example of one of several different ENA detector designs being considered for flight and describe the design and testing of several critical subsystems of this detector.

Introduction

Conventional spacecraft observations of charged particles, electromagnetic fields, and plasmas are essentially single point measurements that sample only a very small region at any one time. Thus while individual components of magnetospheres (for example, the plasma sheet, ring currents, aurora, etc.) have been relatively well surveyed as separate entities, they are poorly understood as interacting parts of a whole. Recent measurements of energetic neutral particle emissions from the magnetospheres of Jupiter and Saturn (Kirsch et al., 1981a,b; Cheng, 1986) and from the Earth's ring current (Roelof et al., 1985) have raised the possibility of an innovative class of instruments devoted to the imaging of energetic neutral atoms (ENAs). These neutrals are created by charge exchange reactions between fast ions and neutral atoms or molecules in the magnetosphere. The resulting ENAs escape rectilinearly from the magnetosphere unaffected by the ambient electric, magnetic, and gravitational fields and can be used in the same way that photons are used in an optical imaging system.

The low-resolution images of the Earth's ring current region constructed by Roelof (1987) using data from the medium energy particle instrument (MEPI) (Williams et al., 1978) aboard the ISEE 1 spacecraft clearly demonstrate the effectiveness of ENA imaging as a tool for the study of magnetospheric plasma dynamics. The ISEE 1 instrument was designed as an energetic charged particle detector and has limited usefulness as an ENA imager since the count rate of the detector is dominated by charged particles over much of the spacecraft orbit. Also, the small geometric factor of the detector allows it to make useful measurements of ENAs only during periods of very high geomagnetic activity. Instrumentation designed specifically for the study of ENAs is clearly needed. To provide the maximum useful information, an ENA imager should possess the following characteristics. It should: (a) provide high-resolution images, (b) have a large geometric factor, (c) provide measurements of the ENA energy over the range of $\lesssim 20$ to $\gtrsim 500$ keV/nucleon, (d) provide identification of the ENA species, and (e) be insensitive to background interferences such as charged particles and extreme ultraviolet (EUV) photons.

Several instrument configurations are being evaluated to fit the needs of several different possible spacecraft configurations. In this report we will describe one instrument configuration and discuss the design and testing of several critical components of the detector, including the EUV and charged particle rejection system, and a position-sensitive microchannel plate (MCP) readout system.

ENA Imager

One ENA imaging instrument designed to meet the requirements defined above is shown in Figure 1. The instrument is designed for use aboard a spinning spacecraft and uses the spin of the spacecraft to scan one dimension of the ENA image. The parallel plates at the entrance of the detector serve the dual purpose of collimating the ENAs in the plane parallel to the spin axis of the spacecraft and eliminating incoming charged particles. The plates in the collimator are charged to alternate potentials of plus and minus several kilovolts so that charged particles are deflected to the plates while the ENAs pass straight through. After entering the detector the ENAs pass through two foils at the front and back of the detector. The secondary electrons that are produced as the ENAs pass through the foils are accelerated toward a central electrostatic mirror where they are reflected through 90° into the two position-sensitive MCP readouts. Special care was

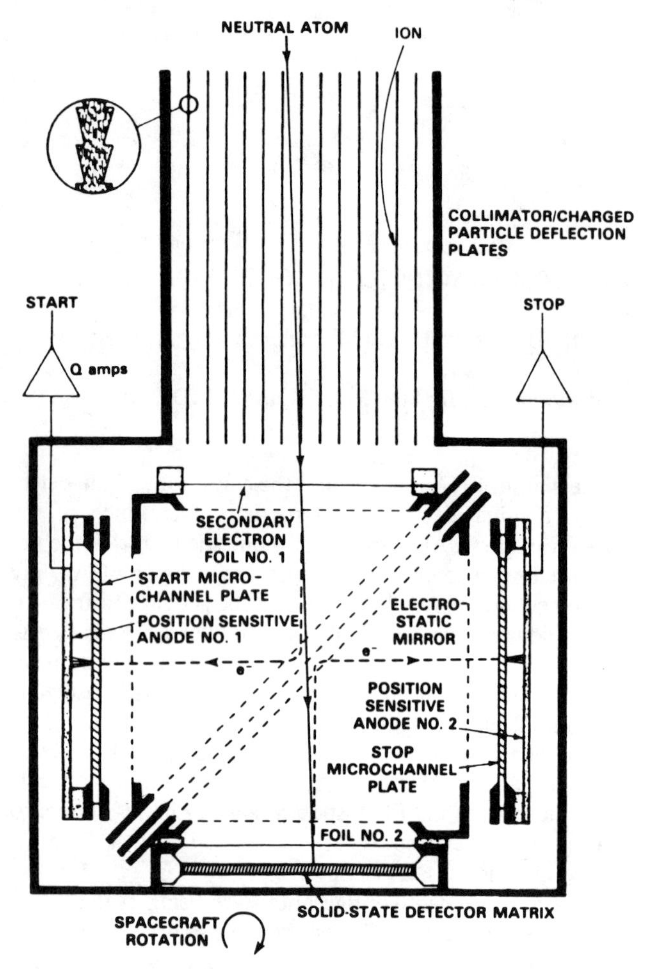

Fig. 1. Schematic diagram of the ENA imager. The unit consists of a collimator/charged particle deflector, an imaging TOF detector for measuring the arrival direction and velocity of the incoming neutral atoms, and a solid-state detector for measuring their energy.

taken in the design of the electrostatic mirrors to ensure a linear mapping between the location of the source of the electrons on the foil and their final destination on the surface of the MCP. By measuring the relative positions at which the ENAs cross the front and rear foils, the arrival direction of the particle in the plane of the collimators can be determined. These data, along with the spin phase of the spacecraft, provide a full description of the arrival direction of the ENAs. Since the total path length of the secondary electrons is independent of the source positions, time of flight (TOF) signals can be generated from the output of the MCPs and can be used to determine the velocity of the incoming ENAs. After they pass through the foils, the final energy of the ENAs is measured by a solid state detector. From the velocity and energy measurements, the mass of the ENAs can be computed. Many components of the detector system described here have been flown successfully in the medium energy particle analyzer (MEPA) detectors aboard the AMPTE Charge Composition Explorer satellite (McEntire et al., 1985).

Charged Particle Rejection Plates

The use of parallel plates at alternate positive and negative potentials to screen charged particles from the detector was motivated by the theoretical effectiveness of this method in screening out high-energy ions and electrons. Given a pair of parallel plates of length L spaced a distance D apart, the minimum energy particle E_m that will pass between the plates, assuming no scattering, is given by $E_m = [1 + (L/4D)^2]eV_p$, where V_p is the potential between the plates and e is the electron charge. For reasonable L/D ratios ($\gtrsim 20$), one might expect rejection of particles with energies of several hundred kiloelectron volts for potentials of only a few kilovolts between the plates. However, in designing a parallel plate system, it is not obvious what the degree of leakage through the plates will be, because of scattering and secondary electron generation following charged particle impacts on the plates. To determine the effectiveness of parallel plates

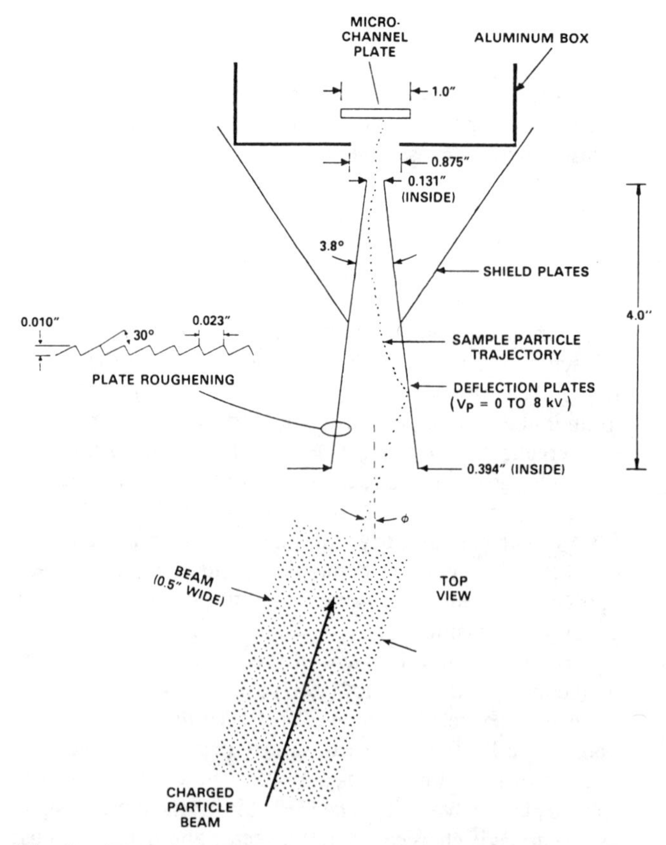

Fig. 2. Experimental setup for measuring the effectiveness of the charged particle rejection plates in screening incoming electrons and ions.

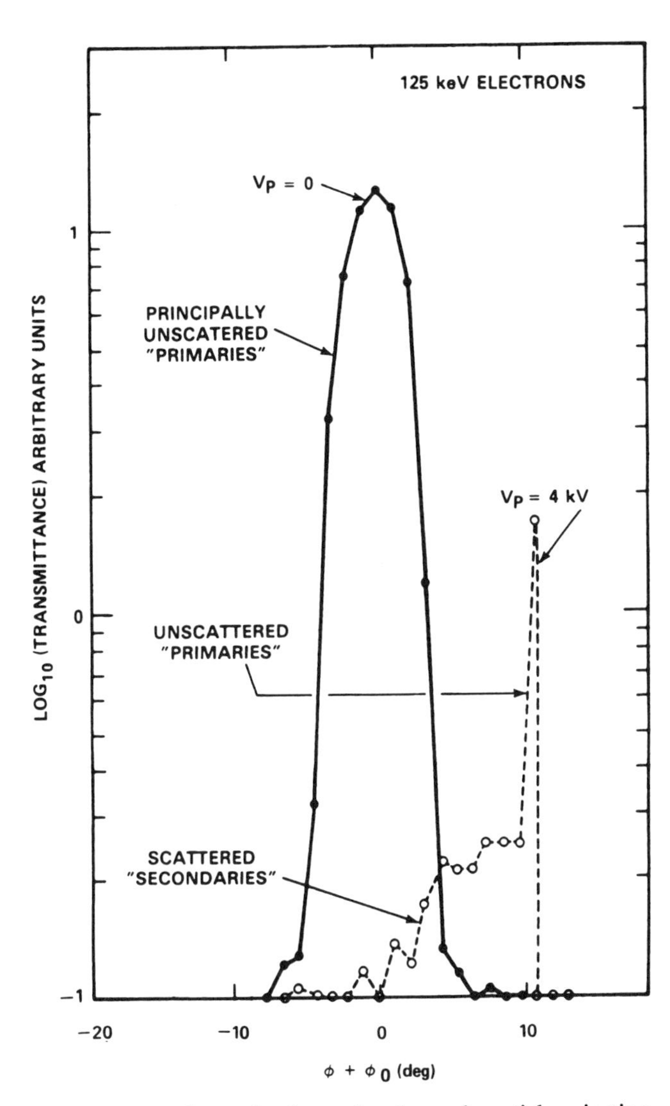

Fig. 3. Sample results from the charged particle rejection experiment shown in Figure 2. Shown is the MCP response as a function of ϕ for a beam of 125-keV electrons with 0 and 4 kV across the deflection plates. For $V_p = 4$ kV, two components appear: a sharp peak resulting from direct unscattered electrons and a more diffuse component corresponding to electrons scattered from the surface of the plates.

in blocking charged particles, a series of tests was run at the NASA Goddard Space Flight Center's accelerator to measure the transparency of the plates to both electrons and protons over a range of energies. A schematic of the test setup is shown in Figure 2. For these tests the plates were not parallel but were flared apart with an ~3.8° angle. Two sets of plates were tested. Grooves were machined in the surface of one set (see inset) to reduce the effects of scattering from the plate surfaces. The surfaces of the second set of plates were smooth. The deflection-plate/MCP assembly was mounted on a rotating table so that the plates could be scanned (angle ϕ) relative to the incoming beam.

Figure 3 shows a plot of the MCP response versus the angle ϕ for a 125-keV beam of electrons with potentials of 0 and 4 keV between the grooved plates. With no voltage ($V_p = 0$) between the plates (solid line), most of the response comes from unscattered electrons passing directly through the system. In the case where the voltage is applied, two components appear: a sharp peak corresponding to direct unscattered electrons, and a diffuse component due to scattered and/or secondary electrons. For a very slightly higher potential, the sharp peak of unscattered electrons disappears.

A summary of the results of the tests for the grooved plates is shown in Figure 4. Each point in the figure was generated by integrating under the transmittance versus ϕ curves illustrated in Figure 3, and then normalizing that integral by the $V_p = 0$ integrated response. The ratios are plotted as a function of the ratio of the particle beam energy per unit charge E/q and the plate potentials V_p. Also shown are the results of a Monte Carlo simulation of the system that includes no scattering or secondary electron effects. Good agreement between the simulation and the data is found, as expected, for high values of E/qV_p. At low values of E/qV_p ($\lesssim 20$), the effects of the secondary and scattered particles become apparent and the measured values deviate sharply from the simulation. To first order, the transmittance appears to follow a power law response at low values of E/qV_p as indicated by the dashed line. Figure 4 also shows the results for the smooth plates. These responses were normalized by the $V_p = 0$

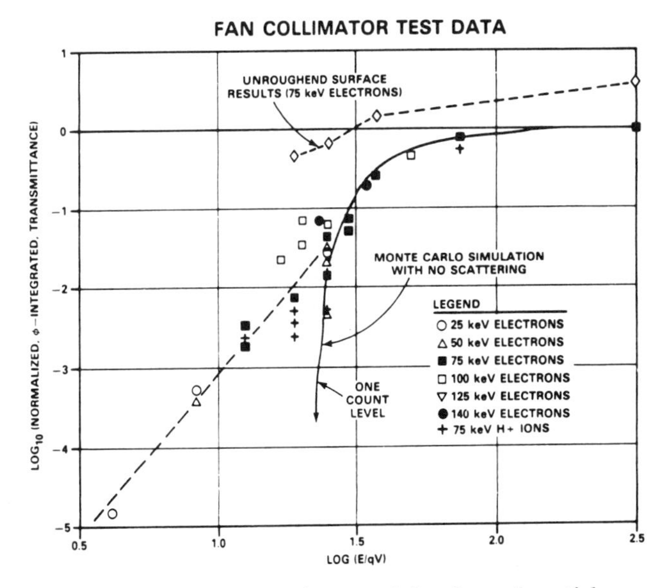

Fig. 4. Measured transmittance of the charged particle rejection plates shown in Figure 2 plotted as a function of the ratio of the particle energy and the potential between the plates. The transmittance exhibits roughly a power law response at low values of E/qV.

grooved runs. The scattered component is dominant (by a factor of 3) for the plates with smooth surfaces, even at $V_p = 0$.

Based on these results, it appears that reasonable potentials across the deflection plates (~10 kV) will provide adequate charged particle rejection in many magnetospheric regions.

EUV Rejection

The front foil of the detector is exposed directly to incoming EUV photons. Hsieh et al. (1980) has shown that EUV photons passing through thin foils induce forward photoelectrons with an efficiency of ~1%. Since Ly-α EUV photon fluxes of up to 2×10^9 photons $cm^{-2}\ sr^{-1}\ s^{-1}$ may be expected in the near-Earth environment (Rairden et al., 1986), the front foil must be thick enough to effectively screen out the incoming photons or the instrument must be able to electronically discriminate against the EUV flux. Since the scattering of ENAs in the front foil reduces the resolution of the imager, increasing the thickness of the foil to reduce the EUV transparency can be done at the expense of the resolution and energy range of the detector. The detector described here does effectively discriminate against EUV photons, since coincidence ($\lesssim$400 ns) outputs from both the front and back MCPs are required to trigger an ENA event while a single EUV photon will produce a response in only one of the MCPs.

To select the optimum foil material and thickness for use in the ENA imager, we have undertaken an evaluation of the mechanical and EUV properties of a variety of foil materials, including foils of Lexan, Lexan and carbon, and Lexan and aluminum. The results of EUV transmittance tests for several of the foils are shown in Figure 5. A schematic of the experimental setup used for testing the transmittance of the thin foils is shown in Figure 5a. Measurements were made at wavelengths of 116.4, 121.6, and 160.8 nm using the strong EUV emission lines from a hydrogen source. Figure 5b shows the transmission of Ly-α EUV as a function of thickness for four of the foils tested, together with similar data for pure carbon (Hsieh et al., 1980). It is clear that the Lexan foils are not as effective as the pure carbon foils. Figure 5c shows the transmittance of the foils as a function of wavelength. The data for pure carbon shown in the figure were also from Hsieh et al. (1980).

To compare the scattering characteristics of Lexan with those of carbon, we have used the approximation $\Psi_{1/2} \propto Z_1 Z_2/(Z_1^{3/2} + Z_2^{3/2})$ (Högberg et al., 1970), where $\Psi_{1/2}$ is the scattering half width, and Z_1 and Z_2 are the atomic numbers of the incident and target materials, respectively. With Z_1 fixed, we can make a rough comparison of the scattering properties of Lexan with those of carbon by using the composition of Lexan ($C_{16}H_{19}O_3$) and its density (1.2 g cm^{-3} vs. 1.35 for carbon) to approximate a given Lexan foil by the linear superposition of appropriate thicknesses of carbon, hydrogen, and oxygen and summing their contribution to $\Psi_{1/2}$. Using arguments by Högberg et al. (1970), we compute that the scattering half width for an incident 40-keV hydrogen atom on Lexan is ~0.6 that of hydrogen on carbon. If this ratio is borne out by experiment, the ratio of scattering to EUV attenuation for Lexan would be slightly more favorable (by ~5%) than that for pure carbon.

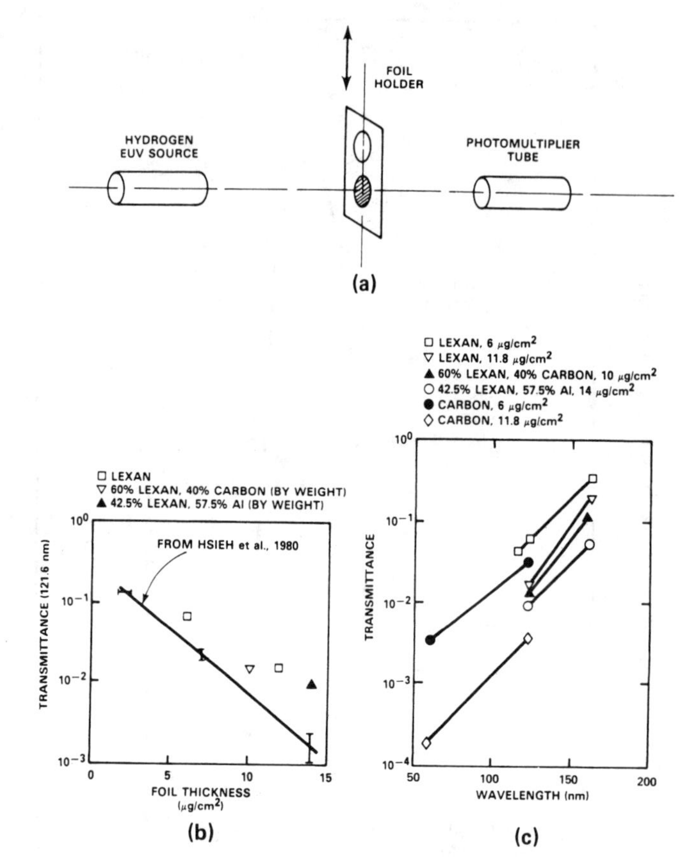

Fig. 5. Transmittance test for thin foils as a function of foil thickness and wavelength.

Position-Sensitive Readouts

The central element of the ENA imager is the position-sensitive detector used in the readout of the distribution of events in the image plane of the instrument. The key requirements for this detector include: (a) high linearity, resolution, and stability to provide accurate representation of the images, and (b) high speed to permit TOF measurements and operating capability in high flux environments. The detector chosen in this study consists of a MCP electron multiplier coupled to a discrete conductor segmented anode. Anodes for both one- and two-dimensional readouts have been fabricated. The anodes are based on the wedge and strip design proposed by Anger (1966) and discussed in detail by Martin et al. (1981).

A photograph of the one-dimensional anode is shown in Figure 6. This anode consists of two interleaved sets of graded-width strips. The individual widths of one set of strips increase linearly in one direction while the widths of the second set decrease in the same direction in such a way that the total width of the pair remains constant. If a charge distribution of 10^6 to 10^7 electrons from the MCP is deposited on the anode, the location of the centroid of the distribution can be obtained from the ratio $(Q_a - Q_b)/(Q_a + Q_b)$ where Q_a and Q_b are the charge signals from the two anodes. The anode has an active area of 12 × 60 mm. The strips vary linearly from 0.1 to 0.93 mm along the length of the anode and have an interelectrode spacing of 0.04 mm.

The inset in Figure 6 is a simplified schematic of the elements used in this design. In addition to the strips (A and B), which provide readouts of the y-coordinates of the events, the wedge pairs (C and D) provide readouts of the x-coordinates of the events. Since the widths of the wedges vary linearly with x, the x-coordinate of the event centroid is given by the ratio $(Q_c - Q_d)/(Q_c + Q_d)$.

The four-electrode design was chosen over the three-element design described by Martin et al. (1981) because it provides high-speed (<1 ns) output on all four electrodes. The four-electrode design does present a problem in connecting all of the elements without using jumper wires. In the present design, connections to the wedges are made to a common electrode on the back of the anode using laser-drilled, 0.13-mm via holes through the ceramic substrate at the wide end of each wedge. The two-dimensional anode has an active area 53 mm in diameter. Each strip and wedge pair is 1 mm wide and the interelectrode spacing is 0.02 mm. To our knowledge, such four-electrode anodes have not been previously constructed.

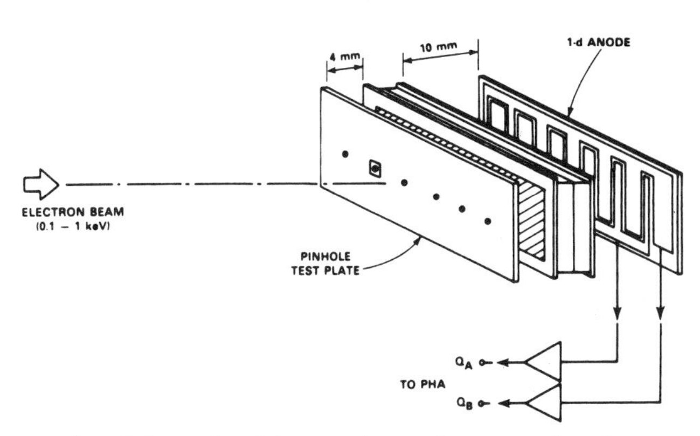

Fig. 7. Schematic of the experimental setup used to demonstrate the resolution of the one-dimensional, position-sensitive anode. The pinhole test plate contains 0.5-mm holes spaced at 4-mm intervals.

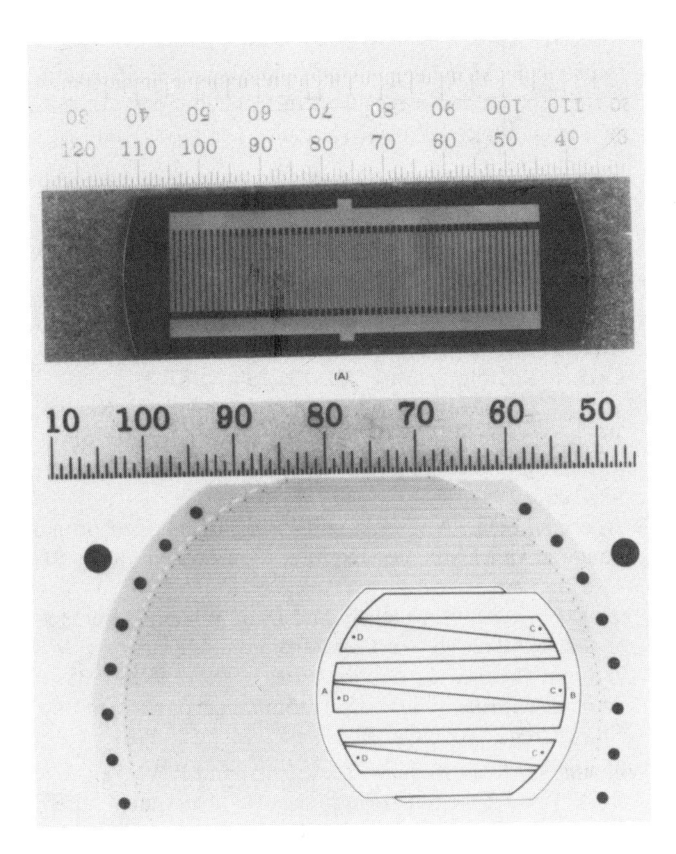

Fig. 6. Photograph of the one- and two-dimensional, wedge and strip, position-sensitive anodes. The one-dimensional anode has an active area of 12 × 60 mm, and the strips vary in width between 0.1 and 0.93 mm along the length of the anode. The inset is a simplified schematic of the elements used in the two-dimensional anode. Connection is made to the wedges through laser-drilled 0.13-mm via holes to common electrodes on the back of the anode.

Figure 7 is a schematic of the test fixture used to demonstrate the resolution of the one-dimensional anode. A pinhole test mask was placed in front of the MCP to screen electrons from an unfocused electron gun positioned 30 cm from the test plate. The nodal gain of the MCP was $\sim 10^6$ and a potential of 200 volts was maintained between the anode and the MCP. The output of the anode was recorded on a dual-parameter pulse-height analyzer.

The results of the test are shown in Figure 8. For comparison, a diagram of the test plate is shown above the plot to indicate the relative positions of the pinholes. Several factors contribute to the full-width/half-maximum of the peaks, including (a) the finite width of the pinholes, (b) digitization errors from the pulse-height analyzer, and (c) the resolution of the anode itself. From these preliminary results, we find

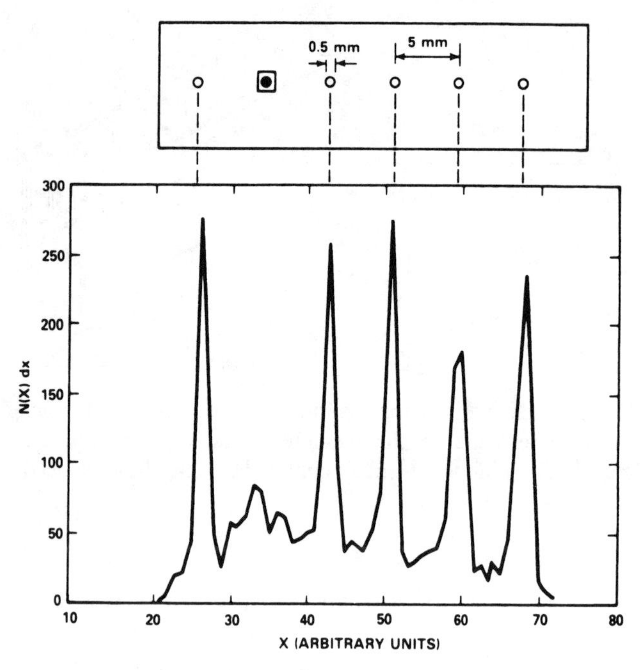

Fig. 8. Results of the pinhole resolution test. Factors contributing to the full-width/half-maximum of the peaks include the finite width of the pinholes, the digitization errors from the pulse-height analyzer, and the resolution of the anode.

that resolution of the order of 0.1 mm should be readily achieved and is sufficient to meet our requirements for an instrument with an angular resolution of 2°. More refined testing would undoubtedly show that the anode is capable of even finer resolution (Martin et al., 1981).

Summary

We have attempted to describe briefly some of our development efforts toward realizing a spacecraft-qualified ENA imager for studying global magnetospheric dynamics. From the studies of three critical elements of the imager, we find no problems with any of the components that would preclude the construction of a high-resolution ENA instrument with effective rejection EUV and charged particles. A prototype of the ENA imager is currently under construction.

Acknowledgment. We wish to thank R. E. McEntire and E. C. Roelof for several useful discussions on the ENA imager. We also wish to thank Steve Brown for his assistance at the Goddard accelerator. This work was supported by NASA grant NAGW-865.

References

Anger, H. O., Beam position identification means, Instrum. Soc. Am. Trans., 5, 311, 1966.

Cheng, A. F., Energetic neutral particles from Jupiter and Saturn, J. Geophys. Res., 91, 4524, 1986.

Högberg, G., H. Nordén, and H. G. Berry, Angular distributions of ions scattered in their carbon foils, Nucl. Instrum. Methods, 90, 283, 1970.

Hsieh, K. C., E. Keppler, and G. Schmidtke, Extreme ultraviolet induced forward photoemission from thin carbon foils, J. Appl. Phys., 5, 2242-2246, 1980.

Kirsch, E., S. M. Krimigis, J. W. Kohl, and E. P. Keath, Upper limits for X-ray and energetic neutral particle emission from Jupiter: Voyager-1 results, Geophys. Res. Lett., 8, 169, 1981a.

Kirsch, E., S. M. Krimigis, W. H. Ip, and G. Glockler, X-ray and energetic neutral particle emission from Saturn's magnetosphere, Nature, 298, 718, 1981b.

McEntire, R. W., E. P. Keath, D. E. Fort, A.T.Y. Lui, and S. M. Krimigis, The medium-energy particle analyzer (MEPA) on the AMPTE CCE spacecraft, IEEE Trans. Geosci. Remote Sensing, GE-23, 230, 1985.

Martin, C., P. Jelinsky, M. Lampton, and R. F. Malina, Wedge-and-strip anodes for centroid-finding position-sensitive photon and particle detectors, Rev. Sci. Instrum., 52, 1067, 1981.

Rairden, R. L., L. A. Frank, and J. D. Craven, Geocoronal imaging with Dynamics Explorer, J. Geophys. Res., 91, 13,613, 1986.

Roelof, E. C., D. G. Mitchell, and D. J. Williams, Energetic neutral atoms (E ~ 50 keV) from the ring current: IMP 7/8 and ISEE 1, J. Geophys. Res., 90, 10,991, 1985.

Roelof, E. C., Energetic neutral atom image of a storm-time ring current, Geophys. Res. Lett., 14, 652, 1987.

Williams, D. J., E. Keppler, T. A. Fritz, and B. Wilken, The ISEE 1 and 2 medium energy particle experiment, IEEE Trans. Geosci. Electron., GE-16, 270, 1978.

A 360° FIELD-OF-VIEW TOROIDAL ION COMPOSITION ANALYZER USING TIME-OF-FLIGHT

D. T. Young, J. A. Marshall, and J. L. Burch

Southwest Research Institute, San Antonio, TX 78284

S. J. Bame and R. H. Martin

Los Alamos National Laboratory, Los Alamos, NM 87545

Abstract. Time-of-flight (TOF) velocity filtering can be used in conjunction with electrostatic energy analysis to provide mass identification. This method has several advantages: all species are measured essentially instantaneously and good mass resolution is obtained over a wide energy range. For example, one can easily maintain sufficient mass resolution using the TOF technique to separate the major magnetospheric ion constituents (H^+, He^+, He^{2+}, O^{2+}, and O^+) at energies per charge from roughly 1 eV/e to 50 keV/e. This energy range and mass resolution are required for many near-Earth plasma investigations, in particular for elements of the International Solar Terrestrial Physics program.

The hot plasma composition analyzer (HPCA) is one example of an instrument that employs the TOF technique, incorporating a 360° field-of-view toroidal electrostatic analyzer and a linear TOF velocity filter. The geometry of the toroidal electrostatic analyzer results in a large geometric factor and allows for focusing within the TOF unit (Young *et al.*, 1988), maximizing transmission and instrument sensitivity. The HPCA is capable of measuring masses from 1–32 amu and energies from 1 to 50,000 eV/e, with $M/\Delta M$ on the order of 5 and $\Delta E/E$ of 0.18. Results of ray tracing and instrument prototype studies are presented along with a discussion of the instrument optics.

Introduction

The basic objective of ion composition measurements on present-day magnetospheric spacecraft is to obtain a complete, well-resolved three-dimensional distribution function for all major ion species in a fraction of the spacecraft spin period. The major constraints on achieving this objective up to now have been: (1) instrument geometric factor, which cannot be arbitrarily large for instruments whose typical mass is 5 kg or less; (2) the field-of-view (FOV) of mass-resolving devices (typically limited to $\lesssim$ 20° whereas 180° to 360° is required); and (3) the speed with which the full range of ion species can be sampled.

During the 1970's, emphasis in satellite-borne mass spectrometry was centered on finding an optical design that could handle a fairly wide range of ion momenta with a fixed magnetic field (electromagnets were not considered feasible due to weight, power, and operational requirements) and yet still produce acceptable mass resolution. Such a design was found in a double-focusing curved $\underline{E} \times \underline{B}$ arrangement that proved quite successful for composition studies from $\sim$ 1 eV/e up to $\sim$ 16 keV/e (see reviews by Balsiger, 1983, and Wilken *et al.*, 1982). (Note that we use eV/e and keV/e to denote the respective units of energy/charge.) Straight-through Wien filter designs have also found some application at energies up to 32 keV/e (Shelley *et al.*, 1972; Johnson *et al.*, 1983).

An alternate approach to satellite-borne mass spectrometry, and the one followed here, has been to apply TOF techniques adapted from energetic particle physics to the problem of magnetospheric composition. TOF analyzers have been studied and constructed by several groups (Gloeckler and Hsieh, 1979; Gloeckler *et al.*, 1985; Gloeckler, 1988, Hovestadt *et al.*, 1978; Fritz and Cessna, 1975; Fritz *et al.*, 1985; McEntire *et al.*, 1985). There are two categories of TOF spectrometers. One type, initially designed for unambiguous determination of charge state and mass composition (as distinct from mass/charge) in solar wind and magnetospheric plasmas, is the triple-coincidence system which measures E/Q, TOF, and total energy. The second type, based on E/Q and TOF analysis, is less complex electronically but is nonetheless quite adequate if charge state and high mass resolution measurements are not required (Fritz *et al.*, 1985; Wilken *et al.*, 1987). The present paper reports on the latter design as applied to the determination of M/Q and E/Q spectra of the major magnetospheric ion species found at energies < 50 keV/e. Species of primary interest are H^+, He^{2+}, He^+, O^{2+}, O^+, and M^+ (M^+ refers to ionospheric molecular species O_2^+, N_2^+, and NO^+ that can be resolved as a group from the atomic species). Note that in principle a resolution $M/\Delta M$ of only 1.5 or less is required. The design presented here is related to somewhat similar devices built for

Giotto (Wilken *et al.*, 1987) and the low-energy magnetospheric ion composition spectrometer (LOMICS) on CRRES (Fritz *et al.*, 1985).

The HPCA TOF analyzer is coupled to a novel 2π-radian toroidal electrostatic analyzer (the 2πTA) described in detail elsewhere (Young *et al.*, 1988) . The 2πTA was designed with focal lengths appropriate for use with a relatively short (5.0 cm) TOF system and in addition has a high geometric factor and, of course, a 360° FOV. The combined E/Q and TOF analyzers are christened the hot plasma composition analyzer (HPCA). The HPCA was constructed and tested in prototype form as part of a proposal effort for the Cluster mission (Johnstone *et al.*, 1987). A second application of this TOF design is the thermal ion dynamics experiment (TIDE) presently under development for the ISTP/Polar satellite. In the case of TIDE, the TOF unit is coupled to a retarding potential analyzer and electrostatic mirror system. We plan to describe the TIDE/TOF analyzer in a separate paper.

Instrument Description

Figure 1 shows a cross-sectional view of the HPCA. The upper portion consists of the 2π-toroidal analyzer (2πTA) which is rotationally symmetric about the central axis. Below the exit of the 2πTA are a set of $\sim 1\,\mu\mathrm{g\,cm}^{-2}$ carbon foils, the TOF housing, and associated microchannel plates (MCPs). The entire TOF section is rotationally symmetric except for evenly spaced pieces of insulators and other mounting hardware.

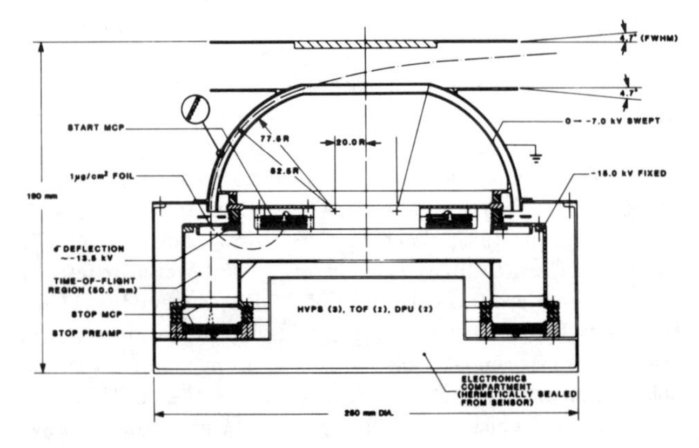

Fig. 1. Cross-sectional view of the HPCA prototype. The entire instrument is rotationally symmetric, providing a 360° field-of-view for E/Q and M/Q analysis from ~ 1 eV/e to 50 keV/e. The TOF foils are $1\,\mu\mathrm{g\,cm}^{-2}$ thick carbon mounted on 93% transmission grids that are 8 mm × 16 mm. Thirty-two of these foils are required for the HPCA. Dimensions in the figure may be scaled by $\sim \pm 20\%$ to adjust the overall size of the HPCA for a particular mission without significantly affecting the optics. Geometric factor at constant energy-angle-mass resolution will scale as R^{-2}.

Ions leaving the 2πTA are accelerated by a fixed potential of -15.0 kV and strike the carbon foil. A smaller potential of, say, -10.0 kV, can also be used with some loss of mass resolution. The ions easily pass through the foil, scattering and losing an amount of energy that depends on initial ion energy as well as on mass and nuclear charge (see below). Passage of the ion out of the foil ejects one or more secondary electrons from the backside of the foil. (There is also a probability of zero electrons being emitted which lowers TOF detection efficiency somewhat). These secondary electrons are emitted at very low energies (averaging on the order of 3 eV) and are easily deflected by the potential of -12.5 kV applied to the small electrode located just below the foil (Figure 1). Care has been taken with the electron deflection optics to produce nearly isochronous trajectories. Ions are virtually unaffected by the electron deflection potential due to their much higher momenta. The electrons are focused onto start-MCP detectors placed in an annular configuration made up of specially cut MCP pieces. These MCPs provide a start pulse for the TOF electronics and, in addition, register the azimuthal sector from which the ion arrived. Although we plan to register start pulses only with a resolution of 22.5°, the 2πTA alone has demonstrated the ability to resolve in polar angle to $< 1°$ FWHM and with the TOF system in place to $< 3°$ FWHM.

After leaving the foil, ions continue over a nearly field-free flight path, striking the MCP at the back of the TOF and giving rise to a stop pulse from the high-voltage stage of the MCP. The complete stop-MCP stack consists of three single MCPs with one at the TOF high voltage and the other two near ground potential (Figure 1). This serves to transfer the stop pulse across the -15.0 kV potential drop without need for high-voltage coupling capacitors.

Because the TOF section is at high negative potential we have paid particular attention to the emission of secondary electrons from internal TOF surfaces. The 2πTA collimator will be baffled and the curved plates (which present a two-bounce system) will be serrated and blackened to reduce both UV and particle scattering. The interior of the TOF system will also be serrated and blackened to reduce secondary emission and scattering (the exterior is polished to reduce field emission).

Electric fields within the TOF region are relatively weak and are designed to cause most of the secondary electrons emitted from the interior to be attracted to the electron deflection electrode rather than back to the start MCP. If required, an electron suppression grid could be placed a few millimeters in front of the first stop MCP (i.e., inside the TOF volume) to further reduce spurious start pulses issuing from the interior of the TOF region.

Primarily because of the care taken with internal TOF optics we have reduced the occurrence of ghost peaks in the mass spectra, i.e., peaks due to internal scattering and resulting spurious timing signals, to an acceptable level ($< 0.1\%$ of main peak height). This is an important consideration for the on-board processing of TOF data in the case that moments are to be computed using pre-assigned mass channels which must therefore be uncontaminated by counts from adjacent TOF peaks.

Results of Instrument Tests

The 2πTA has been tested separately and is reported on elsewhere (Young *et al.*, 1988) and only the principal results will be mentioned. A broad ion beam (5.0 cm diameter FWHM) was used to fully illuminate the 2πTA collimator and establish its transmission characteristics which are reproduced here in Figures 2a and 2b. The image of the ion beam at the position of 2πTA focus is shown as an iso-intensity contour of ion count rate in Figure 2a. This was obtained with an imaging MCP (I-MCP) placed at the focal point, located 30 mm behind the 2πTA exit. In Figure 2b we show the integrated I-MCP count rate plotted as a function of incident ion angle and energy, again with the same broad beam illuminating the 2πTA. Figure 2b shows that the 2πTA transmits about 0.18 in $\Delta E/E$ (FWHM) and 7.6° in angle in the plane of Figure 1. Some of the transmission tails shown in Figures 2a and 2b are due to large impact parameter trajectories (i.e., trajectories well above or below and parallel to the plane of Figure 1) and could be eliminated by placing vertical baffles in the 2πTA collimator. This should result in only a small loss in geometric factor since the offending rays represent a small fraction of the total.

Figure 3 shows a TOF mass spectrum for a mixed species beam obtained by leaking H_2, He, and N_2 into our electron bombardment source. The beam energy was 10.0 keV with a potential of −5.0 kV on the TOF section. The total energy corresponds to the minimum energy ion that the HPCA will encounter with the full −15.0 kV TOF voltage applied. The peak labeled He° results from fast neutral helium formed by the charge exchange of 10.0 keV He^{+} during its transit of the 3-meter long drift tube in our ion acceleration system. Because He° is not affected by the −5.0 kV TOF acceleration it is displaced from the He^{+} TOF peak by a factor of 1.22 corresponding to the ratio of the square root of the total ion and neutral energies. Thus the He° peak is located at the $M/Q = 6$ position for an ion with 15.0 keV total energy, and it is clear that resolution is adequate to separate M/Q = 1, 2, 4, 8, and 16 species even at 1% or less of their peak counting rates.

Test data on mass peak location are shown in Figure 4 for an 8.0-keV ion beam and −7.7 kV TOF acceleration. The open symbols are data; the closed symbols are based on a simple calculation of mean atomic ion flight times given by

$$\bar{\tau}_o = 22.85\, L_o \left(\frac{M}{E_o - \Delta E_s}\right)^{\frac{1}{2}} \times 10^{-9}\,\mathrm{s} \qquad (1)$$

where L_o = 5.0 cm is the length of the TOF drift region, M is ion mass in amu, and E_o is the incident ion energy in keV. The correction for energy straggling in the 1 μg cm^{-2} foil, ΔE_s, is taken from Figure 11 of Ipavich *et al.* (1982). As noted by these authors and by Gloeckler and Hsieh (1979), our quoted foil thickness of 1.0 μg cm^{-2} (specified by the manufacturer) is likely to be in error by as much as $\sim +50\%$; i.e., the foil may in fact be as thick as 1.5 μg cm^{-2}, although there is no evidence for this in the data of Figure 4.

Mass resolution for a TOF instrument is directly related to the TOF resolution

$$\frac{M}{\Delta M} = \frac{1}{2}\frac{\tau}{\Delta \tau} \qquad (2)$$

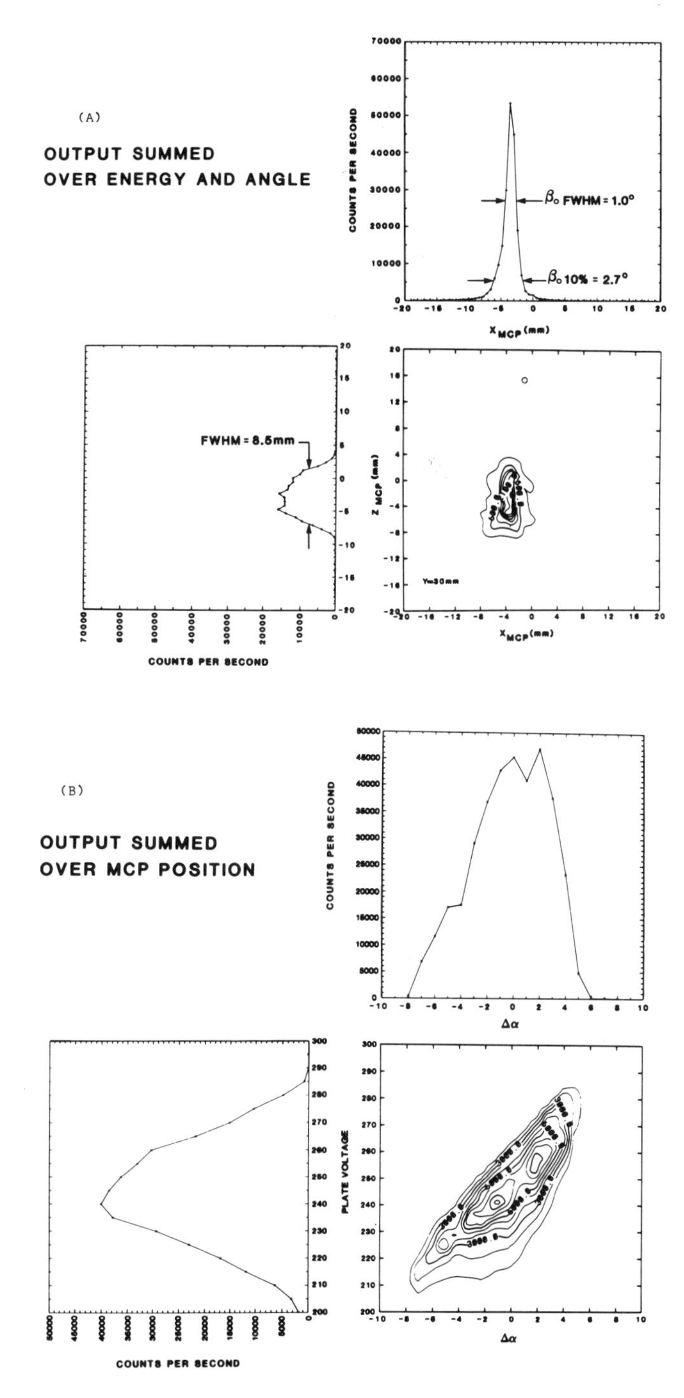

Fig. 2(a) Iso-count rate contours of I-MCP response to broad-beam illumination of the 2πTA entrance collimator including full passband coverage in azimuthal angle and energy. Line plots represent summations over the contour plot to show beam width in the polar and azimuthal directions (from Young *et al.*, 1988). (b) 2πTA response in energy-angle acceptance plotted as iso-count rate contours. Line plots represent summations over the contour plot to show analyzer acceptance in azimuthal angle and energy (from Young *et al.*, 1988).

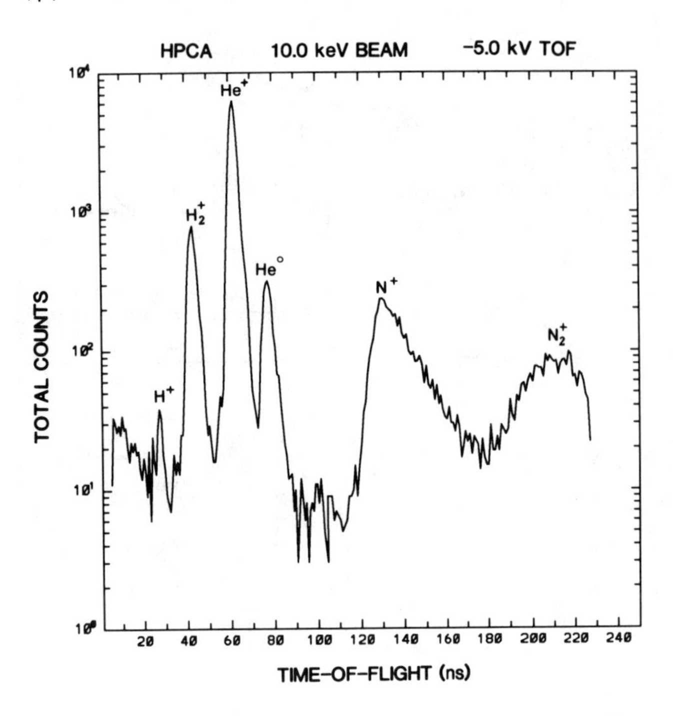

Fig. 3. TOF mass spectrum with full spatial-angle-energy illumination of the HPCA entrance.

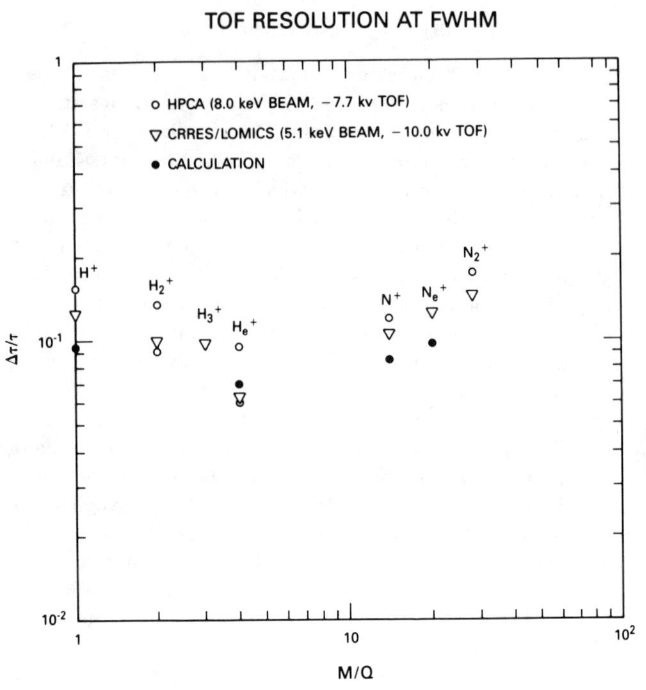

Fig. 5. TOF resolution vs. M/Q for fixed ion energy of 15.7 keV impinging on the foil. Filled points refer to calculation described in text. Mass resolution $M/\Delta M$ is given by one-half the value of $\tau/\Delta\tau$.

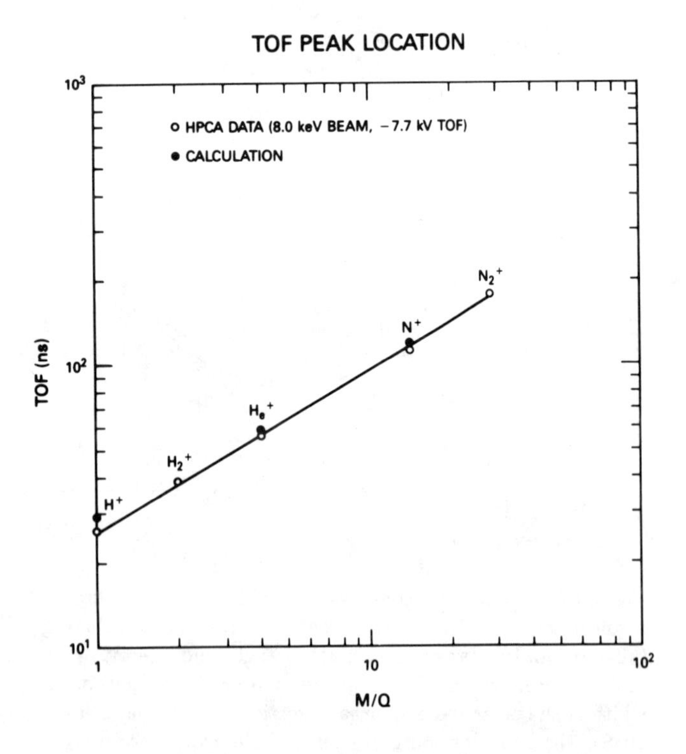

Fig. 4. Mass peak location in TOF vs. M/Q. Filled points refer to calculation of TOF peak location for monatomic species as described in text.

Figure 5 gives values of $\Delta\tau/\tau$ for several ion species measured with both the LOMICS and HPCA instruments. The solid points refer to FWHM resolution calculated by a simple consideration of the several contributions to spreading of ions along their flight path. We define the relative time spread

$$\delta T \equiv \frac{\Delta\tau}{\bar{\tau}_o}$$

where $\bar{\tau}_o$ is the mean TOF and we have the following relations

$$2\delta T = \frac{2\Delta\tau}{\bar{\tau}_o} = \frac{\Delta E}{E} = \frac{2\Delta v}{v} = \frac{2\Delta L}{L_o}$$

where E is the final particle energy after exiting the foil, v is its velocity, and L_o is the nominal path length through the TOF. The contributors to δT are:

(a) Energy spread due to the transmission passband of the 2πTA which is found to be $\Delta E/E_o = 0.18$ (FWHM), then

$$\delta T_{\Delta E} = \frac{0.09}{1 + V_a/E_o} \tag{3}$$

where V_a is the absolute value of the acceleration potential. Note that the larger V_a becomes the smaller is $\delta T_{\Delta E}$, similarly if $\Delta E/E_o$ of the E/Q analyzer decreases so does $\delta T_{\Delta E}$.

(b) Energy spread due to straggling in the foil. We make the approximation that the width of the straggling peak is proportional to the energy lost in the foil (Bernstein *et al.*, 1970). Thus

$$\delta T_{\Delta E\text{foil}} \sim \frac{0.13Z}{m^{\frac{1}{2}}(E_o + V_a)^{\frac{1}{2}}} \tag{4}$$

(c) Angular spread of incident ions due to the transmission passband of the 2πTA. The laboratory prototype HPCA design will accept 18° FWHM at the TOF unit in either polar or azimuthal direction and still hit the stop-MCP. The 2πTA output has a maximum angular divergence of 12° FWHM in the polar direction (Fig. 5, Young *et al.*, 1988) and somewhat less in the azimuthal. Nonetheless we are uncertain about the full spread of exiting trajectories from the 2πTA and will use 18° FWHM for this term which is given by

$$\delta T_{\Delta\alpha} \sim \frac{1}{\cos\Delta\alpha} - 1 \tag{5}$$

with $\Delta\alpha = 9°$.

d) Angular spread due to scattering in the foil is again taken from Bernstein *et al.* (1970) with the approximation

$$\Delta\alpha_{\text{foil}} \sim \frac{15° Z}{E_o + V_a}$$

where Z is the atomic number of the projectile ion. This gives

$$\delta T_{\Delta\alpha\text{foil}} \sim \sec\left(\frac{15° Z}{E_o + V_a}\right) - 1 \tag{6}$$

with E_o and V_a in units of keV.

(e) Dispersion of start electrons due to path differences. This is independent of ion species and energy. Ray tracing shows that this dispersion amounts to < 1.5 ns for the configuration tested with the HPCA. Then

$$\delta T_e = 0.0132 \left(\frac{E_o + V_a}{M}\right)^{\frac{1}{2}} \tag{7}$$

Energy and mass show up in (7) because δT_e is taken relative to the total flight time τ_o which depends on both parameters.

(f) Electronics noise and jitter are quite small ($\ll 1$ ns) when the TOF electronics are tested with laboratory pulsers. However, we have not yet isolated this term using actual MCP pulses. The latter present a variation in pulse shape as a function of amplitude that causes time jitter in the detection of start and stop. We estimate $\lesssim 1$ ns for this in absolute units and have

$$\delta T_{el} \leq 0.0088 \left(\frac{E_o + V_a}{M}\right)^{\frac{1}{2}} \tag{8}$$

Both $\delta\tau_e$ and $\delta\tau_{el}$ are fixed quantities. Therefore the relative values δT_e and δT_{el} increase with shorter flight times (i.e., with higher energies and lower masses) and decrease if a longer flight path (i.e., a larger instrument) is used.

We now assume that all of these terms are statistically independent and therefore add as the root sum of squares. Then the average FWHM value for TOF peak width will be

$$\overline{\delta T} = \left[\sum \delta T_i^2\right]^{\frac{1}{2}}$$

We have calculated $\overline{\delta T}$ using our current best estimates for the HPCA at total ion energies $E_o + V_a = 15.7$ keV. This is plotted in Figure 5 (solid points) as a function of ion M/Q. Up to now we have not tested the HPCA over a range of energies sufficient to show whether our model correctly predicts performance as a function of energy. However it seems to agree reasonably well at 15.7 keV.

Efficiency and Geometric Factor

The HPCA was mounted so that the TOF counting rate could be compared with the counting rate for an identical MCP in an identical mounting configuration–the only difference being that the latter had no foil in place. This gave an efficiency of 0.34 ± 0.026 for the complete TOF relative to the bare MCP. The latter, with a suppression grid in place, should be 60% to 80% efficient. This gives an absolute efficiency for the HPCA system of 0.25 ± 0.05.

Tests of the 2πTA (Young *et al.*, 1988) have shown that the energy-geometric factor of this unit alone is 9.6×10^{-3}cm^2 sr eV/eV for a 22.5° polar angle sector in an instrument with 10.0 and 8.0 cm radii (Figure 1). Here we define energy-geometric factor G_E as

$$G_E = A_e \langle \Delta a \frac{\Delta E}{E_o} \rangle \Delta\beta$$

In order to convert particle flux into count rate one must also specify sensor efficiency, ϵ , such that

$$C(E_o) = G_E \cdot j(E_o) \cdot \epsilon \cdot E_o$$

We might expect some slight dependence of G_E upon incident particle energy E_o due to the focusing caused by the acceleration potential V_a; however this should be negligible since the TOF foil width and MCP detectors are sized to take this into account. With the present data we obtain

$$\epsilon \cdot G_{E(22.5)} = 2.4 \times 10^{-3}\,\text{cm}^2\,\text{sr}\,\frac{\text{eV}}{\text{eV}}\frac{\text{counts}}{\text{ion}}$$

over a 22.5° sector. If one allows roughly 10% dead space for mounting hardware in the full 360° FOV then the total HPCA energy-geometric factor is

$$\epsilon \cdot G_{E(360)} = 3.5 \times 10^{-2}\text{cm}^2\,\text{sr}\,\frac{\text{eV}}{\text{eV}}\frac{\text{counts}}{\text{ion}}$$

To put this in some perspective, in ~ 0.1 s, the HPCA can measure, with an accuracy of $\sim 1\%$ (i.e., with 10^4 total counts), a stationary Maxwellian distribution consisting of H^+ ions having a density of only 1 cm^{-3}. The same measurement for O^+ would of course take 4 times longer.

Summary

We have constructed and successfully tested a 360° FOV plasma mass spectrograph based on the TOF principle. The TOF design is such that it can also be mated to the swept field-of-view analyzers discussed by Bame *et al.* (1989). The HPCA provides a high geometric factor, light-weight, robust design capable of meeting many of the future needs for high time resolution plasma composition measurements onboard magnetospheric spacecraft.

Acknowledgments. This paper is dedicated to Jim Cessna of Los Alamos National Laboratory who designed the TOF electronics and whose invaluable help is gratefully acknowledged. We wish to thank T. Booker and B. Reinhard for assistance in construction and testing of the HPCA. Work at Southwest Research Institute was supported by Internal Research contract 15-9455. Work at Los Alamos National Laboratory was supported by U.S. Department of Energy.

References

Balsiger, H., Recent developments in ion mass spectrometers in the energy range below 100 keV, *Adv. Space Res., 2*, 3, 1983.

Bame, S. J., D. T. Young, R. H. Martin, D. J. McComas, J. L. Burch, and J. A. Marshall, Three-dimensional plasma measurements from three-axis stabilized spacecraft, these proceedings, 1989.

Bernstein, W., A. J. Cole, and R. L. Wax, Penetration of 1-20 keV ions through thin carbon foils, *Nucl. Instrum. Meth., 90*, 325, 1970.

Fritz, T. A., and J. R. Cessna, *IEEE Trans. Aerospace Electron. Systems, AES- 11*, 1145, 1975.

Fritz, T. A., D. T. Young, W. C. Feldman, *et al.*, The mass composition instruments (AFGL-701-11) on *CRRES, in CRRES/SPACERAD Experiment Descriptions*, edited by M. S. Gussenhoven, E. G. Mullen, and R. C. Sagalyn, AFGL-TR-85-0017, Air Force Geophysics Laboratory, Hanscom AFB, Massachusetts, p. 127, 1985.

Gloeckler, G., Measurements of the Charge State and Mass Composition of Hot Plasmas, Suprathermal Ions and Energetic Particles Using the Time-of-Flight vs. Energy Techniques, invited paper, presented 1988 Yosemite Conference on Outstanding Problems in Solar System Plasma Physics: Theory and Instrumentation, Yosemite National Park, California, February 2-5, 1988.

Gloeckler, G., and K. C. Hsieh, Time-of-flight techniques for particle identification at energies from 2 to 400 keV/nucleon, *Nucl. Instrum. Meth., 165*, 537, 1979.

Gloeckler, G., *et al.*, The charge- energy- mass (CHEM) spectrometer for 0.3 to 300 keV/e ions on AMPTE-CCE, *IEEE Trans. Geosci. Remote Sensing, GE-23*, 234, 1985.

Hovestadt, D., G. Gloeckler, *et al.*, The nuclear and ionic charge distribution particle experiments on the ISEE-1 and ISEE-C spacecraft, *IEEE Trans. Geosci. Electron., GE-16*, 166, 1978.

Ipavich, F. M., L. S. Ma Sung, and G. Gloeckler, Measurement of energy loss of H, He, C, N, O, Ne, S, Ar, Fe, and Kr passing through thin carbon foils, University of Maryland Tech. Rept. #82-172, April 1982.

Johnson, R. G., R. J. Strangeway, E. G. Shelley, J. M. Quinn, and S. M. Kaye, Hot plasma composition results from the SCATHA spacecraft in *Energetic Ion Composition in the Earth's Magnetosphere*, edited by R. G. Johnson, p. 287, Terra Sci. Publ. Co., Tokyo, 1983.

Johnstone, A. D., *et al.*, A Plasma Energy Angle Composition Experiment, Part I, Proposal issued by Mullard Space Science Laboratory, Holmbury St. Mary, Dorking, Surrey, UK, 1987.

McEntire, R. W., E. P. Keath, D. E. Fort, A. T. Y. Lui, and S. M. Krimigis, The Medium-Energy Particle Analyzer (MEPA) on the AMPTE CCE Spacecraft, *IEEE Trans. Geosci. Remote Sensing, GE-23*, 230, 1985.

Shelley, E. G., R. G. Johnson, and R. D. Sharp, Satellite observations of energetic heavy ions during a geomagnetic storm, *J. Geophys. Res., 77*, 6104, 1972.

Wilken, B., T. A. Fritz, and W. Studemann, Experimental techniques for ion composition measurements in space, *Nucl. Instrum. Meth., 196*, 161, 1982.

Wilken, B., W. Weiss, W. Studemann, and N. Hasebe, The Giotto Implanted Ion Spectrometer (IIS): Physics and technique of detection, *J. Phys. E: Sci. Instrum., 20*, 778, 1987.

Young, D. T., S. J. Bame, M. F. Thomsen, R. H. Martin, J. L. Burch, J. A. Marshall, and B. Reinhard, A 2π-radian field-of-view toroidal electrostatic analyzer, *Rev. Sci. Instrum., 59*, 743, 1988.

II. Mass, Momentum, and Energy Release and Transfer in Solar System Plasmas

Yosemite Valley from Inspiration Point, Winter, Yosemite National Park.

What is the magnetic energy conversion process in flares?

ENERGY CONVERSION IN SOLAR FLARES

Peter A. Sturrock

Center for Space Science and Astrophysics, Stanford University, Stanford, CA 94305

Abstract. Flares involve a sequence of energy conversion processes beginning with the initial stressing of the magnetic field and ending with radiation. This article is concerned primarily with the processes by which magnetic energy is suddenly converted into other forms. We pay special attention to a sub-class of large flares, for which the flare process appears to begin with the activation of a filament. This activation may be either an MHD instability or a combination of MHD processes and reconnection. The main phase of particle acceleration, that may be either impulsive or gradual, is associated with the sudden release of magnetic energy in a large flux system. This may be attributed to reconnection that is due to, and also generates, fine-scale magnetic fluctuations. This process may be triggered either by the Kelvin-Helmholtz instability of an erupting filament, or by reconnection at a current sheet between adjacent flux systems. As a result of the fine-scale fluctuations, electric field components parallel to the magnetic field develop that can be responsible for the acceleration of large numbers of electrons to keV energies. Acceleration of electrons and ions to relativistic energies may be attributed to stochastic acceleration due to the same MHD fluctuations that are responsible for the E-parallel acceleration.

Introduction

Before the advent of space exploration, solar flares were much simpler than they are today. Thirty years ago, the principal facts to be explained were the total energy (known at that time to be as high as 10^{32} erg) and the time scale for energy release, that was generally taken to be the time scale of the impulsive phase, of order 2 min. Since that time the problem has, for better or worse, become far more complicated. Radio observations now extend from a few kHz (Fainberg and Stone, 1974; Cane and Stone, 1984) (obtained from spacecraft) to almost 100 GHz (Kaufmann et al., 1985). The time resolution of receivers has improved enormously, and we now know that radio bursts of very high frequency sometimes contain structure on a time scale of only a few milliseconds.

Although early observations of flares made by Hα photography were, necessarily, observations of plasma of chromospheric temperature, we now know from X-ray observations that flares typically produce plasma of far higher temperature, certainly 10^7 K and sometimes substantially higher (Lin et al., 1981). In addition, we know that some flares produce hard X-rays and also gamma-rays, both continuum and lines. Observations from spacecraft and from neutron detectors on the Earth's surface show that, on occasion, acceleration up to several BeV occurs during a flare. We know that hard X-ray emission from flares sometimes fluctuates on a time scale as short as 45 ms (Kiplinger et al., 1983). We have also learned, from gamma-ray observations, that acceleration of particles to many MeV typically occurs during the impulsive phase, sometimes on a time scale of order 1 s (Kane et al., 1986). This runs counter to earlier ideas that acceleration to high energies occurs in a "second phase" of acceleration associated with the passage of a shock wave through the corona (Wild et al., 1963).

It was realized many years ago that flares are often preceded by filament disturbances and often associated with filament eruptions (Smith and Ramsey, 1964; Kiepenheuer, 1964). Observations made by means of coronagraphs on spacecraft now show that large flares sometimes occur in association with "coronal mass ejections" (CMEs), the ejection of enormous volumes of plasma through the corona into interplanetary space (Wagner, 1984). These ejections are closely related to filament eruptions.

Given this complexity of the flare phenomenon, it is no longer adequate to look for "the" energy conversion process in flares. There is likely to be a sequence of conversion processes. Some of these processes are the well understood mechanisms of radiation: excitation-radiation, bremsstrahlung, synchrotron radiation, etc. Other radiation mechanisms, such as those involved in radio emission, are more complicated and less well understood. However, we shall focus in subsequent sections on earlier energy-conversion processes that lead to the observed mass motions and to the production of high-temperature plasma and high-energy particles.

It has been emphasized that this article deals only with certain selected energy conversion processes. It must also be emphasized that it deals only with certain types of flares, namely those large flares that involve filament activation and sometimes involve also CMEs. In sum, the goal of this article is to offer a plausible model for one type of flare.

Phases of Solar Flares

Since a flare was originally defined (see, for instance, Smith and Smith, 1963) as a sudden brightening in monochromatic, chromospheric radiation, the start of the flare process was, in consequence, identified with the first Hα brightenings. What happened before these brightenings was therefore termed a "precursor" or "pre-flare activity" (Van Hoven et al., 1980).

However, it has been known for some time that large flares normally occur in regions that involve filaments, and that the filament shows signs of "activation" long before the beginning of the flare. After such activation, the filament may erupt and be associated with a CME. When this occurs, the total kinetic energy associated with this mass motion is likely to be considerably larger than the total energy released during the flare in the form of radiation. When the trajectory of the coronal mass ejection is extrapolated back to the photosphere, it is found to originate with the filament activation, not with the impulsive phase of the flare (Harrison, 1986).

These facts suggest that filament eruption and the production of a CME constitute an important process that is independent of the flare process. This event may or may not give rise to a flare. In fact, most CMEs are not flare related, if the term "flare" is used in its conventional sense (Wagner, 1984). It would probably simplify discussion to extend the term "flare" to include the filament-eruption-CME process, as well as the consequent Hα emission and related emission. With this broader definition, we would say that, when filament activation precedes a flare, the first phase of the flare is that of filament activation.

There is no accepted model for the magnetic field structure associated with (or responsible for) a filament. It is clear, however, that the magnetic field must contain segments that are concave upward, in order to suspend cool gas above the chromosphere. Furthermore, high-resolution photographs show that the striations of a filament run almost parallel to the magnetic inversion line. Hence the magnetic field of the filament may be regarded as some sort of tangled rope of magnetic field running above, and parallel to, the inversion line.

The "activation" process that leads to eruption must be primarily MHD in character. For instance, if the magnetic topology is as shown in Figure 1a, the configuration would become kink unstable (Schmidt, 1966) if the degree of twist of the flux tube exceeds some critical value. It could then lead to an eruption as shown in the sequence (Figures 1b, 1c, and 1d).

Alternatively, the initial activation may involve a combination of MHD processes and reconnection. For instance, the magnetic field of a filament is likely to develop initially from an array of small but highly sheared magnetic loops, as shown in Figure 2. If reconnection occurs at adjacent footpoints, the filament will become detached from the photosphere except at the two ends. This reconnection may give

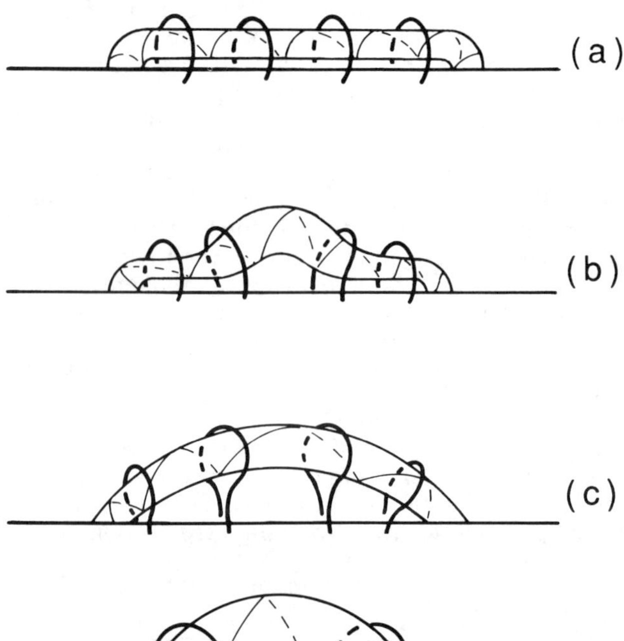

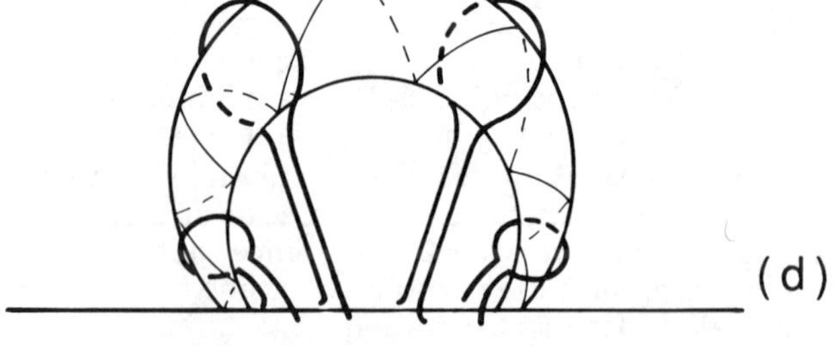

Fig. 1. Schematic representation of possible magnetic field configuration associated with a filament and its surroundings. Initial state (a) may be subject to a kink-type instability that leads to progressive eruption as shown in (b), (c), and (d).

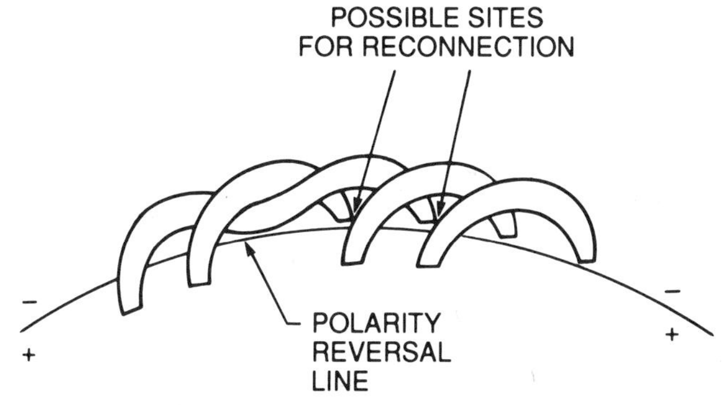

Fig. 2. Alternative schematic representation of possible magnetic field configuration associated with a filament as a rope-like assembly of fine flux tubes. Reconnection at adjacent footpoints of opposing polarities can lead to the disconnection of the filament from the photosphere (except at the ends) and to its eruption.

rise to small brightenings, perhaps primarily in the wings of Hα, but it would also change the forces acting on the filament and so could lead to eruption.

Impulsive and Gradual Phases

Many flares begin suddenly and dramatically with intense brightening in Hα in a small region (sometimes called the "explosive" phase), that rapidly expands to fill a much larger region (the "expansion" phase). This stage is often accompanied by hard X-ray emission and microwave radiation. We also know, from SMM data, that it can also be accompanied by gamma-ray radiation, both continuum emission (due primarily to high-energy electrons) and line radiation (due primarily to high-energy ions) (Forrest and Chupp, 1983). Since gamma-ray observations show unequivocally that ion acceleration occurs during this early phase, it is now believed that this stage of acceleration also contributes to high-energy electrons and ions that are subsequently detected in space.

Bai (1986) has shown that gamma-ray/proton flares may be divided into two categories that he terms "impulsive" and "gradual." Flares with a gradual phase involve larger magnetic loops, produce more intense microwave radiation, and are usually associated with coronal mass ejections. Further comments on these two classes will be made in the section "Late Phase of Flare."

It is frequently proposed that the impulsive phase of a flare is somehow "triggered" by another disturbance, such as the filament eruption. Recent work by Kahler et al. (1986) leads to a suggestion for a possible mechanism for this triggering action. Kahler et al. have studied four flares involving filament eruptions. They find that the impulsive phase of the flare begins when the line-of-sight velocity of the filament reaches a threshold of about 100 km s^{-1}. Since Doppler observations measure only one component of the velocity vector, the actual speed of the eruption may be larger by perhaps a factor of 2.

It is interesting to note that the speed of sound in the corona is of order 150 km s^{-1}. The threshold for onset of the Kelvin-Helmholtz instability is comparable with the speed of sound according to Dobrowolny (1972). This suggests the following scenario. When the speed of the erupting prominence approaches the speed of sound, the Kelvin-Helmholtz instability occurs in the neighborhood of the interface between the prominence and the surrounding corona. The primary spectrum that develops as a result of the instability does not propagate away from the instability region. However, inhomogeneities and nonlinear interactions will couple these disturbances into MHD waves that can propagate away from the shear region into both flux systems. If the coronal flux system comprises a current-carrying (twisted) magnetic loop, these MHD fluctuations will lead to fine-scale fluctuations in the magnetic field configuration. Since the tearing-mode growth rate is very sensitive to the scale of the magnetic field gradients, this could precipitate reconnection. The nonlinear stage of reconnection is itself a source of MHD noise so that reconnection could then propagate, like a flame, across the magnetic loop.

MHD waves are heavily damped if the frequency approaches or exceeds the ion gyro frequency. For protons, the gyro frequency is given by

$$\omega_{g,p} = 10^{4.0} B \qquad (1)$$

Since the Alfvén speed is given by

$$v_A = 10^{11.3} B n^{-1/2} \qquad (2)$$

we see that the inverse of the limiting wave number is given by

$$l_c = 10^{7.3} n^{-1/2} \qquad (3)$$

Hence for coronal densities of order 10^8 cm^{-3}, l_c = $10^{3.3}$ cm so that fluctuations with scales much less than 10^4 cm are likely to be heavily damped.

Let us now suppose that the tearing mode does develop in an extended magnetic loop. If the length of the loop is L (cm), the mean magnetic field strength is B (gauss), the temperature is T (K), and the scale of fluctuations in the magnetic field transverse to the direction of the field is b (cm), then the magnitude of the current density j (emu) is given by

$$j = 10^{-1.1}\, B\, b^{-1} \qquad (4)$$

However, an electric field E (esu) parallel to the magnetic field must develop to carry this current:

$$E = \eta\, j \qquad (5)$$

where the resistivity η is given approximately by

$$\eta \approx 10^{3.6}\, T^{-3/2} \qquad (6)$$

Hence we find that electrons could be accelerated to an energy E_e (eV) given by

$$U_e = 10^{5.0}\, L T^{-3/2}\, B\, b^{-1} \qquad (7)$$

The characteristic growth time of the linear stage of the tearing mode (Furth et al., 1963) is given by

$$\tau_I = \tau_A^{1/2}\, \tau_R^{1/2} \qquad (8)$$

where τ_A is the time for an Alfvén wave to propagate the thickness of the current sheet so that

$$\tau_A = v_A^{-1}\, b \qquad (9)$$

and τ_R is the time required for resistive diffusion over a distance b, so that

$$\tau_R = \frac{4\pi}{c\eta} b^2 \qquad (10)$$

or, numerically,

$$\tau_R = 10^{-13.0}\, T^{3/2}\, b^2 \qquad (11)$$

Hence we see that

$$\tau_I = 10^{-12.1}\, n^{1/4} T^{3/4}\, B^{-1/2}\, b^{3/2} \qquad (12)$$

For a large active-region loop, typical figures might be $L = 10^{10}$, $n = 10^9$, $T = 10^{6.2}$, and $B = 10^2$. We then find from (7) and (12) that $U_e = 10^{7.7}\, b^{-1}$ and $\tau_I = 10^{6.2}\, b^{3/2}$. Hence we find that, if $b \approx 10^{3.5}$, $U_e \approx 10^{4.2}$ and $\tau_I \approx 10^{-1}$. Hence these estimates are consistent with the experimental data that, during the impulsive phase of the flare, electrons are accelerated to some tens of keV to produce the hard X-ray burst and that fluctuations in the hard X-ray emission can exhibit time scales down to 30 ms. (Fluctuations in a millimeter wave radiation have been observed with even shorter time scales (Kaufmann et al., 1980)).

The above calculations indicate that, given the fine scale structure of the order indicated, a sufficiently large electric field develops to accelerate electrons to some tens of keV. However, the above calculation does not indicate how many electrons will be accelerated. To answer this question, we need to estimate how many electrons will "run away" during a typical impulsive phase. The original calculation of the runaway process was made by Dreicer (1959, 1960), but we will use the improved estimates of Kulsrud et al. (1973). According to these calculations, the probability per unit time that an electron will run away may be expressed as

$$\gamma_{RA} = 0.35\, \nu_e \exp\left[-\left(\frac{2}{\varepsilon}\right)^{1/2} - \frac{4}{\varepsilon}\right] \qquad (13)$$

where ν_e is the electron collision rate, given approximately by

$$\nu_e = 10^{2.4}\, n T^{-3/2} \qquad (14)$$

and

$$\varepsilon = \frac{E}{E_D} \qquad (15)$$

where E_D is the "Dreicer" field given approximately by

$$E_D = 10^{-9.6}\, n T^{-1} \qquad (16)$$

We see from (5), (6) and (16) that

$$\varepsilon = 10^{12.1}\, n^{-1}\, T^{-1/2}\, B\, b^{-1} \qquad (17)$$

so that, for the above values of these parameters, $\varepsilon \approx 0.3$. We then find, from (13), that $\gamma_{RA} = 10^{-6.3}$.

The total number of electrons that run away may be expressed as

$$N_{RA} = \gamma_{RA}\, \tau\, n\, L\, A \qquad (18)$$

where τ is the duration of the impulsive phase and A is the cross-sectional area of the tube. If we adopt $\tau = 10^2$ and $A = 10^{18.5}$, we obtain $N_{RA} = 10^{32.2}$. This is lower than the number (10^{36}) produced by a typical large flare. However N_{RA} is very sensitive to the choice of parameters. For instance, if ε is changed from 0.3 to 0.6, the number of runaway electrons increases to $10^{35.5}$.

In order to explain the number of runaway electrons, the required electric field is quite close to the Dreicer field. Moreover, the number of runaway electrons approaches the total number of electrons in the pre-flare flux tube (for the above parameters, this number is $10^{36.5}$). Since the number of runaway electrons depends so sensitively on the parameters of the system, it seems more likely that the electric field is sufficiently high, in some fraction of the volume of the tube, that all electrons in that fraction run away. That is, it seems more likely that, in a small fraction of the tube, the electric field is comparable with or larger than the Dreicer field. We see from (17) that this requires the transverse scale length b to approach, or be less than, the value

$$b_D = 10^{12.1} n^{-1} T^{-1/2} B \qquad (19)$$

for some fraction of the volume of the tube. For the parameters quoted, this becomes $b_D = 10^{3.0}$. This is not unreasonable, since it is still compatible with the limit calculated in (3).

The calculations in this section were directed at the study of a large two-ribbon flare, in which the impulsive phase begins when the speed of the filament is sufficiently large to precipitate the Kelvin-Helmholtz instability. However, it appears that similar considerations may be relevant to other types of flares. For instance, observational data indicate that some flares are associated with newly emerging flux (Heyvaerts et al., 1977). It seems possible, in such a case, that the flare is initiated by reconnection at the current sheet that develops between the new flux system and the old flux system. Equation (11) may be used to estimate the thickness to which a current sheet will grow by diffusion in a given time. If the new flux system has been in place for one day, the thickness of the current sheet will be about $10^{4.3}$ cm. When reconnection occurs in such a sheet, fluctuations of much smaller scale will develop. Hence reconnection in such a current sheet can serve to "trigger" a flare in a large stressed flux tube, just as the Kelvin-Helmholtz instability was invoked to trigger reconnection for the case of a two-ribbon flare.

Late Phase of Flare

The soft X-ray emission from a flare typically lasts longer than the hard X-ray emission that is confined to the impulsive phase. In some flares, such as a simple loop flare, the long duration of soft X-ray emission may be attributed to the energy of the hot, dense "flare plasma" that evaporates from the chromosphere into the corona during the impulsive phase.

In other cases, such as two-ribbon flares, careful analysis (Moore et al., 1980) shows that further release of stored energy (presumably magnetic) is required. An early model of flares (Sturrock, 1968) showed that the topology of flare ribbons and post-flare loop systems could be explained in terms of the progressive reconnection of extended current sheets. Similar ideas were later advanced by Kopp and Pneuman (1976). According to these authors, the formation of the current sheet was attributed to the solar wind.

A modification of this view has more recently been proposed by Sturrock et al. (1984) and Cliver et al. (1986). According to this view, the eruption of a filament deforms the overlying magnetic field structure to form a current sheet extending below the filament. Reconnection of this current sheet then produces high-energy electrons that are responsible for meterwave radio emission, microwave radio emission, soft X-ray emission, and Hα emission.

Finally, we can comment once more on the classification between impulsive and gradual energy release, as discussed by Bai (1986). This division of flares into two classes may be understood if a typical impulsive phase involves energy release in a small magnetic loop below a filament, whereas energy release in a typical gradual phase involves energy release in a magnetic loop that extends above a filament.

Acknowledgments. This work was supported by ONR contract N00014-85-K-0111 and NASA grants NGL 05-020-272 and SE80D7840R.

References

Bai, T., Two classes of gamma-ray/proton flares: Impulsive and gradual, Ap. J., 308, 912, 1986.

Cane, H.V., and R.G. Stone, Type II solar radio bursts, interplanetary shocks, and energetic particle events, Ap. J., 282, 339, 1984.

Cliver, E.W., et al., Solar gradual hard X-ray bursts and associated phenomena, Ap. J., 305, 920, 1986.

Dobrowolny, M., Kelvin-Helmholtz instability in a high-beta collisionless plasma, Phys. Fluids, 15, 2263, 1972.

Dreicer, H., Electron and ion runaway in a fully ionized gas. I, Phys. Rev., 115, 238, 1959.

Dreicer, H., Electron and ion runaway in a fully ionized gas. II, Phys. Rev., 117, 329, 1960.

Fainberg, J., and R.G. Stone, Satellite observations of type III solar radio bursts at low frequencies, Space Sci. Rev., 16, 145, 1974.

Forrest, D.J., and E.L. Chupp, Simultaneous

acceleration of electrons and ions in solar flares, Nature, 305, 291, 1983.

Furth, H.P., J. Killeen, and M.N. Rosenbluth, Finite-resistivity instabilities of a sheet pinch, Phys. Fluids, 6, 459, 1963.

Harrison, R.A. Solar coronal mass ejections and solar flares, Astron. Astrophys., 162, 283, 1986.

Heyvaerts, J., E.R. Priest, and D.M. Rust, An emerging flux model for the solar flare phenomenon, Ap. J., 216, 123, 1977.

Kahler, S.W., E.W. Cliver, H.V. Cane, R.E. McGuire, R.G. Stone, and N.R. Sheeley, Solar filament eruptions and energetic particle events, Ap. J., 302, 504, 1986.

Kane, S.R., E.L. Chupp, D.J. Forrest, G.H. Share, and E. Rieger, Rapid acceleration of energetic particles in the 1982 February 8 solar flare, Ap. J. (Lett.), 300, L 95, 1986.

Kaufmann, P., E. Correia, J.E.R. Costa, A.M. Zodi Vaz, and B.R. Dennis, Solar burst with millimetre-wave emission at high frequency only, Nature, 313, 380, 1985.

Kaufmann, P., F.M. Strauss, and R. Opher, Some characteristics of ultra-fast time structures superimposed on impulsive mm-wave bursts, IAU Symp. No. 86, Radio Physics of the Sun, Reidel, Holland, p. 205, 1980.

Kiepenheuer, K.O., An assembly of flare observations as related to theory, Proc. AAS-NASA Symp. on Physics of Solar Flares, edited by W.N. Hess, NASA SP-50, Washington, DC, p. 323, 1964.

Kiplinger, A.L., B.R. Dennis, A.G. Emslie, K.J. Frost , and L.E. Orwig, Millisecond time variation in hard X-ray solar flares, Ap. J. (Lett.), 265, L99, 1983.

Kopp, R.A., and G.W. Pneuman, Magnetic reconnection in the corona and the loop prominence phenomenon, Solar Phys., 50, 85, 1976.

Kulsrud, R.M., Y.-C. Sun, N.K. Winsor, and H.A. Fallon, Runaway electrons in a plasma, Phys. Rev. Lett., 31, 690, 1973.

Lin, R.P., R.A. Schwartz, R.M. Pelling, and K.C. Hurley, A new component of hard X-rays in solar flares, Ap. J. (Lett.), 251, L109, 1981.

Moore, R.L., et al., The thermal X-ray flare plasma, in Solar Flares, edited by P.A. Sturrock, Colorado University Press, Boulder, Colorado, p. 341, 1980.

Schmidt, G., Physics of High Temperature Plasmas, Academic Press, New York, p. 134, 1966.

Smith, H.J., and E.v.P. Smith, Solar Flares, MacMillan Co., New York, p. 35, 1963.

Smith, S.F., and H.E. Ramsey, The flare-associated filament disappearance., Z. Astrophys. 60, 1, 1964.

Sturrock, P. A., A model of solar flares, IAU Symp. No. 35, Structure and Development of Solar Active Regions, edited by K.O. Kiepenheuer, Reidel, Holland, p. 471, 1968.

Sturrock, P. A., P. Kaufmann, R.L. Moore, and D.F. Smith, Energy release in solar flares, Solar Phys., 94, 341, 1984.

Van Hoven, G., et al., The preflare state, in Solar Flares, edited by P.A. Sturrock, Colorado University Press, Boulder, Colorado, p. 17, 1980.

Wagner, W.J., Coronal mass ejections, Ann. Rev. Ast. Astrophys., 22, 267, 1984.

Wild, J.P., S.F. Smerd, and A.A. Weiss, Solar bursts, Ann. Rev. Ast. Astrophys., 1, 291, 1963.

THE SOFT X-RAY TELESCOPE FOR THE SOLAR A MISSION

M.E. Bruner[1], L.W. Acton[1], W.A. Brown[1], R.A. Stern[1], T. Hirayama[2], S. Tsuneta[2], T. Watanabe[2], and Y. Ogawara[3]

Abstract. The Solar A mission, being conducted by the Japanese Institute for Astronautical and Space Sciences, is a project to study solar flares using a cluster of instruments on an orbiting satellite. It is scheduled to be launched in September or October of 1991. The emphasis of the mission is on imaging and spectroscopy of hard and soft X-rays. The Soft X-Ray Telescope (SXT), one of two major imaging instruments on the satellite, is a joint U.S.-Japan project. It is being prepared at Lockheed under NASA sponsorship. The electronic control system for the SXT is based on microprocessors and is a joint effort between Lockheed and the National Astronomical Observatory of Japan (NAOJ). The SXT uses a glancing incidence telescope of 1.55 m effective focal length to form images in the 0.25 to 3.0 keV range on a 1024 x 1024 virtual phase CCD detector. A selection of thin metallic filters located near the focal plane provides the capability for electron temperature diagnostics. Knowledge of the alignment of soft X-ray images with respect to features observable in visible light is provided by a coaxially mounted aspect telescope which forms its image on the CCD sensor when the thin metallic filter is replaced by an appropriate glass filter A novel mechanical design has permitted a very lightweight structure that remains stiff enough to survive the severe launch environment. Other Solar A instruments include a hard X-ray telescope, a Bragg crystal spectrometer, a wide band spectrometer, and a radiation belt monitor.

[1]Lockheed Palo Alto Research Laboratory, Palo Alto, California, U.S.A. 94304

[2]National Astronomical Observatory of Japan, Mitaka, Tokyo, Japan

[3]Institute for Space and Astronautical Sciences, Sagamihara, Japan

Introduction

The Solar A mission is one of a continuing series of scientific satellites that are being prepared in Japan under the auspices of the Japanese Institute for Space and Astronautical Sciences (ISAS). Solar A is dedicated to the study of solar flares, particularly as characterized by energy radiated in the soft and hard X-ray range. The principal instruments on Solar A will be a pair of X-ray imaging instruments, one for the hard X-ray range and the other for the soft X-ray range. The Hard X-Ray Telescope (HXT) uses an array of modulation collimators to record Fourier transform images of the nonthermal and hot plasmas that are formed during the early phases of a flare. These plasmas are thought to be intimately associated with the sites of primary energy release. The Soft X-Ray Telescope (SXT) uses a glancing incidence mirror to form direct images of the lower temperature (but still very hot) plasmas that form as the solar atmosphere responds to the injection of energy.

The SXT instrument is a joint development effort between the Lockheed Palo Alto Research Laboratory and the National Astronomical Observatory of Japan. The U.S. effort also involves Stanford University, the University of California, Berkeley, and the University of Hawaii who provide support in the areas of theory, data analysis and interpretation, and ground-based observations. The hard and soft X-ray telescopes both have an alignment sensor, operating in white light, to provide co-alignment information. The major properties of the Solar A instruments are summarized in Table 1.

The Solar A spacecraft is 1 x 1 m wide by 2 meters long and is expected to have a total mass of about 420 kg. It is stabilized about all three axes, using inertia wheels and a control moment gyro. Pointing accuracy is expected to be 3 arc min absolute, with a short term stability (over a 5-min interval) of about 5 arc sec. Power is provided by a conventional solar array. Total data rates range from 1 to 32 kbps, depending on the observing mode. A command memory and an onboard computer provide a versatile system for defining observing

TABLE 1. Solar A Mission Instruments

Hard X-Ray Telescope	
PI:	Prof. K. Kai (Nat'l Astron. Obs. of Japan)
Instrument:	Fourier Synthesis Telescope
Energy range:	10 - 100 keV
Angular resolution:	7 arc sec
Effective area:	1.5 cm^2 avg. x 64 elements
Soft X-Ray Telescope	
PI:	Prof. T. Hirayama (Nat'l Astron. Obs. of Japan)
	Japan-U.S. collaboration
	U.S. PI to NASA: Dr. L. W. Acton (Lockheed)
Instrument:	See Table 2
Bragg Crystal Spectrometer	
PI:	Prof. E. Hiei (Nat'l Astron. Obs. of Japan)
Japan-UK collab.	(incl. NRL & NBS in U.S.)
UK PI to SERC:	Prof. J. L. Culhane (MSSL)
U.S. (NRL)	Dr. G. Doschek
U.S. (NBS)	Dr. R. Deslattes
Instrument:	Curved Crystal Spectrometers
Spectral lines:	S XV, Ca XIX, Fe XXV, Fe XXVI
Spectral resolution:	1/3000 - 1/8000
Wide-Band Spectrometer	
PI:	Prof. J. Nishimura (ISAS)
Instrument:	Soft X-Ray / Gas Prop. Counter
	Hard X-Ray / NaI Scint. Counter
	Gamma-ray / BGO Scint. Counter
Energy range:	2 keV - 50 MeV

sequences, including data-dependent control of the observing mode. Onboard data storage is provided by a bubble memory with a capacity of 80 megabits.

The Solar A schedule calls for a launch in August or September of 1991, when the level of solar activity is expected to be at or near maximum. The satellite will be launched from the Kagoshima Space Center in southern Japan, and will be carried by a Japanese-built M3S-II launch vehicle In the remainder of this paper, we will discuss the scientific objectives, design, and capabilities of the SXT instrument as it is currently defined. It is our hope that this description will serve as a useful reference for experiment planning purposes and for potential users of SXT data. We caution the reader, however, that some instrument characteristics given here will probably change as the development and calibration program continues.

Science Objectives

Soft X-ray images provide the opportunity to visualize the distribution of high temperature coronal gas and, thus, the structure of the confining magnetic field. Such information is required to properly define the topological context of solar activity. For flare studies, the highest priority of the mission, the new observational contribution of Solar A, will be to provide simultaneous hard and soft X-ray images with good angular and temporal resolution. In particular, the SXT will provide information on:

- The geometry of the X-ray emitting structures and the inferred coronal magnetic field before, during, and after flaring
- The temperature and density of X-ray emitting plasma
- The spatial and temporal characteristics of flare energy deposition
- The transport of energetic particles and conduction fronts

Hopefully, it will also provide information on:

- The presence of waves or other magnetic field disturbances associated with sprays, filament eruptions, and coronal transients
- The locations of primary energy release and particle acceleration

A possibility exists to use the SXT visible light aspect sensor telescope for helioseismological studies. It is estimated that intensity oscillations as small as $dI/I \approx 10^{-7}$ may be detected if the SXT data stream is dedicated to this experiment for several months.

Expected Results

A correlated study of the observations made with all the Solar A instruments and the simultaneous observations made with ground-based solar radio and optical telescopes will help to answer the following specific questions about solar flares:

- What are the pre-flare conditions which give rise to an energetic flare? What are the fundamental physical differences between flares with strong nonthermal effects such as high energy particle acceleration and mass ejection and those that appear to to be primarily thermal in nature?
- What physical processes are responsible for the energy release? Is the energy released continuously or in discrete pulses (elementary flares)?
- What are the conditions under which the energy released during the impulsive phase will drive the entire flare? What determines if additional energy is to be released during the gradual phase?
- What is the rate of energy release? How does it vary during the flare?
- What is the characteristic time for the acceleration process?
- Are electrons and ions accelerated simultaneously by the same process? Are there multiple phases or steps in the accel-

eration process to cover the wide range of energy (non-relativistic to relativistic) and mass (electrons, protons, and heavier ions)?
• Where does the acceleration occur in relation to the magnetic field structure in the vicinity of the optical flare? What are the dimensions of the acceleration region? Is the acceleration region spatially coincident with the sources of hard X-ray and gamma-ray emission?
• If the acceleration region is not spatially coincident with the hard X-ray and gamma-ray sources, how do the energetic particles propagate from the acceleration region to these sources? Do they diffuse or propagate in well-collimated beams?
• What is the relationship between the energetic particles which escape from the sun into the interplanetary space and those which remain at the sun and produce hard X-ray, gamma-ray, radio, and other emissions?
• How is the energy transported from the site of energy release to the sources of soft X-ray, EUV, and optical emissions? Do the energetic particles play a major role in the energy transport to the chromosphere, especially during the impulsive phase?

SXT Instrument Requirements

The design of the SXT instrument was constrained by the following requirements:

• A dynamic range of $>10^7$ is needed to cover the expected brightness range.
• Time resolution of 2 sec or better is required to cover the evolution of the impulsive phase.
• Angular resolution of 2 to 3 arc sec is required to locate the flare footpoints and observe the filling of loop structures.
• The field-of-view must be large enough to cover substantially the whole solar disk without repointing the spacecraft.
• A spectral diagnostic capability is required to measure plasma properties.
• The capability to make co-aligned images in visible light is needed to determine the locations of the soft X-ray sources with respect to hard X-ray sources and to the magnetic field configuration and other structures observed from the ground.
• Resource constraints include a total mass less than 30 kg, a maximum power of about 20 W, and maximum physical dimensions of 30 x 30 x 170 cm.
• It must withstand a severe launch environment (25 g rms vibration levels). The fundamental mechanical resonance frequency must exceed 100 Hz. The SXT must operate over a temperature range of -10 to +25 °C.

SXT Instrument Overview

The Soft X-Ray Telescope uses glancing incidence optics to form direct images on a CCD detector. The significant properties of the SXT instrument are listed in Table 2.

The optical system, shown schematically in Figure 1, includes an entrance aperture filter, the X-ray mirror, a filter wheel assembly, a rotating shutter, and the CCD camera. A coaxially mounted objective lens assembly allows visible light images to be made on the same CCD detector for aspect determination. The separation of the objectives and the focal plane is maintained by a metering tube of carbon fiber-epoxy composite. The metering tube is designed to compensate for thermal expansion in the other parts of the instrument over the expected range of operating temperatures. The X-ray pass-band of the SXT instrument is determined by the entrance aperture filter, the reflectivity of the mirror, the spectral response of the CCD detector, and by the choice of one of several thin metallic film analysis filters. These filters block

TABLE 2. SXT Characteristics

X-Ray Telescope	
Diameter	230 mm
Geometric area	250 mm^2
Focal length	1550 mm
Wavelength range	4 - 50 Å
Resolution	< 4 arc sec over solar diameter
Mirror	Nariai-Werner Double Hyperboloid
	Single piece, gold on Zerodur
Metering tube	Negative CTE graphite-epoxy
Aspect Telescope	
Aperture	50 mm
Focal Length	1550 mm
Transmission bands	
Wide band filter	4600-4800 Å
Narrow band filter	4293-4323 Å
Filter Wheels	
Two 6-position filter wheels in series	
6 ea. X-ray analysis filters	
1 ea. 10% transmission X-ray mask	
2 ea. optical filters	
1 ea. optical diffuser	
2 ea. open positions	
Detector	
CCD	1024 x 1024 18.3 micron (2.44 ") pixels
Accommodation	
Weight	30 kg
Power	18 watt average, 29 watt peak
Envelope	30 x 30 x 170 cm

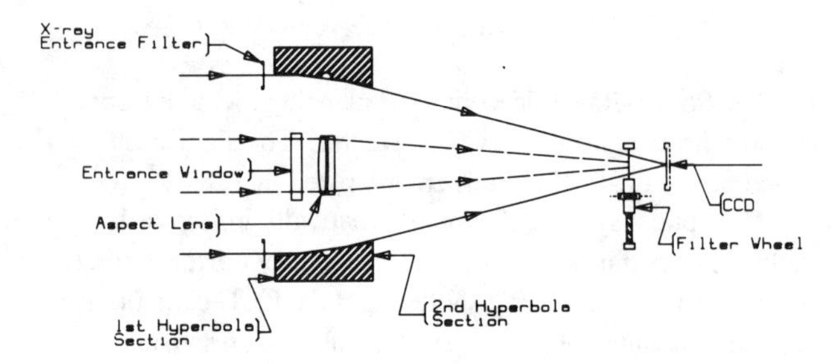

Fig. 1. Schematic of SXT optical system. The X-ray and visible light objectives have equal focal lengths and are co-axially mounted. The setting of the filter wheel determines which image reaches the CCD detector.

the visible light from the aspect telescope aperture so that it does not contaminate the X-ray images. The filter wheel also contains two glass filters that can be used to observe visible images for aspect determination. These filters are opaque in soft X-rays, so that there is no X-ray component in the visible images.

The CCD is cooled to about -20 °C in order to reduce the effect of dark spikes (pixels with abnormally high dark current levels). Exposure times are controlled by a rotary shutter mounted just in front of the focal plane. The CCD camera can also be operated in a pseudo frame transfer mode in the event of problems with the mechanical shutter. A partial frame read-out mode is provided to allow rapid exposure sequences to be made and to reduce the amount of data that must be handled by the spacecraft image processing system.

The SXT electronic system is in two parts: a mechanism control system and a command and data handling system. Both parts of the electronics are microprocessor controlled. The mechanism control system is part of the U.S. development work, while the command and data handling system is part of the Japanese effort.

X-Ray Optics

The X-ray optical system is based on a variant of the Wolter Type I mirror that was recently developed by Nariai (1987, 1988). The Nariai mirror can be understood most readily by comparing it to the Wolter I (Wolter, 1952a, b). The latter consists of a pair of glancing incidence annular mirrors operating in tandem. The first mirror is a section of a paraboloid and the second a section of a hyperboloid. Parallel light incident on the paraboloid is reflected toward its focus. After leaving the paraboloid, the converging beam is intercepted by the surface of the hyperboloid, whose first focus coincides with that of the paraboloid. After reflection from the hyperboloid, the beam converges to its second focus, where the final image is formed on a detector. The Nariai design differs in that both mirror segments have been made hyperbolic in order to gain better off-axis performance at the expense of a slight loss of on-axis resolution. The Nariai design has a lower field curvature than the Wolter I design, a helpful feature when a flat detector such as the CCD must be used. The Wolter I telescope is the glancing incidence analog of the classical Cassegrainian system, while the Nariai design is, in a sense, analogous to the Ritchey-Chre`tien system.

The high intensity levels expected from solar flares, and the high quantum efficiency of the CCD detector have permitted us to use shorter segments for the mirrors than is customary with Wolter I optics. The use of short segments also reduces the effects of field curvature on the angular resolution. The length of each mirror segment is only 2.0 cm and the corresponding width of the annulus is 0.37 mm. Such a highly obscured aperture obviously has an extremely broad diffraction pattern in the visible region of the spectrum, and the effects are noticeable at the 40 Å end of the soft X-ray spectrum. Diffraction effects are smaller than the CCD pixel size for wavelengths below about 20 Å. The expected performance of the Nariai mirror is shown in Figure 2. Resolution is expected to be limited by the CCD pixel size out to about 15 arc min from the axis, and to be no worse than 4 arc sec at the solar limb.

The mechanical design of the SXT mirror system is also somewhat unusual. The mirror system, procured from United Technology Optical Systems, is to be a one-piece design with both optical surfaces formed on the same piece of material.

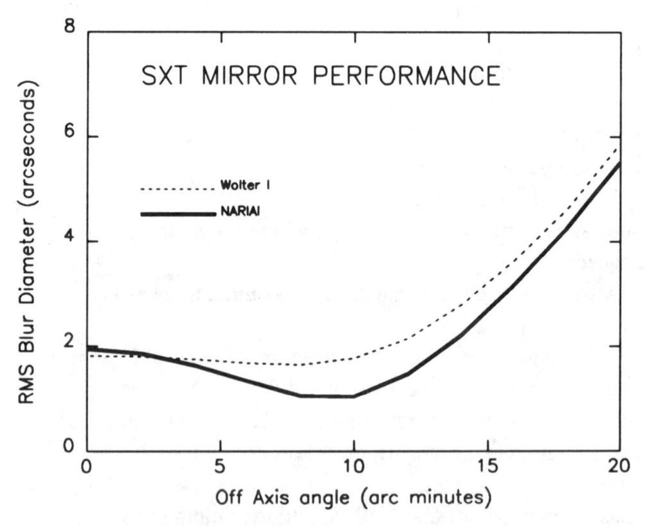

Fig. 2. Calculated X-ray imaging performance. The solid curve shows the expected rms spot diameter as a function of field angle for the Nariai-Werner double hyperboloid mirror system. The focal position has been chosen for optimum performance of an equivalent Wolter I design focussed for the same axial spot size.

The mirror will be made of Zerodur to make it insensitive to temperature gradients and fluctuations. The two hyperbolic sections will be separated by a gap of about 5 mm to facilitate manufacture. This approach obviously places rigorous demands on the optical machining and figuring process but has the great advantage of being absolutely stable in alignment once it has been made. Control of the figuring process is based on precise mechanical metrology methods, with only the final performance evaluation being done optically in soft X-rays. The X-ray mirror is mechanically supported by a set of six titanium fingers that extend forward from a common mounting ring. The mirror support fingers are provided with small invar pads which are bonded to the mirror with epoxy. This ring, in turn, is fastened at three places to the telescope structure. The symmetry properties of this design assure that the mirror will remain centered with respect to the mechanical structure at all temperatures. Lengths and cross sections of the support fingers have been chosen such that the natural frequency of the mirror system avoids known resonances in the instrument, the spacecraft, and the launch vehicle.

Aspect Telescope

The scientific goals of Solar A require that we be able to accurately determine the positions of soft X-ray sources both with respect to the hard X-ray sources and with respect to features visible from the ground. Therefore, both the HXT and SXT are equipped with aspect sensor systems with approximately arc sec accuracy that operate in the visible. The SXT aspect telescope consists of an apochromatic doublet lens mounted on the X-ray telescope axis at the center of the X-ray mirror assembly. Its focal length is to be a close match to that of the X-ray mirror so that their nodal points, and consequently their respective images, coincide. The maintenance of strict co-alignment throughout the launch environment and in orbit is thus reduced to the problem of holding the lens centered in the X-ray mirror. This is accomplished by supporting both lens and mirror from a common mounting plate, and by closely controlling the mechanical tolerance. The lens cell for the aspect telescope is made of titanium, which has a coefficient of thermal expansion that is very close to that of the lens material. The axial symmetry of the design plays an important role in keeping the lens centered with respect to the X-ray mirror throughout the operating temperature range.

The aperture of the aspect telescope is 5 cm and was chosen to match the lens diffraction limit to the CCD pixel size. The telescope operates in two different wavelength ranges, a wide band covering the interval 4600 Å to 4800 Å and a narrow band, 30 Å wide, centered at 4308 Å. The wide band filter is used to produce the white light images on which the aspect determinations are based. The narrow band is designed to observe the CN bandhead which is sensitive to regions of magnetic plage. The 4308 Å filter will be used to assist in locating X-ray sources with respect to the magnetic field, particularly when high quality ground-based magnetograms are not available. These optical bands are potentially useful for spatially resolved helioseismological studies.

The entrance window of the aspect telescope serves two functions: to control the amount of light entering the aperture (only 0.01% of the visible flux in a 30 Å band is required for an adequate CCD signal level), and to act as a passive thermal control element. It consists of a second-surface aluminum mirror, with the thickness of the aluminum chosen to give the desired transmission. During daylight portions of the orbit, the aluminum coating reflects about 90% of the visible radiation, absorbs about 10%, and transmits about 0.1%. The radiation resistant entrance window (fused quartz or Schott BK-7) is black in the infrared and serves as a radiator to dissipate the absorbed visible energy. During orbit night, the glass absorbs infrared radiation from the Earth in an amount that is comparable to to the daytime visible heat load, so that orbital variations are minimized.

As previously discussed, operation of the aspect telescope is controlled by the position of the filter wheel. When one of the visible filters is in position, soft X-rays are blocked by the glass substrates. When one of the X-ray analysis filters is in position, X-rays are transmitted, but visible light is blocked by the metal and carbon contained in the filters. The aspect telescope objective is fitted with a motor-driven cover that can be closed to allow X-ray images to be made without an analysis filter in position or in the event of a visible light leak (pinhole) in an X-ray analysis filter.

CCD Detector

The choice of a detector for the SXT focal plane is crucial to successfully fulfilling the Solar A science and mission requirements. The CCD camera is being developed at the Jet Propulsion Laboratory (JPL) and is based on the Texas Instruments 1024 x 1024 front-illuminated virtual-phase CCD (VPCCD). This device was selected because of its good X-ray quantum efficiency and because its pixel size and format were well matched to the SXT focal length. It has minimal cooling requirements and rapid readout capability.

Unlike other CCDs, the VPCCD is fabricated with a single polysilicon gate of 5000 Å covering 1/2 of a thin (750 Å) oxide layer for each pixel. The remaining half of each pixel is essentially open, being covered only by a protective overcoat (also an oxide layer) about 2500 Å thick. This open gate structure produces relatively little X-ray absorption compared to the overlapping gate structure of a three-phase CCD. Thus the VPCCD can be used in a front-illuminated mode and still provide high soft X-ray quantum efficiency. In addition, diffusion of X-ray created electrons into more than a single pixel is minimized because essentially all of the SXT X-rays are absorbed in the CCD depletion region.

A second major advantage of the VPCCD is its extremely low dark current. The 1024 x 1024 VPCCD has a measured dark rate of less than 50 electrons^{-1} pixel^{-1} s^{-1} at 25 °C. In contrast to CCDs used in optical astronomy, achieving extremely low read noise (i.e., < 10 electrons rms) is not a major concern for the SXT. This is because each interacting X-ray photon produces one electron for each 3.65 eV of energy. For example, at the 18.97 Å O VIII line, roughly 180 electrons per photon are produced. The Solar A CCD camera is expected to have a total system noise of the order of 100 electrons/sec rms. Hence detection of a single photon at the 3 sigma level is possible. Because the dynamic range is limited by the well capacity of the CCD pixels (250,000 electrons full well), the dynamic range of the camera will be about 750-3500 depending upon the X-ray photon energy. This feature, together with exposure times which can range from 0.001 sec to 20 sec and the selection of X-ray filters, will achieve an overall dynamic range above 10^8.

One of the few undesirable characteristics of the VPCCD is the presence of so-called "dark spikes," or pixels with anomalously high dark currents. The number of dark spikes in the current generation of 1024 x 1024 devices relative to the total number of pixels is small, of the order of 3%, and are thus unlikely to have a major impact on the Solar A scientific investigation. The effects of dark spikes can be reduced by cooling the detector and by using short exposures. They can be almost entirely eliminated via dark frame subtraction. Radiation damage manifests itself as an increase in the number of dark spikes which may approach 10% of SXT pixels after 3 years in orbit.

The JPL camera electronics are based on a high-speed, double-correlated sampling technique. The camera requires two inputs: a clock and a serial command message, the length of which depends upon the readout mode desired. The output of the camera is a series of 15-bit parallel words, 12 bits of which are video data from the camera. These 12 bits will be later compressed to 8 bits via a lookup table in the electronic system. The SXT camera has several special design features which are important for Solar A observations, including the ability to sum pixels on-chip in both dimensions (for example, a 2 x 2 or 4 x 4 pixel sum). This allows for data compression on-chip (at the expense of spatial resolution) for low-level signals (e.g., in coronal hole observations) without an increase in readout noise. For readout of selected regions of the solar disk, it also has the ability to select out in one dimension any number of CCD rows of interest. This feature allows for rapid readout of such selected regions, since the CCD can be rapidly clocked in the vertical direction without saving the data until the row of interest is reached, when normal horizontal readout proceeds until the last row of interest is reached. Using this technique, a 2 arc min swath of the Sun's disk can be read out in about 0.5 sec, allowing for rapid timing studies of flaring regions.

Structure

The design of the SXT instrument structure was a major technical challenge because of the severe environment and the tight constraint on weight. The entire instrument assembly could weigh no more than 20 kg, yet had to be stiff enough so that its fundamental resonant frequency was greater than 100 Hz. To meet these limits, the traditional three-point mounting was abandoned in favor of an over-constrained design with elastic averaging.

The elements of the SXT structural system are shown in Figure 3. At the objective end, the X-ray mirror and the aspect telescope objective are fastened to a forward support plate and its mount. The X-ray entrance filter is mounted to a housing assembly (not shown) that surrounds the X-ray mirror. The focal plane group, consisting of the filter wheel assembly, the CCD camera head, and the rotary shutter, are all fastened to a rear support plate and bracket assembly. Separation between the objectives and the focal plane group is maintained by a metering tube made of epoxy-carbon fiber composite manufactured by Fiber Technology Corporation. The inner and outer surfaces of the metering tube are coated with 25 micron thick aluminum foil, which acts as a vapor barrier to prevent water vapor absorption. The fiber properties and winding angles for the metering tube have been chosen such that the overall coefficient of thermal expansion for the metering tube is slightly negative. This was done to compensate for the expansion properties of the X-ray mirror mounting fingers, the filter wheel housing, and the CCD camera housing. With the compensated design, the focus is expected to remain stable to within a few microns over the entire expected operating temperature (-10 to + 25 °C) without the need for active focusing elements.

Careful attention to the geometry of the structure is a key element in the SXT instrument design. As shown in Figure 3, the structure provides a very direct load path between the base of the X-ray mirror mount and the CCD camera to minimize mechanical uncertainties in the focus. All alignment and focus settings are determined by the thicknesses of metal spacers whose dimensions are ground or lapped to size during instrument calibration. This approach allows the instrument to be disassembled for servicing and inspection during the prelaunch phase without loss of alignment or focus, and minimizes alignment changes during the launch. The telescope is designed to be mounted to a flat central wall of the spacecraft in four areas; the two ends of the forward support plate mount, and

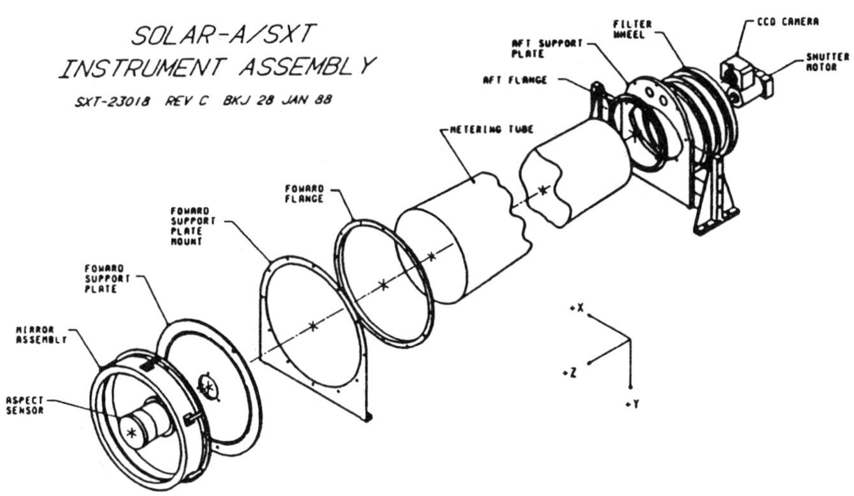

Fig. 3. The SXT Instrument Assembly.

the two support brackets for the focal plane group. Differences in thermal expansion between the spacecraft wall and the instrument are absorbed by the forward support plate mount, which flexes to accommodate the length difference. Twists and departures from flatness of the spacecraft wall are absorbed by the metering tube, which acts as a torsion spring. The completed assembly is expected to have a mass of about 19 kg, and a first mode resonant frequency of about 155 Hz.

X-Ray Filters

The SXT X-ray filters comprise the entrance filter and analysis filters. The X-ray telescope entrance filter excludes visible and UV light while passing X-rays. It consists of a 1800 Å thick Lexan ($C^{16}H^{14}O^{3}$) substrate coated on one side with 900 Å of titanium followed by 800 Å of aluminum. The Al layer faces the Sun to exclude visible light and near UV and to provide appropriate thermal control. The Ti layer blocks the strong He II resonance line at 304 Å.

The analysis filters are contained in a filter wheel assembly at the focal plane end of the telescope. The filter wheel assembly, developed at Schaeffer Magnetics, Inc., contains two wheels mounted in tandem, each accommodating five filters plus an open position. Each wheel is in the form of a pancake stepping motor, with the filters mounted within the motor rotor. The motors are driven from the rim by a rotating magnetic field operating on a ring of permanent magnets mounted on each rotor. Of the10 filter stations ,6 are devoted to X-ray analysis filters, two to visible light bandpass filters, one to a visible light diffuser (for CCD gain measurements), and one for an aperture mask (for reducing the X-ray effective aperture during intense flares). Each motor draws about 8 watts when running, and can reach any arbitrary position in 1 sec or less. Optical encoders (3 bits each) are provided to indicate the position of each rotor. The tentative list of X-ray analysis filters is as follows:

- 100 micron Be (unsupported)
- 12 micron Al on 80% mesh
- 1200 Å Mg on 80% mesh
- 1200 Å Al on 80% mesh
- 3000 Å Al + 2000 Å Mg + 600 Å Mn on 80% mesh
- To be determined

SXT Experiment Control

SXT control and data handling is under the control of electronics and software in the Solar A data processor (DP) computer. As illustrated in Figure 4, it is the function of the DP to control the SXT and to receive, format, and read out the data to telemetry. It is the function of the SXT microprocessor to control the SXT mechanisms and CCD camera as directed by the DP.

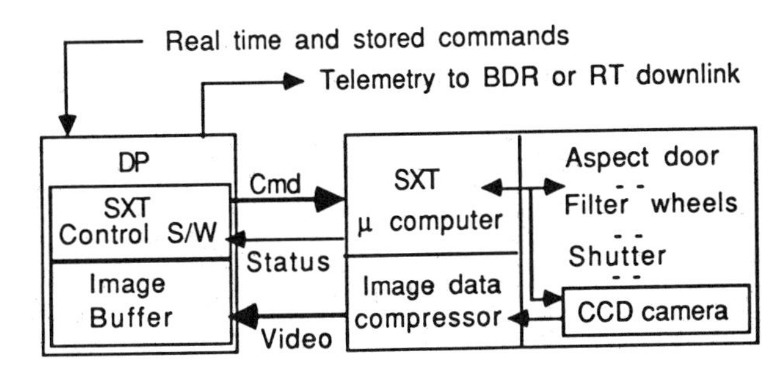

Fig. 4. Schematic of SXT control and data system.

Command and status information is exchanged between the two computers via a block of memory in the SXT microprocessor which serves as a "mailbox." By implementing the "mailbox" approach it was possible to freeze the design of the electrical interface even before the command and status requirements of the system were fully known.

The DP has the capability to examine SXT image data to select regions of interest and to adjust the length of the SXT exposure. It can also monitor data from the hard or soft X-ray spectrometers and issue a flare flag to alter the telemetry mode and SXT observing sequence. These capabilities are not yet fully implemented in software but their intended function will now be briefly described.

The operation and capabilities of the SXT may be described in terms of the parameters which must be specified for each exposure and the constraints on their values. The primary variables for each SXT exposure are:

- Location and size of portion of CCD image to be saved
- On-chip CCD pixel summation of 1 x 1, 2 x 2 or 4 x 4
- Exposure duration
- Filter(s) to be used

In addition, there are secondary instrument functions which may need to be specified:

- Open or shut door behind aspect telescope aperture
- Shutterless (pseudo frame transfer) operation of the CCD
- Transfer image words in compressed or truncated form (either the eight high order or eight low order bits)

The setup of the SXT for each observation is accomplished from tables of parameters stored in the DP. Each table provides all of the information necessary to define an SXT observing sequence, which can be quite complex. The number of such tables is not yet determined. These tables are stored in non-volatile memory and can be modified by ground command. The selection of which observing sequence table to use at any given time may be made from time-tagged stored commands, from real time ground commands, or in response to a flare flag.

Region of Interest Selection

In order to maintain the required time resolution for studies of solar activity with the SXT it will normally be necessary to transmit only a subsection of the full 1024 x 1024 CCD array. As implemented for the SXT such a subsection is limited to a 64 x 64 pixel size and is referred to as a partial frame image (PFI). The field-of-view and time resolution achievable with a PFI is described below. The location of each selected PFI is termed a region of interest (ROI).

It is possible for the DP to locate and track up to 8 regions of interest. The specification of 4 regions is by uplink command. The locations of the other 4 is automatic, based on X-ray brightness. In the automatic mode a special "patrol" exposure is taken and a 42' x 21' section is stored in a dedicated patrol image buffer for search by the DP for the brightest 1, 2, or 4 regions, with a location resolution of 10" (4 pixels). These locations are subsequently used as the center of PFIs for regions of interest. A patrol exposure is normally taken about every 5 min.

Flare Flag

Solar A makes use of an internally generated flare flag to assure that available telemetry capacity is most effectively used for solar observing. Flare observations, especially observations of large flares, normally have priority. Flare alerts are derived from the signals of the non-imaging hard X-ray spectrometer (HXS), soft X-ray spectrometer (SXS), or both instruments through an "or" logic. The flare flag function of the DP is disabled if there is a high signal level from the Solar A radiation belt monitor (RBM).

The operation of the flare flag is controlled by seven settable parameters:

1. Detector (HXS, SXS) selector
2. RBM veto threshold
3. Low flare threshold
4. High flare threshold
5. Flare end threshold
6. Flare mode timeout
7. Preflare data saver

The flare flag software in the DP controls the telemetry mode, the utilization of storage space in the bubble data recorder, and triggers special flare observing sequences in the

During non-flare periods the SXT data rate will be as given in quiet-hi or quiet-med (SXT data rates for different telemetry modes are presented in Table 3) and the SXT may be interleaving full frame images with partial frame images of up to four different solar regions. When the low flare threshold is triggered the SXT enters a pre-defined observing sequence to concentrate on the flare. A special patrol image is taken to locate the flare, assumed to be the brightest X-ray object present. At the same time the telemetry mode is changed to flare-hi and the flare mode timer is started. After the flare mode timeout has elapsed the telemetry mode is switched to flare-med and the flare end threshold is periodically interrogated. When it is found that the flare flag detector signals have dropped below the flare end threshold, the telemetry mode switches back to quiet. For orbits, termed visible orbits,

TABLE 3. SXT Image Readout to Telemetry

Telemetry mode	Image Transfer Time FFI[a]	PFI[b]
Quiet (Hi, FFI:PFI=2:8[c])	1024	2
(Hi, FFI:PFI=8:2)	256	8
Quiet (Med, FFI:PFI=2:8)	8192	16
(Med, FFI:PFI=8:2)	2048	64
Flare (Hi)		2
Flare (Med)		16

[a] A full frame image (FFI) has 1024 x n (n=64, 128, 256 or 1024) pixels. Times given are for 1024 x 512 case.

[b] Partial frame image (PFI) is defined as 64 x 64 pixels.

[c] The FFI:PFI ratio reflects how telemetry is shared between the two images.

in which the bubble data recorder can be dumped to a ground station, the telemetry mode is not switched to the medium rate.

As flare mode data are written to the bubble data recorder they are assigned a write priority. In simplest terms, if the bubble data recorder fills up, flare data can overwrite quiet mode data except for the pre-flare interval specified by the pre-flare data saver. Flares that exceed the high flare threshold have the highest priority; high flare data can overwrite low flare data. This data priority scheme is based on that used on the Hinotori mission which was very successful in capturing data on large flares.

The processing of a flare flag response is terminated by entering orbit night. If continuity of observation of a certain type has priority, SXT response to the flare flag can be inhibited.

SXT Data Handling

SXT operation is constrained to operate in synchronism with the readout of image data, which is done at a fixed rate. That is, during the daylight portion of each orbit the image buffer is always being read out and operation of the SXT is slaved to this readout rate. It is important to keep this constraint in mind as it governs the pacing of scientific observations. It is not possible to take pictures faster than this rate allows and exposures too infrequent or too long to sustain this rate can only be made at the expense of retransmitting old image data. The SXT data rates are given in Table 3. Housekeeping and status data are not included in these telemetry rates.

The variety of data rates deserves a word of explanation. The spacecraft operates with three basic data rates: high (32 Kbps), medium (4 Kbps), and low (1 Kbps). The low rate is for night use and contains no science data. The Hard X-Ray Telescope (HXT) does not produce data in the quiet (QT) mode. In this mode the SXT image buffer control allocates telemetry between two separate images with interleaved data streams. Typically, the full frame image would be read out of the larger section with a 4-min time resolution while partial frame images utilize the smaller section with 8-sec time resolution. In the flare (FL) mode, initiated by internal flare flag or by command, a portion of the data stream is assigned to the HXT and no full frame images are permitted. FL (med) is used to observe flare decay during orbits with no ground station contact.

Ignoring for the moment the fact that SXT operation must be in synchronism with telemetry (telemetry refers to all data transfer, whether it goes to real time downlink, to the bubble data recorder, or both), the time it takes to make one picture comprises the time to position the filters and flush the CCD (nominally 1 sec) plus the exposure time plus the time to clock out the CCD to the image buffer at 131,072 pixels s^{-1}. The CCD camera will select out rows of interest but transfers all 1024 columns. Therefore the readout time for a 1024 x 512 image at full resolution is

$$\frac{1024 \times 512}{131072} = 4.0 \text{ sec}$$

and the time for a partial frame image is 0.5 sec. Thus, for example, to maintain synchronism with telemetry the longest exposure for a partial frame image in the FL (hi) mode is

2.0	sec to read out one PFI to telemetry
− 1.0	sec filter position and CCD flush time
− 0.5	sec CCD clock out time for PFI
0.5	sec maximum exposure duration

Table 4 lists the longest exposures permitted by the telemetry rate for each of a sample of observing formats. The shortest exposure in all cases is 1 ms.

For any given exposure the CCD camera can be commanded to clock out selected rows or groups of rows and to sum the data modulo 2 as it is being read out. On-chip summation is limited to 2 x 2 or 4 x 4 pixels in the Solar-A aplication. Keeping in mind that the angular resolution of the SXT mirror should be 2-4 arc sec over the entire disk it is clear that the scientific trade off between area coverage, pixel size, and time resolution will be challenging. Table 4 provides some typical examples of these choices.

TABLE 4. SXT Observing Trade-Offs

TLM Rate (pix/s)	Pixel Sum	Pixel Size (arc sec)	Partial Frame Image FOV (arc min)	Partial Frame Image Δt (sec)	Partial Frame Image Max. Exp. (sec)	Full Frame Image FOV (arc min)	Full Frame Image Δt (min)	Full Frame Image Max. Exp. (sec)
256	1 x 1	2.4	2.6 x 2.6	16.0	14.5	41.6 x 20.8	34.1	2043
256	2 x 2	4.9	5.2 x 5.2	16.0	14.5	41.6 x 41.6	17.1	1019
256	4 x 4	9.8	10.4 x 10.4	16.0	14.5	41.6 x 41.6	4.3	253
512	1 x 1	2.4	2.6 x 2.6	8.0	6.5	41.6 x 20.8	17.1	1019
512	2 x 2	4.9	5.2 x 5.2	8.0	6.5	41.6 x 41.6	8.5	507
512	4 x 4	9.8	10.4 x 10.4	8.0	6.5	41.6 x 41.6	2.1	125
2048	1 x 1	2.4	2.6 x 2.6	2.0	0.5	41.6 x 20.8	4.3	251
2048	2 x 2	4.9	5.2 x 5.2	2.0	0.5	41.6 x 41.6	2.1	123
2048	4 x 4	9.8	10.4 x 10.4	2.0	0.5	41.6 x 41.6	0.5	29

Sensitivity and Diagnostic Capability

In comparison to the X-ray telescopes on Skylab (which used film), the SXT has a high photometric sensitivity because of the high quantum efficiency of the CCD (typically 30% or higher). The geometric aperture of the X-ray mirror is of the order of 2 cm^2. After accounting for the transmission of the entrance filter, the reflectivity of the X-ray mirror, and the quantum efficiency of the CCD, the effective area of the telescope still approaches 1 cm^2 at 10 Å. The useful range of sensitivity is about 4 to 50 Å, as shown in Figure 5.

A somewhat more meaningful representation of the sensitivity properties of the SXT is shown in Figure 6. Here, we have plotted as functions of temperature, the response in electrons s^{-1} expected for a constant emission measure of 10^{44} cm^{-3}, as viewed through several of the analysis filters. The data for the curves were developed by computing the expected solar spectrum at each temperature from the tables of Mewe et al. (1985) and then integrating the result over the sensitivity function of the SXT instrument. The curve marked "noback" represents the response with only the entrance filter in position and defines the highest sensitivity to low temperature plasmas. The curve marked Ber100 corresponds to the 100 micron Be filter, which is mostly sensitive to high temperatures. The other curves may be identified with other filters

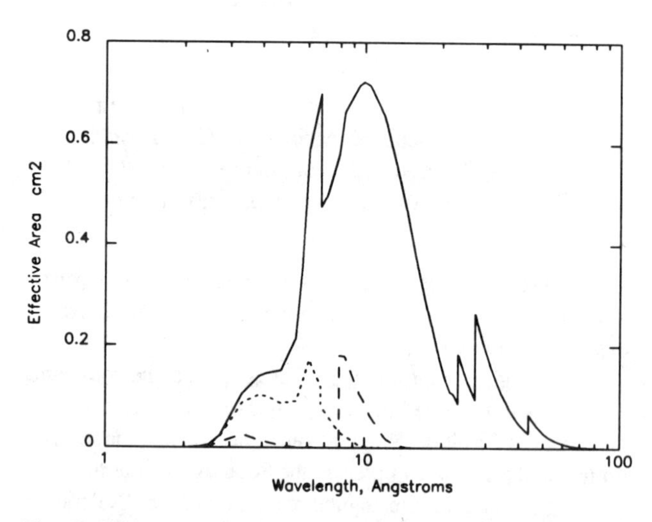

Fig. 5. SXT effective area as a function of wavelength for the basic instrument (solid curve), for the basic instrument plus the thick aluminum filter (long dashed curve), and for the basic instrument plus the thick beryllium filter (short dashed curve).

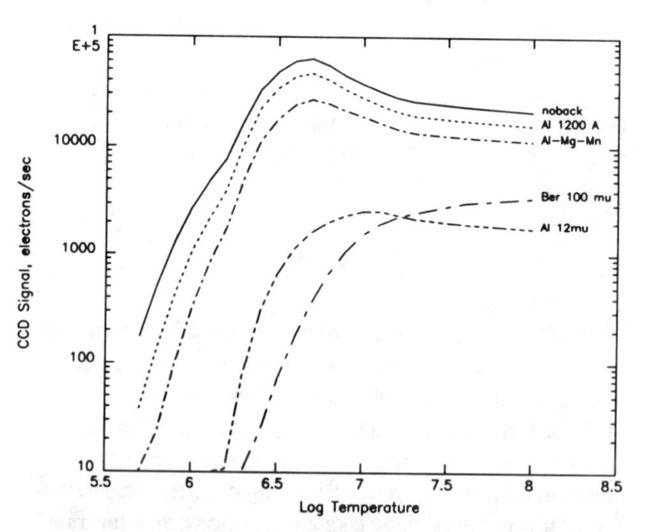

Fig. 6. SXT instrument response to an isothermal plasma of emission measure 10^{44} cm^{-3} at various temperatures. The solid curve represents the basic instrument with no analysis filter in place. The other curves correspond to different analysis filters as indicated in the legend.

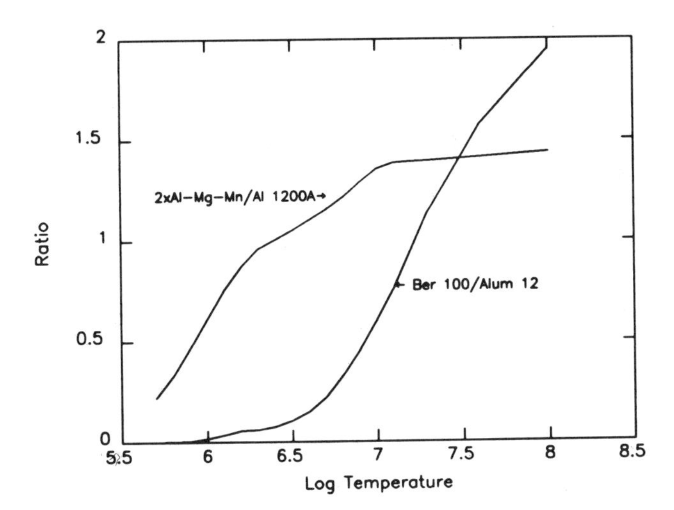

Fig. 7. Plasma-temperature diagnostic signal ratios. The steep curve represents the ratio through the two thickest filters. The other curve represents the ratio through a specially tailored multi-element filter to a thin aluminum filter.

discussed previously. The signal levels represent the total charge accumulation rate for the plasma in the field-of-view and would be divided among the several pixels covering a typical source.

An important consideration in the selection of analysis filters is their potential for plasma diagnostics. Figure 7 shows the dependence of the ratio of signals from pairs of filters as functions of the temperature of an assumed isothermal plasma. The shallow curve in Figure 7 is the ratio comparing the Al-Mg-Mn filter to the Al (1200 Å) filter, and gives a useful temperature diagnostic in the range 5×10^5 to 5×10^6 K. The steep curve compares the Be filter to the Si filter, and can be used from about 4×10^6 to 10^8 K. These examples are encouraging, but the selection of SXT X-ray analysis filters is by no means complete.

In summary, the SXT is the product of the particular requirements of the Solar A mission and advances in X-ray imaging instrumentation. Since the Skylab era, we have seen the advent of new mirror polishing methods that produce lower scatter; of large format CCD detectors with high X-ray quantum efficiency; and new structural materials, space-qualified microprocessors, bubble memories, and a host of other advances. The incorporation of these advances in the SXT will provide a powerful tool for solar research. The SXT, combined with the HXT and the other instruments of Solar A, should put us in a position to unlock some of the elusive secrets of the flare process during the next solar maximum -- an exciting prospect, indeed!

Acknowledgments. The authors are grateful to the Institute for Astronautical and Space Sciences and to NASA for initiating this program and for the effort it took to bring it to fruition. Among the many people who have contributed to the development of the instrument, we would especially like to recognize I. Kondo and T. Sakurai of the who contributed to the electronic system architecture and experiment design; L. Hovland of JPL who played a key role in the CCD camera development; D. Kyrie of Harmony Systems who designed the SXT microprocessor system; R. C. Catura, K. Appert, R. Caravalho, R. Fielder, M. Finch, B. Jurcevich, B. Rix, J. Vieira, and C. J. Wolfson of Lockheed who supported the development work; and John Owens, Paul Schwindt, and Skip Hassler from the Marshall Space Flight Center who provided technical and contractual support. We thank S. Ko for help with the manuscript. The program was supported at Lockheed under contract NAS8-37334 and by the Lockheed Independent Research Program.

References

Mewe, R., E.H.B.M. Gronenschild, and G.H.J. van Oord, Calculated X-radiation from optically thin plasmas. V, Astron. Astrophys. Suppl. Ser., 62, 197, 1985.

Nariai, K., Geometrical aberration of a generalized Wolter type I telescope, Appl. Opt., 26, 4428, 1987.

Nariai, K., Geometric aberration of a generalized Wolter type I telescope. 2: Analytical study, Appl. Opt., 27, 345, 1988.

Wolter, H., Mirror systems with glancing incidence on image-producing optics for X-rays, Ann. Phys., 10, 94, 1952a.

Wolter, H.,Generalized Schwarzschild mirror systems with glancing incidence as optics for X-rays, Ann. Phys., 10, 286, 1952b.

POWER SUPPLY PROCESS FOR SOLAR FLARES

S.-I. Akasofu

Geophysical Institute, University of Alaska, Fairbanks, AK 99775

Abstract. It is argued that a force-free magnetic field must be fed by field-aligned currents $J_\parallel$ and thus that $J_\parallel$ must be induced by divergence of $J_\perp$ which is generated by a dynamo process in the photosphere and below. It is suggested that the kinetic energy of the neutral component flow in the photosphere is the ultimate source of energy for solar flares.

Introduction

There is little doubt that solar flare phenomena are primarily various manifestations of electromagnetic energy dissipation processes. Therefore, there must be a generator process which can supply or regenerate the power needed for solar flares. In the past, solar flare studies have mostly been concentrated on the instability of magnetic configurations which could convert the stored magnetic energy into flare energy; a spontaneous magnetic reconnection has been singled out as the process for conversion. For these reasons, there have been few attempts to understand solar flares in terms of the power supply (generator)-dissipation process.

Similarly, in magnetospheric physics, it has been considered that the magnetotail contains enough magnetic energy for many substorms. Thus, there had been a period of one decade (~1965 - 1975) during which magnetospheric substorms had been discussed only in terms of spontaneous instability of the magnetotail field. This situation has slowly been changed after the discovery of the causal relationship of the polarity of the interplanetary magnetic field (IMF) with the occurrence of substorms. As a result, some aspects of the substorm phenomena have now been discussed in terms of the power supply-dissipation process. The power input is a function of the IMF magnitutde (B), the solar wind speed (V), and the polar angle (θ) of the IMF vector; an example of such a function is $\varepsilon = VB^2 \sin^4 (\theta/2)\ \ell_o^2$ where ℓ_o is a constant (~7 Earth radii); see Akasofu (1981).

Generation of Force-Free Fields

There appears to be a general agreement among many solar physicists that force-free magnetic fields are the configurations which store magnetic energy needed for solar flares (e.g., Nakagawa et al., 1971; Low, 1989). Therefore, let us consider a generation mechanism--the power supply process--of a simple force-free field.

One of the simplest force-free fields can be produced by feeding field-aligned current ($J_\parallel$) along a loop-shaped potential field (Figure 1). In a steady state, the current continuity condition at the feet of the field line is given by

$$\nabla \cdot J_\perp = - \nabla \cdot J_\parallel \qquad (1)$$

The above equation indicates that in order to feed $J_\parallel$ along the magnetic field lines, the divergence of $J_\perp$ must be non-zero. It is quite obvious then that the current component ($J_\perp$) perpendicular to $\underline{B}$ must be generated. Therefore, there must be a dynamo process ($\underline{V} \times \underline{B}$) which can generate $J_\perp$, in which $\underline{J}\cdot\underline{E}$ is negative.

Furthermore, for a given force-free field configuration, there must be a specific velocity field ($\underline{V}(x,y,z)$) and magnetic field ($\underline{B}(x,y,z)$) which can provide the needed distribution of $J_\parallel$

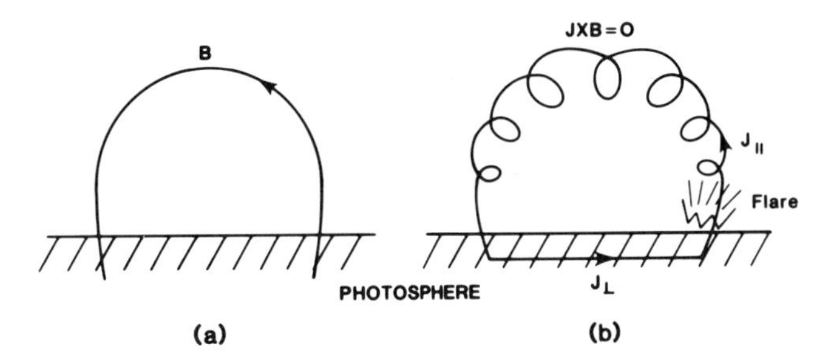

Fig. 1. (a) Loop-like magnetic field line. (b) Force-free field is produced by feeding $J_\parallel$ along the loop. The field-aligned current $J_\parallel$ is fed by $J_\perp$.

(x,y,z). For our simple example, assuming a uniform Pedersen conductivity in the dynamo region, the velocity field may take a circular or vortex motion around the feet of the loop. Actually, this is a very complicated problem even for such a simple geometrical situation.

Importance of the Power Generation Process

There are several reasons why a study of the power supply processes is important and cannot be ignored. As mentioned earlier, a force-free field requires the generation of field-aligned current $J_{\parallel}$ (x,y,z). Furthermore, there are some indications that the location of flare ribbons agrees fairly well with that of upward field-aligned currents ($J_{\parallel} > 0$); see Krall et al. (1982) and Akasofu (1984, p. 1491). Therefore, we must find ways to generate $J_{\parallel} > 0$ along narrow strips. In order to produce $J_{\parallel} > 0$ in such a specific way, there must be a specific distribution of $J_{\perp}$ and thus a specific velocity field under a specific magnetic field configuration for the dynamo process. Thus, the dynamo process cannot be an arbitrary one.

Now, we must find the height in the solar atmosphere, where singificant amounts of $J_{\perp}$ and $J_{\parallel}$ can be generated. The answer to this question will be most important in finding the ultimate source of energy for solar flares. In the coronal height, the Pedersen conductivity Σ_p is very small, while well below the photosphere, there is little relative motion between the gas and the magnetic field. On the other hand, it is interesting to note that the photosphere and the terrestrial ionosphere are remarkably similar in many ways, in terms of the mass density, the degree of ionization ($\sim 10^{-7}-10^{-5}$), etc. One of the main differences is the temperature, 1000 K in the ionosphere and 6000 K in the photosphere; however, both are well below the ionization potential of the main constituents.

Kan et al. (1983) have already shown that the power P of the flare dynamo is given by

$$P = \Sigma_p \underline{E} \cdot (\underline{V}_n - \underline{V}_i) \times \underline{B} \qquad (2)$$

Furthermore, they showed that it is incorrect to assume a priori that there is no differential velocity between the neutral component velocity $\underline{V}_n$ and the ionized component velocity $\underline{V}_i$. This is equivalent to say that one cannot discuss the power supply without knowing the entire circuit. The neutral component can drag the ionized component with exactly the same velocity when there is no dissipation. When the dissipation rate is appreciable, $\underline{V}_i$ should be reduced, so that $\underline{V}_i$ depends on the power dissipation rates during a flare. In fact, a small differential speed ($\underline{V}_n - \underline{V}_i$) ensures the transfer of the flow energy of the neutral gas to that of the ionized component. The assumption of $\underline{V}_n - \underline{V}_i = 0$ occurs when there is no dissipation in the entire flare circuit, namely for an open circuit. From the above consideration, it is not difficult to infer also that the ultimate source of the energy is the kinetic energy of the neutral gas flow which may be maintained by pressure (temperature) gradient or other causes.

One should realize also that by the time a strong force-free field is being built up (meaning to many that the flare energy is sufficiently stored), an intense field-aligned current must already be flowing; the upward current must be carried by downward flowing electrons which could excite or ionize hydrogen atoms in the atmosphere, causing flare ribbons. An auroral curtain is produced by downward streaming electrons, and this is one of the major findings in magnetospheric physics. This consideration suggests also that the force-free field can be built up even more during a flare, although it is tacitly believed that it should relax from a force-free field to a potential during a flare. In fact, observations have not been conclusive whether or not the magnetic fields around flare regions become a potential field. In the case of magnetospheric substorms, the solar wind-magnetosphere dynamo power ε is high and is most often increasing at their onset.

Details of the idea presented here are published in Kan et al. (1983) and Akasofu (1984). The formation of a potential structure is essential in accelerating the downstreaming electrons. This is also the case for the aurora. This structure is considered to be a double layer and requires a threshold current density. It is for this reason that the onset of an auroral substorm is a rather sudden event; in fact, the auroral luminosity can increase 2 orders of magnitude in a few minutes. Indeed, Lin and Schwartz (1987) found that solar electron spectra are similar to those of the aurora; see also Scholer (1988).

Acknowledgments. The work reported here was supported in part by a grant from the National Science Foundation (ATM-85-18512) and a USAF contract (F19628-86-K-0030).

References

Akasofu, S.-I., Energy coupling between the solar wind and the magnetosphere, Space Sci. Rev., 28, 121, 1981.

Akasofu, S.-I., An essay on sunspots and solar flares, Planet. Space Sci., 32, 11, 1469-1496, 1984.

Kan, J. R., S.-I. Akasofu, and L.-C. Lee, A dynamo theory of solar flares, Solar Phys., 84, 153, 1983.

Krall, K. R., J. B. Smith, Jr., M. J. Hagyard,

E. A. West, and N. P. Cumings, Vector magnetic field evolution, energy storage, and associated photospheric velocity shear within a flare-productive active region, Solar Phys., 79, 59, 1982.

Lin, R. P., and R. A. Schwartz, High spectral resolution measurements of a solar flare hard x-ray burst, Ap. J., 312, 412, 1987.

Low, B. C., Magnetic free-energy in the solar atmosphere, these proceedings, 1989.

Nakagawa, Y., M. A. Raadu, D. E. Billings, and D. McNamara, On the topology of filaments and chromosphere fibrils near sunspots, Solar Phys., 19, 72, 1971.

Scholer, M., Acceleration of energetic particles in solar flares, in Activity in Cool Star Envelopes, edited by O. Havnes, B. R. Pettersen, J.H.M.M. Schmitt, and J. E. Solheim, Kluwer Academic Publishers, The Netherlands, 1988.

IMAGING SOLAR FLARES IN HARD X RAYS AND GAMMA RAYS FROM BALLOON-BORNE PLATFORMS

C. J. Crannell

Code 682, Solar Physics Branch, Laboratory for Astronomy and Solar Physics
NASA-Goddard Space Flight Center, Greenbelt, MD 20771

Abstract. Hard X rays and gamma rays carry the most direct evidence available for the roles of accelerated particles in solar flares. An approach that employs a spatial Fourier transform technique for imaging the sources of these emissions is described, and plans for developing a balloon-borne imaging device based on this instrumental technique is presented. This instrument, designed for 15-day, long-duration flights, will provide 1.6-arc sec angular resolution and 10 ms time resolution with a whole-Sun field of view.

Introduction

The solar activity cycle that is expected to peak in 1991 has already begun to produce flares. These events are relatively small and infrequent, as is usual during intervals between maxima in the Sun's 11-year cycle, but they serve as reminders of the opportunities soon to be available. "Opportunities" refer to the new technology, now ready to address the outstanding questions in high-energy solar physics, and to the rich variety of high-energy phenomena to be observed. Major advances in the spatial and spectral resolutions of instrumentation required for high-energy solar physics have been achieved since the previous solar maximum. Now, for the first time, it is possible to observe processes associated with the fundamental acceleration of energetic particles on their intrinsic spatial, spectral, and temporal scales.

Vigorous efforts are underway for a coordinated, multiwavelength campaign during the next solar maximum. A prime thrust of these efforts for MAX '91 is to realize the potential of the new technology in imaging and spectroscopy for high-energy solar-flare research. The success of these efforts is challenged, however, by NASA's serious deficit in flight opportunities and the brief time interval in which development must be accomplished. Flexible and innovative approaches are required. High-altitude, long-duration, scientific ballooning offers just such an approach and also provides the heavy-lift capability that is needed for the high-energy payloads. Flights of approximately 15-days duration can be achieved by circumnavigation of the globe in the southern hemisphere. The author is leading the team of investigators who are working toward the development of a hard X-ray and gamma-ray imaging instrument for such balloon-flight applications. This instrument is known as the Gamma-Ray Imaging Device or GRID on a Balloon.

Objectives

The principal objective of the GRID on a Balloon effort is to advance solar flare science during the next solar maximum using state-of-the-art balloon-borne instruments. GRID on a Balloon would achieve that objective by providing 1.6-arc sec angular resolution and sub-second temporal resolution of hard X-ray and gamma-ray sources in solar flares in the energy range from 20 keV to $\gtrsim$ 0.511 MeV. Because this angular resolution is sufficient to resolve fundamental flare structures, observations with GRID would enable definitive tests of solar flare models.

The flare dynamics that could be observed with GRID for two competing models are illustrated in Figure 1. The hard X-ray source distributions would exhibit distinct differences for the two different cases, particularly during the rise of the impulsive emissions. According to the conduction-front model (Batchelor et al., 1985) represented in Figure 1a, the hard X-ray source is a quasi-thermal population of electrons energized at the top of a loop and confined by conduction fronts that propagate down the loop legs at the ion sound speed, $\simeq 500$ km s^{-1}. The confined energetic electrons produce hard X rays by the thermal bremsstrahlung process in a volume that grows at a rate characterized by this velocity until the front reaches the footpoints (cf. Starr et al., 1988). At the same time, a fraction of the most energetic electrons penetrates the conduction front, travels down the loop leg along the magnetic field lines at relativistic velocities, and produces hard X-ray bremsstrahlung by interactions with the ambient medium at the loop footpoints. In the

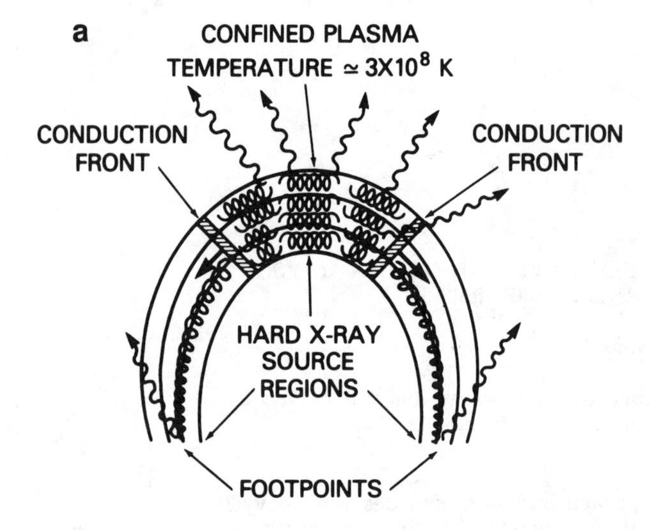

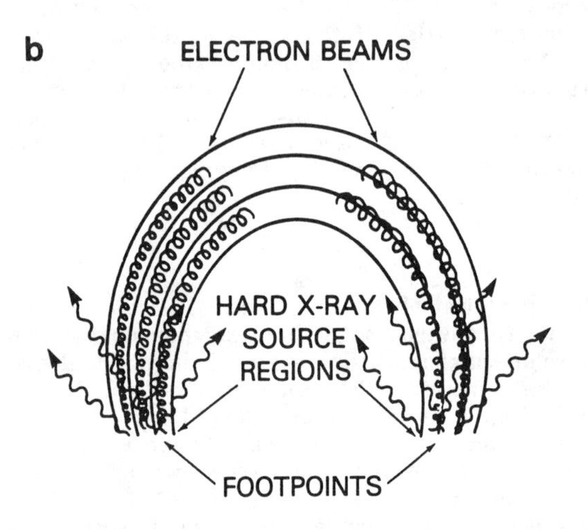

Fig. 1. Schematic representations of flaring loops illustrating (a) the onset of a flare in a coronal arch according to the conduction-front model with the hard X-ray source predominantly at the top of the loop and (b) the onset of a flare as expected for a thick-target beam model with energetic electrons spiraling along magnetic field lines until they penetrate the denser region at the footpoints.

case of this quasithermal model, therefore, hard X-ray sources originate nearly simultaneously at the top of the loop, where most of the emission would be produced, and at the footpoints, where a smaller fraction of the emission would be produced with a harder spectrum. The emission region then grows to fill the loop legs.

In the thick-target beam model, represented in Figure 1b, the energetic electrons responsible for the hard X-ray emission are accelerated as a beam with a power law energy distribution. They spiral along the magnetic field lines of the loop until they reach the footpoints. There, they produce hard X-ray bremsstrahlung by interactions with the dense ambient medium. Most of the energy of the electrons, however, is dissipated by Coulomb collisions. The energy deposited heats the ambient material, which then expands into the loop, so that emission from the remainder of the loop occurs only after its density has had time to increase substantially.

Observations with GRID also would serve to complement spatially resolved observations at microwave frequencies for characterizing physical parameters of the associated flare structures. Hard X-ray emission is predominantly optically thin. Its spectrum depends on the instantaneous spectrum of the interacting energetic electrons that produce it and the density of the medium in which they interact. Optically thin microwave emission depends on the number of emitting electrons, their spectrum, and the strength of the magnetic fields with which they interact. The observed optically thick microwave emission produced by energetic electrons depends on their spectrum and the projected area of the source region. Multiwavelength observations, spectrally and spatially resolved in both hard X rays and microwaves, would provide a capability never before available for investigating the basic physical characteristics of high-energy flare sources.

Instrumental Approach

Hard X rays and gamma rays cannot be reflected or focused by lenses or mirrors. Even grazing incidence reflection, used very effectively in soft X-ray astronomy, is impossible in the photon-energy domain above a few keV. What is required is some varia-

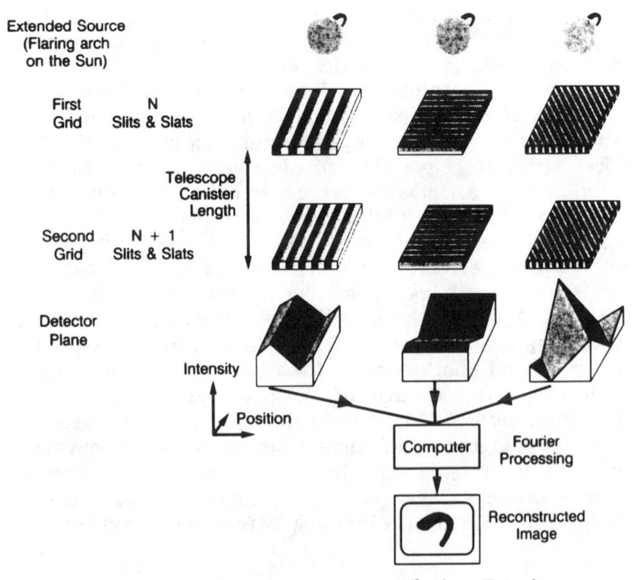

Fig. 2. Schematic representation of the Fourier-transform imaging technique (courtesy of G. Hurford).

tion on a pinhole camera or collimator, consisting of material opaque to this radiation, interspersed with transparent apertures. The mask or collimator casts a shadow on a position-sensitive detector. The GRID telescope is based on a Fourier transform imaging technique employing an array of subcollimators (Makishima et al., 1977; Crannell et al., 1985; Palmer and Prince, 1987; Prince et al., 1988). One Fourier component of the angular distribution being imaged is measured with each subcollimator and its associated position-sensitive detector module. Each subcollimator consists of a pair of grids, one at the top and the other at the bottom of the telescope; each grid is formed from a uniformly spaced distribution of transparent apertures, or slits, and material opaque to the hard X rays and gamma rays of interest, or slats. The array includes subcollimators with a variety of slit/slat orientations and a range of slit/slat spacings corresponding to the range of angular dimensions to be measured.

In the application described here, the number of slits and slats in the top grid of any subcollimator differs by one from the number of slits and slats in the corresponding bottom grid. The resulting X-ray transmission for each subcollimator forms a Moiré pattern with a single intensity contrast cycle extending over the width of the subcollimator, as illustrated in Figure 2. While fine slit/slat spacings are required for fine angular resolution, only modest spatial resolution (approximately 20% the width of any subcollimator) is required of the position-sensitive detector. This is particularly important because it is extremely difficult to make hard X-ray and gamma-ray detectors with spatial resolution much better than 1 cm. Displacements of one grid relative to the other by only a slit width cause shifts in the Moiré pattern of half a subcollimator width. This amplifier effect is readily demonstrated by following the suggestions accompanying Figure 3.

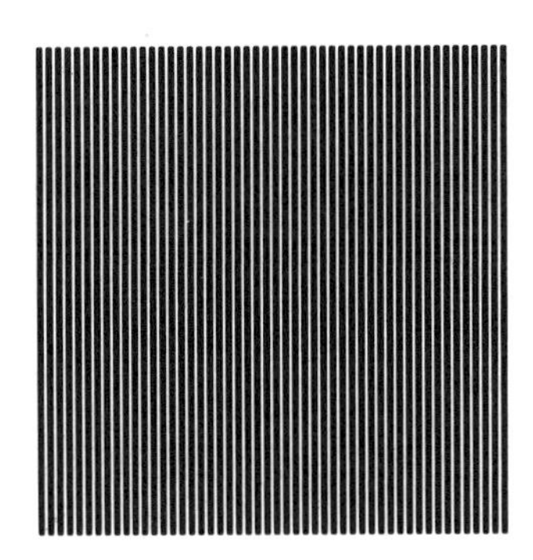

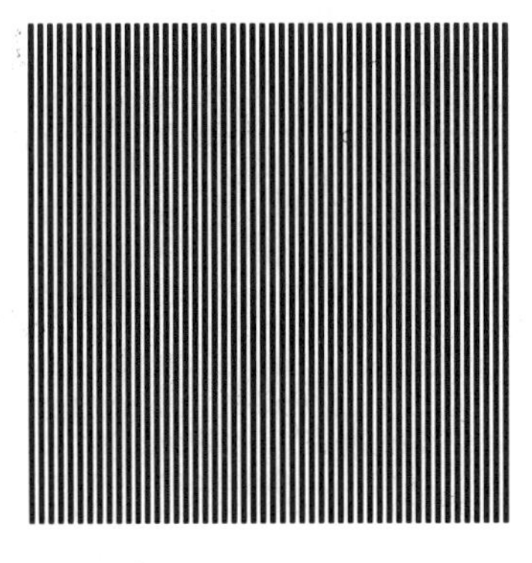

Fig. 3. Schematic representation of a pair of grids. The reader is invited to make a copy of each grid on a transparency and then to superpose the two transparencies. The sensitivity of the Moiré pattern to small shifts of one grid relative to the other illustrates the amplifier effect that obviates the need for detectors with high spatial resolution.

For the subcollimators illustrated in Figures 2 and 3, the slits and slats in one grid are parallel to the slits and slats in the other member of its subcollimator grid pair. A rotation of one grid in the plane of its slits and slats relative to the angular orientation of the other would have the effect of rotating the orientation and spatial frequency of the resultant Moiré pattern. This effect can be employed to enable a great simplification in the construction of the detector modules. The detector modules can all be identical and can be packed with optimally efficient use of the available space if all the Moiré patterns are oriented in a common direction with the same spatial frequency. For each spatial Fourier component to be sampled there is a unique angle of rotation between the two grids and a unique combination of slit/slat spacings (N and N+1 only for a rotation angle of zero) that yields a Moiré pattern with a specific orientation and spatial frequency (cf. Prince et al., 1988). The subcollimators for GRID on a Balloon are designed to enable this effect to be employed in the optimization of the instrument design.

GRID on a Balloon is shown schematically in Figure 4. Because the Fourier-transform technique imposes only modest requirements on the spatial resolution of the GRID detectors, the detectors can be constructed very simply; and, because these requirements are the same for each subcollimator independent of its slit/slat spacing, the detector can be constructed as an assembly of identical modules, one for each subcollimator. A detector module and its placement with respect to a subcollimator are illustrated in the upper left-hand corner of Figure 4. The module consists of five rectangular NaI(Tl) rods, each approximately 2 cm x 2 cm in cross section, a single piece of CsI(Na) for active anticoincidence shielding, and the associated light guide, photomultiplier tubes, and electronics.

All of the information characterizing the angular distribution of a source is contained in the amplitudes of the intensity contrasts of the Moiré patterns and the spatial positions of their maxima. These are the amplitudes and phases of specific, well chosen Fourier components of the X-ray image. This information is the exact mathematical analog of that provided by a pair of radio antennas in a radio interferometer. Because of this direct, one-to-one correspondence, the analytic tools that have been developed and refined for radio interferometry can be used for analysis of the hard X-ray and

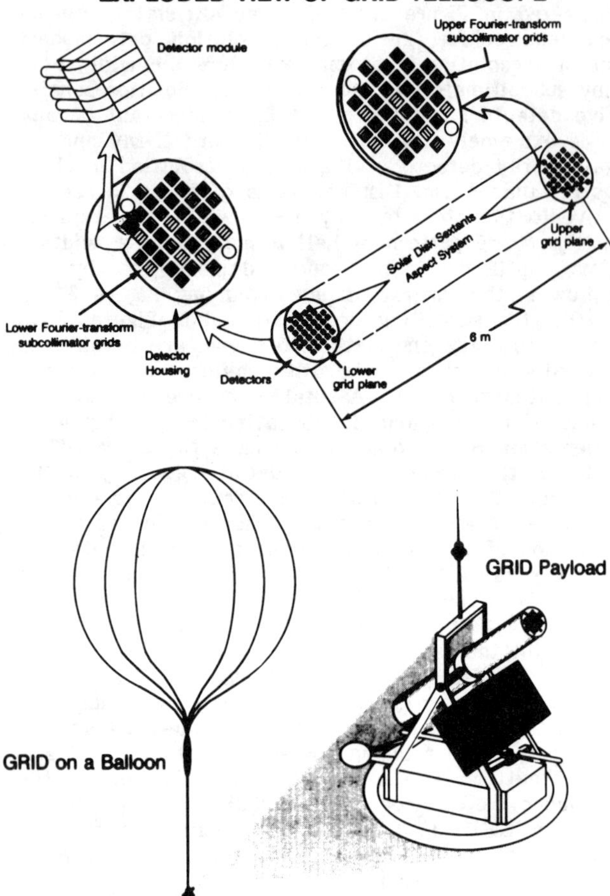

Fig. 4. Schematic representation of GRID showing the telescope integrated into a balloon payload and an exploded view showing the end plates holding the grid subcollimator pairs and the SDS and the canister housing the detector modules.

gamma-ray observations obtained with GRID. Numerical simulations of flaring magnetic loops such as those illustrated in Figure 1 have been constructed for typical flare and instrument parameters. Their images have been reconstructed successfully using these analysis techniques.

The aspect system, indicated by small circles in the telescope endplates, is, in reality, the Solar Disk Sextant (SDS), a sensitive instrument for measuring the Sun's diameter on very short time scales. For GRID, it provides the unique advantage of enabling determination of the precise telescope pointing every 10 ms. The requirements on pointing stability thus permit angular drift rates as great as an arc min per second without compromising the resolution of a reconstructed image. The characteristics and capabilities of GRID are summarized in Table 1.

Observing times of many days are highly desirable for measurements of a number and a variety of solar flares. With the sensitivity afforded by GRID

TABLE 1. Characteristics and Capabilities of GRID

Detector type	Scintillation counters
Effective detector area	500 cm^2
Energy range	20 to $\gtrsim$ 511 keV
Subcollimator material	Tungsten and tantalum
Number of subcollimators	34
Sensitivity	$\sim$100 flares per week
Angular resolution	1.6 arc sec
Field of view	Whole Sun
Required pointing accuracy	3 arc min
Required pointing stability	$\lesssim$ 0.5 arc sec per 10 ms
Expected flight duration	15 days

and a mean flare rate typical of that observed during the maximum of the previous solar cycle, a 15-day balloon flight would provide high-contrast images of about ten flares at energies up to 100 keV and of several of these up to 200 keV. It also would enable measurements of the size and location of more than 30 flares at energies up to 100 keV. The energy input to the scintillator and time of detection will be recorded for each detected photon, so that spectra and high-resolution time/intensity profiles will be obtained for all of these flares and for a large number of smaller flares, as well.

The technology for long-duration ballooning was demonstrated with successful flights of 26 and 22 days in 1981 and 1983. Two flights, one of 7 days and another of 12 days, were launched from Alice Springs, Australia, in 1987. Currently, more long-duration flights are planned for observations of Supernova 1987a, and further development of the technology is being pursued for MAX '91. High-altitude scientific balloons offer unique opportunities for repetitive long-duration flights throughout the next peak in solar activity (1990-1994).

Acknowledgments. The instrumental concepts described here have been stimulated by contributions from numerous individuals throughout the solar physics and astrophysics communities. Extensive discussion and review have come from NASA working groups and study teams, in particular: the Hard X-ray Imaging Instrument (HXII) Facility Definition Team led by L. E. Peterson; the Pinhole/Occulter Facility (P/OF) Science Working Group led originally by E. Tandberg-Hanssen and H. R. Hudson and more recently by J. M. Davis and H. R. Hudson; and the MAX '91 Science Study Committee led by B. R. Dennis. Further concept development was contributed by the members of the team, headed by T. A. Prince, that proposed GRID for the Solar High-energy Astrophysical Plasmas Explorer (SHAPE) mission.

As Principal Investigator for the balloon effort, I am especially indebted to the members of the investigative team for GRID on a Balloon, including the following: B. R. Dennis, A. L. Kiplinger, E. R. Maier, L. E. Orwig, and R. Starr, NASA-Goddard Space Flight Center; J. M. Davis and J. R. Dabbs,

NASA-Marshall Space Flight Center, responsible for the balloon gondola and pointing system; G. J. Hurford and T. A. Prince, Caltech, responsible for the image-reconstruction analysis capability and providing expertise in position-sensitive detectors; F. van Beek, Delft University in The Netherlands, pioneering the development of new technology for fabrication of fine subcollimator grids; H. R. Hudson, UCSD, liaison with the P/OF Science Working Group; S. Sofia, Yale, responsible for development of the SDS; E. E. Fenimore, Los Alamos National Laboratory, providing expertise in image reconstruction for X-ray imagers and investigating grid fabrication technology; and K. S. Wood, Naval Research Laboratory, providing design/performance analysis and contributing to grid technology. Goddard has fabricated the first prototype subcollimator grids and is currently constructing a test model of the position-sensitive detector module. This work was supported in part by NASA RTOPs 682-353-40-10-03 and 682-188-38-51-14.

MAX '91 holds great promise; new technology is ready to enable great advances in high-energy solar physics. The team for GRID on a Balloon is prepared to supply the effort necessary to make these potentials reality.

References

Batchelor, D. A., C. J. Crannell, H. J. Wiehl, and A. Magun, Evidence for collisionless conduction fronts in impulsive solar flares, Ap. J., 295, 258, 1985.

Crannell, C. J., G. J. Hurford, L. E. Orwig, and T. A. Prince, A Fourier transform telescope for sub-arcsecond imaging of X rays and gamma rays, SPIE, 571, 142, 1985.

Makishima, K., S. Miyamoto, T. Murakami, J. Nishimura, M. Oda, Y. Ogawara, and Y. Tawara, in New Instrumentation for Space Astronomy, edited by K. A. van der Hucht and G. Vaiana, Pergamon Press, New York, 1977.

Palmer, D., and T. A. Prince, A laboratory demonstration of high-resolution hard X-ray and gamma-ray imaging using Fourier-transform techniques, IEEE Trans. Nucl. Sci., NS-34, 71, 1987.

Prince, T. A., G. J. Hurford, H. R. Hudson, and C. J. Crannell, Gamma-ray and hard X-ray imaging of solar flares, Solar Phys., in press, 1988.

Starr, R., W. A. Heindl, C. J. Crannell, R. J. Thomas, D. A. Batchelor, and A. Magun, Energetics and dynamics of simple impulsive solar flares, Ap. J., in press, 1988.

IS THE PLASMA TRULY TURBULENT DURING THE IMPULSIVE PHASE OF SOLAR FLARES?

P. L. Bornmann

National Oceanic and Atmospheric Administration, Space Environment Laboratory,
NRC Resident Research Associate, Boulder, CO 80303

Abstract. Solar Maximum Mission observations of line shifts and line broadenings in soft X-ray lines during the impulsive phase of solar flares have generally been accepted as indications of plasma turbulence. In support of this, the rate of decay of the line shifts and line broadenings is consistent with the rate of decay of fluid turbulence. However, the turbulent line broadening interpretation has recently been disputed. They argue that at high time resolution the spectrum consists of discrete, random, wavelength-shifted features and that the apparent turbulent line profiles are the result of integrating these discrete features over long time periods. I consider whether these discrete features are truly random. Preliminary results of this new analysis of these features are presented. I attempt to determine if these discrete features are actually appearing randomly at discrete wavelengths or are moving continuously to new wavelengths. I also consider whether the appearance of these discrete features could be the result of instrumental effects.

Introduction

During the early impulsive phase of solar flares, blueshifts and line broadenings are observed in soft X-ray lines of ions such as Ca XIX and Fe XXV. These line profiles are generally taken as evidence for plasma turbulence (*e.g.* Wu *et al.*,1986). The rate of decay of these line broadenings and line shifts is consistent with the rate of decay of plasma turbulence (Bornmann,1987), further supporting this interpretation.

The existence of this plasma turbulence may play an important role in the storage and conversion of energy during solar flares. Turbulence provides a mechanism for converting the mass motions seen during the impulsive phase of the flare into the thermal energy observed in the later, gradual phase. Thus, in essence, the blueshifted plasma degrades into the turbulent broadening about the rest wavelength. Bornmann (1987) reports that at least 20% of the thermal energy observed during the gradual phase may arise from impulsive phase mass motions that have cascaded through successively smaller turbulent eddy sizes until they are ultimately thermalized.

The blue shifts observed early in the flare are often taken as evidence for chromospheric evaporation (e.g., Antonucci et al., 1984), although this claim has been disputed by Karpen et al. (1986). During chromospheric evaporation, cool chromospheric material is heated and then rises to coronal temperatures and heights. Chromospheric evaporation is believed to be the source of the material that emits the soft X-ray thermal radiation during the gradual phase of the flare.

Recently Doyle and Bentley (1986) have suggested that the turbulent interpretation of the observed line profiles is incorrect. They suggest that at high time resolution, the Ca XIX resonance line consists of many discrete, wavelength-shifted features. Only when the data are integrated over sufficiently long time periods (to give better statistics) do the broadened line profiles appear. Doyle and Bentley conclude that the discrete features indicate the presence of multiple parcels of upflowing material with different velocities rather than randomly oriented turbulent motions. I reexamine their published data for the May 21, 1980 flare.

Data

I consider the observations reported by Doyle and Bentley (1986) during the early impulsive phase of the May 21, 1980 flare, beginning at 20:55:05 UT (Figure 1). This 2B/X1 flare located near disk center at S13 W15 was a two-ribbon flare with a filament eruption.

The data consist of spectra that include the Ca XIX resonance line at 3.176 Å recorded by the bent crystal spectrometer (BCS) onboard the Solar Maximum Mission satellite (described by Acton et al., 1980). This instrument integrates the flux from a 6' x 6' FWHM field-of-view which should include the entire solar flare. The X-rays are diffracted by a Bragg crystal which has been bent to simultaneously record the entire spectrum using a position sensitive detector.

The data considered for this flare consist of a sequence of 11 observations that were integrated for 6 s each. At each observation time the spectrum was simultaneously recorded into 50 wavelength bins. Each bin covers 0.301 mÅ, although the instrumental resolution is 0.6 mÅ. The absolute wavelength scale is

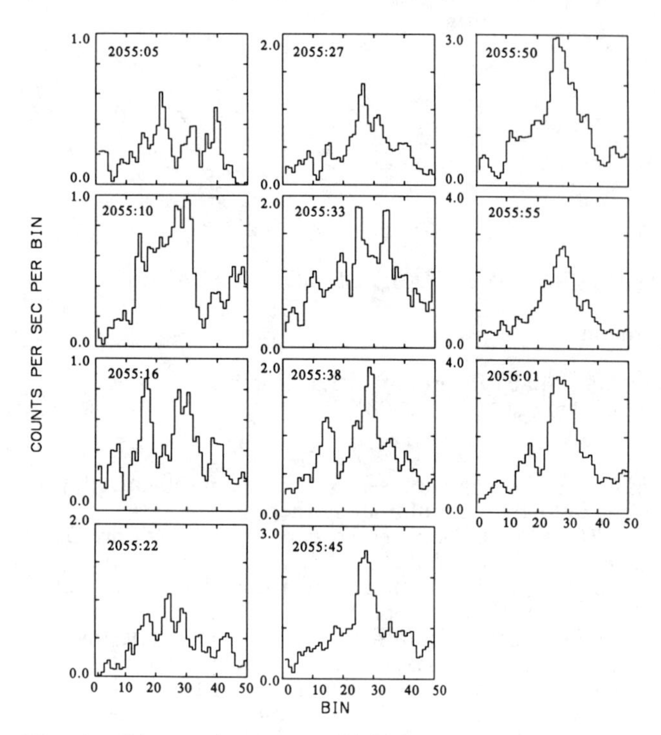

Fig. 1. Observed count rates during the impulsive phase of the May 21, 1980 flare. The rest wavelength of Ca XIX 3.176 Å is at roughly bin 28; each bin covers 0.301 mÅ.

not well determined because offsets of the flare location with respect to the optical axis of the BCS will produce an apparent wavelength shift.

An apparent wavelength shift of one bin or 0.3 mÅ can be caused by either Doppler velocities of 30 km s^{-1}, or by sources separated by a spatial distance of approximately 30,000 km on the solar surface. This distance is large compared with the typical flare size. Based on similar data, Bentley et al. (1986) concluded that the spatial shifts are an unlikely explanation for the observed wavelength shifts.

To determine if the discrete spectral features exhibit systematic or random appearances, I developed and used a "clean" program on the data to identify the location of each feature. This program, in principle similar to the technique used by radio astronomers, sequentially identifies the brightest feature in the spectrum, subtracts its contribution, and examines the remaining spectrum for the next brightest feature. At this stage a crude approximation was used to determine the location and width of each feature. Improvements in the location, width, and flux of each feature were found using a gradient chi-squared search (Bevington, 1969). A representative derived fit is shown in Figure 2. The properties of these derived fits were then used to produce synthetic spectra in which the width and/or the flux of each feature was artificially reduced or set to the same values.

Discussion

I considered the possibility that the appearance of these discrete features might be the result of a wavelength-dependent variation of the reflectivity caused by the bend of the BCS crystal. A normal, unbent crystal will diffract X-rays according to the Bragg relation

$$\sin \theta = n \lambda / 2 d_0$$

where θ is the angle between the incident X-rays and the crystal lattice, n is the order of diffraction, λ is the wavelength of the X-rays, and d_0 is the atomic spacing in the crystal. The BCS instrument uses a Bragg crystal that has been bent. Therefore, for non-radial paths through the crystal, the effective atomic spacing d will vary. However, since the reflectance of the crystal is very small (K. Strong, private communication, 1988), the photons are all diffracted in the first few layers. Therefore, the diffracted photon is not likely to have traveled sufficiently far into the crystal to have experienced a variation of the atomic spacings along its path.

A fixed-pattern variation in the instrumental sensitivity might create the appearance of discrete spectral features at regular spacings. Therefore I performed period analyses on the observed and synthesized data using both individual spectra and the entire observing sequence of spectra. No strong, persistent wavelength separation was found in the data. Therefore it is unlikely that these discrete features are the result of fixed pattern instrumental noise.

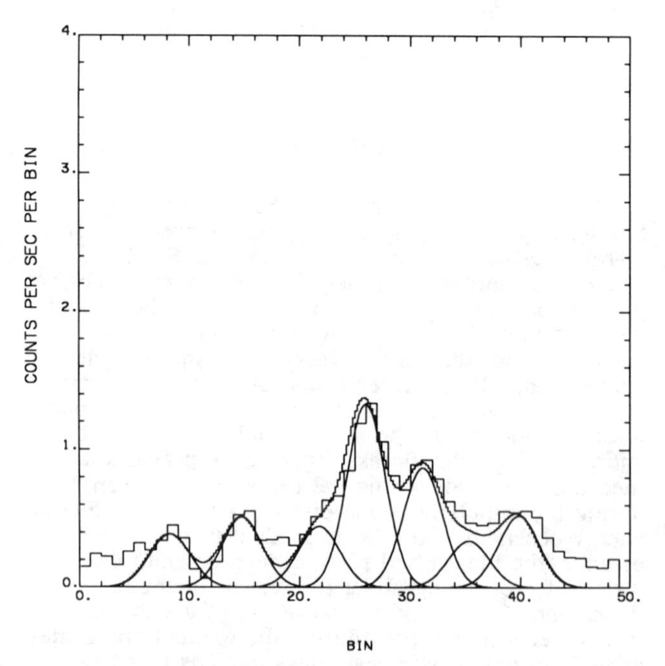

Fig. 2. Example of Gaussian fits to observed data.

I have searched for evidence that the wavelengths of these discrete features are changing, causing the features to move through the spectra. This is in contrast to the conclusion of Doyle and Bentley (1986) that the features occurred randomly and lasted less than one integration time. My study used the synthetic spectra (Figure 3) with features of artificially narrowed widths so that the location of the discrete features could be followed more easily. There may be a subjective impression that these features are moving, but the evidence is not conclusive at this stage in the analysis.

Conclusions

These initial tests have not fully addressed the discrete features interpretation of the observed line profiles. If these discrete features are real, then they should appear in other lines formed at similar temperatures. Such a study would determine whether these discrete features are of solar origin or are simply the result of statistical fluctuations in a turbulently broadened line profile observed at very low count rates.

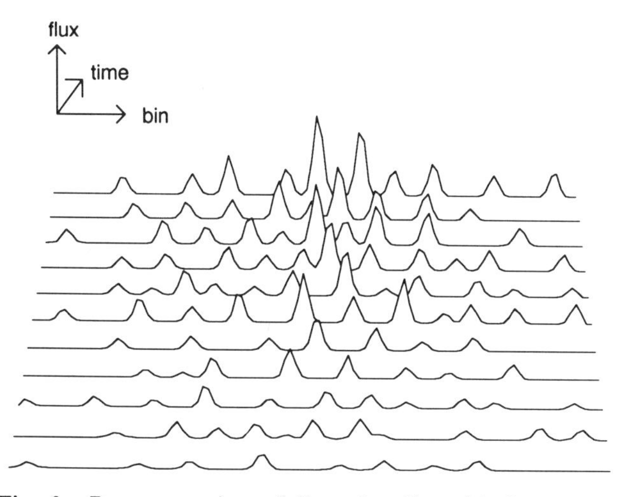

Fig. 3. Reconstruction of Gaussian fits with feature widths artificially reduced by a factor of 4.

There still remain three possible explanations for these discrete features. They may be the result of discrete parcels of solar plasma, they may be the spectral signature of the onset of turbulence in the solar flare plasma, or they may be noise fluctuations.

Acknowledgments. This work was supported by the NOAA Space Environment Laboratory through the National Research Council. The X-Ray Polychromator was built by a consortium of three groups: Lockheed Palo Alto Research Laboratory (L. W. Acton), UCL Mullard Space Science Laboratory (J. L. Culhane), and SERC Rutherford and Appleton Laboratories (A. H. Gabriel).

References

Acton, L. W. et al., The soft X-Ray Polychromator for the Solar Maximum Mission, Solar Phys., 65, 53–71, 1980.

Antonucci, E., A. H. Gabriel, and B. R. Dennis, The energetics of chromospheric evaporation in solar flares, Ap. J., 287, 917-925, 1984.

Bentley, R. D., J. R. Lemen, K. J. H. Phillips, and J. L. Culhane, Soft X-ray observations of high velocity features in the 19 June 1980 flares, Astron. Astrophys., 154, 255-262, 1986.

Bevington, P. R., Data Reduction and Error Analysis for the Physical Sciences, McGraw-Hill Book Co., New York, 1969.

Bornmann, P. L., Turbulence as a contributor to intermediate energy storage during solar flares, Ap. J., 313, 449-455, 1987.

Doyle, J. G., and R. D. Bentley, Broadening of soft X-ray lines during the impulsive phase of solar flares: Random or directed mass motions?, Astron. Astrophys., 155, 278-282, 1986.

Karpen, J. T., G. A. Doschek, and J. F. Seely, High resolution X-ray spectra of solar flares. VIII. Mass upflow in the large flare of 1980 November 7, Ap. J., 306, 327-339, 1986.

Wu, S. T. et al., Flare Energetics, Chapter 5 in Energetic Phenomena on the Sun, NASA Conf. Publ. 2439, edited by M. Kundu and B. Woodgate, pp. 5.1-5.73, NASA, Washington, D.C., 1986.

THE EVALUATION OF ENERGY STORAGE MECHANISMS IN THE GRADUAL PHASE OF SOLAR FLARES

H. A. Garcia

Space Environment Laboratory, NOAA, Boulder, CO 80303

Abstract. Current wisdom provides that the basically thermal character of the gradual phase of a solar flare is a tertiary effect, resulting from the gradual release of energy from the ablated products of chromospheric evaporation. Our studies which quantitatively incorporate the measured bulk flows and the turbulent flows in a comprehensive flare energetics model tend to confirm this theory. The model is a boundary value problem which solves the one-dimensional energy equation semi-empirically for plasma density, utilizing continuous time-referenced electron temperatures, emission measures, and velocity measurements. Observed variations in enthalpy of the flaring plasma are equated to the energy sources and modeled loss mechanisms.

Introduction

A certain sequence of events often observed during the early stages of a flare has lead to the generally accepted theory that chromospheric evaporation is the main intermediate storage mechanism between the primary energy release from stressed magnetic reservoirs and the appearance of hot thermalized plasma during the gradual phase. This theory is supported by the following observations: (1) The period of hot chromospheric upflow appears to coincide with the generation of hard X rays and it generally encompasses the rapid rise time of soft X ray emission and electron temperature. (2) The soft X rays usually reach maximum intensity about the same time that the hard X ray emission ceases and the bulk flows fall to less than observable levels. (3) Nonthermal line broadening appears to diminish to less than threshold intensities approximately coincident with the disappearance of the bulk flows (Antonucci et al., 1982; Antonucci and Dennis, 1983; Antonucci et al., 1984). The well-observed flare on November 5, 1980, is prototypical of this flare sequence (see Figure 1b) and will be discussed in the context of this model. However, there are notable exceptions to this scenario that will be covered in the Discussion section.

Concept

The model attempts to solve a relatively primitive form of the energy balance equation by exploiting a few types of observables to extract quantitative estimates of the main thermodynamic properties of a flare. The model incorporates measured temperatures and emission measures to quantify the response of the flaring plasma to the combined effects of energy gains from chromospheric evaporation and energy losses from radiation and thermal conduction. It also incorporates measured velocities of upflow and turbulence to quantify the energy input.

Because of the great complexity of a solar flare, many simplifying assumptions are necessary to make a modeled description mathematically tractable. In the following we outline the principal assumptions used in the model and include, where appropriate, a statement of how the assumption fits into the overall scheme, how it is implemented, and how the assumption is justified.

The main assumption is that the energy input for the gradual phase comes from the introduction of thermal and kinetic energy derived from the upflowing plasma and from the turbulent energy in the stationary plasma at the loop apex. Although "turbulence" observations in low-β magnetic loops have been interpreted by some authors as the result of line-of-sight velocity gradients, there seem to be few alternatives to the theory that the coherently propagating kinetic energy must pass through the turbulence phase on the way to complete thermalization.

A second major assumption is that the heated plasma is cooled by radiation and thermal conduction. Radiation is a simple function of temperature and density, both of which appear explicitly in the energy equation as dependent variables; however, thermal conduction involves a temperature gradient and therefore a spatial dimension that appears in the form of an independent variable. The dependent variables are either measured or solved for, but the spatial variable in this case is not amenable to either process unless we appeal to some ancillary principle relating the required spatial dimension to parameters that are already a part of the solution. For this purpose we have exploited the scaling law of Rosner et al., 1978, which relates the maximum loop temperature (at the apex), the pressure, and the loop semi-length. An average temperature gradient is then inferred from the observed temperature (assumed to be the maximum temperature) and the semi-length (assumed to be equivalent to the conductance scale length). In the present model we do not address the fact that thermal losses may become thermal gains further down the loop. We anticipate that future versions will consider conductive flowback.

A third major assumption is that the flare consists of two plasma populations: a hot, stationary-but-turbulent plasma at the loop apex that produces most of the soft X ray, and a warm, coherently upward moving plasma that supplies the upper stationary plasma with both mass and energy. It is assumed that each population can be described by one temperature and one density. The density of the stationary plasma is the principal unknown variable of the differential form of the energy equation that we have derived for this purpose. The density and the density rate of the upflowing plasma are solved deterministically with respect to the density and density rate of the stationary plasma. The temperature of the stationary plasma is identified with the observed temperature; the temperature of the upflow is assumed to be proportional to it by a constant factor (Antonucci et al., 1984).

The underlying premise is that it is possible to solve the time-dependent energy equation for an effective, time-dependent plasma density that is responsible for the soft X ray production. In most observational scenarios where the emission measure is observed and the density is the desired unknown, an independent estimate of the emitting volume is inferred by some means, usually from imagery. However in this model, where no volumetric information can be construed from any of the available data, we approach the problem from the opposite direction and obtain the density by solving the energy equation. Volume is derived later from the observed emission measure.

Solution of the One-Dimensional Energy Equation

The energy equation in its most fundamental form is an expression of the variation in heat per unit volume equated to the net effect of all energy sources and sinks introduced into that volume.

A useful form of this equation is given by Priest (equation 2.28g, p. 85, 1982)

$$\rho \frac{d}{dt}(C_p T) - \frac{dP}{dt} = -\mathcal{L} \tag{1}$$

where ρ is the mass density, T is the temperature, is the specific heat at constant pressure, P is the pressure, and $\mathcal{L}$ is the energy loss function. We invoke the equation of state, differentiated with respect to time, and arrive at a form utilized in this model

$$n(mC_p - k)\frac{dT}{dt} - kT\frac{dn}{dt} = -\mathcal{L} \tag{2}$$

where m is the mean particle mass ($\sim$ one-half proton mass), k is the Boltzmann constant, and n is the total number density ($n \sim 2n_e$).

We consider only energy inputs resulting from the dissipation of turbulence in the stationary plasma, the dissipation of coherent blueshifted kinetic energy, and the direct infusion of enthalpy from the blueshifted plasma.

Turbulent energy rate:

$$\mathcal{L}_{tk} = -\frac{d}{dt}\left(\frac{3}{2} m n v_t^2\right) \tag{3}$$

Blueshifted kinetic energy rate:

$$\mathcal{L}_{bk} = -\frac{d}{dt}\left(\frac{1}{2} m n_b v_b^2\right) \tag{4}$$

Blueshifted enthalpy rate:

$$\mathcal{L}_{be} = -\frac{d}{dt}\left(\frac{5}{2} k n_b T_b^2\right) \tag{5}$$

where v_t is the turbulent velocity, v_b is the blueshifted velocity, n_b is the blueshifted density, and T_b is the blueshifted temperature. The principal unknown variable is n; v_t, v_b, and T are observed. This leaves n_b and T_b to be accounted for. As previously discussed, T_b is assumed to be proportional to T by a freely prescribed constant factor, β

$$T_b = \beta T \tag{6}$$

We postulate that all of the new mass introduced into the X ray emitting volume comes from chromospheric evaporation. By invoking the definition of the emission measure

$$\epsilon = n^2 V/4 \tag{7}$$

it is possible to derive the following analytic connection between n_b, dn_b/dt, n, and dn/dt

$$n_b = \frac{4\epsilon}{n^2 \cdot v_b \cdot A}\left[\frac{n}{\epsilon}\frac{d\epsilon}{dt} - \frac{dn}{dt}\right] \tag{8}$$

$$\frac{dn_b}{dt} = \frac{-4}{v_b \cdot A}\left[\frac{1}{n^2}\frac{d\epsilon}{dt} + \frac{2\epsilon}{n^3}\frac{dn}{dt}\right]\cdot\frac{dn}{dt} \tag{9}$$

where ϵ is the observed, time-dependent emission measure $\epsilon(t)$, and A is the freely prescribed flux tube cross-sectional area. Using the above relationships for n_b and dn_b/dt we may express the source functions related to the streaming plasma in terms of measured quantities, one free parameter, and the basic unknown variable, n (see equation (14)).

The radiative loss function has the usual form

$$\mathcal{L}_r = \chi n^2 T^{\alpha} \tag{10}$$

Since n is the total particle density (twice the electron density), $\chi = 3.8 \times 10^{-20}$ erg s^{-1} cm^{-3}. In our investigations, the exponent of temperature α has been found heuristically to be more in accordance with Summers and McWhirter's (1979) value -2/3 for $T > 10^7$ K than the more often quoted value -1/2.

The conductive loss (gain) function is more difficult to deal with since in the classical form it is second order in length. Without knowledge of the detailed spatial temperature structure we take $\partial T/\partial \ell = T/L$ everywhere, where L is the semi-loop length. Antonucci et al. (1982) compute the total conductive losses over the volume of the conducting loop using

$$\mathcal{L}_c \cdot V = \frac{2}{7}\kappa_o T^{7/2} A/L \text{ erg s}^{-1} \tag{11}$$

where $\kappa_o = 9 \times 10^{-7}$ erg s^{-1} cm^{-1} $K^{7/2}$.

Substituting for V with ϵ, we again obtain average unit volume conductive losses

$$\mathcal{L}_c = \frac{1}{14} \kappa_o T^{7/2} A n^2 / L \epsilon \qquad (12)$$

However, the conductive length L remains unspecified. Rosner et al. (1978) have constructed an analytic model of the quiescent, inhomogeneous coronal loop structure, demonstrating that (1) hydrostatic equilibrium obtains only when the maximum temperature plasma resides at the loop apex (as previously noted); and (2) the maximum temperature, pressure, and loop semi-length are functionally related by the scaling law

$$T_{max} = 1.44 \times 10^3 \ (P \times L)^{1/3}$$

Although apparently contrary to the steady state provision of the theory, the scaling law has been demonstrated to hold for prominence-related flare loops (Sylwester et al., 1986). We use this law in the form (Jakimiec et al., 1986)

$$T_{max} = 6.48 \times 10^{-4} \ (n \times L)^{1/2}$$

and write the conductive flux term (utilizing the parameters convenient to this analysis)

$$\mathcal{L}_c = 3 \times 10^{-8} \kappa_o T^{3/2} A \ n^3 / \epsilon \qquad (13)$$

The energy equation may now be written in its final form, expressed as a first-order, polynomial in n

$$\frac{dn}{dt} = (U \cdot n^3 + Q \cdot n^2 + P \cdot n) / W \qquad (14)$$

where

$$U = 3 \cdot 10^{-8} \kappa_o T^{3/2} A/\epsilon$$

$$Q = \chi T^{\alpha}$$

$$P = \left[(mC_p - k) \frac{dT}{dt} - \frac{V}{A} \frac{d v_b}{dt} \frac{m}{\epsilon} \frac{d\epsilon}{dt} - \frac{5}{2} \beta \frac{k}{v_b} \frac{V}{A} \frac{dT}{dt} \frac{1}{\epsilon} \frac{d\epsilon}{dt} \right]$$

$$W = kT + \frac{3}{2} m v_t^2 - \frac{V}{A} m \frac{d v_b}{dt} - \frac{5}{2} \beta \frac{k}{v_b} \frac{V}{A} \frac{dT}{dt}$$

$$- \frac{2}{v_b \cdot A} \left(\frac{1}{n^2} \frac{d\epsilon}{dt} + \frac{2\epsilon}{n^3} \frac{dn}{dt} \right) \left(m v_b^2 + 5 \beta kT \right)$$

The problem as posed is that of a bounded solution where an initial condition must be adjusted to satisfy a specified end boundary condition. Solutions are obtained by numerically integrating equation (14). In this case the end condition is a vanishing temporal density variation, $dn/dt \rightarrow 0$, at the time when X ray flux relaxes to a basically pre-flare state. The initial condition, determined by "cut and try" is the initial density, n_o. In general, n_o is found iteratively, employing the inequality association

$$\Delta n_o \begin{bmatrix} > \\ < \end{bmatrix} 0 \Rightarrow \Delta \left(\frac{dn}{dt} \right)_f \begin{bmatrix} > \\ < \end{bmatrix} 0 \qquad (15)$$

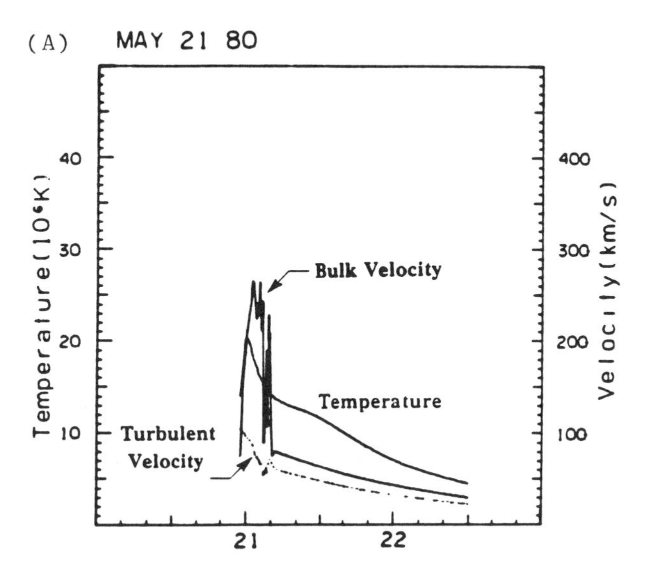

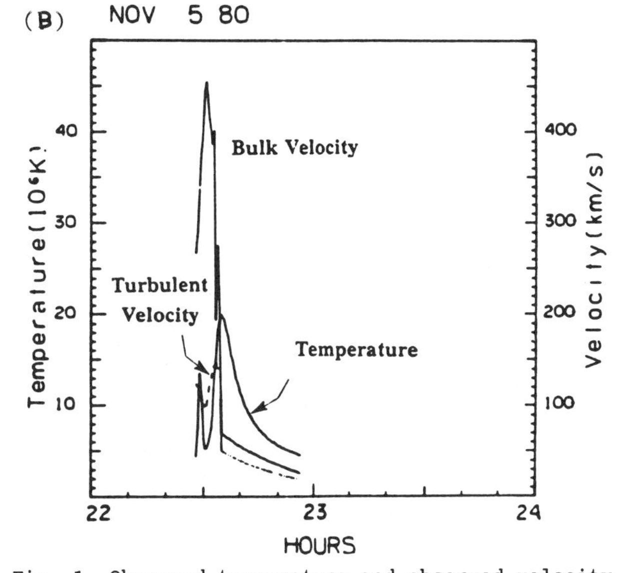

Fig. 1. Observed temperature and observed velocity profiles. (a) May 21, 1980. Two peculiarities distinguish this event. The temperature peak slightly precedes the first velocity maximum, and the blueshifted velocities extend 10 minutes into the decay phase. (b) November 5, 1980. This flare is typical of a chromospheric evaporation energization of the gradual phase, i.e., the flows end concurrent with the maximum temperature.

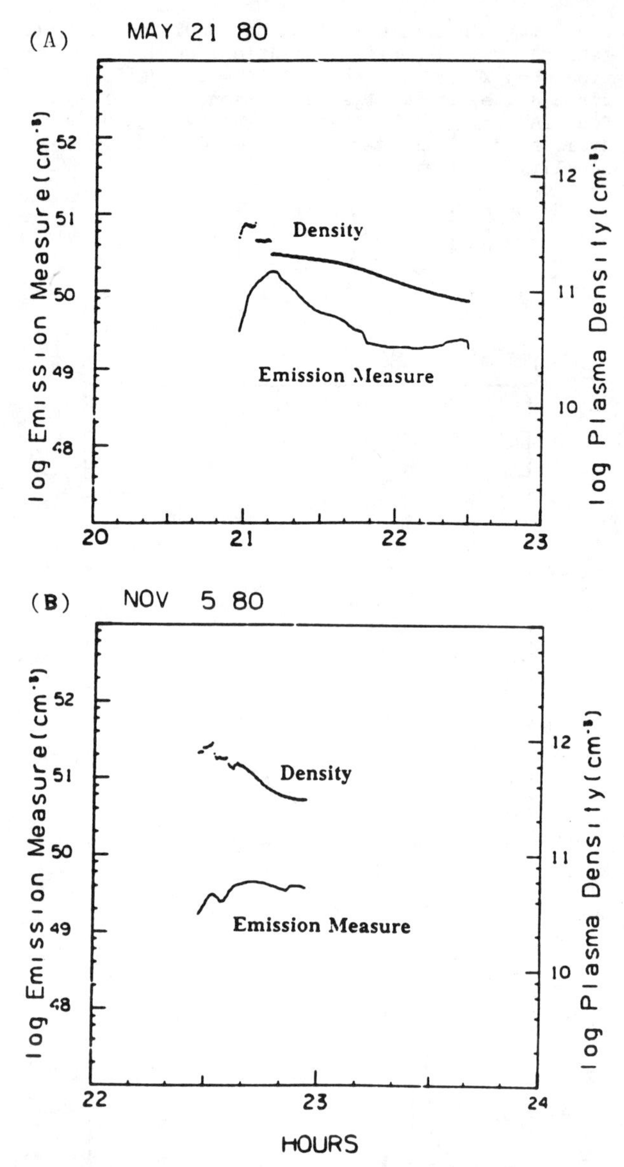

Fig. 2. Computed density and observed emission measure profiles. (a) May 21, 1980. Maximum density is less than that derived by other observation methods, implying that not all energy sources have been accounted for by thermalization of observed kinetic energies. (b) November 5, 1980. Maximum density is slightly greater than obtained by other methods, implying that no other sources are required and that thermalization is less than 100% efficient.

Discussion

In this paper we describe two well-observed flares for which we have the requisite X ray and velocity data with which to execute the model. These flares were selected primarily on the basis of the availability of data from other sources to compare our findings. The flares which occurred on May 21 and November 5, 1980, respectively, apparently were driven by distinctly different storage and transport mechanisms.

Figure 1 contains the superimposed temperature and velocity profiles which are intended to illustrate the temporal correlation between the mechanical energy input and the thermal response of the flaring plasma. In most flares as we have noted, the upwelling motions end coincidentally with the peak temperature. The November 5 flare (Figure 1b) is typical of this type. This accords with the theory that supposes that the thermal flare is the product of dissipating chromospheric upflow.

The November 5 flare shows clear evidence that mechanical energy dissipated into thermal energy. The large (> 450 km s^{-1}) velocities observed in both Fe XXV and Ca XIX lines (E. Antonucci, private communication, 1987) ended abruptly as the plasma temperature peaked. The higher densities computed by this model (Figure 1b), however, slightly exceed that of Antonucci and Dennis (1983), suggesting that no additional unseen energy sources are required for this flare. Moreover, it also suggests that thermal dissipation may at times be less than 100% efficient.

In the context of the mathematical model we have described here, the May 21 flare presents a problem. The blueshifted velocities continued well past the temperature maximum. Note that in Figure 1a the peak temperature occurred before the first velocity maximum. This does not accord with chromospheric evaporation theory since the continued release of mechanical energy into heat would be expected to sustain high temperatures. Batchelor (1986) has cited other evidence that suggests Ca IX blueshifts were more attributable to a filament eruption than to a chromospheric ablation.

The observed mechanical energies are required by this model to dissipate into heat. The upwelling mass associated with these energies is expected to produce the density profiles known to occur for that flare by the other observational methods. The densities for the May 21 flare, shown in Figure 2, are in reasonable agreement with Antonucci and Dennis (1983) as well as Duijveman's (1983) observed densities for this date. If these blueshifts are attributable to an erupting filament as Batchelor (1986) has suggested, then at least a significant part of this energy must have thermalized to produce the observed thermal flare. The alternative is that most of the thermal energy came from unobserved processes that pumped energy directly from stressed magnetic reservoirs.

References

Antonucci, E., B.R. Dennis, Observation of chromospheric evaporation during the Solar Maximum Mission, Solar Phys., 86, 67, 1983.

Antonucci, E., A.H. Gabriel, L.W. Acton, J.L. Culhane, J.G. Doyle, J.W. Leibacher, M.E. Machado, L.E. Orwig, and C.G. Rapley, Impulsive phases of flares in soft x ray emission, Solar Phys., 78, 107, 1982.

Antonucci, E., D. Marocchi, and G.M. Simnett, Origin and location of chromospheric evaporation in flares, Adv. Space Res., 4(7), 111, 1984.

Antonucci, E., B.R. Dennis, A.H. Gabriel, G.M. Simnett, Initial phase of chromospheric evaporation in a solar flare, Solar Phys., 96, 129, 1985.

Batchelor, D.A., Evidence against chromospheric evaporation from the 21 May 1980 flare, Adv. Space Res., 6(6),159, 1986.

Duijveman, A., Heat balance for the high-temperature component of a solar flare, Solar Phys., 84, 189, 1983.

Jakimiec, J., B. Sylwester, J. Sylwester, R. Mewe, G. Peres, S. Seriv, and J. Schrijver, Investigation of flare heating based on X ray observations, Adv. Space Res., 6(6), 237, 1986.

Rosner, R., W.H. Tucker, G.S. Vaiana, Dynamics of the quiescent solar corona, Ap. J., 220, 643,1978.

Summers, H.P., and R.W.P. McWhirter, Radiative power loss from laboratory and astrophysical plasmas, J. Phys. B, 12(14), 1979.

Sylwester, B., F. Farnik, J. Sylwester, J. Jakimiec, B. Valnicek, Flare diagnostics based on Prognoz 9 x ray data, Adv. Space Res., 6(6), 233, 1986.

ON THE MECHANICAL ENERGY AVAILABLE TO DRIVE SOLAR FLARES

A. N. McClymont and G. H. Fisher

Institute for Astronomy, University of Hawaii, Honolulu, HI 96822

Abstract. Where does solar flare energy come from? More specifically, assuming that the ultimate source of flare energy is mechanical energy in the convection zone, how is this translated into energy dissipated or stored in the corona? This question appears to have been given relatively little thought, as attention has been focussed predominantly on mechanisms for the rapid dissipation of coronal magnetic energy by way of MHD instabilities and plasma micro instabilities. We consider three types of flare theory: the steady state "photospheric dynamo" model in which flare power represents coronal dissipation of currents generated simultaneously by sub-photospheric flows; the "magnetic energy storage" model where sub-photospheric flows again induce coronal currents but which in this case are built up over a longer period before being released suddenly; and "emerging flux" models, in which new magnetic flux rising to the photosphere already contains free energy, and does not require subsequent stressing by photospheric motions. We conclude that photospheric dynamos can power only very minor flares; that coronal energy storage can in principle meet the requirements of a major flare, although perhaps not the very largest flares, but that difficulties in coupling efficiently to the energy source may limit this mechanism to moderate sized flares; and that emerging magnetic flux tubes, generated in the solar interior, can carry sufficient free energy to power even the largest flares ever observed.

Introduction

It is generally accepted that solar flares occur only in regions of highly stressed magnetic field, and it is widely supposed that the flare energy is stored in situ in the coronal magnetic field, or at least that the magnetic field serves as a conduit for the energy flux supplying the flare. Although many mechanisms for releasing the energy stored in stressed coronal fields have been investigated, and models have been developed for the ways in which magnetic fields can be stressed by motions of their photospheric footpoints, the fundamental processes responsible for moving the footpoints have received relatively little attention. For instance the steady state "photospheric dynamo" models assume a given photospheric flow field, neglecting the reaction of the flow to generated **J**×**B** forces. Under this assumption, an infinite energy is available! Here we address the question of the origin of flare power, a question which appears to have been considered only crudely in the past: Stenflo (1969) concluded that kinetic energy in the convection zone beneath a sunspot was sufficient to supply a major flare, while Spicer (1982) commented that there was probably insufficient kinetic energy in the convection zone to directly power a flare.

In this paper, our goal is to compute upper limits to the available energy flux in as general a way as possible, avoiding many of the considerations which would have to be resolved in the construction of a realistic model. We examine the coupling of mechanical energy in sub-photospheric flows to the coronal magnetic field. The flows are assumed to differ at the two footpoints of a magnetic arch or arcade of loops, so that the differential motion of the footpoints exerts a twisting or shearing force on the coronal magnetic field. We evaluate the mechanical energy of turbulent motions in the upper convection zone and of large-scale differential rotation as possible sources of flare energy. We also estimate the free energy carried by a newly emerging flux tube, generated in the solar cycle dynamo region, which is independent of near-surface flows.

Observations suggest that magnetic flux at the photospheric level is concentrated into tight bundles or "flux tubes." The amount of mechanical energy which can be intercepted depends on the geometry of the sub-surface magnetic flux tube, the height and time scales over which coherent motions extend, and the time scale for communication of forces to the solar surface. It is believed that isolated flux tubes extend at least part way down through the convection zone and that they are in hydrostatic equilibrium with magnetic pressure plus internal gas pressure equal to external pressure (see Fisher et al., 1989). We will adopt this flux tube model and assume that flux tubes are of roughly circular cross section. The assumptions of pressure confinement and a compact cross section impose the main limiting factor on the power supplied to flux tubes by convective motions.

We calculate the work done on the magnetic field in the following two limits. If currents induced by mechanical stresses are simultaneously dissipated resistively, the plasma slips across the magnetic field, as in an MHD generator, and the maximum power which can be extracted from the flow is a fraction of the kinetic energy flux intercepted by the flux tube. This is the "photospheric dynamo" model. We assume optimum impedance matching, so that the maximum energy flux can be extracted as flare power.

On the other hand, if resistive dissipation is negligible, the flow distorts the magnetic field and the work done by the gas is stored inductively. We presume that in this "energy storage" model the flow couples to the magnetic flux tube through aerodynamic drag. The available power is computed in the same way as for the "dynamo" model. To make use of this power, however, the coronal magnetic field must match the "impedance" (ratio of force to velocity) of the source. We find that optimum power transfer occurs when the opposing force due to bending of the coronal magnetic field lines allows the flux tube to move at 1/3 of the convective driving velocity, and that 4/27 of the kinetic energy flux can be transferred to the corona.

Observed Flare Energies and Time Scales

In a major solar flare, more than $\sim 10^{32}$ erg is dissipated over a solar surface area of 3×10^{19} cm^2 during 10^3–10^4 s (e.g., Svestka, 1976; Lin, 1982). Thus the average energy flux is $\sim 10^9$ erg cm^{-2} s^{-1} and the power is at least $\sim 10^{28}$ erg s^{-1} over the flare area. In the initial, impulsive phase of a flare, higher energy fluxes of order 10^{10}–10^{11} erg cm^{-2} s^{-1} are concentrated in flare kernels of area $\sim 10^{18}$ cm^2 or less. The impulsive phase lasts $\sim 10^2$ s and provides the most severe test of energy release mechanisms.

Major flares are often eruptive, with much of the energy released during mass ejections (e.g., eruptive prominences and surges). Smaller flares (say 10^{30}–10^{31} erg over 10^{19} cm^2 in 10^3 s) are frequently compact with most of the energy being carried away by radiation. There is some evidence that the energy released impulsively during a flare is accumulated over a period of 1 to 2 days prior to the flare. For instance, "homologous flares" recur at the same location with very similar appearance (e.g., Svestka, 1976), and a single active region may give rise to a series of major flares (e.g., Lin and Hudson, 1976). The require ment on power input for such energy accumulation is clearly less severe by at least an order of magnitude compared to the power required to drive a flare directly.

Thus the problem is to supply a power of 10^{27}–10^{29} erg s^{-1} over an area of order 10^{19} cm^2 to a directly driven "dynamo" flare, or, for a "stored energy" flare, to accumulate an energy surplus of order 10^{30}–10^{32} erg over a similar area, which requires an average energy flux of 10^{26}–10^{27} erg s^{-1} over a period of order a day.

Solar Flare Theories

In this section we briefly review the three types of flare models under consideration in this paper. They are the photospheric dynamo model, in which energy of convective motions is converted directly to flare energy on the flare time scale (minutes or hours), the magnetic energy storage model, where pre-existing coronal fields are stressed by convective motions over longer time scales (hours or days), and the emerging flux model, in which new magnetic flux, which may have accumulated free energy from convective motions over solar cycle time scales (months or years), is able to produce flares as soon as it reaches the solar surface.

The Photospheric Dynamo Model

This model, which is based on analogy with magnetospheric substorms, appears in the literature in many forms (e.g., Sen and White, 1972; Heyvaerts, 1974; Kan et al., 1983; Henoux and Somov, 1987). Its principal feature is that the flare is driven directly by the power of photospheric flows. All the above expositions assume that the internal resistance of the photosphere is so high that plasma can slip across magnetic footpoints in a steady state. Such dissipation occurs in the weakly ionized ionosphere of the Earth and is an important component of the theory of magnetospheric substorms. But in the solar context, even the weakly ionized layer of the photosphere is a good conductor. Moreover, the weakly ionized layer of the photosphere is not "heavy" enough to support the required stresses, which on a short time scale must propagate deeper into the atmosphere. This is again contrary to the concepts developed with respect to the Earth's magnetosphere-ionosphere system, where the ionosphere is "heavy," resistive, and in addition has an insulating boundary beneath it. These and other objections to the "photospheric dynamo" flare model have been detailed by Melrose and McClymont (1987).

Nevertheless, the basic concept of the dynamo model (continuous generation in the sub-photospheric region of the power being simultaneously dissipated in a flare) is still valid in principle. Although the photospheric resistance is in fact very low, such resistance must be present in the coronal part of the circuit to account for flare dissipation. Since the weakly ionized layer is really of no relevance, it does not have to support the stresses, which will be provided at a deeper level where mechanical motion can couple more effectively to the magnetic field.

Magnetic Energy Storage in the Corona

It is commonly believed that flares occur in regions of highly sheared magnetic field and that the shear is produced by photospheric flows moving the footpoints of the magnetic field (e.g., Hagyard et al., 1986). Mechanisms for dissipating the field-aligned currents which result from such twisting of the magnetic field have been proposed by, e.g., Alfvén and Carlqvist (1967) and Spicer (1977a,b). We find that sub-photospheric flows can indeed move magnetic footpoints and so stretch the coronal field. However, simple stretching does not necessarily produce coronal currents; production of shear requires relative rotation of the footpoints. In our simple cylindrical model of a sub-surface flux tube, there is clearly not enough mechanical coupling to produce rotation unless the flux tube is fluted or fragmented beneath the photosphere, making it easier for flows to impart torques. Since we want to compute upper limits to the available energy flux we leave aside such difficulties and assume that the entire kinetic energy flux intercepted by the flux tube can be utilized.

Emerging Flux Models

Although major flares seem to be correlated with highly sheared magnetic fields in magnetically complex regions, and are often associated with the eruption of pre-existing prominences, which indicate that the magnetic configuration has existed for some time, many flares and brightenings are observed at points of eruption of new magnetic flux. The energy release could be caused by either reconnection of the new flux with overlying pre-existing magnetic field (Heyvaerts et al., 1977, see also Syrovatskii, 1969; Somov and Syrovatskii, 1977; Uchida and Sakurai, 1977), by conversion of the free energy due to release of the flux tube from pressure confinement in the photosphere (Spicer et al., 1986), or by dissipation of field-aligned currents in the emerging flux tube, as in the ''coronal energy storage'' models. The photospheric shear flows could be a response to the eruption of twisted flux tubes, rather than the generator of them.

As the magnetic field of the pre-emergent flux tube is far from potential, and the flux tube therefore carries a great amount of free energy, it seems unnecessary to assume that the field must be stressed by subsequent footpoint motions before a flare is produced.

The Convection Zone as a Source of Flare Power

The Convection Zone Model

Here we consider first convective motions, then differential rotation as flare energy sources.

We have used the convection zone model of Spruit (1974) as a basis for estimating the available energy in convective motions. This model, which is based on a mixing length theory, provides an estimate of mean convection speeds as a function of depth. These vary from ~0.01 km s^{-1} near the bottom of the convection zone, at a depth of 10^5 km, to ~1 km s^{-1} near the photosphere.

The energy flux available from differential rotation was computed from the relation for the surface rotation rate, $\Omega \approx 14.44 - 3.0 \sin^2\phi$ ° day^{-1}, where ϕ is the latitude (Allen, 1973). The velocity gradient at a latitude of $\phi \approx 15°$ is then $\sim 3 \times 10^{-7}$ s^{-1}. For a magnetic loop spanning a latitude range of $\sim 3 \times 10^9$ cm, the velocity differential between the footpoints is then of the same order as the velocities near the bottom of the convection zone. Since these are much smaller than the convective velocities nearer the surface, it is clear that energy fluxes and forces similar in magnitude to those obtainable from convective motions will only be obtained if it is assumed that the surface rotation rate extends throughout the convection zone, and that flux tubes are able to convey forces from great depths.

Models of Sub-Surface Flux Tubes

We have used two models of flux tubes. In the first case, we assume that the magnetic field strength is independent of depth, so that the flux tube area remains constant and so intercepts the maximum feasible fluxes of energy and momentum. This model is not at all realistic, and although it presents the greatest cross section at depth, the Alfvén speed, and so the ''rigidity'' of the flux tube, drops so rapidly with increasing depth that it is unable to transmit forces to the surface on the relevant time scales from depths of greater than ~5,000 km. In the second case, we use the thin flux tube approximation of Spruit (1981) to compute ''pressure-confined'' model flux tubes with surface magnetic field strengths of 100–2000 G, which are rooted at the base of the convection zone and are in hydrostatic equilibrium (see Fisher et al., 1989). These flux tubes present a much smaller cross-section at depth, but are ''stiff'' enough to transmit stresses from deep in the convection zone to the surface in only a few hours.

Power Available to the Photospheric Dynamo

The maximum power available to a magnetic footpoint down to a depth z beneath the photosphere is computed from

$$P(z) = \int_0^z \tfrac{1}{2}\rho(\hat{z})\, \upsilon(\hat{z})^2\, w(\hat{z})\, \upsilon(\hat{z})\, d\hat{z}$$

where ρ is the density, υ is the mean convective velocity, and w the width of the flux tube. The results of this calculation are shown in Figure 1 for a flux tube of unit width at the photosphere. The solid lines represent the energy flux intercepted by pressure-confined flux tubes with surface magnetic fields of 100–2000 G, while the dashed line is for the "uniform area" flux tube. On the curves open circles mark the greatest depths from which stresses can propagate on a flare time scale of 1 hour. (A field strength of 1000 G was assumed for the uniform field flux tube in this computation.) To estimate flare power levels, we reduce the powers intercepted by 50%, assuming optimum impedance matching.

Thus we find that for a flux tube of surface area 3×10^{19} cm^2 ($w \approx 5 \times 10^9$ cm), say, the maximum power ranges from 8×10^{25} erg s^{-1} for the "uniform" and 100 G flux tubes to 4×10^{26} erg s^{-1} for the 2000 G flux tube. These energy fluxes are at least a factor of 30 too small to directly power major flares. However, a power level of $\sim 10^{26}$ erg s^{-1} is typical of minor subflares. Thus the "photospheric dynamo," while it might play a role in producing the generally enhanced levels of emission in active regions, and even small flares and brightenings, appears to be effectively ruled out as a mechanism for major or even moderate flares.

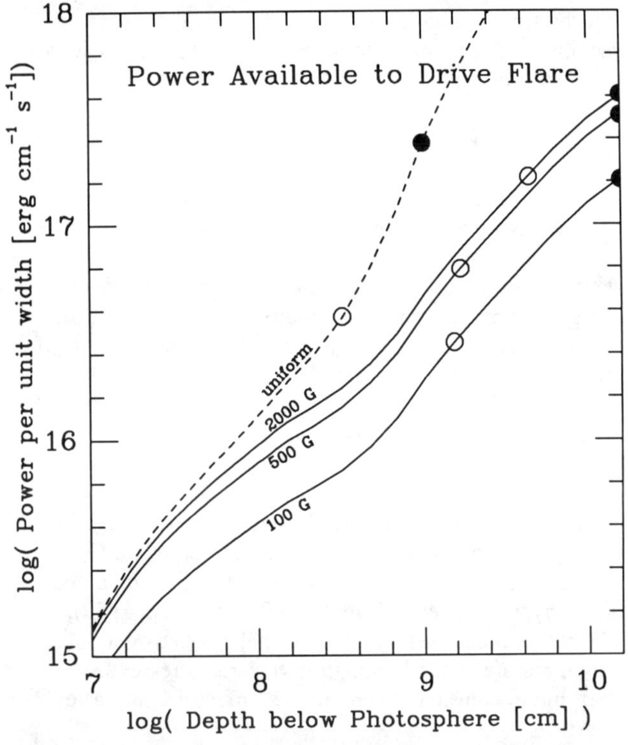

Fig. 1. The kinetic energy flux of convective motions which can be intercepted by a magnetic flux tube above depth z beneath the photosphere for an unrealistic flux tube of uniform magnetic field strength (1000 G) and for more realistic pressure-confined model flux tubes of photospheric field strengths 100, 500 and 2000 G. Open circles mark the maximum depths from which energy can propagate on a flare time scale of 1 hour, and filled circles mark the depths from which stresses can propagate on a coronal energy accumulation time scale of 1 day. (All the pressure-confined flux tubes can propagate stresses from the base of the convection zone in 1 day.)

Such power levels, however, may be sufficient to supply pre-flare buildup of energy over a longer period, even without allowing for the fact that, over a longer time, stresses may propagate from deeper in the convection zone. This topic is discussed further in the next section.

We have also computed the power available from differential rotation, on the assumption that the observed surface rotation rates hold all the way to the base of the convection zone. For a latitude span of 3×10^9 cm between footpoints, the power available is $\lesssim 10^{24}$ erg s^{-1} for all flux tube models. Thus differential rotation certainly cannot drive flares directly, and is very unlikely to contribute significantly to flare energy storage.

Power Available for Coronal Energy Storage

Just as the "dynamo" model requires a matching coronal resistance to extract maximum power from the flow, there is an optimum "spring constant" for the coronal magnetic structure in order to store magnetic energy at the highest rate. To estimate the maximum rate at which convection zone flows can do work on a flux tube, we simplify the problem by assuming that a layer of the convection zone only one scale height thick, over which the convection velocity υ is approximately constant, contributes most of the power. This assumption is consistent with our neglect of the fact that different levels of the atmosphere may be pushing the flux tube in different directions. The rate at which work is done on a flux tube by the drag force is then $P \propto \frac{1}{2}\rho(\upsilon - V)|\upsilon - V|V$, where V is the speed of the flux tube relative to the other footpoint of the coronal flux tube. Taking the other end of the coronal magnetic field line to be fixed, P is maximized when $V = 1/3\,\upsilon$, yielding $P_{max} \approx (4/27)\,\frac{1}{2}\rho\upsilon^3$. Therefore in estimating the available power we have reduced the values from Figure 1 by 4/27.

The filled circles on the curves in Figure 1 indicate the depths from which stresses can propagate in an assumed energy accumulation time of 24 hours. All the pressure-confined flux tubes can transmit stresses from the base of the convection zone in this interval. From Figure 1, the maximum power available to any of the flux tubes of photospheric width 5×10^9 cm (as considered in the previous section) lies in the range 10^{26} erg s^{-1} to 3×10^{26} erg s^{-1}. When multiplied by an energy accumulation time scale of 1 day, these yield flare energies of 10^{31}–3×10^{31} erg, typical of the energy release in a major flare (e.g., Wu et al., 1986), although still an order of magnitude short of the energy release inferred in the very largest flares (Lin and Hudson, 1976).

As mentioned in the previous section, the power levels available from differential rotation are smaller than the above figures by about 2 orders of magnitude and so cannot explain flares.

Because the power transmitted to a flux tube must couple to the coronal magnetic field, it is of interest to consider the forces and displacements corresponding to the above energy fluxes. Convective velocities in Spruit's (1974) model are ~1 km s^{-1} at a depth of 100 km beneath the photosphere, ~0.4 km s^{-1} at 1000 km, ~0.1 km s^{-1} at 10,000 km, and $\lesssim$0.03 km s^{-1} in the lower half of the convection zone, $z \gtrsim$ 100,000 km. Assuming optimum coupling with $V \approx \upsilon/3$, these velocities correspond, respectively, to footpoint displacements of 3×10^9 cm, 10^9 cm, 3×10^8 cm, and $\lesssim 10^8$ cm in one day. The latter two displacements are very small, being much less than the width of the flux tube.

Actual observed speeds of rapidly moving magnetic features in active regions are of order 0.2–0.3 km s^{-1} (e.g., McIntosh, 1981), which suggests that if flux tubes are indeed moved by convective motions, they must be coupled at depths of only a few hundred kilometers below the photosphere, where the energy flux from convective motions falls far short of that required for a major flare. Our canonical flux tube would accumulate only ~3×10^{29} erg in 24 hours, an energy typical of a small but observable subflare (cf. Wu et al., 1986). It is not possible, however, to dismiss rapidly moving sunspots as unconnected with flare energy storage, since Zirin and Lazareff (1975), for instance, find that rapidly moving spots produce many flares: flare activity ceases when the spot motion stops.

It is apparent that motions deep in the convection zone can provide the energy of a large flare (although possibly not the very largest) but would move the footpoints of coronal magnetic field lines at very slow speeds and produce very small displacements. Thus a very "stiff" coronal magnetic field is required to couple efficiently to such a driving force. On the other hand, flows in the subphotospheric layer, which can energize only minor flares, yield much smaller forces but larger velocities and can do significant work only against much weaker coronal fields.

Consider a coronal magnetic arcade of length L and width W between footpoints, with average coronal magnetic field strength B_{CO}. If this structure is sheared by displacing one set of footpoints parallel to the length L, by a distance ΔL, a magnetic field component $\Delta B = B_{CO} \Delta L / W$ is induced, yielding an increase in the magnetic energy of $\Delta\varepsilon = B_{CO}^2 L \, \Delta L^2/(8\pi)$, where we have assumed a volume $W^2 L$, i.e., that the height of the arcade is comparable to its width. Taking $L = 5 \times 10^9$ cm, and using the energies and displacements computed above, we then find that optimum coupling for a minor flare driven by shallow convection ($\Delta\varepsilon = 3 \times 10^{29}$ erg, $\Delta L = 3 \times 10^9$ cm) requires $B_{CO} \approx 15$ G, while a major flare driven by deep convection ($\Delta\varepsilon = 3 \times 10^{31}$ erg, $\Delta L = 3 \times 10^8$ cm) requires $B_{CO} \approx 1250$ G.

While the first case corresponds to a feasible, even low, coronal field strength, it seems unlikely that the average coronal field strength can be over 1000 G. The magnetic potential energy of such an arcade would obviously be huge, since we are requiring a very small footpoint displacement (and therefore very small shear) to account for the energy of a major flare. If such a stiff "spring" is not available, extraction of energy from the deep (as opposed to upper) convection zone becomes difficult and would require at minimum a much longer time scale for energy accumulation. This, together with the fact that magnetic structures observed in flaring regions appear to be highly sheared, suggests that perhaps flare energy cannot be obtained from the lower convection zone, in which case we are led to the conclusion that the "energy storage" model cannot account for major flares, but only moderate ones (e.g., 3×10^{30} erg from $z < 10^4$ km, requiring $B_{CO} \approx$ 120 G for optimal coupling).

Energy Available in an Erupting Flux Tube

A magnetic flux tube erupting through the photosphere contains free energy which it can give up on expanding from its pressure-confined state to a quasi-potential configuration. In addition to converting magnetic energy to kinetic energy of its expansion (see Spicer et al., 1986), it will form current sheets at the interface with pre-existing overlying magnetic fields which can then dissipate with resulting reconnection (see Heyvaerts et al., 1977). Furthermore, micro instabilities resulting in the dissipation of any field-aligned currents it carries could be triggered by the expansion. Thus there seems to be little reason to suppose that magnetic flux tubes must remain quiescent until stressed by photospheric flows.

To estimate the available energy, we note that the magnetic energy in a flux tube, $\varepsilon \approx B^2 A \, L/8\pi$, where A is the cross section and L the length of the flux tube, can be written as $\varepsilon \approx \Phi^2 L/8\pi A$, where Φ is the total magnetic flux. Now, if we assume that a flux tube, freed from the confining pressure of the photosphere, expands upward into the corona so that its area increases from A_0 to A_1, by a much larger factor than its length increases, the change in its magnetic energy is $\Delta\varepsilon \approx (\Phi^2 L/8\pi)(1/A_1 - 1/A_0)$. Assuming $A_1 \gg A_0$, we have $\Delta\varepsilon \approx -B^2 A_0 L/8\pi$, i.e., the bulk of the initial magnetic energy is available as free energy. Taking values of $B_0 = 1000$ G, $A_0 = 10^{19}$ cm^2 (implying a magnetic flux of 10^{22} Mx), and a length of 3×10^9 cm, we find ~10^{33} erg of available energy. This is more energy than we were able to obtain from convective motions and is sufficient to supply the largest flares ever observed.

Of course the magnetic flux tube presumably originated in or near the solar cycle dynamo region, which probably lies at the base of the convection zone (e.g., Gilman, 1986), and so was subjected to roughly the same forces and convective motions as the flux tubes considered in the previous section. The difference is that this flux tube may have been able to accumulate energy over a significant fraction of a solar cycle (11 years), rather than for the 1 day considered in the previous section.

Conclusions

We have considered three modes by which mechanical energy of the solar convection zone could be converted to transient energy release in solar flares. The mechanical energy due to differential rotation was found to be much too small to be significant, as was the energy in near-photospheric convective flows. However, the kinetic energy of motions in the upper convection zone can power moderate flares, while motions deep in the convection zone appear to be able to generate the energy required by large flares.

First, we examined the "photospheric dynamo," which has been proposed by a number of authors in analogy with magnetospheric substorms and MHD generators. Here, flare energy is dissipated in the corona simultaneously with its generation by sub-photospheric flows across the magnetic field lines at the footpoints of coronal magnetic fields. We find that the kinetic energy flux in convective motions is far too small to power flares in this way, except perhaps for minor subflares.

Next we considered "coronal energy storage," in which the kinetic energy of convective motions is intercepted in a similar way to the photospheric dynamo. In this case, however, the energy is stored in distortions of the coronal magnetic field, rather than being dissipated instantaneously. If energy is allowed to accumulate for periods of order 1 day, this mechanism has the potential to fuel even major flares. However, it appears to fall short by an order of magnitude of the energy thought to be released in the very largest observed flares. Moreover, it seems to be difficult to efficiently couple the energy source to the coronal magnetic field in order to extract the energy required for major flares on time scales of order 1 day, suggesting that convective motions may be restricted to powering only moderate sized flares.

Last, we computed the free energy available in an "emerging flux tube" and concluded that a large flux tube should carry enough free energy to account easily for even the largest flares. This theory is less satisfactory than the previous two in that the source of the free energy is the mechanism which generates magnetic flux in the first place, about which little is understood. But it is the only mechanism which appears able to fully account for the energy release in a large flare.

In conclusion, we note that solar flares exhibit an amazing variety of phenomena, and it is quite possible that many different forms of energization can result in a phenomenon which we recognize as a "flare." But the extreme energy requirements of a major flare narrow the range of candidates considerably.

Acknowledgments. This work was supported by NSF under grant ATM86-19853 and by NASA under grant NAGW86-4.

References

Alfvén, H., and P. Carlqvist, Current in the solar atmosphere and a theory of solar flares, *Solar Phys., 1*, 220, 1967.

Allen, C. W., *Astrophysical Quantities*, 3rd edition, Athlone Press, London, 1973.

Fisher, G. H., D.-Y. Chou, and A. N. McClymont, Emergence of anchored flux tubes through the convection zone, this volume, 1989.

Gilman, P. A., The solar dynamo: Observations and theories of solar convection, circulation, and magnetic fields, in *Physics of the Sun*, Vol. I, edited by P. A. Sturrock, T. E. Holzer, D. M. Mihalas, and R. K. Ulrich, D. Reidel, Dordrecht, p. 95, 1986.

Hagyard, M. J., et al., Preflare magnetic and velocity fields, in *Energetic Phenomena on the Sun*, NASA Conf. Publ. 2439, edited by M. Kundu and B. Woodgate, p. 1-16, 1986.

Heyvaerts, J., Coronal electric currents produced by photospheric motions, *Solar Phys., 38*, 419, 1974.

Heyvaerts, J., E. R. Priest, and D. M. Rust, An emerging flux model for the solar flare phenomena, *Astrophys. J., 216*, 123, 1977.

Henoux, J. C., and B. V. Somov, Generation and structure of the electric current in a flaring activity complex, *Astron. Astrophys., 185*, 306, 1987.

Kan, J. R., S.-I. Akasofu, and L. C. Lee, A dynamo theory of solar flares, *Solar Phys., 84*, 153, 1983.

Lin, R. P., Solar flare energetics, in *Gamma Ray Transients and Related Astrophysical Phenomena*, AIP Conf. Proc. No. 77, edited by R. E. Lingenfelter, H. S. Hudson, and D. M. Worrall, p. 419, 1982.

Lin, R. P., and H. S. Hudson, Non-thermal processes in large solar flares, *Solar Phys., 50*, 153, 1976.

McIntosh, P. S., The birth and evolution of sunspots: Observations, in *The Physics of Sunspots*, edited by L. E. Cram and J. H. Thomas, Sacramento Peak Observatory, p. 7, 1981.

Melrose, D. B., and A. N. McClymont, The resistances of the photosphere and of a flaring coronal loop, *Solar Phys., 113*, 241, 1987.

Sen, H. K., and M. L. White, A physical mechanism for

the production of solar flares, *Solar Phys.*, *23*, 146, 1972.

Somov, B. V., and S. I. Syrovatskii, Current sheets as the source of heating for solar active regions, *Solar Phys.*, *55*, 393, 1977.

Spicer, D. S., The thermal and non-thermal flare: a result of non-linear threshold phenomena during magnetic field-line reconnection, *Solar Phys.*, *53*, 249, 1977a.

Spicer, D. S., An unstable arch model of a solar flare *Solar Phys.*, *53*, 305, 1977b.

Spicer, D. S., Magnetic energy storage and conversion in the solar atmosphere, *Space Sci. Rev.*,*31*, 351, 1982.

Spicer, D. S., J. T. Mariska, and J. P. Boris, Magnetic energy storage and conversion in the solar atmosphere, in *Physics of the Sun*, Vol. II, edited by P. A. Sturrock, T. E. Holzer, D. M. Mihalas, and R. K. Ulrich, D. Reidel, Dordrecht, p. 181, 1986.

Spruit, H. C., A model of the solar convection zone, *Solar Phys.*, *34*, 277, 1974.

Spruit, H. C., Motion of magnetic flux tubes in the solar convection zone and chromosphere, *Astron. Astrophys.*, *98*, 155, 1981.

Stenflo, J. O., A mechanism for the build-up of flare energy, *Solar Phys.*, *8*, 115, 1969.

Svestka, Z., *Solar Flares*, D. Reidel, Dordrecht, 1976.

Syrovatskii, S. I., On the mechanism of solar flares, in *Solar Flares and Space Research*, edited by C. de Jager and Z. Svestka, North-Holland, Amsterdam, p. 346, 1969.

Uchida, Y., and T. Sakurai, Heating and reconnection of the emerging magnetic flux-tubes and the role of the interchange instability, *Solar Phys.*, *51*, 413, 1977.

Wu, S. T., et al., Flare energetics, in *Energetic Phenomena on the Sun*, NASA Conf. Publ. 2439, edited by M. Kundu and B. Woodgate, p. 5-1, 1986.

Zirin, H., and B. Lazareff, Sunspot motion, flares and type III bursts in McMath 11482, *Solar Phys.*, *41*, 425, 1975.

What do solar radio bursts tell us about particle beams and wave-particle interactions in flares?

ELECTRON BEAMS AND INSTABILITIES DURING SOLAR RADIO EMISSION

Martin V. Goldman

Department of Astrophysical, Planetary and Atmospheric Sciences,
Campus Box 301, University of Colorado, Boulder, CO 80309

Abstract. We summarize current theoretical ideas concerning two kinds of solar emissions: Type III bursts, and microwave spike bursts, with emphasis on the underlying Langmuir wave and cyclotron maser emission processes, and the relevant sources of free energy in the high-energy electron streams which drive them.

Introduction

A number of different kinds of solar radio emission are thought to be driven by free energy in electron streams.

In this review we will describe two emission mechanisms which arise from instabilities driven by non-thermal features of electron streams. One of these is the electrostatic bump-on-tail instability; the other is the cyclotron maser instability. Both have been studied theoretically in some detail, and are relatively well understood compared to other emission mechanisms.

Our current understanding of Type III solar radio bursts is that they are the result of a multi-stage process in which electron plasma waves are excited by the stream, and convert to photons at the plasma frequency and the second harmonic by nonlinear processes. The electron plasma waves are excited by a "bump" on the tail of the reduced one-dimensional electron distribution function. One can think of this as a parallel energy "maser" process, since there is a local inversion of electron energy associated with the positive slope of the distribution.

It has been proposed that microwave spike bursts are the result of a cyclotron maser instability, which produces electromagnetic radiation near harmonics of the electron cyclotron frequency. This instability is driven by inversion of the energy associated with the component of electron velocity perpendicular to the local solar magnetic field, and hence is a perpendicular maser process.

The plan of this paper is as follows: After briefly describing the observed properties of the two kinds of emission, we discuss how different kinds of electron free energy might develop naturally. We then describe the associated instabilities.

Electron plasma waves (Langmuir waves) and electron streams measured in association with Type III events by spacecraft in the solar wind will be considered next. We briefly review current theoretical ideas about how the unstable Langmuir waves saturate and emit Type III radiation.

Finally, we turn our attention to emission processes near one solar radius, and discuss the linear and nonlinear properties of the cyclotron maser process.

Observations and Theoretical Models

Type III Emissions

Type III emissions are characterized (Goldman, 1984) by a fast frequency drift, down from hundreds of MHz to tens of kHz and lower, as the electron stream propagates out into the solar wind along an open magnetic field line. (Emission can also be from closed field lines, as evidenced by inverted "J" or "U" events. See Stewart, 1975.)

In the lower corona the emissions are often observed to occur in frequency pairs, near the local plasma frequency and its second harmonic (with almost a two-to-one ratio).

Typical brightness temperatures of the high-frequency emissions are 10^{10}-10^{11} K. The degree of circular polarization varies from 10-35%, and is in the O-mode.

The electrons which drive the plasma waves responsible for the Type III's are typically in the 3-100 keV range, and are said to occur in bunches of up to ten bursts, separated by seconds, with 10^{31} to 10^{32} electrons per bunch.

Dulk et al. (1984) investigated whether interplanetary Type III bursts are due to fundamental radiation, second harmonic radiation, or both. They identified several features of the bursts which indicate that the initial part of most emissions, from onset to the peak or beyond, is fundamental radiation; thereafter second harmonic radiation becomes more and more dominant. They suggested that most, but not all, bursts have a fundamental component.

Microwave Spike Bursts

Microwave spike bursts are narrow-band emissions from the lower solar corona, ranging from hundreds of MHz up to about 5 GHz in frequency. The bursts exhibit rapid temporal variation, on time scales as short as milliseconds. They are characterized by very high brightness temperatures of up to 10^{15} K, and exhibit 50-100% circular polarization in the X-mode (rotation in the same sense as electrons). The relevant electrons are in the energy range 100 keV.

According to theory (Wu and Lee, 1979; Melrose and Dulk, 1982a,b), these bursts occur only under conditions when the electron cyclotron frequency is above the plasma frequency (i.e., low in the corona). Emissions, according to this interpretation, are near the electron cyclotron frequency and its harmonics. Fundamental emission is thought to be reabsorbed in flux tubes higher in the corona, so that the observed emission is probably at harmonics of the cyclotron frequency.

Geometry

Figure 1 is a schematic of the geometries associated with both kinds of emissions. Cyclotron maser emission propagates out almost at right angles to the direction of the local (closed) magnetic field line, which is assumed to be low in the corona.

Type III emissions come from a spectrum of beam-driven electron plasma waves. If this spectrum were fairly broad but anisotropic, with symmetry axis parallel to the beam, theory predicts that the radiation at the plasma frequency would propagate in a dipole pattern (peaked at 90° to the magnetic field line), and that the radiation at the second harmonic of the plasma frequency would propagate in a quadrupole pattern (45° cones).

We shall show later that there are theoretical reasons to believe the nonlinearly saturated electron plasma wave spectrum may be rather isotropic (mostly out of resonance with the beam). In that case the electromagnetic emission patterns of both fundamental and second harmonic radiation are more isotropic.

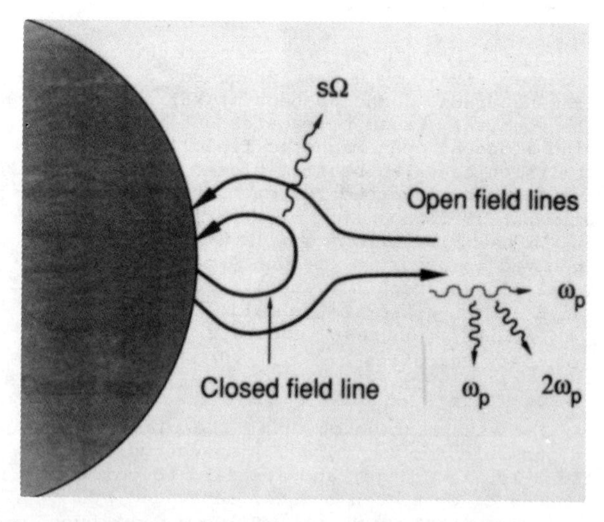

Fig. 1. Sketch of geometry of solar magnetic field lines and two kinds of radio emission. The nearly orthogonal emission at $s\Omega_e$ is due to cyclotron maser emission in the lower corona. The emissions at the electron plasma frequency and its second harmonic are Type III bursts, produced by Langmuir waves (dotted wavy lines).

Electron Stream Free Energy and Instabilities

Energy of Localization

Let us imagine that energetic electrons have been injected in some region of a flaring solar flux tube. Even if this distribution is assumed to be Maxwellian, it will possess free energy associated with spatial localization. Since very little is known about the details of such injected distributions, let us assume an isotropic Maxwellian and follow the evolution of free energy as the electrons move. This is illustrated in Figure 2.

Parallel Energy Inversion

Time-of-flight effects will change the form of free energy downstream. Electrons with higher parallel velocity in the injected distribution arrive at a remote downstream point on the guiding magnetic field line before the slower ones. This leads to local parallel energy inversion and the formation of a "bump" on the tail of the reduced one-dimensional distribution function. Observations of such bumps, and of their temporal drift to lower velocities, are

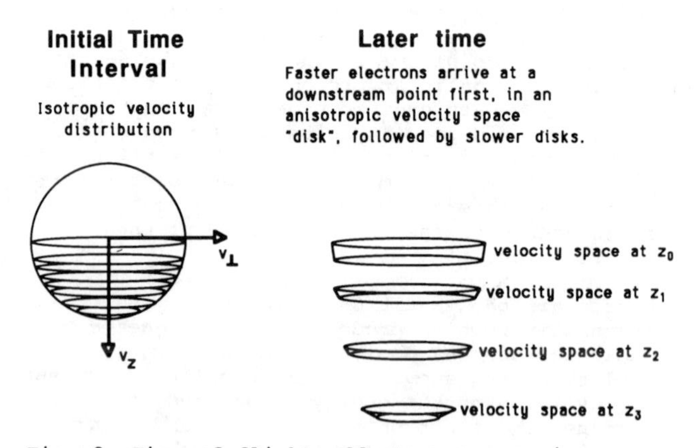

Fig. 2. Time-of-flight effects cause an isotropically-injected distribution of electrons in a localized region of a coronal flux tube to develop parallel energy inversion and velocity anisotropy downstream (the flux tube is assumed to have parallel magnetic field lines).

well documented for Type III events in the solar wind (Lin. et al., 1981). Inversion of parallel electron energy (i.e., a local positive slope on the electron distribution) drives electron plasma waves, and possibly other kinds of waves, such as whistlers. (The bump-on-tail distribution is not the only source of free energy which can drive plasma waves. If the bump is narrow, and the beam density high, the instability changes character, becoming nonresonant instead of resonant. The so-called "cold beam" instability is reactively driven by the entire beam, and is not a "maser" instability.)

Anisotropy Free Energy

Another form of free energy associated with electron propagation has to do with the development of velocity anisotropy. To isolate this effect, let us first consider straight (parallel) magnetic field lines, and neglect collisions. At a remote downstream point, the instantaneous range or "spread" in parallel velocities is determined by the duration and volume of the upstream injection. The spread of perpendicular velocities, on the other hand, is determined only by the perpendicular temperature of the injected distribution. The perpendicular velocity spread can be much larger than the parallel velocity spread, provided the injection time is short. This produces "sliced disks" downstream of the original spherically symmetric injected distribution, as illustrated in Figure 2. The resulting velocity anisotropy represents a form of free energy capable of driving non-resonant electromagnetic instabilities, such as the whistler-beam-mode instability (Goldman and Newman, 1987; Newman et al., 1988b).

Perpendicular Energy Inversion

If we now consider the effects of converging magnetic field lines, yet another form of free energy appears, and, with it a new instability: the cyclotron maser instability. As shown in Figure 3, numerical solutions for the propagation of electrons into converging magnetic field lines (White, 1984; White et al., 1986) reveals that the "slices" are bent into "crescents." This is because electrons with high parallel velocity and moderate pitch angle in the injection region, arrive downstream with larger pitch angle, and slower parallel velocity. After electrons have mirrored or precipitated near the foot of a a flux tube, this effect becomes more exaggerated, and a loss cone distribution develops. The resulting local inversion of perpendicular electron energy may drive whistlers or the cyclotron maser instability, which we will discuss later.

Nonlinear Wave-Induced Diffusion

It is important to bear in mind that the free energy in the stream may be affected by the growing waves it produces. Thus electron plasma waves may cause nonlinear parallel electron diffusion which levels the driving "bump" ("plateau" formation). Electromagnetic waves from maser and whistler instabilities can cause nonlinear electron diffusion which tends to eliminate perpendicular energy inversion and velocity anisotropy.

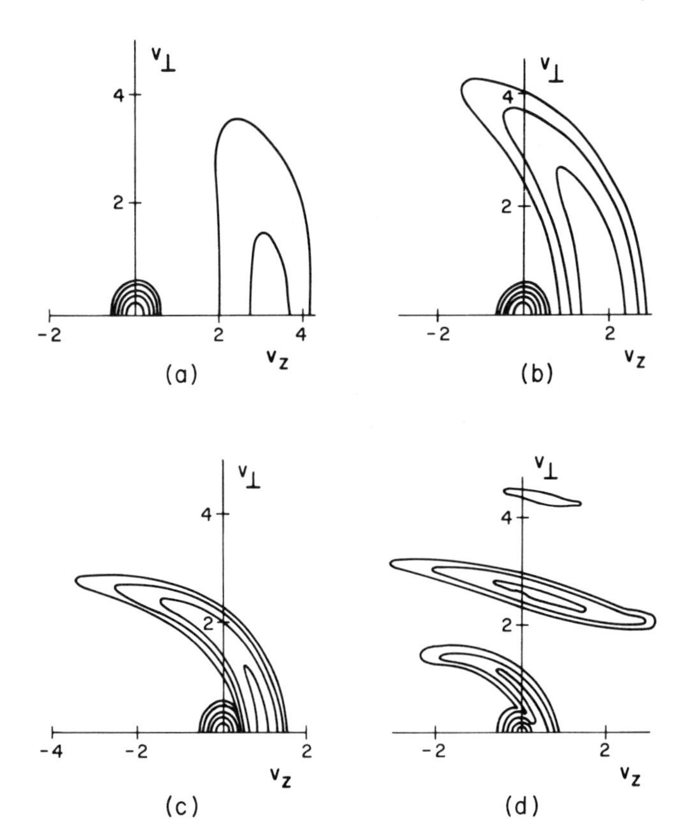

Fig. 3. Time-of-flight effects in converging magnetic field lines cause an isotropic distribution of impulsively- injected electrons to become crescent-shaped and, eventually, to form a loss cone distribution (White, 1984; White et al., 1986). The time sequence in (a)-(d) is t = 0.5T, 1T, 2T, and 4T, where T is the time-of-flight between injection and observation regions for a "thermal" electron in the energetic distribution.

Other Phenomena

The situation is complicated still further by phenomena we have neglected in the simplified picture elucidated thus far. The details of electron injection may be critically important (duration, angular distribution, volume size, etc.). Possible low-velocity return currents may excite ion-acoustic instabilities. Twisted magnetic field lines could alter the evolution of electron streams. On a much longer time scale than it takes for the maser to operate (i.e., some seconds), chromospheric evaporation occurs, due to precipitating electrons near the footpoints of closed flux tubes. This may increase the electron density and hence the plasma frequency in the flux tube, possibly turning off the cyclotron maser instability.

Type III Events in the Solar Wind

Spacecraft Observations

Type III events in the solar wind should be easier to understand than Type III emissions from the low solar corona, since extensive measurements from satellites have provided valuable data on the radiation, the underlying plasma waves, and the properties of the electron stream itself (Lin et al., 1981, 1986; Goldman, 1983).

During Type III events, electron streams are observed to develop bumps on the measured reduced electron distribution function. Positive slopes on these bumps correlate well with observations of the more intense Langmuir waves. (It is appropriate to refer to electron plasma waves as Langmuir waves only when the cyclotron frequency is much less than the plasma frequency, and the beam density is much less than the background plasma density, as in the solar wind.)

Wave fields as high as 10 mV m^{-1} have been measured, but much lower values are more typical.

There are important limitations on present-day spacecraft measurement capabilities.

The magnitude of Langmuir wave electric fields in wave packets smaller than ~10 km cannot be sampled properly because of instrumental limits. This is unfortunate, because theory predicts the strongest wave fields to occur in nonlinear (self-focused) wave packets smaller than 1 km.

The wave growth rate is proportional to the slope of the bump at the resonant velocity: $\gamma \propto \partial_{v_\parallel} f$. With existing instruments the observation time required to reconstruct the velocity distribution function of the electron stream (and hence its slope) is typically from 10 to 100 seconds. Shorter time scale "microstructure" in the velocity distribution cannot presently be resolved. The time to measure the beam is often larger than the inferred growth time for the waves.

A further difficulty concerns the theoretically predicted effect of ambient long wavelength density fluctuations in the solar wind (Muschietti et al., 1985; Celnikier et al., 1987). The measured level of fluctuations should rapidly scatter the growing Langmuir waves out of resonance with the beam, and constitute an effective barrier to wave growth, provided the growth is proportional to the measured values of $\partial_{v_\parallel} f$.

A proposed theoretical solution to this dilemma is that the Langmuir waves possess much larger growth rates, proportional to larger values of $\partial_{v_\parallel} f$ in the unseen, short time-scale microstructure of the beam distribution function (Melrose and Goldman, 1987). R. Lin (1989) has proposed that future satellite experiments (e.g., WIND) be designed with much higher time resolution for particle measurements. This would be a very useful instrumentational response to theory.

Nonlinear Theory of Langmuir Waves

The emission of Type III radiation at the fundamental and second harmonic of the plasma frequency cannot be understood without detailed knowledge of the nonlinearly saturated Langmuir wave spectrum. The nonlinear saturation of the beam-excited Langmuir waves in the solar wind is strongly sensitive to the properties of the beam. Both wave-particle and wave-wave processes may contribute, as well as advection of slower electrons from the remote injection region (as described earlier).

A number of investigators (Magelsson and Smith, 1977; Grognard, 1982, 1985) have studied wave-particle saturation numerically using the "quasi-linear" theory of weak plasma turbulence. In this theory the wave phases are assumed to be random, and the scattering of Langmuir waves off ions or ion-acoustic waves is neglected. The main effect is Langmuir-wave-induced diffusion of electrons, which occurs when the wave energy density becomes a significant fraction of the free energy in the beam. This process tends to level the bump (to cause plateau formation). A competing process, tending to restore the bump, is the advection of slower stream electrons.

A side effect of advection is that portions of the lower-velocity bump can be in resonance with Langmuir waves driven earlier, and can reabsorb them when their phase velocity matches values of $v_\parallel$ where the slope $\partial_{v_\parallel} f < 0$. This is one explanation for how the free energy in the beam is able to survive the long journey from the sun, and why the Langmuir wave energy is relatively low.

Another explanation has to do with the stimulated scattering of Langmuir waves out of resonance with the beam. If this process occurs faster than quasi-linear diffusion, then plateau formation will be inhibited, and again the beam will survive. For conditions appropriate to Type III events in the solar wind, a comparison of the rate of stimulated Langmuir wave scatter with the nonlinear diffusion rate yields the following condition for the nonlinear wave-wave process to dominate

$$16(v_e/v_b)^2 \ll (\Delta v_b/v_b)^3 \tag{1}$$

Here, v_b is the beam velocity, v_e is the thermal velocity of the background plasma, and Δv_b is the spread of the beam (i.e., width of the bump). Measured values of these parameters often show that both sides are of the same order. However, this estimate neglects the role of advection.

Advection quasi-linear diffusion and stimulated scattering of Langmuir waves have been studied numerically by Grognard (1982, 1985) in the random phase approximation of weak turbulence theory, using measured beams as a boundary condition. He has found that stimulated scattering is the faster effect. A cascade of backscatters, each to lower wave number, leads to a pileup of Langmuir wave energy at long wavelengths (compared to the resonant wavelength, $\lambda \approx f_p v_b$, where f_p is the plasma frequency in Hz). Since there is no damping at long wavelengths, Langmuir wave energy continues to build up in this "condensate." Saturation therefore may not occur within the framework of weak turbulence, even though it contains both wave-wave and wave-particle nonlinearities.

The solution to this deficiency appears to lead us into the theory of "strong" Langmuir turbulence (Goldman, 1983, 1984, 1986). This theory is also based on an expansion in the wave energy. However, certain coherent-phase wave-wave nonlinearities are included. In particular, the long wavelength condensate is unstable to a modulational instability, which transfers energy to higher wavenumbers.

In real space, one observes a related process of nonlinear spatial collapse of wave packets. The physics of this collapse is as follows: The ponderomotive force of the wave packet expels electrons from the site of the packet. The electrons drag out the ions, leaving a density cavity. However, a region of lowered density is a region of higher index of refraction for Langmuir waves, so the rays associated with the wave packet are internally reflected. As a result, the wave packet intensifies and the index of refraction increases. The process is unstable, and the wave packet collapses to small dimensions as it intensifies to high energy.

Collapse, modulational instability, and stimulated scatter are all included in the so-called Zakharov equations (Goldman, 1984). Recent numerical solutions to the driven and damped Zakharov equations in three dimensions are shown in Figures 4 and 5 (Robinson et al., 1988). In real space, the collapsing wave packets are visible. For parameters associated with Type III events in the solar wind, scaling laws suggest wave electric fields of order 10 mV m^{-1} at spatial scales of ~10 km, and approaching 1 V m^{-1} in the interior of collapsed wave packets of a few hundred meters (Newman et al., 1988a).

In k-space (not shown), a fairly isotropic wave spectrum is found, with maximum at $k = 0$. This is because many long wavelength plane waves must interfere destructively in order to produce the empty space between intense wave packets. Wave saturation occurs because energy can be extracted by electrons from the wave packets when they have collapsed to wavelengths on the order of a few Debye lengths.

Theory of Plasma Emission

The shape and intensity of the saturated Langmuir wave spectrum controls the emission of Type III radiation at the fundamental and at the second harmonic. We shall discuss two mechanisms for plasma emission, although many others have been proposed.

Coalescence of two Langmuir waves from different parts of the spectrum can produce radiation near twice the plasma frequency. Figure 6 (Newman, 1985) illustrates the kinematical and dynamical requirements on an

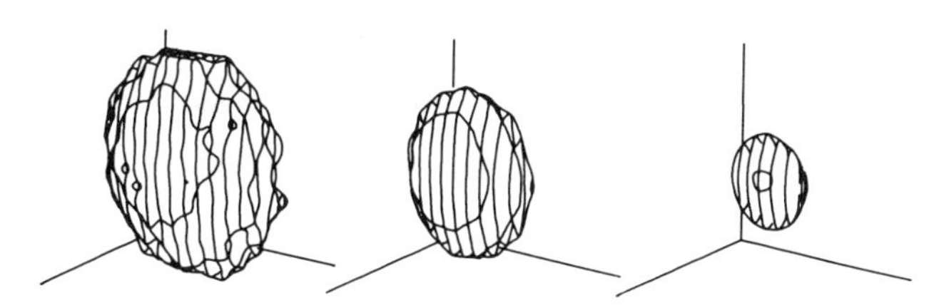

Fig. 4. Time sequence of electric field magnitude level surfaces drawn at 80% of the peak value for a single collapsing wave packet, in a numerical solution to the driven Zakharov equations. The "pancake" shape is characteristic of most collapsing objects in three dimensions. Each frame is 1/8 the simulation volume, centered on the collapsing packet.

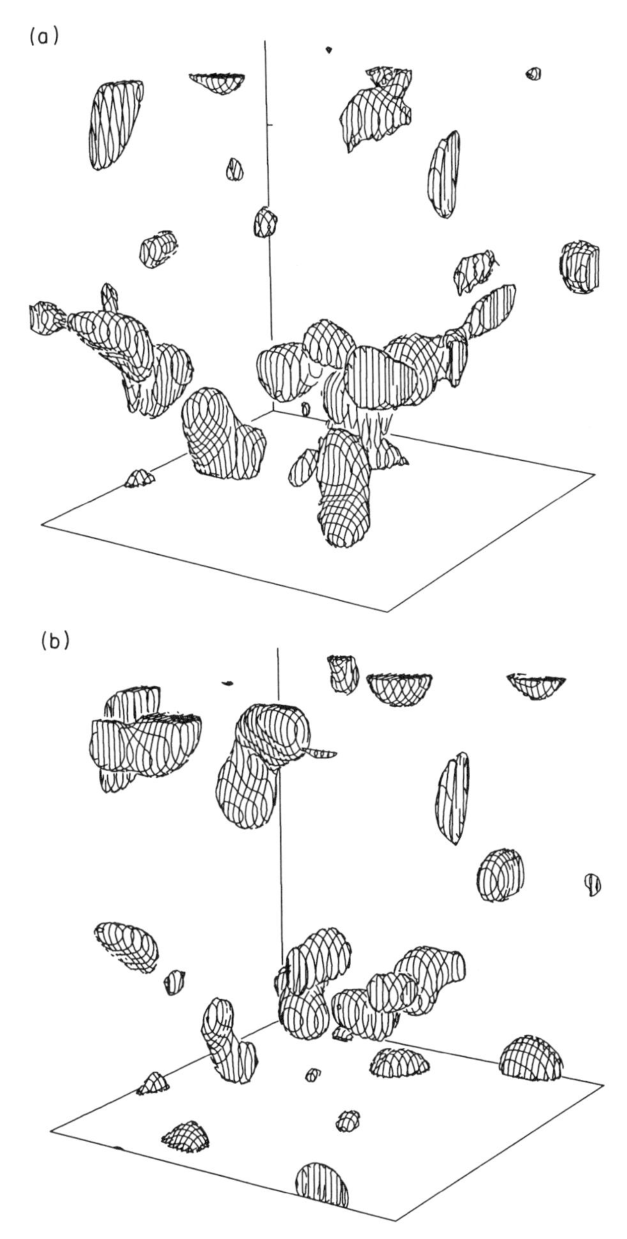

Fig. 5. Three-dimensional numerical solutions to the driven/damped Zakharov model of strong Langmuir turbulence. (a) Surfaces of constant $|E|^2$ are drawn surrounding the cores of collapsing packets at a level of 16% maximum. (b) Surfaces of constant density deviation, $\delta n/n$.

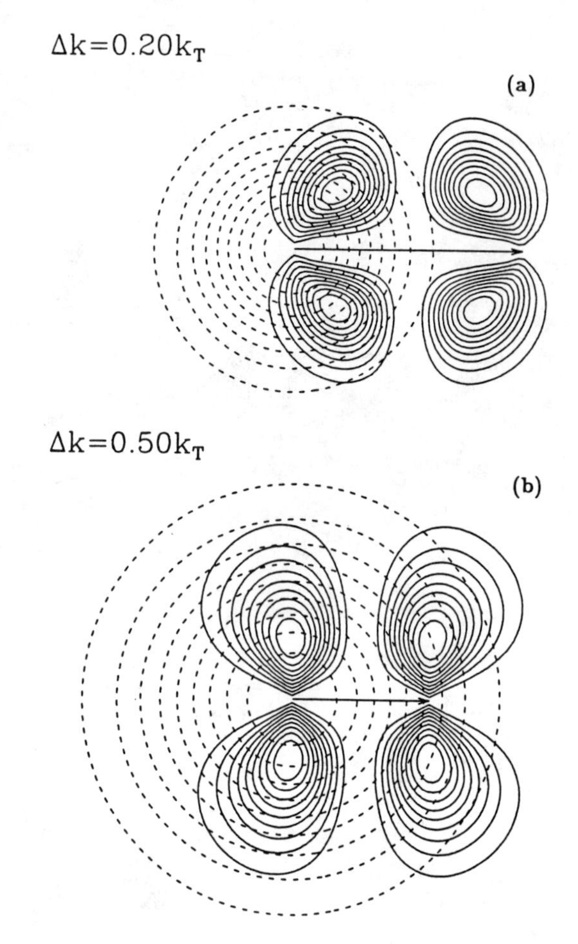

Fig. 6. Assumed contours of Langmuir wave spectrum (dashed), and effective portions of the spectrum (solid) contributing in the coalescence of two Langmuir waves to emit a photon with wave vector $\mathbf{k}_T$ (arrow). Contour intervals are 10%.

assumed Langmuir spectrum centered at $k = 0$, for efficient production of a photon at twice f_p, propagating in the given direction (arrow). The solid line contours must fall within reasonably intense portions of the spectrum (dashed) to efficiently drive this process. The solid contours reveal a quadrupole pattern of source Langmuir waves for the given emission direction. Note, however that emission can occur with equal probability in all directions, since the spectrum is assumed to be isotropic, in accordance with numerical simulations of strong-turbulence theory. The low directivity of harmonic plasma emission is consistent with observation. It is important to recognize that a spectrum entirely in resonance with the beam (i.e., narrowly centered around $k = \omega_p/v_b$) would not emit efficiently at $2\,\omega_p$.

A likely mechanism for the emission of fundamental radiation is the process $L \rightarrow T + i$, in which a Langmuir wave undergoes a parametric decay instability into a photon near the plasma frequency and an ion-acoustic wave (or discrete ions).

There is some evidence for this process from ISEE 3 measurements (Lin et al., 1986) in which ion-acoustic waves with wave number approximately equal to the beam resonant Langmuir wave number were observed coincident in time with the most intense spikes of the Langmuir waves. However, a self-consistent theoretical analysis of this process appears to require a mechanism which boosts the intensity of Langmuir waves at short spatial scales. The postulated short time scale enhancements of the positive slope on the electron bump distribution function (Melrose and Goldman, 1987), together with wave packet collapse, would produce such a result.

Alternative explanations for the ISEE 3 correlation measurements have also been proposed recently (Chian and Alves, 1988). This is one more reason for developing instrumental tools to resolve electron streams on a shorter time scale.

Cyclotron Maser Emission in the Corona

The Langmuir wave instability is driven by beam electron energy inversion in the parallel direction; the growth rate of a mode with wave number k is proportional to $\partial_{v_\parallel} f$, on the resonant "surface" in velocity space determined by the Cerenkov resonance condition: $\omega - k_\parallel v_\parallel = 0$.

The electromagnetic cyclotron maser instability is driven by perpendicular energy inversion, and the growth rate is proportional to $\partial_{v_\perp} f$, on the resonant surface in velocity space determined by the cyclotron resonance condition

$$\omega - s\Omega_e - k_\parallel v_\parallel = 0, \quad s = 1,2,.. \tag{2}$$

where the electron cyclotron frequency, Ω_e, must contain the relativistic mass, γm_0, for the growth rate to be non-zero. Since γ depends on velocity, the resonant surface is an ellipsoid. Electrons on the resonant ellipsoid which have the property $\partial_{v_\perp} f > 0$ provide the maser action which drives the instability.

An unstable situation is illustrated in Figure 7, in which a loss cone distribution is indicated by the shading, and a cross section of the resonant surface is indicated by the curve labeled "resonant ellipse." Most of the electrons on the ellipse satisfy the condition $\partial_{v_\perp} f > 0$, leading to net growth.

In addition, electromagnetic instability requires that the electron plasma frequency be less that the electron cyclotron frequency:

$$\omega_p \lesssim \Omega_e \tag{3}$$

This defines relevant flare parameters (usually only flares, or portions of flares, occurring low in the corona qualify).

A further prediction is that the emission propagates almost at right angles to the solar magnetic field. Neither this directivity nor the constraint of equation (3) has been verified by observations in the context of solar microwave spike bursts.

Recent particle-in-cell simulations (Winglee et al., 1988) show how a loss cone distribution of the type sketched in Figure 7 saturates non-

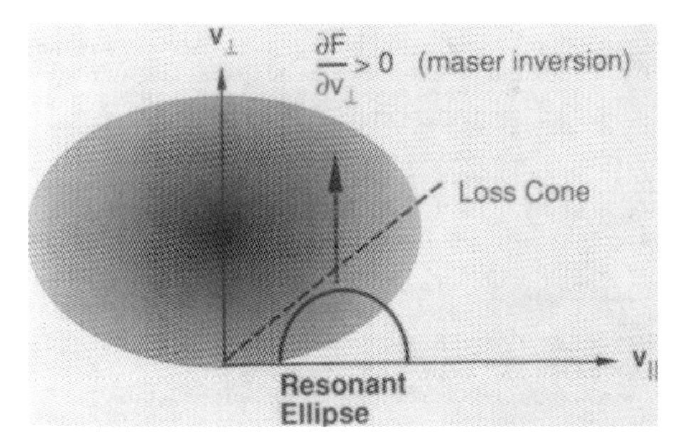

Fig. 7. Illustration of a loss cone electron distribution (shaded) which is unstable to cyclotron maser action by the electrons on the resonant ellipse. Such a process has been proposed as an explanation for microwave spike bursts.

linearly. If the loss cone is assumed to be the initial distribution in an initial value problem, then wave-induced perpendicular velocity-space diffusion of electrons removes the perpendicular free energy and saturates the instability. If, on the other hand, the loss cone distribution is sustained by advecting mirrored electrons ("driven" problem), then the emission recurs intermittently. A number of wave packet bursts occur on a sub-millisecond time scale, each carrying a few percent of the beam energy. These results are shown in Figure 8.

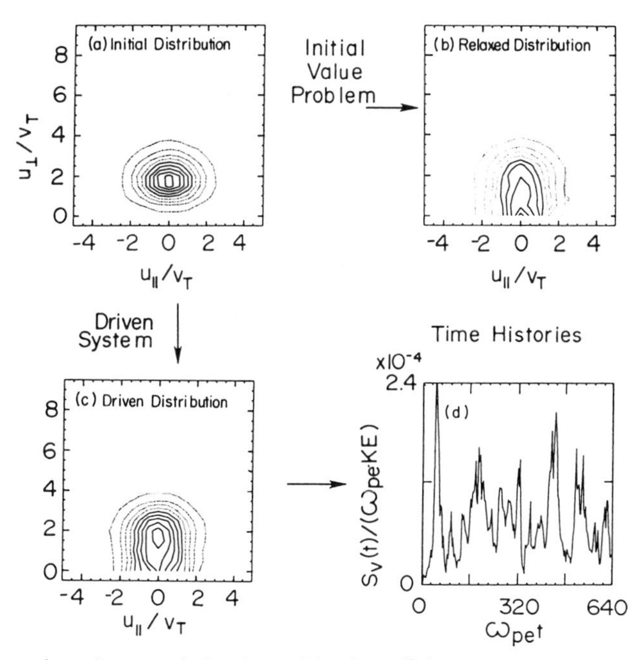

Fig. 8. Particle-in-cell simulations of the electron cyclotron instability, treated both as an initial value problem and as a "driven" simulation, in which the loss cone is sustained by recycling electrons. (a) Initial distribution. (b) Late state developed from initial value problem. (c) Late state developed during a driven simulation. (d) Burst time history. Parameters were chosen so that $\omega_p = \Omega_e/10$ (Winglee et al., 1988).

Conclusion

We conclude by summarizing the current status of theory and observation for the two kinds of emission.

Type III emissions have been studied extensively, both by theorists and by observers. The theory has led us into basic issues concerning Langmuir turbulence which are still under study. There are a number of different theoretical explanations for the survival of beam free energy, for the saturation of Langmuir waves, and for the nonlinear processes of plasma emission which give rise to the observed bursts.

Even in the solar wind, where we have in situ measurements of the beam, the Langmuir waves, and the radiation, we are missing the relevant short-time, short-spatial-scale information which would enable definitive comparisons to be made with theory. Future improvements in satellite instrumentation could prove to be pivotal. The relationship between Type III bursts and the electron streams which produce them is currently incompletely understood in the solar wind context, and is even less clear in connection with the low solar corona.

We have yet to establish a comprehensive model of electron streams in the solar corona. Key issues which remain to be resolved by theory and observation are the following: How, and for how long are electrons injected, and what are the beam modifications associated with propagation (time-of-flight effects)? What are the effects of return currents, twisted magnetic field lines, and chromospheric evaporation? If resonant reabsorption of maser radiation is as strong as calculated, how is the observed radiation produced? What is the role of cyclotron harmonic emission?

Promising new theoretical developments concerning the theory of microwave spike bursts have linked them to cyclotron maser emission. However, this mechanism works only under the special constraint that the plasma frequency be less than or on the order of the cyclotron frequency. A final issue is whether there are competing instabilities which may offer an alternate explanation for some microwave spike bursts.

Acknowledgment. This work was supported by NSF's Solar-Terrestrial Divison and NASA's Terrestrial Theory and Solar Heliospheric Programs under grants ATM-8511906, NAGW-91 and NSG-7287, respectively, to the University of Colorado at Boulder.

References

Celnikier, L.M., L. Muschietti, and M.V. Goldman, Aspects of interplanetary turbulence, Astron. and Astrophys., 181, 138-154, 1987.

Chian, A. C.-L., and M. Alves, Nonlinear generation of the fundamental radiation of interplanetary Type III radio bursts, Ap. J. (Lett.), submitted, 1988.

Dulk, G.A., J.L. Steinberg, and S. Hoang, Type III Bursts in interplanetary space: Fundamental or harmonic?, Astron. Astrophys., 141, 30, 1984.

Goldman, M.V., Progress and problems in the theory of Type III solar radio emission, Solar Phys., 89, 403-442, 1983.

Goldman, M.V., Strong turbulence of plasma waves, Rev. Mod. Phys., 66, 709-735, 1984.

Goldman, M.V., Langmuir wave solitons and spatial collapse in plasma physics, Physica, 18D, 67-76, 1986.

Goldman, M.V., and D.L. Newman, Electromagnetic beam-modes driven by anisotropic electron streams, Phys. Rev. Lett., 58, 1849-1852, 1987.

Grognard, R. J.-M., Numerical simulation of the weak turbulence excited by a beam of electrons in the interplanetary plasma, Solar Phys., 81, 173, 1982.

Grognard, R. J.-M., Propagation of Electron Streams, Chapter 11 in Solar Radiophysics, edited by D.J. McLean and N.R. Labrum, pp. 253-286, Cambridge Univ. Press, Cambridge, 1985.

Lin, R.P. New techniques for charged particle measurements in the interplanetary medium, these proceedings, 1989.

Lin, R.P., D.W. Potter, D.A. Gurnett, and F.L. Scarf, Energetic electrons and plasma waves associated with a solar Type III radio burst, Ap. J., 251, 364, 1981.

Lin, R.P., W.K. Levedahl, W. Lotko, D.A. Gurnett, and F.L. Scarf, Evidence for parametric decay of Langmuir waves in solar Type III radio bursts, Ap. J., 308, 954, 1986.

Magelsson, G.R., and D.F. Smith, Nonrelativistic electron stream propagation in the solar atmosphere and Type III radio bursts, Solar Phys., 55, 211, 1977.

Melrose, D.B., and G.A. Dulk, Electron-cyclotron masers as the source of certain solar and stellar radio bursts, Ap. J., 259, 844, 1982a.

Melrose, D.B., and G.A. Dulk, Radio wave heating of the corona & electron precipitation during flares, Ap. J. (Lett.), 259, L41, 1982b.

Melrose, D.B., and M.V. Goldman, Microstructures in Type III events in the solar wind, Solar Phys., 107, 329-350 1987.

Muschietti, L., M.V. Goldman, and D. Newman, Quenching of the beam-plasma instability by 3-D spectra of large scale density fluctuations, Solar Phys., 96, 181-198, 1985.

Newman, D.L., Emission of electromagnetic radiation from beam-driven plasmas, Ph.D. dissertation, Dept. of Astrophysical, Planetary, and Atmospheric Sciences, Univ. of Colorado, 1985.

Newman, D.L., P.A. Robinson, and M.V. Goldman, Three-dimensional simulation of plasma turbulence in the solar wind, 1988 American Geophysical Union Spring Meeting, Baltimore, Maryland, May 16-20, 1988, 1988a.

Newman, D.L., R.W. Winglee, and M.V. Goldman, Theory and simulation of electromagnetic beam-modes and whistlers, Phys. Fluids, 31, 1515, 1988b.

Robinson, P.A., D.L. Newman, and M.V. Goldman, Three-dimensional strong Langmuir turbulence and wave collapse, Phys. Rev. Lett., in press, 1988.

Stewart, R.T., An example of a fundamental Type IIIb radio burst, Solar Phys., 40, 417, 1975.

White, S.M., Coherent radiation processes in astrophysics, Ph.D. thesis, Univ. of Sydney, 1984.

White, S.M., D.B. Melrose, and G.A. Dulk, Particle propagation effects on wave growth in a solar flux tube, Ap. J., 308, 424, 1986.

Winglee, R.W., G.A. Dulk, and P.L. Pritchett, Fine structure of microwave spike bursts and associated cross-field energy transport by cyclotron maser radiation, Ap. J., 328, 809, 1988.

Wu, C.S., and L.C. Lee, A theory of the terrestrial kilometric radiation, Ap. J., 230, 621, 1979.

SOLAR RADIO BURST SPECTRAL OBSERVATIONS, PARTICLE ACCELERATION, AND WAVE-PARTICLE INTERACTIONS

Dale E. Gary and G. J. Hurford

Solar Astronomy 264-33, California Institute of Technology, Pasadena, CA 91125

Abstract. We consider solar radio bursts throughout the radio spectrum from 30 kHz to 30 GHz. While the range of phenomena over this spectral range is large, the subject can be conceptualized by considering three characteristic frequencies of the plasma: the plasma frequency f_p, the gyrofrequency f_B, and the frequency $f(\tau_{ff}=1)$ at which the plasma becomes optically thick due to bremsstrahlung. We present an overview in terms of these characteristic frequencies to show why each dominates under various physical conditions in the Sun and solar wind. In the broad regime where plasma emission dominates, we discuss the progress that has been made in explaining the observed burst characteristics theoretically, concentrating on bursts of type II (shock wave related) and type III (excited by electron beams). Moving then to microwaves, we show that the microwave spectrum contains considerable diagnostic information, but that present observations cannot access it except in rare cases when the burst is spatially simple. We conclude with a discussion of future prospects for making further progress in the study of solar radio bursts.

Overview

Radio bursts in the Sun and solar wind occur over more than six decades in frequency, ranging from <30 kHz (the typical plasma frequency at Earth) to >30 GHz. Over this spectral range, a variety of emission mechanisms vie for dominance; yet much of the complexity exhibited by solar bursts can be understood in terms of three characteristic frequencies. One such frequency is the plasma frequency (Hz),

$$f_p = 9.0 \times 10^3 \sqrt{n_e}, \qquad (1)$$

which depends only on the ambient electron density n_e (cm^{-3}). The plasma frequency is important for two reasons. (1) Electromagnetic waves with frequency $f < f_p$ cannot propagate in the medium. (2) Burst exciters generate copious radio waves at the local plasma frequency and at its second harmonic $2\,f_p$. This type of emission is called, simply, plasma emission.

A second characteristic frequency is $f(\tau_{ff} = 1)$, at which the optical depth due to free-free emission (bremsstrahlung) reaches unity. This frequency is important because it marks the boundary below which radiation, by whatever mechanism, will be significantly absorbed by the ambient plasma. It is given (in Hz) approximately by

$$f(\tau_{ff} = 1) \approx 0.5 n_e T_e^{3/4} L^{1/2}$$

where T_e is the local electron temperature and L (cm) is a length which, for our purposes, can be taken as the density scale length (assuming constant coronal temperature).

A third characteristic frequency is the electron gyrofrequency (Hz),

$$f_B = 2.80 \times 10^6\, B$$

where B is the magnetic field strength (Gauss). The gyrofrequency is important because various gyromagnetic emission mechanisms operate at harmonics $s = f/f_B$ of the gyrofrequency. Gyroresonance emission, for example, operates at low harmonics ($s \leq 5$), while gyrosynchrotron emission operates at higher harmonics ($10 \leq s \leq 100$).

The values of these three characteristic frequencies as a function of distance from the Sun can be estimated from models of temperature, density, and magnetic field. This has been done in Figure 1, which shows the frequency spectrum along the abscissa, and distance from the Sun along the ordinate, both on a log scale. Density was based on the VAL model B (Vernazza et al., 1981),

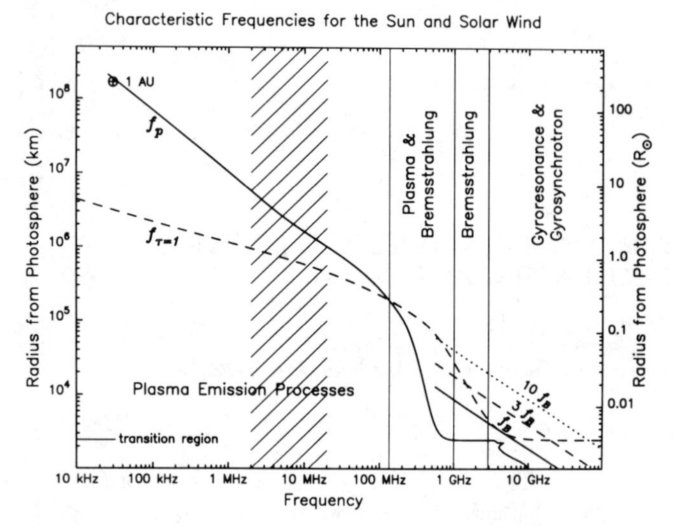

Fig. 1. Plot of three characteristic frequencies of the solar atmosphere, as described in the text. The regions of dominance of various emission mechanisms are noted. The hatching from 2-20 MHz marks an observational gap.

extended to 10^5 km by requiring hydrostatic equilibrium. From 10^5 km to 1 AU, the Saito (1970) model was used, with a scale factor of 5. The VAL model was scaled in density to match the 5 times Saito model at 10^5 km. Temperature was based on the VAL model to about 10^5 K, then extended to 2×10^6 K by a hydrostatic equilibrium model, and assumed constant at that temperature to the Earth. Magnetic field strength was based on the simple active region field model of Dulk and McLean (1978). For the $f(\tau_{ff} = 1)$ curve, a scale height L is needed. The scale height used was $L = H_0(T/T_0)(R/R_\odot)^2$ where $H_0 = 0.1\ R_\odot$ and $T_0 = 2 \times 10^6$ K. Near the Sun, the curves apply to active regions and are meant to be schematic only. In general, the characteristic frequency that appears highest in Figure 1 for a given observing frequency is the one that determines the emission mechanism. Broadly speaking, plasma emission dominates at frequencies from ~200 MHz down to 30 kHz, covering a height range of ~0.2-200 $R_\odot$ above the photosphere. The dominance of plasma emission over this range is simply because the plasma level lies above the $\tau_{ff} = 1$ surface at these frequencies. At frequencies between 200 MHz and 1 GHz, the plasma level is generally beneath the $\tau_{ff} = 1$ surface for active region densities, but plasma emission may still be important for at least three reasons. (1) Plasma emission can reach a brightness temperature $T_b \leq 10^{15}$ K, so that even when $\tau_{ff} \approx 10$, the $e^{-\tau}$ extinction of $\sim 2 \times 10^4$ still allows emission as bright as $T_b \approx 10^{10}$ K. (2) The corona in this height range is highly inhomogeneous, so that emission in relatively dense structures may still be physically adjacent to a less dense structure where $\tau_{ff} > 1$ (i.e., the transverse scale length L is effectively small). (3) Waveducting and scattering of the radiation along underdense flux tubes can allow the emission to escape with little absorption.

Emission in the range 1-3 GHz may be dominated by bremsstrahlung in higher density areas of active regions. Above 3 GHz, gyromagnetic emission becomes more important in active regions, where low harmonics of the gyrofrequency occur at heights overlying the $\tau_{ff} = 1$ level. At still higher frequencies, gyromagnetic emission again falls off in importance, and the observed low brightness temperature emission (outside of flares) arises from bremsstrahlung in the chromosphere.

What we learn about particle acceleration and wave-particle interaction is determined to a large extent by which mechanism generates the observed emission. Having seen where, and to some extent why, the different emission mechanisms operate, we now consider specific frequency ranges in more detail.

What We Have Learned—30 kHz-1 GHz

Although this is a very broad frequency range, the phenomena observed here are unified by the common mechanism of plasma emission. It will be convenient, however, to subdivide this range into three observationally distinct regimes. These are: (1) ground-based metric observations, from ~200 MHz down to the ionospheric cutoff near 20 MHz; (2) space-based observations from 2 MHz down to < 30 kHz; and (3) ground-based decimetric observations from ~ 200 MHz to 1 GHz. Note that there is an observational gap (the hatched area in Figure 1) at 2-20 MHz, spanning the height range ~1-10 $R_\odot$ above the photosphere.

Ground-Based Observations 20-200 MHz

The dominance of plasma emission over other processes in this spectral range yields a relatively small, well classified set of phenomena. Although there are many subclasses, five major types of burst in this range have been identified. The classification of types I, II, and III were made by Wild and McCready (1950) on the basis of their appearance on dynamic spectrograph records. Bursts of type IV were later classified by Boischot (1958) and extended by others, and bursts of type V were classified by Wild et al. (1959). Excellent reviews exist that describe these bursts and their variants in detail (see the recent reviews by Dulk (1985) and McLean (1985), and references therein). In this paper we will limit ourselves to a discussion of types II and III. Although the others have been extensively studied, these two have contributed most to our knowledge of particle acceleration and wave-particle interaction.

Type II bursts. These often appear on a spectrograph record (e.g., Figure 2) as a pair of bands of emission drifting slowly (≤ 1 MHz s^{-1}) to lower frequencies with time. Since the original suggestion by Uchida (1960), it has been accepted that the emission is due to plasma radiation at the fundamental and second harmonic of the plasma frequency, excited by a magnetohydrodynamic (MHD) fast-mode shock. Under this hypothesis, from equation (1), the frequency of the emission depends on the electron density of the medium through which the shock passes. The observed range of frequency drift rates corresponds to shock velocities in the range 200-2000 km s^{-1}.

The emission on a spectrograph record is commonly observed to begin at < 100 MHz some 5-10 min after an associated impulsive flare. A possible reason for this can be seen in Figure 1. The condition for a fast-mode MHD shock to form is that the wave speed exceed the local fast-mode speed. In the case of the solar corona, where the sound speed $c_s \ll v_A$, the fast mode speed reduces to the Alfvén speed, $v_A \propto f_B/f_p$. A wave traveling at ~ 1000 km s^{-1} will exceed the local Alfvén speed when $f_p > 7 f_B$. From Figure 1, extrapolating the $10 f_B$ (dotted) line, this occurs near ~200 MHz. Other effects, such as strength of shock required for emission, push this to somewhat lower frequencies, making this a likely explanation why type II bursts usually are not observed above 100 MHz.

A less commonly recognized but important fact is that, as in Figure 2, type II emission usually ceases before reaching the ionospheric cutoff near 20 MHz. In other words, metric type II bursts typically emit over a relatively small range of frequencies (and, hence, heights). Gary et al. (1984) studied such a type II burst for which simultaneous spatially resolved radio maps

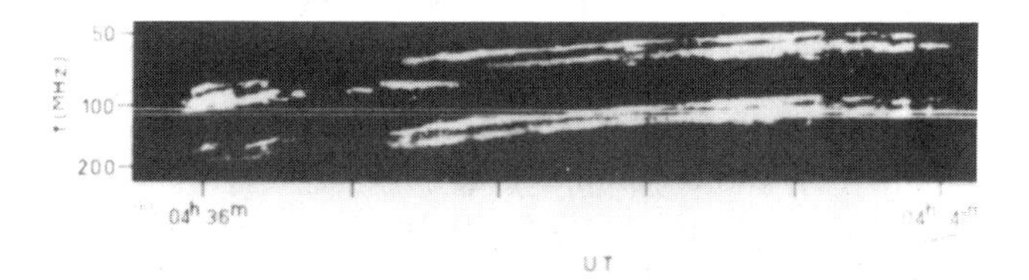

Fig. 2. A dynamic spectrograph record from Culgoora, Australia, showing the slow frequency drift characteristic of type II emission. This burst shows simultaneous emission at the plasma frequency f_p and at the second harmonic $2 f_p$, due to the propagation of a shock into the corona.

(from Culgoora) and white-light coronagraph observations (from the Solar Maximum Mission (SMM) Coronagraph/Polarimeter) were available. To explain why the type II emission ceased in the metric wavelength range, they concluded that the shock, a blast wave initiated by the associated flare, produced the type II emission only while it traversed a (slower) coronal mass ejection (CME) already in progress. A similar explanation was offered by Cane (1983) to explain discrepancies in shock velocities deduced from metric and kilometric type II bursts.

How electrons are accelerated at a fast-mode shock front remains uncertain. (Generation of Langmuir waves by the shock itself, without the intermediary of electrons, has even been proposed, see Nelson and Melrose, 1985.) Two alternative mechanisms for acceleration of electrons are shock drift acceleration, which requires the shock normal to be quasi-perpendicular to the magnetic field (Holman and Pesses, 1983), and diffusive acceleration, which requires a quasi-parallel shock (Achterberg and Norman, 1980). Both mechanisms work best for ions, however, while observationally it is the electrons that are needed. Preferential acceleration of electrons is possible as long as they, and not the ions, have an initial suprathermal component to act as a seed population. The shock drift acceleration mechanism has an advantage over diffusive acceleration in that the former works for a weak shock while the latter requires a strong (Alfvén Mach number $M_A > 5$) shock.

Type III bursts. Figure 3 shows a spectrograph record of an isolated metric type III burst displaying emission at the fundamental and the harmonic of the plasma frequency. Type III bursts are generated by a beam, or stream, of electrons moving rapidly outward in the corona; the rapid drift in frequency corresponds to exciter speeds of 0.2-0.6 c. The starting frequency of type III bursts is typically < 200 MHz, only a little higher than for type II bursts, but occasionally they are seen to begin at very high frequencies (~1 GHz). The starting frequency is probably regulated by the $\tau_{ff} = 1$ level, which overlies the plasma level (as shown in Figure 1) above ~200 MHz.

Type III bursts have received more attention by theorists than any other type of solar burst. The process by which the emission is generated starts with electrons accelerated in a "bump-on-tail" velocity distribution. The positive slope on the low velocity side of the "bump" is a source of energy that goes into plasma (Langmuir) waves in resonance with the electrons to create a nonthermal level of waves, at the expense of the speeding electrons. In weak turbulence theory, some of these Langmuir waves scatter off ions (actually a wave-wave interaction with ion-acoustic waves)

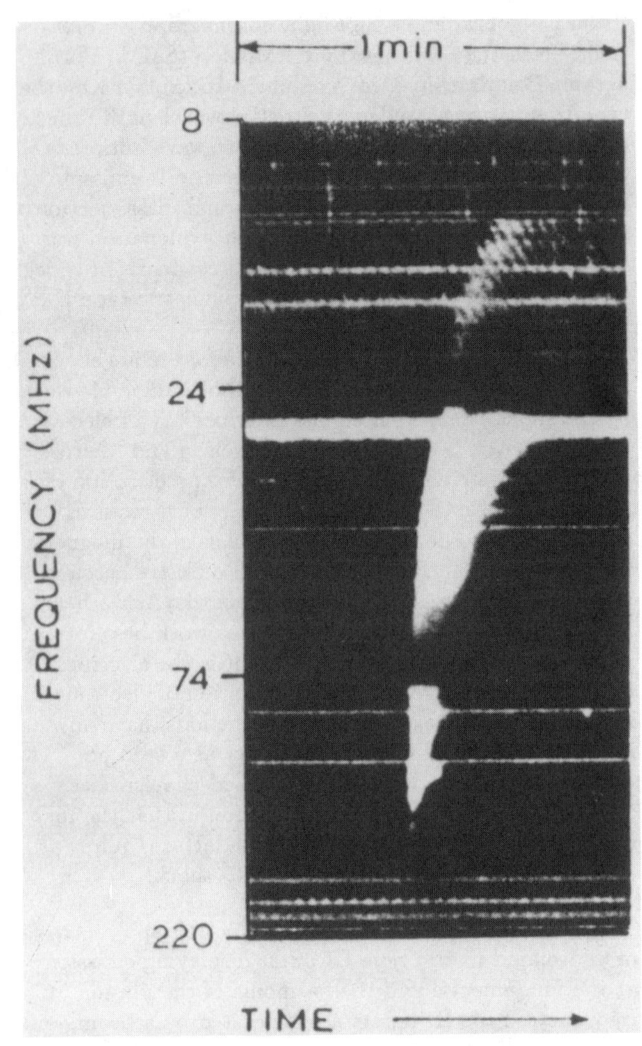

Fig. 3. A type III burst in the same format as Figure 2. Again, fundamental/harmonic structure is clearly visible. The fast frequency drift for the type III is due to the speed of the exciter, a stream of mildly relativistic electrons, as it travels outward in the corona (from Suzuki and Dulk, 1985).

to generate radio emission at the fundamental of the plasma frequency. Others coalesce with oppositely directed Langmuir waves to generate radio emission at $2 f_p$. Most, however, are reabsorbed (Landau damped) by the slower electrons coming along immediately after so that, in essence, the beam cleans up after itself. This reabsorption is essential if the beam is to keep most of its energy and continue to propagate, as observed, to beyond the Earth. Observations that confirm many aspects of this theory will be discussed in the next section. An alternative mechanism for converting Langmuir wave energy into transverse (radio) waves is given by strong turbulence theory (e.g., Goldman 1983). Here the Langmuir wave energy reaches such high levels that nonlinear collapse into "cavitons" is possible. These collapsing cavitons generate radio waves due to currents arising during the collapse.

Space-Based Observations 30 kHz-2 MHz

Analogies to metric bursts of types II and III appear at longer wavelengths. For example, Figure 4 shows spectrograph records from the ISEE 3 spacecraft, in which a type III burst is visible starting at ~1412 UT on August 18, 1979, followed by a much slower type II burst more than 2 hours later (Cane et al., 1982).

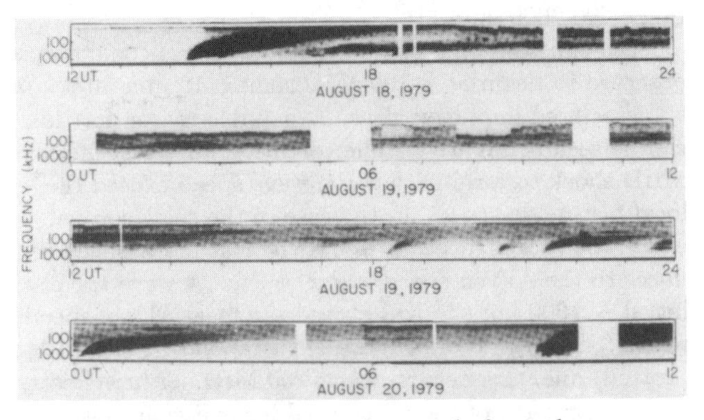

Fig. 4. Dynamic spectrograph records from the radioastronomy experiment onboard the ISEE 3 spacecraft. These show a type III burst associated with a flare at ~1412 UT on August 18, 1979, followed by a more slowly drifting type II burst. The shock passed the spacecraft at ~0550 UT on August 20 (from Cane et al., 1982).

Type II bursts. Kilometric type II bursts remain consistent with the fast-mode shock hypothesis and are observed to continue their drift to lower frequencies until the emission weakens and becomes confused with the galactic background near ~0.7 AU. When the shock is directed at the spacecraft, the shock is detected after an appropriate delay, leaving little doubt as to the association. It should be emphasized, however, that most interplanetary shocks do not produce type II bursts, and some other condition must be necessary to produce the emission.

It has long been assumed that the shock producing the metric type II is the same one that produces the kilometric type II. However, we saw in the previous section that metric type II bursts usually cease before leaving the frequency band observable from the ground. The shock may cease emitting and then begin again at lower frequencies, or there may be two shocks. In fact, Cane (1983) suggests that metric type IIs may sometimes be due to blast waves while the associated kilometric type IIs are due to completely separate piston-driven shocks pushed ahead of fast CMEs. Observations at frequencies in the gap from 2-20 MHz are needed to clarify this important topic.

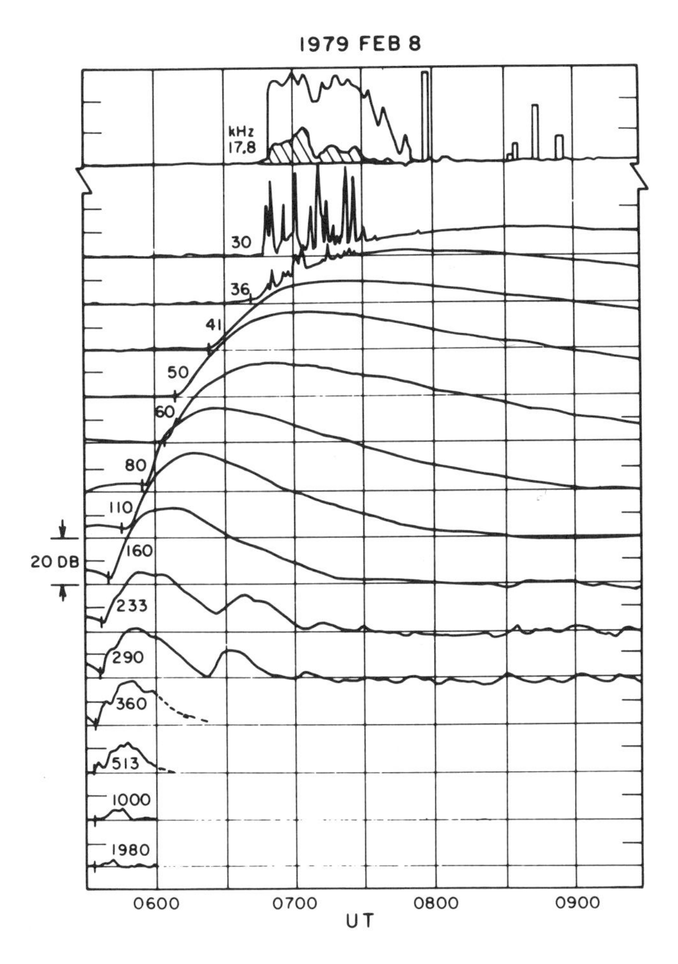

Fig. 5. Time profiles at a number of frequencies (shown in kHz) of a type III burst observed at 1 AU by ISEE 3. The spikes near 30 kHz are due to longitudinal (Langmuir) waves, which occur just when the radio emission reaches the local plasma frequency (26 kHz on this date) (from Melrose et al., 1986).

Type III bursts. Kilometric type III bursts have played an important role in confirming some aspects of the "standard" weak turbulence theory discussed earlier. Figure 5 shows time profiles of a type III burst at a number of frequencies from 1980 to 30 kHz. Also shown at the top of the figure is a plot of the electrostatic noise due to Langmuir waves, measured with the plasma wave experiment on ISEE 3. Note that the spiky electrostatic emission at 30 kHz, due to Langmuir waves generated by the beam (at a time when the local plasma frequency $f_p = 26$ kHz), appears just when the radio burst begins at that frequency, in agreement with theory. The fact that the electrostatic emission began at the start of the 30 and not the 60 MHz emission proves that it begins at the fundamental of the plasma frequency. Dulk et al. (1984) showed that fundamental emission dominates during the rise phase of these bursts, but harmonic emission dominates during the decay.

Additional evidence in favor of weak turbulence theory was obtained by Lin et al. (1981) by direct observation of the velocity distribution of the burst producing electrons. Figure 6 shows parallel velocity distributions deduced from actual observations before and during passage of a type III producing stream of electrons. The curves show that a bump builds on the tail of the original distribution, and a positive slope in the distribution develops at the time that evidence for Langmuir waves first appears in the radio data. This positive slope, a necessary part of the weak turbulence theory, persists throughout the time Langmuir waves are detected, and the two cease together. Many details of observation and theory remain to be reconciled, but the key aspects appear to be well understood.

Ground-Based Observations 200-1000 MHz

The decimetric regime exhibits many of the types of bursts seen in the metric range, such as type Is, type IIIs, and type IVs. During decimetric type IV bursts, short-duration fine structures called blips and spikes (Wiehl et al., 1985; Benz,1986) sometime appear. Some surprises are seen in this regime, such as the fact that reverse frequency-drift bursts are found in similar numbers to forward drift (type III-like) bursts. These are thought to be downgoing streams of electrons. While some knowledge can be gained by observations in this regime (such as limits on bandwidth and duration of type III bursts, and more direct timing comparisons with X rays), the complex plasma physics and geometry of coronal structures make interpretation difficult in many cases.

What We Have Learned—1-30 GHz

As Figure 1 indicates, plasma emission ceases to be important above 1 GHz, and completely new emission mechanisms dominate. During impulsive bursts, the dominant mechanism is some form of gyromagnetic emission, usually either thermal or nonthermal gyrosyn-

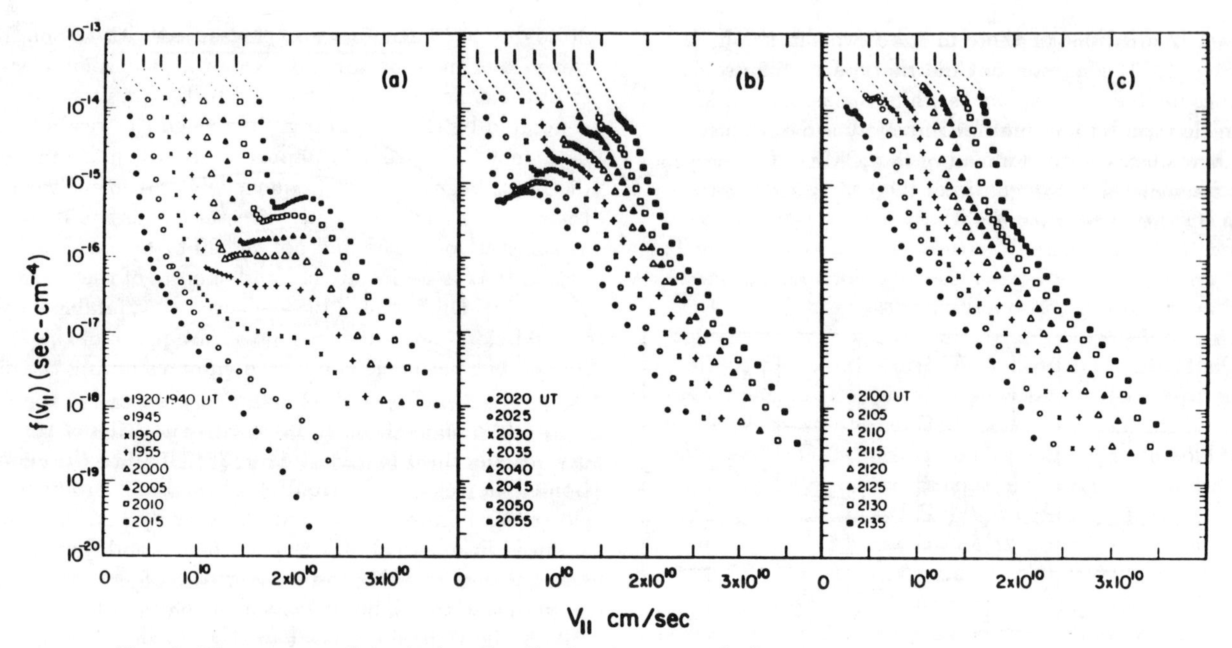

Fig. 6. Electron parallel velocity distribution function $f(v_{\|})$ deduced from ISEE 3 spacecraft observations before and during passage of a type III electron stream. Each succeeding distribution function within a panel is shifted in velocity by 2 x 10^9 cm s^{-1}. Note the development and subsequent decay of a bump-on-tail distribution (from Lin et al., 1981).

chrotron. At times, however, bremsstrahlung and gyroresonance emission may be important at low frequencies, for example in the preflare or decay phase of flares. (Bremsstrahlung and gyroresonance emission are responsible for thermal radio sources in non-flaring active regions.) To understand what can be learned from flares in this frequency range, we must first examine the characteristics of the emission mechanisms. It is most useful to look at the shapes of spectra predicted by theory for different mechanisms when the sources are homogeneous (that is, sources with uniform B, n_e, T_e, etc.). For this purpose, we will use the simplified formulae presented by Dulk and Marsh (1982), which are sufficiently accurate to elucidate the main features of the mechanisms.

Figure 7 shows spectra from homogeneous sources for the three emission mechanisms that are important for bursts. The top panels show normalized brightness temperature spectra, whose measurement requires that the bursts be spatially resolved at each frequency, and the bottom panels show the more usually obtained spectra of normalized flux density. The shapes of the curves are universal (approximately valid for the parameter range of interest); that is, when parameters change, the curves merely shift in the directions indicated by the arrows, without a substantial change in shape. The arrows attached to each curve represent the direction and magnitude of shift when the corresponding parameter increases by a factor of 2. The parameter $\Delta\Omega$ attached to the flux density curves represents an increase in the

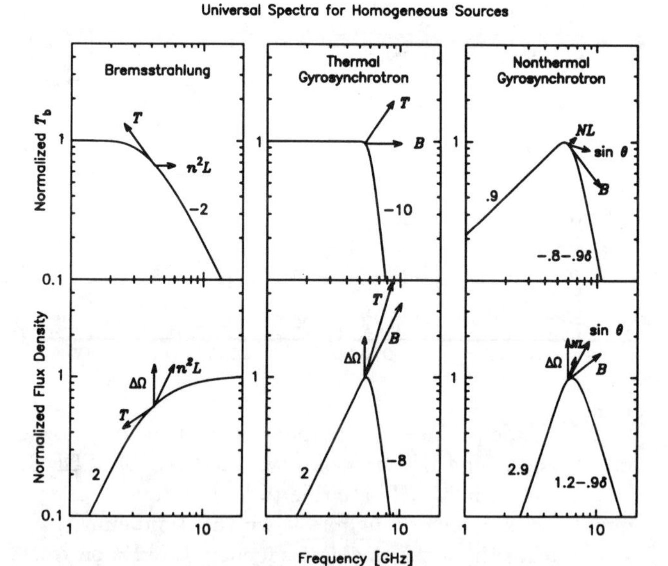

Fig. 7. Universal curves, from theory, for emission by various mechanisms from a homogeneous source. The arrows represent the magnitudes and directions of shift for an increase of parameters by a factor of 2. The top panels show normalized brightness temperature spectra (requiring measurements with spatial resolution) and the bottom panels show the corresponding flux density spectra. The arrow labeled B in the thermal gyrosynchrotron curves actually represents the quantity $B(nL)^{0.11}(\sin\theta)^{2/3}$.

angular size of the source. Note that the shapes are quite different for the three cases.

Figure 7 allows us to see how the microwave spectrum of homogeneous sources can be used as a diagnostic of the plasma parameters in the source. The shape of the spectrum immediately determines the emission mechanism, and the appropriate shifts of the curves to match the peak frequency and brightness temperature determine the desired plasma, B field, and electron parameters. Measurement of brightness temperature spectra is more desirable than flux density spectra for at least two reasons. The first is that flux density spectra have an additional unknown, the source size ($\Delta\Omega$), which widens the ranges of parameters that give acceptable fits. The second is that brightness temperature spectra are obtained from a restricted area of the Sun (depending on the resolution of the observation), so that the condition of homogeneity is more likely to be met.

How well do actual observations fit these curves? Spatially resolved observations of flares, such as from the Very Large Array (VLA), resolve the sources and show their morphology, but only at one or two frequencies. For example, the VLA can observe (at present) at 0.327, 1.45, 5, 8.8, 15, and 23 GHz. Essentially all flares observed so far have been resolved at only one or two frequencies. Figure 8 shows one of the early VLA observations, at 15 GHz by Marsh and Hurford (1982), with the simple result that the emission is confined to a small source located between Hα kernels marking the footpoints of a coronal loop. Subsequent observations, however, have indicated a more complicated situation. Simultaneous observations at two frequencies (e.g., Melozzi et al., 1985; Shevgaonkar and Kundu, 1985; Dulk et al., 1986) generally show that the sources at the two frequencies are at different locations, and their shapes differ. This implies that two different components are being observed, with different spectra. What is needed is imaging with better spectral coverage.

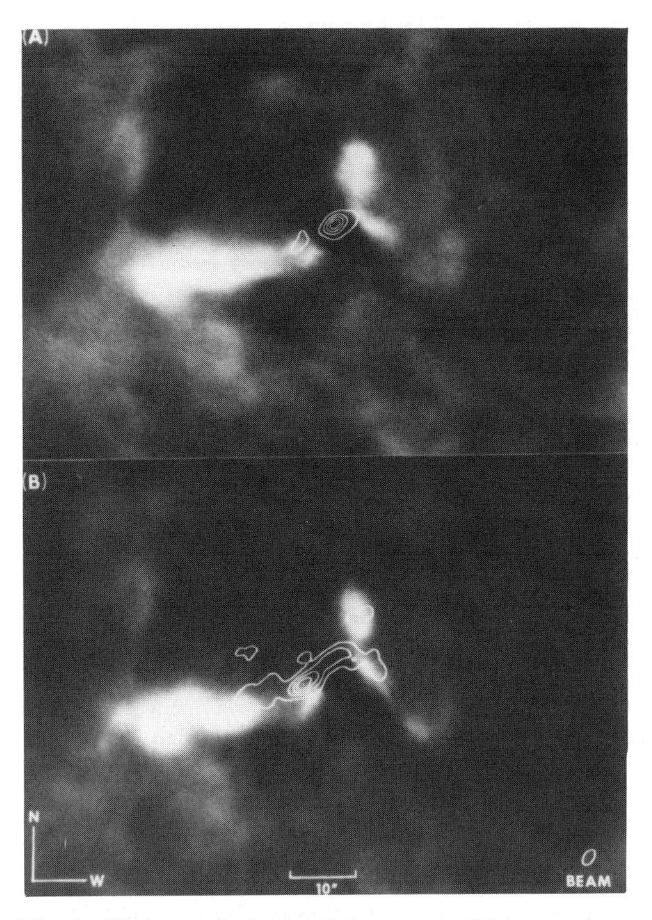

Fig. 8. VLA maps of a small flare at 15 GHz, superimposed on Hα images. (a) Map at the peak of the impulsive phase, showing a radio source located between Hα footpoints. (b) Map during the gradual phase, showing a radio loop spanning the gap between footpoints (from Marsh and Hurford, 1982).

Observations with better spectral resolution have been obtained with the Owens Valley frequency-agile interferometer (Hurford et al., 1984). Figure 9 shows a total power spectrum of a typical flare (1B/M1.0) observed at Owens Valley, with a nonthermal gyrosynchrotron homogeneous source spectrum superimposed. Note that this is a flux density rather than a brightness temperature spectrum. As is typically the case, theoretical spectra are too narrow to fit such observations. This does not mean the theory is wrong, only that the assumption of a homogeneous source is a bad one. Solar flares seen in Hα often are distributed over the flaring active region, and the same is likely true of radio flares. The changes in parameter values in an extended source and/or in multiple sources, when integrated in space across the active region, cause a broadening of the spectrum and, hence, completely mask the diagnostic information inherent in the homogeneous source spectra.

On the rare occasion that a spatially simple flare occurs, however, the observations are more consistent with expectations based on Figure 7. One such case is that of a small flare (SF/C1.0) observed at Owens Valley on February 3, 1986. Evidence for the simple spatial structure of this flare was derived from the interferometer data by plotting the relative visibility vs s^2, where s is the spatial frequency (the inverse of the fringe spacing in arc sec). On such a plot, a single Gaussian source lies on a straight line whose slope depends on the source size. As shown in Figure 10, and independently confirmed by interferometer phase data, the radio source for this flare was consistent with a single Gaussian source.

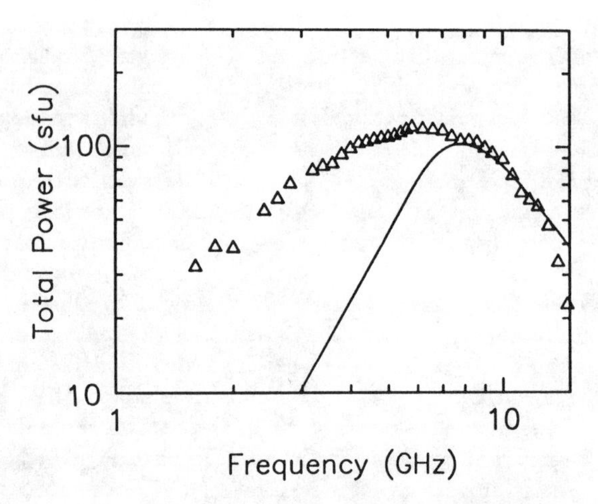

Fig. 9. Total power spectrum of a large burst on April 16, 1987, as observed at Owens Valley Radio Observatory, with a theoretical homogeneous source spectrum superimposed. As is typical of large bursts, the observed spectrum cannot be fit by such a curve because the true radio source structure is complex and highly inhomogeneous.

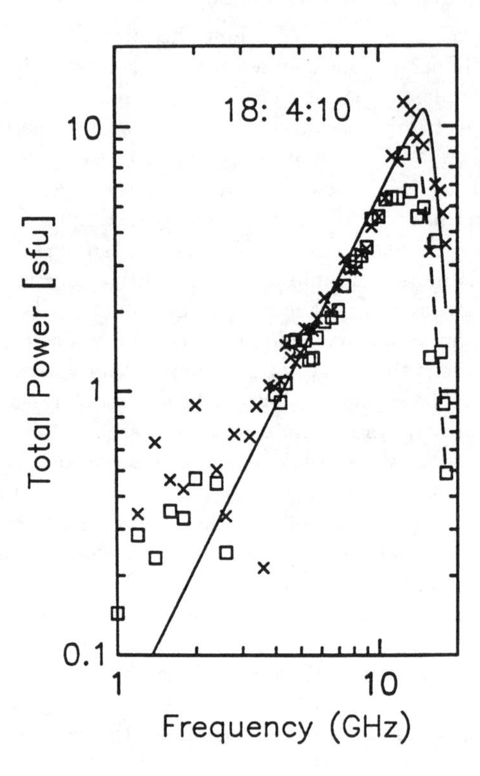

Fig. 10. Total power spectra in right-hand (RH) and and left-hand (LH) polarization for a simple flare. The observed spectra are shown with crosses denoting RH circular polarization and squares denoting LH polarization. The spectra are overlaid with fits of theoretical thermal gyrosynchrotron spectra for a homogeneous source. Both x-mode (solid) and o-mode (dashed) fits are shown. The parameters of the fits are given in the text.

The total power spectrum in both right- and left-circular polarization is shown in Figure 11, along with a superimposed fit of a thermal gyrosynchrotron spectrum (see the bottom center panel of Figure 7). Interpretation of the fit makes use of the interferometrically determined source diameter of 8 arc sec, which we also take to be the depth of the source. The fit then yields an electron temperature of 3×10^7 K and a magnetic field strength of 850 G. (The latter value depends very weakly on an assumed electron density, taken here to be 10^{10} cm^{-3}.) The polarization then indicates that the field was oriented at 75° to the line of sight. For this demonstratively simple source, the measured spectra very much resemble the theoretical flux density spectra, and yield plausible values of the plasma and field parameters. This graphically demonstrates that, in the rare case when the radio burst is spatially simple, one can successfully apply the diagnostic power of the microwave spectrum.

Spike Bursts and Maser Emission

There is another class of bursts in the 1-30 GHz range that does not fit with any of the above emission mechanisms. These rare bursts, called spike bursts because of their short duration, were first reported by Dröge (1977) and Slottje (1978). They most often appear at frequencies < 5 GHz, as highly polarized ($\sim 100\%$), short-duration (milliseconds), small bandwidth bursts superimposed on large, continuum bursts. Observed rise times as short as $\sim$1 ms suggest a source size of <300 km. These characteristics can be explained as due to an electron-cyclotron maser, a coherent emission mechanism that is expected to operate whenever a nearly isotropic population of 10-100 keV electrons is present in a converging/diverging magnetic field. The mechanism is known to operate in the Earth's magnetosphere (Hewitt et al., 1982) and also is thought to account for Jovian decametric and longer wavelength emission and stellar spike bursts (Lang et al., 1983).

An important reason for studying this emission mechanism is that it has been suggested (Melrose and Dulk, 1984) that the mechanism may extract as much as half the total energy in energetic electrons and re-

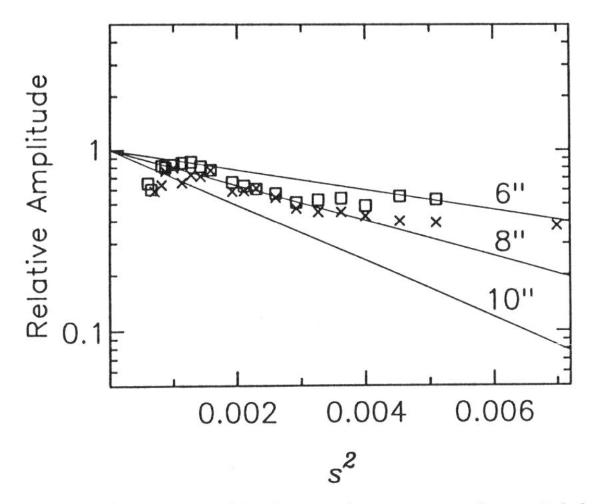

Fig. 11. Relative amplitude vs the square of spatial frequency for the flare shown in Figure 10. The dependence for gaussian sources with sizes (FWHM) of 6, 8, and 10 arc sec lies along straight lines as shown. A size of 8 arc sec was used in fitting the spectrum of Figure 10.

lease it in the form of intense radio waves, most of which are reabsorbed in neighboring loops in the flaring region. This offers an enticing explanation of how flares spread throughout an active region so quickly. Unlike charged particles, the radio radiation can travel across field lines at the speed of light and deposit its energy in remote loops very efficiently.

The main contention between observation and theory at present is that the emission should occur at low harmonics of the gyrofrequency (s = 1 or 2) and will then be strongly absorbed by overlying layers. Perhaps this explains the rarity of spike bursts, but, if so, it is important to obtain spatially resolved observations of spike bursts that do occur to determine what special geometry, if any, is needed to allow the radiation to escape.

Observational Prospects for the Future

We can identify several advances that are needed to make progress in the study and interpretation of solar bursts in the frequency range 30 kHz-200 MHz. The most obvious, perhaps, is to obtain space-based observations spanning the gap from 2-20 MHz. This would allow us to address the important question of whether type II burst producing shocks in the lower corona are the same shocks as those producing the kilometric type IIs. It is also necessary to regularly obtain high dynamic range spatial observations throughout this frequency range as was done in the past at Culgoora and Clark Lake radio observatories.

Study of the frequency range 200-1000 MHz would benefit from spatial resolution to locate the sources relative to solar surface features. The VLA will soon operate all dishes at 327 MHz, helping to bridge the gap between centimetric and metric spatially resolved observations, but at present there are no corresponding spectral data available in the US to provide the spectral context.

At microwaves, we would like to be able to utilize the diagnostic procedures described above for all microwave bursts, not just the rare simple bursts. This will require an imaging spectrometer that combines the imaging ability of the VLA with the spectral coverage of the Owens Valley frequency-agile interferometer. The potential benefits of such an instrument are enormous. The time development of a flare and transport of energy into several sources could be monitored in detail, with full knowledge of the parameters in each source, including the coronal magnetic field strength and direction, which are central to the flare process but which cannot be measured in any other way.

In conclusion, solar bursts throughout the radio spectrum have helped us understand the important processes taking place in the solar atmosphere and solar wind, but have left many questions unanswered. Recent observations and interpretation have shown how those questions might be answered and point the way toward future instruments and observing strategies.

References

Achterberg, A., and C. A. Norman, Particle acceleration by shock waves in solar flares, *Astron. Astrophys.*, **89**, 353, 1980.

Benz, A. O., Millisecond radio spikes, *Solar Phys.*, **104**, 99, 1986.

Boischot, A., Étude du rayonnement radioélectrique solaire sur 169 MHz, à l'aide d'un grand interférometre à réseau, *Ann. Astrophys.*, **21**, 273, 1958.

Cane, H. V., Velocity profiles of interplanetary shocks, Solar Wind Five, *NASA Conf. Publ.*, **2280**, 703, 1983.

Cane, H. V., R. G. Stone, J. Fainberg, J. L. Steinberg, and S. Hoang, Type III solar radio events observed in the interplanetary medium, I: General characteristics, *Solar Phys.*, **78**, 187, 1982.

Dröge, F., Millisecond fine-structures of solar burst radiation in the range 0.2-1.4 GHz, *Astron. Astrophys.*, **57**, 285, 1977.

Dulk, G. A., Radio emission from the Sun and stars, *Ann. Rev. Astron. Astrophys.*, **23**, 169, 1985.

Dulk, G. A., and K. A. Marsh, Simplified expressions for the gyrosynchrotron radiation from mildly relativistic, nonthermal and thermal electrons, *Ap. J.*, **259**, 350, 1982.

Dulk, G. A., and D. J. McLean, Coronal magnetic fields, *Solar Phys.*, **57**, 279, 1978.

Dulk, G. A., J. L. Steinberg, and S. Hoang, Type III bursts in interplanetary space. Fundamental or harmonic?, *Astron. Astrophys.*, **141**, 30, 1984.

Dulk, G. A., T. S. Bastian, and S. R. Kane, Two-frequency imaging of microwave impulsive flares near the solar limb, *Ap. J.*, **300**, 438, 1986.

Gary, D. E., G. A. Dulk, L. House, R. Illing, C. Sawyer, W. J. Wagner, D. J. McLean, and E. Hildner, Type II bursts, shock waves, and coronal transients: The event of 1980 June 29, 0233 UT, *Astron. Astrophys.*, **134**, 222, 1984.

Goldman, M. V., Progress and problems in the theory of type III solar radio emission, *Solar Phys.*, **89**, 403, 1983.

Hewitt, R. G., D. B. Melrose, and K. G. Rönnmark, The loss-cone driven electron-cyclotron maser, *Aust. J. Phys.*, **35**, 447, 1982.

Holman, G. D., and M. E. Pesses, Solar type II radio emission and the shock drift acceleration of electrons, *Astrophys. J.*, **267**, 837, 1983.

Hurford, G. J., R. B. Read, and H. Zirin, A frequency-agile interferometer for solar microwave spectroscopy, *Solar Phys.*, **94**, 413, 1984.

Lang, K. R., J. Bookbinder, L. Golub, and M. Davis, Bright, rapid, highly polarized radio spikes from the M dwarf AD Leo, *Ap. J. Lett.*, **272**, L15, 1983.

Lin, R. P., D. W. Potter, D. A. Gurnett, and F. L. Scarf, Energetic electrons and plasma waves associated with a solar type III radio burst, *Ap. J.*, **251**, 364, 1981.

Marsh, K. A., and G. J. Hurford, High spatial resolution solar microwave observations, *Ann. Rev. Astron. Astrophys.*, **20**, 497, 1982.

McLean, D. J., Metrewave solar radio bursts, in *Solar Radiophysics*, edited by D. J. McLean and N. R. Labrum, p. 37, 1985.

Melozzi, M., M. R. Kundu, and R. K. Shevgaonkar, Simultaneous microwave observations of solar flares at 6 and 20 cm wavelengths using the VLA, *Solar Phys.*, **97**, 345, 1985.

Melrose, D. B., and G. A. Dulk, Radio-frequency heating of the coronal plasma during flares, *Ap. J.*, **282**, 308, 1984.

Melrose, D. B., G. A. Dulk, and I. H. Cairns, Clumpy langmuir waves in type III solar radio bursts, *Astron. Astrophys.*, **163**, 229, 1986.

Nelson, G., and D. B. Melrose, Type II bursts, in *Solar Radiophysics*, edited by D. J. McLean and N. R. Labrum, p. 333, 1985.

Saito, K., A non-spherical axisymmetric model of the solar K corona of the minimum type, *Ann. Tokyo Astron. Obs., Ser. 2,* **12**, 53, 1970.

Shevgaonkar, R. K., and M. R. Kundu, Dual frequency observations of solar microwave bursts using the VLA, *Ap. J.*, **292**, 733, 1985.

Slottje, C., Millisecond microwave spikes in a solar flare, *Nature*, **275**, 520, 1978

Suzuki, S., and G. A. Dulk, Bursts of type III and type V, in *Solar Radiophysics*, edited by D. J. McLean and N. R. Labrum, p. 289, 1985.

Uchida, Y., On the exciters of type II and type III solar radio bursts, *Publ. Astron. Soc. Japan*, **12**, 376, 1960.

Vernazza, J. E., E. H. Avrett, and R. Loeser, Structure of the solar chromosphere. III. Models of the EUV brightness components of the quiet Sun, *Ap. J. Suppl.*, **45**, 635, 1981.

Wiehl, H. J., A. O. Benz, and M. J. Aschwanden, Different time constants of solar decimetric bursts in the range 100-1000 MHz, *Solar Phys.*, **95**, 167, 1985.

Wild, J. P., and L. L. McCready, Observations of the spectrum of high-intensity solar radiation at metre wavelengths—I. The apparatus and spectral types of solar bursts observed, *Aust. J. Sci. Res.*, **A3**, 387, 1950.

Wild, J. P., K. V. Sheridan, and G. H. Trent, The transverse motions of the sources of solar radio bursts, in *Paris Symp. on Solar Radio Astronomy*, edited by R. N. Bracewell, Stanford, Stanford University Press. 1959.

REMOTE SENSING OF PLANETARY MAGNETOSPHERES: IMAGING VIA ENERGETIC NEUTRAL ATOMS

C. C. Curtis and K. C. Hsieh

Department of Physics, University of Arizona, Tucson, AZ 85721

Abstract. Planetary magnetospheres may be imaged using energetic neutral atoms (ENAs) produced by charge exchange between energetic magnetospheric ions and neutral background gas. An imaging technique is described and its potential is demonstrated using an ENA model of the Saturn/Titan plasma system.

Introduction

"A picture is worth a thousand words" might be rephrased to "An image of the magnetospheric ion distribution is worth a thousand in situ measurements." Imaging provides global information, while in situ measurements are parochial and do not permit the separation of spatial from temporal variations. Roelof (1987) has recently described the imaging of magnetospheric ions via energetic neutral atoms (ENAs) resulting from charge exchange between the energetic ions and the neutral background gas. When coupled with mass and energy analysis of ENAs, imaging provides a relatively detailed picture of a magnetosphere through remote sensing. Elsewhere in this volume (Hsieh and Curtis, 1989), we describe the mass and energy analysis of ENAs by ISENA (imaging spectrometer for ENAs). This report concentrates on a process that could be used by ISENA for ENA imaging.

Model of an ENA Object

The Saturn/Titan plasma system is an interesting candidate for ENA imaging. Hsieh and Curtis (1988) constructed a model of this object, incorporating its neutral hydrogen distribution, the L-B dependence of energetic ions and their energy spectrum, and the appropriate charge-exchange cross sections, to estimate its production of energetic hydrogen atoms (EHAs). Figure 1 represents an image of the object, "seen" in EHAs from an observation point 66 R_s away from the planet's center and 25° above the equatorial plane. The image is projected onto a plane through Saturn's center, whose normal passes through the observation point. The plane contains a 75 x 100 element matrix of squares, each 0.5 R_s on a side. The value of each matrix element is the neutral flux generated by all sources within the solid angle subtended by a particular square, as seen from the observation point. Contour lines of constant flux thread the matrix. Summing all the matrix elements gives a total EHA flux of 6 $cm^{-2}\ s^{-1}$ in the energy range of 10 to 100 keV. The model is cylindrically symmetric for simplicity, but an imaging instrument at the hypothetical observation point could easily detect any irregularities in the global image of the Saturn/Titan system.

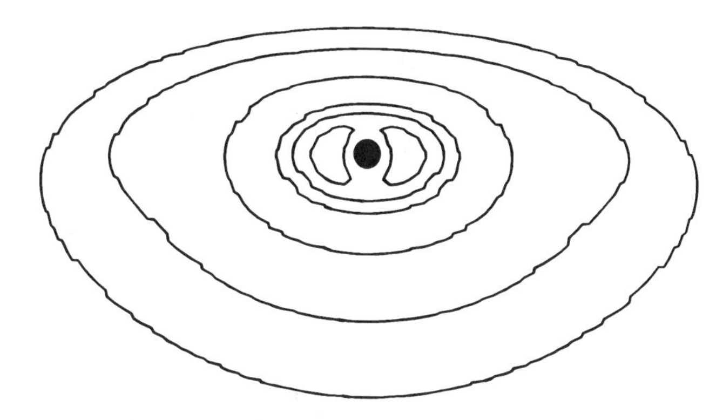

Fig. 1. Contour plots of the Saturn/Titan plasma system "seen" in EHAs from a vantage point 66 R_s away from the planet and 25° above the equatorial plane. Saturn's silhouette has been added. From outermost to innermost, the flux contours range from 3×10^{-5} to 1.8×10^{-2} $cm^{-2}\ s^{-1}$ for 10-100 keV neutrals.

Imaging Technique

A variety of techniques can be used to image neutral particles. Virtually all of these involve either determining two points on the straight line trajectory of each particle being detected, or delimiting a minimum solid angle containing a particular trajectory. Thus, a particle's arrival direction could be determined from the intercept coordinates on two successive planes penetrated by the particle, assuming no scattering, or from identification of the collimator elements through which the particle passed. The technique employed by ISENA is a hybrid of collimator and coded-aperture methods. A fan collimator constrains one of the arrival direction angles and deflects unwanted charged particles. The coded aperture (CA) allows determination of the orthogonal direction angle from the intercept coordinate in a single plane.

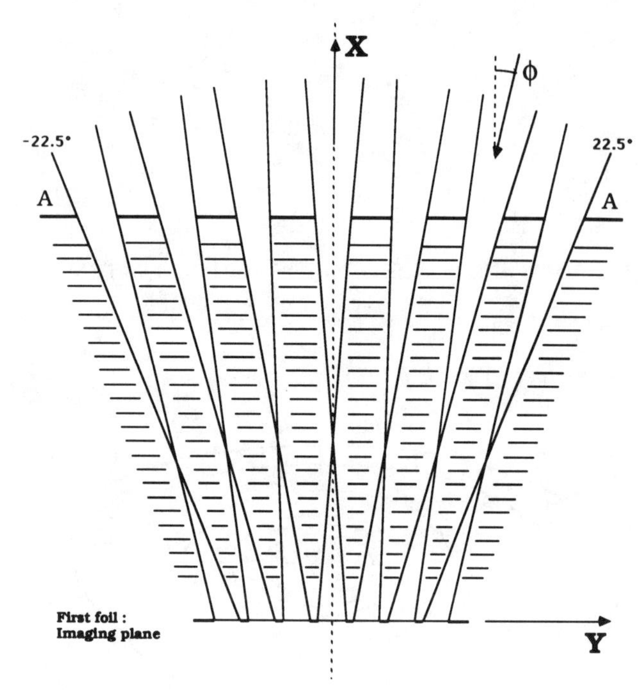

Fig. 2. Cross-sectional view of the ISENA "front end," an ion-trapping collimator that generates directional information in the x-y plane (angle ϕ). Charged particles having energies up to several hundred keV/e are swept away by the biased collimator elements, represented by the horizontal line segments. The crossed rays show the fields-of-view of each of the seven sectors. Coded apertures (Figure 4) are installed normal to the plane of the figure at each of the openings along the line A-A. The carbon foil image plane is at the bottom of the figure along the y-axis.

The sensitivity, the signal-to-noise characteristics, and the collimator of ISENA are described by Hsieh and Curtis (1989); here we discuss the instrument's angular resolution. The rectangular entrances of the collimator shown in Figure 2 are each covered by a one-dimensional (1D) CA. The sectors defined by the collimator provide directional information in the x-y plane (azimuthal angle ϕ), and each CA provides the declination (angle θ) above or below this plane. Together, the two information sets give ISENA a two-dimensional (2D) capability. We will first describe the principle of each technique in 1D and then demonstrate their combined effect in generating a 2D image.

Sectoring

The collimator divides the x-y plane into seven sectors, which allow incident ENAs to reach corresponding sectors on the image plane. (Secondary electrons generated at the foil are mapped onto a position-sensitive detector, thus registering the ENAs' arrival directions in the x-y plane.) This directional information can be refined, at a cost of additional data processing, by subdividing each of the seven sectors into several subsectors, with a detection element for each. Although the subsectors look through the same sector opening, they have slightly different fields-of-view (FOV) in ϕ. Suppose that each of the seven sectors of the collimator shown in Figure 2 is subdivided into three subsectors, thus yielding a total of 7 x 3 overlapping subsectors spanning a 45° total FOV in ϕ. For each of the 21 subsectors, there is a corresponding response curve or kernel for particles arriving from a particular ϕ direction. Figure 3a shows the set of 21 partially overlapping response curves, which are labeled $K_i(\phi)$. Each kernel is about 9° wide.

Accumulation of a number of ENAs passing through each of the sectors over a period of time results in a set of 21 ENA count rates, C_i, where

$$C_i = \int Q(\phi)\, K_i(\phi)\, d\phi \text{ between } \phi_1(i) \text{ and } \phi_2(i) \qquad (1)$$

$Q(\phi)$ is the ENA anisotropic flux in the x-y plane, and $K_i(\phi)$ is the response function or kernel of the ith subsector, which is bounded by $\phi_1(i)$ and $\phi_2(i)$. Figure 3b displays an example of C_i for a hypothetical object distribution in ϕ. The overlapping of the sectors, and the consequent overlapping of the kernels, allows the retrieval of the original angular distribution in ϕ to a precision much finer than the angular width of each sector.

From the known $K_i(\phi)$ and C_i, we employ an iterative nonlinear inversion algorithm described by Twomey (1975) to retrieve $Q(\phi)$. The iteration sweeps through all the subsectors, and within each subsector, through the entire range of angles encompassed by the response function.

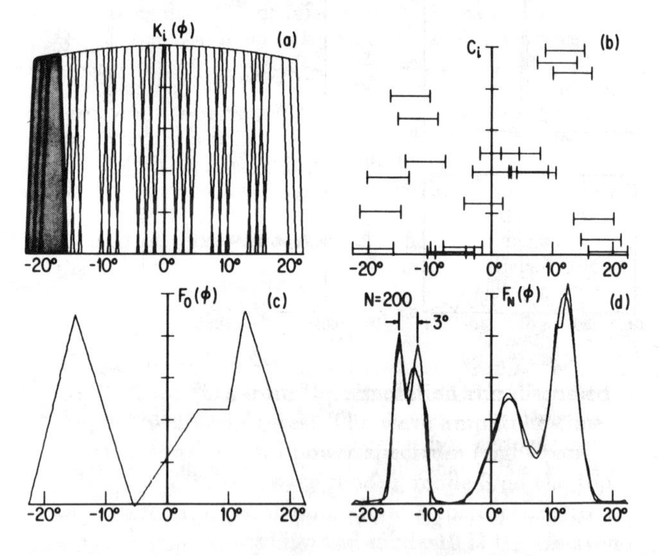

Fig. 3. Directional information provided by the collimator. (a) Overlapping response curves (kernals) of 7 x 3 subsectors. Each subsector is 9° wide. The leftmost response curve is shaded. (b) Relative count rates C_i produced in the 21 overlapping subsectors by an object whose intensity is represented by four contiguous Gaussian distributions in ϕ, shown in Figure 3d. (c) "Guess" intensity distribution of the object based on the count rates in Figure 3b. (d) Superposition of the two-hundredth approximation to the intensity distribution and the four Gaussian curves actually representing the object.

The continuous function $Q(\phi)$ can be replaced conceptually by a sequence of point sources, $F(\phi)$, which are spaced at small increments $\Delta\phi$ and have intensities such that $F(\phi) = Q(\phi)\Delta\phi$. The reconstruction begins by establishing a "guess" sequence, $F_0(\phi)$, which is the zeroth approximation to the actual flux, e.g., see Figure 3c. This guess can be arbitrary (each $F_0(\phi) = 1$, for example) but computation time is saved if the original approximation is more accurate, and the solution to which the interation process converges does depend weakly on the original guess.

A good approximation is $F_0(\phi) = A_1\Sigma_i C_i K_i(\phi)$, where $K_i(\phi) = 0$ for all sources at locations ϕ outside the field-of-view of the detector segment producing the count rate C_i of the ith subsector. Several subsectors thus contribute, with relative strengths $K_i(\phi)$, to the initial guess for the point source located at position ϕ. The normalizing constant A_1 is picked to give $\Sigma_i\Sigma_\phi F_0(\phi)\ K_i(\phi) = \Sigma_i C_i$. The $F(\phi)$ must also meet the requirement that $\Sigma_\phi F(\phi)\ K_i(\phi) = C_i$ for each subsector. This is accomplished through an iterative process in which successive approximations to $F(\phi)$ are obtained by setting

$$F_n(\phi) = F_{n-1}(\phi)\ [1 + (C_i/S_{i,n-1} - 1)\ A_2\ K_i(\phi)] \tag{2}$$

where

$$S_{i,n} = \sum_{\phi} F_n(\phi)\ K_i(\phi) \tag{3}$$

at the nth iterative step, and A_2 is a constant (determined empirically) which controls the speed of convergence. In a single iteration, a given $F_n(\phi)$ is adjusted several times, because the operation described by equation (2) is performed sequentially for each subsector in which $F_n(\phi)$ contributes to the count rate. When the set $S_{i,n}$ matches the set C_i to some precision, no further adjustments are made to $F_n(\phi)$. Figure 3d superposes a reconstructed distribution $F_n(\phi)$, after n = 200 iterations, upon a source distribution $Q(\phi)$ (smooth curve). It resolves two structures, each 2° wide and separated by 3°. (A feature 1 R_s wide subtends 1° at a distance of 57 R_s.)

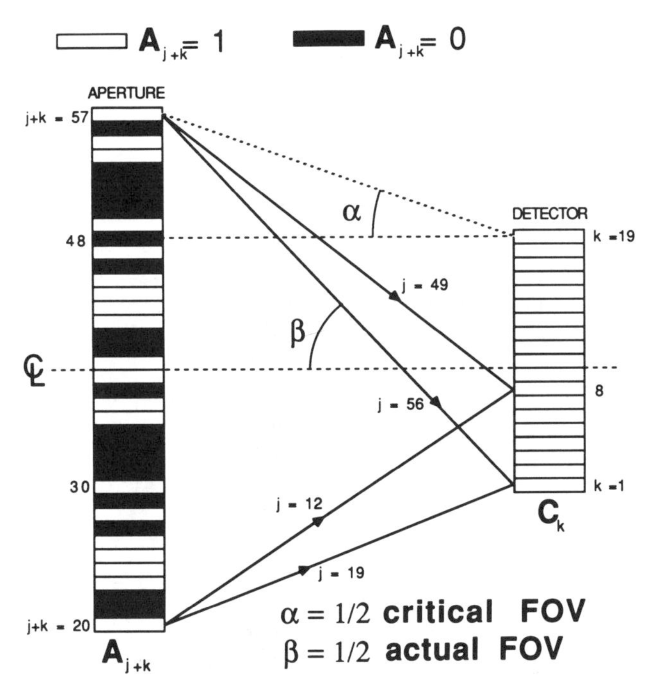

Fig. 4. 1D CA containing two cycles of a 19-element code in front of a 19-element detector. One of these masks is installed with its long axis parallel to Z at each of the openings along line A-A in Figure 2.

1D Coded Aperture

The CA provides directional information in the orthogonal plane (θ). Our description of the technique must necessarily be brief; the reader is referred to Barrett and Swindell (1981), Fenimore and Cannon (1978), and Gunson and Polychronopulos (1976) for a more comprehensive treatment. The principle of imaging by a 1D CA is illustrated in Figure 4, where a 19-pixel detector is placed some distance behind two cycles of a 19-element CA. Black elements on the CA are non-transmitting, and blank ones are open or transmitting. We again treat the source function as a discrete set of point sources. Counts produced in the kth pixel result from flux coming from the jth source point through the (j+k)th element of the aperture

$$C_k = \sum_j O_j\ A_{j+k} + \text{noise} \tag{4}$$

where A_{j+k} is set to 1 when the element is transmitting and to 0 when not. The linear superposition can be inverted to recover the original object by

$$R_j = \sum_k C_k\ G_{k+j} \tag{5}$$

where $G_{k+j} = 1$ when $A_{j+k} = 1$, and $G_{k+j} = -1$ when $A_{j+k} = 0$. That R_j is a reconstruction of O_j is based on the property of the code that $A_{mj}\ G_{jk} = \delta_{mk}$ (Fenimore and Cannon, 1978; Gunson and Polychronopulos, 1976). There are other choices for the post-processing array G, but this one appears to be most suitable for our application.

We note that the code is repeated; i.e., the mask contains two cycles of the 19-element code. Each source point lying within the critical FOV (see Figure 4) illuminates 10 of the 19 pixels through any set of 10 open elements of the 19-element CA. There are objects lying outside the critical FOV that can be "seen" by less than 10 pixels, in which case the sums of equation (4) are truncated and, consequently, side bands are generated that distort even the R_j that lies within the critical FOV. However, if the extended object lies totally within the critical FOV, the reconstruction duplicates the object, except for the effects of noise.

For an extended object such as the Saturn/Titan system, good resolution is needed. The resolution of a CA is determined by the size of the aperture elements and the distance between the aperture and the detector. We demonstrate what a two-cycle, 63-element, 1D CA detector having 0.5° resolution could produce when its

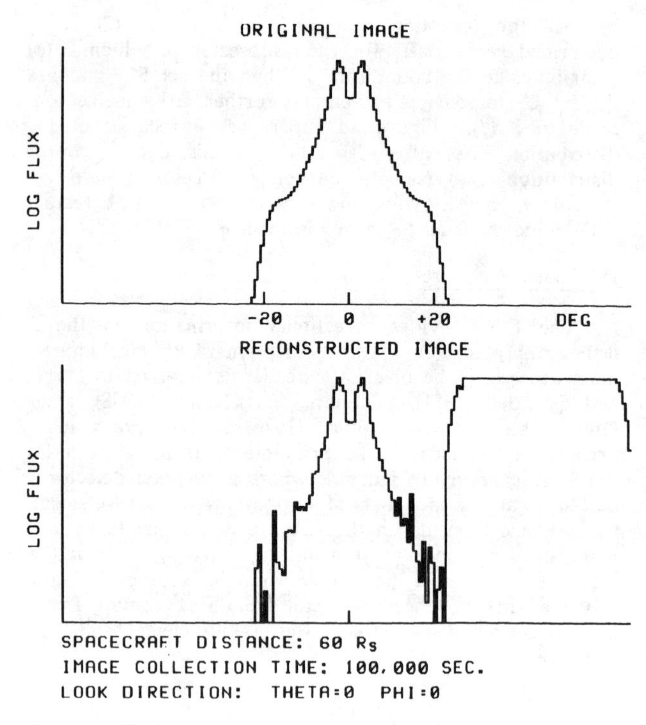

Fig. 5. ENA flux from a 0.5° wide strip along the equatorial plane, and its simulated reconstruction using a 63-element 1D CA.

long axis is parallel to the equator of Saturn. Figure 5a shows the ENA flux in a 0.5° wide strip in the equatorial plane coming from Saturn/Titan at a distance of 60 R_s, and Figure 5b shows the reconstruction of this distribution. Here, we have generated random counts in the detector elements to simulate the effects of noise, which show up most prominently in the low-intensity wings of the reconstruction. Shorter integration times (1000 s, for example) still reproduce the intense central structure quite nicely at this distance, but provide less resolution in the low-intensity regions. As can be seen, the fidelity of the reconstruction is very reasonable.

Combining the Collimators and Coded Apertures

We now consider a 2D object, which can be regarded as a 2D array of point sources, the flux from which is imaged onto a detector by a collimator and CA in combination. The collimator elements are arranged to provide position information in the ϕ direction, and the CA gives θ-direction information. The detector plane contains a 2D array of pixels, in each of which a count rate C_{ij} is recorded. The i index refers to the ith subsector of the collimator, while the j and k indices in the following equations relate to the CA. We have, in the absence of noise

$$C_{ij} = \sum_k O_{ik} A_{j+k} = \sum_\phi \sum_k K_i(\phi)\, Q_k(\phi)\, A_{j+k} \quad (6)$$

where the index k is summed over the full range of θ, and ϕ is summed over the range in ϕ viewed by subsector i. Although each pixel contains flux from point sources in a limited range of ϕ directions and the full range (with gaps) of θ directions, it is possible to separate the recorded image into orthogonal components. We begin by constructing a series of "pseudo detector" elements

$$R_{ik} = \sum_j C_{ij} G_{k+j} = \sum_\phi K_i(\phi)\, Q_k(\phi) \quad (7)$$

Each element R_{ik} contains a count rate equivalent to that produced by point sources lying only in the kth row (i.e., having a fixed θ value) of the object array. Only sources having ϕ values between $\phi_1(i)$ and $\phi_2(i)$ contribute. We then use the process described by equations (2) and (3) to reconstruct the whole sequence of sources along the ϕ direction in the kth row; i.e., we have

$$F_n(k,\phi) = F_{n-1}(k,\phi)\, [1 + (R_{ik}/S_{ik,n-1} - 1)\, A_2\, K_i(\phi)] \quad (8)$$

where

$$S_{ik,n} = \sum_\phi F_n(k,\phi)\, K_i(\phi) \quad (9)$$

S_{ik} is adjusted until the nth iteration matches R_{ij} to some precision. This procedure is then repeated for the next k value. A hypothetical detector count rate set C'_{ij} is generated then from the set $F_n(k,\phi)$ and is compared to the actual detector counts C_{ij}, as shown in Figure 6, which diagrams the entire reconstruction process.

The image formation and reconstruction process is demonstrated by taking the Saturn plasma system shown in Figure 1 and creating a pixel array on a simulated detector using equation (6). Each collimator sector was divided into three subsectors, and a two-cycle, 63-element

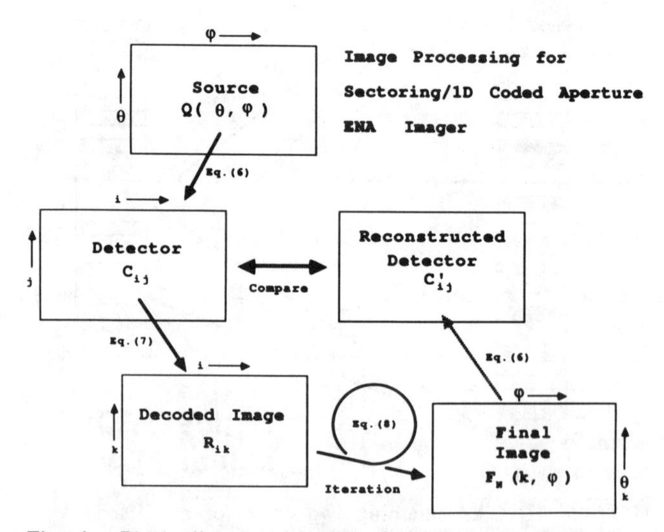

Fig. 6. Block diagram describing the steps employed in obtaining an image using the collimator/1D CA technique.

CA was used along each sector. The reconstructed image, generated in the computer using equations (7) and (8), is shown in Figure 7. Fifty iterations of equation (8) were performed for each k value.

Conclusion

We have modeled the operation of a concept instrument ISENA in producing an ENA image. It features a large field-of-view (45° x 45°), a high-duty cycle with no requirement for scanning motion, and a potential for high angular resolution (0.5° x 0.5°).

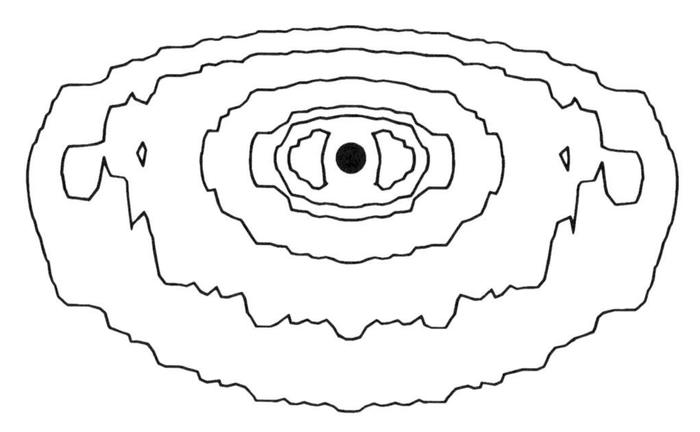

Fig. 7. Simulated image of the model shown in Figure 1 obtained by a 21 × 63 pixel collimator/1D CA camera. The same contour values as in Figure 1 are plotted.

Acknowledgments. We thank A. L. Broadfoot, C. Y. Fan, D. M. Hunten, and S. A. Twomey for helpful discussions and encouragements. We are grateful to D. Hall for assistance with the graphics. This work is supported, in part, by NASA grant NAG5-625.

References

Barrett, H. E., and W. Swindell, Radiological Imaging Academic Press, New York, 1981.

Fenimore, E. E., and T. M. Cannon, Appl. Opt., 17, 337, 1978.

Gunson, J., and B. Polychronopulos, Optimum design of a coded mask x-ray telescope for rocket applications, Mon. Not. R. Astron. Soc., 177, 485-497, 1976.

Hsieh, K. C., and C. C. Curtis, A model for the spatial and energy distributions of energetic neutral atoms produced within the Saturn/Titan plasma system, Geophys. Res. Lett., 15, 772-775, 1988.

Hsieh, K. C., and C. C. Curtis, Remote sensing of planetary magnetospheres: Mass and energy analysis of energetic neutral atoms, these proceedings, 1989.

Roelof, E. C., Energetic neutral atom image of a storm-time ring current, Geophys. Res. Lett., 14, 652, 1987.

Twomey, S. A., Comparison of constrained linear inversion and an iterative nonlinear algorithm applied to the indirect estimation of particle size distributions, J. Comp. Phys., 18, 188-200, 1975.

ENERGETIC NEUTRAL PARTICLE IMAGING OF SATURN'S MAGNETOSPHERE

A. F. Cheng and S. M. Krimigis

Applied Physics Laboratory, The Johns Hopkins University, Laurel, MD 20707

Abstract. Energetic charge exchange neutrals have been detected from the magnetospheres of Jupiter and Saturn by the Voyager spacecraft. Imaging of charge exchange neutrals provides information on the global three-dimensional structure and dynamics of the magnetosphere. Analysis of charge exchange neutrals also allows remote sensing of energetic ion composition and energy spectra. Such analyses will address several critical scientific objectives of the proposed Cassini mission to Saturn, including the ring current, radiation belts, potential substorm activity, and magnetosphere–satellite interactions. On the basis of the Voyager observation it is expected that, in addition to the atomic hydrogen of the Titan torus, there is a water product neutral cloud in Saturn's inner magnetosphere. A model for the Titan torus places an upper limit on the H_2 density consistent with the Voyager upper limit to the thermal H_2^+ density. Hence charge exchanges with H_2 are likely to be relatively unimportant. The charge exchange neutrals from the Saturn system consist of fast H from the Titan torus region and fast heavy atoms from the inner magnetosphere. Simple estimates are given for instrument sensitivities and required exposure times, and possible results are sketched for neutral particle imaging of the Saturn system.

Introduction

The Voyager 1 low energy charged particle (LECP) instrument has detected energetic neutral particle fluxes emitted by the magnetospheres of Jupiter and Saturn (Kirsch et al., 1981a,b). The LECP detectors were scanned in a plane that was divided into eight sectors, each sector being 45° wide. A statistically significant count rate enhancement was observed in the sector facing the planet while adjacent sectors were at background rates, during times when the spacecraft was outside the magnetosphere and the interplanetary magnetic field did not connect the spacecraft to the bow shock. The enhanced count rate from the direction of the planet was therefore attributed to energetic neutral particles or photons, since at these times escaping magnetospheric charged particles could not reach the spacecraft. Furthermore, the count rate enhancement could not be explained as X ray emission from the magnetospheres of Jupiter or Saturn, since magnetospheric electron fluxes are far too low to produce the required X ray fluxes (Kirsch et al., 1981a,b). Hence the Jovian and Saturnian magnetospheres are sources of energetic neutral atoms or molecules. In addition, energetic ions also escape from the Jovian and Saturnian magnetospheres, and these were also detected by Voyager in the upstream regions during different time periods when the spacecraft was magnetically connected to the bow shock (Zwickl et al., 1981; Krimigis et al., 1985; Krimigis, 1986).

The energetic neutral fluxes emitted by the Jovian and Saturnian magnetospheres can be produced by charge exchange reactions between radiation belt ions and ambient neutrals (Kirsch et al., 1981a,b; Ip, 1984; Cheng, 1986). The fast charge exchange neutrals are far above the escape velocity, and they escape into interplanetary space on nearly rectilinear trajectories. Just like photon emissions, the fast charge exchange neutrals carry unique and important information concerning their source region.

Hence charge exchange neutral fluxes from a magnetosphere can be imaged and their composition and energy spectra determined to carry out remote sensing of planetary magnetospheres. Imaging of charge exchange neutrals can reveal the global three-dimensional structure and dynamics of the magnetosphere, allowing a separation to be made between effects of temporal and spatial variations. Species identification of the charge exchange neutrals and measurement of their energy spectra provide further information on the energetic ion and ambient neutral populations of the magnetosphere. Of course, remote sensing of charge exchange neutrals can be best exploited in conjunction with in situ measurements of energetic ion spectra, angular distributions, and composition as well as ultraviolet imaging of ambient neutrals.

Imaging of charge exchange neutrals from Earth's magnetosphere has already been carried out using ISEE 1 data during the main phase of a geomagnetic storm (Roelof, 1987). These images showed a strong enhancement of the ring current near local midnight, as expected. Unfortunately, the Voyager data from Jupiter and Saturn cannot be used for imaging, because in both cases the entire magnetosphere was within a single sector during the times when energetic neutral fluxes could be identified.

Charge Exchange at Saturn

We now focus specifically on Saturn and implications of the Voyager charge exchange neutral observations. The Voyager observations are not well fitted by fast proton charge exchange with hydrogen atoms of the Titan torus only (see Figure 1). The proton on hydrogen charge exchange cross section above 40 keV drops nearly exponentially with increasing proton energy, predicting a charge exchange neutral flux that would decrease with energy more rapidly than was observed. Furthermore, the measured energetic ion fluxes in the outer magnetosphere, where the composition is mainly protons (Krimigis et al., 1983), combined with an atomic H density of 10 cm^{-3} in the Titan torus, yields charge exchange neutral fluxes that are several times smaller than observed (see Figure 1). With an H density of 20 cm^{-3} the discrepancy would be reduced but still exceeds an order of magnitude at the highest energies.

A better fit to the Voyager data is obtained if an additional source of charge exchange neutrals is included. Cheng (1986) suggested that the principal source region lies in the inner magnetosphere, where a water product neutral cloud is maintained by ice sputtering on Saturn's moons (Cheng et al., 1986) and where the energetic ions are mainly heavies (O^+ or N^+; Krimigis et al., 1983). By water products is meant H_2O and dissociation products, such as O, OH, etc. The majority of the charge exchange reactions would then occur between heavy ions and water product neutrals in the inner magnetosphere, and many of the charge exchange neutrals observed by Voyager would be heavy atoms. Alternatively, Ip (1984) has suggested that molecular H_2, of density up to 10 cm^{-3}, is present in addition to H in the Titan torus. In this case the products of proton charge exchanges with H and H_2 in the outer magnetosphere would be fast H atoms.

These two alternative models would be readily distinguished by an instrument that would perform imaging, composition, and spectral analyses of charge exchange neutrals. In the Cheng (1986) picture, the inner magnetosphere is a strong source of fast charge exchange heavy atoms, while the Titan torus region is a weaker source of fast H atoms. On the other hand, the Ip (1984) picture is that the Titan torus is the principal source region of charge exchange neutrals, whose composition could be predominantly fast H atoms.

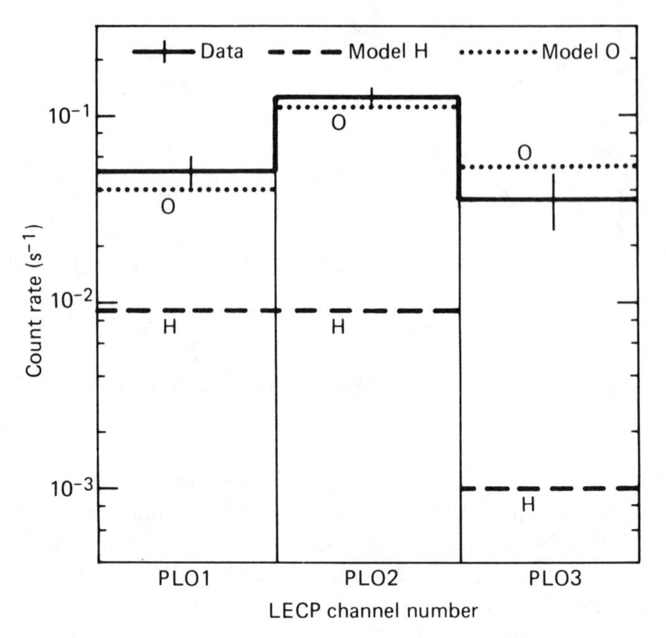

Fig. 1. Energetic neutral count rates as observed by Voyager 1, normalized to 45 R_S, with statistical error bars in three energy channels (crosses). The energy channel boundaries are 40-53 keV, 53-85 keV, 85-139 keV for protons and 100-135 keV, 120-206 keV, 180-260 keV for oxygens. Predicted count rates shown for two models. Model H (dashed lines) includes $H^+ + H$ in the Titan torus only. Model O (dotted lines) includes $H^+ + H$ in the Titan torus and charge exchanges with a water product neutral cloud from 3 to 11 R_S, with average density 20 cm^{-3} and total thickness 1.5 R_S.

Composition-resolved imaging of charge exchange neutrals is an especially important observation to be carried out, because neither the water product neutral cloud of the inner magnetosphere nor the H_2 cloud in the Titan torus has ever been detected in optical or UV emissions. Likewise, the heavy ion plasma of the inner magnetosphere could not be detected by Voyager, and has not been detected from the ground, in any optical or UV emissions (it was detected by in situ plasma experiments). Of course the atomic H cloud of the Titan torus has been mapped in H Lyman-α resonantly scattered emissions (Broadfoot et al., 1981). The absence of independent information from optical or UV spectroscopy of the water product cloud or the H_2 cloud has seriously hampered our understanding of Saturn's magnetosphere. This can be appreciated by considering how much of our current understanding of the Jovian magnetosphere depends on spectroscopy of the Io torus. Thus the density, spatial configuration, and other parameters of the Saturnian water product and H_2 neutral clouds are very poorly known. Their importance for Saturnian plasma sources, energy budgets, etc. are likewise poorly understood.

Despite the absence of UV or optical detections, the existence of a water product cloud in the inner magnetosphere and an H_2 cloud in the Titan torus is firmly predicted theoretically and is not seriously in question (e.g., Frank et al., 1980; Cheng et al., 1982; Johnson et al., 1984, 1988; Hunten, 1977; Hunten et al., 1984). Imaging and composition studies of charge exchange neutrals can be expected to provide important and interesting new information on these and other ambient neutral populations in Saturn's magnetosphere.

There are still other possibilities for charge exchange in Saturn's magnetosphere. Saturn's atmosphere, rather than Titan's atmosphere, may be the principal source for H atoms in the Titan torus (Shemansky and Smith, 1983, 1984). If so, then the H atom density would be expected to decrease monotonically from the vicinity of Saturn's main rings out through the Titan torus, rather than going through a minimum in the inner magnetosphere (the general vicinity of Dione, Tethys, and Rhea). If the H density within the inner magnetosphere is at least as large as that in the Titan torus, there would be a significant contribution to the inner magnetospheric source of charge exchange neutrals. However, Richardson and Eviatar (1988) have recently argued that atomic H densities as large as 20 cm^{-3} in the Dione–Tethys–Rhea regions may not be compatible with observed plasma parameters there, based on theoretical modeling of the ion-neutral chemistry.

Observations of charge exchange neutrals, particularly imaging from far above the orbital plane, may well bear on the issue of whether Saturn's atmosphere is the principal source of H for the Titan torus. Of course, H Lyman-α imaging from such a vantage point would also address this issue more directly. Finally, atomic N is also expected to be present within the Titan torus. Eviatar and Podolak (1983) have predicted an average N density of 0.4 cm^{-3} from a detailed steady state model of the Titan torus. If this result is correct, then charge exchanges with atomic N in the Titan torus do not make a significant contribution to the outer magnetospheric charge exchange source.

In the next section we return to the question of whether there is a large H_2 density in the Titan torus. We consider a Titan torus model in which an H_2 density as large as that proposed by Ip (1984) would yield an observable density of thermal H_2^+, contrary to Voyager observations (e.g., Richardson, 1986).

Molecular Hydrogen in the Titan Torus

Eviatar and Podolak (1983) presented a Titan torus model in which the H_2 density was assumed not to be significant. Their basis for this assumption was the absence of an observable thermal H_2^+ density, which was stated to be at least 50 times smaller than either the proton or heavy ion density. However, Eviatar and Podolak did not make a quantitative estimate of the allowed H_2 density, but we shall do so below. We note parenthetically that H_2^+ and H_3^+ were detected by Voyager at MeV energies (Hamilton et al., 1983).

We consider a steady state, uniform model and write the rate equations describing sources and losses of H^+ and H_2^+. From these two equations, we shall solve for the two unknown densities $n(H_2)$ and $n(H_2^+)$ and impose the observational upper limit $n(H_2^+) < 0.02\, n(H^+)$. We do not write rate equations for neutral densities, avoiding questions such as the importance of Saturn's atmosphere as a neutral source. Instead, we use Voyager observations to fix $n(H^+) = 0.1$ cm^{-3}, $n(N^+) = 0.1$ cm^{-3}, $n_e = 0.2$ cm^{-3}, and $n(H) = 10$ to 20 cm^{-3} (Richardson, 1986; Broadfoot et al., 1981; Eviatar and Podolak, 1983). The electron temperature is taken to be 50 eV; measured values range from several tens to hundreds of eV (Sittler et al., 1983), but the significant rate coefficients are not strongly temperature dependent in this range.

Table 1 lists adopted values of rates and rate coefficients to be used in the model. Most were obtained from Richardson et al. (1986) and Hamilton et al. (1983), but use has also been made of cross section data from Tawara (1978), Lo and Fite (1970), and Kieffer (1969). We have omitted charge exchange reactions that do not change the number of ions of any species, e.g., $H + H^+ \rightarrow H^+ + H$. The H_2^+ photodissociation rate given in the table is an average over the lowest two vibrational levels, weighted according to a Franck–Condon distribution. The rate is essentially infinite ($\gg 10^{-6}$ s^{-1}) for higher levels (Hamilton et al., 1983).

We now write the rate equation for $n(H_2^+)$ as

$$n(H_2^+)\, A_1 = n(H_2)\, B_1 \tag{1}$$

where

$$A_1 = \mathfrak{I}_d^{-1} + n_e\, \alpha_r(H_2^+) + \nu_d(H_2^+) + n_e\, \alpha_d(H_2^+) + \alpha_{c1}(H_2^+)\, n(H_2) \tag{2}$$

Table 1. Rates and Rate Coefficients in the Titan Torus

Rate	Reaction
$\nu_i(H) = 8 \times 10^{-10}\ s^{-1}$	$H + \gamma \rightarrow H^+ + e$
$\nu_{i1}(H_2) = 1 \times 10^{-10}\ s^{-1}$	$H_2 + \gamma \rightarrow H^+ + H + e$
$\nu_{i2}(H_2) = 5.9 \times 10^{-10}\ s^{-1}$	$H_2 + \gamma \rightarrow H_2^+ + e$
$\nu_d(H_2^+) = 5.0 \times 10^{-7}\ s^{-1}$	$H_2^+ + \gamma \rightarrow H + H^+$
$\alpha_i(H) = 2.5 \times 10^{-8}\ s^{-1}\ cm^3$	$H + e \rightarrow H^+ + 2e$
$\alpha_{i1}(H_2) = 7 \times 10^{-9}\ s^{-1}\ cm^3$	$H_2 + e \rightarrow H^+ + H + 2e$
$\alpha_{i2}(H_2) = 4 \times 10^{-8}\ s^{-1}\ cm^3$	$H_2 + e \rightarrow H_2^+ + 2e$
$\alpha_r(H_2^+) = 3.5 \times 10^{-9}\ s^{-1}\ cm^3$	$H_2^+ + e \rightarrow H + H$
$\alpha_r(H^+) = 1.8 \times 10^{-14}\ s^{-1}\ cm^3$	$H^+ + e \rightarrow H$
$\alpha_d(H_2^+) = 1.2 \times 10^{-7}\ s^{-1}\ cm^3$	$H_2^+ + e \rightarrow H^+ + H + e$
$\alpha_{c2}(H^+) = 2 \times 10^{-10}\ s^{-1}\ cm^3$	$H^+ + H_2 \rightarrow H + H_2^+$
$\alpha_{c1}(H_2^+) = 2.1 \times 10^{-9}\ s^{-1}\ cm^3$	$H_2^+ + H_2 \rightarrow H + H_3^+$
$\alpha_{c1}(N^+) = 6 \times 10^{-9}\ s^{-1}\ cm^3$	$N^+ + H \rightarrow N + H^+$
$\alpha_{c2}(N^+) = 1.2 \times 10^{-8}\ s^{-1}\ cm^3$	$N^+ + H_2 \rightarrow N + H_2^+$

ν_d (H_2^+) for lowest two vibrational states (Hamilton et al., 1983); electron temperature 50 eV and ion-neutral relative velocity 100 km s^{-1} (Sittler et al., 1983; Richardson, 1986; Richardson et al., 1986).

$$B_1 = [\nu_{i2}(H_2) + n_e\,\alpha_{i2}(H_2) + n(H^+)\,\alpha_{c2}(H^+) + n(N^+)\,\alpha_{c2}(N^+)]\eta \tag{3}$$

Loss processes for H_2^+ are represented by $n(H_2^+)\,A_1$ and are explicitly (from left to right) radial transport, dissociative recombination, photodissociation, electron impact dissociation, and the ion-molecule reaction $H_2^+ + H_2 \rightarrow H + H_3^+$. The source processes $n(H_2)\,B_1$ are photoionization of H_2, electron impact ionization, charge exchange by H^+ on H_2, and charge exchange by N^+ on H_2. Here η is the Franck–Condon factor.

The Franck–Condon factor η is introduced to take account of the strong dependence of the H_2^+ photodissociation lifetime on the vibrational state (Hamilton et al., 1983). Of the newly created H_2^+, only the fraction η within the lowest two vibrational states has an appreciable lifetime and is included in equations (1)–(3). Any H_2^+ created in a higher excited state is immediately photodissociated and converted to a proton; hence this portion $(1 - \eta)$ of the H_2^+ source will be counted below as a proton source. Given the high electron temperature, electron impact ionization of H_2 yields H_2^+, of which a fraction $\eta = 0.254$ is in either the ground vibrational state or first excited state, assuming a Franck–Condon distribution (Hamilton et al., 1983). We also assume the same distribution for H_2^+ from photoionization and charge exchange, in the absence of better information on the vibrational state distribution.

The rate equation for H^+ is similarly

$$n(H^+)\,A_2 = n(H)\,B_2 + n(H_2)\,B_3 + n(H_2^+)\,B_4 \tag{4}$$

where

$$A_2 = \Im_d^{-1} + n_e\,\alpha_r(H^+) + n(H_2)\,\alpha_{c2}(H^+) \tag{5}$$

$$B_2 = \nu_i(H) + n_e\,\alpha_i(H) + n(N^+)\,\alpha_{c1}(N^+) \tag{6}$$

$$B_3 = B_1\,\eta^{-1}\,(1 - \eta) + \nu_{i1}(H_2) + n_e\,\alpha_{i1}(H_2) \tag{7}$$

$$B_4 = \nu_d(H_2^+) + n_e\,\alpha_d(H_2^+) \qquad (8)$$

Again $n\,(H^+)\,A_2$ represents loss processes, and the remaining terms represent source processes of protons.

We next consider the transport rate $\mathfrak{I}_d^{-1}$ in the Titan torus. Hood (1985) and Paonessa and Cheng (1986) have estimated $\mathfrak{I}_d^{-1}$ from phase-space density analyses of ring current ions. If radial transport must be at least fast enough to balance absorption by Saturn's moons, $\mathfrak{I}_d^{-1} \gtrsim 10^{-7}\ s^{-1}$ near Rhea (8.8 R_S) according to either study. Hence $\mathfrak{I}_d^{-1}$ in the Titan torus should be even greater, at least a few times $10^{-7}\ s^{-1}$.

We now note that $\mathfrak{I}_d^{-1} \gg \alpha_{c1}(H_2^+)\,n(H_2)$ in equation (2) and $\mathfrak{I}_d^{-1} \gg n(H_2)\,\alpha_{c2}(H^+)$ in equation (5), provided $n(H_2) \ll 10\ cm^{-3}$ (as will turn out to be true). We neglect these small terms, so all A's and B's are independent of the unknowns and

$$n(H_2^+) = \frac{n(H^+)\,A_2 - n(H)\,B_2}{A_1\,B_3\,B_1^{-1} + B_4} \qquad (9)$$

The other unknown, $n(H_2)$, is given by equation (1).

For an acceptable solution, $n(H_2^+)$ must be positive and must be less than 0.02 $n(H^+)$ from Voyager observations. Hence

$$0 < n(H^+)\,A_2 - n(H)\,B_2 < 0.02\,n(H^+)\,[A_1\,B_3\,B_1^{-1} + B_4] \qquad (10)$$

We note that $A_2 = \mathfrak{I}_d^{-1}$ for all practical purposes, so equation (10) can be regarded as a constraint on the values of $\mathfrak{I}_d^{-1}$ for which the model yields acceptable $n(H_2^+)$. For the adopted $n_e = 0.2\ cm^{-3}$, $n(H^+) = 0.1\ cm^{-3}$, $n(N^+) = 0.1\ cm^{-3}$, and the values in Table 1, the constraint of equation (10) becomes

$$6.4\left(\frac{n(H)}{10\ cm^{-3}}\right) < \frac{\mathfrak{I}_d^{-1}}{10^{-7}\ s^{-1}} < 6.45\left(\frac{n(H)}{10\ cm^{-3}}\right) + 0.047 \qquad (11)$$

Hence the observed $n(H) = 10$ to $20\ cm^{-3}$ yields $\mathfrak{I}_d^{-1} \approx (6.4-13) \times 10^{-7}\ s^{-1}$, agreeing with estimates from the inner magnetosphere (Hood, 1985; Paonessa and Cheng, 1986). For a fixed $n(H)$, the transport rate $\mathfrak{I}_d^{-1}$ is very tightly constrained, to less than 1%. This is because of the small (observationally imposed) value of $n(H_2^+)/n\,(H^+)$, which forces a near cancellation between $n(H^+)\,A_2$ and $n(H)\,B_2$. Of course the numerical values in equation (11) are very uncertain.

Finally, the condition $n(H_2^+) < 0.02\ n(H^+)$ can be written using equations (1)–(3) as

$$n(H_2) < 0.42\ cm^{-3} + 0.08\left(\frac{\mathfrak{I}_d^{-1}}{10^{-7}\ s^{-1}}\right)\ cm^{-3} \qquad (12)$$

or $n(H_2) < 1.5\ cm^{-3}$ using equation (11). We conclude that H_2 is not an important population in the Titan torus, agreeing with Eviatar and Podolak (1983). If $n(H_2) \ll 10\ cm^{-3}$, energetic proton charge exchange with H_2 in the Titan torus does not make a significant contribution to the charge exchange neutral fluxes observed by Voyager. We confirm the suggestion of Cheng (1986) that the Voyager energetic neutral fluxes imply an additional source in the inner magnetosphere from charge exchanges between heavy ions and water product neutrals. We caution, however, that the Eviatar and Podolak (1983) upper limit on H_2^+ applies only to a few spectra of exceptionally high quality, and that typical spectra yield upper limits several times higher (J. Richardson, private communication, 1988). Nevertheless, we expect H_2^+ to be uniformly distributed, so it is appropriate to use the strongest upper limit obtainable from the data.

Conclusions: Future Observations of the Saturn System

Observations of charge exchange neutrals provide an important new technique for remote sensing of the Saturn system. Even the initial Voyager observations of charge exchange neutrals (Kirsch et al., 1981b) have opened up many interesting scientific issues. If, in the future, composition-resolved imaging and energy spectral analyses of charge exchange neutrals can be carried out, we expect dramatic advances in our understanding of the Saturn system.

Thus far we have focused on inferring properties of ambient neutral populations at Saturn, using Voyager observations of cold plasmas and energetic particles. Both Ip (1984) and Cheng (1986) suggest that charge exchanges between fast protons and atomic H in the Titan torus do not suffice to explain the Voyager observations, and that another source is needed. Ip (1984) suggested that the additional source is proton charge exchange with H_2 in the Titan torus. However, we have argued here that H_2 is not an important population in the Titan torus, based on the Voyager upper limit $n(H_2^+) < 0.02\ n(H^+)$ (Eviatar and Podolak, 1983). Our model for the Titan torus requires $\mathfrak{I}_d^{-1} = (6.4$ to $13) \times 10^{-7}\ s^{-1}$, agreeing with earlier estimates for the inner magnetosphere, and yields $n(H_2) < 1.5\ cm^{-3}$. Of course there are many uncertainties regarding rate coefficients, Franck–Condon factors, etc.

We confirm the suggestion of Cheng (1986) that the Voyager energetic neutral fluxes are best explained if charge exchanges occur in the inner magnetosphere between heavy ions and water product neutrals. Charge exchange neutral observations are especially important because both the water product neutral cloud of the inner magnetosphere and the H_2 cloud of the Titan torus are not observable in optical or UV emissions. Composition-resolved imaging of charge exchange neutrals will readily distinguish between the suggestions of Ip (1984) and Cheng (1986): in the former case, the main source is in the Titan torus and yields fast H atoms, whereas in the latter case there is a significant inner magnetospheric source of fast heavy atoms.

Of course, charge exchange neutral observations also yield information on energetic ion populations, as shown by Roelof (1987), who used independent information on Earth's hydrogen geocorona to obtain images of Earth's ring current. At Saturn, UV imaging of H Lyman-α can determine atomic H distributions, and then charge exchange neutral observations may allow remote sensing of energetic ion densities in regions where atomic H is expected to be the dominant neutral. Atomic H is probably dominant except very close to Saturn and within the water product cloud of the inner magnetosphere.

Perhaps the most exciting prospect is that substorm-like particle injections may be observed at Saturn. Imaging of charge exchange neutrals, by virtue of being a remote sensing technique, may detect substorm-like activity in Saturn's magnetosphere even while the spacecraft lies outside the magnetosphere. A substorm-like particle injection would be manifested as an intense, localized, transient brightening in emissions of charge exchange neutrals.

Table 2 gives a list of science objectives for charge exchange neutral observations on the proposed Cassini mission to the Saturn system. Figure 2 shows an artist's concept of possible results from composition-resolved imaging of charge exchange neutrals from the Saturn system. The bow shock and magnetopause are sketched to orient the reader, but these structures are not expected to be imaged in charge exchange neutrals. However, fast heavy atoms will be imaged from charge exchanges between heavy ions of the ring current and water product neutrals in the inner magnetosphere. There will be localized enhancements near the orbits of each of Saturn's icy moons, resulting from the locally increased ambient neutral density. Fast H atoms will also be imaged from the Titan torus, and in addition, charge exchange neutrals may be imaged from Titan's cometary interaction with Saturn's magnetosphere. Enhanced charge exchange neutral emissions are also expected from the near vicinity of Saturn itself, from interactions between the ring atmosphere/upper exosphere and the innermost radiation belts. Finally, a substorm-like particle injection is sketched in the nightside plasma sheet.

Table 2. Cassini Neutral Imager Science Objectives

- Global imaging of ring current, radiation belts, and neutral clouds in magnetosphere of Saturn
- Remote sensing of energetic ion composition and distribution functions
- Search for and monitor substorm-like activity at Saturn
- Imaging and composition studies of magnetosphere–satellite interactions at Saturn and formation of neutral hydrogen, nitrogen, and water product clouds
- Imaging of Titan's induced magnetosphere and interactions with Saturn's magnetosphere or solar wind
- Global imaging of ring current, radiation belts, and neutral clouds in magnetosphere of Jupiter during flyby
- In situ energetic ion measurements

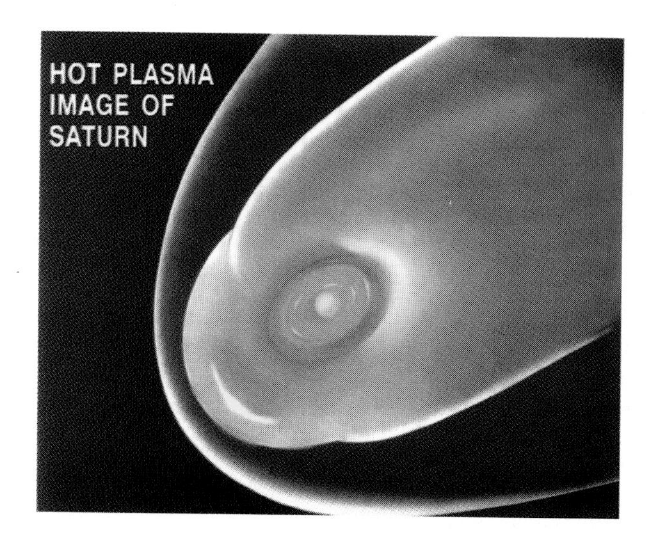

Fig. 2. Artist's concept of charge exchange neutral image of the Saturn system, showing Titan torus, water product neutral cloud, higher density neutral clouds near orbits of Saturn's icy moons and rings, Titan's interaction with the magnetosphere, and a substorm-like particle injection in the nightside plasma sheet.

We conclude with a simple estimate of the instrumental requirements for charge exchange neutral imaging at Saturn. The Voyager observed flux at a distance of 45 R_s was 4.3 $cm^{-2}\ s^{-1}$ from 40 to 139 keV assuming a composition of all H (Kirsch et al., 1981b). The flux was 4.8 $cm^{-2}\ s^{-1}$ for the mixture of H from 40 to 139 keV and O from 100 to 260 keV suggested by Cheng (1986); this difference is not significant for the present estimate. The effective area of the Voyager detector was 0.08 cm^2

in the upper two channels (53 to 139 keV for H) but 0.032 cm^2 in the lowest channel. For a dedicated neutral particle imager, the effective area can be greatly increased, say by a factor of 400. We adopt a nominal distance of 45 R_s for global imaging of the Saturn system, and we suppose the image is resolved into 400 pixels. Then, on the average each pixel should observe a count rate equal to that observed by Voyager and shown in Figure 1. Thus the average count rate per pixel, in each of three energy channels as in the Voyager instrument, would be ~ 0.05 count s^{-1}. An average of 50 counts per pixel would be accumulated in 10^3 s.

Hence imaging of charge exchange neutrals from the Saturn system is a feasible proposition. The estimate of 50 counts per pixel in 10^3 s applies to an instrument steadily pointed at the source. Longer exposure times would be required for the same instrument on a spinning platform, depending on the fraction of the time spent pointed at the source, for images with better counting statistics or higher energy resolution, or for composition-resolved images of minor species. However, this estimate used the fluxes observed by Voyager, which likely correspond to a quiet-time average over the entire Saturn system. Substantially greater fluxes will be emitted from localized regions of high ambient neutral density, as well as from transient particle injection events if substorm-like processes occur. We conclude that neutral particle imaging of the Saturn system will yield important new scientific results. Other instrumental requirements, such as suppression of charged particle and photon backgrounds, are discussed in companion papers.

Acknowledgments. This work was supported by NASA under Task I of contract N00039-87-C-5301 between the Johns Hopkins University and the Department of the Navy and by NASA grant NAGW-865 to the Johns Hopkins University.

References

Broadfoot, A. L., et al., Extreme ultraviolet observations from Voyager 1 encounter with Saturn, *Science, 212,* 206-211, 1981.

Cheng, A. F., Energetic neutral particles from Jupiter and Saturn, *J. Geophys. Res., 91,* 4524-4530, 1986.

Cheng, A. F., L. J. Lanzerotti, and V. Pironello, Charged particle sputtering of ice surfaces in Saturn's magnetosphere, *J. Geophys. Res., 87,* 4567-4570, 1982.

Cheng, A. F., P. Haff, R. E. Johnson, and L. J. Lanzerotti, Interactions of magnetospheres with icy satellite surfaces, in *Satellites,* edited by J. Burns and M. Matthews, pp. 403-436, Univ. of Arizona Press, Tucson, 1986.

Eviatar, A., and M. Podolak, Titan's gas and plasma torus, *J. Geophys. Res., 88,* 833-840, 1983.

Frank, L. A., L. Burek, K. Ackerson, J. Wolfe, and J. Mihalov, Plasmas in Saturn's magnetosphere, *J. Geophys. Res., 85,* 5695-5708, 1980.

Hamilton, D. C., D. Brown, G. Gloeckler, and W. I. Axford, Energetic atomic and molecular ions in Saturn's magnetosphere, *J. Geophys. Res., 88,* 8905-8922, 1983.

Hood, L. L., Radial diffusion of low energy ions in Saturn's radiation belts. A combined analysis of phase space density and satellite microsignature data, *J. Geophys. Res., 90,* 6295-6303, 1985.

Hunten, D. M., Titan's atmosphere and surface, in *Planetary Satellites,* edited by J. A. Burns, pp. 420-437, Univ. of Arizona Press, Tucson, 1977.

Hunten, D., M. Tomasko, F. Flasar, R. Samuelson, D. Strobel, and D. Stevenson, Titan, in *Saturn,* edited by T. Gehrels and M. Matthews, pp. 671-759, Univ. of Arizona Press, Tucson, 1984.

Ip, W. H., An estimate of the H_2 density in the atomic hydrogen cloud of Titan, *J. Geophys. Res., 89,* 2377-2379, 1984.

Johnson, R. E., L. J. Lanzerotti, and W. L. Brown, Sputtering processes: Erosion and chemical change, *Adv. Space Res., 4,* 41-51, 1984.

Johnson, R. E., M. Pospieszalska, E. Sieveka, A. F. Cheng, L. Lanzerotti, and E. Sittler, The neutral cloud and heavy ion inner torus at Saturn, *Icarus,* submitted, 1988.

Kieffer, L. J., Low energy cross section data, *At. Data, 1,* 19-90, 1969.

Kirsch, E., S. M. Krimigis, J. Kohl, and E. Keath, Upper limits for x-ray and energetic neutral particle emission from Jupiter. Voyager 1 results, *Geophys. Res. Lett., 8,* 169-172, 1981a.

Kirsch, E., S. M. Krimigis, W. H. Ip, and G. Gloeckler, X-ray and energetic neutral particle emission from Saturn's magnetosphere, *Nature, 292,* 718-721, 1981b.

Krimigis, S. M., Energetic ions upstream of planetary bowshock: Fermi acceleration or leakage?, in *Comparative Study of Magnetospheres,* CNES, Cepadues Editions, Toulouse, France, pp. 99-124, 1986.

Krimigis, S. M., J. F. Carbary, E. P. Keath, T. P. Armstrong, L. J. Lanzerotti, and G. Gloeckler, General characteristics of hot plasma and energetic particles in the Saturnian magnetosphere: Results from the Voyager spacecraft, *J. Geophys. Res., 88,* 8871-8892, 1983.

Krimigis, S. M., R. Zwickl, and D. N. Baker, Energetic ions upstream of Jupiter's bowshock, *J. Geophys. Res., 90,* 3947-3960, 1985.

Lo, H., and W. Fite, Electron capture and loss cross sections, *At. Data, 1,* 305-328, 1970.

Paonessa, M., and A. F. Cheng, Limits on ion radial diffusion coefficients in Saturn's inner magnetosphere, *J. Geophys. Res., 91,* 1391-1396, 1986.

Richardson, J., Thermal ions at Saturn: Plasma parameters and implications, *J. Geophys. Res., 91,* 1381-1389, 1986.

Richardson, J., and A. Eviatar, Limits on the extent of Saturn's hydrogen cloud, *J. Geophys. Res.,* in press, 1988.

Richardson, J., A. Eviatar, and G. L. Siscoe, Satellite tori at Saturn, *J. Geophys. Res., 91,* 8749-8755, 1986.

Roelof, E. C., Energetic neutral atom image of a stormtime ring current, *Geophys. Res. Lett., 14,* 652-655, 1987.

Shemansky, D., and G. R. Smith, Whence comes the "Titan" torus, *Eos, Trans. AGU, 63,* 1019, 1983.

Shemansky, D., and G. R. Smith, Hydrogen blow-off from Uranus: An extrapolation from Jupiter and Saturn, *Bull. AAS, 16,* 660, 1984.

Sittler, E. C., K. Ogilvie, and J. Scudder, Survey of low energy plasma electrons in Saturn's magnetosphere: Voyagers 1 and 2, *J. Geophys. Res., 88,* 8847-8870, 1983.

Tawara, H., Cross sections for charge transfer of hydrogen beams in gases and vapors in the energy range 10 eV-10 keV, *At. Data Nuc. Data Tab., 22,* 491-525, 1978.

Zwickl, R. D., S. M. Krimigis, J. F. Carbary, E. P. Keath, T. P. Armstrong, D. C. Hamilton, and G. Gloeckler, Energetic particle events ($\geq$ 30 keV) of Jovian origin observed by Voyager 1 and 2 in interplanetary space, *J. Geophys. Res., 86,* 8125-8140, 1981.

NOVEL METHODS FOR ACTIVE SPACECRAFT POTENTIAL CONTROL

R. Schmidt and H. Arends

Space Science Department of ESA/ESTEC, Noordwijk, The Netherlands

K. Torkar

Space Res. Inst. Austrian Academy of Science, TU Graz, Austria

N. Valanvanoglou

Inst. of Communications and Wave Prop., TU Graz, Austria

Abstract. A spacecraft embedded in a plasma is subject to charging of its surfaces to potentials determined by the ambient plasma and solar photons. In a tenuous plasma a sunlit surface may acquire more than +50 V. In eclipse and in the presence of a hot electron plasma negative potentials in excess of several kilovolts are not uncommon. For accurate cold plasma and electric field measurements it is almost mandatory to neutralize the potential. This requires conductive surfaces and active potential control by emission of charged particles. Solar cell powered spacecraft operate mostly in sunlight; thus, large positive potentials in a tenuous plasma must be reduced. We suggest to emit a beam of kilo-electron-volt ions at a current of several microamperes. Two types of emitters are being developed for the INTERBALL, GEOTAIL, and possibly CLUSTER missions. The liquid metal ion emitter allows for a small and compact design. The saddle field emitter provides the advantage that almost any gaseous substance can be used for emission.

Introduction

The principles of electrostatic charging have been extensively discussed in the literature (see Grard, 1973, and Whipple, 1981, for reviews). In the simplest case the equilibrium potential is determined by photoemission from sunlit surfaces and impinging charged plasma particles on the surface. Spacecraft in high plasma density environments charge slightly negative because of the higher thermal velocities of the electrons as compared with the ions. Outside the plasmasphere the current balance is usually between plasma electrons and photoemission as long as the plasma density is very low and electron temperature is less than a few kilo-electron-volts. In extremely low-density plasmas (i.e., lobes) potentials can be as high as +50 V.

For shortness we only treat conductive surfaces that are sunlit and use data from the geostationary spacecraft GEOS 2 for which Knott et al. (1984) reported that the potential rarely exceeded ±10 V. It is an example for a design effort to minimize charging. The detrimental effect of a possible deterioration of the conductive coating on the ISEE 1 solar panels has been reported by Whipple and Olsen (1986); Schmidt and Pedersen (1987), reported possibly eroded wire boom coatings. In both cases the spacecraft potential might have been affected.

GEOS 2 and ISEE 1 have clearly shown that provision of conductive surfaces to suppress differential charging is mandatory. However, reliable electric field and cold plasma measurements require additional measures. Active potential control is the only means to achieve stable potentials very close to the local plasma potential. Several attempts have been undertaken in the past (see, e.g., Olsen, 1985) and solutions have been described (Pedersen, 1983) but no dedicated system has been implemented yet. Two principles look most promising. The cold plasma source is a self-controlling emitter as it embeds the spacecraft in a very dense, cold plasma. The surfaces are neutralized by interaction between the charged surfaces and the artificial plasma cloud. Another approach is to discharge the positive surface by emission of high-energy ions. Typical energies are in the kilo-electron-volt range with an emission of >10 μA for a standard size spacecraft.

We are developing two different energetic ion emitters as we believe that the dense plasma created by a plasma source would interfere with other instruments. A multiple liquid metal ion source (LIMIS), based on indium, and a discharge type saddle field ion emitter have been chosen and are described in more details. Both systems will be integrated into one package and fly on the polar orbiting spacecraft of the INTERBALL mission in 1990 while LIMIS-only based designs will be used on GEOTAIL and possibly on CLUSTER.

Reasons for Satellite Charge Control

Typical satellite floating potentials in the tenuous lobe plasma reach up to 50 V. Such potentials obscure the measurement of the low-energy part of the ion distribution function, which has a thermal energy comparable to the satellite potential. This problem was indicated by discrepancies in density calculations from satellites such as GEOS 1 and 2 (Décréau et al., 1978). Other fundamental problems occur in the measurement of the anisotropic distributions outside the plasmasphere, such as the field-aligned flows which make up the polar wind (Olsen et al., 1986). The bulk on the distribution is lost due to the satellite potential.

With active potential control, it will be possible to measure the low-energy ions under almost all conditions. Electron measurements will also be improved because the low-energy portion of the electron spectra is usually contaminated by photoelectrons from the satellite surface.

Detrimental effects of spacecraft charging extend also to electric field measurements. The double probe technique is sensitive to a spurious electric field induced by the satellite charge. Besides extending the wire booms, the reduction of the positive potential of the satellite and on the supporting wire booms would minimize the local field perturbation.

Derivation of the Equilibrium Potential

The floating potential of a sunlit probe is in first approximation determined by the photoemissivity of the surface and incident ambient plasma electrons and ions. In the outer magnetosphere only the plasma electrons are of relevance. An approximate value for the floating potential V_S can be derived from a simple model. The ambient electron current I_A impinging on the spherical surface A at the positive potential V amounts in the orbital limited mode (i.e., λ_D much larger than the body) to

$$I_A = A\ N\ e\ v\ (\tfrac{1}{2}\sqrt{\pi})\ (1 + eV_S/kT_e) \quad (1)$$

For the retarding regime ($V_S < 0$) we get

$$I_A = A\ N\ e\ v\ (\tfrac{1}{2}\sqrt{\pi})\ \exp(eV_S/kT_e) \quad (2)$$

with e, N, v, and kT_e being, respectively, the plasma electron charge, density, velocity, and temperature.

The photoelectron current I_p is essentially determined by the surface properties. The mean photoelectron temperature, kT_p, is about 1.5 eV (Grard, 1973). Laboratory measurements on various spacecraft materials yielded a photoelectron saturation current density, j_{ps}, of about 1-5 nA cm^{-2} (Feuerbacher and Fitton, 1972). Under certain circumstances it is possible to obtain estimations from in-orbit measurements. Such values were found, at least for a few materials, to be probably more than 6 times higher than values derived on ground.

The photoelectron current density saturates if $V_S < 0$ because all photoelectrons are repelled and it decreases with $V_S > 0$ due to increased attraction of the lower-energy photoelectrons. Ignoring angular variations of the photoelectron yield, the total emitted current from the sunlit, positively charged area S amounts to

$$I_p = S\ j_{ps}\ \exp(-eV_S/kT_p) \quad (3)$$

and if $V_S < 0$

$$I_p = S\ j_{ps} \quad (4)$$

The high-energy ion beam can be considered to be independent of V_S; thus

$$I_i = \text{const} \quad (5)$$

The floating potential is established if the net flow to and from the surface vanishes

$$I_p + I_A + I_i = 0 \quad (6)$$

I_p and I_A, in arbitrary units, are plotted in Figure 1 as functions of the surface potential

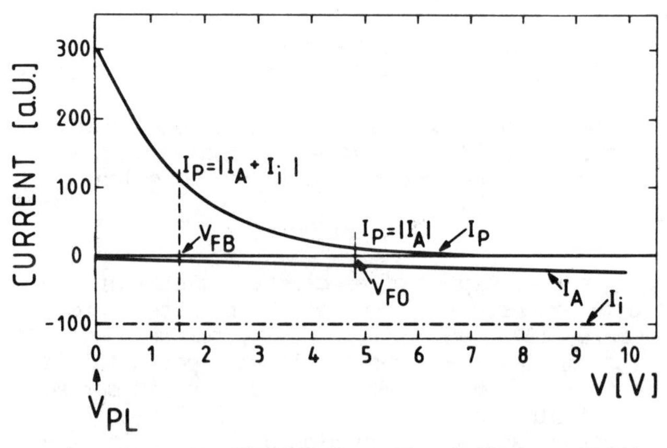

Fig. 1. Currents flowing toward and from a sunlit sphere. I_p, I_A and I_i denote, respectively, the currents due to photoelectrons, ambient electrons, and ion emission. V_{FO} and V_{FB} are the floating potentials for $I_i = 0$ and $I_i > 0$, respectively. V_{PL} is the local plasma potential. Note that arbitrary units are plotted on the vertical axis.

which is taken relative to the plasma potential V_{PL}. With $I_i = 0$, the floating potential V_{Fo} is settled when $|I_p| = |I_A|$. Typical values for V_{Fo} on GEOS 2 would be 6-10 V.

Active potential control aims at clamping the potential in the steep part of the I_p/V relation. This is achievable by emitting the current I_i which forces the equilibrium potential to V_{FB}. With 10 μA < I_i < 20 μA eqs. (1)-(6), together with typical plasma parameters for R = 6.6 R_E, yield for V_{FB} about 1-2 V. V_{FB} should not be too close to V_{PL} in order to avoid that $|I_p| < |I_A+I_B|$ which leads to a negative potential.

Concept of Spacecraft Discharging

The emission of high-energy particles to discharge a spacecraft can be performed in two ways. In the autonomous mode a constant current between 1 - 20 μA will be emitted. This is low enough to avoid a swing over to a large negative potential when the spacecraft suddenly enters a denser plasma region and the emitter remains operative. The autonomous mode keeps the potential low but does not stabilize it. A stable potential can only be achieved if the emitter is connected onboard to the electric field double probe instrument (Figure 2) that provides the actual spacecraft potential by the probe current biasing technique (Pedersen et al., 1983) and maintains the probes very close to the ambient plasma potential. The variable ion emission maintains the potential at a fairly stable value.

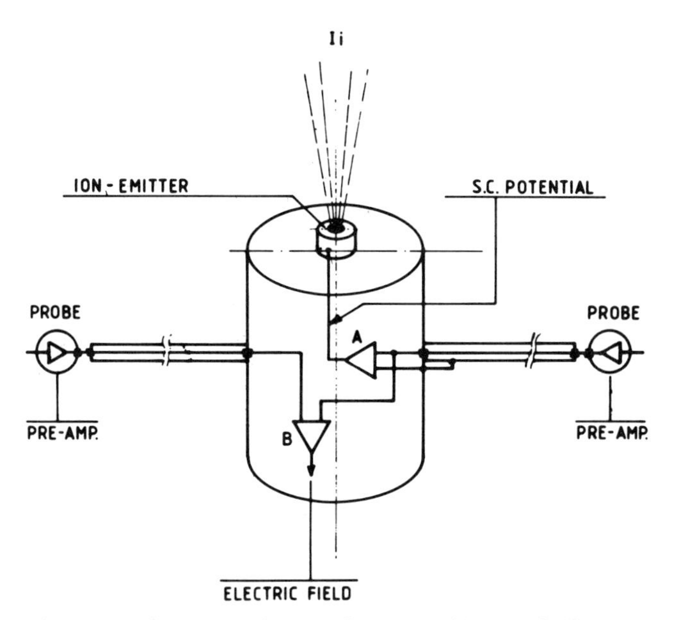

Fig. 2. Interconnection of ion emitter and the electric field instrument to achieve a stable potential.

Technical Realization

Liquid metal ion source. The design is kept simple and essentially consists of a cartridge filled with indium, a needle protruding from it, and a heater to melt the metal (Figure 3). The electrostatic stress at the needle tip pulls the liquid metal toward the extractor electrode (Gomer, 1979). This stress is counteracted by the surface tension forces of the liquid. The liquid surface assumes in the equilibrium a configuration called a Taylor cone (Taylor, 1964) with a total tip angle of 98.6°. The apex of this cone has a diameter of 1 to 5 nm (Kingham and Swanson, 1984). Since the emission zone is in the liquid state, ions leaving the surface can be continuously replenished by the hydrodynamic flow of liquid metal from the reservoir to the needle apex so that a stable emission can be maintained. The beam energy spread amounts to about 15 eV at 3-5 keV beam energy. For reliability reasons five or more emitters will be grouped into one module (Figure 3, lower part). If one emitter cartridge is emp-

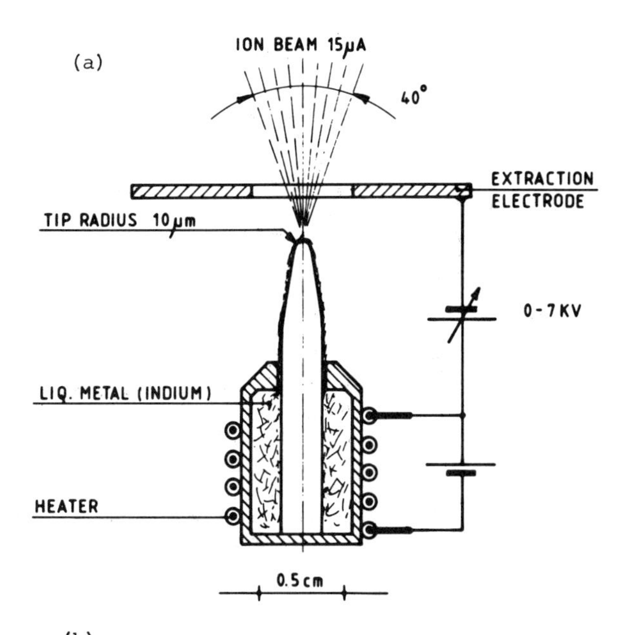

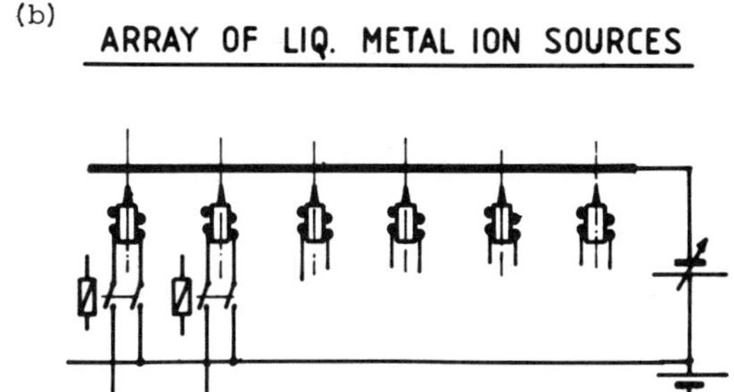

Fig. 3. Individual liquid indium emitter (a) and a schematic depiction of a multi-needle array (b).

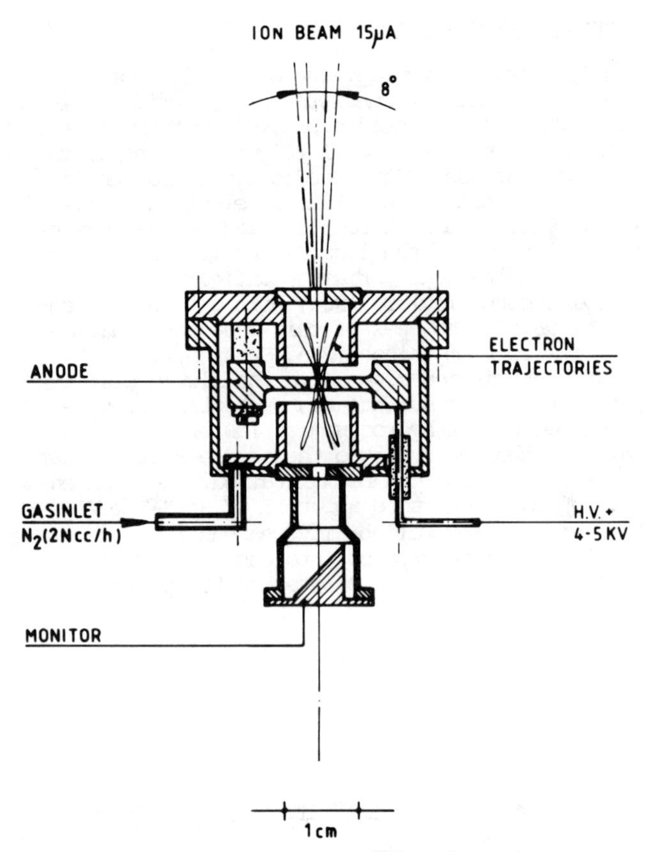

Fig. 4. Saddle field ion emitter. Typical electron trajectories are shown in the center of the source.

tied, the heater of another emitter will be switched on. Such an arrangement also provides better efficiency for currents above 20 μA. The efficiency decreases above 20 μA as droplet emission from the tip occurs. It can be avoided by operating several needles in parallel at a lower individual emission current. The advantages of the LIMIS principle are (a) the low power consumption, (b) high mass efficiency, and (c) its compactness and low mass; one individual emitter has a volume of 170 mm^3 and a mass <5.2 g.

Saddle field ion emitter. A detailed description of this source is given by Franks (1979). It consists of a grounded cylindrical tube that accommodates a ring-shaped anode connected to a positive potential <4 kV. The particular shape of the internal electric field configuration led to its name. Electrons describe longer oscillatory trajectories to prolong their average life time. Two symmetrial ion beams can be extracted from the flat sides of the cylinder. Our design uses only one of these holes to minimize the gas consumption (Figure 4). Extensive laboratory tests were performed to investigate sputter effects produced by the energetic ions (~3 kV). Internal sputter effects make it difficult to achieve the envisaged design life time of 5000 hours, but with proper material selection and surface protection this aim is achievable. While the source itself is rather simple the whole system becomes quite complex as it also needs gas storage, associated pressure reduction, and active flow control. With a gas flow of 2 Ncc/h the source generates about 20 μA of emission current. The associated source power dissipation is about 1.6W. The source operates on most gases but nitrogen was chosen as it is also used for the spacecraft reaction control system. In Figure 5 we give an example of how the emission current, $I_{collector}$, varies as a function of the internal gas pressure. The diagram shows that the source becomes unstable if operated above 7.6 mT and emission currents <3 μA. The onboard computer has to be programmed to combine pressure and current such that the source does not enter this "forbidden" region.

Discussion

After describing the emitters we will briefly discuss possible interferences with spacecraft systems and instruments. Condensation of indium on the spacecraft near to the emitter is very unlikely as the metal is predominantly expelled either in the form of ions or charged droplets. In view of the very low vapor pressure of indium and the high expelling efficiency, surface condensation of indium is not to be expected. First vacuum tests over >200 hours support this. Additional provision of a baffle could give further safety in this respect.

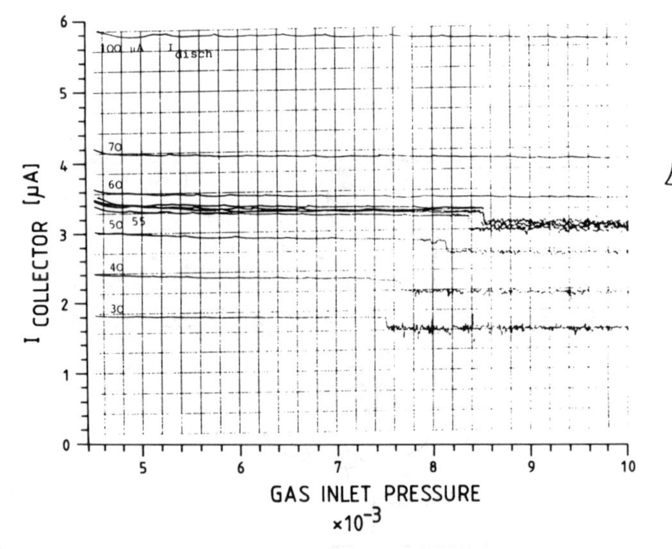

Fig. 5. Current emission of the saddle field ion emitter as a function of the internal gas pressure. The graphs were taken for discharge currents as indicated. The lower right corner indicates the emission instability region.

For both emiters we do not expect beam return to the spacecraft after one or more gyrations in the ambient magnetic field as both mass and injection energies are high. The gyration radius of indium at 3 keV and B = 100 nT is ~800 km. Hence the beam will not perform undisturbed gyrations. Beam scattering by beam/particle interaction, ∇B drift, and energy transfer into waves will take place.

Beams with current densities and energies as we use are practically not treated in the literature. Hence it is difficult to predict whether the beam is capable of triggering any wave activity. Ion beams with 2 orders of magnitude higher current densities were injected into the ionosphere and generated waves around the lower hybrid resonance frequency. Other experiments found increased emission in the magnetic field around the local electron gyrofrequency. The electric field antenna, however, did not measure any effect. Tests and simulations in a plasma chamber will be performed to come to final conclusions.

References

Décréau, P.M.E., J. Etcheto, K. Knott, A. Pedersen, G. L. Wrenn, and D. T. Young, Multi-experiment determination of plasma density and temperature, Space Sci. Rev., 22, 633, 1978.

Feuerbacher, B., and B. Fitton, Experimental investigation of photo-emission from satellite surface materials, J. Appl. Phys., 43, 1563, 1972.

Franks, J., Properties and applications of saddle-field ion sources, J. Vac. Sci. Technol., 16, 181, 1979.

Gomer, R., On the mechanism of liquid metal electron and ion sources, Appl. Phys., 19, 365, 1979.

Grard, R.J.L., Properties of the satellite photoelectron sheath derived from photo-emission laboratory results, J. Geophys. Res., 78, 2885, 1973.

Kingham, D. R., and L. W. Swanson, A theoretical model of a liquid metal ion source, Appl. Phys., 34, 123, 1984.

Knott, K., A. Pedersen, P. Décréau, A. Korth, and G. L. Wrenn, The potential of an electrostatically clean geostationary satellite and its use in plasma diagnostics, Planet. Space Sci., 32, 227, 1984.

Pedersen, A., C. R. Chappell, K. Knott, and R. C. Olsen, Methods for keeping a conductive spacecraft near the plasma potential, in Proceedings of the 17th ESLAB Symposium on Spacecraft/Plasma Interactions and their Influence on Field and Particle Measurements, Noordwijk, The Netherlands, September 13-16, 1983, Eur. Space Agency Spec. Publ., ESA SP-198, 185-190, 1983.

Olsen, R. C., Experiments in charge control at geosynchronous orbit: ATS-5 and ATS-6, J. Spacecraft, 22, 225, 1985.

Olsen, R. C., C. R. Chappell, and J. L. Burch, Aperture plane potential control for thermal ion measurements, J. Geophys. Res., 91, 3117, 1986.

Schmidt, R., and A. Pedersen, Long-term behavior of photon-electron emission from the electric field double probe sensors on GEOS-2, Planet. Space Sci., 35, 61, 1987.

Taylor, G. I., Proc. Royal Soc. (London), A280, 383, 1964.

Whipple, E. C., Potentials of surfaces in space, Rept. Prog. Phys., 44, 1197, 1981.

Whipple, E. C., and R. C. Olsen, High spacecraft potentials on ISEE-1 in sunlight, AGARD symposium, The Hague, The Netherlands, June 1986.

How are coronal mass ejections driven?

DRIVING MECHANISMS FOR CORONAL MASS EJECTIONS

R.S. Steinolfson

Institute for Fusion Studies, The University of Texas at Austin, Austin, TX 78712

Abstract. Mass ejections are new bright features observed in white-light that give the appearance of outward moving material. An acknowledged key ingredient in both the ejection and its progenitor is the magnetic field, although the precise nature of its role, particularily in the driving mechanism, remains unclear. We begin by reviewing analyses of coordinated data sets that establish the relative time sequence and spatial location of individually identified phenomena (such as the flare impulsive phase, eruptive prominence, CME trajectory, etc.) that better define potential drivers. The overwhelming implication from the observations is that a loss of equilibrium or instability in the global magnetic field configuration initiates the subsequent rapid nonlinear evolution that results in the ejection of mass. Some of the models and numerical simulations that have been developed with the intent of determining the physical interactions in the driving mechanism and coronal mass ejection are then considered. Although progress has been made in identifying the primary initiating mechanism, our understanding of the precise physics involved in the nonlinear ejection process is in its infancy. Some future research that may improve our knowledge of this phenomenon through continued analytical work and, perhaps more importantly due to the complex nonlinear processes involved, numerical simulations is also discussed.

Introduction

Coronal mass ejections (CMEs) are observed by coronagraphs as transient changes in the white-light brightness of large-scale structures in the solar corona. Distinguishing features of CMEs are the appearance of new bright features in the coronagraph field of view and temporal changes on time scales of minutes to hours (Hundhausen et al., 1984a). The instruments measure photospheric light that has been Thomson scattered by coronal electrons so bright regions can be interpreted as containing excess mass. During the early analyses of CMEs in the 1970's, they were defined as transient phenomena involving large-scale (> 500 km) motion away from the sun with the understanding that material did not have to escape from the sun to be termed an ejection (Rust et al., 1980). The subclass of CMEs with the largest outward motion (~100 to 1000 km s^{-1}) was referred to as coronal transients. For the purpose of this discussion we will only consider transients in which material appears to escape the solar gravitational field, and these events will be referred to as CMEs.

Since the first observance of CMEs about 15 years ago with the Orbiting Solar Observatory (OSO-7) white-light coronagraph (Tousey, 1973), an appreciable data base has been accumulated with results from three subsequent orbiting instruments. The Skylab coronagraph operated during 1973-1974 and recorded 77 events (Munro et al., 1979). This was followed by the Solwind coronagraph on the P-78 satellite, which obtained in excess of 1200 CME observations during 1979-1985 (Sheeley et al., 1980). The only instrument currently operating is the coronagraph-polarimeter (C/P) on the Solar Maximum Mission (SMM) satellite. It recorded approximately 70 CMEs over a time period of a few months until it failed in 1980, but it has been operational since its repair in 1984. The orbiting coronagraphs individually observe over different portions of the corona varying from a minimum of 1.6 solar radii (C/P on SMM) out to a maximum of 10 solar radii (Solwind). This data set is supplemented by the ground-based High Altitude Observatory K-coronameter at Mauna Loa, Hawaii (Fisher and Poland, 1981), which observes the corona nearer the sun (from 1.2 to 2.0 solar radii), and by the zodiacal light photometers on the Helios spacecraft (Jackson and Leinart, 1985), which detect interplanetary transients.

We will begin in the following section by reviewing coronagraph observations as well as data from other sources in order to determine a typical structure for a CME and the state of the initial atmosphere in which it occurs. Later, additional data (Hα, X-ray) that relate to the possible driving mechanism for CMEs will be considered. Then some of the models of CMEs will be discussed and evaluated with respect to their ability to reproduce characteristic features of the observations. Although our understanding of the entire CME phenomenon has improved substantially since the time of Skylab, there remain several outstanding questions. These will be discussed in the final section along with suggestions for future observations that may assist in resolving them.

CMEs and Helmet Streamers

For the purpose of this paper, we will concentrate on those properties of CMEs determined from coronagraph and related observations that appear to be most useful in identifying the possible driving mechanism. Statistical studies (frequency rate, size, latitude dependence, etc.) based solely on coronagraph measurements will generally not be considered. Furthermore, we will begin by using available data to develop what will be adopted as a generic CME and a typical initial atmosphere through which it propagates. The further discussion will be directed toward this specific subclass of CMEs, although the general results may certainly have wider application.

CMEs are observed to occur in a variety of shapes (Munro et al., 1979; Howard et al., 1984), but those that have the appearance of a radially expanding loop will be emphasized. In the Skylab data set, these comprise about 20% of the total (Sime et al., 1984). Sime et al. (1984) determined that the loop-like CMEs they considered had the following common features, which we adopt as characteristics of our generic CME: (1) The legs or sides of loop ejections are brighter and contain more material than the loop tops. (2) A density depletion forms within the expanding bright loop. (3) Once the loop legs form, they display very little latitudinal motion as the loop tops move radially outward. We will also be primarily concerned with CMEs that appear to originate near the base of, and propagate outward through, pre-existing, well-formed helmet (coronal) streamers. This was the case for all of the loop-like CMEs from Skylab studied by Sime et al. (1984). In addition, the meridional span of all but 1 of 16 SMM CMEs used in the CME onset program (Harrison et al., 1988) included the central position angle of the pre-event streamer (R.A. Harrison, private communication, 1988). It is difficult to determine the pre-event state from Solwind data since this data set ordinarily requires that difference images be used to identify coronal structures.

Coronal streamers are generally believed to lie over polarity inversion (neutral) lines in the line-of-sight magnetic field on the solar surface. Prominences tend to form within the closed-field region of streamers and also lie over the neutral line. Support for the overall picture developed above is provided by the observation that CMEs can frequently be associated with eruptive prominences (Munro et al., 1979; Webb and Hundhausen, 1987). A schematic depicting the pre-event state and the CME is shown in Figure 1. Material flows outward along the open field lines in the ambient atmosphere, while there is not a net outflow from the closed region. The magnetic field in the streamer tends to become more aligned with the prominence as the footpoints approach

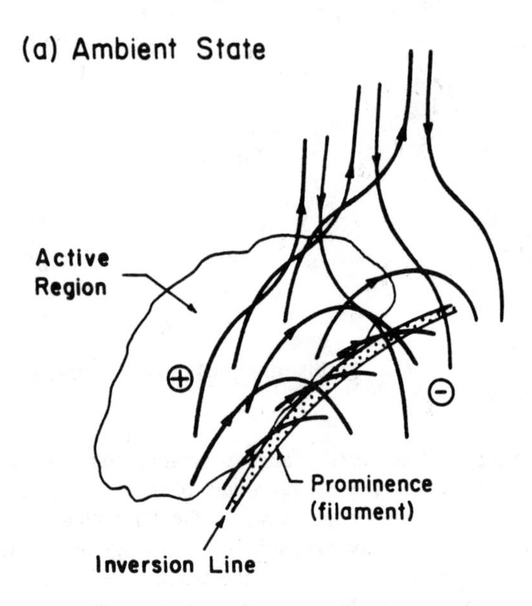

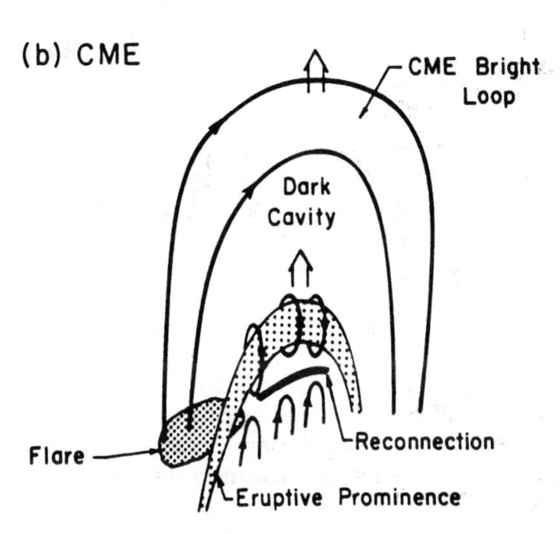

Fig. 1. Schematic representation of (a) a helmet streamer overlying a prominence as may exist in the ambient corona prior to a CME and (b) the disruption of the streamer configuration during a CME.

the inversion line. The initial state, then, contains a highly sheared, nonpotential magnetic field in the low-lying closed-field portion with an overlying open field that may be more nearly potential.

Loop-like CMEs often appear to have a three-part structure consisting of a bright leading loop followed by a dark cavity (the rarefaction region in the Sime et al. study) surrounding a more structured bright inner region, which may or may not appear as a loop (Hundhausen, 1988; Kahler, 1988). An interpretation of this structure is that the preceding bright loop corresponds to compressed ambient corona, the dark cavity is the cavity originally surrounding the prominence in the closed field of the streamer, and the trailing bright portion consists of prominence material. Whether or not the prominence material appears loop-like may be partially determined by the orientation of the event in the plane-of-the-sky. As indicated on Figure 1, the rising prominence probably involves reconnection at some location. Both the flare, if one occurs, and the eruptive prominence may be asymmetric with respect to the CME.

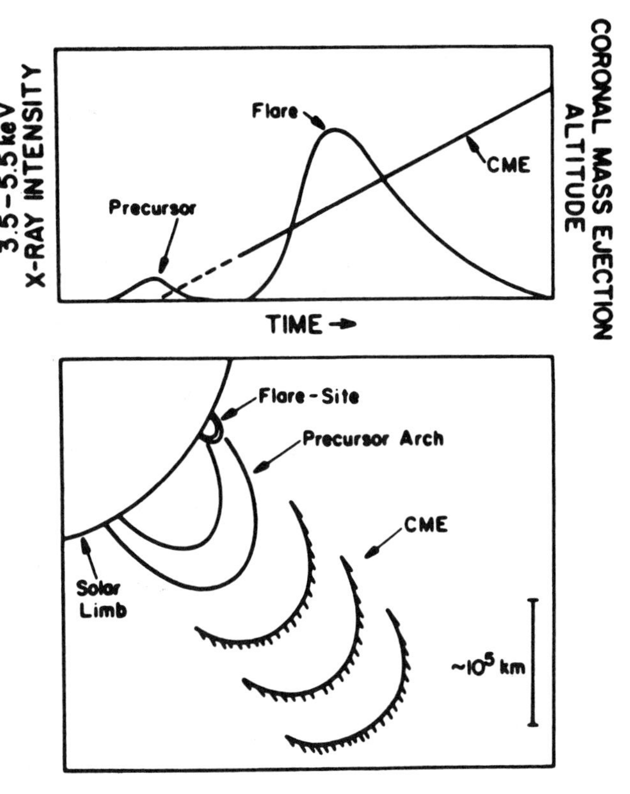

Fig. 2. The relative magnitudes and times of X-ray emission in the precursor and flare impulsive phase are shown on the upper portion along with a typical trajectory of the CME leading edge. The relative spatial location and size of the flare and precursor arch as seen in X-rays and the CME are indicated on the lower part (from Harrison, 1986).

Observations Related to Possible Driving Mechanisms

When studied separately, individual data sets for one particular type of activity (such as Hα, X-ray, line emission, etc.) will not provide a comprehensive view of a phenomenon as complex as one involving a flare impulsive phase, eruptive prominence, and a CME. It is only when several data sets for various forms of activity, and thereby for various physical conditions, are combined does a more complete picture begin to emerge. Unfortunately, the experimentalist seldom has the luxury of having multiple simultaneous data sets for the same event at his disposal. When such information is available, however, the advantages are substantial. Some of the results derived from combined data sets relating to CME drivers are now reviewed.

By combining data from the hard X-ray imaging spectrometer (HXIS) instrument on SMM with coronagraph data, it becomes possible to more clearly define the role of the flare impulsive phase in driving CMEs outward (Simnett and Harrison, 1985; Harrison, 1986). A schematic of the results from this study is shown in Figure 2. The line labeled CME indicates the location of the bright leading edge. If this line is extended back to the surface with no acceleration, the CME onset appears to coincide with a weak precursor some tens of minutes prior to the impulsive flare. In addition to establishing the time sequence of events, the imaging capability of the X-ray instrument provides information on spatial structure. The flare appears to occur at one of the footpoints of the large magnetic arch that brightens in X-rays as the precursor. This study was subsequently extended using data from the X-ray polychromator (XRP) on SMM with similar results (Harrison et al., 1988). Since the CME is well under way before the flare impulsive phase, the flare clearly does not drive the CME. The fact that the flare is asymmetrical with respect to the CME also argues against it being the driving mechanism. As indicated by Harrison (1986), these results seem to support a loss of equilibrium or instability in the large-scale magnetic configuration as the initiating agent.

Another useful data combination was considered by Kahler et al. (1988). They identified the filament location from Hα movies taken at Big Bear Observatory and determined the flare impulsive phase from hard X-ray emission observed by instruments on the ISEE 3 spacecraft. Each of the four events they studied was similar to the one shown in Figure 3. As seen in the figure, the filament eruption begins several minutes before the impulsive phase, and there is no appreciable change in the filament motion during the flare. Kahler et al. (1988) concluded that the evidence indicates the filament eruption is not driven by any pressure pulse associated with the impulsive phase. They suggested that both the filament eruption and the flare are driven by an instability in the global magnetic field configuration. They also found that the flare was initiated when the outward motion of the filament exceeded a threshold of approximately 100 km sec^{-1}. The apparent existence of such a universal threshold certainly requires further study with data from other events to establish that it is not event-dependent and, in fact, to provide more convincing evidence that it is not just coincidental. The concept of the relatively passive role of the impulsive flare in driving transients, however, corroborates the preceding studies by Harrison and others.

There have also been several studies involving the use of coronagraph and Hα data to determine the relative outward motion of CMEs and erupting filaments (e.g., Illing and Hundhausen, 1986; Wagner et al., 1983; Fisher et al., 1981).

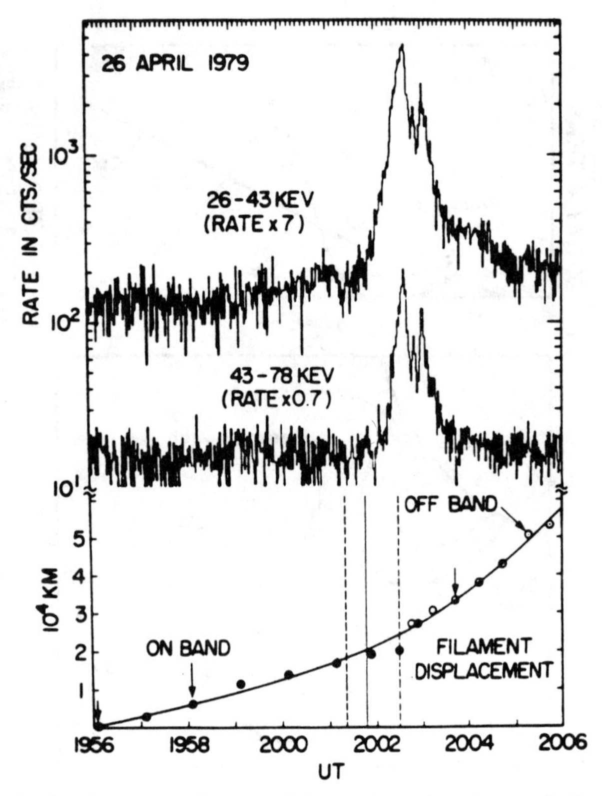

Fig. 3. X-ray counting rates in two energy increments during the April 26,1979, flare are shown on the top graphs. The circles on the bottom portion locate relative filament displacements. The solid vertical line identifies the X-ray onset time of the impulsive phase, while the dashed lines show the earliest and latest possible onset times for the Hα flash phase (from Kahler et al., 1988).

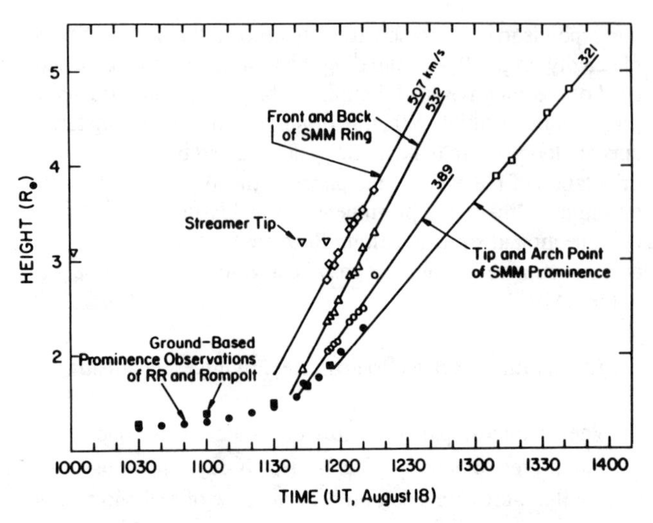

Fig. 4. Height of several features as a function of time during the prominence eruption and CME. Velocities are given next to the corresponding best fit line (from Illing and Hundhausen, 1986).

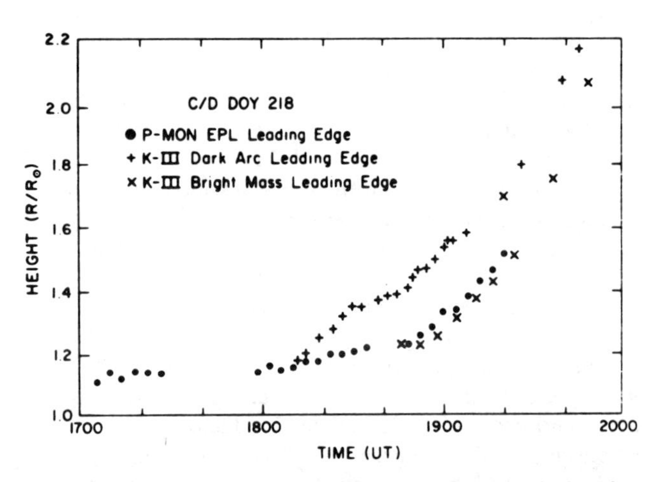

Fig. 5. Height vs. time for three features observed during the August 5, 1980 event (from Fisher et al., 1981).

A representative result is reproduced here as Figure 4 from Illing and Hundhausen (1986). The linearly extrapolated (no acceleration) onset time for the CME is approximately coincident with the time at which the prominence begins moving upward more rapidly. It is also apparent that the CME velocity exceeds that of the eruptive prominence.

The CME location in Figure 4 was taken from SMM C/P observations, and hence its trajectory below the occulting disk cannot be determined directly. The K-coronameter, however, observes down to about 1.2 solar radii and, thereby, has the potential of seeing earlier stages of the CME development and ejection. The trajectory of a CME from the K-coronameter is compared to that of the corresponding eruptive filament in Figure 5 from Fisher et al. (1981). These results suggest that the CME begins its eruption prior to the filament. In this particular case the CME began as a dark arc, indicating a depletion of material, without a preceding bright region. The characteristic bright loop did appear ahead of the darker cavity some tens of minutes later. This example may be somewhat abnormal because, in contrast to this example, the bright loop generally forms immediately (E. Hildner, private communication, 1988). It does, however, display the usual tendency for the CME to lift off prior to the filament. Once again, this behavior appears more consistent with an unstable magnetic configuration than an explosive, thermal-type driver. In fact, Low et al. (1982) were able to model the motion as being due to a buoyant loop system.

To summarize the results discussed above, the observations do not support the flare impulsive phase as the CME driving mechanism primarily because the CME is already under way at the time of the flare and the flare is asymmetrical with respect to the CME. In addition, CMEs often occur without an associated flare impulsive phase. Eruptive prominences or filaments also do not seem to be likely candidates for driving CMEs. This view is substantiated by the observations that the CME onset appears to precede that of the eruptive prominence and the CME travels much faster. The observations do suggest that at least the initiating agent for the driver may be a loss of equilibrium or instability in the large-scale (global) magnetic field configuration. In this scenario, the CME, flare impulsive phase, and eruptive prominence are all secondary effects resulting from the nonlinear evolution in the ensuing unstable state. The nature of the physical processes in this nonlinear stage and the manner in which they produce the observable consequences of the initiating instability remain, for the most part, unresolved.

Models: Theory and Numerical Simulation

The ultimate test of any proposed CME driving mechanism is a direct comparison of the CME produced by the suggested driver with the coronagraph observations discussed earlier. The driver must, of course, also be consistent with the observational results of the previous section. Some recent reviews that discuss theories for CME drivers have been given by Hundhausen et al. (1984b), Steinolfson (1985), Rosner et al. (1986), and Kahler (1987).

A common ingredient of all models of CME drivers is the magnetic field. Some models only include the field in the initial state in which case it may actually act to retard the ejection. In others, however, the field is directly involved in the driving mechanism.

The model that has received the most extensive study with numerical simulations incorporates a localized thermal pressure rise as the driver. These models originated about 10 years ago during the analysis of Skylab data (Steinolfson et al., 1978) and were based on the assumption that thermal energy release during the flare impulsive phase may act as a driver. If the impulsive phase involves energy conversion by reconnection in the nonlinear tearing instability, it has been shown for a particular model that the majority of converted magnetic energy would indeed go into thermal energy (Steinolfson and Van Hoven, 1984). However, this type of thermal driver is in sharp disagreement with the observational results of the previous section that indicate a significant time lapse between the start of the outward progression of the CME and the following flare impulsive phase.

Models using a thermal driver have been able to demonstrate the significant influence that the physical conditions (magnetic configuration, flow velocity, thermodynamics) in the ambient corona have on the CME. For instance, it was shown by Sime et al. (1984) that the CMEs produced in the Dryer et al. (1979) simulations, which assumed a static atmosphere with an open field, clearly did not display any of the observed CME characteristics indicated in the section on CMEs and helmet streamers. Furthermore, Steinolfson and Hundhausen (1988) show that the observed features are similarly not reproduced by thermal pulses in a static atmosphere in a closed field, in agreement with a similar study by Hildner et al. (1986). Steinolfson and Hundhausen also considered an initial helmet streamer configuration in a polytropic atmosphere in which case all CME characteristics were duplicated with the exception that the loop tops were brighter than the legs. With the addition of a coronal heating term in an initial streamer (Steinolfson, 1988), their simulation reproduced all the observed features. A comparison between simulated and observed brightness from the Steinolfson and Hundhausen study is shown in Figure 6. These are actually brightness difference values obtained by subtracting off the brightness in the pre-event state. The dashed circle locates the edge of the occulting disk and the shaded region represents a rarefaction. The time increment between the simulations in the right column is the same as for the observations. Note that the simulated rarefaction is beginning to break up similar to that observed. Consequently, we have a situation in which an obviously inadequate driving mechanism does produce a CME that displays the primary observed features, for this class of loop-like CMEs, as given by Sime et al. (1984).

There is one very important observation that this latter simulation (with the coronal heating term) does not and, in fact, cannot replicate. That is, it does not have the bright inner structure believed to be produced by prominence material, as discussed in the section on CMEs and helmet streamers. An obvious suggestion here is that the state of the ambient corona plays the major role in determining the shape of the outer bright loop and following rarefaction in the CME. On the other hand, the physical nature of the driving mechanism and the ambient conditions within the prominence cavity and the prominence itself may be more important in dictating the structure of the inner bright region as well as the longer-term coronal evolution.

Several other methods that are by nature one-dimensional and concentrate on the driver while neglecting interactions with the ambient atmosphere have also been considered. These models assume that the driving mechanism forms an integral part of the CME and propagates with it at all times. Such assumptions reduce the mathematical complexity and allow a study using a relatively simple force balance in the radial direction. Radially unbalanced magnetic pressure within the CME loop provides the driver in the Mouschovias and Poland (1978) model, while self-induced magnetic forces internal to the loop were used by Anzer (1978). Models by Pneuman (1980) and Anzer and Pneuman (1982) drive the loop by increasing magnetic pressure beneath it. The neglect of thermal pressure forces in these models has been questioned by Yeh and Dryer (1981), and obviously these would play a role in a more realistic treatment. Since all of these models have only been studied in one dimension, it is meaningless to question how well they could reproduce

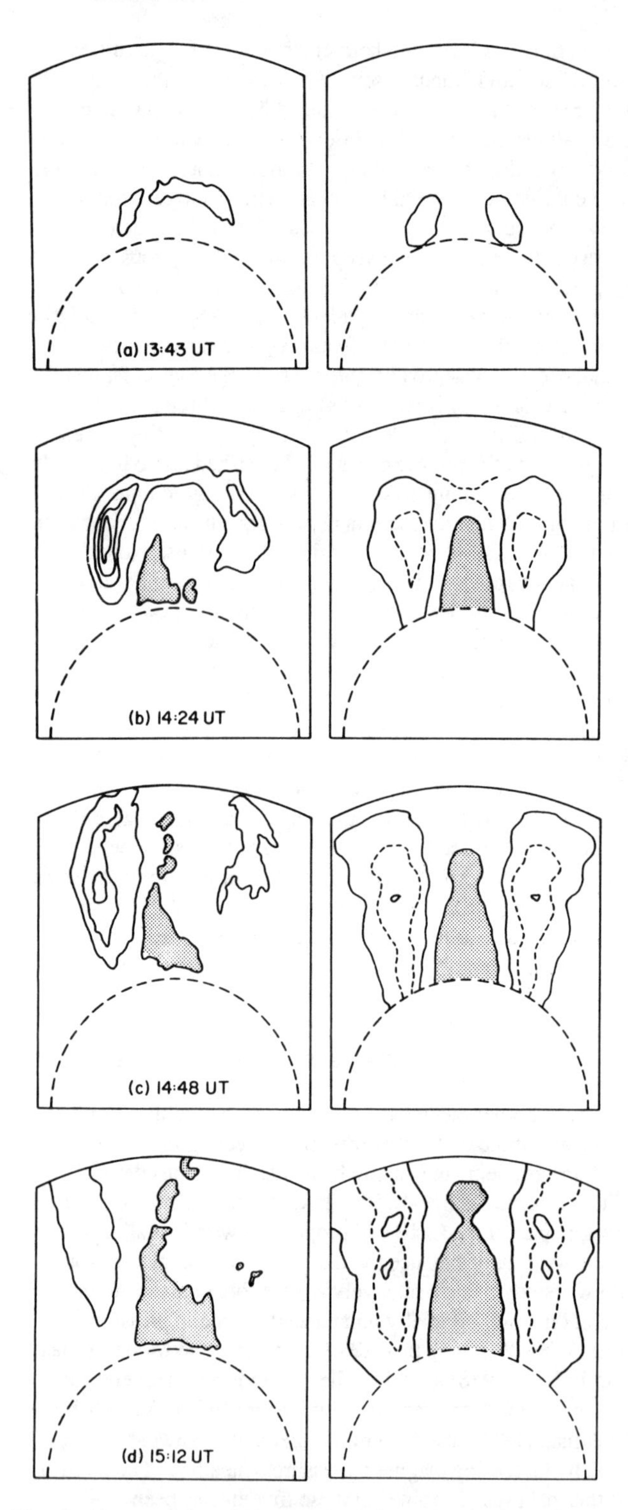

Fig. 6. Comparison of observations from the August 10, 1973, event in the left column with simulation results in the right column. Solid contours enclose an increase in brightness from that in the pre-event corona, while the shaded areas denote a brightness reduction. The outer edge of the occulting disk for the ATM coronagraph is located by the dashed circle (from Steinolfson and Hundhausen, 1988).

coronagraph observations. The only real accomplishment of such models is the demonstration that radially unbalanced magnetic pressures and forces can, for not unreasonable parametric values, produce outward motion of the magnitude observed for some CMEs.

A remaining attractive possibility is that a coronal field configuration may evolve into a highly stressed, nonpotential state as a result of photospheric motion of the field line footpoints. With continued photospheric motion the entire field configuration may reach a state beyond which it either no longer has an equilibrium solution (Low, 1981) or begins evolving much more rapidly (Birn and Schindler, 1981). Theoretical aspects of such unstable configurations are reviewed by B.C. Low (1989).

An example demonstrating how small photospheric changes can produce respectively larger changes in the corona is given in Figure 7 from Low (1981). This behavior was produced by moving the field line footpoints apart. The same four field lines are indicated in each frame, and they all separate equal magnetic flux. This approach assumes quasi-static evolution so the plasma motion that would be induced by the field line expansion is neglected.

The earlier discussion of the observations certainly suggests that the entire flare process may be initiated by a loss of equilibrium in the large-scale field. The subsequent nonlinear driver would then derive its energy from that available in the sheared fields. It must be recognized here that although there may be increasing evidence that a loss of equilibrium is the initiator, there still remains the question of how the sheared magnetic energy actually drives the CME outward. This important issue is just beginning to be addressed using numerical simulations.

The coupling between the magnetic field and velocity in a magnetic arcade in an initially static atmosphere as a result of shear photospheric motion parallel to the arcade axis has been investigated by Mikic et al. (1988). Their analysis is restricted to incompressible fluids with low plasma beta and with no coupling to the thermodynamics. The gravitational force is also neglected. The important new physics here, which is neglected in quasi-static studies, is the interaction between the magnetic field and dynamics mentioned above and the ability to self-consistently simulate the energy build-up phase, as well as the more rapid evolution when some measure of the critical shear is exceeded. The results are shown in Figure 8 for two views initially and at several times during the more rapid phase. The evolution is essentially independent of the z-direction (perpendicular to the plane in the side view). Although the results appear promising, they have the major restriction that symmetry conditions were applied in the y-direction. The several approximations here do not allow a meaningful comparison with coronagraph observations.

Outstanding Problems

Most of the available evidence suggests that the progenitor of the complex, multi-faceted solar phenomena, collectively

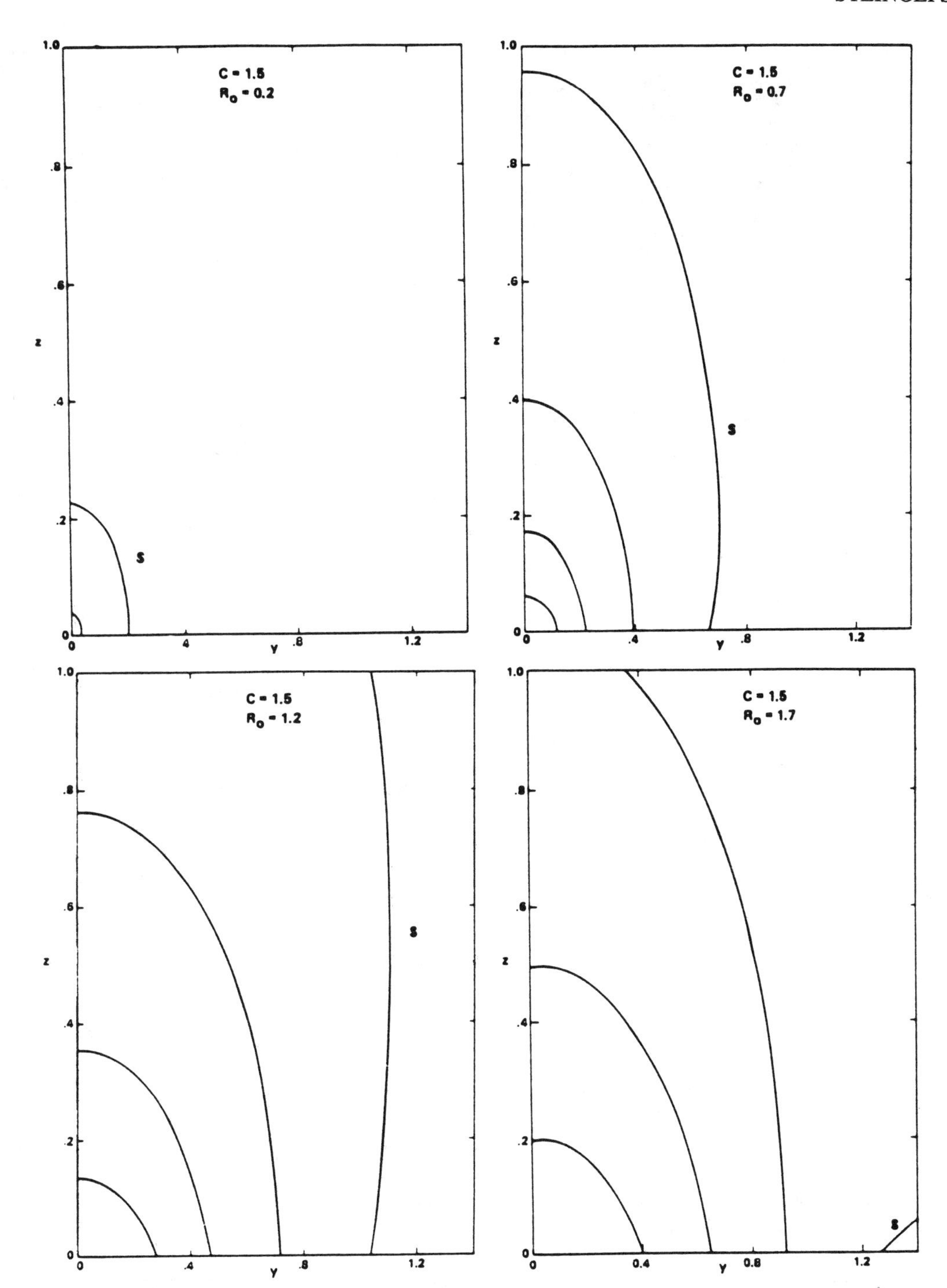

Fig. 7. Change in the coronal field produced by moving the field line footpoints apart (from Low, 1981).

referred to as a two-ribbon flare and often involving an eruptive prominence, flare impulsive phase, CME, and post-flare loops, is a loss of equilibrium or instability of the large-scale or global magnetic field configuration. This realization by itself does not specifically implicate one particular mechanism or physical interaction as the driving mechanism for that phase of the phenomenon in which a CME is propelled outward. However, as discussed above, it does not appear that either the flare impulsive phase or the erupting prominence provides the driver. A convenient simplification here would be to attribute the driver to nonlinear interactions during the time period in which the unstable field seeks out a

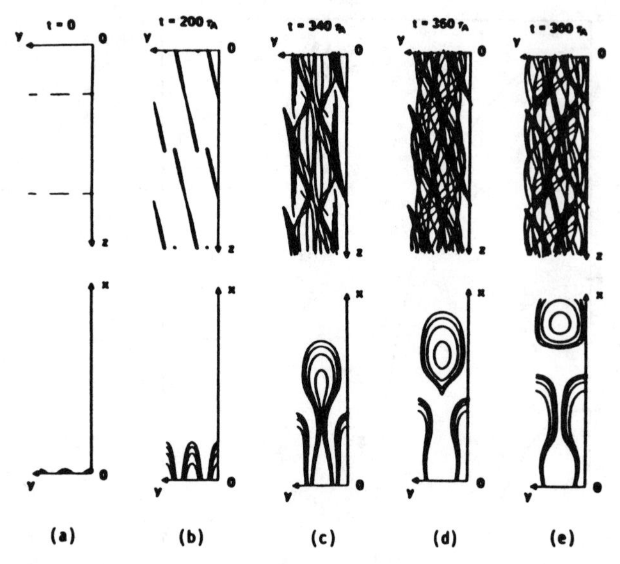

Fig. 8. Projections of field lines in the y-z plane (arcade top view) and in the x-y plane (arcade front view) initially and at several times during the more rapid phase in the evolution (from Mikic et al., 1988).

new equilibrium and let it go at that. Although this is most probably the case, one would like to know the precise nature of the driver and, more specifically, how an unstable magnetic configuration at a given level of magnetic energy evolves to a stable configuration at a lower level of nonpotential magnetic energy. Perhaps equally important is the determination of how the field loses equilibrium, and whether it occurs in the primarily open-field region of the streamer or in the underlying closed field, which tends to be more sheared and, hence, more nonpotential.

The source of energy for driving CMEs has been identified by Moore (1988) with that released by a filament as it rises, expands, and untwists. For this to be the case, one must then explain how the slower erupting prominence transfers energy to a more rapidly moving CME. In addition, not all CMEs appear to be associated with an observed erupting filament. It is certainly possible that the sheared filament field is always present even if the thermodynamics are not such as to produce the characteristic filament Hα signature. The precise nature of the role of the prominence and its surrounding cavity should be better defined.

The emergence of new magnetic flux into an active region has often been invoked as a possible triggering mechanism and as initiating an energy source when the emerging field reconnects with the ambient overlying magnetic structure (Canfield et al., 1974). One process by which new flux may emerge is as a result of the nonlinear Parker instability (Shibata et al., 1988). In this process, magnetic buoyancy has a significant effect on the emerging flux. The importance of such buoyancy in the continued evolution of the emergent flux and also in the relaxation of a field configuration driven beyond a stable limit needs to be further clarified.

The reconnection of magnetic fields provides one obvious process by which the field energy can be converted to other forms. Just where the possible reconnection occurs and how the converted energy acts to drive the CME are poorly understood at present.

A primary result of theoretical research relating to CME driving mechanisms has been to demonstrate that global field configurations can be driven by photospheric motion to the extent that they lose equilibrium or become unstable. This has been shown using both analytic theory and numerical simulations. Although analytic theory will continue to be used to study field equilibrium, the complexity of the subsequent nonlinear evolution dictates that major theoretical progress in improving our understanding of CME drivers will come largely from continued use of numerical simulations.

The observations proposed by Dennis et al. (1987) for the Max '91 program would certainly improve our knowledge of the entire flare-eruptive prominence phenomenon and, in particular, should assist in better defining the CME driver. A major emphasis of this observational program of crucial importance to identifying potential driving mechanisms is the improved spatial and temporal resolution. The important role of the magnetic field also indicates the benefits of more sensitive vector magnetograph observations.

Acknowledgement. This research was supported by NASA with an SMM Guest Investigator Grant.

References

Anzer, U., Can coronal loop transients be driven magnetically?, *Solar Phys.*, *57*, 111, 1978.

Anzer, U., and G.W. Pneuman, Magnetic reconnection and coronal transients, *Solar Phys.*, *79*, 129, 1982.

Birn, J., and K. Schindler, Two ribbon flares: Magnetostatic equilibria, in *Solar Flare Magnetohydrodynamics*, edited by E.R. Priest, pp. 337-338, Gordon and Breach, New York, 337-378, 1981.

Canfield, R.C., E.R. Priest, and D.M. Rust, A model for the solar flare, in *Flare-Related Magnetic Field Dynamics*, edited by Y. Nakagawa and D.M. Rust, NCAR, Boulder, Colorado, p. 361, 1974.

Dennis, B., R. Canfield, M. Bruner, G. Emslie, E. Hildner, H. Hudson, G. Hurford, R. Lin, R. Novick, and T. Tarbell, *Max '91 Flare Research at the Next Solar Maximum*, 1987.

Dryer, M., S.T. Wu, R.S. Steinolfson, and R.M. Wilson, Magnetohydrodynamic models of coronal transients in the meridional plane. II. Simulation of the coronal transient of 1973 August 21, *Ap. J.*, *227,* 1059, 1979.

Fisher, R., C.J. Garcia, and P. Seagraves, On the coronal transient-eruptive prominence of 1980 August 5, *Ap. J. Lett.*, *246*, L161, 1981.

Fisher, R.R., and A.I. Poland, Coronal activity below 2 R_o: 1980 February 15-17, *Ap. J.*, *246*, 1004, 1981.

Harrison, R.A., Solar coronal mass ejections and flares, *Astron. Astrophys., 162*, 283, 1986.

Harrison, R.A., E. Hildner, A.J. Hundhausen, G.M. Simnett, and D.G. Sime, The launch of coronal mass ejections: Results from the coronal mass ejection onset program, *J. Geophys. Res.*, submitted, 1988.

Hildner, E., et al., Coronal mass ejections and coronal structures, Chapter 6 in *Energetic Phenomena on the Sun*, edited by M. Kundu and B. Woodgate, NASA Conf. Publ. 2439, 1986.

Howard, R.A., N.R. Sheeley, Jr., M.J. Koomen, and D.J. Michels, The statistical properties of coronal mass ejections during 1979-1981, *Adv. Space Res., 4*, 307, 1984.

Hundhausen, A.J., The origin and propagation of coronal mass ejections, in *Proceedings of the Sixth International Solar Wind Conference*, edited by V.J. Pizzo, T.E. Holzer and D.G. Sime, pp. 181-214, NCAR/TN-306, 1988.

Hundhausen, A.J., C.B. Sawyer, L. House, R.M.E. Illing, and W.J. Wagner, Coronal mass ejections observed during the solar maximum mission: latitude distribution and rate of occurrence, *J. Geophys. Res., 89*, 2639, 1984a.

Hundhausen, A.J., et al., Coronal transients and their interplanetary effects, in *Solar Terrestrial Physics: Present and Future*, edited by D.M. Butler and K. Papadopoulos, p. 6-1, NASA Ref. Publ. 1120, 1984b.

Illing, R.M.E., and A.J. Hundhausen, Disruption of a coronal streamer by an eruptive prominence and coronal mass ejection, *J. Geophys. Res., 91*, 10,951, 1986.

Jackson, B.V., and C. Leinert, Helios images of solar mass ejections, *J. Geophys. Res., 90*, 10,759, 1985.

Kahler, S., Coronal mass ejections, *Rev. Geophys., 25*, 663, 1987.

Kahler, S., Observations of coronal mass ejections near the sun, in *Proceedings of the Sixth International Solar Wind Conference*, edited by V.J. Pizzo, T.E. Holzer and D. Sime, pp. 215-231, NCAR/TN-306, 1988.

Kahler, S.W., R.L. Moore, S.R. Kane, and H. Zirin, Filament eruptions and the impulsive phase of solar flares, *Ap. J., 328*, 824, 1988

Low, B.C., Eruptive solar magnetic fields, *Ap. J., 251*, 352, 1981.

Low, B.C., Magnetic free-energy in the solar atmosphere, this volume, 1989.

Low, B.C., R.H. Munro, and R.R. Fisher, The initiation of a coronal transient, *Ap. J., 254*, 335, 1982.

Mikic, Z., D.C. Barnes, and D.D. Schnack, Dynamical evolution of a solar magnetic field arcade, *Ap. J., 328*, 830, 1988.

Moore, R.L., Evidence that magnetic energy shedding in solar filament eruptions is the drive in accompanying flares and coronal mass ejections, *Ap. J., 324*, 1132, 1988.

Mouschovias, T. Ch., and A.I. Poland, Expansion and broadening of coronal loop transients: A theoretical explanation, *Ap. J., 220*, 675, 1978.

Munro, R.H., J.T. Gosling, E. Hildner, R.M. MacQueen, A.I. Poland, and C.L. Ross, The association of coronal mass ejection transients with other forms of solar activity, *Solar Phys., 61*, 201, 1979.

Pneuman, G.W., Eruptive prominences and coronal transients, *Solar Phys., 65*, 369, 1980.

Rosner, R., B.C. Low, and T.E. Holzer, Physical processes in the solar corona, in *Physics of the Sun Volume II: The Solar Atmosphere*, edited by P.A. Sturrock, T.E. Holzer, D.M. Mihalas, and R.K. Ulrich, p. 135, D. Reidel, Boston, 1986.

Rust, D.M., and 12 co-authors, Mass Ejections, *Solar Flares*, edited by P.A. Sturrock, p. 273, Colorado Assoc. Univ. Press, Boulder, 1980.

Sheeley, N.R., Jr., R.A. Howard, D.J. Michels, and M.J. Koomen, Solar observations with a new earth-orbiting coronagraph, in *Solar and Interplanetary Dynamics*, IAU Symp. No. 91, edited by M. Dryer and E. Tandberg-Hanssen, p. 55, D. Reidel, Dordrecht, Holland, 1980.

Shibata, K., T. Tajima, R. Matsumoto, T. Horiuchi, T. Hanawa, R. Rosner, and Y. Uchida, Nonlinear Parker instability of isolated magnetic flux in a plasma, *Ap. J.*, submitted, 1988.

Sime, D.G., R.M. MacQueen, and A.J. Hundhausen, Density distribution in loop-like coronal transients: A comparison of observations and a theoretical model, *J. Geophys. Res., 89*, 2113, 1984.

Simnett, G.M., and R.A. Harrison, The onset of coronal mass ejections, *Solar Phys., 99*, 291, 1985.

Steinolfson, R.S., Density and white-light brightness in coronal mass ejections: Importance of the pre-event atmosphere, *J. Geophys. Res.*, in press, 1988.

Steinolfson, R.S., Theories of shock formation in the solar atmosphere, in *Collisionless Shocks in the Heliosphere: Reviews of Current Research*, Am. Geophys. Monogr. Ser., vol. 34, p. 1, edited by B.T. Tsurutani and R.G. Stone, AGU, Washington, 1985.

Steinolfson, R.S., and A.J. Hundhausen, Density and white-light brightness in loop-like coronal mass ejections: Temporal evolution, *J. Geophys. Res.*, in press, 1988.

Steinolfson, R.S., and G. Van Hoven, Nonlinear evolution of the resistive tearing mode, *Phys.Fluids, 27*, 1207, 1984.

Steinolfson, R.S., S.T. Wu, M. Dryer, and E. Tandberg-Hanssen, Magnetohydrodynamic models of coronal transients in the meridional plane, I, The effect of the magnetic field, *Ap. J., 255*, 259, 1978.

Tousey, R., The solar corona, in *Space Research XIII*, edited by M.J. Rycroft and S.K. Runcorn, p. 713, Akademie-Verlage, Berlin, 1973.

Wagner, W.J., R.M.E. Illing, C.B. Sawyer, L.L. House, N.R. Sheeley, Jr., R.A. Howard, M.J. Koomen, D.J. Michels, R.N. Smartt, and M. Dryer, A white-light/Fe X/Hα coronal transient observation to 10 solar radii, *Solar Phys., 83*, 153, 1983.

Webb, D.F., and A.J. Hundhausen, Activity associated with the solar origin of coronal mass ejections, *Solar Phys., 108*, 383, 1987.

Yeh, T., and M. Dryer, Effects of self-induced magnetic force in a coronal loop transient, *Ap. J., 245*, 704, 1981.

CME AND SOLAR WIND STUDIES USING GOES SOLAR X-RAY IMAGERS AND SOHO REMOTE SENSING

W. J. Wagner

Space Environment Laboratory, NOAA/Environmental Research Laboratories, Boulder, CO 80303

Abstract. All inputs to the heliosphere and to planetary magnetospheres and ionospheres are generated in the plasma which constitutes the outer layers of the solar atmosphere. Although the NASA Solar Maximum Mission satellite is expected to re-enter the atmosphere this year, the outlook for observational studies of the solar corona in the mid-1990's looks optimistic. A variety of new instruments will be available for investigations of solar plasma physics processes. Of preeminent importance is the NASA-ESA SOHO satellite, to be stationed at the L1 libration point from 1995. I describe SOHO EUV spectrographic and coronagraphic capabilities for use in identifying these inputs to the magnetosphere as they leave the transition region and high corona. To these data will be added views of launches of coronal mass ejections and the unambiguous locations of soft X-ray coronal holes and magnetic structure in the lower corona. The latter will be recorded by the solar X-ray imagers on the NOAA GOES satellites beginning in 1992 or 1993. I also review the ground-based observational capabilities which are expected to be upgraded in time for the solar cycle maximum campaigns beginning in 1991. The tailoring of the above sensor array to the outstanding problems in mass ejection and coronal research is noted.

Introduction

Space observations of heliospheric phenomena over the past 15 years have provided many surprises, among these being mass ejections in the solar wind, unanticipated planetary magnetospheric X-ray coronal holes, interplanetary magnetic clouds, etc. With a reconnaissance of the heliosphere now accomplished, progress in the next decade will best be made by coordinated, mutually supportive campaigns which will investigate plasma processes found to connect one discipline to another. Shocks launched from the Sun or the outward tracing of surface magnetic fields provide a good example.

The ambitions of humans and their increasingly sophisticated space systems have evolved to the extent that their further economic progress has become ever more dependent on answers from space scientists. NASA engineering and flight offices will more and more find their goals stymied and their schedules disrupted by environmental problems such as premature spacecraft reentry, hazardous solar particle radiation, or satellite anomalies. These are phenomena under the purview of the NASA space science offices. Thus the loop has closed so that today's scientific progress helps to bootstrap along man's facility for work in space.

In the next section, I suggest how progress in coronal mass ejection and solar wind research will look during the coming decade—how answers to specific problems will impact the other solar fields, stellar and galactic astronomy, and the successful operation of NASA itself.

An outline is provided in the third section showing the anticipated programs of the 1990's in coronal and solar wind research. I conclude that the future looks bright if we are ready to support and utilize those programs now on the horizon.

CME and Solar Wind Programs

The programs and missions anticipated for the next decade in coronal mass ejection and solar wind research (discussed in the next section) will pay off in a number of ways. These include both scientific progress in enhancing man's knowledge and also practical benefits in accomplishing national goals.

Open Questions on Coronal Mass Ejections

The Solar and Heliospheric Observatory (SOHO) mission and the GOES solar X-ray imager (SXI) will increase our understanding of coronal mass ejections (CMEs) and processes occurring in the solar wind. Outstanding CME problems have been reviewed by numerous authors (see references in Wagner, 1988). Steinolfson 1989 noted some of these questions and discussed perhaps the most important one, the driving mechanism for CMEs. The anticipated space science missions, especially with full time and complete spatial coverage of X-ray long-duration events and the passage of CMEs through the outer corona, should be able to confirm the sequence of events involving CME initiation. Close coordination with ground-based observatories will be crucial for completing the data sets with H-alpha and visible coronal line records, thus adding a precision to initiation studies not heretofore available.

What is the role of or the requirement for cool prominence material in CME initiation? Do CMEs occur by virtue of processes involving the larger filament cavity only, or must suspended material be present to lead to nonequilibrium? As never before, the morphology of coronal magnetic fields will be visible to the SOHO instruments, upgraded ground-based visible and white-light coronagraphs, and to the SXI.

Related to the initiation and driver problem is the question as to whether CMEs are disconnected from the Sun or instead still remain attached with mass outflow continuing in time. The latter is apparently observed by the photometers on the Helios solar probes. Reconnection seems to be more a consequence of the CME than a cause--if it occurs at all. But without reconnection, should we not expect to see a buildup of interplanetary magnetic flux? Magnetic clouds and bidirectional particle streaming are not regarded today as conclusive evidence for disconnected plasmoids.

We do not understand the role of mass ejections in the acceleration of solar cosmic rays at the Sun or in their shepherding at 1 AU by the sweeping action of CMEs. Like many of the problems discussed here, complete observational coverage in several energy regimes from GOES and SOHO will contribute answers.

The generic signatures of CMEs in the solar wind are still not clear. A variety of effects are seen; how many are the result of misidentifications of CME occurrences is not known. The phenomenon of magnetic field lines draping around an interplanetary CME should be confirmed and any possible influence of this on the CME or on its terrestrial effects should be clarified. As yet, the expected slow-mode shocks have not been revealed to be yet in association with CMEs, but the large 30 solar radii field-of-view of the SOHO coronagraph may show surprises.

SOHO and the NOAA interplanetary scintillation network (in collaboration with U.K. and Indian colleagues) may confirm that a steady state solar wind structure at 1 AU is largely precluded by the busy prior history of CME events. The GOES SXI will provide vital input to efforts in modeling the changing interplanetary medium.

The efficacy of CMEs as generators of geomagnetic storms and their consequent thermospheric heating may well be determinable by an in situ spacecraft at L1 with plasma sensors and coronagraphs onboard. At the present time, however, it is not clear that such measurements will be made from SOHO.

Other Coronal and Solar Wind Problems

In addition to the unsolved problems of the coronal mass ejection phenomenon, SOHO and the GOES solar X-ray imager will help address other aspects of the outer solar atmosphere and the solar wind. With observations in soft X-rays, the series of SXIs will provide a rather complete accounting of the X-ray bright points as they wax and wane over the course of the solar activity cycle. Their apparent anti-correlation with sunspot number will be easily confirmed and their relation to the appearance and disappearance of magnetic flux can be clarified.

This conference agenda does not stress solar wind research to any great degree. The European Cluster mission, a sister to SOHO under the Solar Terrestrial Science Programme, and WIND, a NASA component of the Global Geospace Science Mission (previously known as the International Solar Terrestrial Program), both concentrate on the solar wind. Therefore, I shall not dwell long on SOHO and SXI contributions to the study of the solar wind per se. On the other hand, there does seem to be considerable interest here in the interaction of the solar wind and its disturbances (CMEs, high-speed streams from X-ray coronal holes) with planetary magnetospheres, ionospheres, and comet tails. Hence, my emphasis is in the subsection above, on CME studies.

It may be sufficient to note that one of the prime purposes of the SOHO mission is the study of the fundamental physical processes in the solar corona and the resulting solar wind. For the SXI, it is to monitor and understand the generation of disturbances in the corona and their transmittal to Earth. The most basic input to models of the solar wind is its magnetic structure and initial velocity fields, both visible at sub-Earth longitudes in X-rays. In these respects, the SOHO and SXI will clearly support the other above-mentioned ESA and NASA programs.

The Solar Interior

Coronal mass ejection and solar wind studies also hold insight for other solar specialties. The long-enduring, large-scale structure of the corona and interplanetary medium is commonly believed to be related to processes in the Sun's interior. In some aspects, for example, the double cycle maxima of the Fe XIV green line intensity and possibly the existence of leading indicators of the next solar cycle (e.g., polar plume angles), the corona may be a better indicator of internal dynamics than are surface features or measurable photospheric fields.

When the internal pattern changes episodically in the course of the solar cycle, surface magnetic fields become unsteady; such an epoch is probably marked by changes in the CME frequency and distribution. A few hints already exist (Golub and Vaiana, 1980; Wagner and Wagner, 1984) of a deep-rooted global synchronicity of flux emergence and CME occurrences.

The number of waves in the "ballerina skirt" shape of the heliospheric current sheet is most easily observed in coronal structure and in situ in the solar wind. The current sheet and sector boundaries apparently are also revealing processes deep within the Sun.

Finally, the large-scale X-ray patterns known as coronal holes are believed to reflect circulation modes occurring in the solar interior. So we see that the corona, mass ejections, and the solar wind all can function as indicators of deep-seated processes. Programs of their study will indeed aid our understanding of the solar interior.

Guidance on Stellar and Galactic Processes

The solar-stellar connection of CME's may be dramatized by showing Figure 1, from Dreher and Feigelson (1984), a cosmic ejection of mass. Most interpretations of the engines for such galactic jets involve processes beyond those familiar to solar physics. However, the MHD laws governing the outward plasma flows, the shocks propagated with their interactions and accompanying wave systems, and probably some forms of energetic particle acceleration are similar to the coronal and heliospheric mass ejection phenomena.

To date, astronomers have not enjoyed the luxury of obtaining long time series of X-ray emissions from stellar and cosmic sources. X-ray variability of stars is a field in its infancy, but this will rapidly change with the completion of astronomical X-ray reconnaissance missions. The Roentgen Satellite (ROSAT), the Gamma Ray Observatory (GRO), and the Advanced X-ray Astrophysics Facility (AXAF) satellites will allow a search for cosmic CMEs as long-duration events in X-rays. As in prior studies of stellar chromospheres, activity cycles, and magnetism, solar physicists with their knowledge of CMEs and solar wind behavior will serve to guide and inform the stellar community. The new findings from observational programs involving SOHO, SXI, and the ground-based facilities should be of great interest to stellar astronomers.

A common technique employed in experimental physics is to perturb a steady state system in the laboratory and then monitor the resulting time-dependent behavior. For astronomers, this ave-

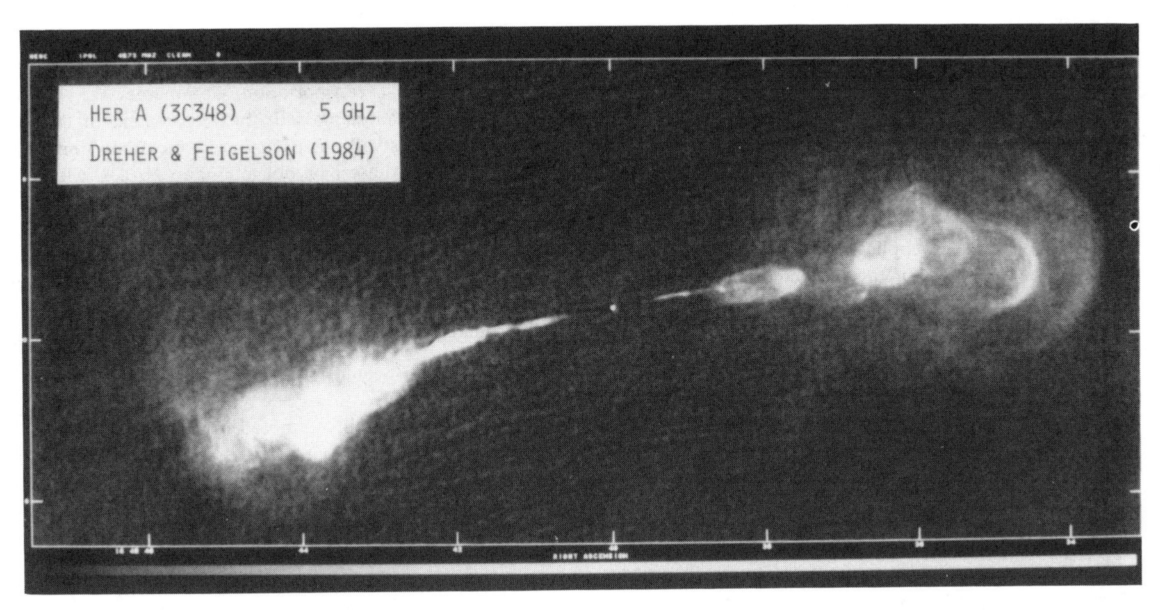

Fig. 1. Cosmic mass ejection from the bright extragalactic radio source Hercules A (3C348) seen in high-resolution mapping at 5 GHz.

nue of investigation is generally precluded. Luckily, our closest star cooperates by performing this experiment for us almost daily. With ingenuity in perceiving the characteristics of the input and with persistence in the provision of monitors (remote sensing from the ground or from space, in situ "weather stations" at the L1 libration point such as is SOHO), we can treat transient CMEs as probes of coronal structure and the solar wind. When CME behavior is understood, the solar corona and interplanetary medium will also have been clarified; these results will be pertinent to the interstellar and intergalactic media as well.

Unrecognized Relevance to NASA Engineering Codes

An asset of NASA which seems unrecognized by the agency lies in its in-house Space Science and Applications Office (Code E). The flight and engineering codes have not been made to fully appreciate just what Code E has and can yet do for expediting the hardware project goals of the agency. The fault lies partly with NASA-supported researchers and partly with other agency scientists (USAF/Geophysics Lab; DOC/NOAA) more conversant with the amelioration of geospace hazards. In the interest of decorum, perhaps, the full story has not been widely advertised.

The cost to the engineering offices of ignoring NASA assets in the science codes is high. Not only is money wasted, but its reputation for excellence tends to be tarnished, a more costly penalty. The re-entry of the nation's first space station, Skylab, was misjudged and this precious resource lost. Little or no effort has been invested since the last epoch to better predict the course of the rising solar activity, of the increasing atmospheric drag induced by EUV and by geomagnetic storms. As a consequence of this lack of investment in improved knowledge of our geospace, shuttle missions have been replanned to rescue the Long Duration Exposure Facility and insure the return of its data. The Solar Maximum Mission and Solar Mesospheric Explorer will also soon require attention, in view of the uncertainty of our forecasts and present lack of EUV monitoring. At the same time, the Hubble Space Telescope is held in storage at millions of dollars per month to insure an adequate on-orbit life.

As in James Michener's *Space*, "damage control" may not be sufficient to quell criticism of NASA in the case of another August 1972 energetic proton event occurring, this time with humans irradiated in space. Not even lip service has been paid to improving our ability to warn against these sporadic radiation hazards. The direct costs of a launch "hold" while clouds cover ground-based USAF observatories during critical periods would be substantial. Courts may see law suits argued concerning the non-provision of warnings and monitoring required under the Occupational Safety and Health Act. These would be especially pertinent for missions to service the International Polar Platform (Earth Observing System), Geostationary Platform , and satellites, or even during extra-vehicular astronaut activity through the South Atlantic Anomaly passages of the Space Station.

Science Code ES has unique contributions. The initiation of a highly visible cooperative program between the engineering offices and the Space Physics Division (Code ES) would provide substantial protection as well as a "best effort" attempt to ameliorate the hazards of space operations. The requisite expertise lies in or is funded by the Space Physics Division. The Offices of Space Flight (Code M), Space Station (Code S), Aeronautics and Space Technology (Code R), and Exploration (Code Z) are all vulnerable to hazards ultimately deriving from the Sun, but also dependent on the Earth's magnetosphere, ionosphere, plasmasphere, and neutral atmosphere. Virtually all these specialties are currently found as branches in Code ES.

Payoffs from Previous Science Missions. The link that is presently lacking is recognition of the success of the previous NASA space science missions in providing answers to some of the basic questions regarding disturbances in the geospace. From revelations of the magnetospheric missions, the OSO's, Skylab, SMM, and DOD missions such as P78-1, we know today what information is needed to provide forecasts and alerts of crucial occurrences of (1) anomalous satellite drag (geomagnetic storms and radiant EUV flux); (2) hazardous energetic solar particles (their generation and acceleration at the Sun, forecasts of impact zones); and (3) satellite anomalies (electron plasma parameters, geomagnetic storms, solar cosmic rays).

Figure 2 envisages one such application, that of X-ray technology from NASA Skylab used for the protection of a NASA manned mission for servicing DOD satellites or the Earth Observing System (International Polar Platform). This application, the NOAA-USAF solar X-ray imager, will be noted further in the next section.

The combination of continued Code ES research missions, together with mutual cooperation of the Space Physics Branch with the mission-oriented agencies such as the U.S. Air Force and NOAA, could provide the NASA engineering and flight codes with the support for operations which they will increasingly need.

Slowing the Trade of Science for Engineering Missions. The tasks imposed on NASA in the 1990's will involve more than the internally-chosen space science programs and engineering challenges.

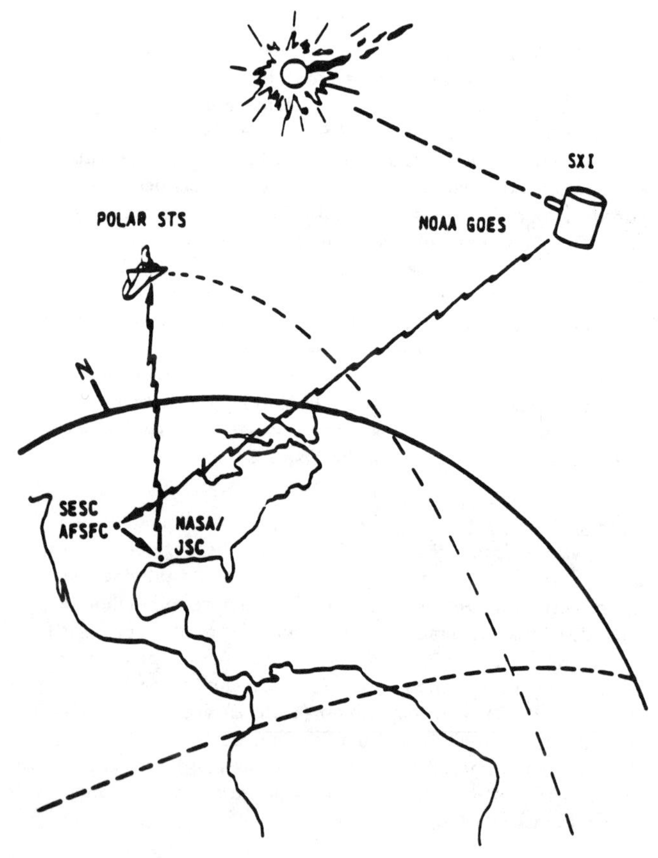

Fig. 2. NOAA's operational satellite as it provides radiation protection for the manned shuttle during a mission to service science experiments on the Earth Observing System of the International Polar Platform.

The agency will be supporting the increasing commercial opportunities sought in space, additional utilization of space by other U.S. civil departments (Transportation, Commerce, etc.), and heavier cost-sharing commitments with the Department of Defense. Non-NASA employees will be space-farers, including non-civil service personnel. With these new obligations to NASA will come the need for the safer and more efficient operation of people and systems in space. With demonstrations of the payback from science missions (mentioned in the subsection above) in the form of smoother expedition of space activities, it would not be surprising to find the NASA engineering offices more amenable to additional heliospheric research missions under the NASA banner.

Specific Observational Programs of the 1990's

The scientific problems noted in the above section, along with the project and agency goals discussed, will be addressed by a number of observational projects in the next decade. These appear to be well designed for accomplishing our objectives, provided that sufficient cooperation can be engendered in bringing all available assets to the tasks.

Two campaigns may be instrumental in bringing together researchers from different agencies and nations. In the U.S., NSF and NASA will sponsor Max '91, whose initiation is this year. On the international scale, the IAU and COSPAR, together with SCOSTEP, will provide coordination under the mantle of FLARES 22. Both programs address flare research and the active Sun, but will also benefit solar wind, interplanetary, and quiet Sun investigators.

Preparations under these programs will be done from now until 1991, at which time coordinated observing campaigns will begin, in time for 1992 which has been designated the International Space Year. Analyses and workshops to report results are planned under Max '91 and FLARES 22, starting in 1994.

Sketched below are the observational programs which should form the primary vehicles for acquisition of coronal and solar wind data during the next decade.

The Japanese SOLAR-A

Bruner et al. (1989) described the exciting technology and plans envisaged for the U.S. instrument on SOLAR-A, the soft X-ray telescope. Consequently, I will only note this project in

TABLE 1. The Japanese SOLAR-A

Objective:	Investigate high-energy phenomena with a coordinated set of instruments; also quiet Sun and helioseismology.
Launch:	By ISAS, September 1991, low-Earth orbit.
Life:	Nominal 2 years; orbit life: 4 years.
Instrumentation:	Hard X-Ray Telescope, 10–100 keV, 7 arc sec
	Soft X-Ray Telescope, 4–50 Angstroms, 3 arc sec
	Bragg Crystal Spectrometer
	Wideband Spectrometer, 2 keV–50MeV

passing and cheer for its potential. Table 1 gives some very basic information on the mission.

The ESA-NASA Solar and Heliospheric Observatory (SOHO)

The European and U.S. space agencies are cooperating in a Solar-Terrestrial Science Programme under whose aegis two missions will be launched in the 1990's (Schmidt et al., 1988). The first is the SOHO which will investigate the solar interior structure using helioseismology and surface radiance variations. SOHO's second purpose is to study the physical processes that cause the formation, heating, and maintenance of the corona and the establishment of the solar wind. SOHO will be stationed at the Earth-Sun L1 Lagrangian point (Table 2). The appearance of this spacecraft is illustrated in Figure 3.

The second component to this program is a constellation of four magnetospheric spacecraft, known as Cluster, which will fly in various elliptical polar orbit formations with apogees swinging from the geotail out beyond the magnetopause into the solar wind. Cluster will study small-scale structures in three dimensions in the Earth's plasma environment, beginning approximately January 1996.

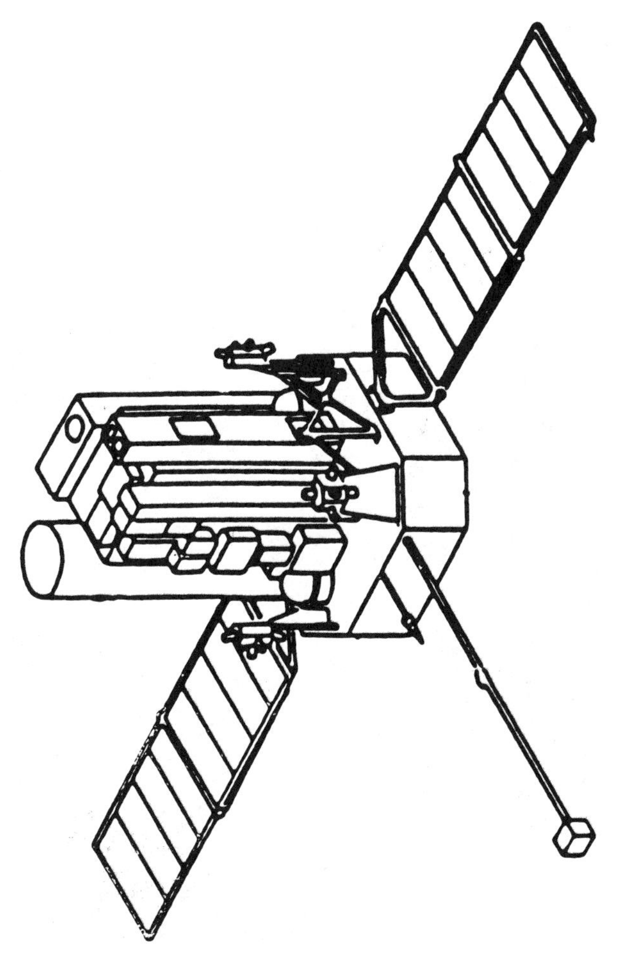

Fig. 3. NASA SOHO spacecraft, as shown in the 1987 Announcement of Opportunity.

TABLE 2. The ESA-NASA SOHO

Objective:	Study fundamental physical processes in the corona and solar wind as well as structure of the solar interior and variability of solar radiance; to investigate the solar wind in situ and its relation to remotely-sensed solar phenomena.
Launch:	March 1995, ESA satellite at L1.
Operations:	To March 1997 (consumables gone by 2001).
Instrumentation:	Solar atmosphere remote sensing. Suprathermal and energetic particle composition of the solar wind. Helioseismology.

TABLE 3. SOHO Payload

Helioseismology:
- Michelson Doppler imager (MDI).
- Global oscillations at low frequencies (GOLF).
- Variability of solar irradiance (VIRGO).

Coronal:
- Solar ultraviolet emitted radiation (SUMER).
- Ultraviolet coronagraph spectrometer (UVSC).
- Extreme ultraviolet imaging telescope (EIT).
- Coronal diagnostic spectrometer (CDS).
- White light and spectrometric coronagraph (LASCO).
- Solar wind anisotropies (SWAN).

Solar Wind in situ:
- Charge, element, and isotope analysis (CELIAS).
- Suprathermal and energetic particle analyzer (COSTEP).
- Energetic particle analyzer (ERNE).

The remote sensing complement of SOHO telescopes (Table 3) is designed for plasma diagnostics. All instruments have spectroscopic capability ranging from the EUV to the visible wavelengths. Taken together, the generation of the corona and solar wind will be studied from the transition region to 30 solar radii.

With Cluster performing its observations, it will be important for SOHO to monitor inputs which are headed toward the magnetosphere. Thus while primarily a quiet Sun mission, SOHO also will have some capability to support the Cluster mission by viewing the active corona.

In the second section of this paper, I noted that observations of the corona and interplanetary medium can indicate processes or changes of state in the Sun's interior. It is thus reasonable to assume that studies of the interior can provide forecasts of phenomena visible later in the outer atmosphere of the Sun and the solar wind.

From the vantage point of SOHO, helioseismology experimenters will be able to observe finer scale modes than ever before. Contaminating side lobes in the frequency spectrum will be removed because of the elimination of day/night cycles and weather gaps in the observations. The usual telescope-Sun relative velocities will be substantially reduced because the SOHO rides its Lagrangian point halo orbit. Measurement accuracy of the spacecraft-Sun velocity is expected to be greater than 1 cm s^{-1}.

The helioseismology sensors may permit the use of oscillation modes as diagnostics for interior circulation, perhaps as reflecting prominent structures such as X-ray coronal holes. With its promise of improved signal to noise, SOHO's MDI and GOLF may also routinely detect incipient sunspots prior to their emergence at the solar surface.

The SOHO will carry no instruments capable of measuring those parameters indicative of the varying state of the bulk flow of the solar wind. Thus no "ground truth" or verification checks of the remote solar sensing will be available from SOHO in situ instrumentation. However, included on spacecraft of the Cluster constellation will be ac and dc magnetometers, electron plasma analyzers, and ion plasma analyzers. Cluster will spend partial orbits several months per year in front of the magnetopause (20 Earth radii). Unfortunately, no plans exist to provide data in real time. Two similar spacecraft will be flown by the Moscow Institute of Space Research (IKI) which will travel upstream of Earth as many as 30 radii; real time data transmittal plans of IKI are not known. Intentions exist for exchanging the complete data sets between Cluster and IKI operations centers.

NOAA-USAF Solar X-ray Imagers (SXIs)

The SXIs (Table 4) will be operational telescopes designed for the civil and defense interests of the U.S. (Mulligan and Wagner, 1987). A series of SXIs are planned for geostationary orbit on NOAA's GOES satellites, beginning about 1992.

Data will be in the form of moderate-resolution, full-disk soft X-ray images (in two passbands of 10–20 Angstroms and 20–60 Angstroms) and extreme ultraviolet images (255–300 Angstroms and 304 Angstroms). The pixel size will be 5 arcsec. Data will be available (at rates up to one image per minute) at NOAA and USAF forecast centers in real time on a continuous basis. Table 5 indicates the parameters of the SXI.

The chief space environment services upgraded by these sensors are in the areas of warnings of energetic solar particle events expected at the Earth, forecasts of geomagnetic storms (both sporadic from coronal mass ejections and quasi-recurrent from X-ray coronal holes), and monitoring/predictions of EUV flux levels as input to atmospheric heating at satellite levels.

TABLE 4. Solar X-Ray Imagers (SXIs)

Objective: Provide continuous real time operational data in the form of soft X-ray and EUV full solar disk images for use as input to warnings of energetic particle events at Earth, forecasts of geomagnetic storms, and monitoring plus predictions of EUV heating of the thermosphere; science, as permitted, with a decade of such frequent, moderate-resolution X-ray data.

First launch: 1992, NOAA's GOES satellites.

Operations: Continuous, from the first SXI launched.

TABLE 5. Parameters of the Solar X-Ray Imagers (SXI)

Inner corona soft X-ray and EUV imaging:

8–20 Angstroms.
20–60 Angstroms.
255–300 Angstroms.
304 Angstroms.

Full solar disk plus 0.5 R_o field-of-view.
Pixel size 5 arc sec.
One image per minute (normal: one per 10 min).
Real-time data link to NOAA/SESC and USAF/SFC.
No day/night (geostationary orbit on NOAA's GOES).

Although not their primary function, SXIs can naturally contribute information for retrospective science analyses, in support of other research programs, and as fiducial data for research telescopes. These data, such as shown in Figure 4, may mitigate limited fields of view, day/night outages, or observing program commitments.

Active experimentation could be initiated by researchers when favorable conditions are indicated by SXIs. Operational instruments such as SXIs will lift some of the burden from future NASA science missions such as the Solar-Terrestrial Observatory, thereby allowing more resources to be applied for STO research goals.

In addition to their space environment service duties, SXIs will be used for research by NOAA and the USAF into the processes causing disturbances in the geospace environment. Particu-

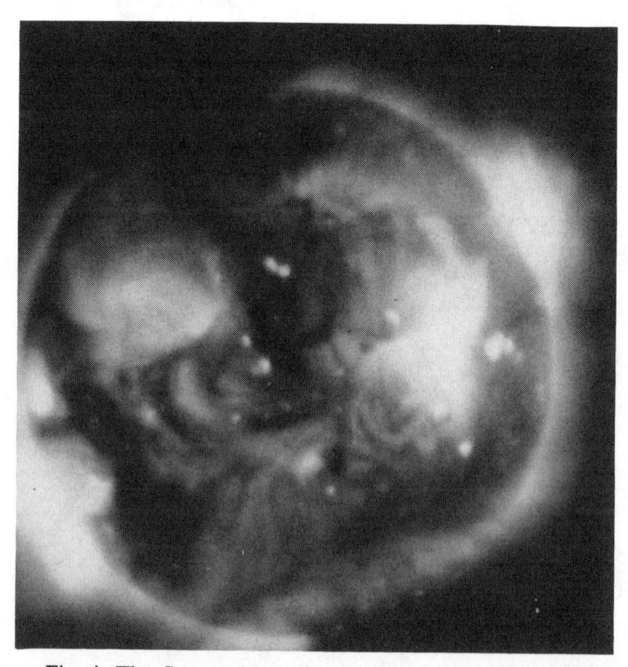

Fig. 4. The Sun on May 28, 1973, seen in X-rays by the American Science and Engineering Corporation S-054 instrument on NASA's Skylab.

larly noteworthy are the SXI capabilities for long-term monitoring of X-ray coronal flares, low corona (magnetic) structure/evolution/rotation, X-ray coronal holes, launches of mass ejections, X-ray bright point populations, and solar oscillations in the EUV.

Enhanced Ground-based Observations

The above describes those space science programs planned for the 1990's to advance our understanding of the solar wind and disturbances in the interplanetary medium such as coronal mass ejections. For solar-terrestrial physics, a suite of ground-based programs will continue to play a crucial role in completing our data sets.

NOAA's Space Environment Laboratory is working with Cambridge University and the British Antarctic Survey to establish a program of interplanetary scintillation observations at the Mullard Radio Astronomy Observatory. Similar cooperation with the Indian Physical Research Laboratory should provide a second daily patrol of maps showing disturbances in the interplanetary medium heading toward the Earth. The possibility also exists for a Japanese component to this program, thereby providing virtually 24-hour coverage. The first data should be available in 1989.

Several projects are underway to construct and operate vector magnetographs during the next solar cycle maximum. These involve, either independently or in consortia, the University of Hawaii, the High Altitude Observatory, the National Solar Observatory, NASA/Marshall Space Flight Center, the USAF Geophysics Laboratory (Sunspot), and the Applied Physics Laboratory of Johns Hopkins University.

Upgrades of mountaintop coronagraph facilities are also in the planning. With its scientists participating in the SOHO mission, the National Solar Observatory (Sacramento Peak) is considering designs for a large-aperture advanced coronagraph.

In 1991 the Global Oscillations Network is scheduled to begin observations. Instrument development is currently underway, after which six telescope sites will be selected. A 3-year program will begin in helioseismology; coarse 8 arcsec magnetic field maps will also be produced by this network on which the Sun will never set.

Other Possible Programs

The NASA WIND spacecraft is currently under construction and is slated for launch in 1992. It will be doing lunar swing-bys and may give upstream measurements of the solar wind.

The high resolution solar optical (HRSO) telescope appears to be regarded by NASA as a potential attached payload for the shuttle or Space Station at this time. The USAF solar activity measurements experiment (SAMEX) may be a free-flyer or, more likely, proposed as an enhanced HRSO strap-on facility for observations of the active Sun.

Other solar-terrestrial programs of NASA will be pursued from Scout and sub-orbital rocket opportunities and from long-duration balloons. These include test beds for the Pinhole Occulter Facility, spaceborne vector magnetographs, and various advanced high-energy imaging systems.

As a supplement to the NOAA interplanetary scintillation network data, a photometer similar to that flown a decade ago on the German Helios solar probes for zodiacal light studies is being proposed for monitoring disturbances in the solar wind. A NOAA operational satellite might serve as its bus.

Conclusion

Progress in coronal and solar wind research will increasingly be dependent on coordinated observing campaigns. An understanding of the plasma processes of coronal mass ejections and the interplanetary medium is important in itself for solar-terrestrial physics. It will guide other astronomers in their studies, and will assist U.S. agencies such as NASA, DOD, and NOAA as they pursue their missions.

Opportunities are being offered to us for cooperative investigations during the next decade. As we have noted, in space these include the Japanese SOLAR–A with its soft X-ray telescope, the ESA-NASA SOHO, and the NOAA-USAF solar X-ray imagers. An active parallel program of groundbased instrumentation may provide even more data than the above space missions.

The outlook is bright. To emerge from the 1990's with answers to the problems enumerated above as outstanding, our theorists should begin now to sharpen their pencils and we observers should begin planning to coordinate mutually supportive measurements.

Acknowledgments. I am grateful to E. Hildner for delivering this paper, with very little advanced warning, when illness prevented my attending the conference.

References

Bruner, M. E., L. W. Acton, W. A. Brown, R. A. Stern, T. Hirayama, S. Tsuneta, T. Watanabe, and Y. Ogawara, The soft X-ray telescope on the SOLAR–A mission, these proceedings, 1989.

Dreher, J. W., and E. D. Feigelson, Rings and wiggles in Hercules A, Nature, 308, 43, 1984.

Golub, L., and G. S. Vaiana, Evidence for globally coherent variability in solar magnetic flux emergence, Ap. J. (Lett.), 235, L119, 1980.

Mulligan, P. J., and W. J. Wagner, Plans for science studies and space environment services from the GOES Solar X-ray Imagers (SXI), Bull. Am. Astron. Soc., 19, 944, 1987.

Schmidt, R., V. Domingo, S. D. Shawhan, and D. Bohlin, Cluster and SOHO: A joint endeavor by ESA and NASA to address problems in solar, heliospheric, and space plasma physics, Eos, 69, 177, 1988.

Steinolfson, R. S., Driving mechanisms for coronal mass ejections, these proceedings, 1989.

Wagner, W. J., and J. J. Wagner, Coronal mass ejection recurrence studies indicating global activity and local suppression, Astron. Astrophys., 133, 288, 1984.

Wagner, W. J., Understanding the structure and dynamics of the inner corona through emission line transients and coronal mass ejections, in Proc. of the Workshop on Solar and Stellar Coronal Structure and Dynamics (edited by R. C. Altrock), Sunspot, New Mexico, 473, 1988.

HELIOSPHERIC REMOTE SENSING USING THE ZODIACAL LIGHT PHOTOMETERS OF THE HELIOS SPACECRAFT

B.V. Jackson

University of California, San Diego, La Jolla, CA 92093

Abstract. Our technique of remote sensing a large portion of the heliosphere using the Helios spacecraft zodiacal light photometers has shown that it is possible to measure the large-scale plasma structures around a spacecraft by electron Thomson scattering. The features observed include coronal mass ejections, corotating density enhancements, shock waves, and any other disturbances that can affect the intensity of the electron-scattered coronal brightness. In addition, it is possible to observe comet gas and dust to the brightness level of the background variations of the electron density. The Helios data show that it is possible to make good measurements of material in and out of the ecliptic plane and to determine the velocities and spatial distributions of the large-scale features which propagate into the heliosphere.

Introduction

Following detections of mass ejections by using the Helios zodiacal light photometers (Richter et al., 1982), a wealth of obervations using these data has followed. The transient phenomena studied thus far have included coronal mass ejections (Jackson and Leinert, 1985; Jackson, 1985b), corotating regions of enhanced density (Jackson, 1988a,b), shocks (Jackson, 1986), and comets (Benensohn and Jackson, 1988).

The Helios data showed that the intensity of Thomson-scattered light in the outer corona is great enough so that the entire heliosphere, at least in terms of variable structure out to 1 AU, can be sensed remotely by suitable photometric systems. Such global observations provide the best possible input into efforts to model and understand the physics of heliospheric dynamics, and they provide the best possible data base upon which to forecast terrestrial effects of heliospheric disturbances. In addition, they also show that background light sources such as stars and the zodiacal light do not vary enough to confuse the faint electron Thomson-scattering signal at distances from the Sun of 1 AU. In this report, I briefly summarize the observations of heliospheric structures (and comets) to date.

Coronal Mass Ejections

A major achievement of past research at the University of California at San Diego (UCSD) has been measurement of interplanetary masses and speeds of coronal mass ejections using the Helios zodiacal light photometer data. These mass ejections have also been observed by coronagraphs, interplanetary scintillation techniques, and by in situ spacecraft measurements. A two-dimensional imaging technique which displays Helios data has been developed at UCSD. Figure 1 gives an example of these images as used to show the coronal mass ejection of May 7, 1979. The combination of these Helios data with others has been used to

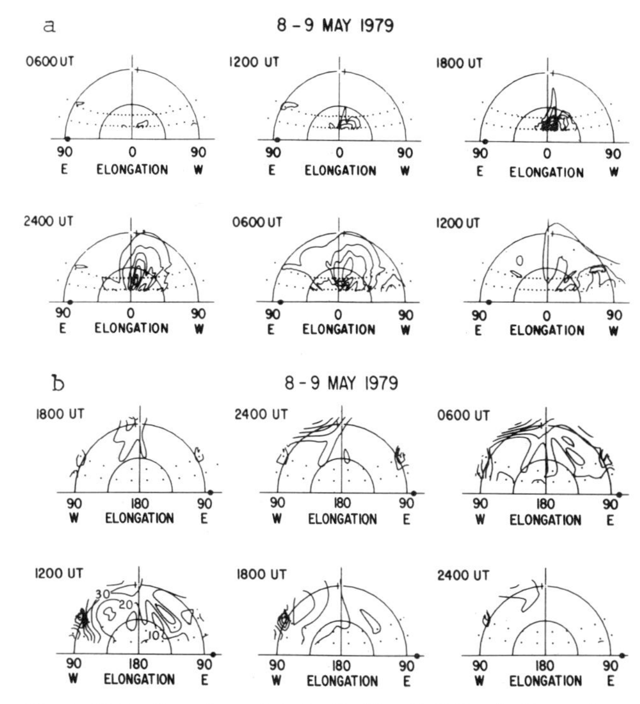

Fig. 1. Helios B contour plots for the May 7, 1979, ejection as it moves outward from the Sun over a period from 0600 UT on May 8 to 1200 UT on May 9. (a) In this presentation, the Sun is centered and various solar elongations labeled on the abscissa form semi-circles above the ecliptic plane (represented by the horizontal line). The vertical line is the great circle to the north of the spacecraft. The position of the Earth is marked as the ⊕ near east 90° and the solar north pole tilt indicated by the short line segment crossing 90° elongation. Positions of the sector centers are marked by dots. Electron columnar density is contoured in levels of 3 $\times 10^{14}$ cm^{-2}. The larger elongations are generally the lowest level contoured. (b) In this presentation the direction 180° opposite the Sun is centered. Electron columnar density is contoured in levels of 10^{14} cm^{-2}. At 12:00 UT contour levels are numbered.

advantage for each ejection studied. These studies form the basis of the papers (Jackson, 1985a; Jackson et al., 1985; and Jackson and Leinert, 1985; as reviewed in Jackson, 1985b).

The masses obtained from these observations indicate that indeed the material of a mass ejection observed in the lower corona continues to move outward into the interplanetary medium. Mass estimates of coronal mass ejections observed by Helios are generally approximately twice those determined by the SOLWIND coronagraph for the same events. This comparison is satisfactory given the measurement errors of both instruments, but it is consistent from ejection to ejection. We interpret the difference, if it is real, as due primarily to the inability of coronagraph images to measure the total mass of an ejection at any given instant in time (e.g., see Jackson, 1985b). The analyses of specific events show that coronal mass ejections supply significant mass to the interplanetary medium, and that the mass flow often extends over times longer than 1 day.

Because the Helios spacecraft move in their own orbits about the Sun, a perspective view of mass ejections also observed from Earth is available. Specifically, the shapes of three loop-like mass ejections observed by coronagraphs were measured as they moved past the Helios photometers in order to determine their edge-on thicknesses. Jackson et al. (1985) find an extent in Helios data for each event studied that is nearly the same as in the coronagraph view.

One coronal mass ejection in particular (that of May 21, 1980) has been studied in detail as to its surface manifestation (McCabe et al., 1986). The perspective view from the Helios spacecraft combined with that for SOLWIND allows a far more accurate mass to be determined for this event. It also indicates a highly non-radial motion at the onset of the ejection. Thus, from this example it is clear that knowing the site of the ejection on the solar surface does not necessarily imply good knowledge of the direction of travel of the major portion of ejected mass.

Elongated Regions of Enhanced Density that Rotate with the Sun

The images have also been used to measure the extent of elongated regions of enhanced density that rotate with the Sun. Figure 2 shows several of these features that can be followed for several days in Helios data. While some of the corotating features can be mapped to streamers observed in SOLWIND images, others cannot. Thus, at this time, the solar surface manifestation of all of these features is not known. Measurement of the position angle motion of these features gives their heliospheric latitude and longitude. Measurement of their curvature with distance from the Sun gives a material speed (Jackson, 1988a) to ~10% accuracy. Careful analysis of more than 30 of these features observed during several 2-month intervals from 1976 to 1979 is currently underway (Jackson, 1988b). The data show that some of the features outline magnetic field reversals near the solar surface (sector boundaries).

Shocks

The density jump behind an interplanetary shock can manifest itself as a brightness increase in the Helios zodiacal light photometers. The shorter duration of these density increases challenges the capability of the 5-hour time cadence of the Helios photometer system. To help circumvent this problem we have developed an analysis program that uses the 10-minute time-cadence samples from a given photometer on Helios. To date

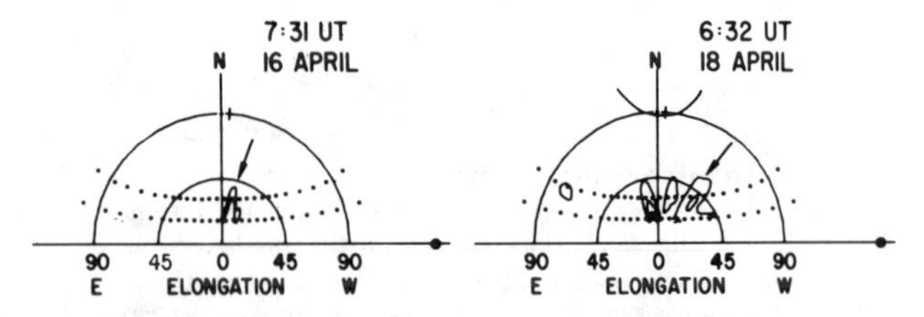

Fig. 2. Helios B contour plots showing a heliospheric corotating structure (arrows mark the feature). In this presentation the Sun is centered. Electron columnar density is contoured in levels of 3 $\times 10^{14}$ cm^{-2} upward from that level. The contours are drawn at times (given) when the 31° photometers measure brightness. Data from the 16° and 90° photometers are interpolated to the times indicated. Movement of the feature from east to west over a period of approximately 2 days from April 16-18, 1979, is shown.

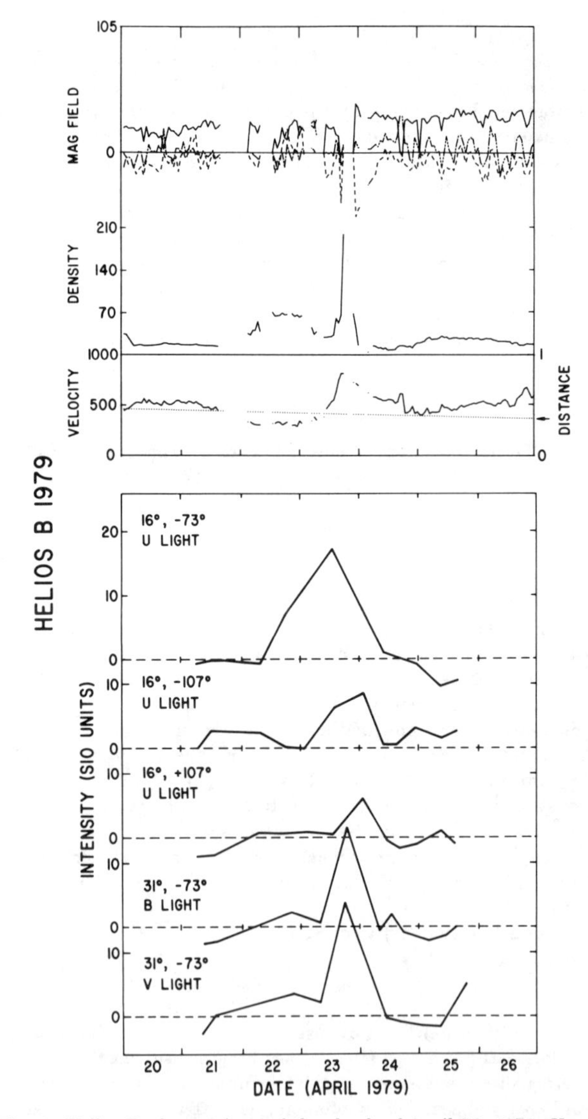

Fig. 3. Helios B observations of the shock of April 23, 1979. Upper panel: In situ magnetic field, proton density, and velocity plasma data. Lower panel: Photometer time series from photometer ecliptic latitudes, longitudes of sector centers relative to the Sun (- to the west), and colors listed.

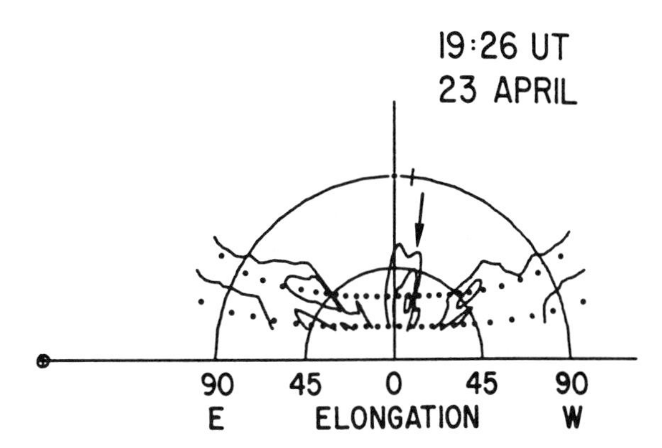

Fig. 4. Helios B photometer view of the density enhancement behind a shock also observed in situ by the spacecraft plasma probes. Electron columnar density is contoured in levels of 3×10^{14} cm^{-2} upward from that level. The columnar density contours to either side of the spacecraft at 0.4 AU ($\geq 45°$ elongation) show the brightness from the enhanced density behind the shock front. The primary structure (arrow) is a corotating region of enhanced density to higher latitude and is not involved with the shock. The shock density enhancement lasts only a few hours in the plasma data. The brightness increase is registered on only one set of photometer observations, but because of the cadence (16° photometer, then 31° photometer) is observed at slightly lesser elongations in the 16° photometer. The enhancement can be observed sweeping past the spacecraft in the three-color cadence observed by the 31° photometer.

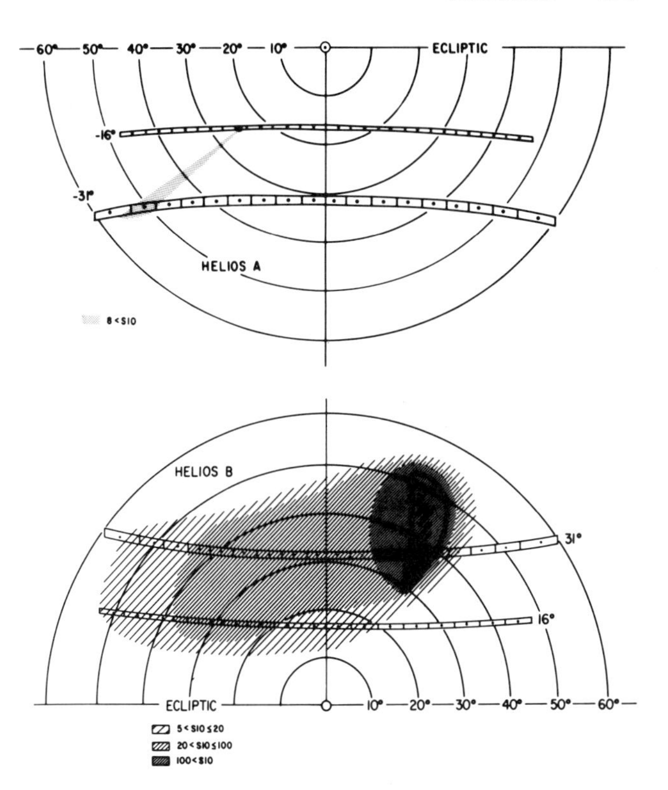

Fig. 5. Comet West images. (a) In the Helios A photometers when the spacecraft is approximately in the comet orbital plane on February 17, 1976. The shaded region indicates the maximum possible longitudinal extent of the Comet West tail on this date. (b) In the Helios B photometers on 1200 UT March 5, 1976. The shaded areas show the different regions of comet dust intensity. Data only exist for the photometer sector patches indicated. A line drawing of the comet shows the trace of a photograph view of Comet West from Earth which has been enlarged by a factor of 1.24 and foreshortened by 5% in longitude to correspond to the Helios B perspective.

we have measured five density enhancements behind shocks observed in situ (Jackson, 1986). While no three-photometer image capability exists for the brightness increases observed in the 10-minute cadence, there is enough information to map the general east-west extent of the shock front (to determine how the density enhancement relates to the magnetic shock normal). In some of the events it is also possible to observe the shock a few hours before it arrives at the Helios spacecraft. Figure 3 gives in situ and corresponding photometer observations of one well-observed shock in April 1979. The brightness increases observed by Helios, like radio observations and in situ density measurements, indicate that interplanetary shocks are fairly extensive heliospheric features.

In addition to mapping the extent of the interplanetary density increases behind shocks, we have attempted to use the UBV photometry and polarization information from the Helios photometers to determine to what extent (if any) entrained dust plays a role in the outward moving shock. To the noise limit of the zodiacal light photometers, there is no evidence that the heliospheric brightness of the material measured in situ as enhanced density following a shock is anything but Thomson scattering from electrons. Figure 4 gives a blurred view of one of these shocks to either side (at 90° elongation) of the Helios spacecraft.

Comets

The Helios photometers can be used to view comets as they pass the spacecraft. The orbit of the Helios spacecraft allows us to view comets in their orbits around the Sun from a different perspective than from Earth. Comet West (1976VI) passed through perhelion on February 25, 1976, and was observed by both the Helios A and B spacecraft. Figure 5 shows images of Comet West from both spacecraft, the first when Helios A was in the plane of the Comet West orbit, the second from Helios B at the approximate time the comet was closest to Earth. Data from the U, B, V photometry show a distinct blueing followed by a slight reddening corresponding to the ion and dust tails, respectively, entering the field of view of each photometer sector. The extent of the tail of Comet West is far greater seen from the Helios spacecraft than seen from Earth, even taking into account their generally closer viewing perspective. As reported in Jackson and Benensohn (1988), a brightness enhancement that preceeds the comet and is thought to be the density enhancement behind the Comet West bow shock can be observed as the comet moves from closer to farther from the Sun. Measurements of this feature show that it is the color of sunlight, and that it changes distance from Comet West with solar distance much as theory would predict (Mendis and Flammer, 1984) for the Comet West bow shock.

As Comet West travels away from the Sun, it can be observed in the zodiacal light fields of view at a solar distance of more than 1.4 AU. The zodiacal light photometers also measure brightnesses of Comet Meier (1978XXI). Comet Meier is far more compact than Comet West, extremely blue, and, unlike Comet West, shows no significant dust tail. Both comets are measured intrinsically several magnitudes brighter in the photometer data than as measured by Earth-based observers.

Conclusions

The Helios spacecraft zodiacal light photometers have been used to image solar mass ejections, corotating heliospheric density enhancements and density enhancements behind shock waves to distances of 1 AU from the Sun. To first order, sunlight scattered from heliospheric electrons provides the measured brightness ascribed to these features. Remote sensing observations of these features indicate extensive structures which propagate throughout the inner heliosphere.

Acknowledgments. I appreciate many helpful discussions with my colleagues about this review, especially those with Ch. Leinert who has graciously supported my analysis of the Helios data supported in part by his grant WRS-0108. Data from the Helios photometers are available at the National Space Science Data Center, Goddard, Greenbelt, Maryland. The work described here was supported in part by National Science Foundation grant ATM 86-09469 to the University of California at San Diego and by a subcontract to the University of California at San Diego through Air Force contract F19628-87-0077 to the Johns Hopkins University Center for Applied Solar Physics.

References

Benensohn, R.M., and B.V. Jackson, Comet West 1976 VI: A view from the HELIOS zodiacal light photometers, in Proc. of the STIP Symposium on Physical Interpretation of Solar/Interplanetary and Cometary Intervals (Huntsville, Alabama, May 12-15, 1987), in press, 1988.

Jackson, B.V., Helios observations of the Earthward-directed mass ejection of 27 November, 1979, Solar Phys., 95, 363, 1985a.

Jackson, B.V., Imaging of coronal mass ejections by the Helios spacecraft, Solar Phys., 100, 563, 1985b.

Jackson, B.V., Helios photometer measurement of in situ density enhancements, Adv. Space Res., 6, 307, 1986.

Jackson, B.V., Helios observations of a coronal streamer during STIP Interval VI, in Proceedings of the STIP Symposium on Retrospective Analyses and Future Coordinated Intervals (Les Diablerets, Switzerland, June 10-12, 1985), in press, 1988a.

Jackson, B.V., Helios spacecraft photometer observation of elongated corotating structures in the interplanetary medium, to be submitted to J. Geophys. Res., 1988b.

Jackson, B.V., and C. Leinert, Helios images of solar mass ejections, J. Geophys. Res., 90, 10,759, 1985.

Jackson, B.V., and R.M. Benensohn, The interplanetary medium near Comet West (1976VI): Possible views of the Comet West bow shock, Earth, Moon and Planets, submitted, 1988.

Jackson, B.V., R.A. Howard, N.R. Sheeley, Jr., D.J. Michels, M.J. Koomen, and R.M.E. Illing, Helios spacecraft and Earth perspective observations of three looplike solar mass ejection transients, J. Geophys. Res., 90, 5075, 1985.

McCabe, M.K., Z.F. Svestka, R.A. Howard, B.V. Jackson, and N.R. Sheeley, Jr., Coronal mass ejection associated with the stationary post-flare arch of 21-22 May 1980, Solar Phys., 103, 399, 1986.

Mendis, D.A., and K.R. Flammer, The multiple modes of interaction of the solar wind with a comet as it approaches the Sun, Earth, Moon, and Planets, 31, 301, 1984.

Richter, I., C. Leinert, and B. Planc, Search for short term variations of zodiacal light and optical detection of interplanetary plasma clouds, Astron. Astrophys., 110, 115, 1982.

DESIGN CONSIDERATIONS FOR A "SOLAR MASS EJECTION IMAGER" ON A ROTATING SPACECRAFT

B.V. Jackson, H.S. Hudson, and J.D. Nichols

University of California, San Diego, La Jolla, CA 92093

R.E. Gold

Johns Hopkins University, Applied Physics Laboratory, Laurel, MD 20707

Abstract. We describe an instrument capable of imaging the time-varying features of the entire outer corona (from near the Sun to beyond 90° elongation) via the Thomson-scattered diffuse solar light. This "all sky" imager works on a spin-stabilized spacecraft, preferably in deep space. The design for such an imager, which can for example study solar mass ejections at great distances from the Sun, must deal with spurious signals from stray light, zodiacal light, and stars to surface brightness levels below 1 S10 unit. The design discussed here envisions a set of three slit apertures, feeding one-dimensional detectors through a lens system; the spacecraft rotation allows a complete sky survey during each spin of the spacecraft. Data clocked into a computer memory complete the "image" of the whole sky. We have analyzed a "median filter" approach to reducing the effects of starlight, in real time, on the statistics of the residual diffuse background. The analysis also included simulations of spacecraft nutation, spin-phase timing error, and image quality in the necessary wide-field optics.

Introduction

This paper describes an instrument capable of remotely sensing heliospheric disturbances via fluctuations in the surface brightness of the Thomson-scattered solar light from electrons in the solar corona and solar wind – the classical "K-corona." This instrument is not a coronagraph, however, and aims instead at imaging the entire heliosphere beyond elongation angles of about 20° (equivalent to about 80 $R_\odot$). Existing literature in this domain comes exclusively from the zodiacal light photometers onboard the Helios spacecraft (Leinert et al., 1981; Richter et al., 1982; Jackson, 1985). The physical processes to be studied include coronal mass ejections, coronal streamers, shock waves, in situ density fluctuations of the solar wind, comet bow shocks and tails, and other phenomena; see Jackson (1989) for a fuller description of the literature.

The design discussed here is for a spin-stabilized spacecraft in deep space, for example at the inner Lagrangian point (L1) of the Earth-Sun system; the model assumed in this description is the WIND spacecraft of the International Solar-Terrestrial Physics (ISTP) program (APL, 1987). A set of three cameras provide one-dimensional images of 60° width along a meridian of the ecliptic coordinate system; the satellite spin motion then scans each pixel of the one-dimensional imaging detector along a small circle of the sky, building up a complete image.

The imaging data in the WIND configuration needed to be compressed to ~50 bits per second, implying coarse angular pixels, onboard processing, and a slow cadence of observation. These compromises necessarily affect image quality, and one of the main purposes of our study has been to prepare for intelligent trade-off decisions in the different instrument parameters.

Description

The instrument consists of three cameras covering 60° ranges of ecliptic latitude, with field optics and detectors giving image-resolved data (128 pixels) in the latitude direction and 2.5° integration in azimuth. A clock divides the azimuth range into 2.5° sectors, or 20.8 ms time bins at a spin rate of 20 rpm. The resulting pixels are approximately 0.5° wide in latitude by 2.5° (FWHM of triangular response) in the azimuth direction. The pixels are then combined electronically into 2.5° × 2.5° "superpixels," which can be integrated in an onboard memory for accumulation periods of up to 1 hour for transmission at low rate through the telemetry. In the numerical simulations described in the next section, we have made slightly different assumptions about the superpixel and pixel geometry. These differences should not affect the general conclusions.

The three cameras view, respectively, from -90° to -30°, -30° to +30°, and +30° to +90° ecliptic latitude. Scattered light from the Sun, and to a lesser degree the Moon, presents problems for the camera at small elongations, and the camera lens and the image it forms must therefore be carefully baffled. A schematic of the camera front end is sketched in Figure 1a along with its configuration on a rotating spacecraft (Figure 1b). Following the basic principles of scattered-light control described by Leinert and Klüppelberg (1974) for the Sun-crossing baffle, we have laid out a baffle as sketched in Figure 2. We calculate that this baffle and a clean lens system can maintain a scattered light level well below that of the background sky beyond ~20° elongation. In order to avoid detector saturation, a "bright object sensor" will actuate a shutter mechanism to block the detector physically from sunlight within about 10° elongation.

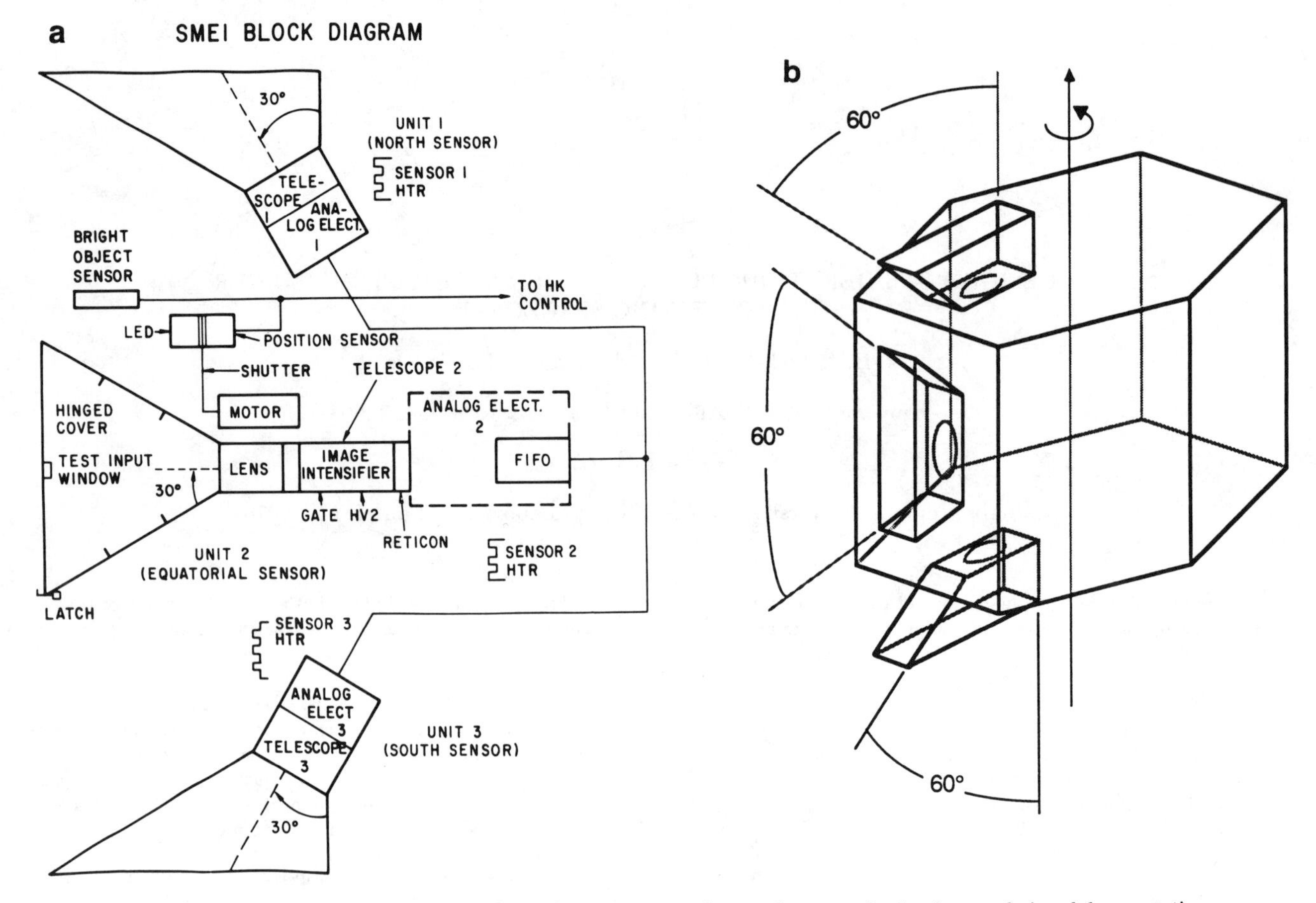

Fig. 1. Camera schematics. (a) The baffle/lens/detector system for a solar mass ejection imager designed for a rotating spacecraft. (b) As configured on a rotating spacecraft.

Signal Levels Expected

The Thomson-scattered coronal light must be detected in the presence of background diffuse light from many sources: scattered light from bright sources such as the Sun, Moon, or Earth; the zodiacal light and Gegenschein; and the stars themselves, either individually as bright point sources or collectively as a contribution to the diffuse sky brightness. The ultimate limit of diffuse-light sensitivity should be set by photon counting statistics; this limit depends upon the optics and scanning configuration, spectral bandpass, and total detector efficiency. The total detected photon count N can be approximated by

$$\log N \; = \; 4.95 \; - \; 0.4 m_V \; + \; 2 \log D \; + \; \log(\Delta t) \qquad (1)$$

where m_V is the stellar visual magnitude, D the aperture (diameter) in cm, and Δt the integration time in seconds; this formula assumes 10% overall quantum efficiency and 20% spectral bandwidth. Astronomical surface brightness can be expressed in "S10 units," equivalent to the flux of one, tenth-magnitude star per square degree; in these terms the background sky brightness varies roughly over the range 100-6000 S10 between the darkest sky and the ecliptic plane at solar elongations $\geq 20°$. This corresponds, according to the above formula, to photon counting rates as low as 1000 counts/s for a 1-s integration 1° pixel through 1-cm optics. This corresponds to a minimum 1-σ photometric error of $\sim$3 S10 in a 1-s direct exposure. Across the darker areas of sky in the spin-sampling environment we expect to gather only a tiny fraction of these number of counts on any given pass. Thus, we conclude that we must count photons with our detectors, and that the signal-to-noise ratio of the instrument will be limited by integration times, viewed area of sky, and the size of the aperture when we view typical heliospheric features (see below).

Table 1 estimates the signal levels expected for various phenomena at 1 AU, on the assumption that the feature in question moves outward at constant velocity without dispersion. The brightnesses of coronal mass ejections (CME's) and streamers were derived from features traced outward from the Naval Research Laboratory (NRL) SOLWIND coronagraph to the Helios photometer field (Jackson, 1989). Shock brightnesses were estimated from the in situ plasma density enhancements behind shocks observed from the Helios spacecraft, and assumed to be viewed at 60° and 90° from the Sun-spacecraft line.

The noise estimates (Table 2) are based upon Helios photometric data taken at 0.85 ± 0.06 AU, supplemented with estimates of brightness fluctuations due to starlight. The final estimates ("all-sky") assume an all-sky imager similar to that described here, but with capability for removal of the effects of stars to different levels: seventh and ninth magnitude.

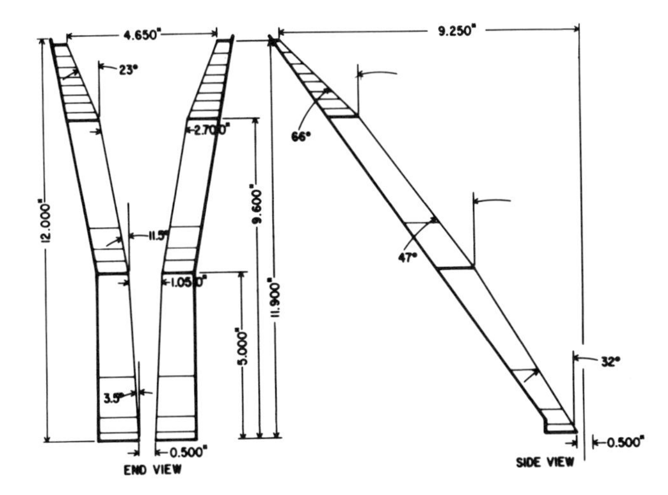

Fig. 2. Baffle layout for the Sun-crossing ecliptic baffle, showing cross-sections in the wide and narrow dimensions and the defining aperture stops. The wide dimension is oriented ecliptic NS and the spacecraft spin scans the narrow dimension, which is imaged on a one-dimensional detector, around the sky. In the narrow dimension, three "stages" of baffling can be identified by the successively narrower angles allowed to incident radiation. For this design, the volume available within the spacecraft did not permit a three-stage design in the wide dimension.

Simulations of Imaging Data

Overview

We have carried out numerical simulations of data from an imager of this design, using selected parts of the sky to provide representative background light levels. The light sources included bright stars (with $m_V \leq 10$) taken from a machine-readable version of the Smithsonian Astrophysical Observatory (SAO) sky catalogue. Diffuse light sources due to zodiacal light, the Gegenschein, and faint galactic stars were taken from tabulated values in Roach and Gordon (1973) and interpolated using a bicubic spline approximation. The actual parameters used in the simulations may differ in some details from those described above, but the results should be generally applicable. Fuller details on the numerical simulations are given by Nichols (1987).

The pixel extent was taken to be 0.5° in latitude, 2.5° in longitude. The electronics are assumed to be able to create 2.5° × 2.5° superpixels composed of five adjacent subpixels. Counts from each pixel are integrated for 1/144 of the spacecraft's 3-s spin period. This means that each pixel sweeps for 2.5°, giving each superpixel a range of 2.5° in latitude and 2.5° (FWHM) in azimuth. The range in ecliptic longitude measured increases with latitude by a factor

$$\arccos\left(\frac{\cos(2.5°)}{\cos(\beta)^2} - \tan(\beta)^2\right) \tag{2}$$

where β is the ecliptic latitude.

A simulation run consisted of making a time series of data from a superpixel viewing a given sector of the sky. The simulated photon count rate was given by equation (1) assuming a 1-cm diameter aperture. Two time scales were relevant: the averaging time and the total measurement time. Each rotation of the spacecraft gives the sample pixel a "flash" view of the sample sky sector. These flash samples are accumulated, with appropriate statistics, over the averaging time to get the total number of counts. An accumulated image is then sent out in the telemetry stream. The total measurement time determines how many of these values are generated in the simulation. The numbers determined for each run include the average brightness of the pixel in S10 units, and the standard deviation of the sample points around that mean. The standard deviation is significant because it measures the sensitivity of the instrument; we will be unable to determine real changes below the level set by statistical fluctuations and the various sorts of uncertainties discussed below.

A number of different superpixels were used for the simulation, in order to include representative sections of the sky containing different levels of diffuse light and bright stars. Table 3 lists these trial regions, giving identification numbers and sky location (β giving the ecliptic latitude, λ the ecliptic longitude).

TABLE 1. Signal Levels Expected at 1 AU

Feature	Elongation degrees	Signal S10	Signal Duration days
Bright CME	60	3	1.5
	90	2	1.5
Bright streamer	60	2	1
	90	1	1
Bright shock	90	1-2	0.5
Major in situ fluctuation	60	3	2
	90	2	2
Comet shock	20	3-10	0.1

TABLE 2. Noise Estimates

Elongation degrees	Bandpass days	Ambient Medium* S10	Helios S + N S10	All-Sky, S10 m = 7	All-Sky, S10 m = 9
60	0.2-4	1.0	1.6	0.5-1	0.1
90	0.2-4	0.5	1.3	0.5-1	0.1

* Assumes an Allen $1/r^2$ heliosphere

TABLE 3. Pixels Used in Simulation Studies.*

Pixel No.	β	λ
1	70.0	267.6
2	50.0	269.3
3	30.0	269.8
4	10.0	270.0
5	−12.5	270.0
6	−32.5	269.8
7	−52.5	269.3
8	−72.5	267.6
11	−72.5	157.6
12	−52.5	159.3
13	−32.5	159.8
14	−12.5	160.0
15	10.0	160.0
16	30.0	159.8
17	50.0	159.3
18	70.0	157.6

*β is the ecliptic latitude and λ is the ecliptic longitude. The values given are the minimum for each 2.5° × 2.5° superpixel. Note that the apparent $\Delta\lambda$ increases at large β by the factor given in equation (2).

The effects of a number of different types of errors were simulated, including: spacecraft nutation, resulting in ecliptic latitude pointing error; timing error, resulting in errors in azimuthal binning of data; and fuzziness in the star images on the detector surface due to imperfect optics. The standard deviation of the brightness is measured for a range of these parameters. The average brightness itself (determined over the total measurement time) seems insensitive to these errors.

Underlying these simulated systematic effects is the unavoidable fluctuation due to photon counting statistics. In order to minimize this effect, most of the the simulation runs were with averaging times of 15 min of spacecraft time (or 300 rotation periods). Some of the test runs were done for 60 min of spacecraft time; Poisson statistics predicts that the fluctuations would be reduced by a factor of $\sqrt{60/15} = 2$, and this ratio was generally observed between 15- and 60-min test run simulations.

An additional feature of the simulation was a test of a "median filter" star rejection scheme, designed to remove subpixels containing bright stars from the averaging. The purpose of this procedure is to reduce the overall telemetry requirement to less than about 50 bps. The median filter scheme simply excludes the brightest 0.5° subpixels from averaging into the 2.5° superpixel sums for telemetry, thus (in principle) excluding the photon statistics due to bright stars. The median filter would be easy to implement electronically as a simple sorting of the subpixel data prior to summation. A possible defect of the median filter approach is that a bright star near a pixel boundary could cause noise-like time variations as a result of pointing errors of various types. The median filter was examined for the range of cases where we took data from only the dimmest pixel, the two dimmest pixels and so on up to all five pixels (filter disabled). The interaction of the median filter with the various errors was also considered.

Nutation Error

Simulations were run for eight pixels with four different values of nutation amplitude (A_ν) at an assumed period of $T_\nu = 20$ s. Results for other values of T_ν should be similar, as long as the nutation period was much shorter than the averaging time and was not an exact multiple of the spin period. The latitude pointing was assumed to be changed by an additive factor of

$$A_\nu \sin(2\pi t/T_\nu) \tag{3}$$

The sensitivity of the standard deviations of the measured brightnesses varied greatly among the pixels, as is seen in Table 4.

TABLE 4. Dependence of Sensitivity on Spacecraft Nutation Amplitude (A_ν) for a Fixed Star Image Radius (R_*)*

Pixel No.	R_*	$\sigma_{0\nu}$	S_ν
11	0.00	0.92 ± 0.01	2.95 ± 0.32
11	0.20	0.50 ± 0.11	6.02 ± 0.81
12	0.00	0.93 ± 0.22	1.87 ± 1.76
12	0.20	0.88 ± 0.04	0.57 ± 1.00
13	0.00	0.70 ± 0.10	2.56 ± 0.53
13	0.20	0.67 ± 0.05	5.10 ± 0.29
14	0.00	0.74 ± 0.10	0.58 ± 0.63
14	0.20	0.85 ± 0.16	-0.20 ± 1.06
18	0.00	0.83 ± 0.12	14.14 ± 1.03
18	0.20	0.71 ± 0.10	3.47 ± 1.12

*Values are the result of a least-squares fit to $\sigma = \sigma_{0\nu} + S_\nu \times A_\nu$.

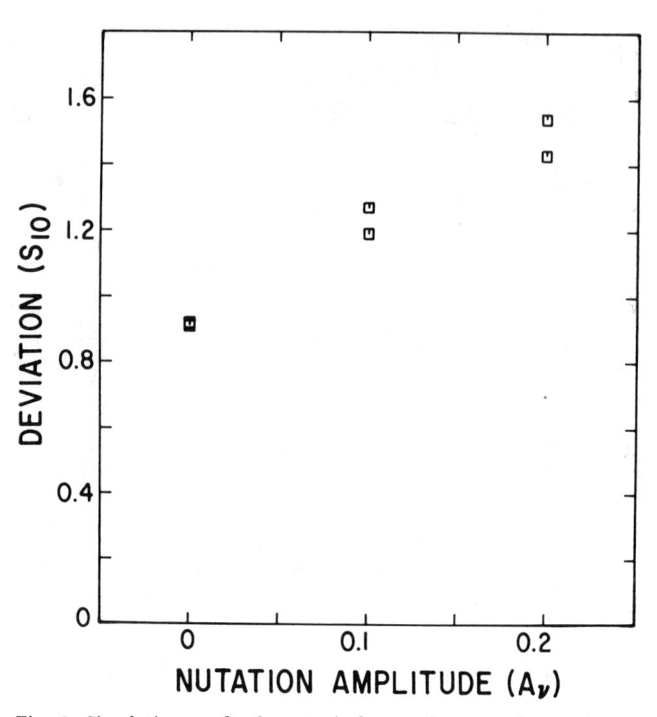

Fig. 3. Simulation results for a typical case of spacecraft nutation, showing the decrease of sensitivity (increase of scatter in S10 units) with nutation angle, as described in the text. The results of linear fits to this relationship for other pixels are given in Table 4. The results show that nutation is an important factor governing sensitivity, and that if it can be kept small enough the measurement error can be below 1 S10 unit (rms) in a 15-min integration.

Figure 3 shows the results for a typical pixel, and Figure 4 shows the worst case observed. Pixel 18 is the worst, with the measured spread increasing by a factor of 4 for a nutation magnitude of 0.2°. This case will be treated later. Linear fits were done for each pixel to the form

$$\sigma = \sigma_{0\nu} + S_\nu \times A_\nu(\text{degrees}) \tag{4}$$

Not all of the curves were linear, but the fits give an indication of the sensitivity of the spread in the brightnesses to the increase in the nutation amplitude. If pixel 18 is excluded, the average slope is 1.9 S10/deg, and all of the slopes are non-negative. As expected, increasing the latitudinal pointing error increases the value of σ. If we want to limit the deviation due to nutation to ~0.1 S10, then the nutation must be limited to ~0.05°.

The distinctive feature (at least among the pixels considered) of pixel 18 is a bright star (magnitude 4.90) near a subpixel boundary. (The star is at $\beta = 72.03°$, and the subpixel boundary is at 72.0°.) When the nutation amplitude becomes greater than 0.03°, the deviation measured jumps abruptly from about 0.6 to 2.2 S10. This suggests an explanation in terms of count truncation. The loss of one count per flash due to the fact that counts come only in integer values would give a change of 2 S10 for a subpixel, which is the correct order of magnitude.

Effects of Timing Error

Timing error would result from the Sun pulse being read inexactly, so that the azimuthal angle when a given pixel is enabled would differ slightly from one spin to the next. The effects of the

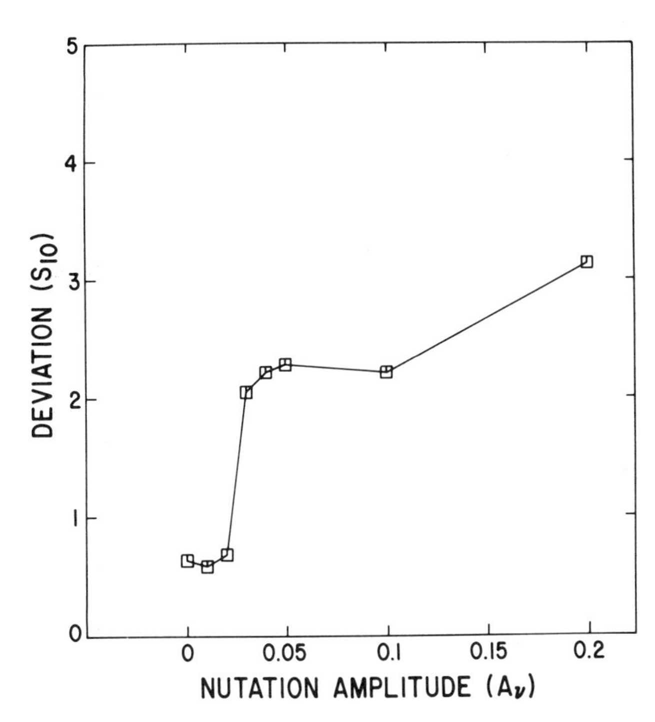

Fig. 4. Nutation results for the worst-case pixel, no. 18 in Table 1, showing that nutation-induced error can become intolerably large.

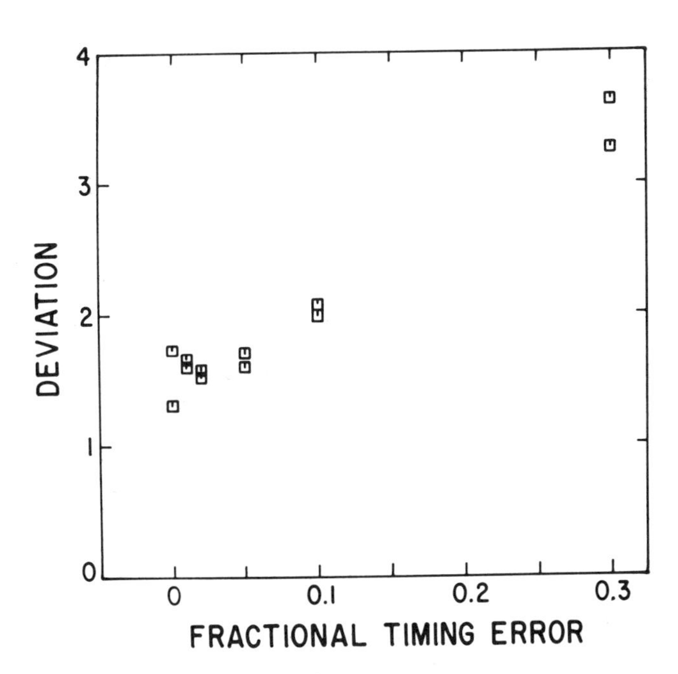

Fig. 5. Timing error results for pixel 5, showing the decrease of sensitivity (increasing scatter in S10 units) with increasing fractional timing error. This pixel was the worst in the simulation list (Table 1), and a more typical pixel did not show a significant sensitivity to timing error. See Table 5 for the simulation results of the other pixels studied.

TABLE 5. Dependence of Instrument Sensitivity on the Fractional Timing Error (ϵ).*

Pixel No.	σ_{0t}	S_t
1	0.718 ± 0.032	-0.049 ± 0.210
2	0.900 ± 0.017	0.390 ± 0.141
3	0.981 ± 0.036	-0.183 ± 0.314
4	1.127 ± 0.006	0.086 ± 0.033
5	1.491 ± 0.033	5.537 ± 0.614
6	0.981 ± 0.027	-0.021 ± 0.227
7	0.728 ± 0.007	0.531 ± 0.174
8	0.755 ± 0.004	0.264 ± 0.046

*The values are a result of a least-squares fit of $\sigma = \sigma_{0t} + S_t \times \epsilon$.

timing error were measured for eight pixels assuming perfect optical sharpness and no spacecraft nutation. The errors were considered to be normally distributed, and ranged from 0 to 30% of the flash (6.2 ms or 0.75°). The worst pixel is shown in Figure 5, and Table 5 details other results.

The results were in general analogous to those for nutation. The scatter in pixel brightness determinations is not very sensitive to errors in the initial azimuth of a pixel. The few extreme cases such as pixel 5, with a bright star near the boundary, could be masked out and disregarded in data analysis.

Optical Fuzziness

The effects of imperfection in the optics were also considered. As above, linear fits were made to

$$\sigma = \sigma_{0f} + S_f \times R_*(\text{degrees}) \tag{5}$$

for four values of R_*, the radius of an assumed disk-shaped image of uniform brightness. The selected maximum value for R_* of 0.2° meant that no star fell in more than two pixels at once. The light from a star falling in multiple subpixels was divided according to the areas of the intercepted parts of the images. Figure 6 gives the dependence of the scatter in brightness determination on optical fuzziness, for a range of nutation amplitudes, for a typical pixel. Table 6 shows the results for some of the pixels. Note that these are the same data as in Table 4, except that now A_ν is held constant and σ is considered a function of R_*. If we again exclude pixel 18 the slopes are generally small, and of mixed signs. The average slope is −0.2 S10/deg, so a decrease in optical precision actually increases the sharpness in our brightness measurement. Generally, it seems that optical precision is not that critical, which is not surprising, since only items which are point sources (e.g., stars) create a nuisance, and smearing them out should decrease their bad effects. This can be seen acting strongly in pixel 18: if A_ν = 0.2° then σ = 3.6 S10. Smearing the stars out to 0.2° radii decreases σ to ~1.2, which is nearing usability. Perhaps one could redeem a number of otherwise useless pixels by defocussing without harming the resolution on the others. We have not actually counted the pixels that would fall in this category.

The Median Filter

All of the above tests were run with the median filter engaged, so that results were obtained for five sets of filters, where the i th filter result is given by the average of the i dimmest subpixels.

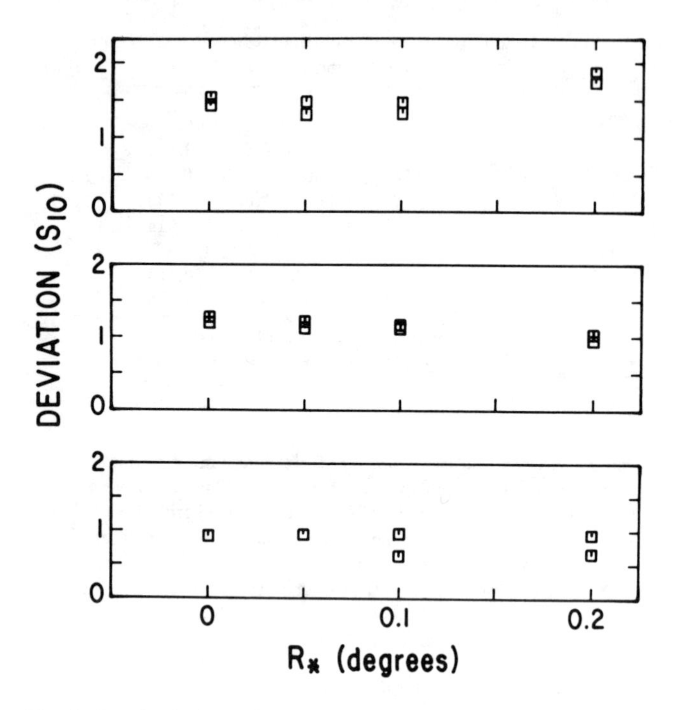

Fig. 6. Results for imperfect ("fuzzy") optics, simulated as disk-shaped stellar images of radius R_* degrees, for three different nutation amplitudes from bottom to top of 0.0°, 0.1° and 0.2° for pixel 11. See Table 6 for further results. As expected, the sensitivity does not depend strongly upon image quality, because fuzzy images reduce the contrast across the detector.

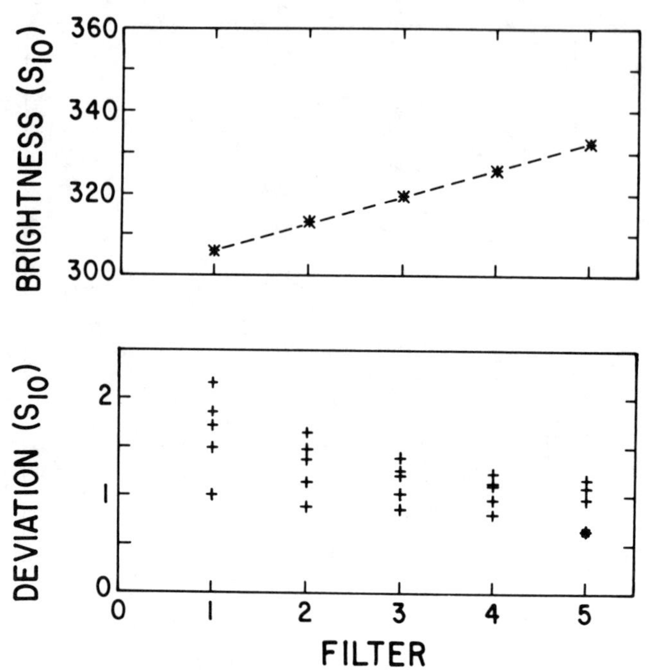

Fig. 7. The median filter at work: five simulation runs on a typical pixel, no. 11. The upper panel shows the estimated surface brightness in the pixel simulated with number of subpixels included by the median filter. As expected, the median filter biases the measurement towards low surface brightnesses. The dashed line shows the trend that would be expected for a constant-brightness sky, leading to an unbiased measurement for the full number of subpixels (5) constituting one pixel in our simulation. The lower panel shows the scatter in the brightness estimates, which decreases with number of subpixels included in the filter because of improved photon counting statistics.

The model was run on the 18 pixels described in Table 1 and the brightness and the standard deviation of the brightness were examined as a function of filter number. Five different runs were done for each case, and the means and standard deviations of the brightness and its standard deviation were computed for each superpixel and filter.

In many of the cases the filter had no effect. Figure 7 shows a typical case (pixel 11) ; the brightness increases in a linear fashion as more subpixels are used, while the standard deviation decreases.

TABLE 6. Dependence of Sensitivity on Star Image Radius (R_*) for a Fixed Spacecraft Nutation Amplitude (A_ν).*

Pixel No.	A_ν	σ_{0*}	S_*
11	0.00	0.92 ± 0.01	-0.31 ± 0.71
11	0.20	1.40 ± 0.07	1.61 ± 0.56
12	0.00	0.87 ± 0.03	0.03 ± 0.29
12	0.20	1.29 ± 0.20	-1.49 ± 1.61
13	0.00	0.69 ± 0.06	-0.09 ± 0.46
13	0.20	1.05 ± 0.02	2.64 ± 0.19
14	0.00	0.77 ± 0.09	0.61 ± 0.98
14	0.20	0.85 ± 0.06	-0.21 ± 0.71
18	0.00	0.77 ± 0.09	-0.19 ± 0.73
18	0.20	3.37 ± 0.13	-9.91 ± 1.29

*Values are the result of a least-squares fit to $\sigma = \sigma_{0*} + S_* \times R_*$.

This is what would be predicted from the increasing number of photon counts. Bright stars are not important here, and the median filter is of little use.

However, in a few of the cases the effect of the bright stars was clearly seen. Figure 8 (pixel 12) shows an example of this case. The case where all five subpixels are used clearly deviates from the linear relation expected. In these cases a median filter which rejects the brightest subpixel and keeps the other four is clearly better at measuring the diffuse light than a simple average. The increased variance in the standard deviation of the brightness also indicates that the brightest pixel is one that we can well do without.

Simulation Results: Instrumental Errors Included

Figure 9 (pixel 11) shows the effects of the median filter on a superpixel heavily affected by spacecraft nutation. It seems that the standard deviation of the brightness is heavily dependent upon the total number of photon counts, which means that the median filter is exactly the wrong approach to take here, since it throws away a portion of the counts. Another effect which could be operating here is that nutation, for example, could systematically change the list of subpixels selected by the median filter. Since the individual subpixels always contain intrinsically different levels of

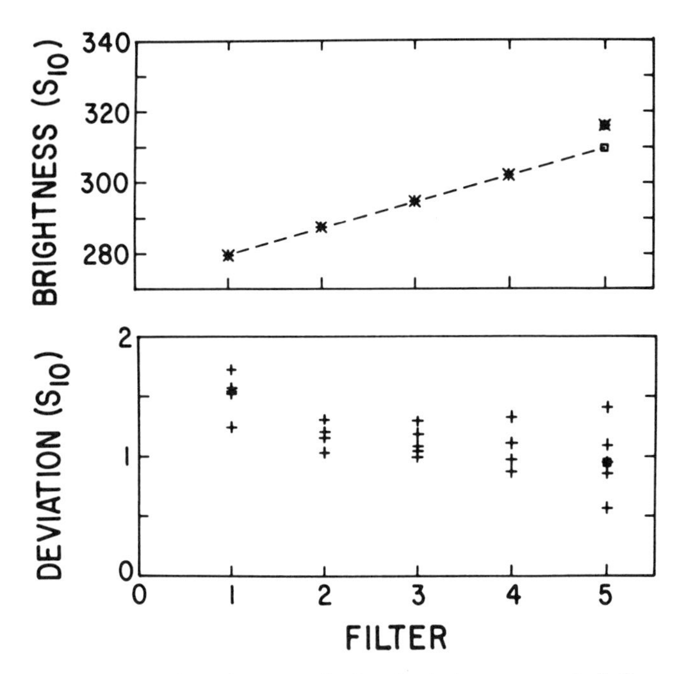

Fig. 8. The median filter at work: five simulation runs on pixel 12. This pixel contains a bright star in one subpixel. Thus, the final estimate (all five subpixels passed by the median filter) is in excess of the true value of surface brightness due to diffuse sources.

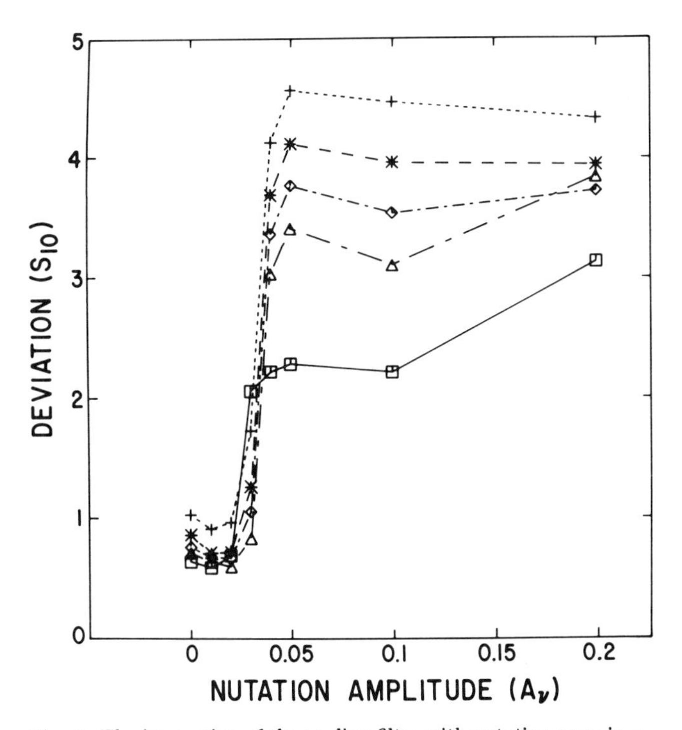

Fig. 9. The interaction of the median filter with nutation error in a case where the latter is highly significant. The + are the dimmest subpixels only, the ∗ the dimmest two subpixels, the △ three, the ◇ four, and the □ all five subpixels. This is pixel 18, to be compared with Figure 4 above.

diffuse brightness (including faint stars), the result of the median filter reselection would be to add unwanted variance. For example, when a bright star causing trouble is in a subpixel, then obviously that subpixel is not the dimmest, but when it is out, then it is the dimmest. This would definitely increase the standard deviation of the measured brightness.

Because the effects due to timing error and fuzzy star images did not seem to be much of a problem, they were not investigated in depth for their interaction with the median filter.

Conclusions

We have carried out a preliminary design for a heliospheric imager intended for a spin-stabilized spacecraft. Our analysis has shown that a rather low-resolution imager can achieve sensitivities ≥1 S10, if the stability of the spin motion can be maintained at a reasonable level. This is sufficiently good for studies of a number of interplanetary phenomena. We have not dealt in depth yet with scattered light problems, concentrating instead on the errors due to the background light sources and their effects on signal-to-noise ratio.

We have shown that it is possible to reject light from bright stars using an onboard algorithm (a "median filter"), allowing the data to be averaged in a meaningful way and transmitted over a narrow telemetry bandwidth. We would prefer a much higher telemetry rate than the one envisioned here, which would allow us to transmit all the data to the ground for processing. Shorter accumulation times would also allow us to study rapidly changing phenomena with time scales on the order of minutes near the Sun where there are enough photons to get accurate values. The median filter approach, although simple and in some cases effective, necessarily compromises data quality and complicates the analysis. Thus, it is to be avoided if sufficient telemetry is available to transmit all subpixel information (about 500 bps for all-sky imaging on this cadence).

Acknowledgments. The work described here was supported in part by a subcontract to the University of California at San Diego through Air Force contract F19628-87-0077 to the Johns Hopkins University Center for Applied Solar Physics.

References

Applied Physics Laboratory, All-Sky Heliospheric Imager (ASHI) Interface Control Document, JHU/APL SDO 8416, 1987 .

Jackson, B.V., Imaging of coronal mass ejections by the Helios spacecraft, Solar Phys., 100, 563, 1985.

Jackson, B.V., Heliospheric remote sensing using the zodiacal light photometers of the Helios spacecraft, these proceedings, 1989.

Leinert, C., E. Pitz, H. Link, and N. Salm, Calibration and in-flight performance of the zodiacal light experiment on Helios, Space Sci. Instrum., 5, 257, 1981.

Leinert, C. and D. Klüppelberg, Stray light suppression in optical space experiments, Applied Optics, 13, 556, 1974.

Nichols, D., Data Simulations for a Rotating Heliospheric Imager, UCSD-SP-87-23, 1987.

Richter, I., C. Leinert, and B. Planc, Search for short term variations of zodiacal light and optical detection of interplanetary plasma clouds, Astron. Astrophys., 110, 115, 1982.

Roach, F.E, and J.L. Gordon, The Light of the Night Sky, Reidel, 1973.

FLUX ROPE DYNAMICS FOR LOOP PROMINENCES, CORONAL MASS EJECTIONS AND INTERPLANETARY MAGNETIC CLOUDS

Tyan Yeh

Cooperative Institute for Research in Environmental Sciences, University of Colorado
Boulder, CO 80309

Abstract. Many solar and interplanetary phenomena involve magnetic flux ropes immersed in magnetized media. Notable examples are loop prominences, loop-like coronal mass ejections, and interplanetary magnetic clouds. These objects have the common feature that they are magnetically separated from their surrounding media. Their temporal evolutions can be described by flux rope dynamics. A flux rope is subjected to external and internal forces. The external forces consist of the gravitational force exerted on the distributed mass of the flux rope by the massive Sun and the hydromagnetic buoyancy force exerted by the surrounding medium. The hydromagnetic buoyancy force includes hydrostatic, hydrodynamic, and diamagnetic buoyancy forces. The diamagnetic force amounts to the magnetic force exerted on the distributed current in the flux rope by external currents. The internal forces consist of the magnetic force on various parts of the internal current by other parts and the pressure gradient force resulting from the pressure difference between internal and external thermal pressures. The external forces drive the translational motion of the flux rope, while the internal forces drive its expansional motion. The translational motion determines the temporal displacement of the centroidal axis of the flux rope, and the expansional motion determines the displacements of various mass elements relative to the axis. The sum of translational and expansional motions is the resultant motion.

Introduction

Coronal mass ejections (MacQueen et al., 1978; House et al., 1981; Howard et al., 1982) and interplanetary magnetic clouds (Burlaga et al., 1981; Klein and Burlaga, 1982) are two important discoveries that were made from observations with space-borne instruments in recent years. Very likely, magnetic clouds are interplanetary manifestations of coronal mass ejections (Burlaga et al., 1982; Wilson and Hildner, 1986). Moreover, many coronal mass ejections occurred in conjunction with eruptions of solar prominences (Hildner, 1977; Wagner, 1984).

Some of the prominences and coronal mass ejections appear loop-like morphologically. So do magnetic clouds. It is desirable to have a unified description of the motions of this class of loop-like magnetic objects. Flux rope dynamics provides such a framework. Its theoretical basis is the notion of hydromagnetic buoyancy. The theory of hydromagnetic buoyancy force was originally advanced to provide an explanation for loop-like coronal mass ejections (Yeh, 1982). Recently, with suitable modifications, that theory was applied to magnetic clouds (Yeh, 1988a) and loop prominences (Yeh, 1988b).

In this paper, we shall give a non-mathematical description of the flux rope dynamics for loop prominences, coronal mass ejections, and interplanetary magnetic clouds. Since no analytical tractability is needed, our discussion will not be restricted to flux ropes with circular cross sections.

Flux Rope as an Extraneous Body

A flux rope is composed of mass elements constricted to field lines that wind toroidally. It is immersed as an extraneous body in a vast magnetized medium. The immersion is an instantaneous state of continual intrusion, because, in general, a flux rope keeps moving through the background medium. The pre-existing conditions are perturbed by the presence of the flux rope.

The presence of a flux rope precludes the pre-existing magnetic field from the region occupied by the flux rope. The polarization current induced on the interface between the flux rope and the surrounding medium separates the internal and external magnetic fields. The internal field is produced solely by the internal current that circulates toroidally inside the flux rope, because the external current is shielded by the peripheral current in the sense that together they produce a null field in the interior region. On the other hand, the external field is produced jointly by the external, peripheral, and internal currents. The external current produces the pre-existing part. The perturbation part produced by the peripheral current makes the external field tangential at the interface. The internal current is so distributed that the magnetic field produced by it is tangential at the peripheral boundary.

Because the electrical conductivity of the plasma is very high, the peripheral layer in which the polarization current is induced is very thin. This thin layer has a negligible mass. Both the boundary magnetic field of the internal field on the inner surface of the peripheral layer and the ambient magnetic field of the external field on the outer surface are tangential. Their difference is accommodated by the sheet current formed by the polarization current.

External and Internal Forces

The masses and currents in the flux rope are subjected to forces exerted by masses and currents at a distance. These are gravitational and magnetic forces, respectively. In addition there are thermal forces, caused by molecular collisions of the plasma particles.

First, consider gravitational forces between masses. The external gravitational forces are by and large due to the massive Sun. Contributions from other masses are negligible. The internal gravitational forces, whose magnitudes are trivially small, sum up to zero. This is so because action and reaction are equal in magnitude but opposite in direction. The point we want to make is that internal forces are not capable of causing the displacement of a body's centroid. This holds even for internal magnetic forces.

Next, consider magnetic forces between currents. The magnetic forces exerted on the internal and peripheral currents by the external current are external forces. The magnetic forces exerted on various parts of the internal and peripheral currents by other parts of them are internal forces. Again, the sum of the internal magnetic forces is zero.

By its nature of transmitting stresses from one region to another, thermal force may arise from transformation from gravitational and magnetic forces. They may also be caused by the thermal pressure exerted by the surrounding medium. Thus, the thermal forces may be regarded as comprising two parts: externally-caused and internally-caused. The sum of the internally-caused thermal forces is zero.

Hydromagnetic Buoyancy Force

By the relationship between force density and stress, the volume integral of magnetic force density is equal to the corresponding surface integral of magnetic stress, and the volume integral of thermal force is equal to the corresponding surface integral of thermal pressure. Accordingly, because the volume integral of the internal magnetic force is zero, the volume integral of the external magnetic force is simply equal to the surface integral of the ambient magnetic stress. Since the magnetic field has no normal component at the periphery, only the magnetic pressure, not the magnetic tension, will contribute to this surface integral of magnetic stress. This provides a way to determine the resultant external magnetic force without knowing the spatial distributions of the external current. What we have to do is the calculation of the perturbation to the pre-existing magnetic field so that the resulting external magnetic field is tangential at the interface between the flux rope and the surrounding medium. This is a boundary value problem of magnetism. The surface integral of the ambient magnetic stress yields the diamagnetic force (Yeh, 1983), which is nothing but the sum of the external magnetic forces exerted on the flux rope by the external current that sustains the pre-existing external magnetic field.

Likewise, the surface integral of the ambient thermal pressure yields a thermal force exerted on the flux rope by the external medium. This force includes the well-known hydrostatic buoyancy force, because part of the ambient thermal pressure is hydrostatic. The inhomogeneity of hydrostatic pressure is caused by the effect of gravity. The remaining part of the ambient thermal pressure is caused by the flow motion of the external medium. This additional hydrodynamic buoyancy force is very much like the aerodynamic lift force on an aerofoil. Calculation of the perturbation to the pre-existing pressure that results from realignment of the flow to become tangential to the flux rope is a boundary value problem of hydrodynamics.

Altogether, the surface integral of the hydromagnetic pressure (viz., the sum of thermal and magnetic pressures) yields the hydromagnetic buoyancy force (Yeh, 1985), which includes hydrostatic, hydrodynamic, and diamagnetic buoyancy forces. The hydromagnetic buoyancy force is the external force exerted by the magnetized medium in which a flux rope is immersed as an extraneous body. This external force acts on the mass elements of the flux rope through the gradient of internal stress that matches the ambient stress.

Hydromagnetic Stress in a Flux Rope

Stress is spatially continuous. By spatial transmission and transformation of stress, thermal pressure in one region can be matched by magnetic stress in a neighboring region, and vice versa. In this manner, thermal force cushions the spatial unevenness of magnetic force.

First, consider the stress in the peripheral layer of polarization current. Since this thin layer is essentially massless, it is impossible to have gravitational force or inertial force to counterbalance or bear any unbalanced hydromagnetic forces. Hence the magnetic force density must be balanced exactly by the thermal force density everywhere in the peripheral layer. Consequently, the hydromagnetic pressure is invariant across the thin peripheral layer. In other words, the hydromagnetic pressures on the two sides of the peripheral current sheet have the same distributions. The normal transmission of the circumferential inhomogeneity in hydromagnetic pressure means that the inhomogeneity in ambient magnetic pressure on the outer surface is transformed into inhomogeneity in boundary thermal pressure on the inner surface. This is so because the boundary magnetic pressure on the inner surface is rather homogeneous circumferentially.

Now, since the boundary thermal pressure is matched pointwise by the internal thermal pressure, the inhomogeneity in boundary thermal pressure is mapped into inhomogeneity in internal thermal pressure. A part of this internal inhomogeneity gives rise to a force density that accounts for the spreading of the external hydromagnetic force. The remaining part gives rise to a force density which, when combined with the Lorentz force density associated with the internal magnetic stress, accounts for the internal hydromagnetic force. The externally-caused thermal pressure gradient manifests the hydromagnetic buoyancy force in the interior of the flux rope. The internally-caused thermal pressure gradient sustains the pressure difference between internal and external thermal pressures.

Translational and Expansional Motions

Before we discuss the motion of a flux rope, we shall make a general remark about translational and expansional motions of a non-rigid body. The motion of a non-rigid body involves its displacement as a whole and the change of its volume due to the changes of distances among its mass elements. The former can be described by the translational motion of its centroid and the latter by the expansional motion of its mass elements relative to the centroid.

An extraneous body immersed in a magnetized medium is subjected to external and internal forces. The gravitational force exerted by the Sun and the hydromagnetic buoyancy force exerted by the surrounding medium are external forces. The internal forces are entirely hydromagnetic. The Lorentz force density inside an immersed body is an internal force, because its magnetic stress is entirely due to the internal current. It does not involve the external current, whose effect is nullified by the peripheral current.

Internal forces, which sum up to zero, are not capable of causing the displacement of the centroid. Translational motion is driven by external forces, not by internal forces. Internal forces drive the expansional motion.

Motions of a Flux Rope

Topologically, a flux rope is a doubly-connected body, exemplified by a toroid. Flux ropes near the solar surface have their lower portions submerged beneath the solar surface. Hence as far as prominences and coronal mass ejections (and perhaps also interplanetary magnetic clouds) are concerned, we are dealing with upper portions of flux ropes, each with two legs embedded in the solar surface.

A flux rope in the astrophysical context is rather slender. Its axial length is much greater than its cross-sectional lengths. Therefore, it makes more sense to consider a centroidal axis instead of a centroidal point for a whole flux rope or for its portion above the solar surface. Accordingly, a flux rope in our consideration may be regarded as a stack of thin frustum sections. Each frustum has a centroid. These contiguous centroids constitute a centroidal axis.

Each frustum section of a flux rope has its own translational motion. The centroid of each frustum has transverse and longitudinal motions. The hydromagnetic pressure on its lateral surface drives the transverse motion, the hydromagnetic pressure on its two end surfaces drives the longitudinal motion, whereas the magnetic tension on the end surfaces and the gravitational force drive both motions. In this treatment, the forces exerted by the currents in other frustums are external forces for the frustum under consideration. The differential motion between neighboring centroids will cause the stretching of the centroidal axis of the flux rope.

As to the expansional motion of a mass element relative to the pertinent centroid, it is driven by a force density equal to the total force density less the inertial force density of the translational motion. The latter inertial force density is uniform in a frustum, given by the sum of external forces divided by the sum of masses.

Conclusion

Movement of a flux rope or its portion above the solar surface toward or away from the Sun involves the displacement of its centroidal axis. It is a translational motion. Translational motion is driven by external forces, not by internal forces. The viable external force that can overcome solar gravity, which is itself an external force, is the hydromagnetic buoyancy force. Therefore, for a flux rope to move away from the Sun, it is utterly necessary to have sufficiently large hydromagnetic buoyancy force exerted by the surrounding medium to overcome inward solar gravity and provide outward acceleration. Self-propulsion (Mouschovias and Poland, 1978; Anzer, 1978) is not attainable, even just to defy solar gravity.

In this non-mathematical description of flux rope dynamics, the discussion has been focused around Newton's law of motion. To give a mathematical description, it is necessary to include all equations of magnetohydrodynamics. In addition to obeying Newton's law of motion, flux rope dynamics must also satisfy the physical laws for conservation of mass, energy and magnetic flux.

Acknowledgment. The author thanks Ernest Hildner for reading the manuscript. This work was supported by Space Environment Laboratory, National Oceanic and Atmospheric Administration.

References

Anzer, U., Can coronal loop transients be driven magnetically?, Solar Phys., 57, 111, 1978.

Burlaga, L. F., E. Sittler, F. Mariani, and R. Schwenn, Magnetic loop behind an interplanetary shock: Voyager, Helios, and IMP 8 observations, J. Geophys. Res., 86, 6673, 1981.

Burlaga, L. F., L. W. Klein, N. R. Sheeley, Jr., D. J. Michels, R. A. Howard, M. J. Koomen, R. Schwenn, and H. Rosenbauer, A magnetic cloud and a coronal mass injection, Geophys. Res. Lett., 9, 1317, 1982.

Hildner, E., Mass ejections from the solar corona into interplanetary space, in Study of Travelling Interplanetary Phenomena, edited by M. A. Shea, D. F. Smart, and S. T. Wu, p. 3, Reidel Publishing Co., Dordrecht, 1977.

House, L. L., W. J. Wagner, E. Hildner, C. Sawyer, and H. U. Schmidt, Studies of the corona with the Solar Maximum Mission Coronagraph/Polarimeter, Ap. J. Lett., 244, L117, 1981.

Howard, R. A., D. J. Michels, N. R. Sheeley, Jr., and M. J. Koomen, The observation of a coronal transient directed at earth, Ap. J. Lett., 263, L101, 1982.

Klein, L. W., and L. F. Burlaga, Interplanetary magnetic clouds at 1 AU, J. Geophys. Res., 87, 613, 1982.

MacQueen, R. M., J. A. Eddy, J. T. Gosling, E. Hildner, R. H. Munro, G. A. Newkirk, Jr., A. I. Poland, and C. L. Ross, The outer solar corona as observed from Skylab: Preliminary explanation, Ap. J., 220, 675, 1978.

Mouschovias, T. Ch., and A. J. Poland, Expansion and broadening of coronal loop transients: A theoretical explanation, Ap. J., 220, 675, 1978.

Wagner, W. J., Coronal mass ejections, Ann. Rev. Astron. Astrophys., 22, 267, 1984.

Wilson, R. M., and E. Hildner, On the association of magnetic clouds with disappearing filaments, J. Geophys. Res., 91, 5867, 1986.

Yeh, T., A magnetohydrodynamic theory of coronal loop transients, Solar Phys., 78, 287, 1982.

Yeh, T., Diamagnetic force on a flux rope, Ap. J., 264, 630, 1983.

Yeh, T., Hydromagnetic buoyancy force in the solar atmosphere, Solar Phys., 95, 83, 1985.

Yeh, T., A model structure of magnetic clouds at 1 AU, to appear in Proceedings of Symposium on Physical Interpretation of Solar/Interplanetary and Cometary Intervals (held Huntsville, Alabama, May 12-15, 1987), 1988a.

Yeh, T., A dynamical model of loop prominences, Solar Phys., to be submitted, 1988b.

What macroscopic and microscopic processes are responsible for particle acceleration in the solar wind and in planetary magnetospheres?

ACCELERATION AND TRANSPORT IN THE PLASMA SHEET BOUNDARY LAYER

Maha Ashour–Abdalla and David Schriver

Institute of Geophysics and Planetary Physics, University of California, Los Angeles, CA 90024
Department of Physics, University of California, Los Angeles, CA 90024

Abstract. Observations in the plasma sheet boundary layer have indicated the presence of field–aligned beams, ionospheric plasma and intense broadband electrostatic noise. In this paper we use analytic theory and numerical simulations to investigate the linear properties and nonlinear consequences of instabilities caused by two different free energy sources. The free energy sources considered are an ion beam and electron current. It is found that for the case of an ion beam free energy source, the presence of a cold electron background allows the electron acoustic instability to be excited. This instability rapidly heats the cold electrons. Once the cold electrons are heated, the interaction of the ion beams with the ionospheric ions permits the ion–ion mode to be unstable. The ion–ion instability saturates by heating the background ions and, to a lesser extent, the beam ions. The ion–ion instability is oblique when the ion drift speed is greater than the sound speed ($U > C_s$) and occurs in the direction parallel to the ambient magnetic field for $U \leq C_s$. Considering the other free energy source, the electron current, the electron acoustic instability as well as a lower frequency electron–ion instability are excited. The high–frequency electron acoustic heats both the hot and cold electrons whereas the electron–ion instability heats the ions as well. This paper discusses the implications of these results for plasma transport and makes suggestions for additional experimental measurements that need to be made in the plasma sheet boundary layer.

1. Introduction

It has become apparent over the past few years that the plasma sheet boundary layer (PBSL) is an enduring feature that exists continuously in the magnetotail (Eastman et al., 1984; Huang et al., 1987). Although the spatial location and extent of the PSBL may vary in time, its relative position remains fixed as a moderately thin wedge that lies between the lobe (a volume of relatively low plasma density and high magnetic field intensity) and the central plasma sheet (the domain of hot, high density plasma and a less intense magnetic field).

The PSBL appears to be a permanent feature of the Earth's magnetotail; it is found during both quiet and active magnetic periods and at all times represents a major region of plasma transport through the tail (DeCoster and Frank, 1979; Lui et al., 1983; Eastman et al., 1984; Huang et al., 1987). In this highly dynamic region of space, fast plasma flows, field–aligned beams, plasma inhomogeneities, and ionospheric plasma are all observed. A prominent feature of the PSBL is the presence of intense broadband electrostatic noise (BEN). These waves were first noted by Scarf et al. (1974) who observed them in IMP 7 data and are characterized by their wide range of frequencies, from about 10 Hz all the way to the electron plasma frequency near 10 kHz, and large amplitudes in the range of mV/m (Gurnett et al., 1976).

The cause of BEN is not completely understood, although as suggested by Cattell and Mozer (1986), it is probably a combination of instabilities that generates the entire frequency spectrum. The peak in wave power is at low frequencies (< 50 Hz) and is near the lower hybrid frequency (Cattell and Mozer, 1986). Ashour–Abdalla and Thorne (1986) suggested that BEN is generated by current–driven ion cyclotron waves, but the instability excites frequencies that are too low to account for the entire BEN spectrum. Another possible cause is the lower hybrid drift instability (Huba et al., 1978) driven by density gradients. This instability, however, is not broadbanded enough to be responsible for the entire BEN spectrum, which is observed to extend up to the electron plasma frequency, near 10 kHz (Gurnett et al., 1976).

More recently, Grabbe and Eastman (1984) used ISEE data to show that the presence of BEN is correlated with that of ion beams. This important observation prompted their extensive analytic study of ion beam instabilities in the geomagnetic tail. Grabbe and Eastman (1984) showed that an ion beam streaming through a background of hot plasma sheet electrons and ions can excite waves with a broad range of frequencies, provided the ion beam is sufficiently cold.

The work of Grabbe and Eastman (1984) generated considerable interest in BEN and resulted in several studies. First, Omidi (1985) and Akimoto and Omidi (1986) correctly identified the two unstable wave modes for the above plasma configuration as being the ion acoustic and ion–ion instabilities. For PSBL conditions the ion acoustic instability excites waves with a frequency near $0.1\omega_{pe}$ while the ion–ion mode drives lower frequency waves of $\sim 0.01\omega_{pe}$.

Dusenbery and Lyons (1985) noted that the theoretical analysis of Grabbe and Eastman (1984) did not explain the association of intense BEN with warm ion beams in the PSBL. Dusenbery and Lyons (1985) therefore investigated the wave generation, including both warm and cold ion beams in their study. They found that, in agreement with Grabbe and Eastman (1984), warm beams alone were insufficient to generate the waves. On the other hand, wave growth was enhanced significantly when both warm and cold ion beams were present simultaneously. Their model predicted that the wave intensity of the broadband noise would peak in the plasma sheet boundary layer, whereas observations of the less intense electrostatic waves in the lobes and plasma sheet were attributed to an absence of hot ion beams and large ion temperatures, respectively, which result in smaller growth rates.

The above authors concentrated on the mechanism whereby linear electrostatic instabilities excite broadband electrostatic noise. In order to understand the effects of the waves on the particle distribution as well as to obtain a wave spectra directly comparable with satellite data, Ashour-Abdalla and Okuda (1986a) undertook a simulation study of broadband electrostatic noise driven by ion beams. They found that, by varying the temperature of the beam and the drift velocity of the beam, they could excite ion acoustic waves, ion–ion waves, or ion cyclotron waves. When the temperature of the ion beam T_b was much lower than that of the plasma sheet ions, both the ion acoustic waves and the ion–ion waves, were unstable. When the beam temperature was comparable to that of the plasma sheet ions, ion cyclotron waves grew in the frequency range $\omega = n\omega_{ci}$, where ω_{ci} is the ion cyclotron frequency. However, BEN cannot be attributed to ion cyclotron wave growth, since its growth rate cannot account for broadband noise extending to high frequencies.

To overcome the difficulty of the observed ion beams being too warm to be unstable, Grabbe (1985) prompted by the observations of cold ionospheric electrons (Etcheto and Saint-Marc, 1985) in the PSBL included them in the dispersion relation. Not only did Grabbe (1985, 1987) find that the upper frequency cutoff of the ion beam instability in the presence of cold electrons approaches ω_{pe}, but it was also shown by Schriver and Ashour-Abdalla (1987) that warm ion beams continue to be strongly unstable over the broadbanded frequency range. The addition of cold electrons drastically alters the wave dispersion relation by allowing new wave modes to be unstable in the presence of ion beams. These wave modes are the electron acoustic and beam resonant instabilities, both of which require cold electrons to exist and are unstable in the presence of warm ion beams (Ashour-Abdalla and Okuda, 1986b; Grabbe, 1987; Schriver and Ashour-Abdalla, 1987).

In a recent study, Schriver and Ashour-Abdalla (1988) carried out a number of simulation runs from various regions of the PSBL in an attempt to understand the relationship between ion beams and BEN, and to form a global picture of how the PSBL serves as a heating region in which cold ionospheric plasma passes through to become part of the hot central plasma sheet. They found that at the edge of the PSBL, electron acoustic ion beam resonant instabilities were excited which led to rapid heating of the cold electrons. Nearer to the central plasma sheet, the ion–ion mode was excited which resulted in both perpendicular ion heating ($U > C_s$) and parallel ion heating ($U \leq C_s$).

In keeping with the theme of the conference and defining what measurements are required to advance the theory, in this paper we will examine several free energy sources for the generation of BEN. Specifically we consider field–aligned currents (Frank et al., 1981; Frank, 1985) as well as warm and cool ion beams (Eastman et al., 1984; Sharp et al., 1981). Where appropriate, these instabilities are considered in both the absence and presence of cold ionospheric plasma. In conducting such a systematic study of free energy sources and the resultant instabilities we have been able to identify what measurements are needed to further our physical understanding of transport in the PSBL.

In section 2 we will consider instabilities due to ion beams in the presence of cold electrons. Section 3 concentrates on wave–particle interactions for the case of a source of ion beams in the absence of cold electrons. Last, we consider the effect of the electron current energy source in section 4. For all these different configurations we discuss both linear theory and results from particle simulations. In section 5 we summarize our findings and make suggestions for future experimental studies.

2. Instabilities Due to Ion Beams in the Presence of Cold Electrons

In this section we consider instabilities due to an ion beam free energy source in the presence of ionospheric plasma. A schematic of the distribution function is shown in Figure 1a, where the ion beam is assumed to be a drifting Maxwellian with temperature T_b and drift speed U, whereas the cold electrons are modeled as an isotropic Maxwellian with temperature T_c. Such a configuration may exist near the lobeward edge of the PSBL. This type of configuration would be the most commonly occurring one in the PSBL, since over most of the length of the tail there is thermal mixing of cold lobe plasma with the hot plasma sheet. Energetic ion beams are created in the tail, streaming earthward and mirror reflecting near the Earth as shown by the black arrows in Figure 2. As discussed by Takahashi and Hones (1988), plasma convection ($\overline{E} \times \overline{B}$ drift) may carry lobe plasma across the boundary layer into the central plasma sheet in the general direction shown by the large open arrows. In addition, cool ionospheric plasma is streaming along magnetic field lines out from the auroral zone directly into the PSBL (small black arrows in Figure 2), and, as it encounters the warm ion beams, it may be heated as it convects into the central plasma sheet. Observations have shown that upon entering the PSBL from the lobe, the satellite first encounters hot electrons and then, somewhat deeper in the PSBL toward the central plasma sheet, encounters the first (and fastest) earthward streaming beams. Further into the PSBL, the earthward streaming speed decreases gradually, after which counterstreaming beams are observed. This structure is shown graphically in Figure 3 from Takahashi and Hones (1988).

Schematic of Magnetotail

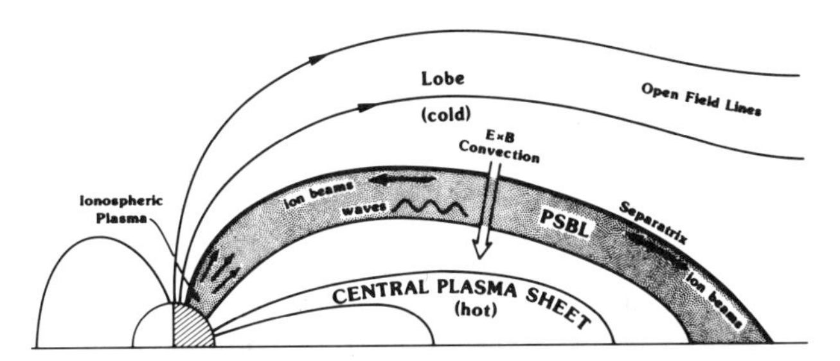

Fig. 2. A schematic of the Earth's magnetotail in the north–south plane showing the relative position of the lobe, plasma sheet boundary layer (PSBL), and the central plasma sheet.

Ion Beam Configurations

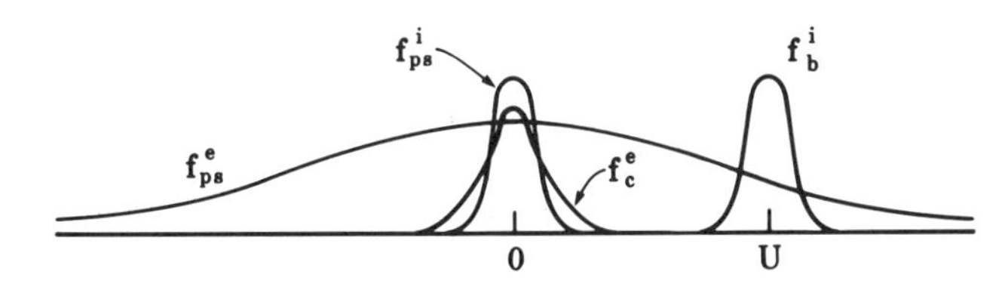

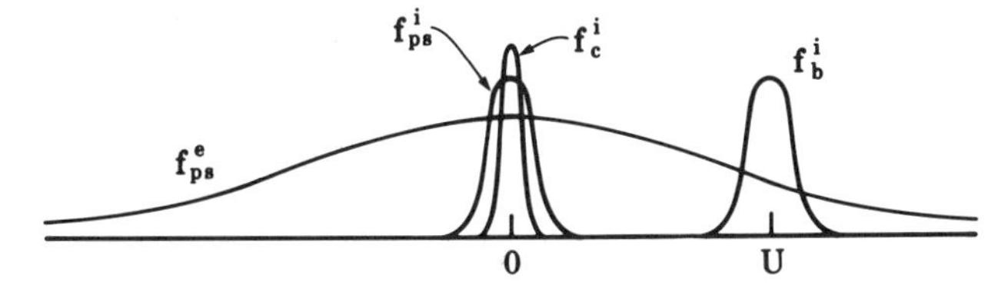

Fig. 1. A schematic model used for the linear theory and simulation. In panel (a) we consider a four-component plasma: an ion beam, warm plasma sheet ions and electrons and cold ionospheric electrons. In panel (b) we consider an ion beam, warm plasma sheet ions and electrons, and cold ionospheric ions.

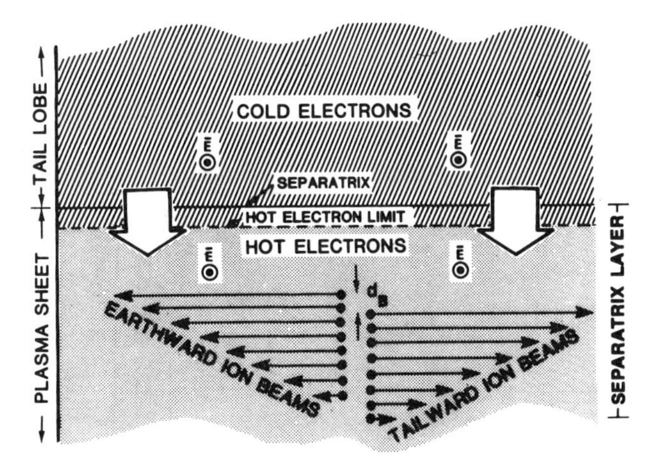

Fig. 3. Morphology of ion beams in the plasma sheet boundary layer from Takahashi and Hones (1988). The gap in vertical distance (labeled d_B) between the earthward and tailward ion beams is due to the fact that the tailward beams are a result of near-Earth magnetic reflection of the earthward beams. Note the gradual decrease of the beam speed in plasma sheet. The large white arrows indicate the direction of plasma convection.

Since cold lobe plasma convects across the separatrix into a region of hot plasma and ion beams, a thermally mixed background plasma must be considered. This type of configuration, found over most of the length of the PSBL, permits the excitation of a number of instabilities. The parameters most critical in determining which instability will be excited are the temperature and drift speed of the ion beam. Ion beam temperatures measured in the PSBL are usually on the order of hundreds of electron volts (Eastman et al., 1984; Takahashi and Hones, 1988). This is considered as the high (warm) beam temperature limit, and the two beam instabilities present are the electron acoustic and beam resonant instabilities (Ashour–Abdalla and Okuda, 1986b; Schriver and Ashour–Abdalla, 1987). The growth rate of the beam resonant instability will be lower in the presence of hot electrons than in the case of an ion beam and cold electrons only. This is because the hot electron velocity distribution overlaps the ion beam distribution such that the hot electron negative slope will tend to compete with the ion beam positive slope. Since the cold electrons are a minority species in this case, we will concentrate on the electron acoustic instability.

The electron acoustic instability is a normal mode of a plasma that has two electron components of different temperatures and is present regardless of whether or not the beam is there (Watanabe and Taniuti, 1977). In the presence of an ion beam, an electron acoustic mode may become unstable when the electron acoustic phase velocity is in resonance with the positive slope of the beam distribution (Ashour–Abdalla and Okuda, 1986a). Figure 4 also shows a diagram of growth rate maximized over wave number and angle versus real frequency for three different cold electron to total density ratios (n_c/n_o). Note that the maximum unstable frequency for the electron acoustic waves driven by ion beams is near ω_{pc}/ω_{pe}. Figure 4 shows a broadband wave spectrum with peak growth above $0.1\omega_{pe}$.

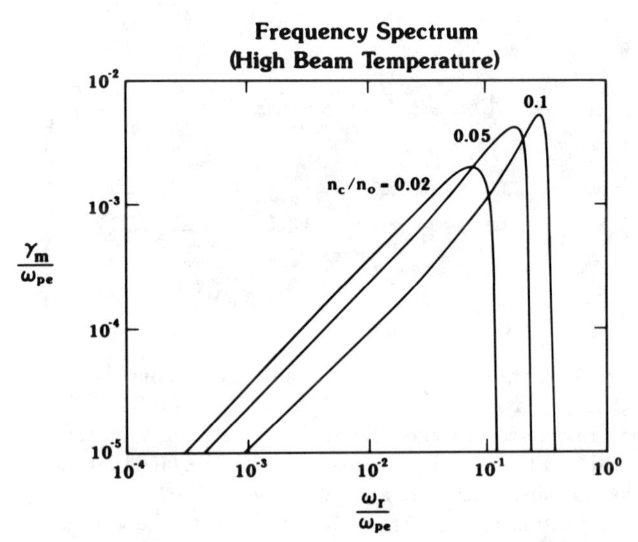

Fig. 4. Growth rate maximized over all wave numbers plotted versus real frequencies for three different cold electron to total density ratios n_c/n_O.

Numerical simulations have been carried out using a one–dimensional electrostatic particle model in the presence of a uniform external magnetic field. As shown in Figure 1, four different components of the particles are considered: hot and cold electrons and thermal and beam ions. The simulation parameters are mass ratio, $m_i/m_e = 400$ (which is reduced for simulation tractability since the full dynamics of both ions and electrons must be followed), hot to cold electron temperature ratio, $T_h/T_c = 100$, hot electron to ion temperature ratio, $T_h/T_i = 1$, and hot electron to ion beam temperature ratio, $T_h/T_b = 1$. The choice of $T_h = T_i = T_b$ guarantees that the ion acoustic wave will be damped. The system length was chosen as $L = 256\Delta$ where Δ is the unit grid scaled to $\Delta = 1/8\lambda_h$. Other parameters used are $n_H = 0.8n_O, n_c = 0.2n_O, n_i = n_b$, and $U/V_b = 12.5$. We assume the wave number k is in the direction of B_O. Figure 5 shows the time history, $e\phi_k/T_e \sim 0.02$ near $\omega_{pe}t = 150$ and decreases in amplitude thereafter. The corresponding power spectrum clearly indicates the presence of a peak around $\omega/\omega_{pe} = 0.3$.

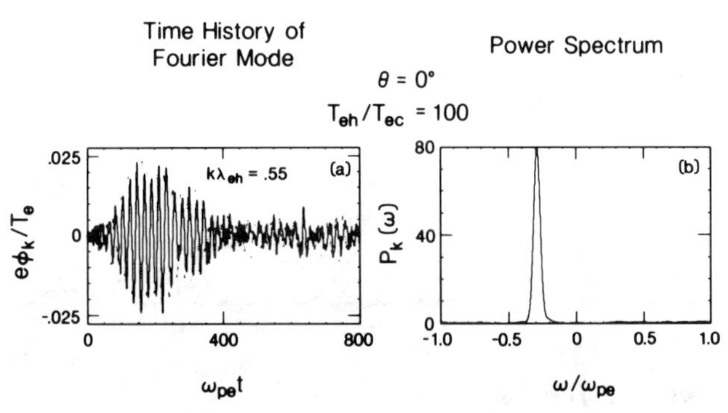

Fig. 5. Time history, electrostatic potential versus time (left panel), and the power spectrum (right panel). In this case, $n_H = 0.8n_O$, $n_c = 0.2n_O$, $n_i = n_b$, and $U/V_b = 12.5$.

Now this was the power spectrum for an individual Fourier mode, but to allow comparison of the simulation results with satellite observations we must sum up all the wavelength modes from the simulation data. Figure 6 shows the result of this summing up; one obvious difference is a much broader distribution than in Figure 5. This is reminiscent of the broadband electrostatic waves observed on ISEE 1 and extends all the

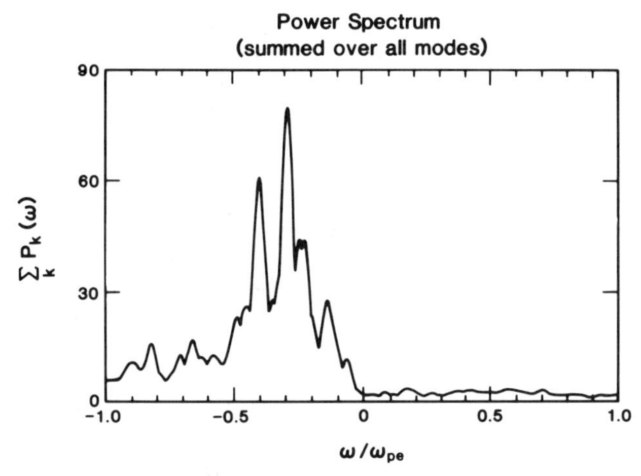

Fig. 6. Power spectrum summed over all wave numbers. Taking into account all Fourier modes illustrates the broadbanded nature of the instability.

way to the cold electron plasma frequency. It should be noted that while the amplitude of the electrostatic potential is smaller than that for the ion–acoustic waves (Ashour-Abdalla and Okuda, 1986a), the fastest growing modes for the electron acoustic waves occur at large wave numbers resulting in an electric field $E \sim mV/m$, in agreement with satellite observations.

What are the nonlinear saturation mechanisms? Figure 7 shows the phase space plots for the cold electrons at different times. This figure shows that the cold electrons are trapped first by the unstable waves which are forming holes in phase space at $\omega_{pe} = 200$. The deterioration of this coherent structure seen at $\omega_{pe}t = 400$ results in the strong heating of cold electrons.

Figure 8 confirms this statement. In panel (a) we plot the electric field energy, (b) the hot electron temperature, (c) the cold electron temperature, and (d) the cold electron velocity distribution function. It is interesting to note that the hot electron temperature changes slightly, whereas the cold electron temperature increases significantly. This can also be seen in panel (d). In fact the hot electrons are heated during the unstable phase as shown for $\omega_{pe}t < 300$, but they lose energy afterward. It is clear that during the unstable phase both cold and hot electrons are heated as they absorb energy from the drifting ion beam. When the instability saturates, cold electrons continue to be heated, albeit at a slower rate, whereas the hot elec-

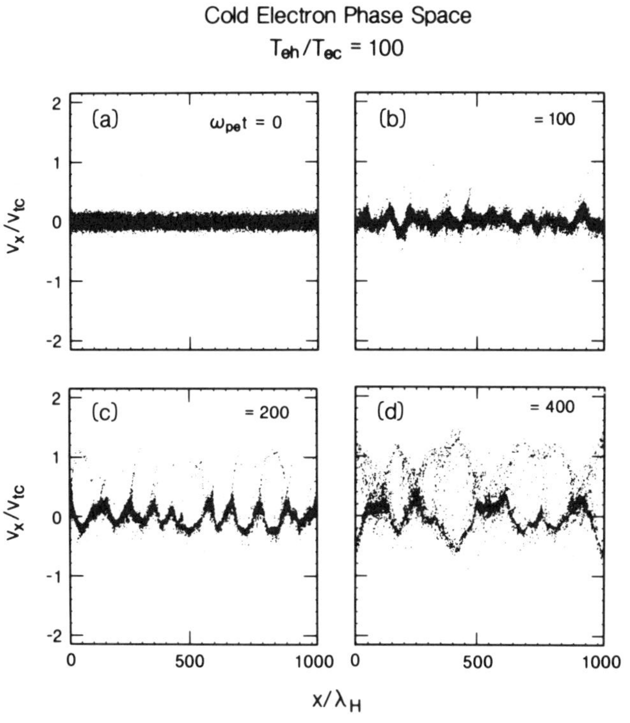

Fig. 7. Cold electron phase space evolution for the same simulation run as in the previous two figures. Electron phase trapping occurs in panel c, and as the wave amplitude decreases after separation, the electrons show significant heating (panel d).

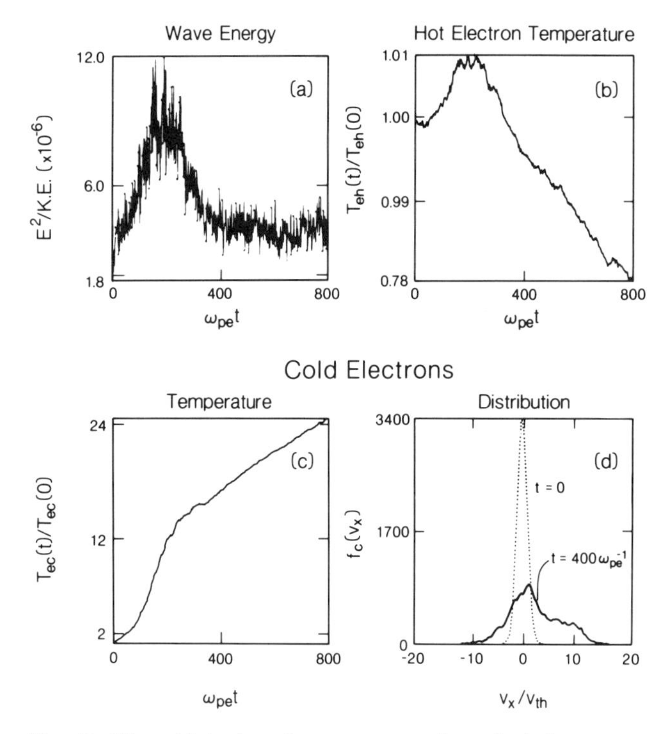

Fig. 8. Time histories of wave energy (panel a), hot electron temperature (panel b), cold electrons (panel c), and the cold electron velocity distribution (panel d). Note that after the wave saturates, the cold electrons continue to heat while the hot electrons lose energy.

trons lose energy. This heat loss is caused partly by collisional coupling between hot and cold electrons, which transfers hot electron energy to cold electrons. However the energy gained by the cold electrons when $\omega_{pe}t > 300$ is more than the energy loss from the hot electrons. Obviously the additional energy gained by the cold electrons is taken form ion kinetic energy, which is transferred in the form of marginally unstable electron acoustic waves. The linear theory and simulation results clearly show how the addition of cold electrons significantly enhances wave growth by a warm ion beam. Without cold electrons, no wave growth is possible. With cold electrons, a broadbanded, intense wave spectrum is excited which in turn heats the cold electrons. The heating of the cold plasma may be indicative of the process by which ionospheric plasma gains energy to become ultimately part of the hot plasma sheet itself.

As shown above, the electron acoustic instability will heat the cold electrons rapidly, so as we look deeper in the interior of the PSBL or closer to the central plasma sheet, the ionospheric electrons will be heated and only cold background ions will remain. This leads us to consider our next plasma configuration, ion beams in the presence of ionospheric ions and hot electrons.

3. Instabilities Due to Ion Beams in the Absence of Cold Electrons

The setup for this case is similar to the original plasma configuration by Grabbe and Eastman (1984) (see Figure 1 b). They considered a single ion beam streaming relative to an ion and electron background, but, if a counterstreaming beam is included, the physics is essentially the same. Akimoto and Omidi (1986) first identified the instabilities for this configuration as the ion acoustic and ion–ion instabilities. The ion acoustic instability is a negative energy wave that results from the interaction between the ion beam and hot electron background (Fried and Gould, 1961). It is a resonant instability driven by the slope of the hot electrons. Thus it is very sensitive to the hot electron and ion beam temperature. The ion–ion instability is a streaming instability that results from the net drift between the two ion species (Fried and Wong, 1966; Gary and Omidi, 1987).

When the relative drift speed between the two ion distributions is greater than the sound speed, both instabilities are driven. In this case the ion acoustic mode is unstable at parallel propagation and the ion–ion mode is unstable at an oblique wave propagation angle given by $\cos\theta \simeq C_s/U$, where $C_s = (T_e/m_i)^{1/2}$ = sound speed, U = relative ion drift speed, and θ = propagation angle (Omidi, 1985). The ion acoustic instability excites higher frequencies, while the ion–ion instability has larger growth rates at lower frequencies, and thus the two instabilities can combine to excite a broadbanded range of frequencies. This is seen in Figure 9, which shows growth rate maximized over all wave numbers versus real frequency for various beam temperatures. The parameters for this figure (9) are $U = 0.2V_{th}, \Omega_e/\omega_{pe} = 0.05, T_h = 4T_i$, and $n_b = 0.6n_h$. The species subscripts are h (or e) for the hot background electrons, i for background ions, and b for beam ions. For $T_h/T_b > 10$, two peaks in the growth rates appear, one at about $0.1\omega_{pe}$. The low–frequency peak is caused by the ion–ion instability and occurs at oblique wave propagation. The angle at maximum growth for the ion–ion instability is shown on Figure 9 and occurs at $\theta \geq 76°$, depending on T_h/T_b. This agrees with the expected angle given by the expression $\theta \simeq \arccos(C_s/U)$. The second, smaller peak in growth rate at high frequencies is due to the ion acoustic instability: the peak growth is at parallel propagation. As the ion beam temperature is increased such that T_h/T_b decreases, the growth rates of both instabilities decrease. However, the ion–ion mode is unstable for warmer beams, whereas the ion acoustic instability vanishes for $T_h/T_b < 10$.

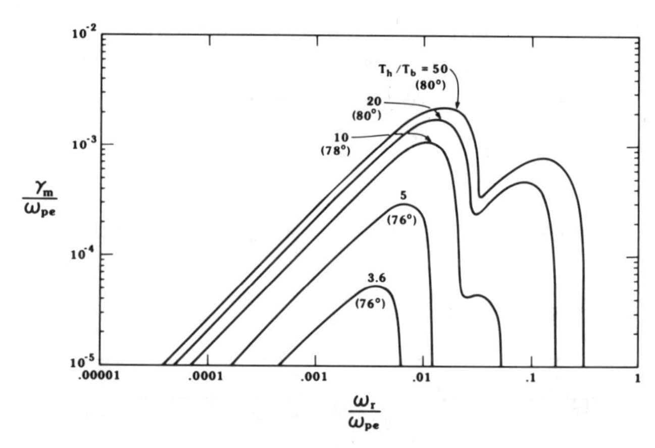

Fig. 9. Growth rates versus frequency for the case of an ion beam drifting with respect to hot electrons and background ions. The beam temperature is varied (T_h held fixed), and when $T_h/T_b < 10$ only the ion–ion mode is unstable at $\omega \sim 0.01\omega_{pe}$. For colder beams $T_h/T_b > 10$, the ion acoustic instability grows at frequencies near $0.1\omega_{pe}$.

Since the ion acoustic and ion–ion instabilities are driven by interactions among the different plasma species, different parameters determine the instability criteria. For the ion acoustic instability T_h/T_b determines stability. For the warm ion beams streaming through hot electrons in the PSBL, $T_h = T_b$, and the ion acoustic mode is stable. For the ion–ion mode, however, T_i/T_b is the important parameter, and for cold background ions, this ratio is about 0.002 to 0.01.

Now turning to simulations and concentrating on the cold ions entering the PSBL from the lobe, the ions first encounter fast moving beams, which in general will have a relative drift speed greater than the sound speed. The ion–ion instability will be oblique to the magnetic field (according to cos $\theta \sim C_s/U$), and the instability is saturated by ion heating. The heating is along the direction of wave propagation and is illustrated in Figure 10 in terms of velocity phase space which shows the positions of the ion beam and ion background particles in velocity space ($v_{\parallel}$ vs. $v_{\perp}$). The left two panels show the initial distribution at $t = 0$ (top) and at $t = 2000\omega_{pe}^{-1}$ (lower panel) for a propagation angle of 70°. The beam speed for this run is $U = 0.14V_{th} = 2.8C_s$, and heating maximizes at this angle (70°) as expected from cos $\theta = C_s/U$. Ion heating does occur at other angles, but gradually decreases as the measurements change from 70°. Note from Figure 10 that the cold ions are heated the most, by a factor of nearly 100, but the beam is heated somewhat as well. As the ions continue convecting into the PSBL, the beam speed decreases and the heating will shift to less oblique angles. This is shown in the right two panels of Figure 10 where the beam speed is $U = 0.05V_{th} = C_s$ and the instability is parallel propagating. The lower panel on the right shows the heating in the parallel direction, and the two distributions merge to form one heated distribution. In both of these simulation runs, the background hot electrons are not affected at all.

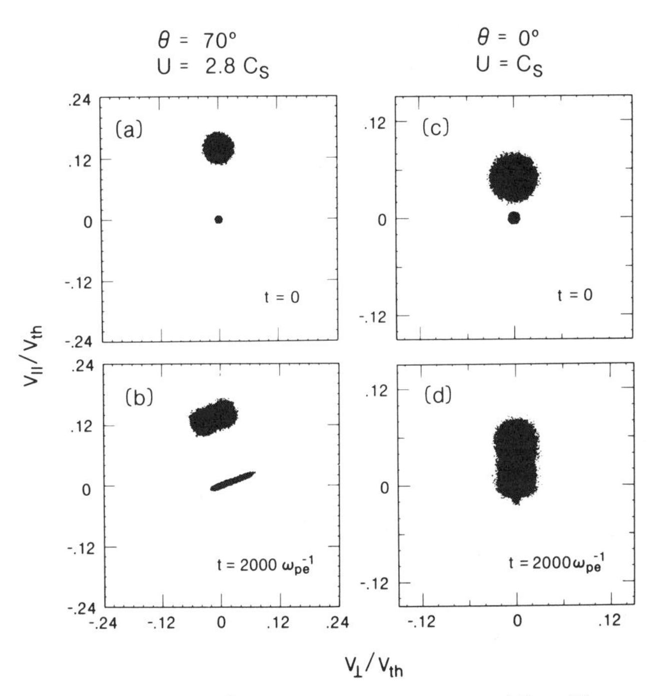

Fig. 10. Velocity phase space $v_{\perp}$ versus $v_{\parallel}$ with cold ion background centered at (0,0) and the warmer ion beam drifting parallel to the magnetic field. The left two panels are for a case with $U > C_s$. The upper panels show the initial distribution for each case. For the case with $U > C_s$ heating is oblique to the magnetic field at an angle $\theta = 70° = \arccos(C_s/U)$ (lower left) and parallel when $U = C_s$ (lower right).

4. Electron Beam Free Energy Source

Having examined ion beams as a free energy source of plasma instabilities in the PSBL in the previous sections, the emphasis in this section will be on an alternative free energy source, electron drifts (currents). In contrast to the numerous observational studies reporting ion beams in the PSBL, there are fewer studies of electron beam observations in the PSBL (Frank et al., 1981; Frank, 1985). Observations from ISEE 3 indicate the presence of energetic field-aligned streaming electrons (Scholer et al., 1986). These observations took place much deeper in the tail ($\geq$ 100 R_E) in the PSBL type structure found tailward of the reconnection region. The relative scarcity of electron beam observations appears to be due to more technical difficulties than from a lack of beams themselves. Also there is the problem of detecting cold electrons below instrument cutoff. Since electron beams drive instabilities differently than do ion beams, a linear theory and computer simulation study is presented in this section to describe the wave spectrum and heating that can result from electron currents in the PSBL.

The initial plasma configuration for modeling the electron beams in the spatial region near 20R_E in the PSBL is shown schematically in Figure 11. As shown in

Electron Beam Configuration

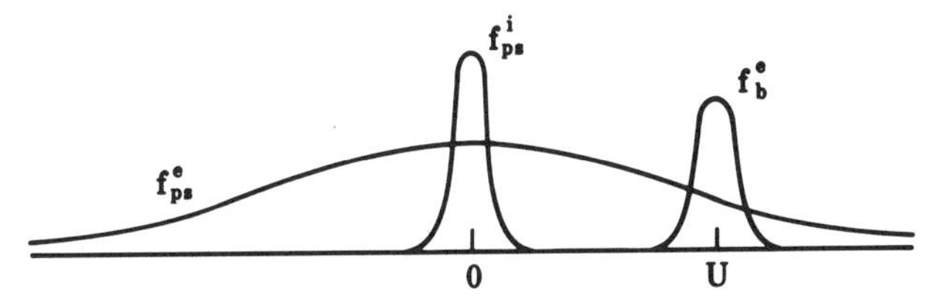

Fig. 11. Schematic plot of the plasma configuration. Here the cold electrons are drifting with respect to a background of hot ions and hot electrons.

Figure 11, we consider a cold drifting Maxwellian electron distribution with temperature T_c and drift speed U, streaming relative to a background of hot ions with temperature T_i and hot electrons with temperature T_h.

Considering the linear theory first, two instabilities are possible. The first type is a fluid–type instability resulting from the interaction between the drifting cold electrons and background ions. This instability is similar to a Buneman–type instability (Buneman, 1958); however, the presence of the hot electron background modifies the wave properties somewhat. The other instability possible is the electron acoustic instability, resulting from the interaction between the hot and cold electrons. The electron acoustic instability here differs from that found in section 2, in that wave growth is due to a negative energy wave resonant with the slope of the hot electron distribution, rather than a positive energy wave driven unstable by the slope of an ion beam. The two instabilities (electron–ion and electron–acoustic) can combine to form a broadbanded instability with low–frequency waves excited by the electron–ion instability and higher frequency waves by the electron acoustic instability. Figure 12 illustrates the dispersion properties of the instabilities at parallel propagation for three different cold electron beam densities with other parameters held fixed and given by the beam speed, $U = 0.64V_{th}$, and temperatures, $T_h = T_i = 500T_c$. The real frequency is seen as solid curves which increase steadily with $k\lambda_e$. This increase occurs since the real frequency is Doppler shifted by the beam speed and is given by

$$\omega = k_{\|}U - kC_{ea}$$

where C_{ea} = electron acoustic speed = $V_{th}\sqrt{n_c/n_h}$ and V_{th} = hot electron thermal velocity. At large k, $\omega/k_{\|}$ approaches U. The growth rates in Figure 12 show the presence of the two instabilities when the electron densities are equal. At lower k values the electron–ion instability is unstable with lower growth rates. When no hot electrons are present, as expected, only the cold electron–ion instability can exist (seen for the curve marked with $n_c/n_e = 1$).

The presence of the two instabilities is illustrated more clearly in Figure 13, which shows growth rates plotted versus real frequency at parallel propagation using the same parameters as the previous figure with various cold electron densities. For high cold electron densities, the electron–ion instability is dominant at low frequencies. As n_c is lowered, the higher frequency electron acoustic instability is enhanced. When hot electrons are more abundant than cold, the electron acoustic instability dominates. As the cold electron density is diminished further, thereby decreasing the amount of free energy available, growth rates eventually decrease.

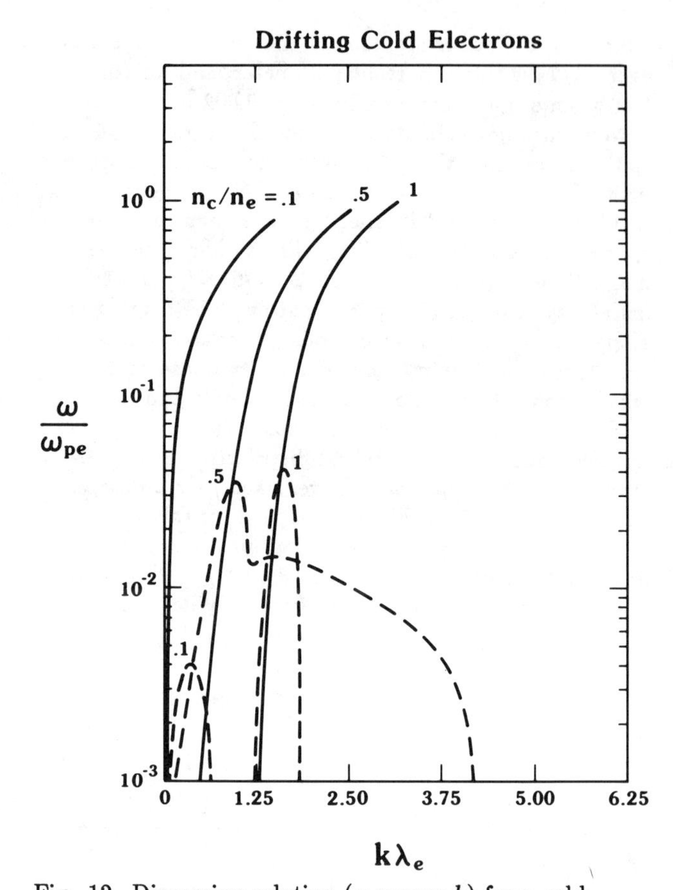

Fig. 12. Dispersion relation (ω versus k) for a cold electron beam drifting relative to a background of hot ions and electrons. The solid lines are real frequency and dashed lines are the growth rates. Three different values of the cold electron density are shown with other parameters held fixed at $U = 0.64V_{th}$, $T_i = T_h = 500T_c$, and $\theta = 0°$.

Analogous to the ion–ion and ion acoustic instabilities (Akimoto and Omidi, 1986), the cold electron drift speed determines at which propagation angle the electron–ion instability will occur. When the drift speed is equal to or greater than the electron acoustic speed ($U \geq C_{ea}$), the electron–ion mode instability has maximum growth at oblique wave angles. The electron acoustic instability on the other hand always has maximum growth at parallel propagation. This is seen in Figure 14 for $U = 1.44C_{ea}$ with $n_c = n_h$ and the temperatures held fixed at the same values as in Figures 12 and 13. At $\theta = 0°$, the electron acoustic growth rates are largest, but, as the angle is increased, the electron acoustic growth decreases and the electron–ion mode becomes most unstable at lower frequencies.

To understand the nonlinear properties of these instabilities, computer simulations have been run using the same plasma particle code as that discussed in the previous section, only here the electrons are allowed to drift and the ions are part of the background. By varying initial parameters such as cold-to-hot electron density ratio and initial drift speed, a number of runs have been performed simulating the instabilities due to a drifting cold electron distribution. In all the cases we have run, the results are qualitatively the same. It is found that the cold electron beam loses kinetic energy due to the beam's slowing down, which allows the fastest growing unstable wave modes to increase in amplitude. As the wave energy increases at the expense of the cold electron drift energy, the waves interact with the cold and hot electron distributions, heating both. This is illustrated in Figure 15, which shows the results from a particular simulation run with $U = 0.64V_{th}, n_c = n_h$, and $T_H = T_i = T_c$. Figure 15 shows the time histories of the wave energy (panel a), cold electron kinetic energy (panel b), hot electron temperature (panel c), and cold electron temperature

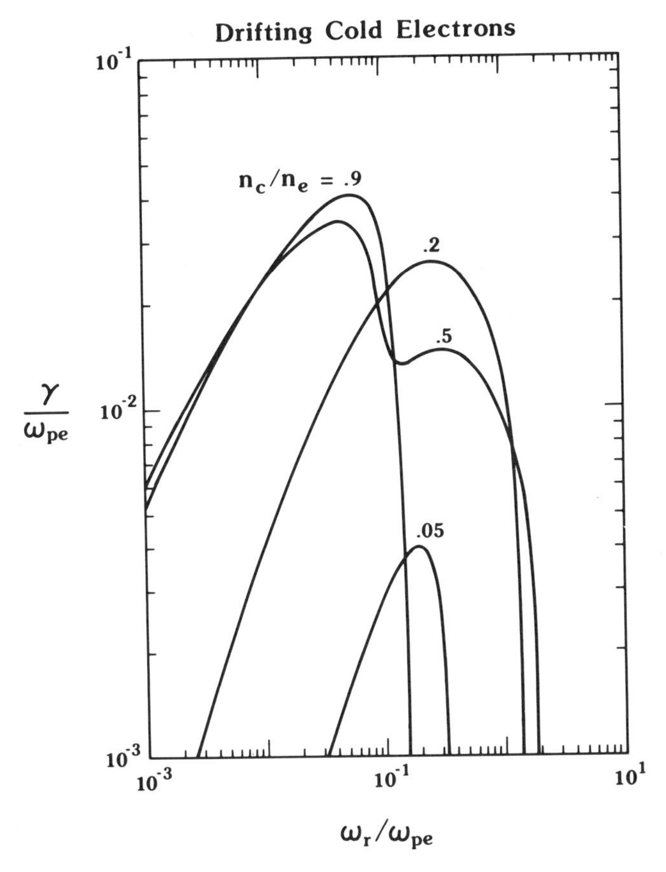

Fig. 13. Growth rate versus real frequency for the cold electron drift case with the same parameters as in the previous figure. The cold electron density is varied, and when $n_c = n_h$, the two instabilities are seen as peaks at different frequencies.

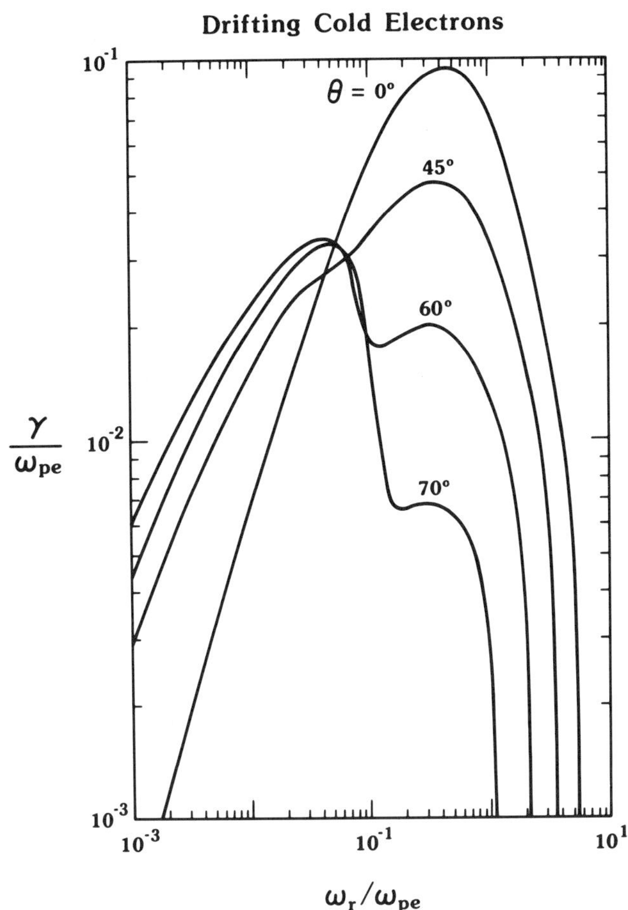

Fig. 14. Same format as Figure 11 except here the angle of wave propagation is varied. In this figure $n_c = n_h$ and $U = 1.44V_{th}$, and only when $\theta \geq 60°$ does the low-frequency electron–ion instability appear. At higher frequencies the electron acoustic instability is present with maximum growth when $\theta = 0°$.

(panel d). These features are enhanced by increasing the beam speed, which in turn increases the available free energy.

The cold and hot electron velocity distributions are shown in Figure 16 which illustrates the heating of each component. The left panel (a) shows the combined electron distribution at $t = 0$ and $t = 400\omega_{pe}$ while the center (panel b) and right panels (panel c) show the hot and cold electron distributions independently. The hot electron distribution (panel b) goes from Maxwellian initially to a flat top type distribution indicative of wave–particle interactions due to the resonant electron acoustic instability which grows off the hot electron slope. The cold electrons lose energy and heat up due to both the electron–ion and electron acoustic instabilities eliminating the positive slope that

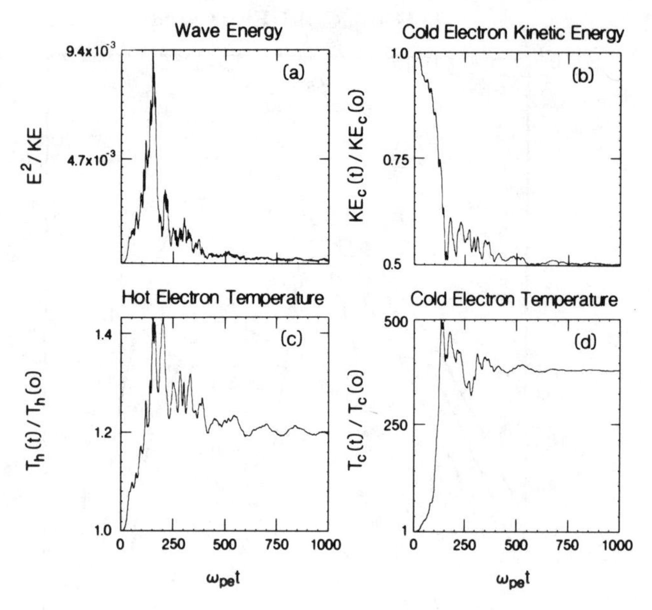

Fig. 15. Simulation results for the cold electron beam configuration with $U = 0.64V_{th}$, $n_c = n_h$, $T_i = T_h = 500T_c$, and $\theta = 0°$. Shown are the time histories of the wave energy, cold electron kinetic energy, hot electron temperature, and cold electron temperature. Note that the energy in the cold electron beam is released and converted into wave energy which in turn heats both electron species.

exists between the cold electrons and the background species. The background ions (not shown) also heat up somewhat, probably due to the electron–ion instability.

Since the particle diagnostics indicate both instabilities are active, it would be instructive to obtain information about the waves. This can be obtained from the simulation by examining the wave spectra. Figure 17 shows the time history (left panels) and corresponding power spectrum for two wave modes that can be identified from linear theory by their values of $k\lambda_h$ and ω/ω_{pe} as being the electron–ion (mode 3, top panels) and electron acoustic (mode 10, lower panels) instabilities. It can be seen that the electron–ion mode reaches a much higher amplitude and is a lower frequency ($\omega/\omega_{pe} = 0.07$) as expected from linear theory. The electron acoustic mode has peak power at just less than $0.2\omega_{pe}$. Since the electron–ion mode has the most wave power and reaches the largest amplitude, this mode would be expected to have the strongest influence in heating the cold electrons, but the electron acoustic mode reaches its peak amplitude at nearly the same time (comparing panels a and c), and so the nonlinear effects of each mode cannot be conclusively determined. The total power spectrum summed over all wave modes (not shown) has peak power at low frequencies ($\omega < 0.1\omega_{pe}$) due to the electron–ion instability and a general decline in intensity as frequency increases up to about $0.5\omega_{pe}$. This is in qualitative agreement with observations of BEN in the magnetotail (Gurnett et al., 1976).

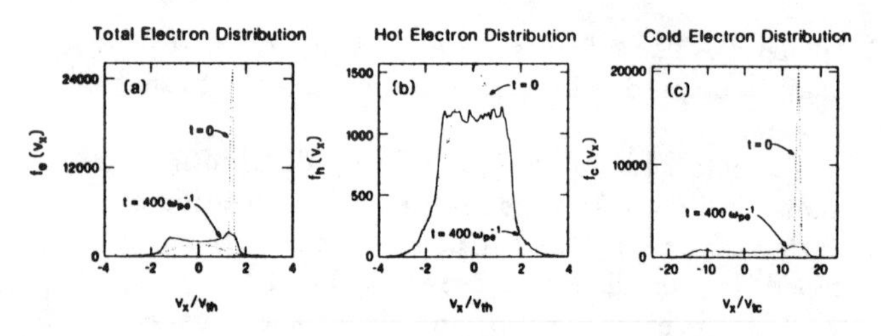

Fig. 16. Velocity distribution of the cold, hot, and total electrons from the simulation run discussed in the previous figure. The distributions at $\omega_{pe}t = 400$ are superimposed on the distributions at $t = 0$. The total electron distribution shows at $t = 400\omega_{pe}^{-1}$ that the positive slope existing between the hot and cold species is virtually eliminated. The hot electrons deform into a flat top type distribution.

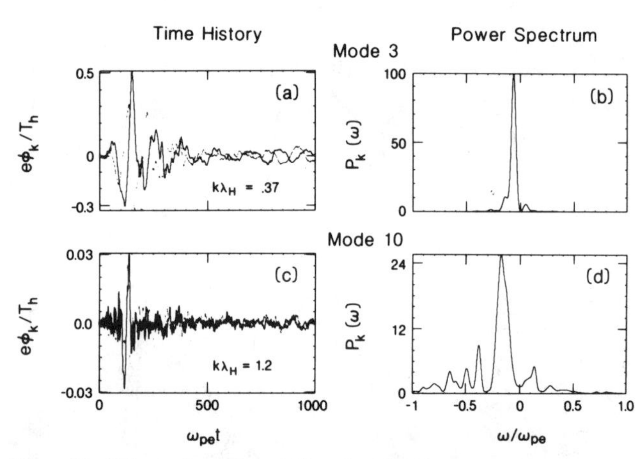

Fig. 17. Wave data from the simulation run discussed in the previous two figures. The wave amplitude time history (left panels) and power spectrum (right panels) are shown for two wave modes, mode 3 on the top and mode 10 on the bottom. Mode 3 corresponds to the electron–ion instability and mode 10 is the electron acoustic instability.

5. Summary and Discussion

In this paper we have considered the linear growth and nonlinear consequences of broadband electrostatic noise in the plasma sheet boundary layer. Studies of different free energy sources with varying plasma backgrounds show that it is possible to excite broadband waves that have different spectral characteristics and varying nonlinear consequences depending on the initial plasma configuration. For the case of an ion beam in the presence of cold ionospheric electrons, the dominant instability is the electron acoustic instability which peaks at frequencies $\omega \sim 0.1\omega_{pe}$. Although the electron acoustic wave peaks at around $\omega \sim 0.1\omega_{pe}$, the power spectrum is reminiscent of the broadband electrostatic waves observed by ISEE 1 and extends all the way to the cold electron plasma frequency. This broadbanded wave spectrum significantly heats the cold electrons, whereas the hot electron temperature and ion kinetic energy are hardly affected. The amplitudes of these waves are in general agreement with the observations ($\sim mV/m$). If we assume the presence of cold electron sources and model the cold electrons constantly entering the PSBL (Schriver and Ashour-Abdalla, 1986), then the wave growth persists much longer since waves saturation is no longer due to cold electron heating, but instead is due to ion beam heating, which is a much slower process. With this cold electron recycling Schriver and Ashour-Abdalla (1988) have shown that the wave spectrum is much more broadbanded, with the highest level of wave power at low frequencies ($\sim 0.05\omega_{pe}$) and a steady falloff in power up to ω_{pe}.

In the absence of cold electrons the electron acoustic instability does not exist. Instead for the case of an ion beam streaming relative to an ion background and hot electron background it is found that the ion acoustic and ion–ion instability grows. As the ion beam temperature is increased such that T_h/T_b decreases, the growth rates of both instabilities decrease, and the ion–ion mode is unstable for warmer beams, whereas the ion acoustic mode is stable. The wave spectrum produced by the ion–ion instability is quite different than that produced by the electron acoustic instability. The wave spectrum in the former case is more narrowbanded and peaks at lower frequencies $\sim 0.01\omega_{pe}$. The ion–ion instability saturates by heating the background ions and to a lesser extent the beam ions. The amount of heating depends on the beam speed and can be greater than one hundred times that for the cold ionospheric ions, whereas the corresponding temperature increase for the beam is only a factor of 2 to 5. This heating is oblique for the ion drift speed of $U > C_s$ and occurs in the parallel direction for $U \leq C_s$.

In summary, ion beam instabilities in the PSBL result in heating ionospheric plasma that enters the plasma sheet region (Figure 18). This ionospheric plasma enters in the plasma sheet region either through the lobe or along the auroral field lines. The heating of the cool plasma takes place through wave–particle interactions whereby the free energy in the energetic ion beams is transferred to the cold plasma through electrostatic waves. The background plasma can be heated using only a fraction of the ion beam kinetic energy, thus allowing the ion beams to travel the extent of the magnetotail and be observed frequently. Over most of the length of the PSBL, cold lobe plasma entering the PSBL encounters ion beams and hot electrons and is rapidly heated to become part of the central plasma sheet as it convects deeper in the plasma sheet. Electrons are heated first by the electron acoustic and beam resonant instabilities, and the constant incoming flux of cold electrons allows a broadband wave spectrum to be excited from ω_{pe} down to $0.05\omega_{pe}$. Once the electrons are heated, then the plasma consists of ion beams, cold ionospheric ions, and hot electrons. The ion beam velocity is layered in the PSBL with speed decreasing into the PSBL toward the central plasma sheet (Figure 3, Takahashi and Hones, 1988). Therefore, the ion heating is first oblique to the magnetic field and then, as slower beams interact with the cold ions as they convect deeper into the PSBL, the heating becomes parallel. Then a single isotropic ion distribution if formed by the combination of the beam and background ions. This is entirely consistent with the observations of Eastman et al. (1984) that show anisotropic ion distributions ($T_\perp/T_\parallel > 1$) closest to the lobe

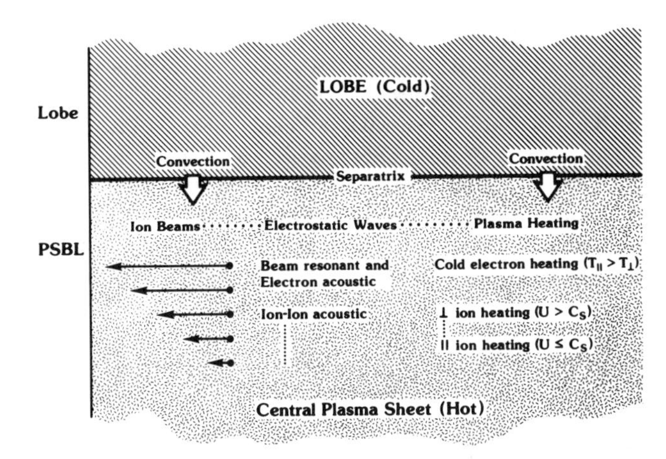

Fig. 18. Summary of the wave particle interactions due to ion beams, occurring in the plasma sheet boundary layer. As cold plasma convects from the lobe into the PSBL, the beam resonant and electron acoustic instabilities are excited by the interaction between the cold electrons and the ion beams. As a result the cold electrons are heated. Once the cold electrons are heated, the cold ions and ion beams can excite the ion–ion instability. This instability heats the ions obliquely to the magnetic field when the beam speed is greater than the sound speed, and parallel to the magnetic field when the beam speed is less than the sound speed.

(in the PSBL) and then heated isotropic distributions deeper in the PSBL toward the central plasma sheet. The frequency spectrum excited in this process is narrowbanded and intense with frequency near $0.01\omega_{pe}$.

For the case of free energy carried by the electron drifts the resultant broadband spectra is different. It was found that if the cold electrons are allowed to drift with respect to the hot electrons and background ions, two instabilities can occur, the electron–ion and electron acoustic instability. The electron acoustic frequency is higher here than for the case of ion beam free energy since the real frequency is Doppler shifted by the beam speed. Simulation studies show that the low–frequency electron–ion instability has the largest growth rates and is the dominant instability causing the peak in wave power to occur at $\leq 0.1\omega_{pe}$. The electron acoustic instability occurs at higher frequencies ($\sim 0.3\omega_{pe}$) and has less power so that the total wave spectrum shows a peak at lower frequencies with a general decline in intensity as frequency increases. This is in good qualitative agreement with observations of BEN (Gurnett et al., 1976). These, when combined, result in the heating of both the cold and hot electron distributions and to a lesser extent the background ions.

In summary, instabilities driven by electron beams in the PSBL heat the cold and hot electrons. Due to the excitation of a low–frequency wave, the ion–electron instability, the ions are also heated but to a lesser extent. In contrast to the ion beams the free energy source here is quickly destroyed due to wave growth, so that observations of electron beams must be made on a very rapid time scale. This may explain the scarcity of electron beam observations in the PSBL. When the temperature of the drifting electrons is increased the instability is limited to an electron acoustic wave and the ion–electron wave is stable. In this case the ions are not affected since the electron acoustic instability does not involve the ions. The lack of observations of electron currents makes it difficult for us to analyze further the transport and acceleration of plasma in the PSBL due to electron currents.

The theory of cold plasma heating and broadbanded electrostatic wave generation pushes the limits of experimental observation. First as we have seen the cold plasma plays a very important role in determining which instability will be excited. Since the cold plasma that enters the plasma sheet boundary is below most instrument energy cutoffs the true density and initial energy are still unknown. Remembering that the initial cold temperature of the lobe plasma will determine the final temperature due to interactions with the ion beam or electron beam driven instabilities, it is not clear how much the cold plasma will be heated by this process. Theoretical studies that determine the mechanisms causing the large plasma outflow from the ionosphere to the lobe would be useful since this type of study may also determine the temperature and density of the plasma when it reaches the plasma sheet boundary layer. Moreover more rapid measurements of the free energy are required. The present temporal resolution for the ion beams does not seem to be a problem since the ion beams are not destroyed by the wave–particle interaction. On the other hand, the drifting electron distributions tend to be flattened rather quickly as a result of wave growth. The only way of identifying the latter free energy source is with instruments that have a faster time resolution. Although the temporal resolution does not present a problem for the case of ion beams, the upper energy cutoff of the instruments may be a cause of concern. At present the upper cutoffs of the instruments are 29keV ($\sim$ 2300 km s^{-1}) for the fast plasma instrument (Takahashi and Hones, 1988) and 45 keV ($\sim$ 3000 km s^{-1}) for LEPEDEA (Frank et al., 1976). Only ion beams with speeds up to instrument cutoff have been observed so the actual upper limit on beam speed is not known and could be much higher (K. Takahashi, private communication, 1987). A very important and essential component in the transport and heating of plasma in the PSBL is the waves. Although we know that waves can extend all the way to the electron plasma frequencies with amplitudes of $\sim 1mVm^{-1}$ the wave modes have not been identified. To advance the theory further, it is essential for us to have wave polarization measurements as a function of frequency. Last, but certainly not least, simultaneous observations of waves, the free energy, and background plasma would greatly enhance our understanding of the physical processes that occur in the plasma sheet boundary layer.

The study of broadband electrostatic noise and the related transport and acceleration of plasma in the plasma sheet boundary layer provides a fine example of theory reaching a level of sophistication sufficient for its direct comparison with observations.

Acknowledgments. This work was supported by NASA Solar Terrestrial Theory Program grant NAGW–78, Air Force contract F16928-88-K-0011, and National Science Foundation grant ATM 85-13215. Computing was performed at the San Diego Supercomputer Center (SDSC).

References

Akimoto, K., and N. Omidi, The generation of broadband electrostatic noise by an ion beam in the magnetotail, Geophys. Res. Lett., 13, 97, 1986.

Ashour–Abdalla, M., and H. Okuda, Theory and simulations of broadband electrostatic noise in the geomagnetic tail, J. Geophys. Res., 91, 6833, 1986a.

Ashour–Abdalla, M., and H. Okuda, Electron acoustic

instabilities in the geomagnetic tail, Geophys. Res. Lett., 13, 366, 1986b.

Ashour-Abdalla, M., and R.M. Thorne, Toward a unified view of diffuse auroral precipitation, J. Geophys. Res., 83, 4755, 1978.

Buneman, O., Instability, turbulence, and conductivity in current carrying plasma, Phys. Rev. Lett., 1, 8, 1958.

Cattell, C.A., and F.S. Mozer, Experimental determination of the dominant wave mode in the active near-Earth magnetotail, Geophys. Res. Lett., 13, 221, 1986.

DeCoster, R.J., and L.A. Frank, Observations pertaining to the dynamics of the plasma sheet, J. Geophys. Res., 84, 5009, 1979.

Dusenbery, P.B., and L.R. Lyons, The generation of electrostatic noise in the plasma sheet boundary layer, J. Geophys. Res., 90, 10,935, 1985.

Eastman, T.E., L.A. Frank, and W. Peterson, The plasma sheet boundary layer, J. Geophys. Res., 89, 1553, 1984.

Etcheto, J., and A. Saint-Marc, Anomalously high plasma densities in the plasma sheet boundary layer, J. Geophys. Res., 90, 5338, 1985.

Frank, L.A., Plasmas in the earth's magnetotail, in Space Plasma Simulations, edited by M. Ashour-Abdalla and D.A. Dutton, p. 211, D. Reidel Publ. Co., Dordrecht, Holland, 1985.

Frank, L.A., K.L. Ackerson, and R.P. Lepping, On hot tenuous plasmas, fireballs and boundary layers in the Earth's magnetotail, J. Geophys. Res. 81, 5859, 1976.

Frank, L.A., R.L. McPherron, R.J. DeCoster, B.G. Burek, K.L. Ackerson, and C.T. Russell, Field-aligned currents in the Earth's magnetotail, J. Geophys. Res., 86, 687, 1981.

Fried, B.D., and R.W. Gould, Longitudinal ion oscillations in a hot plasma, Phys. Fluids, 4, 139, 1961.

Fried, B.D., and A.Y. Wong, Stability limits for longitudinal waves in ion beam-plasma interaction, Phys. Fluids, 9, 1084, 1966.

Gary, S.P., and N. Omidi, The ion/ion acoustic instability, J. Plasma Phys., 37, 35, 1987.

Grabbe, C.L., New results on the generation of broadband electrostatic waves in the magnetotail, Geophys. Res. Lett., 12, 483, 1985.

Grabbe, C.L., Numerical study of the spectrum of broadband electrostatic noise in the magnetotail, J. Geophys. Res., 92, 1185, 1987.

Grabbe, C.L., and T.E. Eastman, Generation of broadband electrostatic noise by ion beam instabilities in the magnetotail, J. Geophys. Res., 89, 3865, 1984.

Gurnett, D.A., L.A. Frank, and R.P. Lepping, Plasma waves in the distant magnetotail, J. Geophys. Res., 81, 6059, 1976.

Huang, C.Y., L.A. Frank, and T.E. Eastman, Observations of plasma distributions during the coordinated data analysis workshop substorms of March 31 to April 1, 1979, J. Geophys. Res., 92, 2377, 1987.

Huba, J.D., N.T. Gladd, and K. Papadopoulos, Lower-hybrid-drift wave turbulence in the distant magnetotail, J. Geophys. Res., 83, 5217, 1978.

Lui, A.T.Y., T.E. Eastman, D.J. Williams, and L.A. Frank, Observations of ion streaming during substorms, J. Geophys. Res., 88, 7753, 1983.

Omidi, N., Broadband electrostatic noise produced by ion beams in the Earth's magnetotail, J. Geophys. Res., 90, 12,330, 1985.

Scarf, F.L., L.A. Frank, K.L. Ackerson, and K.P. Lepping, Plasma wave turbulence at distant crossings of the plasma sheet boundaries and neutral sheet, Geophys. Res. Lett., 1, 189, 1974.

Scholer, M., D.N. Baker, G. Gloeckler, B. Klecker, F.M. Ipavich, T. Terasawa, B.T. Tsurutani, and A.B. Galvin, Energetic particle beams in the plasma sheet boundary layer following substorm expansion: Simultaneous near-Earth and distant tail observations, J. Geophys. Res., 91, 4277, 1986.

Schriver, D., and M. Ashour-Abdalla, Generation of high-frequency broadband electrostatic noise: The role of cold electrons, J. Geophys. Res., 92, 5807, 1987.

Schriver, D., and M. Ashour-Abdalla, Cold plasma heating in the plasma sheet boundary layer: Theory and simulations, J. Geophys. Res., in press, 1988.

Sharp, R.D., D.L. Carr, W.K. Paterson, and E.G. Shelley, Ion streams in the magnetotail, J. Geophys. Res., 86, 4639, 1981.

Takahashi, K., and E.W. Hones, Jr., ISEE 1 and 2 observations of ion distributions at the plasma sheet-tail lobe boundary, J. Geophys. Res., 93, 8558, 1988.

Watanabe, K., and T. Taniuti, Electron acoustic mode in a plasma of two-temperature electrons, J. Phys. Soc. Japan, 43, 1819, 1977.

MACROSCOPIC MAGNETOSPHERIC PARTICLE ACCELERATION

B. H. Mauk and C.-I. Meng

Applied Physics Laboratory, The Johns Hopkins University, Laurel, MD 20707

Abstract. Selected topics of recent theoretical developments in the study of macroscopic magnetospheric particle acceleration mechanisms are reviewed. As such mechanisms are fairly well understood as isolated entities, it is in the incorporation of the mechanisms into the complex workings of real plasma systems that new findings have emerged. In terms of illuminating the outstanding problems in solar system plasma physics, these discussions are used to illustrate the importance of considering the reciprocal interactions between the macroscopic kinetic acceleration mechanisms and the processes of coupling between different macroscopic regions (e.g., magnetosphere/ionosphere) as opposed to the point of view that macroscopic kinetic energization is simply a consequence of coupling. Full particle distributions, as well as accurate distribution moments, are needed to experimentally characterize such interactions.

Introduction

It would be difficult to identify "new" fundamental mechanisms within the recent scientific literature on macroscopic particle energization within planetary magnetospheres. It is in the incorporation of generally well understood mechanisms into the complex working of real magnetospheric environments that new findings have arisen. For example, the standard guiding center motions, fully utilized, can, in the presence of convection electric fields, give rise to complex energization patterns that have only very recently been applied to the middle and inner magnetospheric regions. These complex energization patterns are used here as a basis for an extended discussion of kinetic energization throughout planetary magnetotail environments.

In particular, it is shown that such middle and inner magnetospheric guiding center energizations are virtually identical to the "non-guiding center" energizations that occur within the narrow neutral sheet geometries of the distant magnetotail regions. The role of these respective kinetic particle energization processes in enhancing ionospheric/magnetospheric coupling is then emphasized in terms of the generation of field-aligned electric fields and the consequential auroral-like field-aligned discharges. Based on the similarities between near and distant magnetotail kinetic energization processes, it is argued that at least two spatially separate regions of auroral-like discharges are likely to occur, one mapping to the middle and inner magnetospheric regions and one mapping to more distant magnetotail regions. This assertion is shown to be supported by observations.

What is emphasized in these discussions is the role of the kinetic energizations in initiating and/or augmenting the magnetosphere/ionosphere coupling. The discussions are used to illuminate some outstanding problems in solar system plasma physics by emphasizing that the reciprocal interactions between the kinetic acceleration mechanisms and the processes of coupling between different macroscopic regions (e.g., between the magnetosphere and ionosphere) must be considered to understand the dynamics of the magnetospheric plasma system. The kinetic accelerations are not simply a consequence of the couplings but are active partners in the macroscopic interactions. The signatures of such interactions can be hidden in the details of the particle distributions. Thus, there is no short cut to the experimental characterization of these macroscopic interactions.

Guiding Center Acceleration

Within the guiding center approximation one may write (Northrup, 1963)

$$\frac{mdv_{\parallel}}{dt} = mg_{\parallel} + qE_{\parallel} - \mu\frac{d|\mathbf{B}|}{ds} + m\mathbf{V}_c \cdot \frac{d\hat{b}}{dt} \qquad (1)$$

and

$$\mu = \frac{\frac{1}{2}mv_{\perp}^2}{|\mathbf{B}|} = \text{constant} \qquad (2)$$

where $v_\parallel$ and $v_\perp$ are the velocities parallel and perpendicular to **B**, m is particle mass, q is particle charge, $g_\parallel$ is the parallel component of gravity, $E_\parallel$ is the parallel component of the electric field **E**, μ is magnetic moment, s is distance along **B**, $\mathbf{V}_c$ is the convection velocity $c\mathbf{E} \times \mathbf{B}/B^2$, and $\hat{b}$ is the unit vector in the direction of **B**. It is the last term of (1), dubbed here the "centrifugal term," that will be the focus of the present discussions. This term can be understood heuristically by rewriting equation (1) into a field-aligned energization equation. Considering only the term of interest one may write

$$\frac{d(\frac{1}{2}mv_\parallel^2)}{dt} \sim \left(mv_\parallel^2 \frac{\hat{R}_c}{R_c}\right) \cdot \mathbf{V}_c + \left(mV_c^{\,2} \frac{\hat{R}_1}{R_1}\right) \cdot \mathbf{v}_\parallel \quad (3)$$

where $\mathbf{R}_c$ and $\mathbf{R}_1$ are explained by reference to Figures 1a and 1b, respectively. One sees that two different types of terms have emerged, each with the form of a centrifugal force times a velocity. The two terms have a symmetry in that the particle's parallel velocity and the convection velocity exchange rolls.

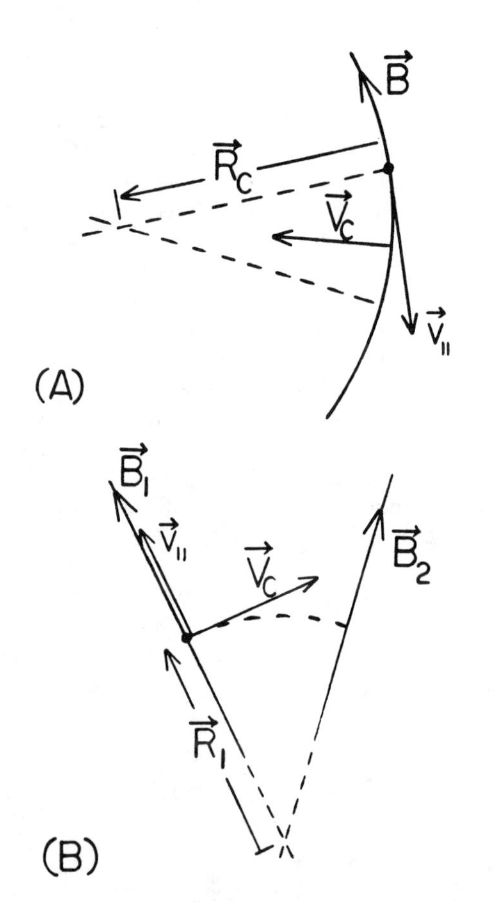

Fig. 1. Illustrations of the "centrifugal" character of the energizations that result from the last term of equation (1). See text for explanations.

The first term of (3) can be understood heuristically by referring to Figure 1a. There is a centrifugal force associated with the motion of the particle's guiding center along a curved field line, with radius of curvature R_c. Energization results when the magnetic force that balances out the centrifugal force (i.e., or that causes the centipetal acceleration) acts in the direction of the convection velocity $\mathbf{V}_c$. This energization is analogous to the energization that results to a stone that is whirling around in a circle on the end of a rope, when the length of the rope is shortened.

The second term of (3) can be understood, again heuristically, by referring to Figure 1b. There is a centrifugal force associated with the convection of the particle from one field line to another field line that is pointing in a different direction, with an effective moment arm R_1 defined within an inertial frame of reference. Energization results when that centrifugal force (this "force" is not balanced in this case) acts against the particle's parallel velocity. This energization is analogous to the energization that results to a bead sliding loosely along a straight length of metal rod when the rod is, say, whirled overhead in a whip-like motion.

Particle energization within electromagnetic systems must ultimately be expressible in the form of an electric field acting on a velocity. The parallel energization that results from the centrifugal term of equation (1) arises simply from the action of the convection electric field on the so-called "magnetic curvature drift" (e.g., Quinn and Southwood, 1982). Care must be taken, however, to include all contributions to that drift, including some contributions that are often ignored (Northrup, 1963). The perpendicular acceleration that results from the conservation of μ can be correspondingly understood as the action of the same electric field on the so-called "magnetic gradient drift."

Until recently, the centrifugal acceleration described above has not been applied to the middle and inner magnetospheric regions (say $R < 10\ R_E$, where R_E = the radius of the Earth) because it has commonly been assumed that parallel energization can be calculated by invoking the conservation of the so-called longitudinal or second-adiabatic invariant ($\oint v_\parallel\, ds$ = constant). Recent interest in the centrifugal acceleration term for these regions has been sparked by (a) the focus of the magnetospheric community's interest on the evolution of very low-energy particles (<50 eV) of ionospheric origin, and (b) the realization of the importance of fast dynamical processing on the transport of particles within these middle and inner regions. Additionally, Chapman and Cowley (1984) used the centrifugal acceleration term to predict particle acceleration and transport for particle clouds released within the near magnetotail during the active experiment Active Magnetospheric Particle Tracer Explorers (AMPTE) program.

Evolution of Ionospheric Ions

There has been substantial recent interest in the ways in which the polar cap and cusp ionospheres can serve as source

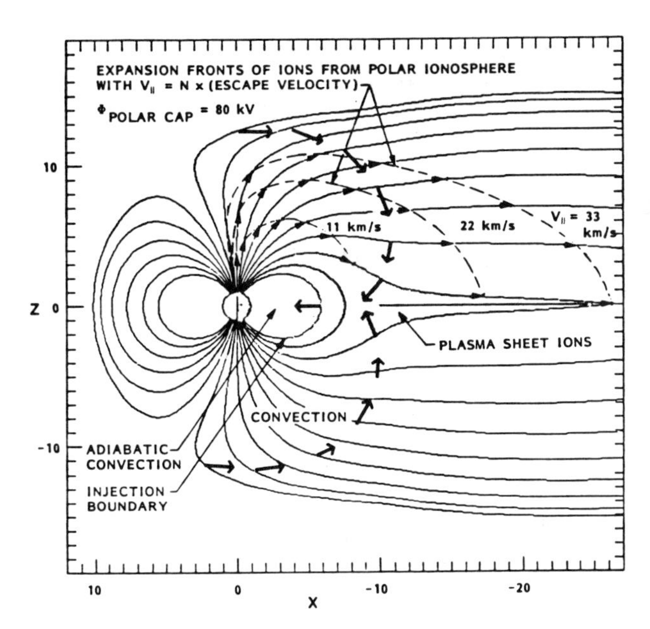

Fig. 2. Schematic of the evolution of ionospheric ions as they populate the middle and inner magnetospheric regions. The noon-midnight meridional plane of the Earth's magnetosphere is shown (from Cladis and Francis, 1985).

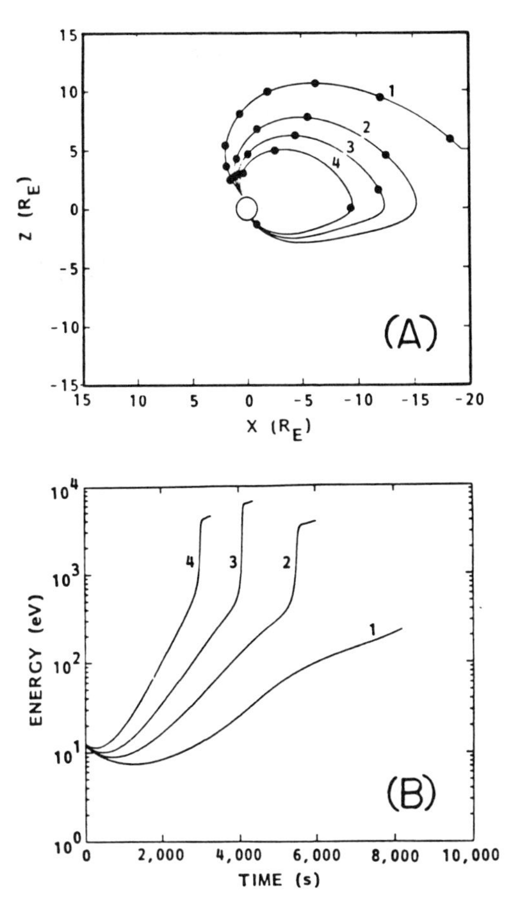

Fig. 3. (a) Sample model guiding center trajectories of initially 12 km s^{-1} (30° pitch angle) O^+ ions starting out from four different polar positions. The noon-midnight meridional plane of the Earth's magnetosphere is shown. The dots show time steps of 1000 s in real time. (b) Energy vs. time for the four trajectories shown in part (a) (from Cladis, 1986).

plasmas for populating the inner and middle magnetospheric regions. This interest was sparked particularly by the findings of the Dynamics Explorer (DE) spacecraft concerning the so-called "Cleft Ion Fountain" (see review by Horwitz, 1987). It was Cladis (1986) who recognized the important role that the centrifugal acceleration term would play in this process. If one hypothesizes that there exist pre-energized (~ 10 eV) populations within and above the polar cusp and/or the polar cap ionospheric regions, the problem is to evaluate the evolution of the ions in the presence of a magnetospheric convection electric field. Figure 2 (from Cladis and Francis, 1985) shows a noon-midnight meridional cross section of the Earth's magnetosphere illustrating the escape of ions out of the polar ionosphere and the transport, due to the convection, of the ions into the inner magnetosphere. It is clear that both types of centrifugal acceleration discussed in reference to equation (3) will participate. Over the polar cap the ions convect from one relatively straight field line to another, pointing in a different direction, giving rise to the "bead-on-a-rod" type of acceleration. Elsewhere, particularly in the vicinity of the equatorial plane, the ions convect in the direction of the field line curvature vector ($\mathbf{R}_c$ in Figure 1a) giving rise to the "stone-on-a-rope" type of acceleration.

Figure 3a shows the trajectories of oxygen ions starting at four different polar positions, calculated using realistic magnetic and electric configurations by Cladis (1986). Figure 3b shows the ions' energizations. Energizations from ~ 10 eV to as high as 7 keV are shown. Energization up to ~ 0.5 keV occurs even away from the highly curved regions close to the equatorial plane. Cladis (1986) argued that the behavior represented on Figures 3a and 3b suggests that the ionospheric ions can become important constituents of the plasma sheet and ring current populations without invoking any additional anomalous energization processes. It should be noted that whether or not the ionospheric ions behave in this fashion depends strongly on the particles' initial conditions. There are many other classes of orbits, as has been shown by the more detailed modeling efforts of, e.g., Horwitz and Lockwood (1985), Horwitz et al. (1986), and Sauvaud and Delcourt (1987).

Dynamical Centrifugal Acceleration

The calculations of Cladis (1986) were performed for steady state conditions. Also, the most dramatic energization occurs

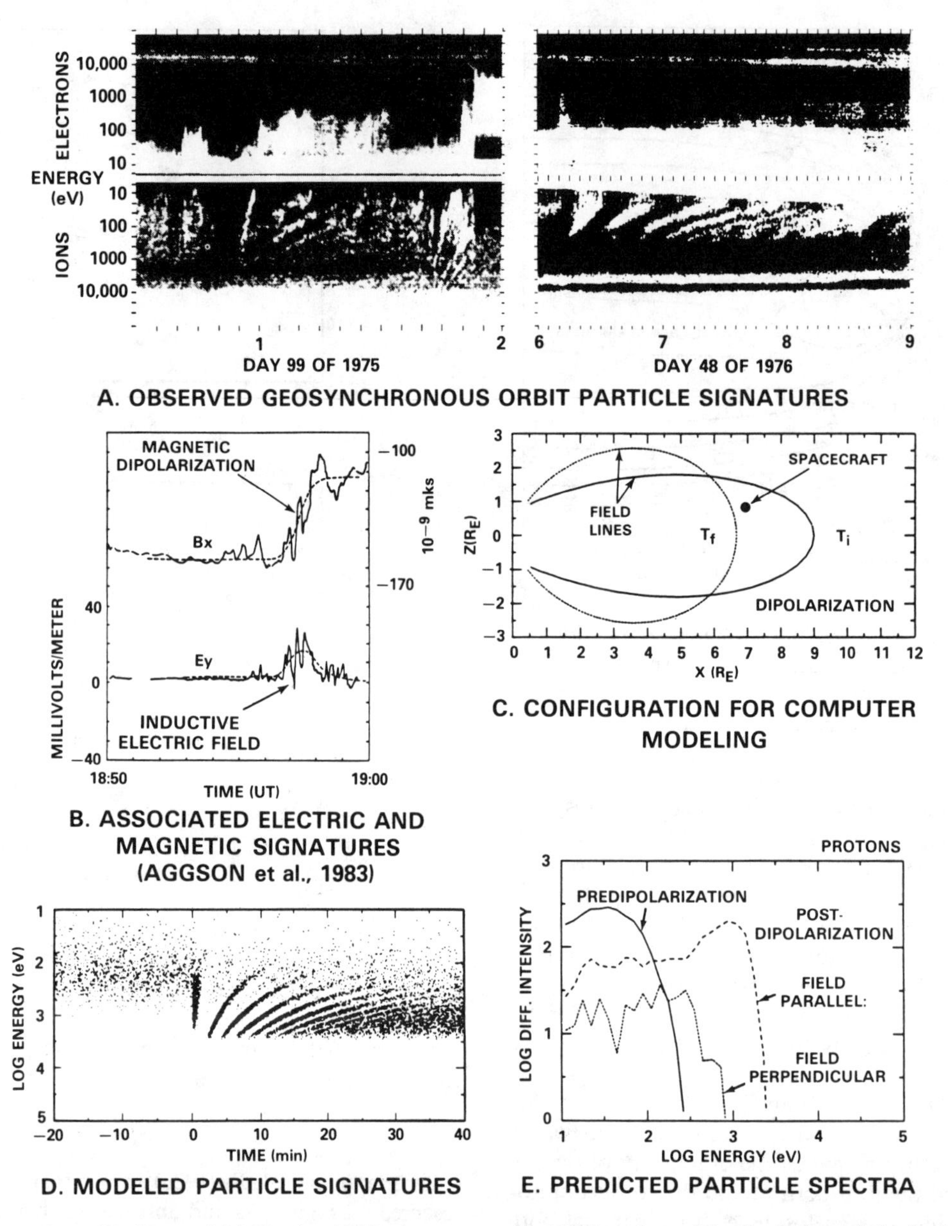

Fig. 4. Observations (Quinn and McIlwain, 1979; Aggson et al., 1983) and model calculations (Mauk, 1986) of the convection surge/dipolarization process occurring within the near geosynchronous regions of the Earth's magnetosphere.

just in the vicinity of the highly curved field line regions of the equatorial plane. One of the unanswered questions concerning the configuration of the global electric configurations is the degree to which the electric field can penetrate to these highly curved regions in a steady state fashion (e.g., Mauk and Zanetti, 1987; Birn and Schindler, 1985; Rostoker and Boström, 1976). Strong penetrations do occur, however, during active dynamical time periods, and it is the investigation of the centrifugal acceleration mechanism during dynamical periods that will be the focus of the present section.

These investigations were motivated by observations rather than by theoretical considerations. Panel A of Figure 4 shows the observations in question. These panels, taken from Quinn and McIlwain (1979), show ion and electron data sampled by the University of California at San Diego (UCSD) electrostatic analyzers on the ATS 6 geosynchronous satellite. The data were sampled near the midnight local time region, and only low pitch angle particles (traveling approximately along the local field line direction) are represented on the figures.

The features of interest are the distinct, repetitive dispersive streaks within the ion displays occasionally reaching up to energies as high as 20 keV. (Two dispersive "events" are seen on the left panel.) Quinn and McIlwain (1979) interpreted the streaks as resulting from a sudden temporal perturbation causing spatially bunched clusters of ions to bounce back and forth between the northern and southern hemispheres along the quasi-dipolar field lines. Quinn and Southwood (1982) proposed that the "bunching" was caused by the centrifugal acceleration mechanism acting in association with sudden magnetic reconfigurations, termed "convection surges."

Panel C of Figure 4, showing a meridional view of quasi-dipolar field lines, shows the concept of the convection surge. At the initial time T_i, the field configuration is quasi-dipolar, but stretched somewhat into a tail-like configuration. In association with an intense, transient, east-to-west, near equatorial electric field, the equatorial (or high field line curvature) region of the field line will move suddenly earthward to the T_f position on the figure. Quinn and Southwood (1982) proposed that the centrifugal acceleration occurring close to the equator will give rise to the observed "bounce-phase-bunched" ions.

This "convection surge" or "dipolarization process" was modeled in detail by Mauk (1986). The electromagnetic field quantities were constrained by the Aggson et al. (1983) observations shown in panel B of Figure 4. For instance, the bottom portion of this panel shows a transient, azimuthal electric field occurring in association with the dipolarizations sensed by the magnetic field component shown. A single flux tube in the model was then populated initially with an isotropic, Maxwellian distribution of ions, and the evolution of the distribution was followed. Panel D of Figure 4 shows a simulated ion spectrogram (low pitch angles only) to be compared with the data in panel A. The simulated spectrogram compares favorably to the data, strongly supporting the Quinn and Southwood (1982) suggestion. Note that the ions have been very substantially energized in bulk.

One of the surprises of these dynamical simulations is the ease by which strongly magnetic field-aligned ion distributions are generated within the geosynchronous regions (field-aligned distributions were anticipated for particles interacting with the more distorted field configurations in the more distant tail regions; Chapman and Cowley, 1984). The modeled predipolarization (angularly isotropic) and post-dipolarization model intensity spectra are shown in panel E of Figure 4. The final distribution is clearly strongly field-aligned. Field-aligned distributions with energies up to several tens of kiloelectron volts can be generated by the model, depending on mass species. The mass species sensitivity of the convection surge mechanism can be seen by comparing the H^+ simulated results shown in panel E of Figure 4 with the O^+ simulated results shown in Figure 5 obtained using identical field configurations. Field-aligned energization to tens of kiloelectron volts is seen for the O^+ ions. As a result of observations that show field-aligned ions predominating within

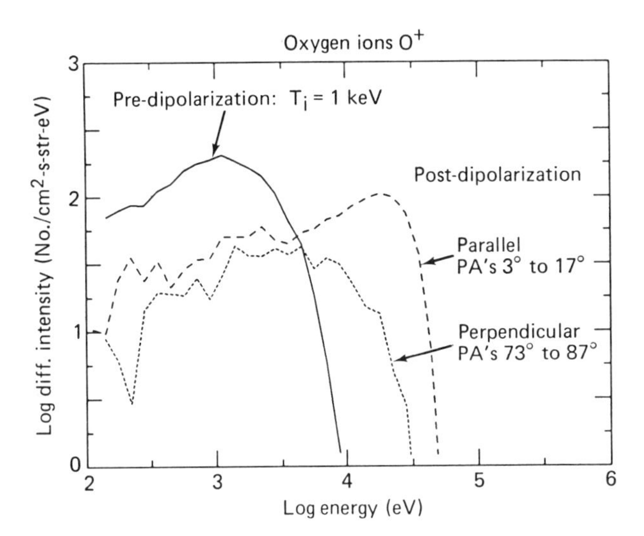

Fig. 5. Similar to Figure 4e, but for O^+ ions rather than protons.

geosynchronous regions at energies below a transition that commonly ranges between ~1 to 10 keV (Mauk and McIlwain, 1975; Kaye et al., 1981), it has been suggested that the convection surge/dipolarization process is a fundamental component of the processes by which plasmas are transported to the middle and inner regions of the Earth's magnetosphere.

Neutral Sheet Acceleration

The focus of these discussions will now shift to the distant magnetotail regions in the vicinity of the neutral sheet. The motivation for this shift is to address the similarities and differences between the centrifugal accelerations discussed above and the neutral sheet accelerations discussed most persistently by Speiser (1965, 1967, 1987) and Lyons (1984, 1987; Lyons and Speiser, 1982). Our conclusions will be that these ostensibly different mechanisms can be quite similar, as has been alluded by Speiser (1987). We will suggest that these similar behaviors have important consequences to magnetospheric dynamics by augmenting the coupling between the magnetosphere and ionosphere in both the near as well as the very distant tail regions.

Figure 6 shows a meridional cross section of a model magnetotail borrowed from Lyons (1984). Up to this point we have been addressing those regions, labeled A on the figure, where it is clear that the guiding center approximations are valid. In the deep tail regions, labeled C, in the vicinity of a very narrow "neutral sheet" structure, the figure correctly states that the guiding center approximations are no longer valid. One certainly anticipates that the physics of these very thin neutral sheets (region C) will be very different from the physics of the inner regions (region A).

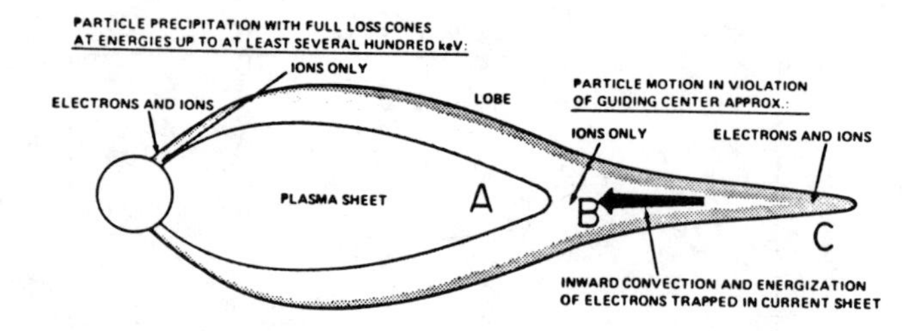

Fig. 6. Schematic meridional cross section of the Earth's magnetotail (after Lyons, 1984).

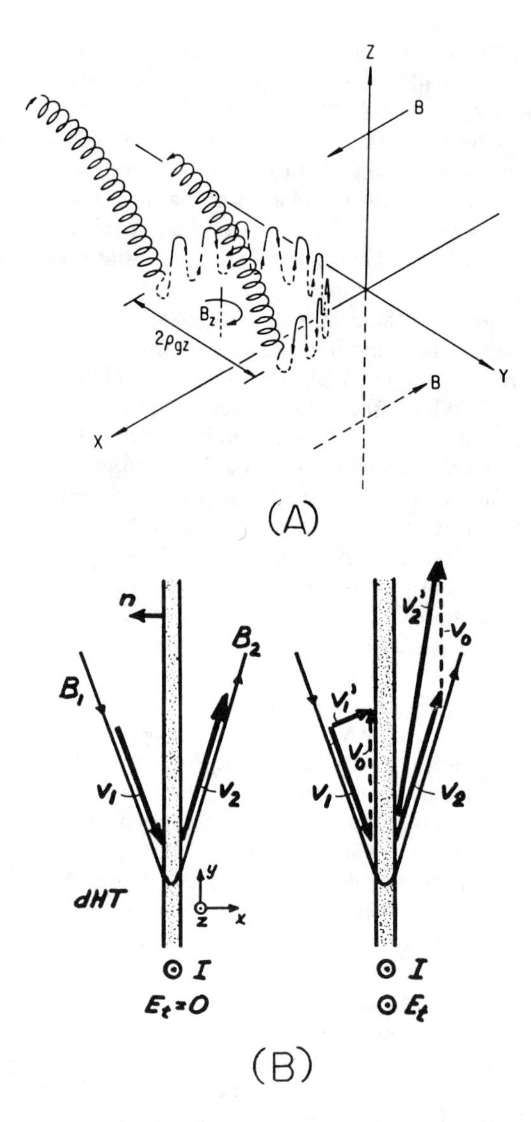

Fig. 7. (a)Trajectory of particle on interacting with a narrow magnetic neutral sheet, with a small normal magnetic field component, such as the sheet that might be found in the distant magnetotail. The velocity frame of reference is the one moving rapidly earthward at the plasma convection velocity cE/B (from Speiser, 1965). (b) Schematic of energization that results from the interaction of a particle with a narrow magnetic boundary structure (the magnetopause in this case). The energization is calculated using a velocity frame of reference change argument (from Sonnerup, 1984). See text.

The interaction between charged particles and thin neutral sheets has been studied extensively by various authors (Speiser, 1987; Lyons, 1987). If one jumps into the velocity frame of reference defined by the $c\mathbf{E} \times \mathbf{B}/B^2$ convection velocity moving toward the Earth (see Figure 6), the behavior of a single particle interaction with a very narrow neutral sheet looks like that shown in Figure 7a. The x-axis of the figure points toward Earth and the neutral sheet is centered on the x-y plane. As the particles enter the sheet from point 1, they get trapped within the sheet for a time, executing a meandering circular motion, corresponding to the gyromotion of the particle about the weak normal component of the magnetic field. After the particle executes a semicircle about that normal component, the particle is ejected from the neutral sheet. Within the moving frame of reference chosen for the display, the particle's energy is conserved. However, when one transforms back into the Earth stationary frame of reference, one finds that the particle energy has gained a substantial amount in the magnetic field-aligned direction.

This energization can be understood even more simply by considering the sketches shown in Figure 7b, borrowed from Sonnerup (1984). Sonnerup was concerned with the particle energization associated with magnetic reconnection at the dayside magnetopause interface between the solar wind and the magnetosphere. At that interface a narrow boundary exists analogous to our thin neutral sheet. The left-hand sketch of Figure 7b is drawn in the velocity frame of reference, the so-called de Hoffman-Teller frame, moving rapidly along the boundary such that the tangential component of the electric field has zero magnitude. In that frame of reference the particle's energy (parallel plus perpendicular) is preserved as the particle enters and leaves the vicinity of the boundary structure. In the Earth stationary frame of reference, drawn in the right panel of Figure 7b, the initial v_1' and final v_2' velocity vectors are very different, and the particle has been very substantially energized in the field-aligned direction.

Our reasons for belaboring this simple discussion involving the de Hoffman-Teller frame is that it has close similarities to the centrifugal acceleration discussed earlier. Reexamining Figure 1a, for instance, one could, as with the Sonnerup discussion, jump into the $\mathbf{V}_c$ frame of reference and allow the particle to move a differential distance along the field line. Within the moving frame, the particle energy does not change. Jumping back into the Earth stationary frame of reference after the differential motion, the particle's energy will have gained in the field-aligned direction.

The problem is that in the complex magnetic geometries to which the centrifugal acceleration has been applied, there is no one de Hoffman-Teller frame of reference to which one can transfer in order to perform a calculation. Within such geometries, the de Hoffman-Teller concept is only useful in a differential sense. Crucial to actually being able to calculate the particle's energization is the existence of the first-adiabatic invariant, in these cases the particle's magnetic moments. In solving equation (1) for complex geometries, the magnetic moment is a crucial player.

We have made the point that the guiding center approximations are violated deep within the magnetotail, and one anticipates that the magnetic moment will no longer be an invariant. However, it has been shown by Speiser (1970) and Sonnerup (1971) for special configurations, and now by Whipple et al. (1986) for more general conditions, that in the vicinity of narrow neutral sheets a first-adiabatic invariant exists that is different from but analogous to the particle's magnetic moment. This first-adiabatic invariant reduces to the magnetic moment in the regions of nearly uniform magnetic field.

As a particle enters the neutral sheet from point 1 on Figure 7a, it has a specific value of the magnetic moment. As the particle meanders about its great semicircle, its magnetic moment (however defined in this environment) can change rapidly. However, the invariant defined generally by Whipple et al. (1986) remains invariant. Then, as the particle pops out of the neutral sheet, to point 2 on Figure 7a, the energy that had been gained in the field perpendicular direction gets converted to field-aligned energy in just the right proportion so that the final magnetic moment at point 2 approximately equals that value at point 1. Thus, contrary to intuition, the existence of a first-adiabatic invariant in these very narrow neutral sheet regions has guaranteed that, in terms of macroscopic kinetic particle energization, the physics of the deep magnetotail regions (region C on Figure 6) is very similar to the physics of the middle, guiding-center-valid magnetospheric regions during dynamical periods.

The applicability of these kinds of simple acceleration processes within the deep magnetotail has, in fact, been demonstrated by Lyons and Speiser (1982). Figure 8 shows one of their comparisons between measured and modeled two-dimensional phase-space density plots. The comparison is quite favorable. Our point (again as alluded to by Speiser, 1987) is that these same favorable results would be obtained if the model sheet were thickened to the extent that guiding center approximations were valid and then the centrifugal acceleration term were employed.

These discussions have left the impression that in terms of kinetic particle energization, the magnetotail behaves the same irrespective of distance down the tail. That impression is, of course, incorrect. We have made no mention of regions like region B on Figure 6, where the neutral sheet is of intermediate thickness. The conditions that exist deep in the tail that allowed Whipple et al. (1986) to define a generalized first-adiabatic invariant are no longer valid in these intermediate regions. What one finds in these regions is that the particles are scattered by the small radii of field line curvature. Propp and Beard (1984), for instance, found that within thick neutral sheet geometries the particles, on interacting with the sheet, can come out with essentially random pitch angles. Based on the Propp and Beard (1984) results, Beard and Cowley (1985) found that they could define a "gyro-phased averaged" first-adiabatic invariant, but such an invariant is quite distinct from the kind of particle-by-particle invariant that concerns us in the present discussion, particularly with regard to the consequences of adiabatic behavior discussed below.

The relative behaviors of thick versus thin neutral sheets have also been examined by Martin (1986a,b), who has taken a somewhat different approach. He has argued, in the fashion of Chen and Palmadesso (1986), that the equations of motion for particles within the magnetotail are formally non integrable such that the motions of particles within the neutral sheet are chaotic in a formal sense. He finds that the motions are chaotic irrespective of whether the sheet is narrow or broad. (However, Whipple et al., 1986, shows that the adiabatic invariant becomes an exact invariant for a zero thickness neutral sheet.) Martin also finds, however, that the time scale for chaotic behavior (the time for two points closely spaced in phase space to substantially separate from each other in an exponential fashion) is a function of the sheet's narrowness. It turns out that in the narrow sheet the time scale for the onset of chaos is longer than the particle's neutral sheet occupation time. In contrast, for a thick neutral sheet the particle occupation times are longer than the chaos onset times. We would suggest that a statement concerning the relative comparison of occupation times and chaos onset times may

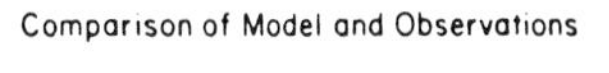

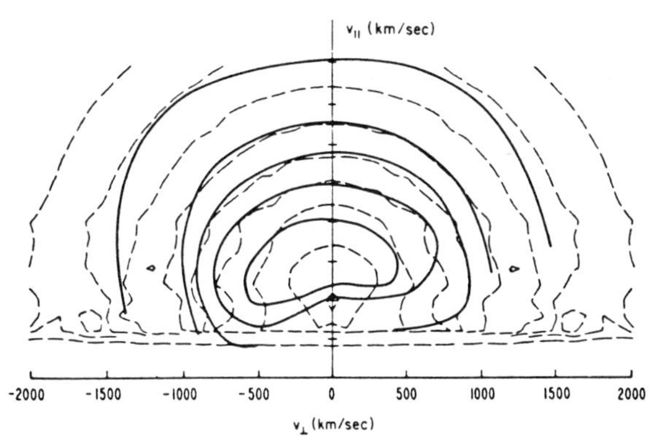

Fig. 8. Comparison of measured (solid lines) and modeled (dashed lines) distributions measured in the magnetotail, ostensibly generated by the neutral sheet acceleration process discussed in the text (from Lyons and Speiser, 1982).

well be equivalent to a statement concerning the existence of an adiabatic invariant.

In any case, in following single particle trajectories, Martin (1986a) indeed finds that the motion of particles within a narrow neutral sheet is well behaved, and that within a thick neutral sheet has the appearance of disorder, as is shown in Figure 9 (top and bottom, respectively). In generating these figures, Martin considered the motions in the vicinity of a reconnection x-line; however, the general conclusion drawn from these figures would be valid for other neutral sheet environments as well.

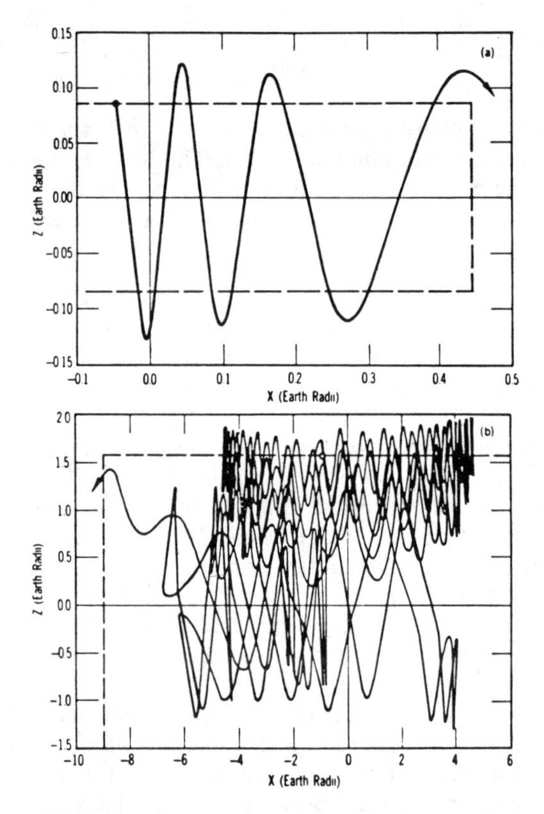

Fig. 9. Trajectories of particles near a narrow (top panel) and thick (bottom panel) magnetic neutral sheet that contains (at $X = 0$) a magnetic reconnection x-line (from Martin, 1986a).

All of the discussions of this section suggest a somewhat different ordering for the physics of the magnetotail than has been stated explicitly in the past. In particular, we suggest that the degree to which the guiding center approximations are valid is not the most useful way of ordering the magnetotail. Rather, the most useful ordering is the degree to which the motions can be described as adiabatic in the sense that a first-adiabatic invariant can be defined. Under this organizational scheme there would be two separated regions of adiabatic behavior where one might expect the physics to be similar in terms of kinetic particle acceleration (see Figure 10). Separating these regions would be a nonadiabatic region where the physics would be quite distinct. Clearly there would be no sharp separation between the regions shown in Figure 10, and the regions would depend on energy and species. Lyons (1984), for instance, took advantage of some nonadiabatic scattering within generally adiabatic regions in order to confine some electrons to the neutral sheet so that they receive more energization than they otherwise would.

Field-Aligned Electric Fields and Ionospheric Coupling

One reason for belaboring the distinction between the adiabatic and nonadiabatic behaviors is that there are important consequences of the adiabatic behaviors that we have discussed in terms of overall magnetospheric dynamics. Among these consequences is the generation of magnetic field-aligned electric fields and, we speculate, enhanced ionospheric/magnetospheric coupling. Within this context we will return to the convection-surge/dipolarization modeling efforts discussed earlier. In particular, it was noted, in reference to Figure 4e, that the mechanism generates strong magnetic field-aligned ion distributions. Hence the average ion mirrors at a much lower altitude than it did prior to the dipolarization. The electrons, on the other hand, will not be affected by the centrifugal term to nearly the degree that are the ions. Hence the electrons will largely be confined to the higher altitude regions relative to the ion positions. In other words, the action of the centrifugal acceleration term during the convection surges tries to separate charges along the field lines. The system will respond to restore quasi-charge-neutrality by setting up magnetic field-aligned electric fields (e.g., Whipple, 1977).

This process has been modeled (B. H. Mauk, 1988, manuscript submitted to the *J. Geophys. Res.*) by expanding the previously described modeling efforts into an electrostatically self-consistent simulation. Approximated forms for the massless electron distributions were assumed (e.g., bi-Maxwellians), and the technique of quasi-neutrality was used to derive the field-aligned electric fields. The ions were forced to respond self-consistently to these electric fields.

Figure 11 shows one example of these modeling efforts. The figure shows magnetic latitude versus normalized potential for various times (labeled by seconds) during the convection surge process. For this particular simulation the pre-dipolarization populations consisted of isotropic 0^+ ions and electrons, each with Maxwellian temperatures of 1 keV. The figure shows that during the initial stages of the 60-sec dipolarization, a relatively sharp potential discontinuity forms near the magnetic equator. As this field-aligned electric field front propagates from the equator to the higher latitude/ionospheric regions, it evolves into a much broader structure, finally settling down into a net equator-to-ionosphere potential drop that is several times the electron temperature. For the simulation shown, the initial front involves a potential drop of

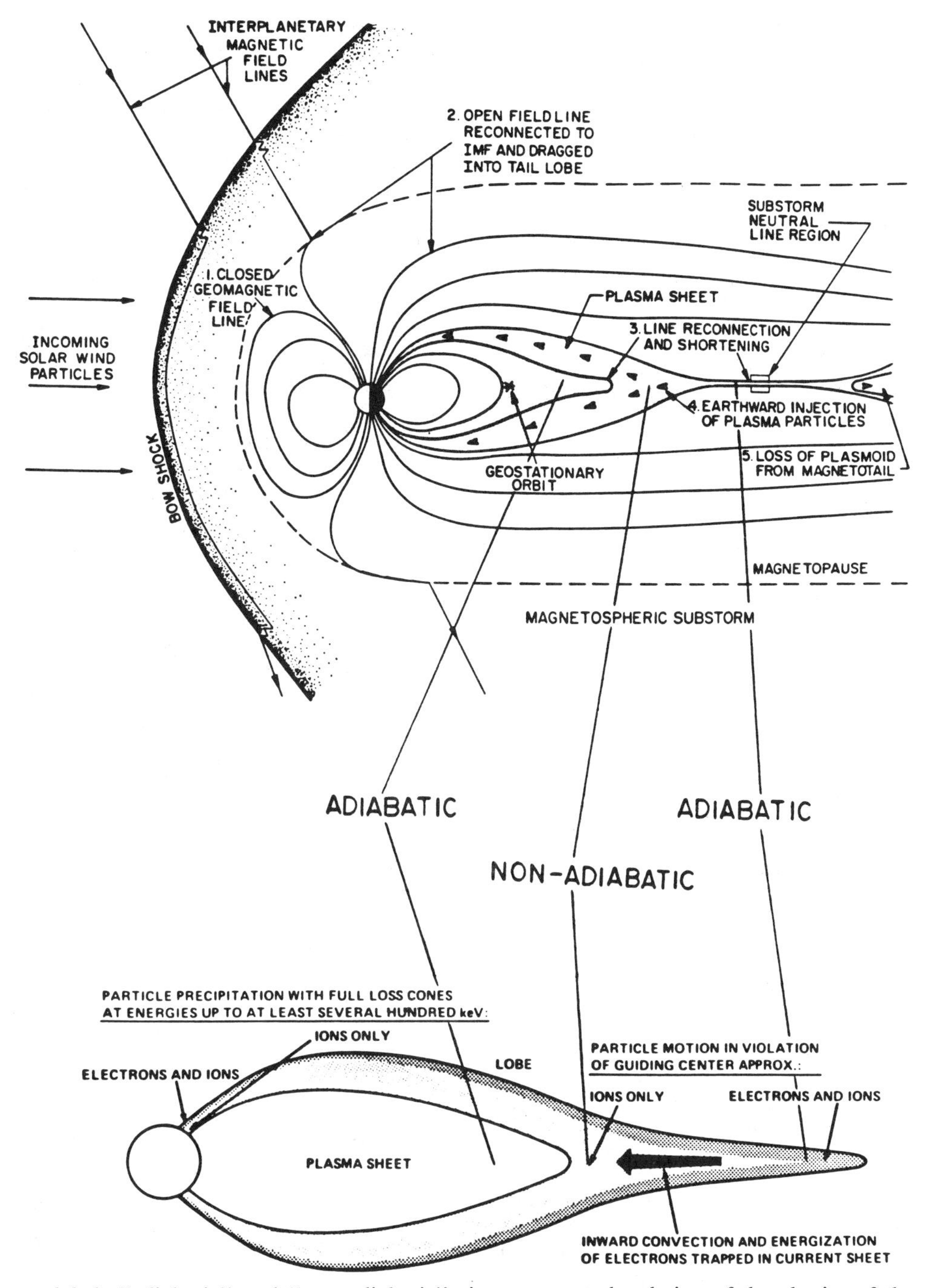

Fig. 10. The new labels "adiabatic" and "non-adiabatic" give a suggested ordering of the physics of the near neutral sheet regions of the magnetotail in terms of kinetic particle acceleration. This ordering may be more useful than an ordering based on the validity of the guiding center approximations (after Baker et al., 1987, top and Lyons, 1984, bottom).

$\sim$1.3 kV, and the equator-to-ionosphere potential drop at times reaches as high as 3 kV. These values are, of course, dependent on initial conditions. Potentials as high as $\gtrsim 10$ kV have been reached for these geosynchronous regions using higher initial temperatures ($\sim$10 keV). At even higher temperatures, the dramatic effects of the centrifugal acceleration mechanism on the ions relative to the electrons begin to fade (the second or longitudinal adiabatic invariant becomes valid for still higher initial energies). Thus there is a limit to the potential drops that can be reached.

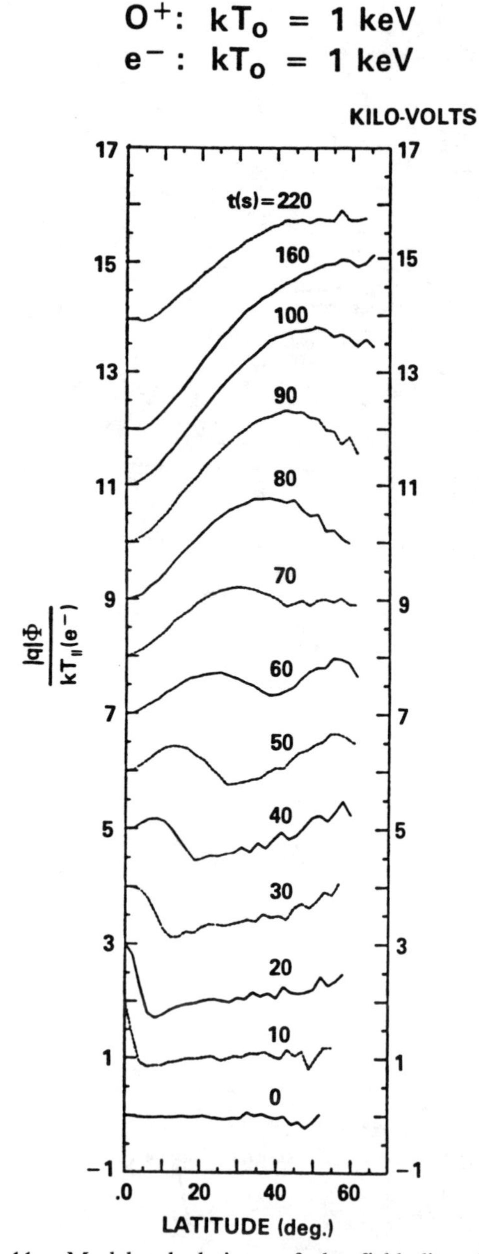

Fig. 11. Model calculations of the field-aligned electric potentials generated by the convection surge/dipolarization process near the geosynchronous regions of the Earth's magnetosphere. The different profiles, corresponding to different times into the process (seconds), are displaced vertically from each other for clarity. The initial particle distributions were O^+ and e^- each at initial Maxwellian temperatures of 1 keV.

One of the things that has been ignored in the self-consistent modeling efforts just described is the possible ionospheric interactions at high latitudes. Immediately into the simulation (e.g., see the Φ profile at $t = 20$ sec) a potential barrier forms which excludes the ionospheric electrons from streaming out of the ionosphere and neutralizing the near equatorial potential structures. However, at later stages, it is clear that the ionospheric interaction is likely to become important. Some idea of what that ionospheric interaction might consist of can be gotten by considering the laboratory studies of Stenzel et al. (1981) on the generation of electrostatic shock/double layer structures within converging magnetic field-line geometries.

Stenzel et al. (1981) performed the experiment diagrammed in Figure 12a. An ion beam was fired from the right of the figure toward one pole of a dipolar magnetic field structure. The crucial feature of the pole of the magnet was the converging nature of the field lines. The potential of the leading face of the magnet generating the dipolar configuration could

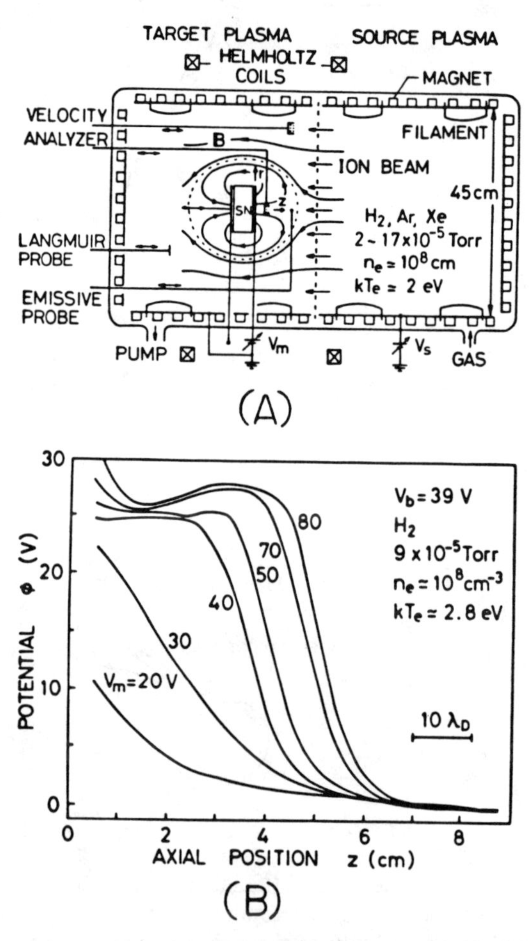

Fig. 12. (a) Laboratory setup for an experimental study of the generation of electric potential double layers within a magnetic geometry of strongly converging field lines. (b) Field-aligned electric potentials measured along the symmetry axis of the dipole magnetic configuration shown in (a). V_m is a potential impressed on the face plate of the magnet (from Stenzel et al., 1981). See text.

be set to arbitrary values V_m relative to the system ground. Also, this face could draw a current.

What Stenzel et al. (1981) found was that they could form the field-aligned potential structures shown in Figure 12b. This figure shows the electric potential as a function of distance from the magnet face along the symmetry axis of the magnetic pole. Note that a detached double layer/electrostatic shock does not form for the relatively low values of the magnet face potential V_m, despite the fact that $\Delta\Phi$ along the axis is $>kT_e/e$ (i.e., satisfying the so-called Bohm criterion). Stenzel et al. (1981) emphasized that a crucial aspect of forming the detached double layer is the use of a potential V_m high enough to reflect the streaming ions so that an electron-ion counter streaming situation is set up. These results have been modeled theoretically by Kan (1975).

The motivation here in showing the Stenzel et al. (1981) results is to emphasize that the field-aligned adiabatic accelerations that have been discussed in previous sections generate pitch angle collimated ion beams which propagate toward the field line converging regions above the atmosphere, in direct analogy with the Stenzel et al. (1981) experiment. We suggest that the field-aligned potentials obtained by Stenzel et al. (1981) for low values of V_m are fully analogous to the potential structures that develop at later times in the simulations shown in Figure 11. The question then becomes: how are detached double layers formed from this situation? It is an obvious speculation that the need for a large repelling potential V_m might well disappear if one were to add an ionosphere at the surface of the magnet in the Stenzel et al. (1981) experiments. Cold ionospheric ion being accelerated out into the magnetosphere by the broad field-aligned potential drops could then provide the counter-streaming condition required to generate the detached double layer structure and the resulting auroral-like discharges.

And indeed it does appear that such auroral-like discharges do occur within the adiabatic inner regions where the guiding center calculations are valid. McIlwain (1975), for instance, has measured intense field-aligned electron beams within the geosynchronous regions in association with auroral expansions. These beams are observed in association with geosynchronous magnetic signatures (Mauk and Meng, 1987) identical to the magnetic signatures associated with the ion bounce-phase-bunched distributions modeled in Figure 4 (Quinn and McIlwain, 1979). An example of the geosynchronous electron beams is shown in Figure 13. The solid dots and open circles show the field-aligned and field-perpendicular components, respectively. A field-aligned beam was present, peaking near 6 keV, that is suggestive of discrete, auroral-like discharge phenomena. Other such phenomena, for instance, transient magnetic field-aligned currents, have also been observed within these regions (Kelly et al., 1984; Robert et al., 1984; Roux et al., 1986). [Field-aligned currents is another requirement for double layer formation of the sort reported by Stenzel et al. (1981). Such currents can be drawn in the vicinity of sharp spatial boundaries of the sort inferred within the inner regions by McIlwain, (1974) and Mauk and Meng (1983, 1986). The mechanism of current generation within the inner regions, in association with spatial boundaries, may well be that suggested by Roux et al. (1986)]. Finally, there is some evidence that the geosynchronous discharge phenomena observed by McIlwain (1975) has a low altitude, auroral ionospheric counterpart (Mauk et al., 1981).

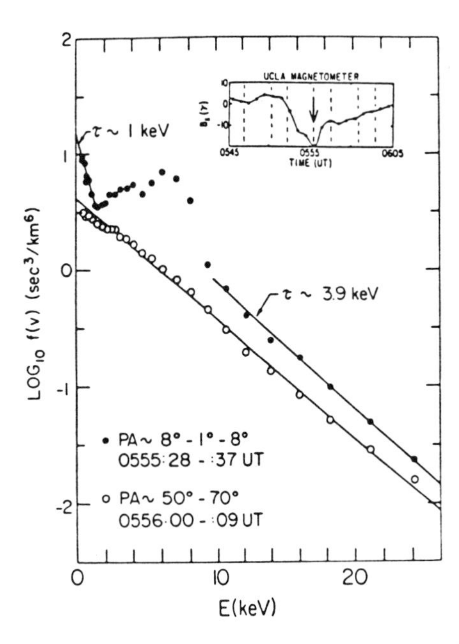

Fig. 13. Field-aligned (dots) and field perpendicular (open circles) electron distributions measured in the geosynchronous orbit near midnight in close association with a substorm particle injection. In this region where the guiding center approximations are valid, the field-aligned distribution has some features associated with discrete auroral-like discharges (after McIlwain, 1975).

The crucial point is that these geosynchronous auroral-like discharges occur in regions where the guiding center approximations for the particle motions are valid. Lyons and Evans (1984) have provided evidence that those auroral emissions, most commonly called "discrete," map along field lines to the narrow neutral sheet regions examined by, e.g., Speiser (1965, 1987) and Lyons and Speiser (1982). Thus phenomena occur in both regions that have similarities. We suggest that in terms of the generation of auroral-like discharge phenomena and the corresponding modification of the coupling between the magnetosphere and ionosphere, it is the "adiabatic" nature of the field-aligned ion acceleration that unifies the inner, guiding-center-valid regions and the outer, non-guiding-center-valid regions. The maintenance of the auroral coupling is presumably dictated by the ability of the different regions to continuously generate the necessary conditions. Stable arc structures will not be generated by the inner region phenomena discussed here. The discrete-auroral-like phenomena occurring within the inner regions are likely to be more transient in nature than they are in the outer regions (e.g., Mauk et al., 1981).

Summary and Discussion

The following points can be abstracted from the foregoing discussion:

A. The importance of the so-called guiding center centrifugal acceleration to the inner and middle ($r \lesssim 10\ R_E$) Earth magnetospheric regions has been demonstrated with regard to the magnetic field-aligned kinetic acceleration of ions.

B. In terms of kinetic acceleration, the consequences of the guiding center centrifugal acceleration are virtually identical to the consequences of the non-guiding center acceleration that occur in the vicinity of narrow neutral sheet regions of the distant magnetotail. The similarities result from the existence of different but analogous first-adiabatic invariants in the two regions.

C. Based on this similarity, an organizational scheme for the magnetotail has been suggested whereby two "adiabatic" regions, one in the near magnetotail and one in the distant magnetotail, are separated by a region of nonadiabatic behavior.

D. One of the consequences of the "adiabatic" behavior is the generation of field-aligned electric fields, and perhaps, through the ionospheric interaction, the formation of field-aligned double layers and the corresponding auroral-like discharges.

E. Evidence is available that auroral-like discharges, and the associated modification of the ionosphere-magnetosphere coupling, does indeed occur in both of the "adiabatic" regions described above, although it is anticipated that the nature of these discharges will be very different due to the probable inability of the inner regions to maintain the appropriate conditions for extended periods.

One popular technique for examining macroscopic particle acceleration processes is to follow the development of kinetic particle trajectories within the dynamical electromagnetic environments established by magnetohydrodynamic calculations (e.g., Scholer and Jamitzky, 1987; Zelenyi, et al., 1984; Matthaeus et al., 1984; Goldstein et al., 1986; Sato et al., 1982). The philosophy behind these calculations is that the primary coupling interactions between different macroscopic regions of the magnetosphere can be described on the basis of fluid-like or "localized" interactions. The particle distribution moments are thus viewed as being the fundamental experimental quantities needed to characterize the macroscopic interactions.

In the systems described in this paper, the kinetic particle acceleration processes appear to be acting as critical components in the self-consistent dynamics of the systems, including the macroscopic electromagnetic couplings between different regions (ionosphere/magnetosphere in the systems considered). The kinetic accelerations are not simply side effects of the electromagnetic interactions. We consider the degree to which kinetic particle acceleration processes moderate the fundamental electromagnetic interactions on macroscopic scales to be one of the outstanding problems in solar system plasma physics. Clearly, full particle distributions as well as accurate distribution moments will be needed over macroscopic scales to experimentally address this problem.

Acknowledgments. We thank D. G. Mitchell for helpful discussions. This work was supported by the Atmospheric Science Division, National Science Foundation grant ATM-8315041, by the Air Force Office of Scientific Research ground 84-0049, and by NASA and the Office of Naval Research support to The Johns Hopkins University Applied Physics Laboratory and Department of the Navy under task I2UOS1P of contract N00024-85-C-5301.

References

Aggson, T. L., J. P. Heppner, and N. C. Maynard, Observations of large magnetospheric electric fields during the onset phase of a substorm, *J. Geophys. Res.*, *88*, 3981, 1983.

Baker, D. N., S. J. Bame, D. J. McComas, R. D. Zwickl, J. A. Slavin, and E. J. Smith, Plasma and magnetic field variations in the distant magnetotail associated with near-earth substorm effects, in *Magnetotail Physics*, edited by A. T. Y. Lui, p. 137, The Johns Hopkins University Press, Baltimore, Maryland, 1987.

Beard, D. B., and S. W. H. Cowley, Electric and magnetic drift of non-adiabatic ions in the earth's geomagnetic tail, *Planet. Space Sci.*, *33*, 773, 1985.

Birn, J., and K. Schindler, Computer modeling of magnetotail convection, *J. Geophys. Res.*, *90*, 3441, 1985.

Chapman, S. C., and S. W. H. Cowley, Acceleration of lithium test ions in the quiet-time geomagnetic tail, *J. Geophys. Res.*, *89*, 7357, 1984.

Chen, C., and P. J. Palmadesso, Chaos and non-linear dynamics of single-particle orbits in a magnetotail-like magnetic field, *J. Geophys. Res.*, *91*, 1499, 1986.

Cladis, J. B., Parallel acceleration and transport of ions from polar ionosphere to plasmasheet, *Geophys. Res. Lett.*, *13*, 893, 1986.

Cladis, J. B., and W. E. Francis, The polar ionosphere as a source of the storm time ring current, *J. Geophys. Res.*, *90*, 3465, 1985.

Goldstein, M. L., W. H. Matthaeus, and J. J. Ambrosiano, Acceleration of charged particles in magnetic reconnection: Solar flares, the magnetosphere, and the solar wind, *Geophys. Res. Lett.*, *13*, 205, 1986.

Horwitz, J. L., Core plasma in the magnetosphere, *Rev. Geophys.*, *25*, 579, 1987.

Horwitz, J. L., and M. Lockwood, The cleft ion fountain: A two-dimensional kinetic model, *J. Geophys. Res.*, *90*, 9749, 1985.

Horwitz, J. L., M. Lockwood, J. H. Waite, Jr., T. E. Moore, C. R. Chappell, and M. O. Chandler, Transport of accelerated low-energy ions in the polar magnetosphere, in *Ion Acceleration in the Magnetosphere and Ionosphere, Geophys. Monogr. Ser.*, vol. 38, edited by T. Chang, p. 56, AGU, Washington, D.C., 1986.

Kan, J. R., Energization of auroral electrons by electrostatic shock waves, *J. Geophys. Res.*, *80*, 2089, 1975.

Kaye, S. M., E. G. Shelley, R. D. Sharp, and R. G. Johnson, Ion composition of zipper events, *J. Geophys. Res.*, *86*, 3393, 1981.

Kelly, T. J., C. T. Russell, and R. J. Walker, ISEE-1 and -2 observations of an oscillating outward moving current sheet near midnight, *J. Geophys. Res.*, *89*, 2745, 1984.

Lyons, L. R., Electron energization in the geomagnetic tail current sheet, *J. Geophys. Res.*, *89*, 5479, 1984.

Lyons, L. R., Processes associated with plasma sheet boundary layer, *Physica Scripta*, *T18*, 103, 1987.

Lyons, L. R., and D. S. Evans, An association between discrete aurora and energetic particle boundaries, *J. Geophys. Res.*, *89*, 2395, 1984.

Lyons, L. R., and T. W. Speiser, Evidence for current sheet acceleration in the geomagnetic tail, *J. Geophys. Res.*, *87*, 2276, 1982.

Martin, R. F., Jr., The effect of plasma sheet thickness on ion acceleration near a magnetic neutral line, in *Ion Acceleration in the Magnetosphere and Ionosphere, Geophys. Monogr. Ser.*, vol. 38, edited by T. Chang, p. 141, AGU, Washington, D.C., 1986a.

Martin, R. F., Jr., Chaotic particle dynamics near a two-dimensional magnetic neutral point with application to the geomagnetic tail, *J. Geophys. Res.*, *91*, 11,985, 1986b.

Matthaeus, W. H., J. J. Ambrosiano, and M. L. Goldstein, Particle acceleration by magnetohydrodynamic reconnection, *Phys. Rev. Lett.*, *53*, 1449, 1984.

Mauk, B. H., Quantitative modeling of the "convection surge" mechanism of ion acceleration, *J. Geophys. Res.*, *91*, 13,423, 1986.

Mauk, B. H., and C. E. McIlwain, UCSD auroral particles experiment, *IEEE Trans. Aerosp. Electron. Syst.*, *AES-11*, 1125, 1975.

Mauk, B. H., and C.-I. Meng, Dynamical injections as the source of near geostationary quiet time particle spatial boundaries, *J. Geophys. Res.*, *88*, 10,011, 1983.

Mauk, B. H., and C.-I. Meng, Macroscopic ion acceleration associated with the formation of the ring current in the earth's magnetosphere, in *Ion Acceleration in the Magnetosphere and Ionosphere, Geophys. Monogr. Ser.*, vol. 38, edited by T. Chang, p. 351, AGU, Washington, D.C., 1986.

Mauk, B. H., and C.-I. Meng, Plasma injection during substorms, *Physica Scripta*, *T18*, 128, 1987.

Mauk, B. H., and L. J. Zanetti, Magnetospheric electric fields and currents, *Rev. Geophys.*, *25*, 541, 1987.

Mauk, B. H., J. Chin, and G. Parks, Auroral X-ray images, *J. Geophys. Res.*, *86*, 6827, 1981.

McIlwain, C. E., Substorm injection boundaries, in *Magnetospheric Physics*, edited by B. M. McCormac, p. 143, D. Reidel, Hingham, Massachusetts, 1974.

McIlwain, C. E., Auroral electron beams near the magnetic equator, in *The Physics of Hot Plasma in the Magnetosphere*, edited by B. Hultquist and L. Stenflow, p. 91, Plenum, New York, 1975.

Northrop, T. G., *The Adiabatic Motion of Charged Particles*, Interscience, New York, 1963.

Propp, K., and D. B. Beard, Cross-tail ion drift in a realistic model magnetotail, *J. Geophys. Res.*, *89*, 11,013, 1984.

Quinn, J. M., and C. E. McIlwain, Bouncing ion clusters in the earth's magnetosphere, *J. Geophys. Res.*, *84*, 7365, 1979.

Quinn, J. M., and D. J. Southwood, Observations of parallel ion energization in the equatorial region, *J. Geophys. Res.*, *87*, 10,536, 1982.

Robert, P., R. Gendrin, S. Perraut, A. Roux, and A. J. Pedersen, Geos 2 identification of rapidly moving current structures in the equatorial outer magnetosphere during substorms, *J. Geophys. Res.*, *89*, 819, 1984.

Rostoker G., and R. Boström, A mechanism for driving the gross Birkeland current configuration in the auroral oval, *J. Geophys. Res.*, *81*, 235, 1976.

Roux, A., S. Perraut, A. Pedersen, R. DeViness, and D. Rodgers, Instability of the plasma sheet in relation to substorms, paper presented at Sixth International Symposium on Solar-Terrestrial Physics, Toulouse, France, June 30-July 5, 1986.

Sato, T., H. Matsumoto, and K. J. Nagai, Particle acceleration in time developing magnetic reconnection process, *J. Geophys. Res.*, *87*, 6889, 1982.

Sauvaud, J. A., and D. Delcourt, A numerical study of suprathermal ionospheric ion trajectories in three-dimensional electric and magnetic field models, *J. Geophys. Res.*, *92*, 5873, 1987.

Scholer, M., and F. Jamitzky, Particle orbits during the development of plasmoids, *J. Geophys. Res.*, *92*, 12181, 1987.

Sonnerup, B. U. Ö., Adiabatic particle orbits in a magnetic null sheet, *J. Geophys. Res.*, *76*, 8211, 1971.

Sonnerup, B. U. Ö., Magnetic field reconnection at the magnetopause: An overview, in *Magnetic Reconnection in Space and Laboratory Plasmas, Geophys. Monogr. Ser.*, vol. 30, edited by E. W. Hones, Jr., p. 92, AGU Washington, D.C., 1984.

Speiser, T. W., Particle trajectories in model current sheets, 1, Analytical solutions, *J. Geophys. Res.*, *70*, 4219, 1965.

Speiser, T. W., Particle trajectories in model current sheets, 2, Applications to auroras using a geomagnetic tail model, *J. Geophys. Res.*, *72*, 3919, 1967.

Speiser, T. W., Conductivity without collisions or noise, *Planet. Space Sci.*, *18*, 613, 1970.

Speiser, T. W., Processes in the magnetotail neutral sheet, *Physica Scripta*, *T18*, 119, 1987.

Stenzel, R. L., M. Ooyama, and Y. Nakamura, Potential double-layers in strongly magnetized plasmas, in *Physics of Auroral Arc Formation Geophys. Monogr. Ser.*, vol. 25, edited by S.-I. Akasofu and J. R. Kan, p. 226, AGU, Washington, D.C., 1981.

Whipple, E. C., T. G. Northrop, and T. J. Birmingham, The signature of parallel electric fields in a collisionless plasma, *J. Geophys. Res.*, *82*, 1525, 1977.

Whipple, E. C., T. G. Northrop, and T. J. Birmingham, Adiabatic theory in regions of strong field gradients, *J. Geophys. Res.*, *91*, 4149, 1986.

Zelenyi, L. M., A. S. Lipatou, D. G. Lominadze, and A. L. Taktakishvili, The dynamics of the energetic proton bursts in the course of the magnetic field topology reconstruction in the earth's magnetotail, *Planet. Space Sci.*, *32*, 313, 1984.

NEW TECHNIQUES FOR CHARGED PARTICLE MEASUREMENTS IN THE INTERPLANETARY MEDIUM

R. P. Lin

Space Sciences Laboratory, University of California, Berkeley, CA 94720

Abstract. The interplanetary medium may be the ideal environment for in situ studies of particle acceleration processes of astrophysical interest and for studies of basic physical processes in collisionless plasmas. New instrumentation, on the upcoming GGS and the Cluster and SOHO missions of the International Solar-Terrestrial Physics (ISTP) program, will have the sensitivity and dynamic range to obtain the first comprehensive measurements of the three-dimensional distribution of suprathermal particle populations in the interplanetary medium from solar wind to cosmic ray energies. The temporal resolution of the planned particle measurements, although much improved over previous instrumentation, is still far short of time scales for evolution of wave-particle phenomena in the interplanetary medium. Wave-particle correlation techniques, however, can detect perturbations of the particle distribution functions by large amplitude plasma waves, on time scales relevant to wave-particle interactions.

Introduction

For many years the solar wind has been regarded as a regime essentially independent from the energetic particle populations found in interplanetary space. The particles in the intervening energy range, from above solar wind plasma to ~100 keV, are referred to here as suprathermal particles. They have not yet been adequately surveyed, much less quantitatively measured. Yet they must play a key role in the varied plasma and energetic particle phenomena observed to occur in the interplanetary medium, and upstream from the Earth's magnetosphere. A wide variety of very interesting plasma phenomena are observed in the interplanetary medium (IPM). These include: collisionless shocks and associated particle acceleration and gyro-phase bunching of particles; a multitude of wave-particle instabilities leading to Langmuir waves, whistlers, ion acoustic waves, etc.; various nonlinear plasma phenomena such as parametric decay, soliton collapse, and electromagnetic wave generation; as well as a broad spectrum of plasma turbulence. A close relationship between waves and particles has been demonstrated in many of these phenomena.

Figures 1 and 2 show the typical energy spectra of various interplanetary and distant magnetospheric particle phenomena. Because of dynamic range considerations, up to now instruments flown to measure the solar wind plasma ions and electrons lack the sensitivity to detect suprathermal particles from just above solar wind energies to a few tens of kiloelectron volts, except during highly disturbed times. For ions, only upper limits are known for the quiet suprathermal population, while for electrons a quiet-time power law component is observed at energies above 2 keV which is orders of magnitude more intense than the extrapolation of the solar wind halo population (Lin, 1985).

New Measurements

The charged particle detectors currently planned for the WIND spacecraft of the Global Geospace Science (GGS) mission and proposed for the SOHO and Cluster spacecraft of the International Solar-Terrestrial Physics (ISTP) program will provide a very significant advance over previous suprathermal particle measurements. These detector systems range from electrostatic analyzers of novel design, to solid state detector telescopes, to time-of-flight systems. Since ion composition detectors were covered in other review papers at this conference, they will not be discussed here. The planned new instrumentation will measure the three-dimensional distribution of plasma and energetic electrons and ions with high sensitivity and good energy, angular and temporal resolution, over the entire energy range ~10 eV to $\gtrsim$several hundred kiloelectron. Some of the scientific objective for these instruments are: *(a)* the first detailed exploration of the interplanetary particle population in the suprathermal energy range between solar wind plasma energies and ~100 keV; *(b)* the study of particle acceleration at the Sun, in the IPM, and upstream from the Earth; *(c)* the study of the transport of particles in the IPM in the critical transition energy range between solar wind plasma and cosmic rays, *(d)* the study of basic plasma processes occurring in the IPM, such as wave-particle interactions, the production of radio emission by beam-plasma processes (Type III bursts), shock waves, nonlinear processes such as soliton collapse, and solar wind heat flux; *(e)* the measurement of the particle and plasma input to and output from the Earth's magnetosphere.

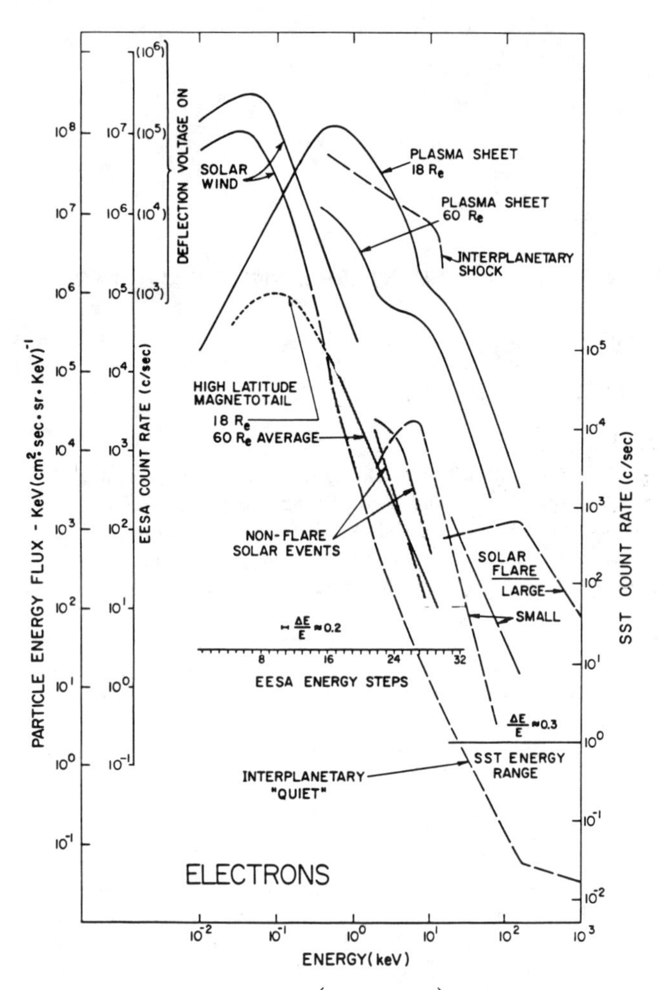

Fig. 1. Electron energy flux $\left(E\,(dJ/dE)\right)$ in the interplanetary medium and outer magnetosphere. Approximate counting rates per channel for the electron electrostatic analyzer (EESA) and semiconductor detector telescopes (SST) planned for the WIND spacecraft are indicated on the left and right axes, respectively.

Plasma Processes

The IPM may be ideal for the study of particle acceleration processes of astrophysical interest. Collisionless shock waves, both those associated with solar flares and those originating in solar wind stream-stream interactions, provide acceleration of particles over spatial scales of ~1 AU or more. The shock waves themselves, however, have thicknesses of ~10^2 km. Multiple scattering of the particles, by waves upstream and downstream of the shock, appears necessary to attain the high energies observed. The "seed" population from which the particles are accelerated is at present still uncertain.

High sensitivity at suprathermal energies is required to identify the seed population, whether solar wind thermal particles or background quiet suprathermal population or residual solar flare particles. High time resolution measurements with good energy and angular resolution and full coverage are needed to study the microscopic interaction of the particles with shock waves, and particle scattering by upstream and downstream waves. The planned WIND instruments will provide full three-dimensional measurements of the electrons and ions with 1.5 sec time resolution and extremely high sensitivity (see Figures 1 and 2). In addition, there will be detailed ion composition measurements—elemental, isotropic, and charge state—which can provide probes of different charge-to-mass ratios, and which can also identify the source of the accelerated particles.

The IPM also may provide a near-ideal environment for the study of basic physical processes in collisionless plasmas. Boundary effects and perturbations by the measurement probes are generally negligible. The low IPM plasma densities and magnetic field strengths lead to typical time scales

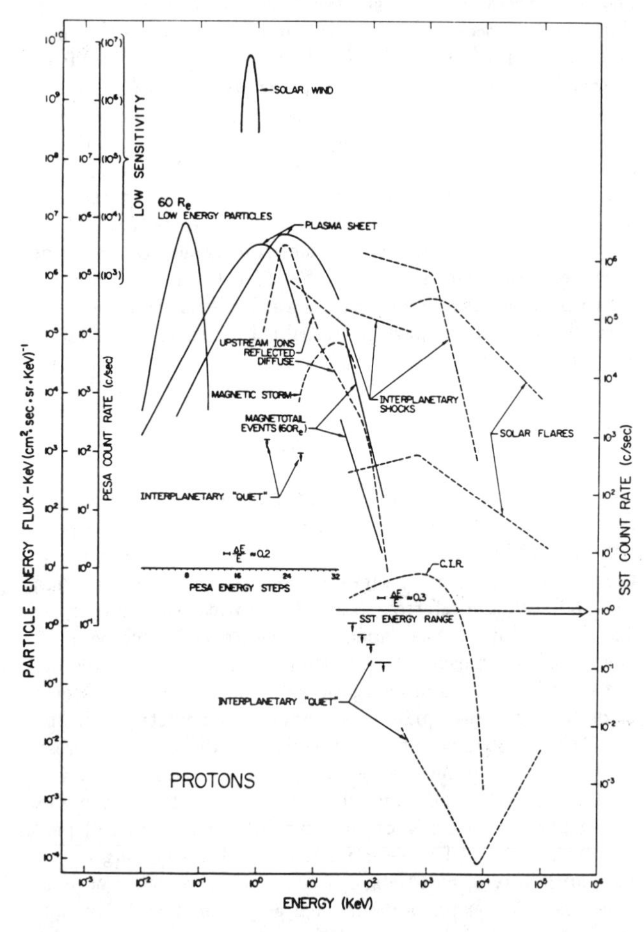

Fig. 2. Proton energy flux $\left(E\,(dJ/dE)\right)$ in the interplanetary medium and outer magnetosphere. Approximate counting rates per channel for the proton electrostatic analyzer (PESA) and semiconductor detector telescopes (SST) planned for the WIND spacecraft are indicated on the left and right axes, respectively.

for evolution of plasma phenomena that are very long compared to those for laboratory plasmas. For example, the fastest plasma wave time scale is set by the electron plasma frequency which is typically of order ~20 kHz in the IPM, but is in the megaHertz range in laboratory or ionospheric plasmas. Perhaps most important, the instruments for measurements of waves and particles in space plasmas have now advanced to the point where particle distribution functions and wave fields can be determined to the detail required for quantitative comparisons with theory.

Particle measurements have typically been obtained from electrostatic analyzers that provide a two-dimensional distribution function on time scales of a few seconds to a minute or more, and spectrum analyzers have provided wave frequency/power spectra on somewhat faster time scales, while the natural time scales for the growth and saturation of these phenomena are tens to thousands of wave periods, i.e., $1-10^2$ ms for Langmuir waves. Thus the measurements only give the time-averaged or quasi-equilibrium state after the instability has grown, evolved nonlinearly, and saturated.

As an example, unstable electron beams (see Figures 3 and 4) have been observed in close association with Langmuir waves in solar Type III radio bursts (Lin et al., 1981) and in the upstream electron foreshock (Fitzenreiter et al., 1984). The positive slopes in the reduced parallel distribution functions of the electrons should be highly unstable and should rapidly (<1 sec) plateau, yet they are observed apparently to persist for minutes. On the other hand, the Langmuir waves occur in bursts of a few hundred millisecond duration and there is evidence (Figure 5) that these waves parametrically decay (Gurnett et al., 1981; Lin et al., 1986). The theoretical interpretation of these phenomena appears to require either microstructures where the unstable electron distribution function evolves on a time scale of $\lesssim 10^2$ ms (Melrose and Goldman, 1987), or density fluctuations which trap the waves (Levedahl, 1987), or various nonlinear plasma phenomena such as modulational instability and soliton collapse (see Goldman, 1983, for review), but the observations are inadequate to pinpoint the process.

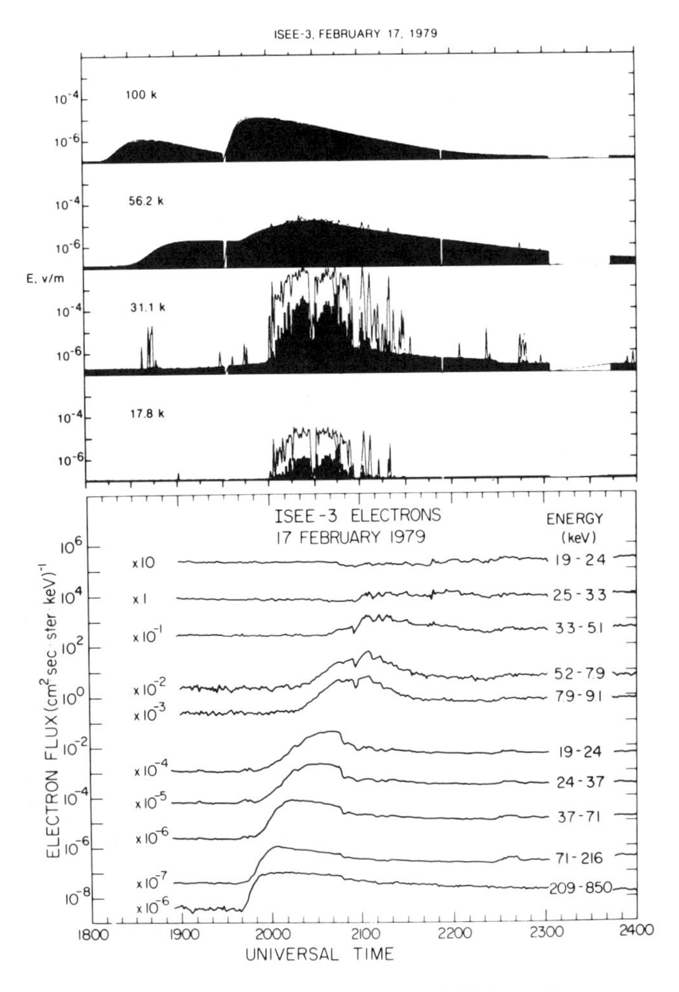

Fig. 3. The top panel shows the electric field intensity measured from ISEE 3 on February 17, 1979, in four broad frequency bands for the event of interest. The black areas show the intensity averaged over 64 s. The solid lines give the peak intensity measured every 0.5 s. The smoothly varying profiles in the 100 and 56.2 kHz channels show two Type III radio bursts. The second one is of interest here. The intense, highly impulsive emissions observed in the 31.1 and 17.8 kHz channels are the plasma waves. The lower panel shows the omni-directional electrons from 2 keV to >200 keV. The velocity dispersion is clearly evident (from Lin et al., 1981).

Fast Correlation Techniques

Ideally, for the study of these phenomena, both wave and particle measurements should be obtained down to the time scales of the electron plasma period, ~50 μs for a typical IPM plasma frequency of 20 kHz. This goal is still not accessible to present-day space experiments but may be possible in the next generation. Fast wave-particle correlation techniques, however, can measure the nonlinear and linear perturbations of the electron distribution function near the phase resonant velocity of large amplitude waves, and thus provide a means for exploring the effects of particle trapping by these waves.

Figure 6 (taken from Chen, 1984) shows that when a plasma wave is present, charged particles will be slowed down or speeded up by the wave electric field, depending on the direction of the electric field relative to the particle motion. For particles traveling much faster or much slower than the wave, the wave-particle interaction averages to nearly zero and little energy is exchanged between the particle and the wave. Particles traveling at speeds near the wave speed can be accelerated or decelerated by the wave and significant energy exchange can occur. If the number of slightly slower particles is fewer than the number of slightly faster particles, then the wave will grow (and vice versa). In the process the particles exchange energy with the wave and are bunched or trapped by the wave.

Correlation techniques can be used to detect this bunch-

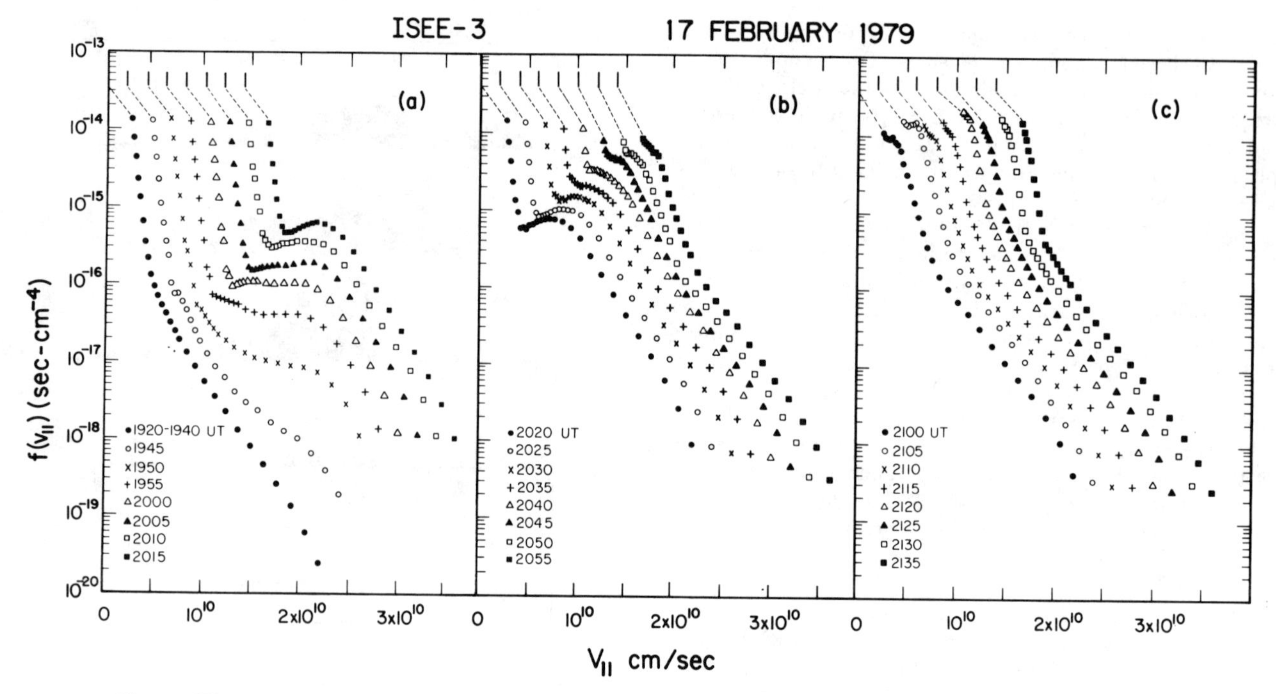

Fig. 4. The comprehensive measurements over energy and angle taken from ISEE 3 on February 17, 1979, have been used to construct the one-dimensional velocity distribution function of the electrons every 64 s. The distribution averaged over 20 minutes prior to the event onset is indicated by the solid dots in panel (a). The 64 s measurements of the distribution during the event are shown every 5 minutes. Successive distributions are offset to the right (see top of each panel) to provide separation. Note the positive slope which develops and moves to lower velocities with time (from Lin et al., 1981).

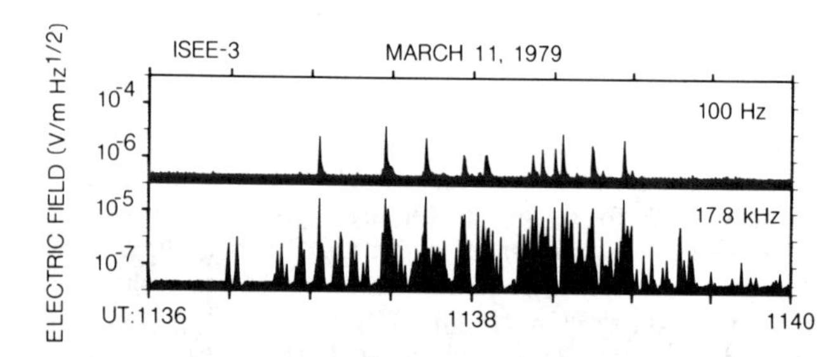

Fig. 5. High time resolution (0.5 s) plots of Langmuir waves (17.8 kHz) and long wavelength ion acoustic waves (100 Hz) for the March 11, 1979, Type III solar radio burst event (from Lin et al., 1986). The close correspondence between the most intense Langmuir wave spikes and the ion acoustic spikes indicates that some nonlinear processes such as parametric decay are occurring. Note the impulsive nature of the Langmuir waves.

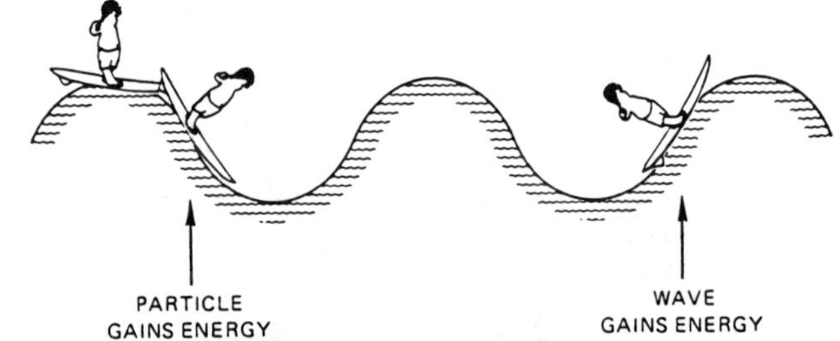

Fig. 6. Customary picture of wave-particle interaction in Landau damping (from Chen, 1984).

ing of the resonant particles by the wave, and thereby provide the first direct observation of the wave-particle coupling. The idea is to obtain essentially continuous, very high sensitivity measurements of electrons in a limited region of phase space, and to provide timing of these electron counts to a small fraction of a wave period. The electric field wave form would be supplied by a plasma wave experiment in real time. The wave period is divided into a number of phase bins and the electrons sorted into those bins to detect phase bunching.

Preliminary estimates, using a Monte Carlo test particle simulation, indicate that the coupling between strong ($\gtrsim$1 mV m^{-1}) Langmuir waves and electrons in solar Type III radio bursts and in the Earth's upstream electron foreshock should be detectable (R. Ergun, private communication, 1988). The simulation program chooses particles with random initial velocity (biased to the desired initial distribution function) at random (no bias) phase of an infinitesimally small Langmuir wave. The wave is constrained to grow exponentially for a set number of wave periods. The motion of the particles is followed as the wave grows to the desired amplitude (1 mV m^{-1}), and afterward up to the typical duration of a wave burst (~200 ms). Using the parameters of the solar Type III radio burst of March 11, 1979 (see Table 1 and Figure 5), the simulation indicates that an electron spectrometer with $\Delta E/E \approx 10\%$ measuring at the resonant velocity will see a correlation of $3.0\% \pm 0.8\%$. This is much larger than the expected linear correlation (at a velocity 10% out of resonance) of 0.25%. For an electron count rate of 5000/s, as few as ten wave bursts would need to be captured to obtain statistically significant ($\geq 3\ \sigma$) results. About 10^2–10^3 bursts occur during a solar Type III radio burst. It should be noted that this test particle simulation does not generate self-consistent wave and particle evolution.

Although much of this discussion has been limited to Langmuir wave phenomena (the fastest plasma wave in the IPM), we expect that fast wave-particle correlation techniques will provide new information on essentially the entire range of electron plasma and wave phenomena observed in the upstream, shock, and magnetosheath region of the Earth as well as in the IPM. Other waves, such as whistlers and ion acoustic waves, have time scales which are much longer and thus can easily be accommodated.

Particle Correlation Detectors

A limitation on present instrumentation for charged particle detection is that the field of view is typically a small fraction of the total 4π steradian solid angle. Instruments are usually designed to utilize the spin of the spacecraft to obtain full three-dimensional coverage but the temporal resolution is then limited to a spin period or half a spin period. For example, an electrostatic analyzer of the symmetric quadrispheric type (Figure 7) provides a 360° field of view in a plane with imaging in the polar angle direction (Carlson et al., 1983). If the spacecraft spin axis is in the plane of the field of view, then full three-dimensional coverage with excellent angular resolution can be obtained in half a spin. This type of instrument can provide measurements of the full electron distribution every half spin, and thus identify unstable regions in phase space.

There may be a number of ways to obtain continuous coverage of a limited region of phase space on a spinning spacecraft. Essentially 2π steradian coverage could be obtained using an onion-like set of nested focusing toroidal analyzers (see Figure 8). We have also considered electrostatic steering of the detector field of view (Figure 9). With

TABLE 1. Solar Type III Burst Plasma, Beam, and Wave Parameters (from Lin et al., 1986)

Parameters	Values: March 11, 1979	Values: February 8, 1979
Solar wind plasma:[a]		
Solar wind density, n	2 cm^{-3}	7
Solar wind velocity, V_{SW}	480 km s^{-1}	350
Angle of magnetic field to solar wind, θ_B	139°	63
Electron temperature, T_e	2×10^5 K	1.7×10^5
Ion temperature, T_i	4×10^4 K	6×10^4
Debye length, λ_D	2.2×10^3 cm	1.1×10^3
Electron plasma frequency, f_{p-}	13 kHz	24
Ion plasma frequency, f_{p+}	3×10^2 Hz	5.6×10^2
Fast electrons:		
Beam velocity, v_b	$\sim 3.5 \times 10^9$ cm s^{-1}	$\sim 3.5 \times 10^9$
Beam density, n_b	$\sim 7 \times 10^{-5}$ cm^{-3}	$\sim 2 \times 10^{-5}$
Positive slope, $\partial f/\partial v_\parallel$	$\sim 10^{-25}$ cm^{-3} s^2	$\sim 10^{-25}$
Beam width, $\Delta v_b/v_b$	~0.1–0.2	~0.1–0.2
Langmuir pump waves:		
Beam resonant wave number, k_0	2.3×10^{-5} cm^{-1}	4.3×10^{-5}
Maximum wave amplitude, $E_{0_{max}}$	~1 mV m^{-1}	~0.15
Maximum normalized energy density, $W_{max} = E_0^2/8\pi n \kappa T_e$	8×10^{-7}	6×10^{-9}
Long wavelength ion acoustic waves:		
Wavenumber, k_I (typical)	1.8×10^{-5} cm^{-1}	4×10^{-5}
Ion acoustic speed, c_s	5.2×10^6 cm s^{-1}	5.5×10^6
Ion acoustic frequency, f_I	15 Hz	35
Maximum electric field, $E_{I_{max}}$	~40 μV m^{-1}	~4

[a] Solar wind parameters provided by J. Gosling and W. Feldmen, and magnetic field by B. Tsurutani, private communication (1984).

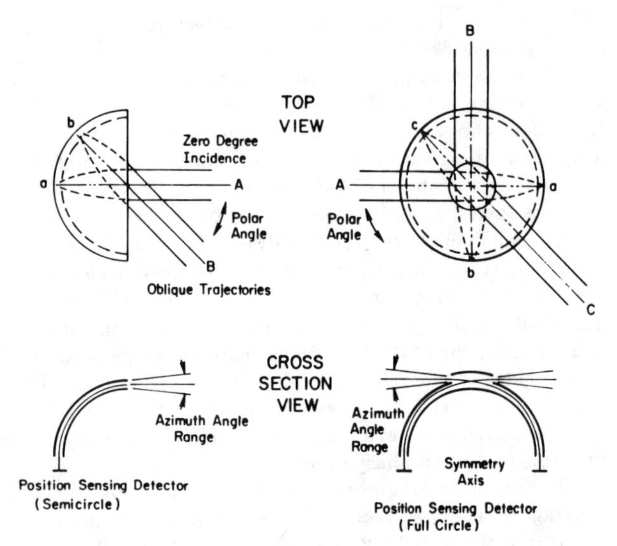

Fig. 7. A comparison of normal and symmetrical quadrisphere geometries. With normal quadrisphere the response varies with polar angle. Symmetrical analyzer has cylindrical symmetry with complete 360° field of view. Typical trajectories illustrate focussing characteristics (from Carlson et al., 1983).

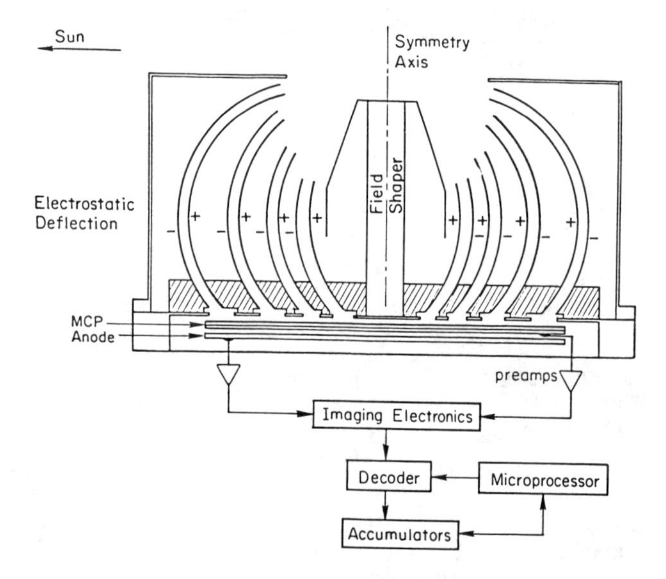

Fig. 8. Electrostatic analyzer with a 2π steradian field of view. A nested set of four toroidal analyzers image particle trajectories onto the microchannel plate (MCP) detector. Charge pulses produced by the MCP are collected by the anode and decoded to produce angular information about particle distribution functions.

input from an onboard magnetometer, such steering could be controlled to include the magnetic field direction and thereby provide continuous coverage of field-aligned beams. Either a conventional electrostatic analyzer with electrostatic steering

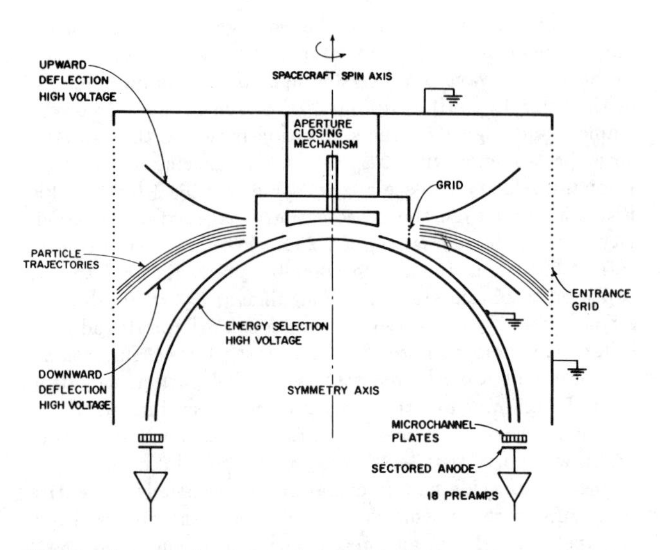

Fig. 9. A normal symmetric quadrisphere analyzer with electrostatic deflection plates to provide up to ±45° deflection from the normal planar field of view. The field of view could be steered to include the magnetic field direction. Trajectory simulations are shown for 45° deflection.

or an "onion" could provide the continuous high sensitivity measurements of a limited region of electron phase space needed for fast wave-particle correlations.

The same type of detector can also be used to detect rapidly evolving microstructures in the particle distributions. In Type III solar radio bursts the unstable positive slope region of the electron distribution is relatively narrow ($\Delta E/E \lesssim 0.3$), so it can be scanned very rapidly (on time scales of $\sim$10–10^2 ms), and the rapid evolution of that part of the distribution can be followed. In addition, this type of detector can provide very high time resolution (down to microseconds, limited only by count statistics) measurements of fluctuations in the thermal or suprathermal electron population for study of electrostatic waves and density fluctuations.

Acknowledgments. I am pleased to acknowledge extensive discussions with C. W. Carlson, R. E. Ergun, J. P. McFadden, P. Kellogg, L. Bougeret, and M. V. Goldman and colleagues. This research was supported in part by NASA contract NAS5-26855.

References

Carlson, C. W., D. W. Curtis, G. Paschmann, and W. Michael, An instrument for rapidly measuring plasma distribution function with high resolution, *Adv. Space Res.*, *2*, 67, 1983.

Chen, F. F., *Introduction to Plasma Physics and Controlled Fusion*, 2nd ed., Plenum Press, New York, 1984.

Fitzenreiter, R. J., A. J. Klimas, and J. D. Scudder, Detection of bump-on-tail reduced electron velocity distributions at the electron floreshock boundary, *Geophys. Res. Lett.*, *11*, 496, 1984.

Goldman, M. V., Progress and problems in the theory of Type III solar radio emission, *Solar Phys., 89*, 403, 1983.

Gurnett, D. A., J. E. Maggs, D. I. Gallagher, W. S. Kurth, and F. L. Scarf, Parametric interaction and spatial collapse of beam-driven Langmuir waves in the solar wind, *J. Geophys. Res., 86*, 8833, 1981.

Levedahl, W. K., Effect of wave localization on plasma instabilities, Ph.D. thesis, University of California, Berkeley, 1987.

Lin, R. P., Energetic solar electrons in the interplanetary medium, *Solar Phys., 100*, 537, 1985.

Lin, R. P., D. W. Potter, D. A. Gurnett and F. L. Scarf, Energetic electrons and plasma waves associated with a solar Type III radio burst, *Astrophys. J., 251*, 364, 1981.

Lin, R. P., W. K. Levedahl, W. Lotko, D. A. Gurnett, and F. L. Scarf, Evidence for nonlinear wave-wave interactions in solar Type III radio bursts, *Astrophys. J., 308*, 954, 1986.

Melrose, D. B., and M. V. Goldman, Microstructures in Type III events in the solar wind, *Solar Phys., 107*, 329, 1987.

A TECHNIQUE FOR FULLY SPECIFYING PLASMA WAVES

Paul M. Kintner

School of Electrical Engineering, Cornell University, Ithaca, NY 14853

Abstract. A propagating wave is described by its frequency and its wave vector. With a few rare exceptions all spacecraft measurements of plasma waves were designed to characterize wave frequency only. In this paper we discuss a series of experiments designed to characterize wave vector and wave frequency. The basic technique is to measure the phase shift between spaced sensors. The phase shift is a measure of the projection of a wave vector onto the axis of an interferometer. Two examples of interferometer data are presented. The first example is of plasma waves within a transverse ion acceleration region and the second example is of spatial density irregularities.

Introduction

Since the mid-1960's spacecraft have successfully conducted measurements of waves within space plasmas. These measurements have opened entire new areas of research. They have proven invaluable in our study of the physics of space plasmas. However they have been limited. With a few rare exceptions (Kelley and Mozer, 1972; Temerin, 1978; Fejer and Kelley, 1980; Bahnsen et al., 1978; Anderson et al., 1982; Fuselier and Gurnett, 1984; Gallagher, 1985), the measurements have all been made in the time domain; only the wave frequency was measured. A rigorous description requires measurement of frequency and wavelength (or wave vector). In this paper we will describe a technique for measuring frequency and wavelength simultaneously. The technique also applies to discrete structures such as double layers where, instead of wavelength, measurements of velocity are desired.

The advantages of measuring wavelength are obvious. Even a crude measurement of wavelength separates long wavelength electromagnetic modes from short wavelength electrostatic modes. More precise measurements of wavelength yield precise phase velocities which in turn can be used for testing theories of Landau resonance. Many theories of wave-particle interactions are sensitive to the ratio of gyroradius to wavelength. In general the measurements of wavelength complete the physical picture for testing theories of wave-particle interactions.

The device used to measure wavelength is termed the plasma wave interferometer. Even though the need to measure wavelength is obvious, the plasma wave interferometers are difficult experiments to perform. They require multiple sensors with spacing the order of the wavelength to be measured. Their signals must be telemetered in unaltered form using wideband channels. Even when a spacecraft can support the hardware and telemetry requirements, the interferometer data can only be interpreted when each frequency is associated with only one wave vector. Despite these obstacles several plasma wave interferometers have been successfully flown on sounding rockets and one has been successfully flown on a satellite.

This paper will be organized into several sections. The second section is a description of space plasma wave dispersion surfaces. The third section discusses signal processing techniques for interpreting plasma wave interferometers. The fourth section presents interferometry data taken in an ion acceleration region. The fifth section discusses spatial irregularities measured with the Viking spacecraft.

A "Simple" Example of a Plasma Wave Dispersion Relation

In the linear regime all plasma waves can be characterized by a dispersion relation which relates the complex wave frequency to the real wave vector, i.e., $D(\omega,\mathbf{k})=0$. From an appropriately general formulation the dispersion relation contains all the physical information to fully characterize a linear wave. The group velocity, phase velocity, polarization, and the relations between wave electric field, wave magnetic field, and wave density field are all contained within the formulation of a dispersion relation. Hence by measuring both the wave frequency and the wave vector, a linear wave can be fully specified through comparison with the dispersion relation.

Formulating a dispersion relation is not typically a straightforward process. Usually one begins with the Vlasov equation and then appeals to a host of approximations that limit the applicable range of the result. A more general scheme has been developed by Rönnmark (1983) and applied by Andre (1985). The process assumes a set of Maxwellian distributions and then numerically solves the Vlasov equation. A variety of multi-ion, multi-temperature, and drifting species can be included. Here we wish to present a "simple" example which is typical of the quiescent ionosphere at 500 km altitude. The plasma is composed of 99% O^+ and 1% H^+ and all the temperature ratios are 1. Except for magnetic field no other anisotropies or inhomogeneities are introduced.

The resulting dispersion relation is shown in the two panels of Figure 1. This is a rather complex surface which requires some effort to explain. The electron modes are in the

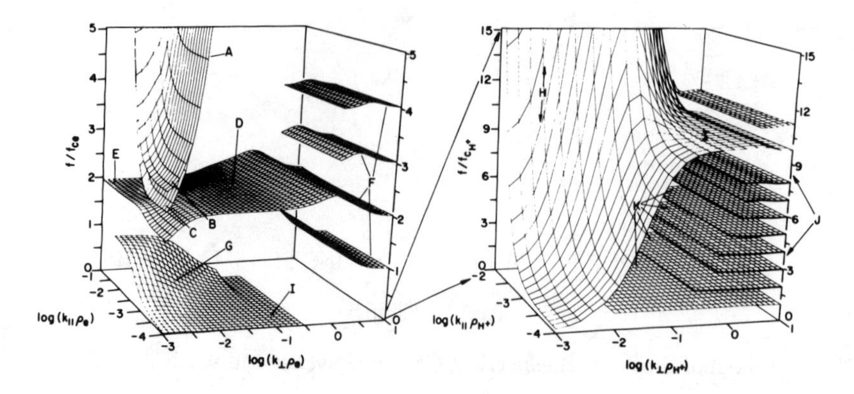

Fig. 1. Dispersion relation for a quiescent ionosphere at 500 km altitude. The left panel shows the electron modes and the right panel shows the ion modes. See text for a complete explanation.

left panel and the hydrogen modes are in the right panel. The oxygen modes are not shown. For both panels the vertical axis is frequency normalized to the gyrofrequency and the horizontal axes are $k_{||}$ and $k_\perp$ normalized to the gyroradius. Beginning with the electron modes in the left-hand panel, the free space mode, where $\omega >> \omega_{pe}$ and ω_{ce}, is labeled A. As we decrease in frequency the free space mode separates near the plasma frequency into the right circular polarized/extra-ordinary mode (B) and the left circular polarized/ordinary mode (C). At still lower frequencies are the Z-mode (E) and the upper hybrid mode (D) which extends into the electron Bernstein modes (F) for large $k_\perp$. Near the bottom of the panel is the whistler mode (G) and the lower hybrid mode (I).

The hydrogen modes are shown in the right panel. The whistler mode in this panel is labeled H and the lower hybrid mode is labeled I. It can be seen that the lower hybrid mode extends into an ion Bernstein mode (J) for large $k_\perp$. The electrostatic hydrogen cyclotron modes are labeled K and roughly refer to modes with $k_\perp\rho \cong 1$ and $k_\perp/k_{||} < 10$. Not shown are the electromagnetic ion cyclotron mode, the Alfvén modes, and the heavily damped ion acoustic modes.

These dispersion surfaces should be regarded as an incomplete road maps. They are useful for orientation but for a specific situation, such as a drifting electron population, they should be redrawn to examine areas of growth and damping. For more complex environments such as ion or electron beams, ion conics, inhomogeneities, loss cones, etc., these surfaces become substantially more complex.

Our goal here is simply to demonstrate that measurement of frequency is inadequate to determine the physical properties of a wave. For example, in the right-hand panel of Figure 1 at $f/f_{cH^+} \cong 7$ the possible modes are parallel propagating whistler waves, mostly perpendicularly propagating lower hybrid waves, or perpendicularly propagating hydrogen Bernstein modes. For more complex environments interpretation becomes even more difficult. The obvious solution is to measure the wave vector. In the next sections we discuss several efforts to measure components of the wave vector and the physical implications of those measurements.

Interferometric Analysis

The classical method of measuring wavelength is an interferometer composed of two or more sensors in a wave field. By comparing the relative phases of the signal at each sensor, the wavelength may be determined. If enough sensors are employed, the full wave vector may be determined. In this paper we shall describe only the simplest situation where two sensors are employed.

The sensors are typically either density or electric field sensors although in one case magnetic field sensors were successfully used (Kintner et al., 1983). Electron density sensors ($\delta n/n$) have the advantage of mechanical simplicity because they measure a scalar and a sensor can be composed of a single probe. However, density probes are also sensitive to electric field fluctuations which makes their interpretation more problematical. In many cases an unambiguous interpretation can be achieved but caution is necessary. Electric field sensors are not sensitive to density fluctuations but they are more complex. For each electric field sensor two probes are required. For an electrostatic wave, when $\mathbf{k}\mathrm{x}\mathbf{E} = 0$, the largest measured electric field signal corresponds to the largest possible phase shift between sensors.

A simple interferometer experiment produces two signals, one from each sensor (S_1 and S_2). There are two ways of obtaining wavelength or velocity information from the two signals. The first method is to create a cross correlation function of the form

$$\Gamma_{12}(\tau) \sim \int S_1(t)\, S_2(t-\tau)\, dt$$

The time for a phase front (or time-of-flight for a discrete structure) to traverse the distance between two sensors is the time of the maximum in the cross correlation function. This approach requires some caution and the reader is referred to LaBelle et al. (1986) for an example of the cross correlation technique.

The second method of data analysis is the cross spectral technique. This method is more powerful and more general than the cross correlation technique although it is more numerically intensive. The cross spectrum is given by

$$C_{12}(\omega) = \frac{<S_1(\omega)S_2^*(\omega)>}{\sqrt{<S_1^2><S_2^2>}}$$

where < > implies an ensemble average. The cross spectrum has the advantage of being a function of frequency so that the wavelength dependence on frequency can be discerned. If the two interferometer sensors are immersed in a wave field of the form $F = F_0\exp[i(\mathbf{k}\cdot\mathbf{r}-\omega t)]$ and the uncorrelated noise signals at sensors 1 and 2 are N_1 and N_2 then the cross spectrum becomes

$$C_{12}(\omega) = \gamma\exp(i\mathbf{k}\cdot\mathbf{d})$$

$$\gamma^2 = \{[1+(N_1/F_0)^2][1+(N_2/F_0)^2]\}^{-1}$$

where **d** is the vector describing the separation between the two interferometer sensors. The coherency γ represents the percentage of signal which is coherent to both sensors. Hence for each frequency ω, the cross spectrum yields a complex number whose amplitude is the coherency and whose argument is $\mathbf{k}\cdot\mathbf{d}$. The quantity $\mathbf{k}\cdot\mathbf{d}$ represents the measured phase shift (θ) between the sensors.

Two points are worth noting here. First the phase shift is a measure of the wave vector projected onto the interferometer axis. To measure the full wave vector instantaneously, an interferometer must be composed of three orthogonal axes. If the wave field is stationary for times long compared to a spacecraft spin period, then two components of **k** can be measured with a single axis interferometer. Second, the phase shift at frequency ω can only be determined when one wave vector is associated with that frequency. Two or more uncorrelated signals at frequency ω will produce a cross spectrum with low coherency. For example, high coherency is produced by a point source or by a random wave field which is completely Doppler shifted into the observing reference frame. For a more detailed description of the cross spectral technique and its applications see Kintner et al. (1987) and LaBelle et al. (1986).

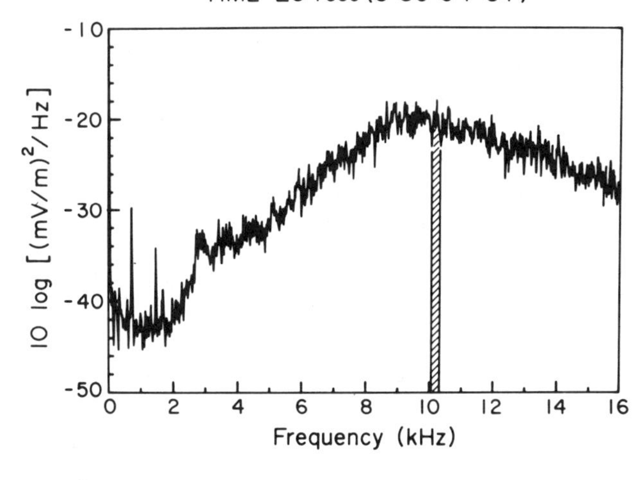

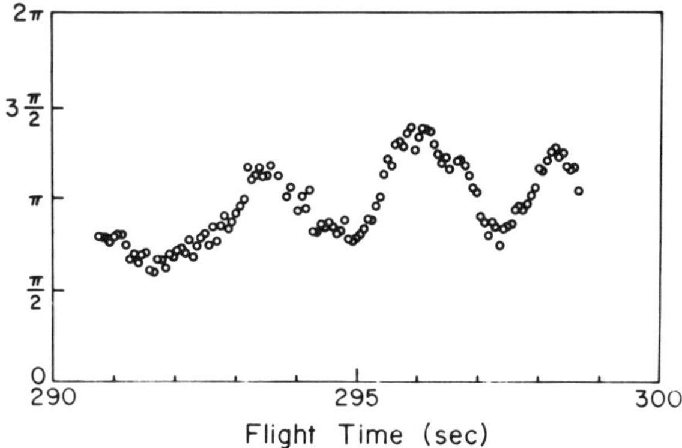

Fig. 2. Upper panel: power spectrum of VLF waves on the border of a transverse ion acceleration region. Lower panel: phase shift of the 10 kHz signal as a function of time.

Short Wavelength "LHR" Waves

These measurements were made onboard the sounding rocket MARIE launched from Ft. Churchill Canada on February 15, 1985 into an auroral break up. Between 500 and 600 km altitude the payload encountered a region of transverse ion acceleration or heating and a region of large amplitude electric fields around the lower hybrid frequency. The ions were observed up to 300 eV and they were perpendicular to the geomagnetic field to within the instrumental resolution ($\cong 1°$). The electric field power spectrum exhibited a broad spectrum above 4 kHz with a peak at roughly 10 kHz. Near 4 kHz there was structure in the electric field spectrum ordered by the hydrogen cyclotron frequency and suggestive of ion Bernstein modes. As the payload departed this region, the interferometer indicated a coherent cross spectrum for about 10 s which we shall discuss next. A more general discussion of this event is found in Yau et al. (1986) and Kintner et al. (1986).

The interferometer on the MARIE payload was composed of three probes. Two of the probes were small spheres separated by 5.5 m and the other probe was the payload skin was located halfway between the two spheres. Two electric field measurements were made; each measurement was between a sphere and the payload skin. The wideband signals up to 16 kHz were telemetered to the ground for detailed analysis.

The top panel of Figure 2 shows the electric field power spectrum as the payload was exiting the region of ion acceleration. The broadband electric field amplitude was about 10 mV m^{-1} (rms). The lower panel of Figure 2 illustrates the phase as a function of time between the two sensors for the cross hatched frequency range in the upper panel (10 kHz). That is, cross spectra were calculated for a period of about 10 s and then the phase at 10 kHz of each cross spectrum was plotted as a function of time. To make the plot more meaningful the phase was only plotted when the coherency was greater than 0.5. Clearly observable in this plot is a quasi-sinusoidal signal at the payload spin rate (3 s). At 10 kHz the quantity $\mathbf{k}\cdot\mathbf{d}$ appeared to be roughly a sinusoid in the payload reference frame. From the amplitude of the phase sinusoid the wavelength can be estimated to be roughly 20 m. Hence this electric field signal has a phase velocity of about 2 x 10^5 m s^{-1} and is located on the lower hybrid resonance cone (surface I of Figure 1).

The appeal of the cross spectral technique is that entire spectrum may be examined to determine its wavelength properties. In Figure 3 we have plotted phase data for five different frequencies and for the same period as Figure 2. The upper two frequencies (10 and 14 kHz) are roughly sinusoids, the middle frequencies (5 and 7 kHz) are flat, and the lowest frequency (4.2 kHz) is roughly sinusoidal. The interpretation is that at the two highest frequencies the power spectrum is

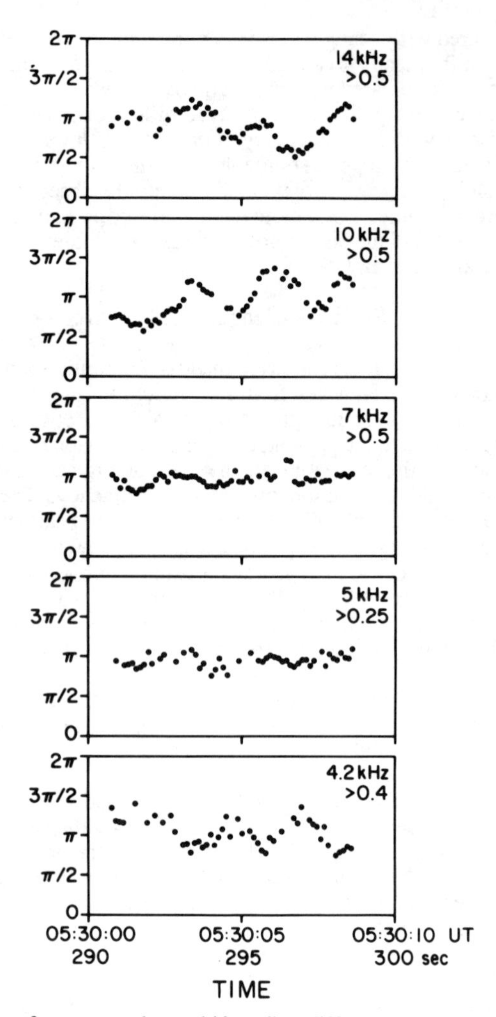

Fig. 3. Interferometer phase shift at five different frequencies for the time period of the lower panel of Figure 2.

dominated by short wavelength modes (20 m) and that the wave vectors are approximately parallel. These waves are most likely propagating on the lower hybrid resonance cone. At 5 and 7 kHz the absence of sinusoidal modulation implies that the wavelengths are long compared to the interferometer length. Their wavelengths are longer than 100 meters ($\mathbf{k}\cdot\mathbf{d} \cong 0$) and most likely these are whistler mode waves. At 4.2 kHz the sinusoidal feature implies wavelengths of about 20 m. Since 4.2 kHz is probably below the lower hybrid frequency and since the electric field power spectrum has features ordered by the hydrogen cyclotron frequency between 3 kHz and 5 kHz, the waves at 4.2 kHz are probably ion Bernstein modes.

At this point we could speculate on the possibility of these short wavelength modes transversely accelerating ions but that is the subject of another paper. Instead we will simply note that for a wavelength of 20 m, $k\rho \cong 1$ which is a condition for optimal exchange of energy between the waves and the ions.

Spatial Density Irregularities

In the previous section we demonstrated how a plasma wave interferometer could be used to distinguish waves on different surfaces of a "simple" dispersion relation even though the waves had the same frequency. In this section we will illustrate how a mode, which at first examination appears to described by the "simple" dispersion relation of Figure 1, is not related to any of the surfaces. This "new" mode is best described by the name spatial density irregularities. Its existence was first suggested by electric field measurements on the OV1-17 and S3-3 spacecraft (Kelley and Mozer, 1972; Temerin, 1978). Motivated by these measurements the Viking spacecraft was instrumented with a plasma wave interferometer composed of two small spheres separated by 80 m which detected electron density fluctuations. Viking was launched in 1986.

Previous measurements of the spatial irregularities indicated that they could be found over the altitude range 400 to 8000 km (the S3-3 apogee) within the auroral oval and polar cap. In these regions Viking regularly encountered large areas of fluctuations at frequencies up to 100 Hz. Given the frequency range these fluctuations might well be a mixture of Alfvén waves, whistler mode waves (surface G), and electrostatic ion cyclotron waves (surface K). They are none of these modes.

Viking was oriented in a cartwheel mode with its spin axis perpendicular to the orbital plane. For illustration we may assume that the Viking velocity vector was perpendicular to the geomagnetic field and that the Viking spin axis was perpendicular to both the geomagnetic field and the velocity vector. The interferometer axis was perpendicular to the spin axis so as the spacecraft rotated the interferometer axis was alternately parallel to the geomagnetic field (perpendicular to the velocity vector) and perpendicular to the geomagnetic field (parallel to the velocity vector). In Figure 4 the raw wideband output for the two sensors is plotted as 0.3 s snapshots for the spin orientations 0°, 90°, 180°, and 270° during one spin of the spacecraft. In panel 1, the interferometer axis was parallel to geomagnetic field and both sensors responded nearly identically. One quarter spin period later in panel 2 the interferometer axis was perpendicular to the geomagnetic field and the two sensors continued to respond nearly identically with one exception. The $\delta n/n$ (1) sensor led the $\delta n/n$ (2) sensor by about 1.4 x 10^{-2} s. In panel 3 the interferometer axis was parallel to the geomagnetic field and the two sensors responded identically. Finally in panel 4 the interferometer

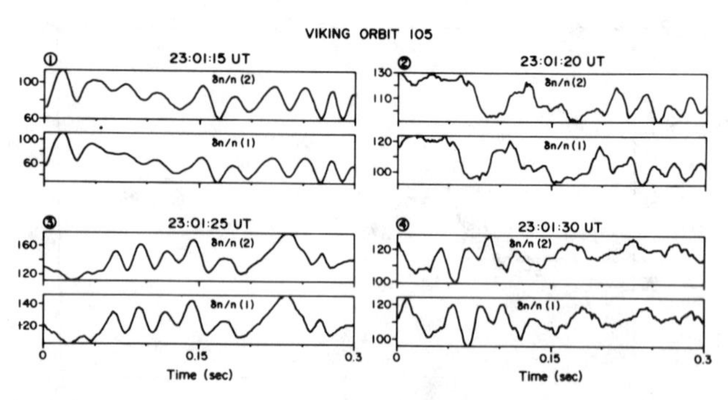

Fig. 4. Four 0.3 s snapshots of "raw" wave data from the two $\delta n/n$ sensors on Viking. The snapshots are taken during one spin period when the interferometer axis was parallel, perpendicular, anti-parallel, and perpendicular to the geomagnetic field. Panels 1, 2, 3, and 4 correspond to the orientations 0°, 90°, 180°, and 270°.

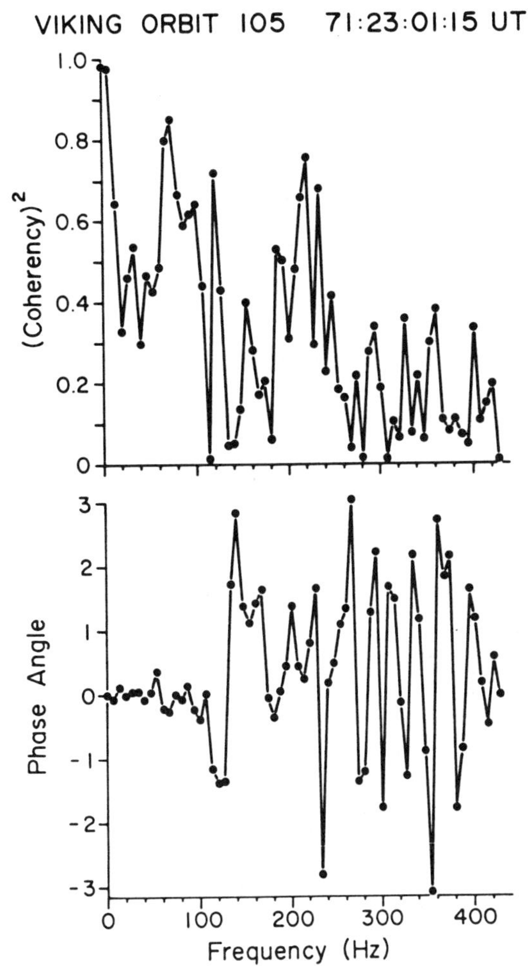

Fig. 5. Cross spectrum of panel 1 in Figure 4. The upper panel shows the coherency spectrum and the lower panel shows the phase spectrum.

axis was again perpendicular to the geomagnetic field but the $\delta n/n$ (1) sensor lagged the $\delta n/n$ (2) sensor by about 1.4×10^{-2} s. This behavior can be explained if the magnetosphere contains highly field-aligned structures (flute mode) with slow phase velocities perpendicular to the geomagnetic field. The signal is then Doppler shifted into the spacecraft reference frame. Using a lead/lag time of 1.4×10^{-2} s. and 80 m for the interferometer length, the phase velocity is roughly 5.7 km s^{-1} which is roughly the spacecraft velocity.

This data may be examined more quantitatively using cross spectra. In Figure 5 the cross spectrum for the data in panel 1 of Figure 4 is plotted. Below 100 Hz, there was a region of enhanced coherency and the phase below 100 Hz was nearly zero radians. Hence below 100 Hz the signals from both sensors were virtually identical. In Figure 6 is the cross spectrum for the data in panel 2 of Figure 4. In this case the region of enhanced coherency extended up to 150 Hz and the phase was a linear function of frequency which passed through the origin (i.e., at zero frequency the phase angle was zero radians). For the situation when the interferometer axis (**d**) is perpendicular to the geomagnetic field and parallel to the spacecraft velocity (**v**), the phase angle for spatial irregularities is given by $\theta = \omega d/v$. From the slope of the phase spectrum in Figure 6 we obtain $v = 5.1 \pm 0.3$ km s^{-1}. This should be compared with the spacecraft velocity perpendicular to the geomagnetic field which was 4.9 km s^{-1}. Within the error the phase velocity of the irregularities in the spacecraft reference frame was the spacecraft velocity in an Earth-fixed frame.

A full interpretation of these irregularities is beyond the scope of this paper but a few comments are nonetheless appropriate. First the observed irregularities are completely Doppler shifted into the spacecraft reference frame. It is somewhat difficult to specify an upper bound for their plasma frame phase velocities but certainly it is less than 300 m s^{-1}. Ideally one would prefer to check the interferometer measurement of velocity with an electric field instrument to obtain **ExB**. However, the necessary electric field component is parallel to the spacecraft spin axis and this component is notoriously difficult to measure. The interferometer measurement is probably much more accurate than any existing electric field measurement. Second, the requirement of phase velocities less than 300 m s^{-1} puts severe constraints on the possible wave modes. The H^+ thermal speed is the order of 15 km s^{-1} so all acoustic modes including the

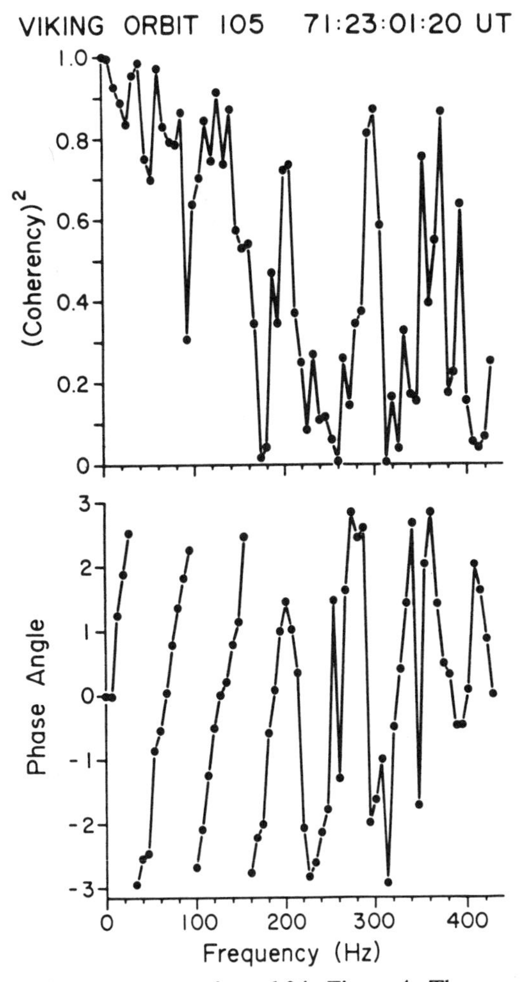

Fig. 6. Cross spectrum of panel 2 in Figure 4. The upper panel shows the coherency spectrum and the lower panel shows the phase spectrum.

electrostatic cyclotron modes are eliminated. The remaining interpretations are any of the various drift modes or turbulent structuring with density acting as a passive scalar.

Conclusions

It is the purpose of this paper to demonstrate that measurement of just wave frequency is inadequate to determine the physical properties of a wave within a magnetoplasma. At a single frequency there may be many different modes, each with different wavelengths, and different phase velocities. On the other hand, the structures may be Doppler shifted or Doppler broadened and thereby lose all of their frequency structure in the instrumental reference frame. These ambiguities may be resolved with the measurement of wavelength or phase velocity using interferometers. Since wavelength is invariant with respect to reference frame, it does not suffer from the difficulties associated with frequency measurement.

However, the measurement of wavelength is not simple. In this paper we only presented single-axis measurements. Three-axis measurements are preferable but carry the penalty of added weight and complexity. Furthermore, the interferometer data can only be interpreted when the signal is coherent, that is, when each frequency is associated with a single wave vector in the spacecraft reference frame.

To end we would like to present one additional example of the utility of the interferometer. Mozer and Temerin (1983) presented time series data suggesting the existence of ion acoustic double layers and solitons. By employing the Viking interferometer Holbach et al. (1986) were able to confirm the result of Mozer and Temerin and to demonstrate unambiguously the ion acoustic properties of the double layers. The interferometer resolved an ambiguity in the electric field direction, established that the structures were propagating upward and that they were propagating at the ion acoustic speed. The curious observation of Mozer and Temerin became established fact when examined with a plasma wave interferometer.

Acknowledgments. The experiments described in this paper were made possible by the collaboration of R. Arnoldy, R. Böstrom, G. Holmgren, H. Koskinen, J. LaBelle, B. Whalen, and A. Yau. The sounding rocket experiments were supported by NASA grant NAG5-601 and the Viking experiment was supported by ONR grant N00014-81-K-0018.

References

Anderson, R. R., C. C. Harvey, M. M. Hoppe, B. T. Tsurutani, T. E. Eastman and J. Etcheto, Plasma waves near the magnetopause, J. Geophys. Res., 87, 2087, 1982.

Andre, M., Dispersion surfaces, J. Plasma Phys., 33, 119, 1985.

Bahnsen, A., E. Ungstrup, C-G. Fälthammar, U. Fahleson, J. K. Olesen, F. Primdahl, F. Spangslev, and A. Pedersen, Electrostatic waves observed in an unstable polar cap ionosphere, J. Geophys. Res., 83, 5191-5197, 1978.

Fejer, B. G., and M. C. Kelley, Ionospheric irregularities, Rev. Geophys. Space Phys., 18, 401-454, 1980.

Fuselier, S. A., and D. A. Gurnett, Short wavelength ion waves upstream of the earth's bow shock, J. Geophys. Res., 89, 91-103, 1984.

Gallagher, D. L., Short-wavelength electrostatic waves in the earth's magnetosheath, J. Geophys. Res., 90, 1435, 1985.

Holbach, B., R. Böstrom, G. Gustafsson, H. Koskinen, G. Holmgren, and P. Kintner, Propagating solitary plasma density structures observed by the Viking spacecraft, Eos, 67, 1156, 1986.

Kelley, M. C., and F. S. Mozer, A satellite survey of vector electric fields in the ionosphere at frequencies of 10 to 500 Hertz, 1. Isotropic, high-latitude electrostatic emissions, J. Geophys. Res., 77, 4158, 1972.

Kintner, P. M., R. Brittain, M. C. Kelley, D. L. Carpenter, and M. J. Rycroft, In situ measurements of transionospheric VLF wave injection, J. Geophys. Res., 88, 7065-7073, 1983.

Kintner, P. M., J. LaBelle, W. Scales, A. W. Yau, and B. A. Whalen, Observations of plasma waves within regions of perpendicular ion acceleration, Geophys. Res. Lett., 13, 1113-1116, 1986.

Kintner, P. M., M. C. Kelley, G. Holmgren, H. Koskinen, G. Gustafsson, and J. LaBelle, Detection of spatial density irregularities with the Viking plasma wave interferometer, Geophys. Res. Lett., 14, 467-470, 1987.

LaBelle, J., P. Kintner, and M. Kelley, Interferometric phase velocity measurements in the auroral electrojet, Planet. Space Sci., 34, 1285, 1986.

Mozer, F. S., and M. Temerin, High-Latitude Space Plasma Physics, edited by B. Hultqvist and T. Hagfors, p. 453, Plenum Publ. Co., New York, 1983.

Rönnmark, K., Computation of the dielectric tensor of a Maxwellian plasma, J. Plasma Phys., 25, 699, 1983.

Temerin, M., The polarization, frequency, and wavelengths of high-latitude turbulence, J. Geophys. Res., 83, 2609-2616, 1978.

Yau, A. W., B. A. Whalen, and P. M. Kintner, Low-altitude transverse ionospheric ion acceleration, in Ion Acceleration in the Magnetosphere and Ionosphere, Geophys. Monogr. Ser., vol. 38, edited by T. Chang, pp. 39-42, AGU, Washington, D.C., 1986.

FLUX TRANSFER EVENTS : A THEORETICAL OVERVIEW

J. R. Kan

Geophysical Institute, University of Alaska Fairbanks, Fairbanks, AK 99775

Abstract. The elbow-shaped flux transfer events (FTEs) proposed by Russell and Elphic (1978, 1979) have changed our concept of the dayside magnetopause reconnection from the traditional two-dimensional configuration to a specific three-dimensional one. Theoretical response to the challenge of three-dimensional reconnections has been slow in coming. Two general approaches to the three-dimensional reconnection problem have been pursued: one assumes the three-dimensional reconnection process to be laminar, the other assumes it to be turbulent. It is too early to tell which approach is on the right track. At the present stage, it is important to encourage new and innovative theoretical ideas. Several such ideas have been proposed in the last few years; these will be summarized and compared. A number of basic issues of FTEs are singled out for further examination.

Introduction

This paper presents an overview of theoretical ideas proposed for the magnetospheric flux transfer events (FTEs) deduced from observations by Russell and Elphic (1978, 1979). It should be made clear from the outset that some of the relevant topics of FTEs were painfully excluded from our discussion due to the page limitation. The FTEs are characterized by a bipolar oscillation in the normal field component, B_n. Poleward motion of a reconnected flux tube in the northern hemisphere can lead to the standard B_n signature of FTEs, a positive B_n followed by a negative B_n when observed in northern latitudes, as noted by Rijnbeek et al. (1982) and Paschmann et al. (1982). The normal field B_n reversal signature observed near the magnetopause can be interpreted as evidence for two possible modes of reconnections: (1) uniform but intermittent reconnections; and (2) patchy and intermittent reconnections. Mode (1) is two-dimensional, while mode (2) is three-dimensional. The elbow-shaped FTEs (Russell and Elphic, 1979) are consistent with the patchy and intermittent reconnection of mode (2). The single x-line intermittent reconnection proposed by D. J. Southwood (private communication, 1988) and the multiple x-line reconnection proposed by Lee and Fu (1985) belong to the uniform but intermittent reconnection of mode (1).

The brilliant interpretation of the B_n reversal signature with the elbow-shaped FTEs by Russell and Elphic (1978, 1979) forced us to take the three-dimensional reconnection at the magnetopause seriously. This conceptual change from the traditional two-dimensional to the more realistic three-dimensional magnetopause reconnection is a major step in advancing the reconnection paradigm of space research.

The objective of this paper is to identify and discuss outstanding theoretical issues of FTEs. It is hoped that the comparative overview presented in this paper will shed some light on the right track to guide our search for the definitive FTE theory.

Models of Magnetopause FTEs

There are three distinct models of FTEs characterized by the flux tube or flux rope configuration. The FTE model proposed by Russell and Elphic (1979) is characterized by isolated elbow-shaped flux ropes of reconnected field lines. This model is commonly known as the elbow-shaped FTEs. The second FTE model proposed by Lee and Fu (1985) is characterized by straight flux ropes parallel to the multiple x-lines on the dayside magnetopause (Sonnerup, 1987). This model will be called the multiple x-line FTEs. The third FTE model is characterized by single x-line intermittent reconnection (D. J. Southwood, 1988, private communication; Elphic, 1988). The multiple x-line and the single x-line FTE models (Lee and Fu, 1985; Southwood, 1988) are two-dimensional models, in contrast to the elbow-shaped FTE model (Russell and Elphic, 1979) which is inherently three-dimensional.

Theories of Magnetopause FTEs

Theories of FTEs are still in a very early stage of development. An overview of these theoretical ideas can provide a proper perspective for future studies of FTEs.

Lee and Fu (1985) recognized that the reconnected field lines in the multiple x-line reconnection are inherently helical when the y-component of the interplanetary magnetic field (IMF) is non-zero. The two-dimensional nature of this model is embodied in the fact that the flux ropes of reconnected field lines in the multiple x-line FTE model are parallel to the x-lines which are straight, extending across the dayside magnetopause (Sonnerup, 1987). The currents associated with the helical field lines in the straight flux ropes are the magnetopause currents.

Galeev et al. (1986) proposed the idea of a stochastic percolating model of FTEs in which the magnetopause current sheet is in a state of turbulence due to tearing instability. However, the coupling process responsible for combining the tearing turbulence with driven reconnection was not described specifically.

Kan (1988) presented a theory for the elbow-shaped FTE model (Russell and Elphic, 1979). He recognized the fact that the dayside magnetopause current sheet is three-dimensional, with current density decreasing as a function of the distance from the subsolar point, so that the normal mode of tearing must also be three-dimensional. Figure 1 shows the current and field configurations: (a) for the two-dimensional tearing, (b) and (c) for a specific normal mode of three-dimensional tearing which are identified by Kan (1988) as the seed for initiating three-dimensional reconnections on the magnetopause. Thus, the three-dimensional reconnection model proposed by Kan (1988) is based on a combination of the component reconnection hypothesis (Sonnerup, 1974) and the specific normal mode of three-dimensional tearing on the dayside magnetopause shown in Figures 1b and 1c, resulting in the three-dimensional magnetopause reconnection configuration shown in Figure 2. He also identified the IMF draping over the dayside magnetopause as a configuration in favor of simultaneous occurrence of the elbow-shaped FTEs (Russell and Elphic, 1979) and the quasi-steady reconnection (Sonnerup, 1979) on the dayside magnetopause.

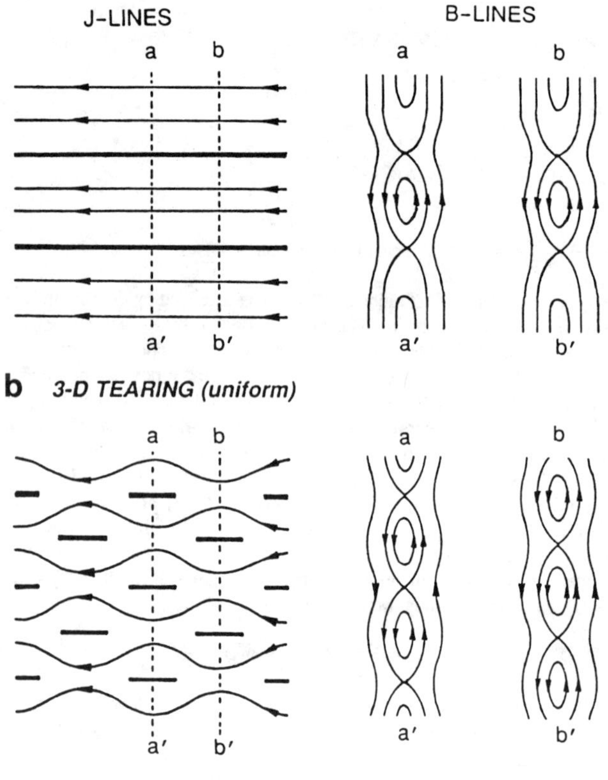

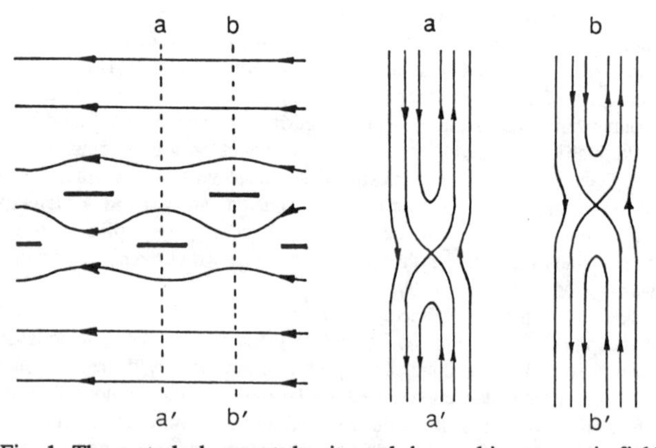

Fig. 1. The perturbed current density and the resulting magnetic field configuration in (a) a two-dimensional tearing of one-dimensional current sheet, (b) a uniform three-dimensional tearing of two-dimensional current sheet, and (c) a confined three-dimensional tearing of a three-dimensional current sheet. The bold lines (without arrows) in the left-hand panel of (a) are the x-lines. The bold line-segments in the left-hand panel of (b) and (c) are the x-segments for the localized reconnection sites produced by the three-dimensional tearing.

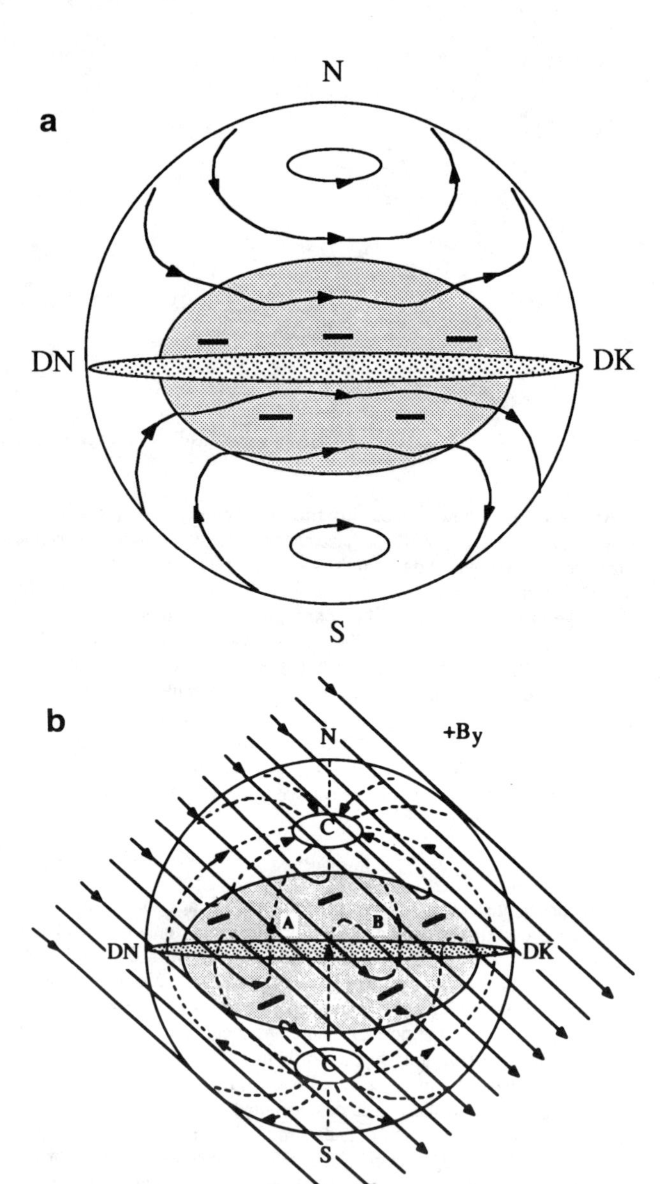

Fig. 2. Snapshots of reconnection sites on the dayside magnetopause viewed from the sun. (a) A snapshot illustrating the relation between the perturbed magnetopause current density and the distribution of the localized reconnection sites. The reconnection sites are indicated by short and bold line segments. The wavy curves (with arrows) indicate the perturbed magnetopause current density lines. Each reconnection site contains two holes in which the magnetopause current density exhibits local minima. The vorticities of the current density in the two holes are opposite to allow the polarity of the non-zero normal magnetic field component to reverse once within a reconnection site. (b) A different snapshot illustrating the elbow-shaped flux ropes produced at the reconnection sites due to the proposed three-dimensional tearing reconnection for an IMF with $+B_y$ and $-B_z$ polarities. The solid curves indicate the IMF field lines. The dashed curves indicate geomagnetic field lines. The densely shaded region delineates the FTE source region in which the three-dimensional tearing can occur. The thinly shaded strip near the equator delineates the quasi-steady reconnection region in which the three-dimensional tearing cannot occur.

Song and Lysak (1987) presented a turbulence model of FTEs. Unlike the tearing turbulence model proposed by Galeev et al. (1986), the mode of turbulence was not identified in the model proposed by Song and Lysak (1987). They suggested that the spatial scale of the FTE rope might be determined by the ionosphere which preferentially damps smaller scale electric fields. This is in contrast with the three-dimensional tearing FTE theory (Kan 1988) in which the dimension of elbow-shaped flux ropes is determined by the normal mode of the three-dimensional tearing which scales to about a few tens of the magnetopause thickness.

La Belle-Hamer et al. (1988) proposed that the Kelvin-Helmholtz (K-H) and the tearing instabilities can operate simultaneously on the dayside magnetopause so that the multiple x-line reconnection may become patchy. The relevant shear flow for producing patchiness should be perpendicular to the magnetic field, if the K-H instability can indeed bring the oppositely directed magnetic fields closer to each other at some locations but further apart at other locations to produce current density holes on the dayside magnetopause as in the three-dimensional tearing (Kan, 1988). Since the transverse flow speed can be expected to increase from the subsolar point toward the dawn and dusk flanks of the dayside magnetosphere, the proposed coupling between the tearing and the K-H instabilities may be operative near the flanks of the dayside magnetopause, rather than near the subsolar point. This is in contrast with the 3-dimensional tearing theory of FTEs [Kan, 1988].

Outstanding Issues of FTEs

In this section, a number of outstanding issues of FTEs are singled out for further discussion.

FTE Flux Ropes

A fundamental feature of FTEs is the polarity reversal of B_n on the dayside magnetopause. This feature can be produced by elbow-shaped FTEs, multiple x-line FTEs, or any time-dependent reconnection so that the flux rope of reconnected field lines is of finite dimension in the direction of the convection. The single x-line intermittent reconnection can also produce B_n reversal as emphasized by Elphic (1988). Thus, as far as observations are concerned, FTE flux ropes can be either two-dimensional or three-dimensional, although the original elbow-shaped FTE model proposed by Russell and Elphic (1979) is definitely three-dimensional.

The three-dimensional tearing reconnection proposed by Kan (1988) can produce the isolated elbow-shaped FTE flux rope if there is only one localized reconnection site on a given geomagnetic longitude during a given period of time. The possibility of a few looped flux ropes on the magnetopause may occur occasionally if more than one localized reconnection sites are initiated simultaneously by the three-dimensional tearing on a given geomagnetic longitude.

Figure 3 shows three types of flux ropes of reconnected field lines. Figure 3a illustrates an elbow-shaped flux rope of helical field lines. Figure 3b illustrates a looped flux rope of helical field lines. Figure 3c illustrates a straight flux rope of helical field lines produced by the multiple x-line model of FTE. Figures 3a and 3b are produced by three-dimensional patchy reconnections, while Figure 3c is produced by either multiple or single x-line intermittent reconnection, which is a two-dimensional configuration.

The distinction between looped flux ropes and helical field lines needs to be discussed. A flux rope can be straight while the field lines in it are helical, as shown in Figure 3c. Field lines in a flux rope are helical due to a net current flowing parallel to the axis of the rope. Without the tube-aligned current, field lines in a flux rope must be exactly parallel to the axis of the rope. The simple question of "What can cause a flux rope to loop?" is not an easy one to answer. We suggest that a flux rope is forced to loop by boundary conditions or forces outside the flux rope rather than

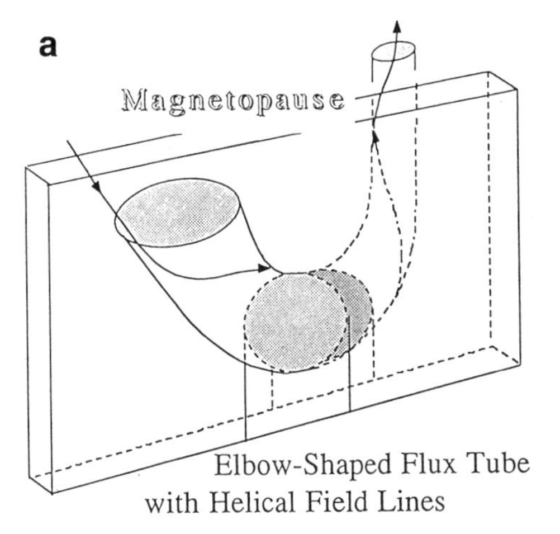

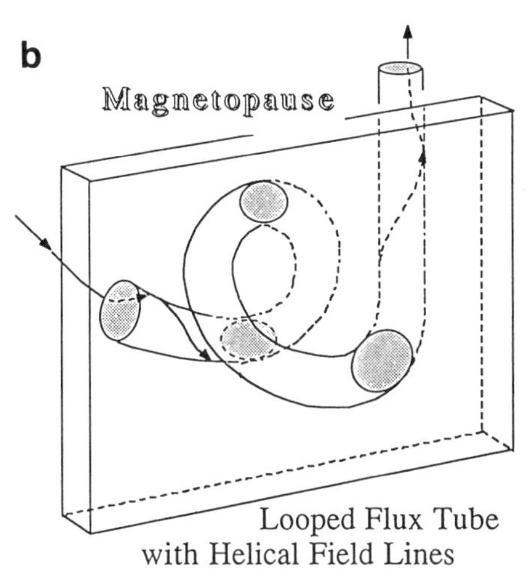

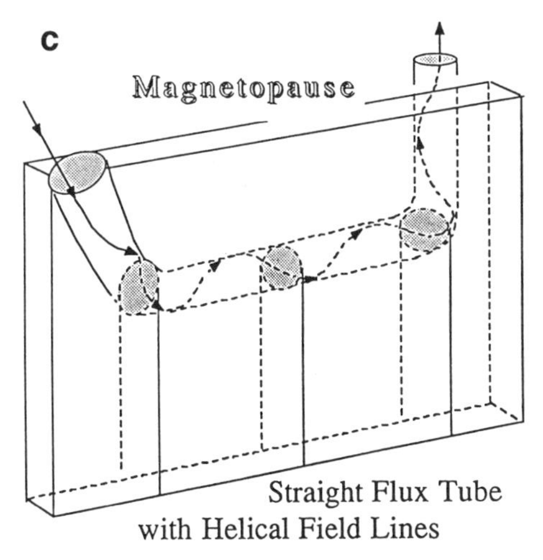

Fig. 3. (a) An elbow-shaped FTE flux rope. The field lines in the rope are helical if a net current is flowing along the rope. (b) A looped flux rope which occurs if there are more than one three-dimensional reconnection sites. (c) A straight flux rope of the multiple x-line FTE model. The field lines in the straight flux rope are helical because magnetopause current is flowing along the rope.

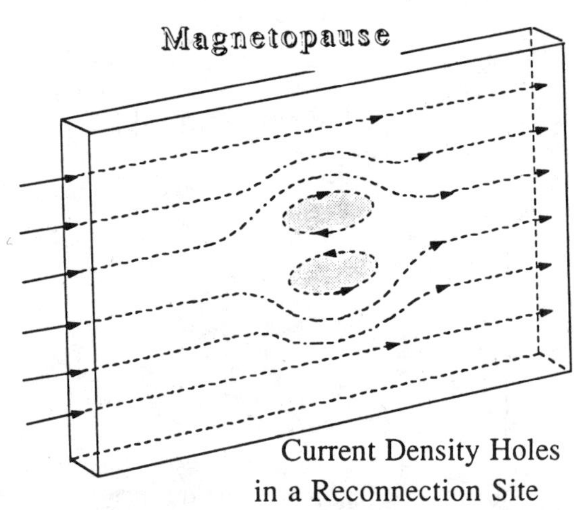

Fig. 4. A current density hole on the magnetopause due to the combined effects of the three-dimensional tearing and the component reconnection process. The current density can be expected to be lower in the hole than the ambient magnetopause current density. Each hole contains a pair of eddy currents flowing in the opposite direction to produce the antiparallel normal field components for the paired elbow-shaped flux ropes at each three-dimensional reconnection site on the dayside magnetopause.

by field-aligned currents inside the rope. The looped flux rope in Figure 3b is produced because of the three-dimensional tearing creating a boundary condition permitting it to occur, independent of the presence of field-aligned currents. It is not at all clear if a straight flux rope can be forced to loop by increasing the field-aligned current flowing along the rope. This problem may be analyzed as a stability problem of a cylindrical current along a uniform magnetic field.

Current Density Holes on the Magnetopause

The three-dimensional tearing in Figures 1b and 1c creates holes in the magnetopause current. As each of these current density holes deepens, it splits into two holes of eddy currents with opposite sense of rotation to form a reconnection site for a pair of elbow-shaped FTE flux ropes. Figure 4 shows one half of the paired elbow-shaped flux ropes produced at each reconnection site. The splitting of a current density hole, originally created by the three-dimensional tearing, into two holes of eddy current requires an "external force" in addition to the "internal force" of tearing. The "external" force can be identified with the solar wind ram pressure which is believed to be responsible for the driven reconnection on the dayside magnetopause. The current density pattern shown in Figure 4 corresponds to an FTE with very large normal field components because the current densities in the two holes form closed loops flowing in the opposite direction. Thus, a three-dimensional reconnection site consists of two current density holes so that the non-zero normal field component can change sign in a reconnection site to produce a pair of elbow-shaped FTE flux ropes.

Field-Aligned Currents Associated with FTEs

As has been mentioned earlier, field lines in a flux rope are helical if a net current is flowing along the axis of the flux rope. The issue of field-aligned currents associated with FTEs is directly related to the issue of magnetosphere-ionosphere (M-I) coupling. However, it should be noted that it is possible for a current to flow along a portion of a flux rope and then leave the flux rope. Such a current will not reach the ionosphere and therefore does not contribute to the M-I coupling field-aligned current. For example, the current flowing along the straight portion of the flux rope in the multiple x-line model of FTEs (Lee and Fu, 1985) is the magnetopause current because the straight flux rope is east-west aligned on the dayside magnetopause. As depicted by Lee and Fu (1985), the helical field lines are confined in the straight portion of the flux rope. From the location where the flux rope turns away from the magnetopause, the helicity of field lines is much reduced in their diagram. It is questionable if a significant amount of the magnetopause current flowing along the straight FTE flux rope will follow the rope as the rope turns away from the magnetopause. The amount of field-aligned current flowing along an FTE flux rope to the ionosphere is determined by the ionospheric line-tying effect. These field-aligned currents could be closed partially in the magnetosphere, at the magnetopause, in the magnetosheath and even further out in the solar wind. This brings up the important issue of the generation of field-aligned currents in collisionless space plasmas, which is beyond the scope of the present paper. The above discussion shows that the field-aligned current cannot be deduced by helicity of field lines in a flux rope when the flux rope is aligned with the magnetopause current or with any other cross-field currents in space. To deduce field-aligned

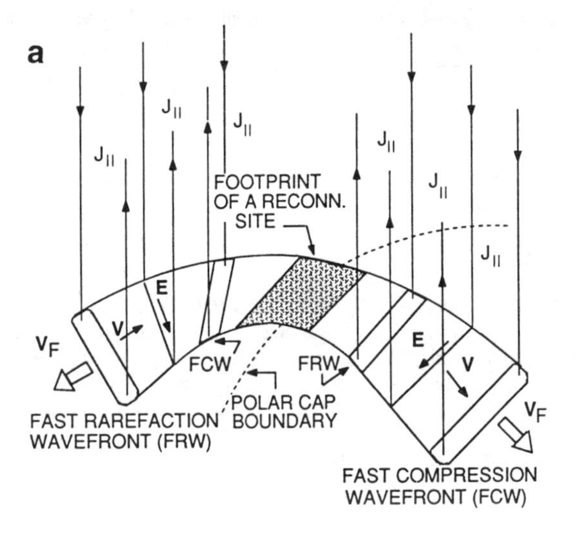

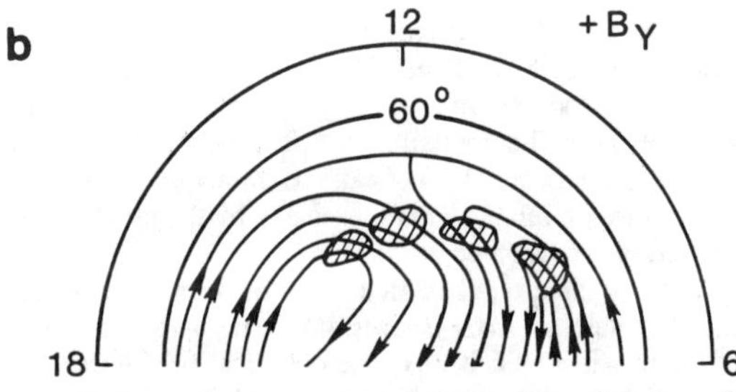

Fig. 5. (a) Sketch of an isolated convection channel projected on the ionosphere driven by the proposed three-dimensional tearing reconnection on the magnetopause. The cross-shaded region is the footprint of a reconnection site on the magnetopause. Fast-mode MHD waves play an important role in initiating and terminating the flow driven by the proposed three-dimensional reconnection on the dayside magnetopause. The open arrows indicate the velocity of fast-mode waves. The solid arrows indicate the convection velocity along the channel. Field-aligned currents are indicated by vertical lines with arrows. (b) Dayside ionospheric convection pattern driven by the proposed three-dimensional reconnections on the dayside magnetopause. The curves with arrows are convection streamlines.

currents associated with an FTE flux rope by helicity of the field lines, one must stay away from these cross-field current regions.

The field-aligned currents in an FTE flux rope due to the ionospheric line-tying effect have been discussed by Southwood (1987), leading to a pair of equal but oppositely directed field-aligned currents on opposite halves of the cylindrical shell of an FTE flux rope. Kan (1988) argued that the oppositely directed currents need not be of equal magnitude due to the ionospheric line-tying effect so that a net field-aligned current can flow along an FTE flux rope from the ionosphere to contribute to the Region I and the cusp field-aligned current systems. Figure 5 shows the ionospheric footprint of a convection channel produced intermittently by FTEs on the dayside magnetopause. The upward field-aligned current can be more intense than the downward current so that a net field-aligned current can flow along FTE flux ropes. This means that the field lines in an elbow-shaped FTE flux rope can be helical due to the ionospheric line-tying effect. The sign of helicity depends on whether the FTE flux rope is connected to the upward or downward part of the Region I or the cusp field-aligned current systems. Although the line-tying effect is believed to play an important role in the production of field-aligned currents, the details of the ionospheric signature of FTEs are still open.

Electric Field Signatures of FTEs

Electric fields produced by FTEs are necessarily nonuniform due to the finite cross section of an FTE flux rope. On the basis of Figure 5, Kan (1988) proposed that the FTE electric fields can expected to be spiky, peaking at the center of the flux rope. A series of intermittently produced FTEs having longitudes in close proximity can lead to spiky electric fields of variable intensity along an enhanced convection channel. Due to the propagation effects of the MHD fast mode, Kan (1988) argued that the spiky electric field signature is more observable on the equatorward side than on the poleward side of the cusp region.

Southwood (1987) proposed that the FTE electric field is dipolar. This electric field model presumably is in a steady state. The main difference between the dipolar electric field model (Southwood, 1987) and the spiky electric field model (Kan, 1987) is the presence of the fast-mode wave fronts shown in Figure 5a, where are not in a steady state. Other electric field signatures of FTEs have been discussed by Cowley (1986).

Quite different from the spiky FTE electric field (Kan, 1988) and the dipolar FTE electric field (Southwood, 1987) is the radial FTE electric field proposed by Saunders et al. (1984) and Lee (1986). The radial field is associated with the cylindrical field-aligned current system with return current flowing along the surface of cylindrical flux rope. As has been discussed earlier, the existence of such a cylindrical current system is questionable because it is not clear why the magnetopause current would want to follow the straight flux rope of the multiple x-line FTE as the rope turns away from the magnetopause toward the ionosphere. The only field-aligned currents that must close in the ionosphere are those that are initiated by the ionospheric line-tying effect as discussed by Southwood (1987) and Kan (1988).

The observed high-latitude convection electric field profiles are often spiky (Burch et al., 1976; Heppner and Maynard, 1987). Figure 6 shows several examples of the observed electric field profiles provided by Heppner and Maynard (1987). The electric spikes on the dayside near the peak of the convection field might be produced by FTEs, for example around 12.5 MLT in the top panel and around 7.7 MLT in the bottom panel. The electric spikes inside the convection reversal region in Figure 5 could be associated either with the FTEs or with vortices produced by the Kelvin-Helmholtz instability driven by the velocity shear (e.g., Miura, 1984). The FTE electric field is spiky because it must fall off toward the edge of the isolated FTE reconnection site; the spiky field must be in the same direction as the average convection electric field. These FTE electric spikes should be most observable in and around the cusp region. An FTE flux rope of $1R_E$ in diameter at the magnetopause (B_m = 50 nT) scales to about 200 km at the ionospheric altitude (B_i = 50,000 nT) which is comparable to the widths of the electric spikes in Figure 5. Further study is required to ascertain whether the electric field spikes on the dayside in and around the cusp region are indeed produced by FTEs.

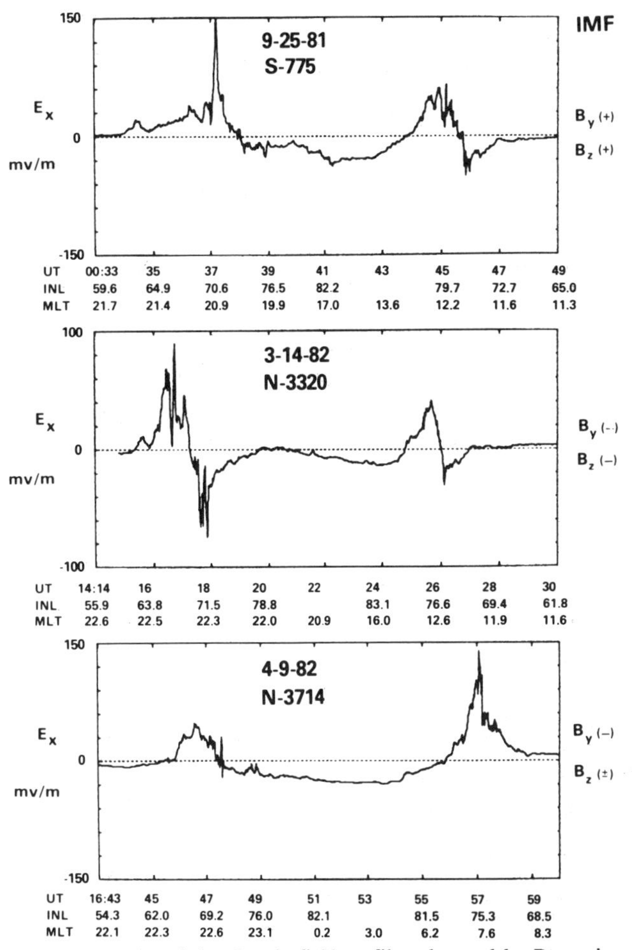

Fig. 6. Examples of the electric field profiles observed by Dynamics Explorer 2 satellite (courtesy of Heppner and Maynard, 1987).

Prediction of the Three-Dimensional Tearing Reconnection Theory

The theory proposed by Kan (1988) for the elbow-shaped FTE is based on the three-dimensional tearing combined with the component reconnection hypothesis. This theory contains a number of specific predictions as summarized below:

(a) A specific normal mode of three-dimensional tearing on the magnetopause is identified in Figures 1b and 1c for producing the elbow-shaped FTE flux ropes in the three-dimensional reconnection configuration on the dayside magnetopause as depicted in Figure 2.

(b) FTEs can be expected to occur simultaneously with the quasi-steady reconnection when the IMF field drapes over the dayside magnetopause. The draping should occur preferentially when the solar wind speed is sufficiently high.

(c) Magnetospheric convection is enhanced intermittently along convection channels driven by the three-dimensional reconnection on the magnetopause, resulting in spiky electric fields. The spikes on the convection electric field in and around the cusp region are proposed to be produced by FTEs. The ionospheric scale length of the FTE (~ 200 km) is comparable to the widths of the observed electric field spikes.

(d) A pair of field-aligned current sheets is associated with each FTE. The upward current intensity need not be equal to the downward current intensity in each pair. The net field-aligned current produces helical field lines in an elbow-shaped FTE flux rope. The current sheets are of finite extent due to the intermittent nature of FTEs. The Region I and the cusp field-aligned currents can be expected to be striated into current sheets of finite extent by FTEs driven by the three-dimensional reconnections on the dayside magnetopause. The latitudinal width of the current sheets is on the order of the convection channel, which is about 100 to 200 km.

(e) Cusp auroral arcs of variable intensity fanning out from the cusp region (Akasofu and Kan, 1980; Meng and Lundin, 1986) may be associated with FTEs. If they are indeed produced by FTEs, cusp auroral arcs should be aligned with the intermittently enhanced convection channels because FTE field-aligned current sheets must be aligned with the FTE convection channels in the dayside ionosphere where the conductivity is fairly uniform.

Summary

The elbow-shaped FTE proposed by Russell and Elphic (1978, 1979) has changed our concept of magnetopause reconnection from the traditional two-dimensional model to the more realistic three-dimensional configuration. Several outstanding issues of FTEs have been discussed. The most crucial issue is to determine by observation whether FTE flux ropes are three-dimensional (elbow-shaped) or two-dimensional (single or multiple x-line) structures. The distinction between helical field lines near the magnetopause and field-aligned currents which are closed in the ionosphere should be kept in mind in future observational study of FTEs. The mechanism responsible for producing a pair of eddy currents in the current density hole on the magnetopause at the FTE reconnection site should be examined quantitatively, because it may reveal the secrete of how the 3-D tearing is related to the driven reconnection. The proposed spiky electric field signature of FTEs is more likely to be observed on the equatorward side of the cusp than on the poleward side. It is too early to tell whether any one of the proposed theoretical ideas of FTEs is clearly on the right track. At this early stage of theoretical development, it is important to keep our minds open to new ideas. It is essential that the basic processes responsible for the three-dimensional reconnection on the dayside magnetopause are identified before constructing any elaborate theoretical or simulation models of FTEs.

Acknowledgments. This work was supported in part by the NSF grant ATM 85-21194 under the Division of Atmospheric Sciences, Solar Terrestrial Section.

References

Akasofu, S.-I., and J. R. Kan, Dayside and nightside auroral arc systems, *Geophys. Res. Lett., 7*, 753, 1980.

Burch, J. L., W. Lennartsson, W. B. Hanson, R. A. Heelis, J. H. Hoffman, and R. A. Hoffman, Properties of spikelike shear flow reversals observed in the auroral plasma by Atmospheric Explorer C, *J. Geophys. Res., 81*, 3886, 1976.

Cowley, S. W. H., The impact of recent observations on theoretical understanding of solar wind-magnetosphere interactions, *J. Geomag. Geoelec., 38*, 1223, 1986.

Elphic, R. C., Magnetic field and electric current measurements of critical phenomena in solar wind interactions, these proceedings, 1989.

Galeev, A. A., M. M. Kuznetsova, and L. M. Zeleny, Magnetopause stability threshold for patchy reconnection, *Space Sci. Rev., 44*, 1, 1986.

Heppner, J. P., and N. C. Maynard, Empirical high latitude electric field models, *J. Geophys. Res., 92*, 4467, 1987.

Kan, J. R., A theory of patchy and intermittent reconnections for magnetospheric flux transfer events, *J. Geophys. Res., 93*, 5613, 1988.

La Belle-Hamer, A. L., Z. F. Fu, and L. C. Lee, A mechanism for patchy reconnection at the dayside magnetopause, *Geophys. Res. Lett., 15*, 152, 1988.

Lee, L. C., and Z. Fu, A theory of magnetic flux transfer at the Earth's magnetopause, *Geophys. Res.Lett., 12*, 105, 1985.

Lee, L. C., Magnetic flux transfer at the Earth's magnetopause,in *Solar Wind-Magnetosphere Coupling*, edited by Y. Kamide and J. Slavin, p. 297, Terra Scientific Publ. Co., Tokyo, 1986.

Meng, C.-I., and R. Lundin, Auroral morphology of the midday oval, *J. Geophys. Res.*, 91, 1572, 1986.

Miura, A., Anomalous transport by magnetohydrodynamic Kelvin-Helmholtz instabilities in the solar wind-magnetosphere interaction, *J. Geophys. Res., 89*, 801, 1984.

Paschmann, G., G. Haerendel, I. Papamastorakis, N. Sckopke, S. J. Bame, J. T. Gosling, and C. T. Russell, Plasma and magnetic field characteristics of magnetic flux transfer events, *J. Geophys. Res., 87*, 2159, 1982.

Rijnbeek, R. P., S. W. H. Cowley, D. J. Southwood, and C. T. Russell, Observations of "reverse polarity" flux transfer events at the Earth's dayside magnetopause, *Nature, 300*, 23, 1982.

Russell, C. T., and R. C. Elphic, Initial ISEE magnetometer results: Magnetopause observations, *Space Sci. Rev., 22*, 681, 1978.

Russell, C. T., and R. C. Elphic, ISEE observations of flux transfer events at the dayside magnetopause, *Geophys. Res. Lett., 6*, 33, 1979.

Saunders, M. A., C. T. Russell, and N. Sckopke, Flux transfer events: Scale size and interior structure, *Geophys. Res. Lett., 11*, 131, 1984.

Song, Y., and R. L. Lysak, turbulent generation and evolution of FTE flux tubes, *Eos Trans. AGU, 68*, 1442, 1987.

Sonnerup, B. U. O., Magnetopause reconnection rate, *J. Geophys. Res., 79*, 1546, 1974.

Sonnerup, B. U. O., Magnetic field reconnection, *Solar System Plasma Physics*,edited by C. F. Kennel, L. J. Lanzerotti and E. N. Parker, Vol. III, p. 45, North-Holland Publ. Co., 1979.

Sonnerup, B. U. O., On the stress balance in flux transfer events, *J. Geophys. Res., 92*, 8613, 1987.

Southwood, D. J., The ionospheric signature of flux transfer events, *J. Geophys. Res., 92*, 3207, 1987.

THE SOLAR WIND INTERACTION WITH NON–MAGNETIC BODIES AND THE ROLE OF SMALL–SCALE STRUCTURES

T. E. Cravens*

Space Physics Research Laboratory, Department of Atmospheric, Oceanic and Space Sciences, The University of Michigan, Ann Arbor, Michigan 48109-2143

Abstract. The solar wind interacts directly with the ionospheres and/or neutral atmospheres of non-magnetic bodies in the solar system. The solar wind with its associated interplanetary magnetic field is usually excluded from the highly conducting ionosphere of Venus due to currents induced in the ionospheric plasma. These currents are largely confined to the narrow ionopause layer during conditions of relatively low solar wind dynamic pressure. The ionosphere remains free of large-scale magnetic fields in this case, except for small-scale structures called magnetic flux ropes which have been observed throughout most of the dayside ionosphere by the magnetometer on the Pioneer Venus Orbiter. The formation and evolution of these flux ropes and their role in the momentum and energy balance of the ionosphere are not yet understood. Other small-scale plasma/field structures have been observed in the nightside ionosphere of Venus. During periods of high solar wind dynamic pressure, currents flow throughout the dayside ionosphere which is then permeated by large-scale magnetic fields. The solar wind interaction with comets is characterized by the mass loading of the solar wind with heavy cometary ions which are produced by the ionization of neutrals in the extensive cometary coma. This mass loading slows down the solar wind and ultimately leads to the formation of a magnetic barrier and a magnetotail. The Giotto magnetometer has observed a diamagnetic cavity surrounding the nucleus, which is a consequence of an outward ion-neutral drag force associated with the flow of cometary neutrals past plasma frozen onto field lines in the magnetic barrier. In addition to these large-scale structures, many small-scale plasma/field structures have been observed in comets including the bow shock, tail rays and kinks, and plasma pile-ups and depletions in the barrier. Theoretically, the existence of a very narrow layer of enhanced plasma density just inside the contact surface has been predicted. The role of these small-scale structures in the overall solar wind interaction scenario is not yet understood.

1. Introduction

The solar wind is an almost collisionless plasma consisting mainly of protons and electrons flowing outward from the sun. This flow is supersonic and super-Alfvénic. The solar wind carries with it the interplanetary magnetic field. The nature of the solar wind interaction with an object or body in the solar system depends on the characteristics of that body, such as the type of atmosphere it has, its heliocentric distance, and the strength of its intrinsic magnetic field. A somewhat arbitrary categorization of the types of solar wind interactions in the solar system is given by:

(1) Lunar Type. The moon has no atmosphere to speak of and no large-scale intrinsic magnetic field. The solar wind directly impacts the surface of the moon and is absorbed. A bow shock does not form in the upstream solar wind flow because there is no significant pressure perturbation. However, there is a wake. Examples of this type of interaction are the moon, asteroids, and inactive comets.

(2) Earth Type. The Earth has a core which is electrically conducting and which is rotating. The resulting dynamo magnetic field is approximately dipole within a few Earth radii of the surface. The solar wind plasma cannot penetrate this large intrinsic magnetic field and a magnetosphere is created which diverts the solar wind at large distances from the Earth (i.e., the subsolar magnetopause is located at about 10 Earth radii). A bow shock is formed as the solar wind flows around the magnetospheric obstacle. The full description of this type of interaction constitutes a whole field of investigation and includes such concepts as magnetic merging of the terrestrial and interplanetary fields, convection electric field, and magnetic storms. Other examples of a terrestrial type of solar wind interaction are Mercury, Jupiter, Saturn, Uranus, and probably Neptune as well.

(3) Venus Type. Venus has virtually no intrinsic magnetic field (see Russell and Vaisberg, 1983, and the references therein) which can serve to divert the solar wind around it. However, Venus possesses a dense neutral atmosphere, and therefore it also possesses an ionosphere because the neutrals are photoionized by solar extreme ultraviolet radiation. The ionospheric plasma is a very good electrical conductor and tries to exclude the solar wind plasma with its associated interplanetary magnetic field--basically this is the diamagnetic effect. Hence, the ionosphere acts an obstacle to the solar wind (see Figure 1). A bow shock exists because the solar wind is supersonic and super-Alfvénic and must become subsonic in order to flow around the obstacle. Other possible examples of a Venus-type interaction are Mars and Titan.

(4) Comet Type. A typical cometary nucleus is a mixture of ice and dust several kilometers in extent. Naturally, they do

*Now at Department of Physics and Astronomy, The University of Kansas.

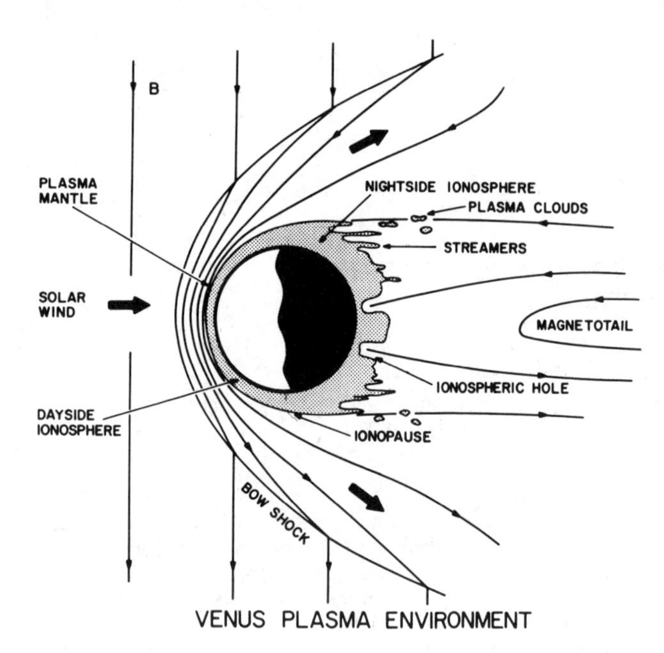

Fig. 1. Schematic of the plasma environment of Venus.

not have significant intrinsic magnetic fields. Cometary nuclei are inactive for most of their lives, which are largely spent in the outer solar system far from the sun, where the solar wind interaction is of the lunar type (see Mendis et al., 1985). But when the heliocentric distance of a nucleus becomes less than a few AU, its surface is heated by solar radiation and consequently water vapor (and other volatiles) is produced at the surface. The outflowing gas carries dust with it. The gravitational acceleration associated with such a small object is negligible, and consequently the neutral atmosphere, or coma, extends millions of kilometers. Photoionization of these neutrals produces both heavy and light cometary ions which are picked up by the solar wind and mass-load the flow. A weak bow shock exists. The interaction of Jovian satellite Io with the Jovian magnetospheric plasma also has some cometary-type interaction characteristics. And mass loading plays some part in the solar wind-Venus interaction.

An overview of the solar wind interaction with non-magnetic bodies will be given in this paper, as exemplified by the interactions with Venus and comets. The emphasis will be on: (1) the ionospheric aspects of the interaction process, and (2) outstanding problems in our understanding of these interactions. Section 2 is a general overview of the solar wind interaction with Venus, section 3 will focus on the ionospheric aspects of this interaction, and section 4 is about magnetic flux ropes. An overview of the plasma environments of comets will be given in section 5. Section 6 considers the plasma environment of the inner coma. Section 7 is a very brief summary. It will be suggested throughout the paper that large-scale, or global, plasma/field structures in the plasma environment of non-magnetic bodies are usually better understood than small-scale structures. The conclusion will be that multi-point, high spatial and temporal resolution plasma and field measurements are required to resolve the outstanding problems in this field.

2. An Overview of the Solar Wind Interaction with Venus

All our knowledge of the Venus plasma environment comes from the data gathered by the Mariner, Venera, and Pioneer Venus missions, plus the theoretical work stimulated by this database. For a detailed review of the literature in this area, the reader is referred to the Venus book including chapters by Russell and Vaisberg (1983), Brace et al. (1983), Nagy et al. (1983), Cloutier et al. (1983), and Gringauz (1983). More recently, an excellent review by Luhmann (1987) was published.

Consider the schematic of the plasma environment of Venus given in Figure 1. The overall shape, location, and structure of the bow shock (i.e., large-scale features) are rather well (albeit not perfectly) understood in terms of basic hydrodynamic and magnetohydrodynamic theory and calculations (Spreiter and Stahara, 1980a,b; Tátrallyay et al., 1983, 1984; Slavin et al., 1983; Breus et al., 1987; Alexander et al., 1985). However, the structure of the shock itself (small-scale/microscopic properties) is not as well understood, although much can be understood in terms of collisionless shock theory as applied to the terrestrial bow shock (Scarf et al., 1980; Omidi et al., 1986; Luhmann, 1987). The foreshock region is even less well understood (Fontheim, 1988).

Both the observational picture and the theory are in better shape for the large-scale aspects than for the small-scale aspects for each of the regions shown in Figure 1 (i.e., ionopause, magnetosheath, plasma mantle, dayside and nightside ionospheres, magnetotail). The shocked solar wind slows down as it approaches the planet, and as a consequence, the magnetic field strength builds up, resulting in the formation of a magnetic barrier located outside the dayside ionopause. The ionopause delineates the upper altitude limit of the cold ionospheric plasma. The ionopause usually is coincident with the bottom of the magnetic barrier. The thermal pressure in the magnetic barrier is much less than the magnetic pressure, which thus accounts for almost all of the upstream solar wind dynamic pressure. The electron distributions observed in at least the lowest part of the barrier (called the plasma mantle) by the Pioneer Venus retarding potential analyzer (RPA) (Spenner et al., 1980) are a mixture of magnetosheath and ionospheric (photoelectron) populations. There is no theoretical explanation for these electron distributions in the mantle. This is just the sort of microscopic/small-scale property that very little is known about. Some O^+ pick-up by the flow also appears to be taking place in this region (see the discussion in Luhmann, 1987).

The global structure of the wake and magnetotail of Venus has been studied observationally (see the review articles in the Venus book and Luhmann, 1987) but our theoretical understanding remains remarkably sketchy. In the next section, outstanding problems in the ionospheric part of the solar wind interaction with Venus will be discussed.

3. The Ionosphere of Venus

The reader is referred again to the Venus book and to the paper by Luhmann (1987) for detailed reviews. Kliore et al. (1985) provided another good review. Pioneer Venus magnetometer observations indicate that the observed ionospheres can be sorted rather well into two categories, or states, with the following characteristics:

(1) Unmagnetized Ionospheres

* Large-scale magnetic fields are not present in the ionosphere.
* Observed during time periods when the solar wind dynamic pressure is less than the maximum thermal pressure in the ionosphere.
* The ionopause exists at high altitudes (z > 300 km) and is a rather narrow ($\Delta z \approx 25$ km) layer in which the ionospheric density decreases by more than a factor of 100 from ionospheric values ($n_e \approx 10^4$ cm^{-3}).
* A substantial nightside ionosphere exists.
* Small-scale (diameter ≈ 10 km) magnetic structures called flux ropes are often present.

(2) Magnetized Ionospheres

* Large-scale magnetic fields (B ≈ 50-150 nT) are present throughout the dayside ionosphere.
* Observed when the solar wind dynamic pressure is about the same as, or greater than the maximum ionospheric thermal pressure.
* The ionopause is located at low altitudes (z<300 km) and is thick ($\Delta z \approx 80$ km).
* The nightside ionosphere is weak or absent.
* A low-altitude (z between 150 and 175 km) layer, or belt, of enhanced magnetic field strength is always present.

The current state of our understanding of the unmagnetized ionosphere of Venus will be described first. Pioneer Venus Orbiter (PVO) magnetometer and Langmuir probe data (Russell and Vaisberg, 1983) are displayed in Figure 2. All the characteristics listed above are apparent for the unmagnetized ionosphere represented by Orbit 186. The magnetic field strength just outside the ionopause is 70 nT, which corresponds to a magnetic pressure of $B^2/8\pi = 2 \times 10^{-8}$ dynes cm^{-2}. The ionospheric pressure just inside the ionopause is given by $p = n_e k (T_e + T_i)$, where n_e is the electron density, k is Boltzmann's constant, T_e is the electron temperature, and T_i is the ion temperature. Langmuir probe and RPA data indicate that $T_e + T_i \approx 6000$ K at these altitudes, which with $n_e = 2.5 \times 10^4$ cm^{-3} from the figure, gives a pressure $p \approx 2 \times 10^{-8}$ dynes cm^{-2}, which is equal to the external magnetic pressure. A static pressure balance exists across the ionopause for Orbit 186

$$n_e k \left(T_e + T_i\right) + B^2 / 8\pi = \text{constant} \tag{1}$$

Statistical studies support the general validity of equation (1) for unmagnetized ionospheres and also indicate that the magnetic pressure in the barrier is almost the same as the solar wind dynamic pressure when adjusted for solar zenith angle effects (cf. Russell and Vaisberg, 1983; Brace et al., 1983). The height of the ionopause was observed to increase with increasing solar zenith angle and this can be understood in terms of the pressure balance relation (and a thermal pressure

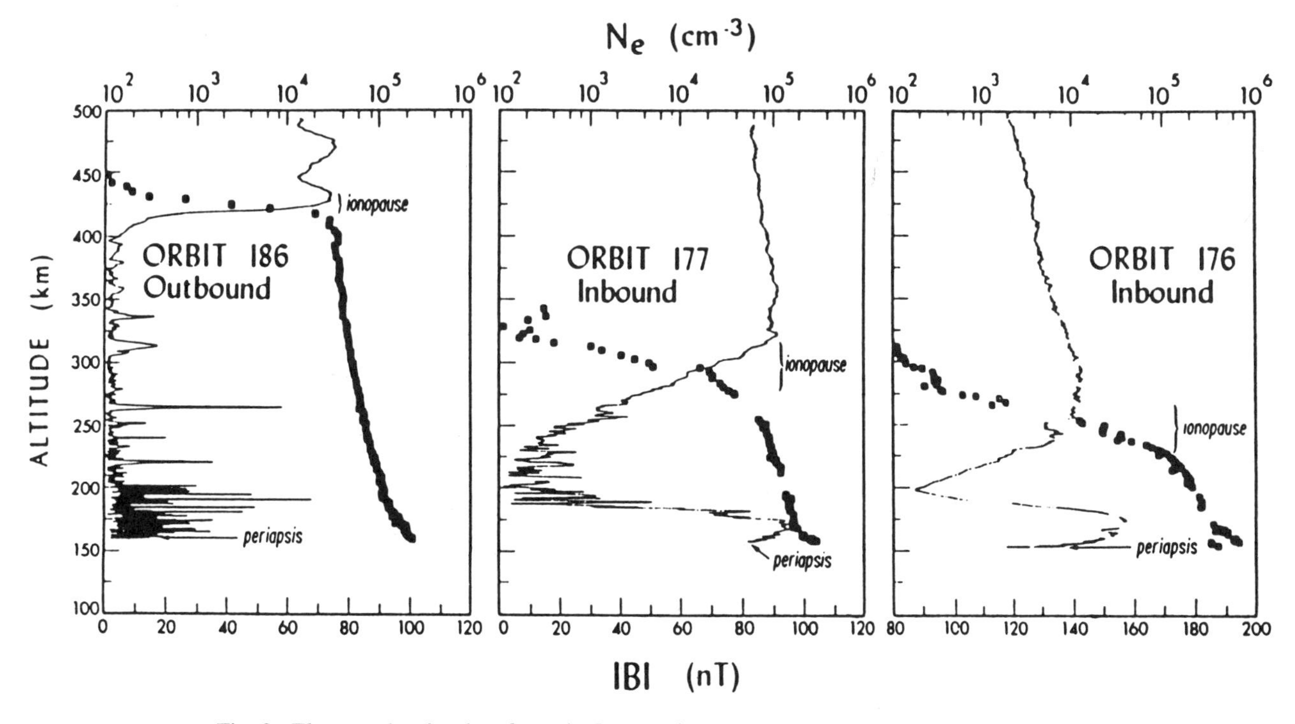

Fig. 2. Electron density data from the Langmuir probe on the PVO--top scale. Magnetic field strength from the PVO magnetometer--bottom scale. Orbit 186 is an example of an unmagnetized ionosphere. Orbits 176 and 177 are examples of a highly magnetized and a moderately magnetized ionosphere, respectively. This figure was adopted from Russell and Vaisberg (1983).

in the ionosphere which decreases with increasing altitude z). Thus the large-scale features of the ionopause are understood. However, the small-scale structure of the ionopause current layer is not understood theoretically, although its thickness is known observationally to be only a couple of O^+ gyroradii (Elphic et al., 1980).

The observed ionospheric density structure for electrons and for individual ion species as a function of altitude and solar zenith can be explained reasonably well theoretically on the dayside (e.g., Nagy et al., 1980, 1983; Cravens et al., 1981; Fox, 1982). And two-dimensional models have been used successfully to explain the overall behavior of the nightside ionosphere (Cravens et al., 1983; Whitten et al., 1984; Elphic et al., 1984; Knudsen et al., 1987). It is now thought that the nightside ionosphere is maintained primarily by horizontal transport of ions from the dayside, although low-energy electron precipitation also makes some contribution to the maintenance of the lower ionosphere (Gringauz et al., 1983). However, both large- and small-scale features are found deep on the nightside, such as ionospheric holes, regions of enhanced ion temperature, streamers, etc. that are only poorly understood.

Most of the unresolved issues concerning the dayside ionosphere involve small-scale structures such as flux ropes and microscopic transport processes such as electron heat conduction. These topics will be taken up in the next section.

Consider now the status of our understanding of the magnetized state of the ionosphere. The observed characteristics were first described by Russell et al. (1979) and Luhmann et al. (1980). Russell and Vaisberg (1983) and Luhmann (1987) also reviewed this topic. The large-scale magnetic fields are nearly horizontal and appear most frequently when the solar wind dynamic pressure is high. A mildly magnetized ionosphere was found during Orbit 177, as shown in Figure 2. Notice that the ionopause is located near 280 km and is quite broad. The magnetic field strength drops to almost zero near 200 km, but the peak magnetic field strength in the low-altitude layer is almost 100 nT. Orbit 176 is an example of a highly magnetized ionosphere. The magnetic pressure just outside the ionopause is 8×10^{-8} dynes cm^{-2} which is considerably in excess of the maximum ionospheric thermal pressure of about 6×10^{-8} dynes cm^{-2}. The peak field in the low-altitude layer is almost 160 nT, and even the minimum field at 200 km is 85 nT. The ionopause is located between 225 and 275 km.

Considerable advances in our theoretical understanding of magnetized ionospheres have been achieved over the last several years (Cloutier et al., 1981, 1983; Russell and Vaisberg, 1983; Luhmann et al., 1981, 1984; Cravens et al., 1984; Phillips et al., 1985; Shinagawa et al., 1987; Cloutier, 1987; Krymskii and Breus, 1988; Shinagawa and Cravens, 1988). The following one-dimensional magnetic induction, or diffusion-convection equation has figured prominently in all recent models of the large-scale magnetic field in the ionosphere of Venus

$$\frac{\partial B}{\partial t} = -\frac{\partial}{\partial z}[w B] + \frac{\partial}{\partial z}[D_B \frac{\partial B}{\partial z}] \qquad (2)$$

$D_B = m_e \nu_e / n_e e^2 \mu_0$ is the magnetic diffusion coefficient, where ν_e is the electron collision frequency, m_e is the electron mass, e is the electron charge, and μ_0 is the permeability of free space. w is the vertical plasma velocity. The first term in equation (2) represents the effects of the convection of magnetic flux which is "frozen-in" the plasma, and the second term represents the diffusion of magnetic flux due to the Ohmic dissipation of the currents responsible for the field.

Cravens et al. (1984) calculated the vertical plasma velocity in the dayside ionosphere, w, and numerically solved equation (2) using this velocity profile (which is shown in Figure 3). The plasma moves downward and attains a maximum speed of 60 m s^{-1} at an altitude of 200 km. Phillips et al. (1984) also used this velocity profile and solved equation (2) with a variety of upper boundary conditions applied at the ionopause. The typical observed shape of the magnetic field profile was correctly reproduced by these various model calculations. The minimum in the magnetic field profile at 200 km is associated with the maximum downward velocity. Magnetic flux frozen into the plasma is moved from the upper ionosphere down into the part of the ionosphere located below 200 km, where the magnetic belt is located. Ohmic resistivity becomes significant for altitudes below 160 km, where the neutral density becomes large. Consequently, the frozen-in assumption becomes invalid below about 160 km and the field "diffuses" away from the magnetic layer.

These earlier calculations neglected the effects of the magnetic field on the velocity and density profiles. Recently, Shinagawa et al. (1987) and Shinagawa and Cravens (1988) constructed a one-dimensional, time-dependent multi-species model of the ionosphere of Venus, solving the continuity equations for CO_2^+, O_2^+, O^+, and H^+ ions, momentum equations for O_2^+, O^+, and H^+, together with equation (2). Shinagawa et al. (1987) used the diffusion form of the momentum equations, and Shinagawa and Cravens (1988) used standard momentum equations and also included a quasi-two-dimensional divergence terms in the continuity equations.

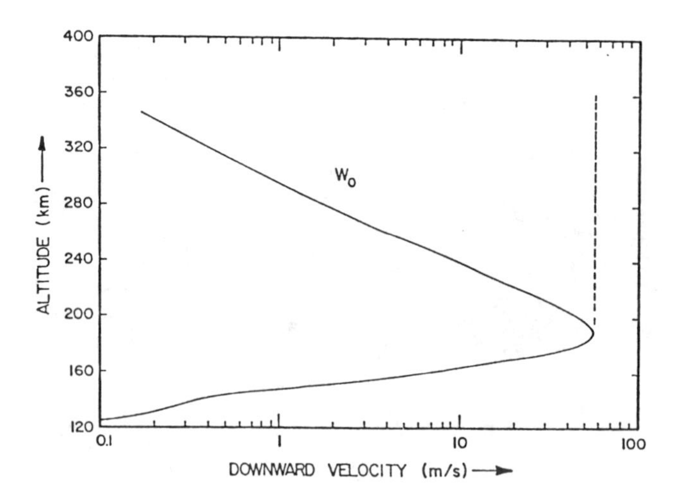

Fig. 3. W_0 is a calculated downward plasma velocity for an unmagnetized ionosphere of Venus. This velocity profile was used to calculate the time evolution of the magnetic field. The dashed line pertains to another calculation and should be ignored (from Cravens et al., 1984).

The calculated plasma velocity profiles from this model are qualitatively similar to the profile shown in Figure 3, although the maximum downward velocity (still located at 200 km) is reduced by about a factor of 2 due to the effects of the upward magnetic pressure gradient force. It should be noted that Cloutier et al. (1987) solved a similar set of equations, but in steady state. However, in order to reproduce a typical observed field profile, Cloutier et al. (1987) required vertical velocities near 160 km which were about a factor of 10 larger than the ones shown here. Calculated magnetic field profiles are shown in Figure 4 for one case from Shinagawa and Cravens (1988). The upper boundary condition chosen for this case was such that magnetic pressure (i.e., solar wind pressure) decreased with time. The initial ionosphere for this case is clearly highly magnetized and the calculated field profile (Figure 4) has the characteristic low-altitude layer structure (Figure 2). An unmagnetized ionosphere, including a relatively high and narrow ionopause, develops over the course of a few hours.

A consensus is beginning to develop (with a dissenting note from Cloutier and colleagues) that we now have a reasonably good, but far from perfect, understanding of the magnetized state of the ionosphere of Venus. The large-scale structure of both the magnetized and unmagnetized dayside ionospheres of Venus is understood reasonably well. The nightside ionosphere is not well understood.

4. Magnetic Flux Ropes

One outstanding problem concerning the ionosphere of Venus and its interaction with the solar wind is the formation and evolution of magnetic flux ropes. Flux ropes are narrow regions of enhanced magnetic field strength ($B \approx 30$ - 100 nT) which exist within an otherwise unmagnetized ionosphere

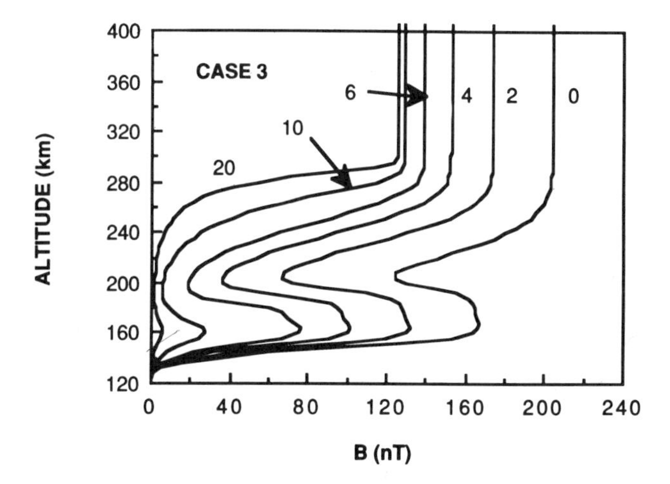

Fig. 4. Calculated magnetic field profiles from the time-dependent model of Shinagawa and Cravens (1988). Each profile is labeled with time in hours. The case shown here is for a solar wind dynamic pressure which is decreasing with time.

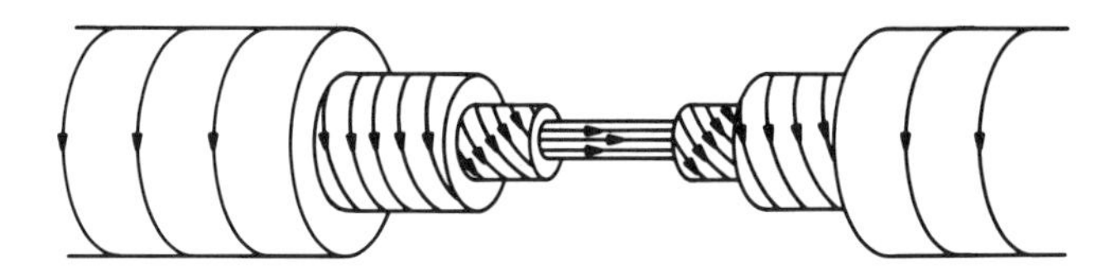

Fig. 5. Schematic of the magnetic field structure in a flux rope in the ionosphere of Venus as deduced from PVO magnetometer measurements. The field at the center of the rope is axial and the field toward the outer part of the rope is azimuthal (from Elphic et al., 1980).

(Figure 2). Considerable observational knowledge of the structure of these flux ropes is available, due to the analysis and interpretation of PVO magnetometer and Langmuir probe data by R. Elphic and others (Russell et al., 1979; Elphic et al., 1980; Russell and Vaisberg, 1983; Brace et al., 1983; Elphic and Russell, 1983). Statistical analyses of the data indicate that the diameter of a flux rope is typically $\approx$ 20 km in the upper ionosphere and $\approx$ 10 km for altitudes less than 200 km. Flux ropes are more abundant at lower altitudes.

An important observed property of flux ropes is that the plasma pressure across a flux rope is observed to be almost constant, particularly below 200 km. In other words, the thermal pressure gradient is almost zero

$$\nabla p \approx 0 \quad \text{with} \quad p = n_e k \left(T_e + T_i\right) \tag{3}$$

This implies that the Maxwell stress on the plasma must be almost zero

$$\mathbf{J} \times \mathbf{B} \approx 0 \tag{4}$$

$\mathbf{J} = \nabla \times \mathbf{B}/\mu_0$ is the current density. In other words, magnetic flux ropes are force-free structures (almost). The current density is proportional to the magnetic field intensity for a force-free structure (see Priest, 1982, or Low, 1989, for a more complete discussion of force-free structures and their role in solar physics). The following relation holds for force-free structures

$$\nabla \times \mathbf{B} = \alpha \mathbf{B} \tag{5}$$

where the parameter α is a function only of the radial distance from the axis of the structure.

The structure of a typical rope as deduced from magnetometer data is shown in Figure 5. The field is aligned with and is largest along the axis of the rope. The field has an increasing azimuthal component with increasing distance from the axis. The observed field structure of a rope largely conforms to the force-free relation (5).

Although much is now known about the structure of individual ropes, we understand very little else about them:

(1) Very little is known about the global configuration of flux ropes. We know that the "ends" of a rope must be

anchored somewhere out in the solar wind and that the middle is "buried" in the ionosphere where it is observed, but we do not know what goes on in-between. Flux ropes found deeper in the ionosphere appear to meander a lot and seem to be subject to the helical kink instability (Elphic and Russell, 1983).

(2) No quantitative and widely accepted model of the formation, or creation, of flux ropes exists, although numerous suggestions have been put forth over the last decade. One school of thought holds that flux ropes are formed at the ionopause due to the Kelvin-Helmholtz instability, or related instabilities (Johnson and Hanson, 1979; Dubinin et al., 1980; Wolff et al., 1980; Krymskii and Breus, 1988). Once formed, the ropes are thought to convect downward into the ionosphere. On the other hand, Cloutier et al. (1983) suggested that flux ropes are formed due to the Kelvin-Helmholtz instability within an initially magnetized ionosphere, which then "breaks up" into flux ropes. And a kinematic-dynamo mechanism was proposed by Luhmann and Elphic (1985) in which weak large-scale background fields are twisted and redistributed by waves and turbulence in the ionospheric plasma.

(3) The transport and evolution of magnetic flux ropes is not understood. However, Krymskii and Breus (1988) presented an excellent qualitative discussion of the forces on a flux rope, which include drag, tension, and buoyancy forces. They suggested that flux ropes which are created near the ionopause during low solar wind dynamic pressure conditions cannot be transported to the lower altitude region where they are observed due to either the existence of weakly upward background plasma convection when the ionopause height is large, or due to an upward buoyancy force. However, when the solar wind dynamic pressure is large, then the background plasma moves both downward (as in Figure 4) and toward the terminator, carrying with it newly-created flux ropes. This is a reasonable scenario, although it is far from being verified.

(4) The effects of flux ropes on electron and ion heat transport are only poorly understood. Cravens et al. (1980) constructed a numerical model of the electron and ion energetics, including heating and cooling processes and heat transport via thermal conduction. One conclusion of this study was that the calculated electron temperatures are very sensitive to the electron conductivity coefficient used. It was assumed that the small-scale magnetic field in the ionosphere could be represented as turbulent magnetic fluctuations which scatter the electrons with a mean free path λ. The calculated temperatures were inversely correlated with the value of λ. But obviously, flux ropes are not merely random fluctuations in the magnetic field, and a more sophisticated model of the energetics will eventually be required to improve our understanding of this aspect of the ionosphere of Venus.

What must be done to improve our knowledge of small-scale structures in the ionosphere of Venus in general, and of flux ropes in particular? More advanced theoretical models are certainly necessary, but these alone will not be sufficient. First, as in all areas of physics, experimental data are required to determine what the relevant physical problems really are, and then the relevant parameters of the models must be set. Second, sophisticated models produce complex sets of results, which require the availability of detailed experimental information for their verification or modification. Thus in order to resolve the outstanding problems concerning flux ropes, we need improved measurements with the following criteria:

(1) High spatial and temporal resolution plasma measurements are required. The magnetometer data from Pioneer Venus (PV) have a time resolution of 0.2 s, or a spatial resolution of 2 km (given the 10 km/s orbital speed of PVO near periapsis). The PV electron density data have a similar resolution. However, the ion velocities measured by the RPA on PV (Knudsen et al., 1980) have a spatial resolution of only ≈ 200 km and are only accurate to within ≈ 200 m s^{-1}. These velocity measurements are adequate for determining the average day to night flow near the terminator, but vector plasma velocities with a minimum accuracy of a few meters per second and a spatial resolution of 1-2 km are needed in order to understand the plasma behavior in the vicinity of a flux rope. Furthermore, the ion composition measurements made by either the RPA (Knudsen et al., 1980) or the ion mass spectrometer on PV (Taylor et al., 1980) have spatial resolutions greatly in excess of 10 km, the typical size of a flux rope.

(2) High-resolution plasma and field measurements by themselves are not sufficient for really understanding phenomena like flux ropes. Ideally, one would like to have "images" of flux ropes and their global configuration; however, this is not feasible. Alternatively, this information could be obtained by simultaneous, multi-spacecraft measurements of a flux rope. Obviously, it would be inadvisable to hold our breath waiting for this to happen, given the current state of the space program, but at least it is technically possible. The problems faced in determining the nature of these force-free structures in the ionosphere of Venus are the converse of the problems faced by solar physicists in understanding force-free structures in the solar atmosphere. In the latter case, remote observations of the global configuration of various coronal loops and filaments are available, but no in situ measurements of the fields in these structures are available (see other papers in these proceedings). Conversely, at Venus in situ measurements are available, but we have no remote observations to guide us concerning the global aspects of flux ropes.

5. The Cometary Plasma Environment

A schematic overview of the cometary plasma environment is given in Figure 6. The nucleus of a typical comet is a chunk of dusty ice several kilometers in extent. When the nucleus approaches within several of astronomical units of the sun, its surface heats up and water vapor is produced (as well as other volatiles). The water vapor atmosphere, or coma, expands millions of kilometers into space since the gravitational attraction of such a small object is negligible. The solar wind interaction with comets is characterized by the mass loading of the solar wind with cometary ions which are produced by the ionization of neutrals in the extensive cometary coma. This mass loading slows down the solar wind and ultimately leads to the formation of a magnetic barrier and a magnetotail, as shown in the schematic.

A very large literature now exists on this subject and a large amount of experimental data have become available from the many recent spacecraft missions to Comet Halley and Comet Giacobini-Zinner (G-Z). Only a brief and limited overview will be given in this paper, which will make the point that the large-scale structure of the cometary plasma environment is now beginning to be understood, but that our understanding of the small-scale structure remains rather poor. The reader is referred to Galeev et al. (1986) for a review of

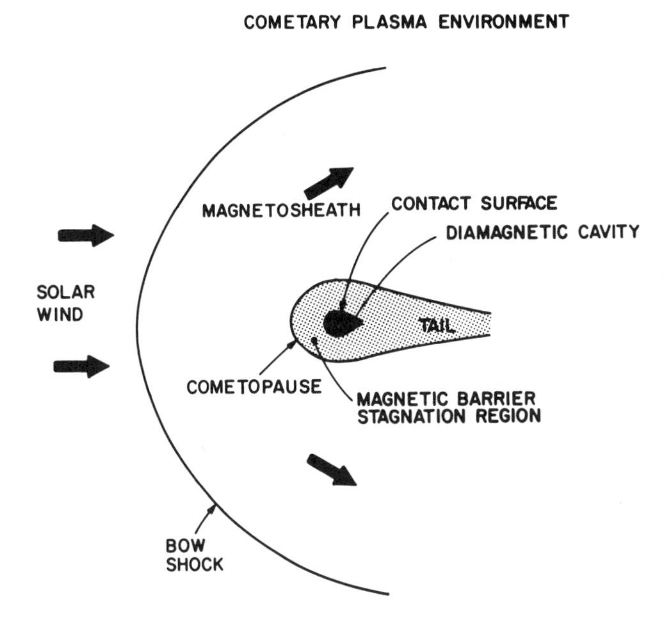

Fig. 6. Schematic of the plasma environment of active comets. The scale depends on the degree of activity. For Comet Halley near 1 AU, the subsolar distance to the bow shock was about 3.5 x 10^5 km, whereas for Comet Giacobini-Zinner, this distance was only 5 x 10^4 km.

recent developments in the solar wind interaction area. Also see the various articles in the book, Comets, or the monograph by Mendis et al. (1985) for general pre-1985 review of comets and the interaction of the solar wind with them.

The outer plasma environment (cometocentric distances exterior to the cometopause--r greater than about 10^5 km) will be discussed first. When a cometary neutral is ionized, the resulting ion is almost at rest with respect to the cometary frame of reference. A newly ionized particle is initially accelerated by the motional electric field of the solar wind and then follows a cycloidal trajectory in the E x B drift direction (cf. Ip and Axford, 1982; Cravens, 1986a). The newly-created ions drift parallel to the magnetic field in the solar wind reference frame, unless the solar wind and interplanetary magnetic field (IMF) directions are orthogonal. The resulting ion distribution is a ring-beam distribution, which is highly unstable and results in wave growth via an ion-cyclotron instability (e.g., Wu and Davidson, 1972; Gary et al., 1986; Winske et al., 1985; Sagdeev et al., 1986). Intense magnetic fluctuations were observed upstream of comet G-Z by the International Cometary Explorer (ICE) spacecraft and upstream of Comet Halley (Tsurutani and Smith, 1986; Johnstone et al., 1986). These waves or fluctuations then leads to pitch angle scattering of the pick-up ions which causes the ion distribution function to isotropize, or at least to partially isotropize. The magnetic fluctuations also lead to the stochastic acceleration of the ions via the second-order Fermi mechanism (Sagdeev et al., 1986; Ip and Axford, 1986; Gribov et al., 1986). Accelerated ions were detected in the vicinity of both Comet Halley and Comet G-Z (e.g., Somogyi et al., 1986; Kecskemety et al., 1988; McKenna-Lawlor et al., 1986; Ipavich et al., 1986; Hynds et al., 1986).

The ion pick-up process described above has been investigated using a variety of theoretical tools: linear stability theory, test particle calculations, the semikinetic approximation, quasi-linear theory, and plasma simulation codes. Overall, the basic processes are beginning to be understood although some problems, such as some nonlinear aspects of the very large amplitude waves ($\delta B/B \approx 1$), remain unresolved.

The cometary bow shock (Figure 6) is rather weak (Mach number $M \approx 2$) and is located far from the nucleus ($r_{shock} \approx 4$ x 10^5 km for Comet Halley and ≈ 5 x 10^4 km for Comet Giacobini-Zinner) (Thomsen et al., 1986). The shock thickness is of the order of a heavy ion gyroradius ($\Delta r_{shock} \approx 10^4$ km) (see Galeev, 1986). The location and strength of the shock is not highly dependent on the details of the cometary ion distribution function (although it does depend on the degree of isotropy) and its subsolar location can be predicted reasonably well using simple semikinetic theories (Wallis and Ong, 1975; Wallis, 1973; Galeev et al., 1986). Its shape and position are well described by global three-dimensional numerical magnetohydrodynamic (MHD) models which include cometary mass addition (Schmidt and Wegmann, 1982; Ogino et al., 1988), although these models do not properly describe the details of the cometary ion distribution function.

Cometary bow shocks are much broader and weaker than the bow shocks found at the planets. A great amount of experimental and theoretical work has been done on the structure and nature of collisionless shocks in plasmas, but the parameter regime relevant to cometary shocks is only now beginning to be studied theoretically (Omidi et al., 1986; Galeev and Lipatov, 1984) or experimentally (e.g., Klimov et al., 1986; Fuselier et al., 1986; Neugebauer et al., 1987). Although it was correctly expected, even prior to the first in situ measurements, that the cometary shock would be quite thick and that there might be a viscous subshock (Galeev et al., 1985; Galeev and Lipatov, 1984), the actual measured shock structure (Neugebauer et al., 1987) is much more complicated than originally envisioned and will take a considerable effort to properly understand.

The overall large-scale dynamics of the solar wind flow downstream of the shock is well described by global, three-dimensional, numerical MHD models (Schmidt and Wegmann, 1982; Boice et al., 1986; Ogino et al., 1988; Fedder et al., 1987; Schmidt et al., 1986). The shocked solar wind moves slower as it approaches closer to the nucleus, due to the increasing addition of mass to the flow. The magnetic field strength increases as the flow slows. The global MHD models indicate that the flow is almost stagnant for distances less than about 10^5 km for Comet Halley, or less than $\approx 10^4$ km for Comet G-Z. This region is often referred to as the stagnation region, and in Figure 6 its outer limit is called the cometopause. A region of enhanced magnetic field (the magnetic barrier) exists in the vicinity of the stagnation region. The development along the sun-comet axis of the stagnation region and the nature of the cometary ion distribution function in this region was also studied using the semikinetic approximation prior to the spacecraft encounters with the comets (Galeev et al., 1985, Ip and Axford, 1982, and Wallis and Ong, 1975) and after the encounter (Galeev, 1988; Gombosi, 1987).

The in situ data showed that the structure of the stagnation region was more complicated than originally envisioned. For instance, the solar wind proton component rapidly disappears as the cometary ion component of the flow rapidly grows near

10^5 km. Gringauz et al. (1986) called this transition the cometopause). Goldstein et al. (1986) also presented ion data for this region and found a less rapid cometopause transition than Gringauz et al. (1986a). Gombosi (1987) explained the gross behavior of the plasma in this region using a two-fluid semikinetic model along the sun-comet axis.

The magnetic field lines which are anchored in the magnetic barrier on the sunward side of the comet get "draped" around the head of the comet and form a tail. The region of enhanced cometary plasma density within the cometopause extends down into the tail. Cometary ion tails have been remotely observed for many decades (cf. Brandt, 1982), but the only in situ measurements are from the encounter of the ICE spacecraft with Comet G-Z. ICE crossed the ion tail of this comet at a distance ≈ 10^4 km downstream of the nucleus. Both plasma (Zwickl et al., 1986; Meyer-Vernet et al., 1986; Ogilvie et al., 1986) and magnetic field measurements (Smith et al., 1986; Slavin et al., 1986) were made. Global MHD models reproduce the gross structure (as determined by remote observations by by the ICE data) of cometary ion tails reasonably well (Schmidt and Wegmann, 1982; Schmidt et al., 1986; Fedder et al., 1986). The plasma found in the center of the tail was observed to be cold (electron energies less than ≈ 1 eV) and ionospheric in nature (Meyer-Vernet et al., 1986). An ionospheric model was able to explain many of the characteristics of this plasma (Marconi and Mendis, 1986), and even simple MHD concepts can explain the overall dynamics of the flow in the tail (Siscoe, 1986). Both the theoretical models and the data show that the greatest plasma densities are found, naturally, in the plasma/neutral sheet (Slavin et al., 1986).

The theoretical understanding which exists for large-scale aspects of the cometary plasma environment does not extend to the small-scale plasma structures seen in the vicinity of comets. The first sentence of the abstract of a review paper by Brandt (1982) states: "Photographs of comet tails reveal a constantly changing array of structural features such as rays, streamers, knots, kinks, helices, condensations, and disconnected tails." Similarly, in situ observations of the plasma and magnetic field show many small-scale variations which are superimposed on the overall large-scale variations (e.g., Yeroshenko et al., 1986; Réme et al., 1986; and many other articles). Theoretical models have attempted to explain a few types of small-scale structures such as tail-rays (Schmidt and Wegmann, 1982), but the results have been inconclusive.

6. Plasma Inside the Cometopause

Now consider the inner region of the cometary plasma environment—that is, the region within the cometopause. A large amount of experimental information is now available in this region due to the Vega and Giotto encounters with Comet Halley in 1986. We know that the plasma in this region is: (1) almost stagnant (velocities less than 20 km s^{-1} or so), (2) relatively cold (ion temperatures less than ≈ 10^4 K), (3) is dense (electron densities $n_e \approx 10^2 - 10^4$ cm^{-3}), (4) almost entirely of cometary origin, and (5) quite complex chemically (Gringauz et al., 1986a; Balsiger et al., 1986; Krankowsky et al., 1986; Lämmerzahl et al., 1986; Schwenn et al., 1986; Korth et al., 1986; Mitchell et al., 1986). Both large- and small-scale structures are evident in both the plasma data and in the magnetometer data.

The magnetic field strength throughout most of the inner region of Comet Halley is 30-65 nT (Riedler et al., 1986; Neubauer et al., 1986; Neubauer, 1986). The magnetic field strength measured by the Giotto magnetometer near closest approach is shown in Figure 7 (Neubauer et al., 1986). The most striking feature in the observed magnetic field profile is the presence of a diamagnetic cavity in the near vicinity of the nucleus (also shown schematically in Figure 6). The boundary of this cavity is located at a cometocentric distance of 4500 km. This boundary has been called the ionopause, the contact surface, or tangential discontinuity, depending on the author. Notice that there is also some small-scale structure in the field.

The plasma both inside and outside the contact surface is ionospheric in its characteristics (Cravens, 1986b), and for cometocentric distances less than about 10^4 km, the densities of the various ion species can be explained photochemically. The overall behavior of the major ion species in this photochemical region can be explained using the physical and chemical processes studied prior to the recent comet encounters (see the review by Mendis et al., 1985). For example, the major ion is H_3O^+, which is formed when the ion species produced most prolifically by photoionization, H_2O^+, reacts with the major neutral constituent, H_2O. And the electron density was correctly predicted to vary inversely with cometocentric distance ($n_e \approx 1/r$). The Giotto ion mass spectrometer data indicated that near closest approach the total plasma density did vary as $1/r$ and that the major ion was indeed H_3O^+. However, these pre-encounter models did not do so well on some of the "details" found in the data outside the contact surface such as the ion pile-up region observed near 1-2 x 10^4 km (Balsiger et al., 1986).

The pre-encounter models also did not correctly predict the large-scale structure of the magnetic field immediately outside the diamagnetic cavity, although they did correctly predict the field pile-up and draping. It was correctly recognized prior to the encounters that a field-free region or cavity would exist around the nuclei of active comets and that ion-neutral coupling would play an important role in the formation of this cavity (cf. Mendis et al., 1985; Ip and Axford, 1982). However, the pre-encounter consensus (including that of this

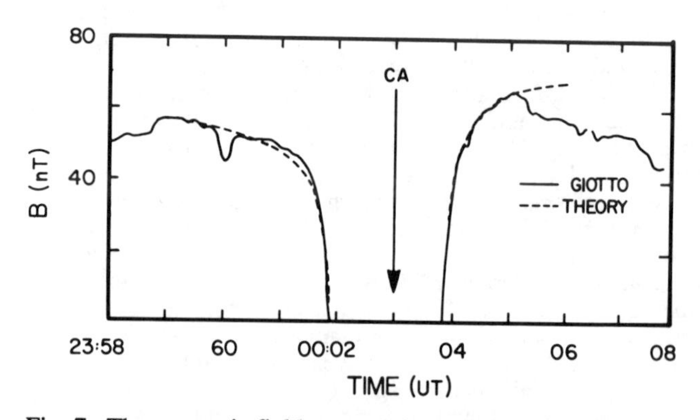

Fig. 7. The magnetic field strength in the inner coma of Comet Halley. The solid line is the magnetic field measured by the Giotto magnetometer (Neubauer, 1986) as a function of time near closest approach. CA designates the time of the closest approach of the spacecraft to the nucleus. The diameter of the diamagnetic cavity is ≈ 9000 km. The dashed line is from the theoretical calculations of Cravens (1986b,c).

author) was that a Venus-like ionopause would coexist with the cavity boundary, at which the plasma density would rapidly decrease and the magnetic field would rapidly increase so as to maintain pressure balance (as in equation (1)). The main role of ion-neutral drag was thought to be the enhancement of the thermal plasma pressure just inside the contact surface. Ip and Axford (1982) suggested that an inward magnetic curvature force must be balanced by an outward ion-neutral drag force in the magnetic barrier itself. The in situ data clearly show that a Venus-like ionopause does not exist at Comet Halley (see the discussion in Cravens, 1986a). The observed electron density continues to vary as 1 / r right through the contact surface (Balsiger et al., 1986), and a simple pressure balance relation like equation (1) does not hold. Furthermore, the magnetic field strength gradually builds up over a couple of thousand km, and not over a 10-60 km-thick layer as it does at the Venus ionopause.

With the benefit of 20/20 hindsight, theoretical modelers are now able to explain the large-scale magnetic field in the inner region of Comet Halley. Analytical expressions for the field strength, B(r), have been obtained by integrating the plasma momentum equation with a variety of simplifying assumptions (Ip and Axford, 1987; Cravens, 1986b,c; Haerendel, 1987; Wu, 1987; Eviatar and Goldstein, 1988; Cravens, 1988; Houpis, 1988). For example, Cravens (1986b,c) derived a simple expression for B(r) by making the following assumptions: (1) a planar geometry (that is, neglecting the magnetic curvature force), (2) stagnated plasma (v = 0), and (3) a 1 / r photochemical variation of the plasma density. The expression for B(r) thus obtained is given by

$$B(r) = B_0 \left[1 - \frac{r_{cs}^2}{r^2} \right]^{1/2} \qquad (6)$$

where the distance to the contact surface, r_{cs}, was shown to vary as $Q^{3/4} / B_0$, where Q is the cometary neutral gas production rate and B_0 is the maximum field strength in the magnetic barrier.

Expression (6) reproduces the observed radial variation of B quite well for the inbound leg of Giotto (Figure 7). The agreement with the outbound data is not as good. For the outbound leg, the magnetic curvature force must be included in order to obtain a maximum in the field profile (see Ip and Axford, 1987, or other references listed above).

Wu (1987) used similar methods to determine the geometrical shape of the contact surface. More elaborate numerical models are now being developed. Boice et al. (1986) applied their three-dimensional MHD model to this region and obtained a diamagnetic cavity, albeit with low ($\Delta r \approx 10^3$ km) spatial resolution. And Roatsch et al., 1986 have developed a higher resolution ($\Delta r \approx 300$ km) quasi two-dimensional MHD model which generates not only the magnetic structure (like the analytical theories), but also plasma velocities and densities. Cravens (1988) has developed a high-resolution ($\Delta r \approx 4$ km) one-dimensional MHD model in order to study the small-scale structure associated with the contact surface.

The basic physical explanation for the formation of the diamagnetic cavity which underlies all these theoretical models is that the inward magnetic pressure gradient force is balanced by an outward ion-neutral drag force on the plasma. In the field-free region, the ions and neutrals flow outward together from the nucleus at a speed of ≈ 1 km s^{-1}. But the plasma in the magnetic barrier is virtually stagnant and is "tied" to the field lines. The neutrals are obviously unaffected by the field and therefore move relative to the ions.

A fair degree of understanding of the physics and chemistry (at least for the major ions) now exists for the large-scale aspects of the plasma environment of the inner region of comets. Considerable work remains to be done on the large-scale aspects of the plasma velocities. The minor ion chemistry is quite complex and is discussed by Huebner and Boice (1989).The energetics for the cometary plasma environment also requires much more work, although Haerendel (1987) and Cravens (1987) have theoretically explained some aspects of the observed ion temperatures inside the cometopause. There has also been some theoretical work on the energetics inside the diamagnetic cavity (Mendis et al., 1985; Körösmezey et al., 1987), but big gaps in our knowledge will remain until electron temperatures are measured in that region.

The small-scale structure in the inner plasma environment, such as that seen in the observed field in Figure 7, remains almost completely unexplained. However, there is one possible explanation for the dip in B observed at 23:60 UT. Russell et al. (1987) have suggested that the mirror instability can produce this type of field perturbation. Another example of small-scale structure is that associated with the contact surface. In this case, there has been some theoretical modeling (Cravens, 1988) but the experimental data are inadequate. A brief discussion of some of the results of this model is now given.

The large-scale features of the model are not significantly different than what has just been discussed, and hence only results for the region near the contact surface will be shown. Figure 8 displays calculated profiles of the plasma density and velocity and of the magnetic field in the immediate vicinity of the contact surface. A 40 km wide transition layer exists just inside the contact surface (defined where B = 0). In this transition layer, the velocity decreases from the neutral outflow speed (1 km s^{-1}) to almost zero. The most noticeable feature of this layer is the density enhancement of a factor of 3 over the background density. The background density varies as 1 / r and is clearly photochemically controlled. The magnetic field gradient is enhanced in the outer part of the transition layer and is related to the thermal pressure bulge associated with the density peak.

The density peak is a by-product of the accommodation of the outward flowing plasma in the field-free region to the presence of the obstacle that the magnetic barrier represents to this flow. The widely accepted pre-encounter conception of this accommodation process (Houpis and Mendis, 1980; Mendis et al., 1985) was that an inner shock formed at a distance about 1000 km inside the contact surface, and that the outward flow was diverted tailward in the region of shocked flow between the inner shock and the contact surface (which was thought of as a Venus-like ionopause). The Cravens (1988) results suggest that this inner shock concept must be significantly modified. The transition layer evident in Figure 8 is indeed shock-like, but it is very narrow--$\Delta r \approx 40$ km rather than ≈ 1000 km. And the plasma flowing into this layer is not diverted into the tail (to any significant extent), but is destroyed in the layer locally via electron-ion recombination.

The above idea of a transition layer has not yet been widely accepted. One problem is that the predicted large ion density enhancement was not observed by the Giotto ion mass

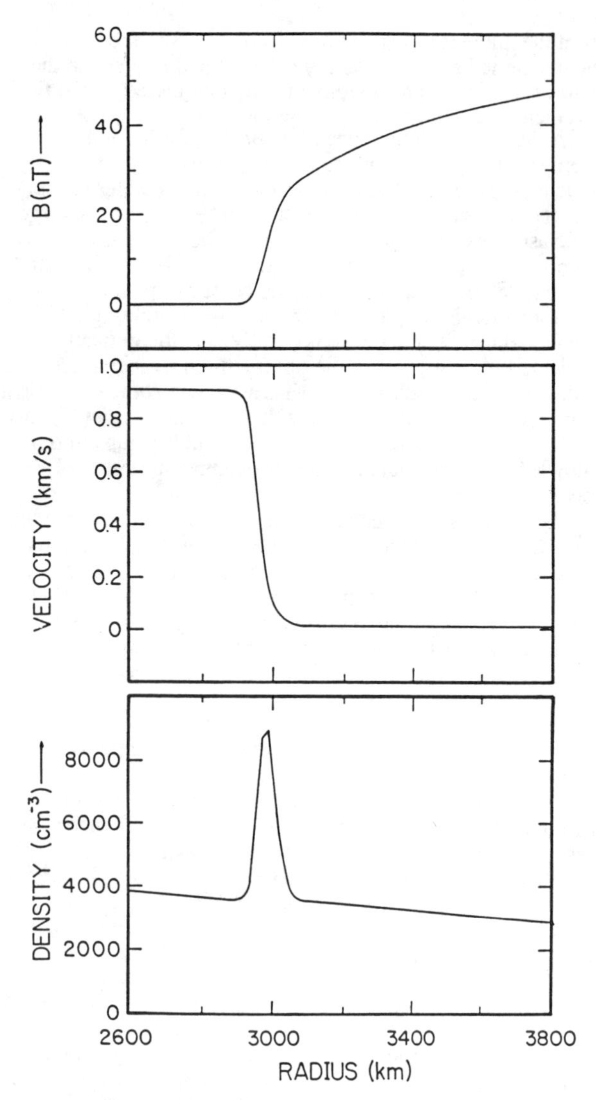

Fig. 8. Results from the theoretical model of Cravens (1988) for the region near the boundary of the diamagnetic cavity (i.e., contact surface). The upper panel displays the calculated magnetic field profile as a function of cometocentric distance. The middle panel shows the plasma/ion velocity, and the bottom panel shows the total ion density.

spectrometer (IMS). The observed total density profile appears to "hesitate" near the contact surface but certainly does not have a factor of 3 enhancement. And none of the individual ion species observed by the IMS had large enhancements either, with the exception of S^+. However, the published cycle time of the IMS instrument was 4 s; this is the time required to sample all the mass channels (Balsiger et al., 1986). The spacecraft encountered the comet at a speed of 68 km s^{-1}, which translates into a spatial resolution for a given ion species of $\Delta r \approx 300$ km. Thus it is possible that the peak was not detected due to undersampling since the predicted width of the peak is only about 40 km.

There is another source of experimental information on the transition region. The Giotto magnetometer had a temporal resolution of 0.03 s which gives a spatial resolution of 2 km. Neubauer (1988) reported that there is a 25 km wide region right at the contact surface in which the magnetic field gradient is enhanced. The observed B profile appears similar to the calculated profile and is shown in Figure 8. Another possible explanation for the enhanced gradient in B profile is an enhanced electron temperature inside the diamagnetic cavity (Neubauer, 1988). This is not likely in my opinion because the electrons are strongly coupled thermally to the neutrals in this region, which will keep the electron temperature close to the neutral temperature of a couple hundred degrees Kelvin (Körösmezey et al., 1987; Cravens et al., 1987).

In order to really understand the small-scale structure of the contact surface region, or small-scale structure in general, throughout the inner plasma environment, plasma measurements with a spatial resolution of at least a few kilometers are required. These measurements should include mass analysis and three-dimensional determinations of the velocity vector of each ion. Neugebauer (1989) discusses the instrumental characteristics of ion mass spectrometers which are capable of making measurements of this type.

7. Summary

A brief overview of the plasma environments of Venus and comets, and the solar wind interaction with these non-magnetic bodies was given in this paper. The point was made that our physical understanding of the large-scale structure of these plasma environments, although far from perfect, is rapidly improving. However, our understanding of most small-scale features, and their role in the overall solar wind interaction, remains undeveloped. One major obstacle to progress in understanding these features is the absence of high-resolution plasma measurements. A related difficulty is the absence of accurate and high-resolution measurements of the velocity vectors of ions (as well as measurements of just densities and temperatures). Another difficulty is our ignorance of the global structure of small-scale features, such as magnetic flux ropes at Venus, which are small-scale in the sense that they are very narrow, albeit very long. Multi-spacecraft measurements are needed to remedy this last deficiency.

Acknowledgments. This work was supported by NASA grants NAGW-15 and NGR-23-005-015 and NSF grant ATM 8417884.

References

Alexander, C. J., and C. T. Russell, Solar cycle dependence of the location of the Venus bow shock, Geophys. Res. Lett., 12, 369, 1985.

Balsiger, H., et al., Ion composition and dynamics at Comet Halley, Nature, 321, 330, 1986.

Boice, D. C., W. F. Huebner, J. J. Keady, H. U. Schmidt, and R. Wegmann, A model of Comet P/Giacobini-Zinner, Geophys. Res. Lett., 13, 381, 1986.

Brace, L. H., H. A. Taylor, Jr., T. I. Gombosi, A. J. Kliore, W. C. Knudsen, and A. F. Nagy, The ionosphere of Venus: Observations and their interpretation, in Venus, edited by D. Hunten, T. Donahue, L. Colin, and V.

Moroz, p. 779, Univ. of Ariz. Press, Tucson, Ariz., 1983.

Brandt, J. C., Observations and dynamics of plasma tails, in Comets, edited by L. L. Wilkening, p. 519, Univ. of Ariz. Press, Tucson, Ariz., 1982.

Breus, T. K., A. M. Krymskii, and V. Ye. Mitnitskii, Interaction of the mass-loaded solar wind flow with blunt body, Planet. Space Sci., 35, 1987.

Cloutier, P. A., Formation and dynamics of large-scale magnetic structures in the ionosphere of Venus, J. Geophys. Res., 89, 2401, 1984.

Cloutier, P. A., Steady state flow/field model of solar wind interaction with Venus: Global implication of local effects, J. Geophys. Res., 92, 7, 289, 1987.

Cloutier, P. A., T. F. Tascione, and R. E. Daniell, Jr., An electrodynamic model of the electric currents and magnetic fields in the dayside ionosphere of Venus, Planet. Space Sci., 29, 635, 1981.

Cloutier, P.A., T. F. Tascione, R. E. Daniell, Jr., H. A. Taylor, Jr., and R. S. Wolff, Physics of the interaction of the solar wind with the ionosphere of Venus: Flow/field models, in Venus, edited by D. Hunten, T. Donahue, L. Colin, and V. Moroz, p. 941, Univ. of Ariz. Press, Tucson, Ariz., 1983.

Cravens, T. E., The physics of the cometary contact surface, in 20th ESLAB Symposium on the Exploration of Halley's Comet, ESA SP-250, edited by B. Battrick, E. J. Rolfe, and R. Reinhard, vol. I, p. 241, ESTEC, Noordwijk,. The Netherlands, 1986a.

Cravens, T. E., Theory and observation of cometary ionospheres, Presented at XXVI COSPAR Meeting, Toulouse, France, 1986b.

Cravens, T. E., Ion distribution functions in the vicinity of Comet Giacobini-Zinner, Geophys. Res. Lett., 13, 275, 1986c.

Cravens, T. E., Ion energetics in the inner coma of Comet Halley, Geophys. Res. Lett., 14, 983-986, 1987.

Cravens, T. E., A one-dimensional magnetohydrodynamical model of the inner coma of comet Halley, J. Geophys. Res., submitted 1988.

Cravens, T. E., T. I. Gombosi, J. U. Kozyra, A. F. Nagy, L. H. Brace, and W. C. Knudsen, Model calculations of the dayside ionosphere of Venus: Energetics, J. Geophys. Res., 85, 7778, 1980.

Cravens, T. E., A. J. Kliore, J. U. Kozyra, and A. F. Nagy, The ionospheric peak on the Venus dayside, J. Geophys. Res., 86, 11,323, 1981.

Cravens, T. E., S. L. Crawford, A. F. Nagy, and T. I. Gombosi, A two-dimensional model of the ionosphere of Venus, J. Geophys. Res., 88, 5595, 1983.

Cravens, T. E., H. Shinagawa, and A. F. Nagy, The evolution of large-scale magnetic fields in the ionosphere of Venus, Geophys. Res. Lett., 11, 267, 1984.

Cravens, T. E., J. U. Kozyra, A. F. Nagy, T. I. Gombosi, and M. Kurtz, Electron impact ionization in the vicinity of comets, J. Geophys. Res., 92, 7341, 1987.

Dubinin, E. M., P. L. Izrailevich, S. M. Shkolnikova, and I. M. Podgorny, Nature of flux ropes in the Venus ionosphere, Pis'mu V A. J., 6, 253, 1980.

Elphic, R. C., and C. T. Russell, Magnetic flux ropes in the Venus ionosphere: Observations and models, J. Geophys. Res., 88, 58-72, 1983.

Elphic, R. C., C. T. Russell, J. A. Slavin, and L. H. Brace, Observations of the dayside ionopause and ionosphere of Venus, J. Geophys. Res., 85, 7679, 1980.

Elphic, R. C., C. T. Russell, and J. G. Luhmann, The Venus ionopause current sheet: Thickness length scale and controlling factors, J. Geophys. Res., 86, 11,430, 1981.

Elphic, R. C., H. G. Mayr, R. F. Theis, L. H. Brace, K. L. Miller, and W. C. Knudsen, Nightward ion flow in the Venus ionosphere: Implication of momentum balance, Geophys. Res. Lett., 11, 1007, 1984.

Eviatar, A. and B. E. Goldstein, A unidimensional model of comet ionosphere structure, J. Geophys. Res., in press 1988.

Fedder, J. A., J. G. Lyon, and J. L. Giuliani, Jr., Numerical simulations of comets: Predictions for comet Giacobini-Zinner, Eos, Trans. Am. Geophys. Union, 67, 17, 1986.

Fontheim, E. G., Venus effects on upstream solar wind, paper presented at 1988 Yosemite Conference on Outstanding Problems in Solar System Plasma Physics: Theory and Instrumentation, Yosemite National Park, California, February 2-5, 1988.

Fox, J. L., The chemistry of metastable species in the Venusian ionosphere, Icarus, 51, 248, 1982.

Fuselier, S. A., W. C. Feldman, S. J. Bame, E. J. Smith, and F. L. Scarf, Heat flux observations and the location of the transition region of Giacobini-Zinner, Geophys. Res. Lett., 13, 247, 1986.

Galeev, A. A., in 20th ESLAB Symposium on the Exploration of Halley's Comet, ESA SP-250, edited by B. Battrick, E. J. Rolfe, and R. Reinhard, vol. I, pp. 1-17, ESTEC, Noordwijk, The Netherlands, 1986.

Galeev, A. A., and A. S. Lipatov, Plasma processes in cometary atmospheres, Adv. Space Res., 4, 229, 1984.

Galeev, A. A., T. E. Cravens, and T. I. Gombosi, Solar wind stagnation near comets, Ap. J., 289, 807, 1985.

Galeev, A. A., et al., Position and structure of Comet Halley bow shock: Vega-1 and Vega-2 measurements, Geophys. Res. Lett., 13, 841, 1986.

Gary, S. P., S. Hinata, C. D. Madland, and D. Winske, The development of shell-like distributions from newborn cometary ions, Geophys. Res. Lett., 13, 1364, 1986.

Gloeckler, G., D. Hovestadt, F. M. Ipavich, M. Scholer, B. Klecker, and A. B. Galvin, Cometary pick-up ions observed near Giacobini-Zinner, Geophys. Res. Lett., 13, 251, 1986.

Goldstein, B. E., et al., GIOTTO-IMS observations of ion flow velocities and temperatures outside the contact surface of comet Halley, in 20th ESLAB Symposium on the Exploration of Halley's Comet, ESA SP-250, edited by B. Battrick, E. J. Rolfe, and R. Reinhard, vol. I, pp. 229-233, ESTEC, Noordwijk, The Netherlands, 1986.

Gombosi, T. I., Charge exchange avalanche at the cometopause, Geophys. Res. Lett., 14, 1174–1177, 1987.

Gombosi, T. I., Preshock region acceleration of implanted cometary H^+ and O^+, J. Geophys. Res., 93, 35-47, 1988.

Gribov, B. E., et al., Stochastic Fermi acceleration of ions in the pre-shock region of comet Halley, in 20th ESLAB Symposium on the Exploration of Halley's Comet, ESA SP-250, edited by B. Battrick, E.J. Rolfe, and R. Reinhard, vol. I, pp. 272-275, ESTEC, Noordwijk, The Netherlands, 1986.

Gringauz, K. I., The bow shock and the magnetosphere of

Venus according to measurements from Venera 9 and 10 orbiters, in Venus, edited by D. M. Hunten, L. Colin, T. M. Donahue, V. I. Moroz, p. 980, Univ. of Ariz. Press, Tucson, Ariz., 1983.

Gringauz, K. I., et al., First in situ plasma and neutral gas measurements at Comet Halley, Nature, 322, 282-285, 1986a.

Gringauz, K. I., et al., Detection of a new "chemical" boundary at Comet Halley, Geophys. Res. Lett., 13, 613, 1986b.

Haerendel, G., Plasma flow and critical velocity ionization in cometary comae, Geophys. Res. Lett., 13, 255, 1986.

Haerendel, G., Plasma transport near the magnetic cavity surrounding Comet Halley, Geophys. Res. Lett., 14, 673, 1987.

Houpis, H. L. F., and D. A. Mendis, Physiochemical and dynamical processes in cometary ionospheres 1. The basic flow profile, Ap. J., 239, 1107, 1980.

Houpis, H. L. F., The global interaction of comets with the solar wind: Heliocentric variation, in Symposium on Physical Interpretations of Solar/Interplanetary and Cometary Intervals, edited by M. A. Shea, in press, 1988.

Huebner, W. F., and D. C. Boice, Polymers and heavy ions in comet comae, these proceedings, 1989.

Hynds, R. J., S. W. H. Cowley, T. R. Sanderson, K.-P. Wenzel, and J. J. VanRooijen, Observations of energetic ions from Comet Giacobini-Zinner, Science, 232, 361, 1986.

Ip, W.-H., and W. I. Axford, Theories of physical processes in the cometary comae and ion tails, in Comets, edited by L. L. Wilkening, p. 588, Univ. of Ariz. Press, Tucson, Ariz., 1982.

Ip, W.-H., and W. I. Axford, The acceleration of particles in the vicinity of comets, Planet. Space Sci., 34, 1061-1065, 1986.

Ip, W.-H., and W. I. Axford, The formation of a magnetic field free cavity at Comet Halley, Nature, 325, 418, 1987.

Ip, W.-H., R. Schwenn, H. Rosenbauer, H. Balsiger, M. Neugebauer, and E. G. Shelley, An interpretation of the ion pile-up region outside the ionospheric contact surface, in 20th ESLAB Symposium on the Exploration of Halley's Comet, ESA SP-250, edited by B. Battrick, E. J. Rolfe, and R. Reinhard, vol. I, pp. 219-223, ESTEC, Noordwijk, The Netherlands, 1986.

Ipavich, F. M., A. B. Galvin, G. Gloeckler, D. Hovestadt, B. Klecker, and M. Scholer, Comet Giacobini-Zinner: In situ observations of energetic heavy ions, Science, 232, 366, 1986.

Johnson, F. S., and W. B. Hanson, A new concept for the daytime magnetosphere of Venus, Geophys. Res. Lett., 6, 581, 1979.

Johnstone, A., et al., Ion flow at Comet Halley, Nature, 321, 344-347, 1986.

Kecskemety et al., Pick-up ions in the unshocked solar wind at Comet Halley, J. Geophys. Res., submitted, 1988.

Klimov, S., et al., Extremely-low-frequency plasma waves in the environment of Comet Halley, Nature, 321, 292-293, 1986.

Kliore, A. J., V. I. Moroz, and G. M. Keating (editors), The Venus International Reference Atmosphere, Adv. Space Res., 5, 1985.

Knudsen, W. C., A. J. Kliore, and R. C. Whitten, Solar cycle changes in the ionization sources of the nightside Venus ionosphere, J. Geophys. Res., 92, 13,391, 1987.

Knudsen, W. C., K. Spenner, K. L. Miller, and V. Novak, Transport of ionospheric O^+ ions across the Venus terminator and implications, J. Geophys. Res., 85, 7803, 1980.

Korth, A., et al., Radial variations of flow parameters and composition of cold heavy ions within 50,000 km of Halley's nucleus, in 20th ESLAB Symposium on the Exploration of Halley's Comet, ESA SP-250, edited by B. Battrick, E. J. Rolfe, and R. Reinhard, vol. I, pp. 199-201, ESTEC, Noordwijk, The Netherlands, 1986.

Körösmezey, A., T. E. Cravens, T. I. Gombosi, A. F. Nagy, D. A. Mendis, K. Szegö, B. E. Gribov, R. Z. Sagdeev, V. D. Shapiro, and V. I. Shevchenko, A comprehensive model of cometary ionospheres, J. Geophys. Res., 92, 7331, 1987.

Krankowsky, D., et al., In situ gas and ion measurements at Comet Halley, Nature, 321, 326-329, 1986.

Krymskii, A. M., and T. K. Breus, Magnetic fields in the Venus ionosphere, J. Geophys. Res., in press, 1988.

Lämmerzahl, P., et al., Expansion velocity and temperatures of gas and ions measured in the coma of Comet Halley, in 20th ESLAB Symposium on the Exploration of Halley's Comet, ESA SP-250, vol. I, 179-182, ESTEC, Noordwijk, The Netherlands, 1986.

Low, B. C., Magnetic free-energy in the solar atmosphere, these proceedings, 1989.

Luhmann, J. G., and R. C. Elphic, On the dynamo generation of flux ropes in the Venus ionosphere, J. Geophys. Res., 90, 12,047, 1985.

Luhmann, J. G., C. T. Russell, and R. C. Elphic, Time scales for the decay of induced large-scale magnetic fields in the Venus ionosphere, J. Geophys. Res., 89, 362, 1984.

Luhmann, J. G., R. C. Elphic, and L. H. Brace, Large-scale current systems in the dayside Venus ionosphere, J. Geophys. Res., 86, 3509, 1981.

Luhmann, J. G., R. C. Elphic, C. T. Russell, J. D. Mihalov, and J. H. Wolfe, Observations of large scale steady magnetic fields in the dayside Venus ionosphere, Geophys. Res. Lett. , 7, 917, 1980.

Luhmann, J. G., The solar wind interaction with Venus, Space Science Reviews, 44, 241, 1986.

Marconi, M. L., and D. A. Mendis, The electron density and temperature in the tail of Comet Giacobini-Zinner, Geophys. Res. Lett., 13, 405, 1986.

McKenna-Lawlor, S., E. Kirsch, D. O'Sullivan, A. Thompson, and K.-P. Wenzel, Energetic ions in the environment of Comet Halley, Nature, 321, 347-349, 1986.

Mendis, D. A., H. L. F. Houpis, and M. L. Marconi, The physics of comets, Fund. Cosmic Phys., 10, 1, 1985.

Meyer-Vernet, N., P. Couturier, S. Hoang, C. Perche, J. L. Steinberg, J. Fainberg, and C. Meetre, Plasma diagnosis from thermal noise and limits on dust flux or mass in Comet Giacobini-Zinner, Science, 232, 370, 1986.

Mitchell, D. L., et al., Derivation of heavy (10–210 AMU) ion composition and flow parameters for the Giotto PICCA instrument, in 20th ESLAB Symposium on the Exploration of Halley's Comet, ESA SP-250, edited by B. Battrick, E. J. Rolfe, and R. Reinhard, vol. I, pp. 203-205, ESTEC, Noordwijk, The Netherlands, 1986.

Nagy, A. F., T. E. Cravens, and T. I. Gombosi, Basic theory and model calculations of the Venus ionosphere, in Venus, edited by D. Hunten, T. Donahue, L. Colin, and V. Moroz, p. 841, Univ. of Ariz. Press, Tucson, Ariz., 1983.

Nagy, A. F., T. E. Cravens, S. G. Smith, H. A. Taylor, Jr.,

and H. C. Brinton, Model calculations of the dayside ionosphere of Venus: Ionic composition, J. Geophys. Res., 85, 7795, 1980.

Neubauer, F. M., Giotto magnetic field results on the magnetic field pile-up region and the cavity boundaries, in 20th ESLAB Symposium on the exploration of Halley's Comet, ESA SP-250, edited by B. Battrickl, E. J. Rolfe, and R. Reinhard, vol. I, pp. 35-41, ESTEC, Noordwijk, The Netherlands, 1986.

Neubauer, F. M., et al., First results from the Giotto magnetometer experiment at Comet Halley, Nature, 321, 352-355, 1986.

Neubauer, F. M., Giotto magnetic field results on the boundary of P/Halley's magnetic cavity region, Presented at XIX General Assembly I.U.G.G., Vancouver, B.C., 1987.

Neubauer, F. M., The ionopause transition and boundary layers at Comet Halley from GIOTTO magnetic field observations, J. Geophys. Res., 93, 7272, 1988.

Neugebauer, M., Ion spectrometers for studying the interaction of the solar wind with nonmagnetic bodies, these proceedings, 1989.

Neugebauer, M., et al., The pick-up of cometary protons by the solar wind, in 20th ESLAB Symposium on the Exploration of Halley's Comet, ESA SP-250, edited by B. Battrick, E. J. Rolfe, and R. Reinhard, vol. I, pp. 19-23, ESTEC, Noordwijk, The Netherlands, 1986.

Neugebauer, M., F. M. Neubauer, H. Balsiger, S. A. Fuselier, B. E. Goldstein, R. Goldstein, F. Mariani, H. Rosenbauer, R. Schwenn, and E. G. Shelley, The variation of protons, alpha particles, and the magnetic field across the bow shock of comet Halley, Geophys. Res. Lett., 14, 644, 1987.

Neugebauer, M., et al., The variation of protons, alpha particles, and the magnetic field across the bow shock of Comet Halley, Geophys. Res. Lett., 14, 995, 1987.

Ogilvie, K. W., M. A. Coplan, P. Boschler, and J. Geiss, Ion composition results during the International Cometary Explorer with Giacobini-Zinner, Science, 232, 374-377, 1986.

Ogino, T., R. J. Walker, and M. Ashour-Abdalla, A three-dimensional MHD simulation of the interaction of the solar wind with Comet Halley, preprint, 1988.

Omidi, N., D. Winske, and C. S. Wu, The effect of heavy ions on the formation and structure of cometary bow shocks, Icarus, 66, 165, 1986.

Phillips, J. L., J. G. Luhmann, and C.T. Russell, Growth and maintenance of large-scale magnetic fields in the dayside Venus ionosphere, J. Geophys. Res., 89, 10,676, 1984.

Phillips, J. L., J. G. Luhmann, and C. T. Russell, Dependence of Venus ionopause altitude and ionospheric magnetic field on solar wind dynamic pressure, Adv. Space Res., 5, 173, 1985.

Priest, E. R., Solar Magnetohydrodynamics, D. Reidel, Dordrecht, 1982.

Réme, H., et al., Comet Halley--Solar wind interaction from electron measurements aboard Giotto, Nature, 321, 349, 1986.

Riedler, W., K. Schwingenschuh, Ye. G. Yeroshenko, V. A. Styashkin, and C. T. Russell, Magnetic field observations in Comet Halley's coma, Nature, 321, 288-289, 1986.

Roatsch, Th., K. Sauer, and K. Baumgärtel, Gasdynamic interpretation of ICE and VEGA/GIOTTO measurements, in 20th ESLAB Symposium on the Exploration of Halley's Comet, ESA SP-250, edited by B. Battrick, E. J. Rolfe, and R. Reinhard, vol. I, pp. 60-69, ESTEC, Noordwijk, The Netherlands, 1986.

Russell, C. T., and O. Vaisberg, The interaction of the solar wind with Venus, in Venus, edited by D. Hunten, T. Donahue, L. Colin, and V. Moroz, p. 873, Univ. of Ariz. Press, Tucson, Ariz., 1983.

Russell, C. T., R. C. Elphic, and J. A. Slavin, Initial Pioneer Venus magnetic field results: Dayside observations, Science, 203, 745, 1979.

Russell, C. T., and R. C. Elphic, Observations of magnetic flux ropes in the Venus ionosphere, Nature, 279, 616, 1979.

Russell, C. T., R. C. Elphic, and J. A. Slavin, Initial Pioneer Venus magnetic field results: Dayside observations, Science, 203, 745, 1979.

Russell, C. T., W. Riedler, K. Schwingenschuh, and Ye. Reroshenko, Mirror instability in the magnetosphere of Comet Halley, Geophys. Res. Lett., 14, 644, 1987.

Sagdeev, R. Z., V. D. Shapiro, V. I. Shevchenko, and K. Szego, MHD turbulence in the solar wind--Comet interaction region, Geophys. Res. Lett., 13, 85, 1986.

Sanderson, T. R., K.-P. Wenzel, P. Daly, S. W. H. Cowley, R. J. Hynds, E. J. Smith, S. J. Bame, and R. D. Zwickl, The interaction of heavy ions from Comet P/Giacobini-Zinner with the solar wind, Geophys. Res. Lett., 13, 411-414, 1986a.

Scarf, F. L., W. W. L. Taylor, C. T. Russell, and R. C. Elphic, Pioneer Venus plasma wave observations: The solar wind-Venus interaction, J. Geophys. Res., 85, 2599, 1980.

Schmidt, H. U. and R. Wegmann, Plasma flow and magnetic fields in comets, in Comets, edited by L. L. Wilkening, pp. 538-560, Univ. of Ariz. Press, Tucson, Ariz., 1982.

Schmidt, H. U., R. Wegmann, and F. M. Neubauer, MHD-model for Comet Halley, in 20th ESLAB Symposium on the Exploration of Halley's Comet, ESA SP-250, edited by B. Battrick, E. J. Rolfe, and R. Reinhard, vol. I, pp. 43-46, ESTEC, Noordwijk, The Netherlands, 1986.

Schwenn, R., W.-H. Ip, H. Rosenbauer, H. Balsiger, F. Bühler, R. Goldstein, A. Meier, and E. G. Shelley, Ion temperature and flow profiles in comet Halley's close environment, in 20 ESLAB Symposium on the Exploration of Halley's Comet, ESA SP-250, edited by B. Battrick, E. J. Rolfe, and R. Reinhard, vol. I, pp. 225-227, ESTEC, Noordwijk, The Netherlands, 1986.

Shinagawa, H., and T. E. Cravens, A one-dimensional multi-species magnetohydrodynamic model of the dayside ionosphere of Venus, J. Geophys. Res., in press, 1988.

Shinagawa, H., T. E. Cravens, and A. F. Nagy, A one-dimensional time-dependent model of the magnetized ionosphere of Venus, J. Geophys. Res., 92, 7317, 1987.

Siscoe, G. L., J. A. Slavin, E. J. Smith, B. T. Tsurutani, D. E. Jones, and D. A. Mendis, Statics and dynamics of Giacobini-Zinner magnetic tail, Geophys. Res. Lett., 13, 287, 1986.

Siscoe, G. L., Solar System Magnetohydrodynamics, in Solar Terrestrial Physics, edited by R. L. Carovillano and J. M. Forbes, D. Reidel Co., Dordrecht, 1983.

Slavin, J. A., R. C. Elphic, C. T. Russell, F. L. Scarf, J. H. Wolfe, J. D. Mihalov, D. S. Intriligator, L. H. Brace, H. A. Taylor, Jr., and R. E. Daniell, Jr., The solar wind interaction with Venus, Pioneer Venus observations of bow shock location and structure, J. Geophys. Res., 85, 7625, 1980.

Slavin, J. A.,R. E. Holzer, J. R. Spreiter, S. S. Stahara, and D. S. Chausee, Solar wind flow about the terrestrial planets 2. Comparison with gas dynamic theory and implications for solar-planetary interactions, J. Geophys. Res., 88, 19, 1983.

Slavin, J. A., E. J. Smith, B. T. Tsurutani, G. L. Siscoe, D. E. Jones, and D. A. Mendis, Giacobini-Zinner magnetotail: ICE magnetic field observations, Geophys. Res. Lett., 13, 283, 1986.

Smith, E. J., B. T. Tsurutani, J. A. Slavin, D. E. Jones, G. L. Siscoe, and D. A. Mendis, International Cometary Explorer encounter with Giacobini-Zinner: Magnetic field observations, Science, 232, 382, 1986.

Somogyi, A. J., et al., First observations of energetic particles near Comet Halley, Nature, 321, 285-288, 1986.

Spenner, K., W. C. Knudsen, K. L. Miller, V. Novak, C. T. Russell, and R. C. Elphic, Observation of the Venus mantle: The boundary region between solar wind and ionosphere, J. Geophys. Res., 85, 7655, 1980.

Spreiter, J. R., and S. S. Stahara, A new predictive model for determining solar wind–terrestrial planet interactions, J. Geophys. Res., 85, 6769, 1980a.

Spreiter, J. R., and S. S. Stahara, Solar wind flow past Venus: Theory and comparisons, J. Geophys. Res., 85, 7715, 1980b.

Tátrallyay, M., C. T. Russell, J. D. Mihalov, and A. Barnes, Factors controlling the location of the Venus bow shock, J. Geophys. Res., 88, 5613, 1983.

Tátrallyay, M., C. T. Russell, J. G. Luhmann, A. Barnes, and J. D. Mihalov, On the proper Mach number and ratio of specific heat for modeling the Venus bow shock, J. Geophys. Res., 89, 7381, 1984.

Taylor, H. A., H. C. Brinton, S. J. Bauer, R. E. Hartle, P. A. Cloutier, and R. E. Daniell, Global observations of the composition and dynamics of the ionosphere of Venus: Implications for the solar wind interaction, J. Geophys. Res., 85, 7765, 1980.

Thomsen, M. F., S. J. Bame, W. C. Feldman, J. T. Gosling, D. J. McComas, and D. T. Young, The comet/solar wind transition region at Giacobini-Zinner, Geophys. Res. Lett., 13, 393, 1986.

Tsurutani, B.T., and E. J. Smith, Strong hydromatic turbulence associated with Comet Giacobini-Zinner, Geophys. Res. Lett., 13, 259, 1986.

Vaisberg, O. L., and L. M. Zeleny, Formation of the plasma mantle in the Venusian magnetosphere, Icarus, 58, 412, 1984.

Wallis, M. K., Weakly shocked flows of the solar wind plasma through atmospheres of comets and planets, Planet. Space Sci., 21, 1647, 1973.

Wallis, M. K., and R. S. B. Ong, Strongly cooled ionizing plasma flows with application to beams, Planet. Space Sci., 23, 713, 1975.

Whitten, R. C., P. T. McCormick, D. Merritt, K. W. Thompson, R. R. Brynsvold, C. J. Eich, W. C. Knudsen, and K. L. Miller, Dynamics of the Venus ionosphere: A two-dimensional model study, Icarus, 60, 317-326, 1984.

Winske, D., C. S. Wu, Y. Y. Li, Z. Z. Mou, and S. Y. Guo, Coupling of newborn ions to the solar wind by electromagnetic instabilities and their interaction with the bow shock, J. Geophys. Res., 90, 2713, 1985.

Wolff, R. S., B. F. Goldstein, and C. M. Yeates, The onset and development of Kelvin-Helmholtz instability at the Venus Ionopause, J. Geophys. Res., 85, 7697, 1980.

Wu, C. S., and R. C. Davidson, Electromagnetic instabilities produced by neutral-particle ionization in interplanetary space, J. Geophys. Res., 77, 5399, 1972.

Wu, Z.-J., Calculation of the shape of the contact surface at Comet Halley, Annales Geophysique, submitted, 1987.

Yeroshenko, Ye. G., V. A. Styashkin, W. Riedler, K. Schwingenschuh,and C. T. Russell, Magnetic fine structure in comet Halley's coma, in 20th ESLAB Symposium on the Exploration of Halley's Comet, ESA SP-250, edited by B. Battrick, E. J. Rolfe, and R. Reinhard, vol. I, pp. 189-192, ESTEC, Noordwijk, The Netherlands, 1986.

Zwickl, R. D., D. N. Baker, S. J. Bame, W. C. Feldman, S. A. Fuselier, W. F. Huebner, D. J. McComas, and D. J. Young, Three component plasma electron distribution in the intermediate ionized coma of Comet Giacobini-Zinner, Geophys. Res. Lett., 13, 401, 1986.

MAGNETIC FIELD AND ELECTRIC CURRENT MEASUREMENTS OF CRITICAL PHENOMENA IN SOLAR WIND INTERACTIONS

R. C. Elphic

Institute of Geophysics and Planetary Physics, University of California, Los Angeles, CA 90024

Abstract. Magnetic fields play a crucial role in every case of solar wind/planetary interaction observed to date. In upstream waves, in bow shock phenomenology, in magnetosheath turbulence characteristics, and in the stability of the magnetopause, the magnetic field mediates critical processes. It is through the magnetotail magnetic field that solar wind energy is stored and released in terrestrial substorms. In the case of non-magnetic bodies, the solar wind magnetic field is central to the ion pickup process, and at Venus (and perhaps Mars) the solar wind magnetic field influences much of the planetary ionospheric behavior. Here we discuss new magnetometer techniques needed to answer some outstanding questions on the stability and structure of planetary magnetopauses and magnetotail dynamics. In particular we focus on using the four spacecraft Cluster configuration as a single instrument to make measurements of currents in flux transfer events and for sounding near-tail magnetic structure. Outstanding issues in the solar wind interaction with unmagnetized bodies include the emplacement and evolution of solar wind magnetic field in a planetary (or cometary) ionosphere and the formation mechanism of magnetic flux ropes. Finally, we note that a mission to study the lunar interaction would not only provide crucial insight on how atmosphereless, unmagnetized bodies such as asteroids (and the Martian moons Phobos and Deimos) affect the surrounding medium, but would also serve double duty as a monitor of the solar wind and terrestrial magnetotail at 60 R_E.

Introduction

The purpose of this paper is to review issues concerning magnetic field measurements in critical processes in solar wind/planetary interactions and to postulate what advances in instrumentation or mission approach are needed to address these issues. The potential topics range widely throughout space plasma phenomenology: ion pickup and solar wind/current sheet interactions with the foreshock and bow shock; the structure and stability of the magnetopause, ionopause, or cometopause; global current systems in magneto-spheres and ionospheres; and the dynamics of intrinsic and induced magnetotails, to name but a few. Because there are so many diverse outstanding questions it is impossible to cover them all; instead we shall focus on only a few selected areas. In general the advances in magnetometry required to attack these selected areas also benefit studies in other areas.

We begin with a brief discussion of the problem of making local magnetic field measurements and touch on some "generic" advances in magnetometry that will directly benefit future measurements. We then review briefly a subset of critical problems in solar wind/magnetosphere interactions. These include magnetic reconnection, specifically flux transfer events at the magnetopause (see Sonnerup, 1984) and magnetotail reconnection phenomena such as plasmoids in the distant tail (Hones, 1979) and near-tail impulsive flow events in the plasma sheet. We next investigate what advances can be expected from the Cluster and International Solar-Terrestrial Program (ISTP) missions on these topics. Mission and measurement scenarios for other planetary magnetospheres are also suggested.

On the topic of solar wind interaction with unmagnetized bodies, we review some observations of magnetic field behavior near obstacles with gravitationally-bound atmospheres (taking Venus as the principal example), with unbound atmospheres (comets), and finally with atmosphereless bodies. The topic of how solar wind magnetic field penetrates and evolves in an ionosphere is still controversial (see Cloutier et al., 1987). There are several competing formation mechanisms proposed for magnetic flux ropes in the Venus ionosphere. The diamagnetic boundary between a cometary ionosphere/atmosphere and the solar wind magnetic field has been observed in only one case. Is the boundary stable to interchange? Finally, our knowledge of the solar wind interaction with an atmosphereless, unmagnetized obstacle comes only from the Earth's moon, based on data acquired using technology now nearly 25 years old. The details of the lunar wake boundary are still unresolved.

Space-Borne Magnetic Field Measurements

Increasingly refined and ambitious space flight missions have placed ever greater demands on magnetic field measurements. Magnetometers have been required to measure fields both larger and several orders of magnitude smaller than at the Earth's surface, with increasing resolution in both field intensity and in time. At the same time the requirements for spacecraft magnetic cleanliness have also become stricter in order to accomplish these more refined measurements. Even under the best circumstances, however, observations on a spacecraft are only instantaneous measures of the local field. A time series of these measures constitutes a set of samples taken along an essentially unknown trajectory through configuration space. Only by knowing the magnetic field's temporal and spatial behavior can we infer currents and thereby make a connection to the observed plasma dynamics.

In laboratory plasma measurements, detailed knowledge of the magnetic field is obtained on some ordered grid of points in space. From this description of the field, one can calculate the approximate current density $\vec{J} = \vec{\nabla} \times \vec{B}/\mu_o$, $\vec{J} \times \vec{B}$ forces acting on the plasma as well as induction electric fields from $-\partial\vec{B}/\partial t = \vec{\nabla} \times \vec{E}$. With a grid of plasma vector flow field and the scalar density and temperature measurements, the state of the plasma is well specified, and we can better understand the salient plasma physics.

In the spirit of laboratory plasmas then, we wish to obtain such detailed knowledge of the magnetic field in the magnetospheric system. Such an idealized measurement scheme is illustrated in Figure 1. Here samples of the Tsyganenko-Usmanov magnetospheric field model (Tsyganenko and Usmanov, 1982) are shown in the noon-midnight meridian plane, for Kp = 0 and about 1640 UT on June 21. The samples are taken every 2 R_E throughout the dayside and partly into the nightside. If we were able to measure the true instantaneous field on such an array, then we could infer an approximate global field $\vec{B}(\vec{x},t)$ shown by the magnetic field lines drawn in the figure. In such an idealized system, currents and induction fields could be determined.

In practice we have much less information about the system we wish to measure and are forced to make assumptions about the spacecraft trajectory through plasma configuration space. However, simultaneous multi-point measurements constrain the various scenarios put forward to explain observations. Thus, our study of the physics benefits most from missions comprised of multiple spacecraft; failing that, however, single orbiting spacecraft can make long-term synoptic measurements which build up an average picture of the field configuration.

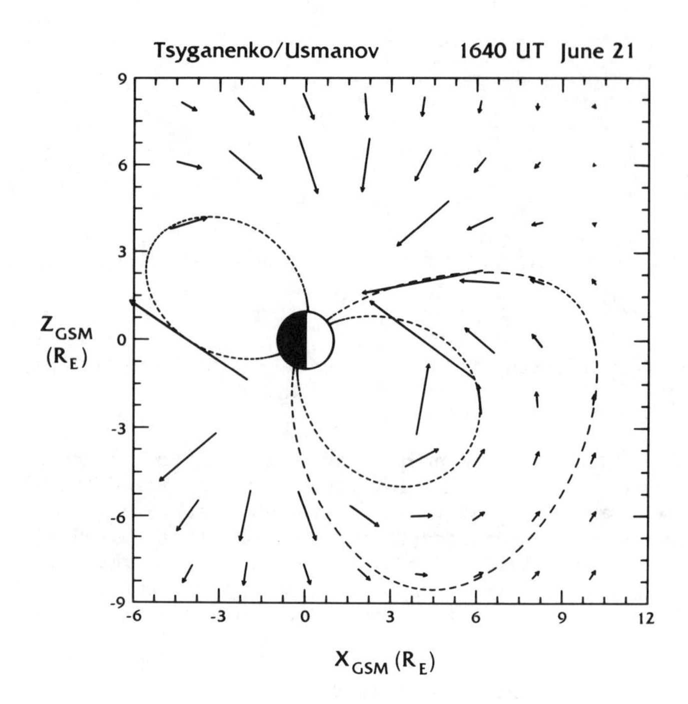

Fig. 1. Magnetic field vectors of the magnetospheric model of Tsyganenko and Usmanov (1982), sampled on a grid of points separated by 2 R_E. The model corresponds to about 1640 UT on June 21, and Kp = 0. Simultaneous multi-point measurements of the global field can yield local current densities, hence $\vec{J} \times \vec{B}$ forces acting on the plasma.

There are several areas where field measurements can make advances. Observations of extremely small-scale magnetic structure require high sampling rates, and magnetometer sampling rates are largely limited only by available spacecraft telemetry. One means of measuring at higher frequencies is to use burst memories which are filled at high rate and then unloaded later at normal telemetry rates to ground. This burst mode must be triggered either by ground command or by an instrument microprocessor with event-recognition software. It is not clear why the telemetry rates of currently planned magnetospheric missions are not much different from those developed more than 10 years ago; spacecraft downlink design should keep pace with scientific demands and advances in experiment technology.

Data compression is another means of increasing magnetometer sampling bandwidth. Various noise-free schemes based on first-differences can provide compression factors of 2 or 3. For spinning spacecraft, onboard despinning reduces the dynamic range needed for compression. Both tasks require significant microprocessor power.

Detection of small-amplitude waves in large background fields on spinning spacecraft requires a wide dynamic range with fine resolution. At the time of this writing, there are no off-the-shelf flight-qualified, radiation-hard, low mass and power 16-bit analog-to-digital converters needed for such resolution. Measurements of, for example, 100-m-scale structures in the low-altitude auroral ionosphere at burst mode (100 Hz) sampling frequencies, require such a device.

Critical Phenomena in Magnetospheres

The magnetospheric field model of Tsyganenko and Usmanov (1982) shown in Figure 1 is the result of assembling a large number of individual spacecraft measurements into a global model. It is the most well-studied model of the intrinsic planetary magnetospheres. For our purposes here we shall focus our attention on the terrestrial magnetosphere, though many important plasma-physical problems await to be investigated in the magnetospheres of other planets.

Over the past 10 years, it has become clear that some important terrestrial magnetospheric processes take place on less-than-global scales. For example, reconnection at the dayside magnetopause can occur in an impulsive, bursty manner, over what appear to be small length scales (Russell and Elphic, 1978), in addition to the quasi-steady merging process (Paschmann et al., 1979). In the magnetotail brief, intense plasma flows are seen associated with small-scale magnetic structures suggestive of impulsive reconnection (Sergeev et al., 1986; Elphic et al., 1986). On much larger scales, deep tail observations suggest the formation of plasmoids, islands of reconnected plasma undergoing ejection from the magnetotail. Tied in with all these phenomena is the global magnetospheric current system, which is also connected to the terrestrial ionosphere. We focus here on some of the issues related to flux transfer events and impulsive reconnection in the near-tail.

Flux Transfer Events

Haerendel et al. (1978) first suggested that certain brief disturbances in the field and plasma seen in the HEOS data at the high-latitude magnetopause could be signatures of transient or time-unsteady reconnection. Russell and Elphic (1978) investigated similar structures in the International Sun-Earth Explorer (ISEE) 1 and 2 data, and surmised that the signature in the magnetic field could be explained by patchy or temporally limited episodes of reconnection. These signatures were dubbed flux transfer events or FTEs. Subsequently, other investigators also found evidence that FTEs are indeed impulsive reconnection [for a review of FTEs, see Paschmann (1984), Sonnerup (1984), and Russell (1984)].

Some of the typical observed features of FTEs are: a bipolar excursion in the magnetic field component normal to the magnetopause boundary; occasionally the field strength will rise above both the magnetosheath and magnetospheric levels; within the FTE there is a mixture of magnetosheath and magnetospheric populations (Thomsen et al., 1987); the total (plasma and field) pressure within a FTE rises above that outside the FTE, but is balanced by magnetic tension in field lines wrapped around the structure (Paschmann et al., 1982); often electron heat flux is observed at the outer boundary of a FTE (Scudder et al., 1984); energetic magnetospheric ions within the FTE have anisotropies suggesting that they are being lost to the magnetosheath (Daly et al., 1984); the thermal plasma in FTEs is often flowing faster than the adjacent magnetosheath. Some of the magnetic field features in FTEs can be seen in Figure 2, showing data from ISEE.

The ISEE 1 and 2 data definitively showed that FTEs are three-dimensional structures, that they preferentially occur when the interplanetary magnetic field is southward (Berchem and Russell, 1984; Rijnbeek et al., 1984), and that they are about 1 R_E in scale size normal to the boundary (Saunders et al., 1984).

In the spirit of Figure 1, a multi-point investigation of the FTE phenomenon would help us understand their occurrence and structure. In particular, several widely-spaced spacecraft can check whether or not FTEs are observed simultaneously at different points on the magnetopause. Such a study has been reported by Elphic and Southwood (1987). The ISEE 1 and 2 and active magnetospheric particle tracer experiment (AMPTE) spacecraft crossed the magnetopause within 30 min of each other on September 19, 1984, and all observed FTEs. The two spacecraft pairs were at nearly the same magnetic local time, but the AMPTE spacecraft were above the equator while ISEE 1 and 2 were far below the equator. Both AMPTE and ISEE observed FTEs within 1 min of each other. Elphic and Southwood (1987) conclude that FTEs thus form pairwise at the equator and convect away, north and south. This study illustrates the value of multi-point measurements by widely separated spacecraft even when studying relatively small-scale processes.

Since the original Russell and Elphic (1978) FTE picture was introduced, several further incarnations have appeared but have retained the essential single reconnected flux tube notion (see Kan, 1989). Important recent modifications include those of Scholer (1988) and Southwood et al. (1988), who attribute FTE signatures to bursty Petschek reconnection. In contrast, Lee and Fu (1985) suggested that FTEs arise naturally if tearing mode reconnection occurs at the magnetopause when the interplanetary field is southward and has a B_y component. The topologies suggested by the two pictures are different, and the local time extent of a FTE is potentially greater than in the original Russell and Elphic (1978) picture. These differences bear directly on the question of how important FTEs are in the

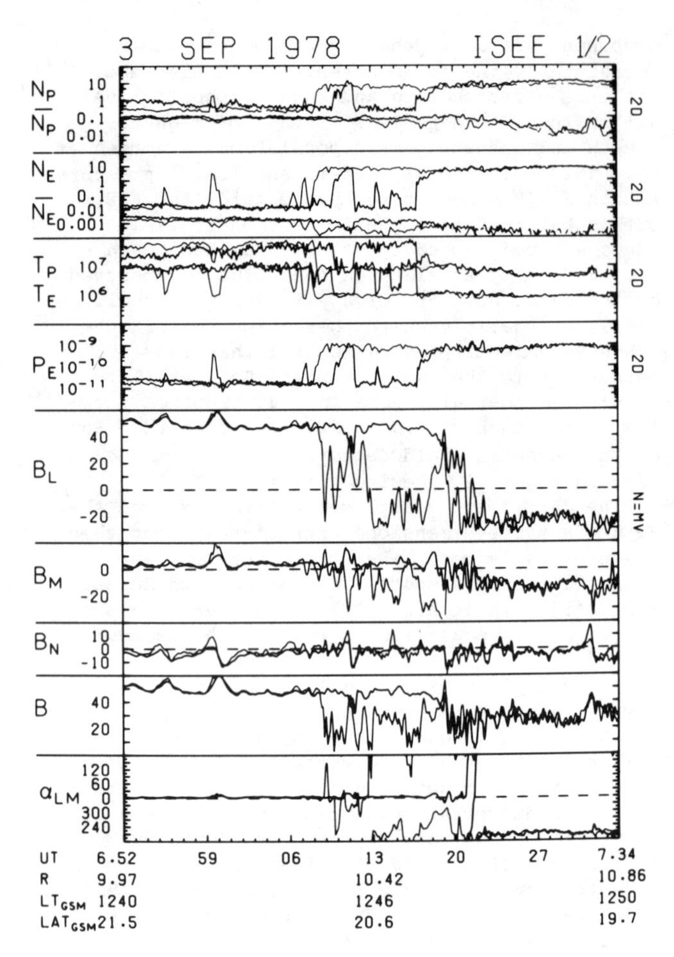

Fig. 2. Plasma and magnetic field data from ISEE 1 (heavy trace) and ISEE 2 (light trace), showing FTEs and magnetopause crossings. The FTEs can be identified by their characteristic bipolar signature in B_N, the component normal to the magnetopause current sheet surface. The plasma in FTEs is a mixture of the thermal magnetosheath plasma and the hot magnetospheric plasma. Not shown is the plasma flow speed enhancement which often accompanies FTE passage.

overall solar wind/magnetosphere energy transfer budget. Consequently, we must understand the exact magnetic structure of FTEs and determine what sort of instability produces them, and we must learn the local time extent of FTEs. Are they only 1 R_E wide or do they span a large fraction of the dayside magnetopause?

Magnetotail Field Structure

One of the more striking results of the ISEE 3 passages through the deep tail of the terrestrial magnetosphere was the observation of plasmoids. These structures were observed to consist of rapidly tailward-flowing plasma accompanied by field variations suggestive of an island-like magnetic configuration. Hones et al. (1976) and Hones (1979) have suggested a mechanism whereby such structures might arise in the magnetotail. This mechanism rests on the existence of two magnetic merging regions, one near Earth and the other far down the tail. The rapid reconnection postulated to occur at the near-Earth site eventually "disconnects" the plasma, sending the island structure and plasma down the tail. Since this model depends completely on the formation of a near-Earth merging region, we must investigate the detailed near-Earth magnetic topology [see reviews on the magnetotail and plasma sheet by Fairfield (1984) and Huang (1987)].

ISEE 1 and 2 allowed us to observe small-scale local changes of field and plasma in the near-Earth plasmasheet. On occasion brief, intense plasma flows are observed in association with crossings through the current sheet separating the northern from the southern plasmasheet. Figure 3, taken from Elphic et al. (1986), shows the plasma flows in the X GSM direction together with the observed field variations on ISEE 1 and 2. The spacecraft were separated by roughly 0.5 R_E at this time. The two spacecraft observations show that the current sheet is quite thin, about 2000 km. Energetic particle sounding also sug-

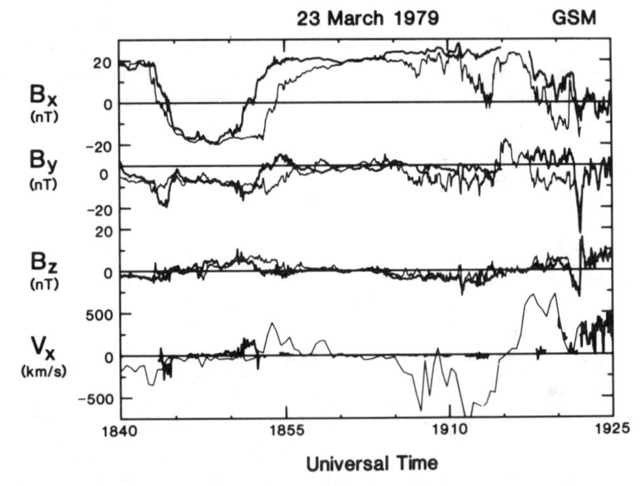

Fig. 3. Magnetic field components and plasma flow speeds from ISEE 1 and 2 during several current sheet crossings in the near-Earth magnetotail. The brief plasma flow events observed during current sheet crossings are earthward (tailward) when the B_z field component is northward (southward). This behavior is consistent with episodes of transient reconnection in the central plasmasheet. Although brief, the flow events yield net transport distances $\int V_x dt$ of several to a few tens of Earth radii, the larger values being comparable to the Earth-satellite distance. Also seen are FTE-like bipolar signatures in B_z (from Elphic et al., 1986).

gests a thin plasmasheet at this time (D. G. Mitchell, personal communication, 1986). The flows are observed within the current sheet crossings and are earthward or tailward for $B_z > 0$ or < 0, respectively. This is consistent with the existence of a near-Earth merging line, but the process is transient. Sergeev et al. (1986) liken these events to solar flare-like phenomena. Other ISEE studies (e.g., McPherron and Manka, 1985) also document near-Earth current and plasmasheet thinning. Fairfield (1984) reviews the evidence for plasmasheet thinning from earlier missions.

Several researchers have begun studying and simulating transient reconnection; their results may apply to FTEs and to the magnetotail as well. Figure 4 shows an example of transient reconnection simulated by Forbes and Priest (1987) with field lines anchored to the lower boundary, just the case for plasmasheet magnetic fields. The field lines and plasma streamlines are shown; high-speed earthward and tailward flows develop and are largely confined to the current sheet.

The ISEE spacecraft were never separated by more than about 1 R_E and were rarely in an orientation favorable for studying the plasmasheet and current sheet. Structure seen at the spacecraft could be local, and investigation of the large-scale variations of magnetic topology and plasma dynamics in the near-Earth tail could not be done easily. As a result, one question still being asked is: Are substorms the result of near-Earth reconnection? The ISEE observations, as well as results from other missions and ground observations, suggest that the answer is yes. However, the large-scale magnetic field change accompanying the sequence of events suggested by Hones (1979) has not been observed directly at the crucial widely separated sites.

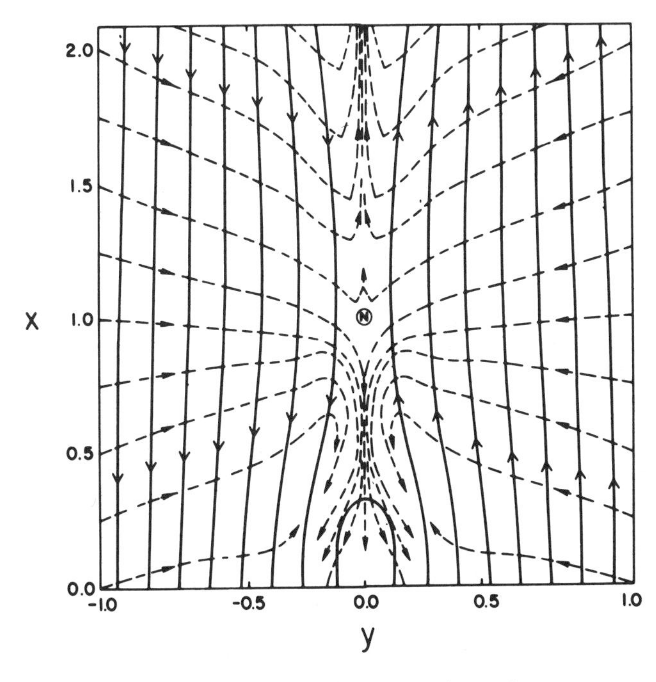

Fig. 4. A two-dimensional simulation of transient reconnection by Forbes and Priest (1987) showing streamlines (dashed) and magnetic field lines (solid). The field lines are tied to the lower boundary; this condition applies to magnetotail field lines connected to the Earth. N marks the position of the neutral line. The convergence of streamlines shows where strong earthward and tailward flow occurs.

Magnetospheric Measurements: The Future

Because the magnetic field is related to our physical understanding of plasmas only through knowledge of its spatial and temporal behavior, multi-point measurements are essential for understanding magnetospheric processes. The multipoint missions of ISEE, Dynamics Explorer (DE), and AMPTE have made that clear. For this reason the two major near-term space plasma physics missions proposed by NASA and ESA involve multiple spacecraft. Here we discuss briefly some aspects of these missions relating to magnetic field studies.

The joint ESA/NASA cornerstone mission, Solar and Heliospheric Observatory (SOHO)/Cluster, envisages one spacecraft at the forward Lagrange point monitoring solar wind conditions and remotely sensing the solar surface and corona (SOHO); in an eccentric polar orbit, four closely-spaced, identical spacecraft would form a basic measurement network for spatially sounding small-scale plasma structures (Cluster). Two representative Cluster orbits are shown in Figure 5. When apogee is on the dayside, the Cluster network can investigate FTEs, structure in the cusp, magnetosheath waves, the bow shock, foreshock, and upstream waves. At lower altitudes currents in the auroral zone can be studied. When apogee is in the nightside the spacecraft can sound the plasmasheet boundary layer, the central plasmasheet, and dynamics such as plasmasheet thinning and expansion. During the nightside apogee season the low-altitude cusp can also be examined.

Figure 5 also illustrates how important the inter-satellite spacing is in these investigations. The separation distance essentially governs what scale lengths can be effectively studied with Cluster; much larger separations are required on the nightside to study plasmasheet thinning than to study, for example, the internal structure of the bow shock. The lessons learned on ISEE and AMPTE will be valuable in Cluster separation strategy.

The Cluster spacecraft will be used to infer local electric currents from the four vector magnetic field measurements. One way to approach this is to expand the magnetic field in a truncated Taylor series about some reference position, for example, the position of one of the

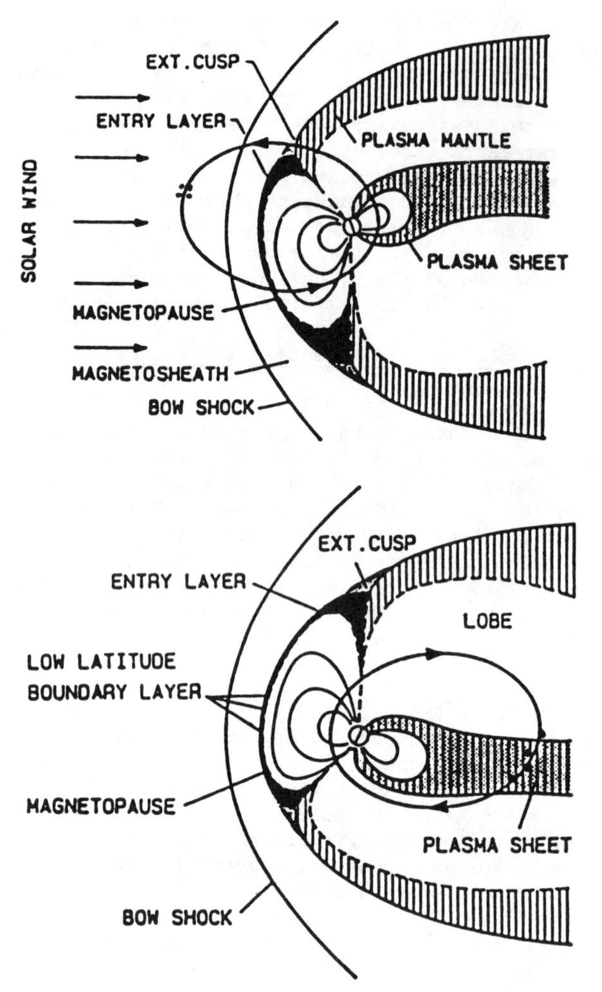

Fig. 5. Proposed orbits for the ESA/NASA Cluster mission. When apogee is on the dayside (left) the foreshock, bow shock, magnetosheath, magnetopause, and cusp can be investigated; when apogee is on the nightside (right) the dynamics of plasmasheet thinning and expansion can be observed, as well as the central current sheet.

spacecraft. Then the field observed at the ith spacecraft will be

$$\vec{B}_i = \vec{B}_o + \Delta\vec{r}_i \cdot \overleftrightarrow{\nabla B} \quad (1)$$

where $\vec{B}_o$ is the observed field at the reference spacecraft, and $\Delta\vec{r}_i$ is the separation vector from the reference spacecraft. The set of three of these equations can be represented in tensor form as

$$\overleftrightarrow{\Delta B} = \overleftrightarrow{\Delta r} \cdot \overleftrightarrow{\nabla B} \quad (2)$$

where the rows of $\overleftrightarrow{\Delta B}$ are the $\vec{B}_i - \vec{B}_o$ difference vectors, and the rows of $\overleftrightarrow{\Delta r}$ are the separation vectors. Then the dyad $\overleftrightarrow{\nabla B}$ can be expressed as

$$\overleftrightarrow{\nabla B} = (\overleftrightarrow{\Delta r})^{-1} \cdot \overleftrightarrow{\Delta B} \quad (3)$$

The errors and uncertainties inherent in $\overleftrightarrow{\Delta r}$ and $\overleftrightarrow{\Delta B}$ can be carried through to give the uncertainties in the elements of $\overleftrightarrow{\nabla B}$. From this calculation we can obtain $\vec{\nabla} \cdot \vec{B}$ and $\vec{\nabla} \times \vec{B}$ directly; the diagonal elements of $\overleftrightarrow{\nabla B}$ are terms in the divergence and the off-diagonal elements are terms in the curl. The trace of $\overleftrightarrow{\nabla B}$ should thus vanish within the errors and uncertainties of $(\overleftrightarrow{\Delta r})^{-1} \cdot \overleftrightarrow{\Delta B}$; if not, the current calculation is also doubtful. A significantly non-vanishing value of $\vec{\nabla} \cdot \vec{B}$ suggests other problems exist. One possibility is spatial aliasing; current structures smaller than the spacecraft separation can lead to spurious values of the curl and divergence.

The ISTP also relies on multi-point measurements, but from widely-separated spacecraft. In its present form the program entails space-borne measurements from four spacecraft: the U.S. WIND spacecraft will monitor the solar wind's energy input to the magnetosphere; auroral imaging and high latitude, intermediate altitude plasma and field measurements will be made on the U.S. POLAR spacecraft; magnetotail observations from the ISAS/NASA GEOTAIL spacecraft and near-Earth equatorial measurements on the Air Force/NASA chemical release and radiation effects satellite (CRRES) will elucidate the processes by which the magnetosphere sheds the solar wind energy. In addition, ground measurements will provide knowledge of magnetospheric behavior at the ionospheric boundary. This system is aimed at monitoring the global response of the magnetosphere to solar wind input, complementing the studies of micro-scale or local phenomena to be investigated on Cluster.

These planned missions were designed to address many outstanding issues in magnetospheric physics. However, there are some crucial questions which they will be unable to answer. We know now that FTEs are seen north and south of the magnetic equator simultaneously, but we do not know their local time extent and, hence, their contribution to magnetospheric convection. Likewise in the magnetotail, the crucial near-Earth region outside geosynchronous orbit will not be routinely monitored during much of the ISTP. To determine the spatial extent of sub-storm phenomena in the near tail we need spacecraft separated in local time across the tail.

A proposed Explorer-class mission, the Twin Equatorial Explorer (TWEQUE) would place two identical spacecraft in an eccentric, equatorial orbit, with apogee at roughly 12 R_E and perigee just outside geosynchronous. As Figure 6 shows, when apogee is on the dayside, the spacecraft can investigate magnetopause and boundary layer phenomena separated widely in local time. This configuration would establish the local time extent of FTEs and quasi-steady magnetopause reconnection phenomena, hence the contribution to the

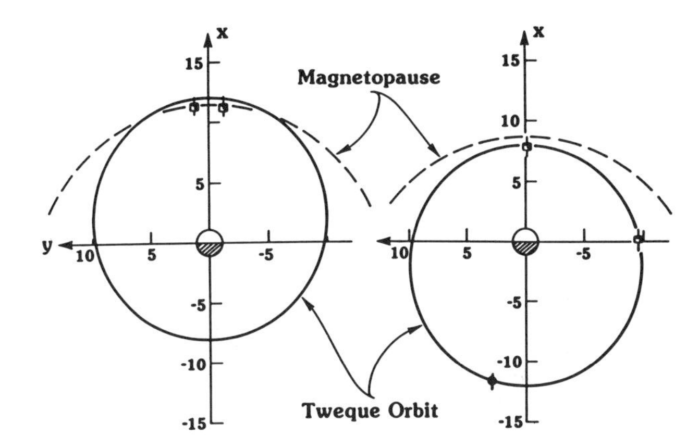

Fig. 6. A proposed Explorer-class mission called TWEQUE. With apogee at roughly 12 R_E, this two-spacecraft mission can investigate the local time extent of magnetopause and near-Earth magnetotail processes.

total polar cap potential drop. Similarly, when apogee is on the nightside, near-Earth plasma-sheet dynamics and substorm current wedge development can be investigated by the TWEQUE spacecraft at different local times.

Major progress in future studies of planetary magnetospheres relies on new spacecraft missions to orbit other planets. Our knowledge of the Mercury magnetosphere comes from only two flybys of that planet nearly 15 years ago. A Mercury orbiter would provide first-order synoptic measurements of the magnetosphere of an atmosphereless (or nearly so) body and would help us to understand the role that ionospheres, or the lack of one, play in magnetospheric physics. Does Mercury, for example, have substorms? The question of whether Mars possesses an intrinsic magnetic field, still unanswered after nearly 25 years of Mars missions, will be addressed by the orbiting spacecraft of the Soviet Phobos and U.S. Mars Observer missions.

Regarding the outer planets, the Galileo orbiter mission will make long-term observations of the Jovian magnetosphere, as will the Cassini mission of Saturn's magnetosphere; it will also investigate the Titan interaction. Are the magnetospheric dynamics of these planets driven entirely by rapid rotation? Do "substorms" occur, not because of solar wind energy input, but because of loading and breaking of magnetotail field lines by satellite-derived plasmas? Our first glimpse of the magnetosphere of Neptune awaits the Voyager encounter with that planet in 1989 (cf. Connerney, 1988).

Critical Phenomena at Non-Magnetic Bodies

The nature of the solar wind interaction with unmagnetized bodies depends crucially on the presence or absence of an atmosphere/ionosphere. We discuss here unmagnetized bodies with two classes of atmospheres, those that are gravitationally-bound (Venus) and those that are freely escaping (comets). We also take up the subject of atmosphereless, unmagnetized bodies in the solar wind, our own moon being the primary example.

Unmagnetized Bodies With Atmosphere

Gravitationally-bound atmospheres. The premiere example of the solar wind interaction with bodies possessing gravitationally-bound atmospheres and ionospheres is Venus. We know more of this case than any other in part because of several flybys of Soviet and U.S. spacecraft, but primarily because of the nearly 10 years of Pioneer Venus orbital operations [see Luhmann (1986) for a review]. However, there is still controversy. Is the whole Venus ionosphere electrodynamically coupled directly to the solar wind (Cloutier et al., 1987), or do photochemical processes dominate the interaction (Luhman, 1986; Cravens, 1989). Related to this interaction are small-scale magnetic flux ropes in the ionosphere, the origin and evolution of which are still unknown. The Pioneer Venus data have shown that massloading of the solar wind by planetary oxygen ions must be important in the interaction, but the magnitude of this effect on the formation of the Venus magnetotail is still unknown. Recently, controversy has arisen over whether there is lightning on Venus, and magnetic fields play an important role in the propagation of lightning signals out into space [see Taylor et al. (1987) and Scarf et al. (1987) for recent work on this topic].

The Pioneer Venus observations indicate that the Venus ionosphere is often free of any large-scale field; instead, small-scale (~10 km diameter), twisted field filaments called flux ropes are found (Russell and Elphic, 1979). But there are times when large-scale magnetic fields are found in the ionosphere. How the solar wind causes the spatial and temporal evolution of the field is the subject of ongoing study and controversy. Cloutier (1984) has suggested that finite absorption of the solar wind drives currents and convection throughout the ionosphere, resulting in large-scale fields and rapid flows; characteristic ionospheric response time scales are minutes or less. Luhmann (1986) and Shinagawa et al. (1987) argue that photochemical processes prevail and lead to the downward diffusion/ transport of magnetic flux on time scales of roughly an hour. The field evolution in both scenarios is strongly dependent on the solar wind dynamic pressure which in turn controls the ionopause altitude.

Since the Pioneer Venus orbital period is roughly 24 hours, each pass through the ionosphere (lasting perhaps 20 min) constitutes a snapshot of the ionospheric state at that time. The time scales for evolution of the ionosphere are several tens of minutes to hours, but less than 24 hours. Consequently it has not been possible to monitor this evolution directly using Pioneer Venus measurements.

To study the Venus ionosphere/solar wind interaction mechanism, it is vital to investigate the spatial and temporal evolution of magnetic fields in the ionosphere. Here again, in the spirit of Figure 1, it would be ideal to know the temporal behavior of the field on a grid of points within the Venus ionosphere. However, the only practical means of studying the field evolution is to sample the ionosphere with an orbiting spacecraft on a synoptic time scale commensurate with characteristic evolutionary time scales, i.e., many tens of minutes to a few hours.

Gravitationally-unbound atmospheres. Comets represent the case of the solar wind interaction with a gravitationally-unbound ionosphere. In this case the interaction between the cometary plasma and the solar wind results in a shock-like structure upstream and a broad interaction region containing plasma of cometary origin together with mass-loaded solar wind plasma. Closer to the nucleus, within the cometopause, the plasma is entirely of cometary origin. The innermost boundary, the contact surface, separates a diamagnetic cavity containing field-free cometary plasma from the almost stagnant inner coma plasma carrying the impressed solar wind field.

Figure 7 shows the magnetic field strength profile observed by the Giotto spacecraft as it passed through closest approach to Halley's Comet (from Neubauer et al., 1986). Superimposed on this profile is the theoretical profile of Cravens (1986); Haerendel (1987) found a similar profile. The distinctive diamagnetic cavity observed during the central 2 min of the pass represents a region of field-free, purely cometary plasma from which the solar wind field is entirely excluded. The boundary of this cavity results from pressure balance between the compressed and draped solar wind magnetic field and ion-neutral drag force exerted by the out-rushing neutrals on the essentially stagnant ions (Ip and Axford, 1987). The boundary thus forms where the collisional coupling between ions and neutrals is just sufficient to stand off the compressed solar wind magnetic field. The ions cannot pass through the magnetic barrier, and so must pile up on the inner edge of the contact surface creating a density enhancement. Here the ions and electrons may recombine and leave the system as neutrals. Thus, tailward flow of the piled-up cavity plasma may not be needed to remove it from the dayside contact surface region (cf. Cravens, 1989).

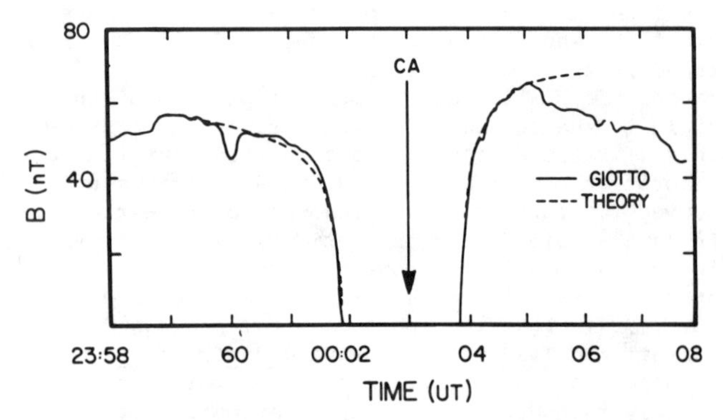

Fig. 7. Total magnetic field observed by the Giotto spacecraft as it passed by Halley's Comet, showing the diamagnetic contact surface boundary (from Neubauer et al., 1986). Superimposed on this profile is the theoretical profile of Cravens (1986), based on the balance of the magnetic pressure of the compressed solar wind field and the outward force of ion-neutral drag.

While the basic pressure balance at the contact surface is understood, the stability of this boundary has yet to be investigated. Interchange instabilities such as Kelvin-Helmholtz (K-H) depend crucially on the detailed flow of plasma along the contact surface; if recombination limits the magnitude of the tailward flow of cavity plasma, the boundary may be K-H stable. Because the contact surface separates a field-free plasma from inwardly curved field lines draped about the inner coma, the flute instability is another logical mode to investigate. In this case restoring forces associated with ion-neutral drag may stabilize the boundary. The collisional plasma beyond the contact surface may also be subject to reconnection when the solar wind fields draped around the comet have opposite orientations. Such a scenario is postulated to explain disconnection events (Niedner, 1984).

Atmosphereless Unmagnetized Bodies

The solar wind interaction with unmagnetized, atmosphereless bodies is perhaps the most basic of all. Objects in this class may be the most numerous in the solar system, ranging from asteroids with diameters of only a kilometer or so, through objects the size of the Martian moons Phobos and Deimos, up to bodies the size of the Earth's moon. While this class of interaction may at first glance seem trivial, it can in fact yield insight on fundamental plasma physical processes. In particular, the physics of plasma wakes can be investigated on length scales ranging from the order of a solar wind electron gyro radius up to many times the solar wind ion gyro radius. What we know of this class of interaction, and hence what questions still wait to be answered, comes from our limited studies of the lunar interaction using Explorer 35 and the Apollo 15 and 16 subsatellites [see Schubert and Lichtenstein (1974) for a review].

The moon absorbs nearly all the solar wind plasma incident on it. However, the solar wind magnetic field is able to diffuse rapidly through the low conductivity lunar interior. Since the solar wind is flowing supersonically with respect to the ion thermal speed, plasma flow cannot close immediately behind the obstacle. Solar wind electrons have thermal speeds much greater than the solar wind flow speed, but they cannot enter the wake unless there are sufficient ions to allow charge balance. Consequently the wake is comprised of essentially vacuum magnetic field. The wake field strength is higher than the upstream field because solenoidal currents flow on the surface of the wake boundary. These currents are just the self-consistent drift currents of the particles in the pressure gradient at the edge of the boundary, i.e., $\vec{J} = (-\vec{\nabla} p \times \vec{B})/B^2$, and they produce a field contribution which reinforces the background field. Additional features of the interaction include an expansion fan and possibly limb shocks.

Average magnetic field intensities as a function of location around the moon for a variety of plasma regions are shown in Figure 8 [from Schubert and Lichtenstein (1974)]. The top trace shows that in the very low β environment of the tail lobes, there is little field variation from lunar dayside to nightside. When the moon is in the magnetosheath, a large nightside enhancement of field strength is observed; this is presumably due to the high β sheath plasma. The enhancement is also present when the moon is in the lower β solar wind and there is also some evidence for enhancements off the limbs of the moon. Finally, when the moon is in the plasmasheet, the field tends to be highest in the sunward sector, indicating that on average plasmasheet flow is earthward at 60 R_E. There is no evidence of limb shocks in the plasma data.

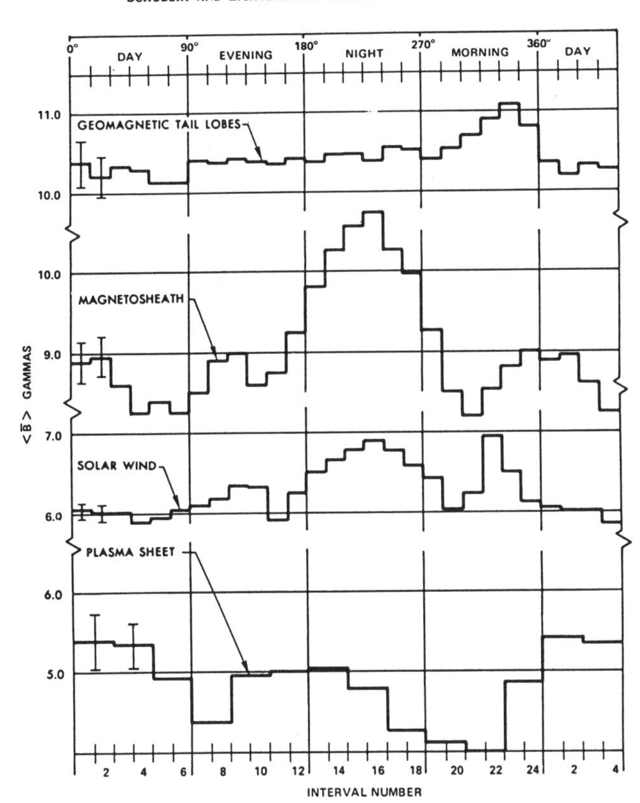

Fig. 8. Average field intensities around the moon for various plasma regimes. Little field variation is seen in the very low β magnetotail lobes. In the high β magnetosheath, a very strong nightside enhancement is seen; the lower β solar wind has a smaller nightside enhancement. In the plasmasheet the field enhancement is bimodal, but the larger peak on the dayside indicates plasma flow is earthward on average at 60 R_E (from Schubert and Lichtenstein, 1974).

Future Measurements of Unmagnetized Bodies

To study some of the critical phenomena related to the solar wind interaction with unmagnetized planetary ionospheres, a low-altitude orbiter is required. For Venus, an orbital period of no more than a few hours is needed to monitor major changes in the magnetic state of the ionosphere. In effect, a single orbiter can perform the desired multi-point function because the phase velocities and transport speeds within the ionosphere are much smaller than the spacecraft orbital speed; the spacecraft makes a snapshot of the ionosphere along its trajectory. As the orbiter revisits the ionosphere every few hours along similar trajectories, the evolution of both large- and small-scale field structures can be monitored. Estimates of the induction electric field due to $-\partial\vec{B}/\partial t$ could be made and the resultant energy input to the ionosphere calculated.

A low-altitude orbiter could also search for possible Venus lightning signatures with a radio receiver or search coil magnetometer system. Better still would be to add a stratospheric balloon carrying similar instrumentation. Then simultaneous signals could be sought both in space and in the atmosphere. A balloon magnetometer could also monitor the magnetic field leaking through the bottomside ionosphere on the dayside.

There are other possibilities for investigating the solar wind interaction with an ionosphere. If Mars has no intrinsic field, Mars Observer may be able to investigate a Venus-like interaction from just such a low-altitude orbit. Cassini, once in orbit around Saturn, will study the interaction of Titan's ionosphere with the magnetospheric plasma. Nearly 30 Titan encounters are planned for this mission.

The Comet Rendezvous and Asteroid Flyby (CRAF) mission promises to address many questions about comet/solar wind interactions. This mission proposes to bring a spacecraft into rendezvous with a comet, allowing measurements of the development of the coma during approach to the sun, during its most active and declining phases. It would allow detailed study of all the interaction regions, including the contact surface. The spacecraft could even presumably be stationed to observe disconnection events as they happen on the dayside or in the tail.

If we wish to understand the lunar interaction better, we must investigate the cis-lunar environment with modern instrumentation and with a sensible orbit. A proposed Explorer-class mission, the Lunar Solar Terrestrial Explorer (LUSTER), would study the moon's interaction with a variety of plasma regimes, as shown in Figure 9. It could sound the deep lunar wake in search of the flow closure region where the expansion fans meet. It would also act as a solar wind and magnetotail monitor at 60 R_E and would sound the distant magnetopause and bow shock. Finally, it would serve as a monitor of external conditions for a low-altitude lunar orbiting spacecraft.

The (presumably) unmagnetized Martian moon Phobos will be visited by the Soviet Phobos spacecraft mission. Based on the little we know of Phobos, this may also be our first glimpse of the solar wind interaction with an asteroid. While enroute to their destinations Galileo, CRAF, and Cassini are all slated to fly by asteroids. Whether or not the spacecraft will be targeted for wake passages will depend on trade-offs between in situ and remote sensing science.

Summary and Conclusions

We have seen that multi-point measurements of the vector magnetic field are central to understanding the plasma physics of solar wind interactions with magnetized and unmagnetized bodies. Many outstanding questions can be answered by this approach. In the case of the terrestrial magnetosphere, the field topology changes associated with reconnection can be sounded in detail. The role and importance of FTEs in the solar wind energy input to the magnetosphere can be established; direct observations of near-Earth plasmasheet thinning and reconnection in the magnetotail can be made. Both the Cluster and the ISTP/GGS missions will add immeasurably, and in much different ways, to our understanding of magnetospheric plasma processes.

Multi-spacecraft magnetic field observations in other planetary magnetospheres are not likely in the near future. However, single orbiting spacecraft can go far in establishing an average, global description of the magnetic field by repeated and comprehensive sampling of interesting regions of the magnetosphere. In short, a flyby is good, an orbiter much better, and more than one orbiter better still. For certain objects about which we know very little, such as Mercury and Mars, an orbiter can establish first-order understanding of the salient plasma physics.

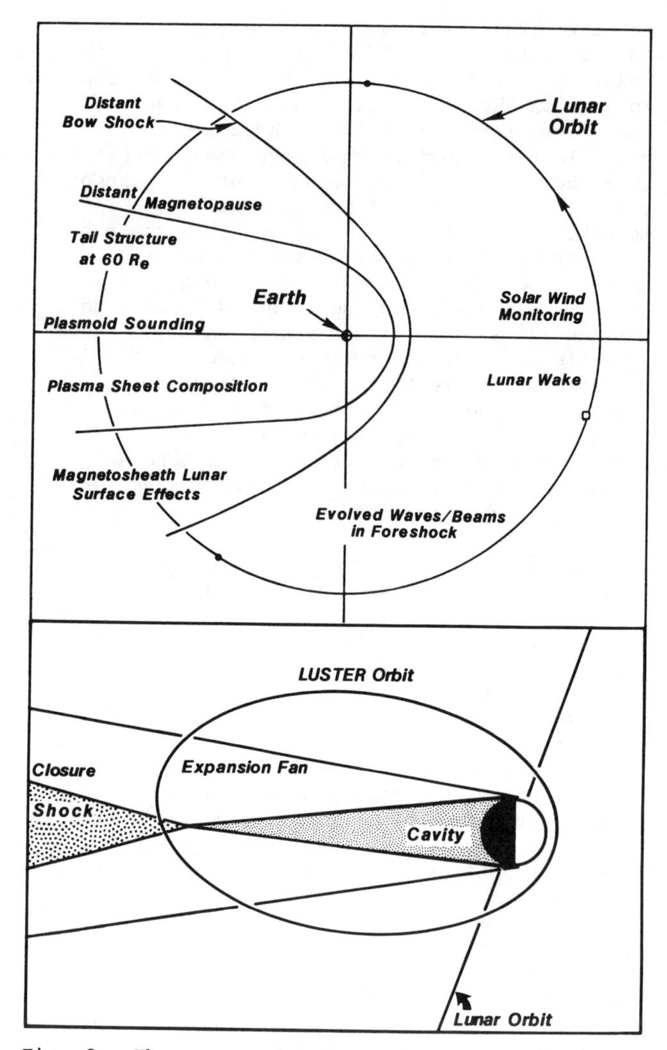

Fig. 9. The proposed LUSTER mission. This lunar orbiter would study many of the features of the lunar interaction with the solar wind, including the expansion fan, compression region, the plasma wake or void, and possibly the reclosure shock downstream. In addition, it would provide solar wind monitoring, studies of the distant foreshock and bow shock, and the distant magnetopause, and would monitor the magnetotail at 60 R_E.

Missions to unmagnetized bodies in the solar system will have to depend on the single orbiter sampling approach. For the solar wind interaction with a planetary ionosphere, the high spacecraft orbital speed relative to plasma drift speeds ensures that the field measurements are essentially a snapshot of the ionospheric state along the trajectory. With a short orbital period, changes in the magnetic state of the ionosphere can be monitored. Comets, because of their extremely low mass, require spacecraft to rendezvous and station-keep; the spacecraft can be positioned in interesting regions such as the contact surface, cometopause, and tail.

Finally, a case can be made for revisiting the moon to study the lunar/solar wind interaction. A lunar orbiter equipped with modern instruments could examine the wake of this field-free, atmosphereless body while also serving as a solar wind monitor and a magnetotail monitor at 60 R_E.

Acknowledgments. This work benefitted from discussions with numerous colleagues, particularly T. E. Cravens, J. R. Kan, J. G. Luhmann, C. T. Russell, and D. J. Southwood. The work was supported by NASA grants NAG5-536 and NAS5-28448.

References

Berchem, J., and C. T. Russell, Flux transfer events on the magnetopause: Spatial distribution and controlling factors, J. Geophys. Res., 89, 6689-6703, 1984.

Cloutier, P. A., Formation and dynamics of large-scale magnetic structures in the ionosphere of Venus, J. Geophys. Res., 89, 2401-2405, 1984.

Cloutier, P. A., H. A. Taylor, Jr., and J. E. McGary, Steady state flow/field model of solar wind interaction with Venus: Global implications of local effects, J. Geophys. Res., 92, 7289-7307, 1987.

Connerney, J.E.P., Magnetic field measurements, invited paper presented at Yosemite 1988 Conference on Outstanding Problems in Solar System Plasma Physics: Theory and Instrumentation, Yosemite National Park, California, February 2-5, 1988.

Cravens, T. E., The physics of the cometary contact surface, in Proc. 20th ESLAB Symposium on the Exploration of Halley's Comet, ESA SP-250, pp. 241-246, 1986.

Cravens, T. E., The solar wind interaction with non-magnetic bodies and the role of small-scale structures, these proceedings, 1989.

Daly, P. W., M. A. Saunders, R. P. Rijnbeek, N. Sckopke, and C. T. Russell, The distribution of reconnection geometry in flux transfer events using energetic ion, plasma and magnetic data, J. Geophys. Res., 89, 3843-3854, 1984.

Elphic, R. C., and D. J. Southwood, Simultaneous measurements of the magnetopause and flux transfer events at widely separated sites by AMPTE UKS and ISEE 1 and 2, J. Geophys. Res., 92, 13,666-13,672, 1987.

Elphic, R. C., C. A. Cattell, K. Takahashi, S. J. Bame, and C. T. Russell, ISEE-1 and 2 observations of magnetic flux ropes in the magnetotail: FTE's in the plasmasheet?, Geophys. Res. Lett., 13, 648-651, 1986.

Fairfield, D. H., Magnetotail energy storage and the variability of the magnetotail current sheet, in Magnetic Reconnection in Space and Laboratory Plasmas, Geophys. Monogr. Ser., Vol. 1., 30, edited by E. W. Hones, Jr., pp. 168-177, AGU, Washington, D.C., 1984.

Forbes, T. G., and E. R. Priest, A comparison of analytical and numerical models for steadily driven magnetic reconnection, Rev. Geophys., 25, 1583-1607, 1987.

Haerendel, G., Plasma transport near the magnetic cavity surrounding Comet Halley, Geophys. Res. Lett., 14, 673-676, 1987.

Haerendel, G., G. Paschmann, N. Sckopke, H. Rosenbauer, and P. C. Hedgecock, The frontside boundary layer of the magnetosphere and the problem of reconnection, J. Geophys. Res., 83, 3195-3216, 1978.

Hones, E. W., Jr., Transient phenomena in the magnetotail and their relation to substorms, Space Sci. Rev., 23, 393, 1979.

Hones, E. W., Jr., S. J. Bame, and J. R. Asbridge, Proton flow measurements in the magnetotail plasma sheet made with IMP 6, J. Geophys. Res., 81, 2270, 1976.

Huang, C. Y., Quadrennial review of the magnetotail, Rev. Geophys., 25, 529-540, 1987.

Ip, W. H., and W. I. Axford, The formation of a magnetic field free cavity at comet Halley, Nature, 325, 418, 1987.

Kan, J. R., Flux transfer events: A theoretical overview, these proceedings, 1989.

Lee, L. C., and Z. F. Fu, A theory of magnetic flux transfer at the Earth's magnetopause, Geophys. Res. Lett., 12, 105-108, 1985.

Luhmann, J. G., The solar wind interaction with Venus, Space Sci. Rev., 44, 241-306, 1986.

McPherron, R. L., and R. H. Manka, Dynamics of the 1054 UT March 22, 1979, substorm event: CDAW 6, J. Geophys. Res., 90, 1175-1190, 1985.

Neubauer, F. M., et al., First results from the Giotto magnetometer experiment at comet Halley, Nature, 321, 352-355, 1986.

Niedner, M. B., Magnetic reconnection in comets, in Magnetic Reconnection in Space and Laboratory Plasmas, Geophys. Monogr. Ser., vol. 30, edited by E. W. Hones, Jr., pp. 79-89, AGU, Washington, D.C., 1984.

Paschmann, G., The earth's magnetopause, in Achievements of the International Magnetospheric Study (IMS), ESA SP-217, pp. 53-64, ESTEC, Noordwijk, 1984.

Paschmann, G., B. U. O. Sonnerup, I. Papamastorakis, N. Sckopke, G. Haerendel, S. J. Bame, J. R. Asbridge, J. T. Gosling, C. T. Russell, and R. C. Elphic, Plasma acceleration at the earth's magnetopause: Evidence for reconnection, Nature, 282, 243-246, 1979.

Paschmann, G., G. Haerendel, I. Papamastorakis, N. Sckopke, S. J. Bame, J. T. Gosling, and C. T. Russell, Plasma and magnetic field characteristics of magnetic flux transfer events, J. Geophys. Res., 87, 2159-2168, 1982.

Rijnbeek, R. P., S. W. H. Cowley, D. J. Southwood, and C. T. Russell, A survey of dayside flux transfer events observed by the ISEE 1 and 2 magnetometers, J. Geophys. Res., 89, 786-800, 1984.

Russell, C. T., Reconnection at the Earth's

magnetopause: Magnetic field observations and flux transfer events, in Magnetic Reconnection in Space and Laboratory Plasmas, Geophys. Monogr. Ser., vol. 30, edited by E. W. Hones, Jr., pp. 124-138, AGU, Washington, D.C., 1984.

Russell, C. T., and R. C. Elphic, Initial ISEE magnetometer results: Magnetopause observations, Space Sci. Rev., 22, 681-715, 1978.

Russell, C. T., and R. C. Elphic, Observation of magnetic flux ropes in the Venus ionosphere, Nature, 279, 616, 1979.

Saunders, M. A., C. T. Russell, and N. Sckopke, Flux transfer events: Scale size and interior structure, Geophys. Res. Lett., 11, 131-134, 1984.

Scarf, F. L., K. F. Jordan, and C. T. Russell, Distribution of whistler mode bursts at Venus, J. Geophys. Res., 92, 12,407-12,411, 1987.

Scholer, M., Magnetic flux transfer at the magnetopause based on single X line bursty reconnection, Geophys. Res. Lett., 15, 291, 1988.

Schubert, G., and B. R. Lichtenstein, Observations of moon-plasma interactions by orbital and surface experiments, Rev. Geophys. Space Phys., 12, 592, 1974.

Scudder, J. D., K. W. Ogilvie, and C. T. Russell, The relation of flux transfer events to magnetic reconnection, in Magnetic Reconnection in Space and Laboratory Plasmas, Geophys. Monogr. Ser., vol. 30, edited by E. W. Hones, Jr., pp. 153-154, AGU, Washington, D.C., 1984.

Sergeev, V. A., A. G. Yahnin, R. A. Rakhamatulin, S. I. Solovjev, F. S. Mozer, D. J. Williams, and C. T. Russell, Permanent flare activity in the magnetosphere during periods of low magnetic activity in the auroral zone, Planet. Space Sci., 34, 1169-1188, 1986.

Shinagawa, H., T. E. Cravens, and A. F. Nagy, A one-dimensional time-dependent model of the magnetized ionosphere of Venus, J. Geophys. Res., 92, 7317-7330, 1987.

Sonnerup, B. U. O., Magnetic field reconnection at the magnetopause: An overview, in Magnetic Reconnection in Space and Laboratory Plasmas, Geophys. Monogr. Ser., vol. 30, edited by E. W. Hones, Jr., pp. 92-103, AGU, Washington, D.C., 1984.

Southwood, D. J., C. J. Farrugia, and M. A. Saunders, What are flux transfer events?, Planet. Space Sci., 36, 503-508, 1988.

Taylor, H. A., Jr., P. A. Cloutier, and Z. Zheng, Venus "lightning" reinterpreted as in situ plasma noise, J. Geophys. Res., 92, 9907-9919, 1987.

Thomsen, M. F., J. A. Stansberry, S. J. Bame, S. A. Fuselier, and J. T. Gosling, Ion and electron velocity distributions within flux transfer events, J. Geophys. Res., 92, 12,127-12,136, 1987.

Tsyganenko, N. S., and A. V. Usmanov, Determination of the magnetospheric current system parameters and development with experimental geomagnetic field models based on data from IMP and HEOS satellites, Planet. Space Sci., 30, 985, 1982.

PLASMA OBSERVATIONS OF FLUX TRANSFER EVENTS: PRESENT AND FUTURE MEASUREMENTS

M. F. Smith

Southwest Research Institute, San Antonio, TX 78229

A. D. Johnstone

Mullard Space Science Laboratory, Holmbury St. Mary, Dorking, Surrey, England

Abstract. Flux Transfer Events (FTEs) are widely regarded as an important feature of the interaction of the solar wind with the dayside magnetopause. Recent observations show the plasma structure of FTEs to be complex. To provide further information on the formation and development of this phenomenon a new generation of plasma instruments will be required. We review possible new instrumentation in the light of the Cluster mission.

Introduction

In recent years, since the initial observations in ISEE 1 and 2 (Russell and Elphic, 1978) and HEOS 2 data (Haerendel et al., 1978), much experimental evidence has been accumulated to support the notion of patchy reconnection. Figure 1 shows an idealized sketch of the present understanding of the flux tube model (Russell, 1984). The coordinate system used is that of boundary normal coordinates (Elphic and Russell, 1979). The $\hat{n}$ component is perpendicular to the plane of the magnetopause, while $\hat{l}$ is roughly northward. Briefly, reconnection occurs for a short period of time in a limited spatial region. The reconnected flux tube is moved across the magnetopause by the magnetosheath flow and the tension in the bent field lines. The magnetosheath field lines near the tube drape over it producing the magnetic field draping signature (Farrugia et al., 1987). A field-aligned current flowing inside the tube produces the twisted field lines that are required for the continuous B_N signature (Saunders et al., 1984a). ISEE plasma data (e.g., Paschmann et al., 1982; Daly et al., 1984) show hot electrons and ions streaming out of the magnetosphere, while the low-energy plasma flows with the magnetosheath plasma. Notice that this scenario says little about the detailed interior plasma structure of FTEs.

An alternative to the Russell and Elphic (1978) model is the approach of Lee and Fu (1985) who suggested that multiple x-line reconnection leads to the formation of flux tubes (see Figure 2). The model requires either forced reconnection on several different field lines or the tearing mode instability (Podgorny et al., 1978, 1980; Quest and Coriniti, 1981; Greenly and Sonnerup, 1981) to occur. Lee and Fu (1985) showed that the tearing mode growth rate is large enough to form small magnetic tubes which may coalesce into larger tubes (FTEs). The small tubes may be smaller scale FTEs as suggested by Smith et al. (1986). The distinction between the Russel and Elphic (1978) and Lee and Fu (1985) models remains one of the most important in the study of FTEs.

In this paper we will show some of the latest observations of FTEs from the Active Magnetospheric Particle Tracer Explorer UK Satellite (AMPTE UKS). In describing the data we will highlight the limitations of the observations. It is not the aim, nor would it be possible in a paper of this sort, to make a thorough review of past observations or to provide a complete theoretical background. Interested readers requiring further information are referred to papers by Cowley (1982), Paschmann et al. (1982), Rijnbeek et al. (1984), Saunders et al. (1984b), Sonnerup (1984, 1987), and the papers referenced therein. From the limitations of previous observations a list of desirable characteristics for future instruments will be presented. These characteristics will then be discussed in terms of the ESA Cluster mission.

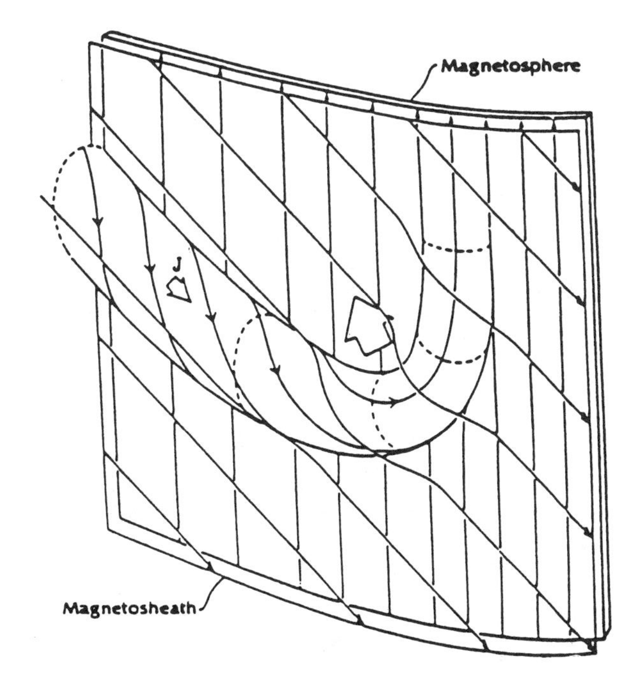

Fig. 1. Russell and Elphic (1978) flux tube model showing draped field lines.

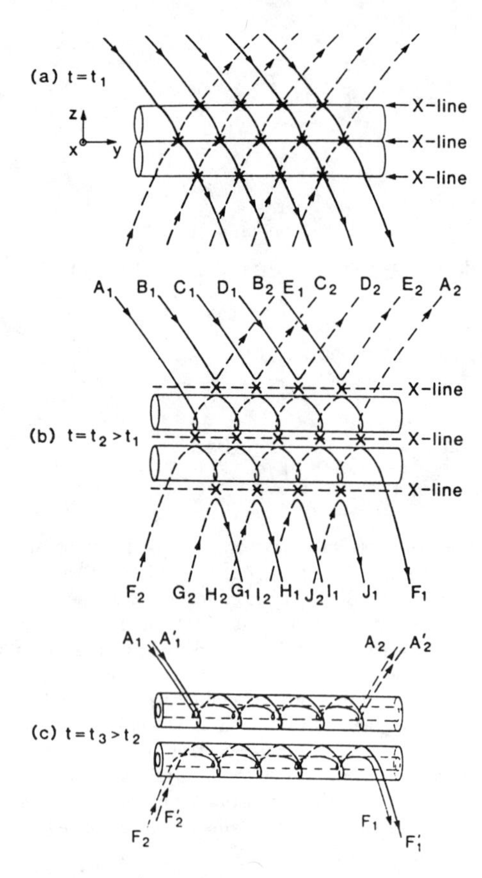

Fig. 2. Lee and Fu (1985) flux tube model. Reconnection occurs along the points marked x (t = t_1). A pair of twisted flux tubes is formed (t = t_2).

Plasma Measurements of FTE Structure

Originally FTEs were defined solely by their bipolar B_N magnetic field signature and much of the work has concentrated on magnetic field measurements. Few detailed plasma observations have been reported. A recent paper on ISEE observations by Thomsen et al. (1987) has provided more detailed observations of the plasma structure. However, the ISEE plasma instruments do not have full three-dimensional coverage with sufficient time resolution to look at the detailed structure of FTEs. Recent AMPTE Ion Release Module (IRM)/UKS observations (e.g., Rijnbeek et al., 1987; Farrugia et al., 1988; Paschmann et al., 1986) have shown that full three-dimensional plasma observations with fast time resolution are required to study the plasma structure in detail. In this section we will concentrate on plasma data from the AMPTE UKS ion (Coates et al., 1985) and electron (Shah et al., 1985) sensors. Both these instruments consist of a pair of 270° electrostatic analyzers. In its usual mode the ion instrument measure ions from 10 eV/q to 20 keV/q with an angular resolution of 45° by 45° measuring the full three-dimensional distribution every spin (≈ 5 sec). The electron sensor measures from 12 eV to 18 keV with an energy sweep made every 1 or 2 sec.

Figures 3 and 4 show a swathe of data taken by the AMPTE UKS magnetometer (Southwood et al., 1984) and ion and electron sensors. This event, October 28, 1984, was the center of an intensive study of magnetospheric FTEs by the AMPTE UKS science team (Farrugia et al., 1988). Here we briefly present the main findings of the study. The third panel of Figure 3 shows the magnetometer B_N component. Clearly this event meets the usual criteria for an FTE of a peak-to-peak amplitude of greater than 10 nT and a duration in excess of 1 min (Rijnbeek et al., 1984). Figure 4 shows energy-time spectrograms for the ion and electron sensors, as well as a frequency-time spectrogram from the UKS wave instrument (Darbyshire et al., 1985). The ion spectrogram is color-coded for count rate while the electron spectrogram is color-coded for intensity. The data have been integrated over all angles for the purposes of this paper. The angular characteristics of these data are discussed in Farrugia et al. (1988). The bipolar magnetometer signature clearly extends beyond that of the plasma signature due to field line draping around the flux tube. Before 1045:00 UT the plasma population is that of the magnetosphere. The spacecraft is outside the reconnected flux tube although within the draping region. A region then appears that is intermediate between magnetospheric and magnetosheath populations. In the Russell and Elphic model (1978) this feature is associated with entry into the reconnected region. This type of distribution is seen until 1045:36 UT, when a region of plasma similar to the magnetosheath occurs. However, this plasma is more isotropic and less dense than the magnetosheath-like plasma seen between 1045:54 UT and 1046:30 UT. This region of pure magnetosheath plasma is the open flux tube region. A similar series of events is seen on the exit of the event except in reverse order. It is important to notice that there is no mixing of magnetosheath and magnetospheric plasma in the open flux tube region at the center of the event. It is likely that the high energy magnetospheric population has already leaked out of the tube. A feature of these data is the layering first reported by Rijnbeek et al. (1987). This feature can be clearly seen in Figure 5, which shows the plasma β. In the draping region the magnetic pressure dominates, and in the wings the plasma pressure dominates. In the layer adjacent to the open region the magnetic field is again dominant, while in the open flux tube itself the plasma pressure is larger. To see this structure clearly, it is essential for the sensors to be fully three-dimensional and have good time resolution. Even with the AMPTE UKS sensors many features are poorly resolved. For example, at the trailing edge of the FTE a flow burst is seen for a single spin of data (1047:00 UT). The burst does not seem to

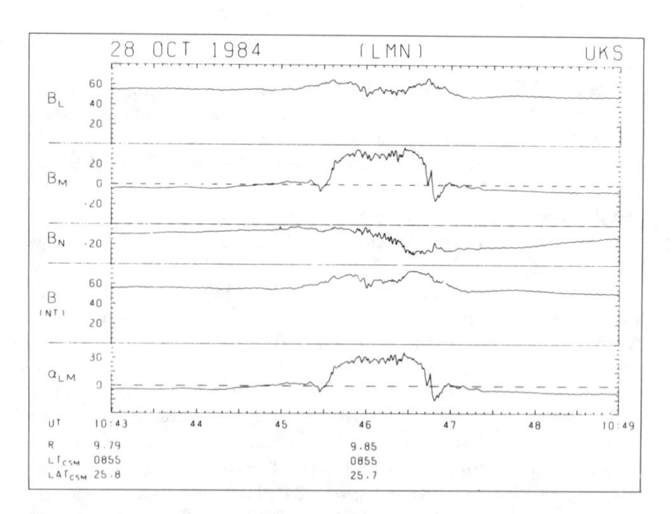

Fig. 3. AMPTE UKS magnetometer data for a magnetospheric FTE on October 28, 1984. The boundary normal coordinate system has been used. The angle, α_{LM}, is the angle in the LM plane.

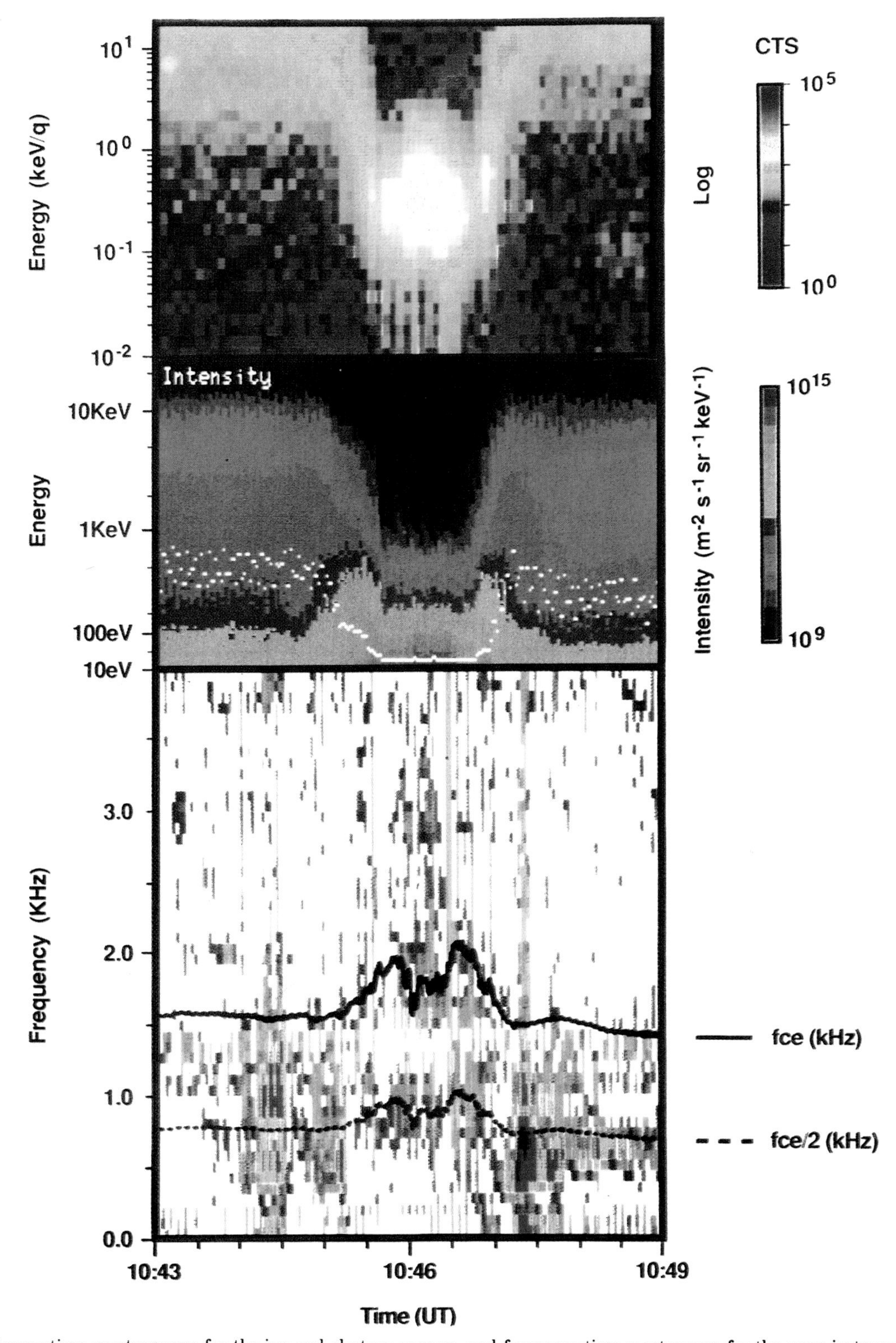

Fig. 4. Energy-time spectrograms for the ion and electron sensors and frequency-time spectrogram for the wave instrument. The top panel is an ion energy-time spectrogram for the October 28, 1984, event, color-coded for count rate. The middle panel is for electrons, color-coded for intensity. The bottom panel is the frequency-time spectrogram from the wave instruments, color-coded for wave power. In both energy spectrograms the data have been summed over the full angular distribution.

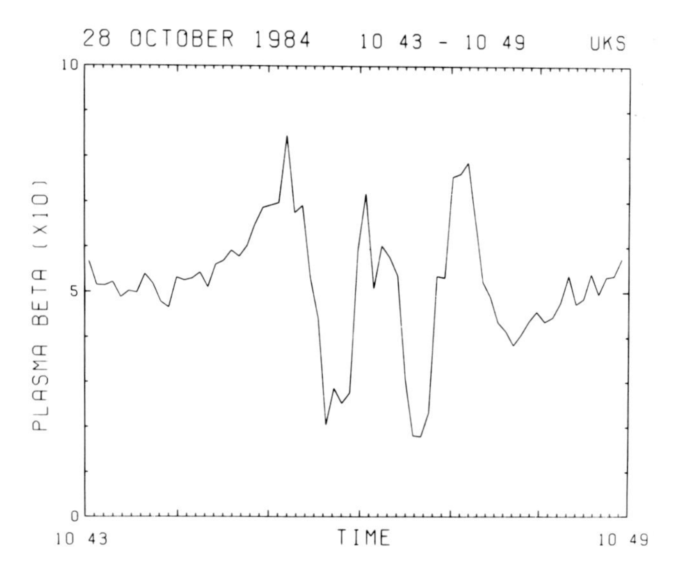

Fig. 5. Plasma β for the October 28, 1984, event.

flow in a direction aligned with any other feature and is possibly the signature of ongoing reconnection. Similar bursts are seen in other events, but with so little data it is impossible to determine much about these phenomena. From this data it can be seen that a requirement of future sensors is that they be fully three-dimensional with good time resolution.

Traditionally FTE identification was done solely by looking at the magnetic signatures. For FTEs with large bipolar excursions and of long duration this method is sufficient. Many possible signatures are seen with smaller bipolar amplitude and of shorter duration than normally associated with FTEs. Whether these are FTEs or not can only be distinguished by using the plasma signature (e.g., Smith et al., 1986). The plasma signature is also the only way to determine whether the open flux tube itself is entered as opposed to the draping region. Thus time resolution in the plasma instruments is important in identifying FTEs. The existence of many small-scale FTEs will have a significant effect on the reconnection rate and hence provide even more of a contribution to the cross tail potential.

Another aspect of FTEs which has attracted attention concerns the Walén relation (Walén, 1944). That is, is there a relationship between the variation in the magnetic field components and the plasma bulk velocity components which implies that the perturbation is a propagating Alfvén wave? As we have seen, an FTE cannot be treated as a region of uniform plasma. Does the Walén relation hold across the whole event? If not, does it hold in a specific region? Observations by Saunders et al. (1984a) suggest that it may indeed hold across the whole event. It is likely that this study looked at the draping field where the Walén relation is expected to hold. Analysis of many AMPTE UKS events by Smith et al. (1987) shows that it generally does not hold across the whole event, nor in the open flux tube region. They found a single event (out of 17) that had the relation hold in the intermediate plasma regime on the edge of the FTE. Figure 6 shows a plot of ΔV against ΔB for the two components perpendicular to the average background field direction. The points should lie on a straight line, with a slope of about 4. This region is where the field line twisting occurs and so it is reasonable to expect that if the Walén relation was to hold anywhere in an FTE, it would be here. As this region is generally small compared to the rest of the event, few data points were obtained in this region (normally less than 4) and it is not clear whether either set of results is definitive. Until plasma instruments with higher time resolution are available it is likely that the question of the Walén relation will still be open.

Recent papers by Saunders et al. (1987, 1988) of high time resolution magnetometer data from AMPTE UKS show data which are consistent with the tearing mode instability at the magnetopause. As we have seen above, this may be a mechanism for producing reconnected flux tubes. Figure 7 shows high time resolution magnetic field data from AMPTE UKS including the angle the magnetic field makes out of the plane of the magnetopause, α. Figure 8 shows the associated ion data from AMPTE UKS along with 5-sec averaged magnetometer data. The top panel shows the ion density while the next four show the boundary normal velocity components and the total velocity V. Before 1419:39 UT and after 1421:35 UT the spacecraft is in the magnetosphere. The region between the two inner dotted lines is the magnetosheath. Tearing-island-like field signatures are observed within the magnetopause layer (marked by MP). This phenomenon is clearly seen in the large values of α that occur during this time. The increase in V at about 1421:00 UT to 1422:30 UT is suggestive of reconnection; however, there are no variations in the plasma parameters associated with the tearing mode. This observation is explained by the fact that the period of the oscillations in the field is of the order of 5 sec, which is roughly the spin

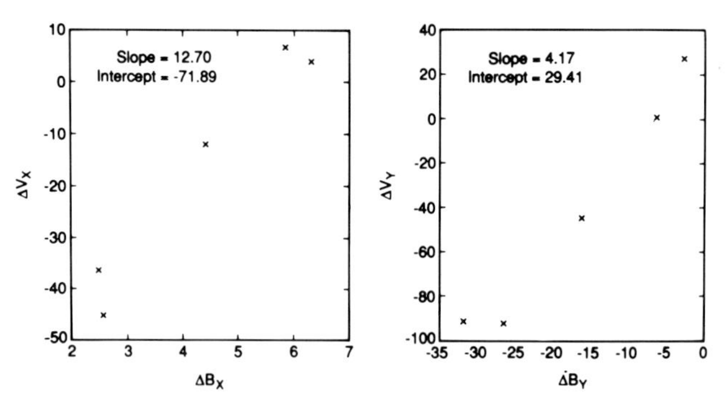

Fig. 6. Plots of Δ V against Δ B for the wings of an FTE. The components are perpendicular to the ambient field direction.

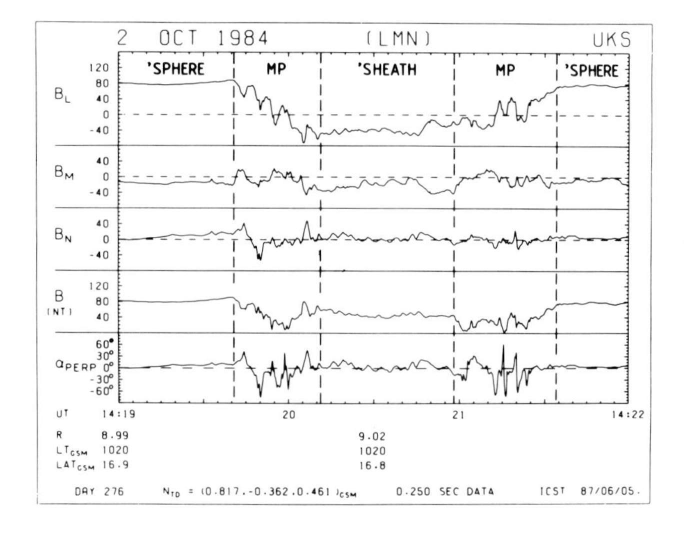

Fig. 7. High resolution magnetometer data for the magnetopause crossings on October 2, 1984. The data are in boundary normal coordinates. The angle, α_{perp}, is the angle of the field out of the plane of the magnetopause.

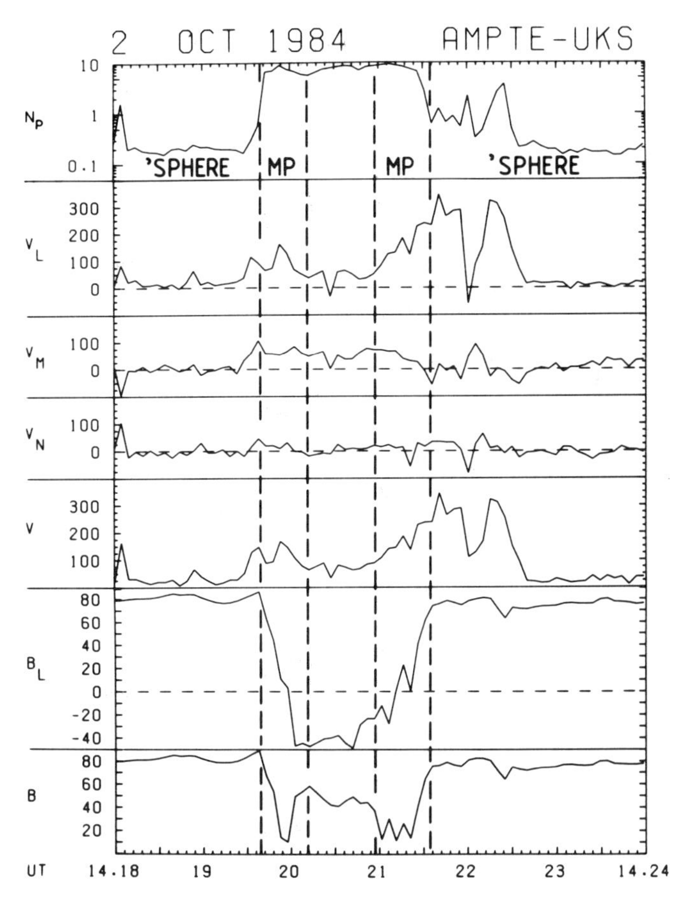

Fig. 8. Ion density and velcoities in boundary normal coordinates for the October 2, 1984, event. The bottom panels show 5-sec averaged magnetometer data.

period. To obtain useful plasma data on this sort of event, i.e., observations of structure in the accelerated flows, thus requires full three-dimensional data within a fraction of a period ($\approx$ 1 sec).

As we have seen temporal resolution is a critical parameter in observing FTEs, however, spatial resolution is also important. What resolution is required to look at the interior of an FTE? This requirement can be estimated by looking at the ratio of the thermal velocity, $V_{thermal}$, to the bulk flow velocity, V_{flow}. The top panel of Figure 9 plots arctan ($V_{thermal}/V_{flow}$) for the duration of the FTE in Figure 3. The minimum angle inside the event is of the order of 45°. So a resolution of half this, 22.5°, should be sufficient to observe the sheath-like plasma. This assumes that there are no features on a scale smaller than that detected by AMPTE UKS. In fact, observations have shown that higher angular resolution is required for some features, e.g., ionospheric ion upwelling in the wake of an FTE (Lockwood et al., 1988).

So far we have concentrated on problems involving a lack of spatial and temporal resolution. Another area in which measurements are needed is that of ion species. The time resolution of mass discriminating instruments flown on spacecraft looking at the magnetopause has been of the order of the typical FTE duration. For example, the AMPTE IRM mass analyzer, SULICIA (Mobius et al., 1985) had a sampling time of 80 sec as compared with a typical FTE duration of 60-120 sec. Although data within FTEs have been obtained, the layered nature of FTEs precludes too much interpretation. A single data sample would cover many of the regions, each of which may have different species present. Good mass data (i.e., M/Δ M of $\approx$ 2) from each of the regions within an FTE would provide information on the sources of plasma making up each of the layers within the FTE.

Up to this point we have been concerned with the instruments individually; however, some measurements require the use of multi-spacecraft missions. For example, to measure gradients, to separate spatial and temporal effects, and to determine the ion flow structure and geometry of an FTE requires the use of more than one spacecraft. The two spacecraft ISEE mission lacked plasma instruments with sufficient time resolution as well as ideal spacecraft separations to make good FTE measurements. An attempt has been made by Smith et al. (1987) to characterize the flows using AMPTE UKS data; however, without multi-point measurements the true three-dimensional nature of FTEs is difficult to ascertain.

Multi-point measurements put constraints on the accuracy of measurements when gradients are to be determined. The third panel of Figure 9 shows the accuracy with which measurements of the ion density are required through the FTE in Figure 3. We assumed two spacecraft 1000 km apart, which is one of the proposed separation distances for the Cluster mission. Using the measured bulk velocity flow and density we derived the difference in percentage of density that would be measured by the two spacecraft. At maximum the difference is 40% dipping to about 2% at the center of the event. So on average the gradients which need to be measured are of the order of 10%. The sensors are required to be accurate within a small fraction of this value, say 1%. This accuracy sets the minimum number of counts required per integration period to be $\approx$ 10,000 to give the required signal-to-noise ratio. An increase in the angular

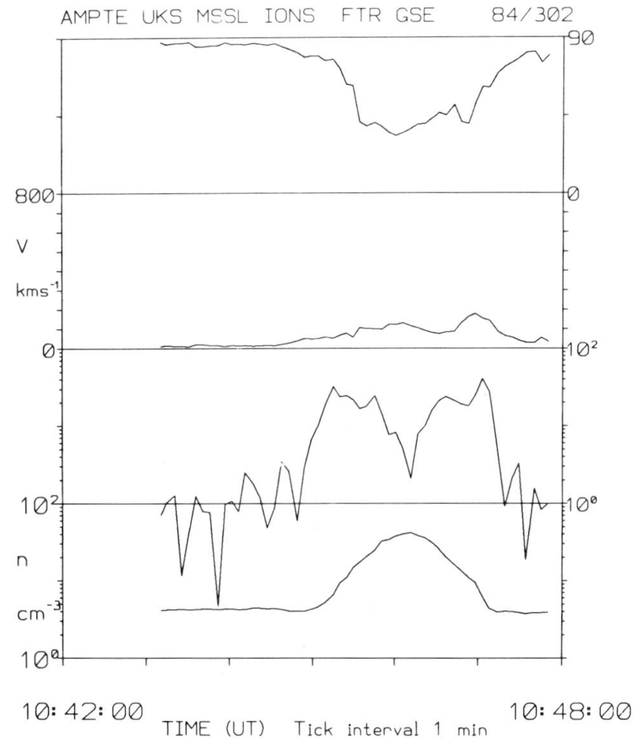

Fig. 9. Angle between $V_{thermal}$ and V_{flow} for the FTE on October 28, 1984 (top panel). The second panel shows the total ion flow velocity, while the bottom panel is the total ion density through the event. The third panel shows the accuracy requirement for density measurements for two spacecraft 1000 km apart.

resolution also has the effect of increasing the signal-to-noise ratio. These requirements are fed back into the choice of geometric factor and angular resolution for future instruments.

Among other areas of interest in FTE research are those of low-energy electron and low-altitude observations. Observations of low-energy electrons (< 10 eV) have not been obtained for FTEs. As the current is thought to be mostly carried by these electrons, observations to determine the current structure of the FTE is important. It is a requirement that these measurements be accurate.

To date no unambiguous low-altitude plasma observations of FTEs have been published. However, some of the Dynamics Explorer (DE) observations (Saflekos et al., 1988; Smith et al., 1988; Winningham and Smith, 1988) may be relevant. Due to the small size of FTEs at low altitudes ($\approx$ 50 km), a sample time of the order of 0.5 sec is required because of the large relative spacecraft velocity ($\approx$ 7 km s^{-1}). Again mass resolution at a high rate would identify the sources of particle populations. As at the magnetopause multi-spacecraft measurements would provide details of the geometry.

The Future

In the above section we have shown the limitations of previous observations of FTEs. Table 1 is a summary of the sensor requirements obtained from the preceding discussion. It can be seen that the requirements are quite difficult to meet. In this section we detail a set of instruments which will meet the requirements for the ESA Cluster mission. This set is not unique and other designs exist. To accommodate the requirements three sensors can be used. The characteristics of these instruments are detailed in Table 2. The low energy electron analyzer (LEEA) and high energy electron analyzer (HEEA) instruments cover electrons while the hot plasma composition analyzer (HPCA) covers positive ions.

The HPCA is designed to make accurate measurements of the major mass species within half a spin. Unlike previous spacecraft instrumentation this sensor combines the role of fast plasma sensor and mass sensor. The instrument consists of a 2π-radian field-of-view toroidal electrostatic analyzer (Young et al., 1988) combined with a time-of-flight (TOF) velocity analyzer of the type developed for the Combined Release and Radiation Effects Spacecraft (CRRES) (Fritz et al., 1985) (see Figure 10). The outer surfaces of the electrostatic analyzer are at ground potential, the inner plate having a negative potential. The analyzer is a mild toroid, which gives energy and azimuthal angle characteristics similar to previously developed spherical section analyzers. An important feature of this design is that the focus of the ion beam is beyond the ends of the

TABLE 1. Details of Sensor Requirements for a Cluster Mission

Time resolution (full three-dimensional distribution)	1 sec
Angular resolution	22.5°
M/Δ M	2
Accuracy	1%
Geometric factor	> 5 x 10^{-4} cm^2 sr eV /eV

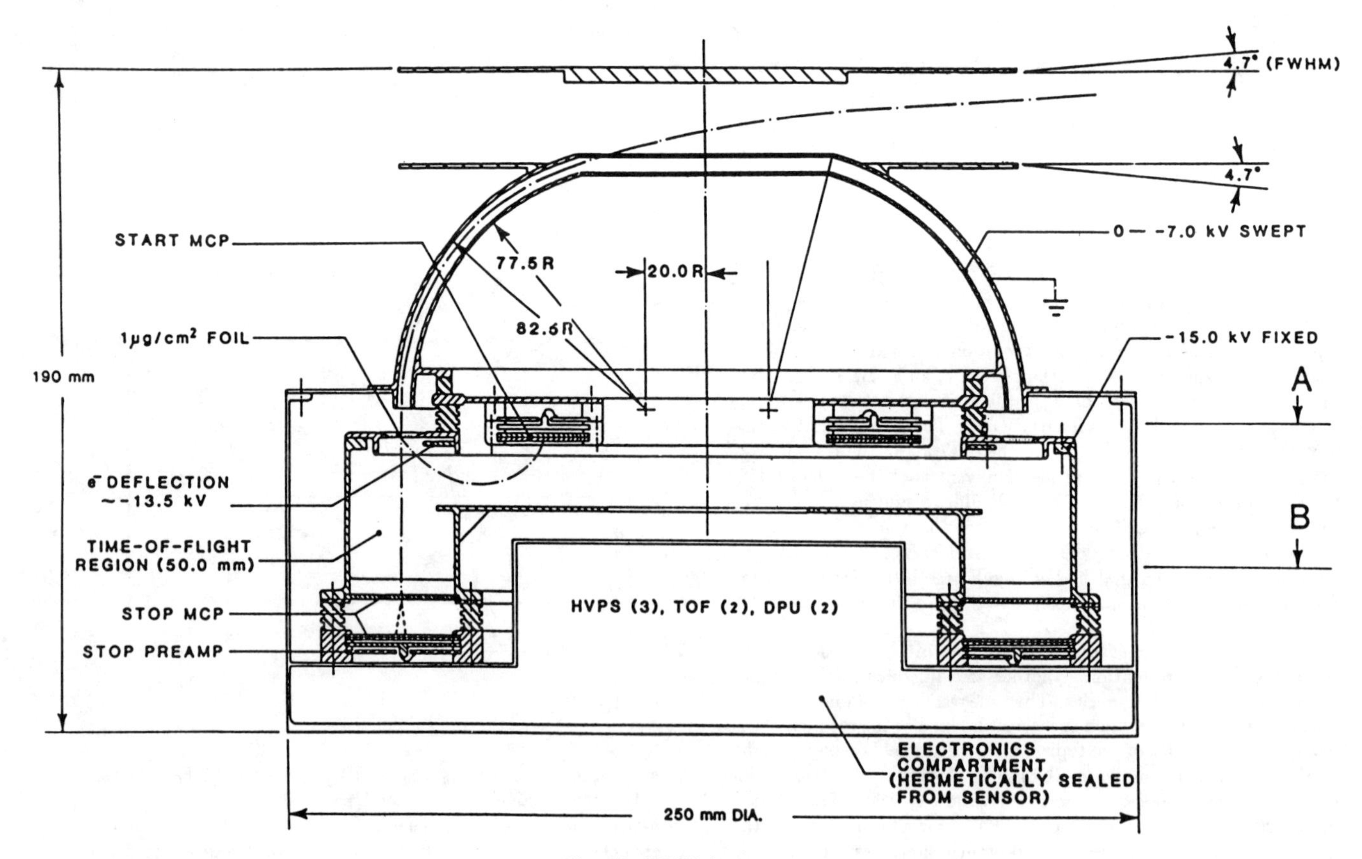

Fig. 10. HPCA instrument.

TABLE 2. Instrument Characteristics

	LEEA	HEEA	HPCA
Energy range	0-500 eV	100-30,000 eV	10-55,000 eV
Energy resolution	0.10	0.07	0.04
FOV	$180^o \times 3.6^o$	$360^o \times 4.6^o$	$360^o \times 5^o$
Angular resolution	5°	10°	2.8^o
Mass range	—	—	1-40 amu
Geometric factor	2.25×10^{-4}	1.1×10^{-3}	3.4×10^{-4}

electrostatic plates. This feature allows for optimum transmission into the TOF optics. On exiting the electrostatic analyzer the ions are accelerated to strike the carbon foil. The secondary electrons from the collision are then focussed by a potential onto one of the 16 start micro channel plate (MCP) detectors. The ions continue to the rear MCP detector producing a stop pulse in one of the 22.5° sectors. The design of the sensor is such as to avoid ghost peaks. It does this by eliminating spurious start electrons, allowing for field-free drift of the ions, and by reducing scattering off the internal walls of the analyzer.

The advantage of this type of analyzer is its ability to image all mass values at one time without the need for scanning. The sensor allows for flexibility in designing modes of operation, enabling either accurate or quick-look mass analysis to be undertaken. The data can also be integrated for ion species with low fluxes. In addition Young et al. (1988) have shown that although a spherical top-hat analyzer can be built with comparable resolution to that of HPCA, it would be of the order of 20% heavier. The toroidal analyzer thus has a considerable advantage in space-borne applications.

The LEEA instrument (Figure 11) is a conventional spherical analyzer mounted radially to reduce the incoming fluxes of photoelectrons and secondary electrons. This instrument thus obtains one distribution per spin. HEEA (Figure 11) has the same geometry as HPCA, although scaled down, to facilitate easier comparison between the two instruments. This sensor is mounted conventionally. Due to the large variation in fluxes that will be encountered, the HEEA instrument has a geometric factor larger by about a factor of 5. For LEEA to be successful requires the spacecraft potential to be accurately controlled. Additionally, the cleanliness requirements will be strict. Indeed the sensors themselves are designed in such a way as to reduce secondary electron emission from inside the instrument itself.

To make full use of these sensors all the spacecraft of the Cluster mission along with their instrumentation should be identical to enable accurate multi-point observations to be undertaken. This approach in turn will allow for the accurate calculation of gradients. An additional point is that any onboard processing of the data must preserve the accuracies required for the science, which is particularly difficult unless high telemetry rates are available. Thus the data processing unit required will be particularly complex.

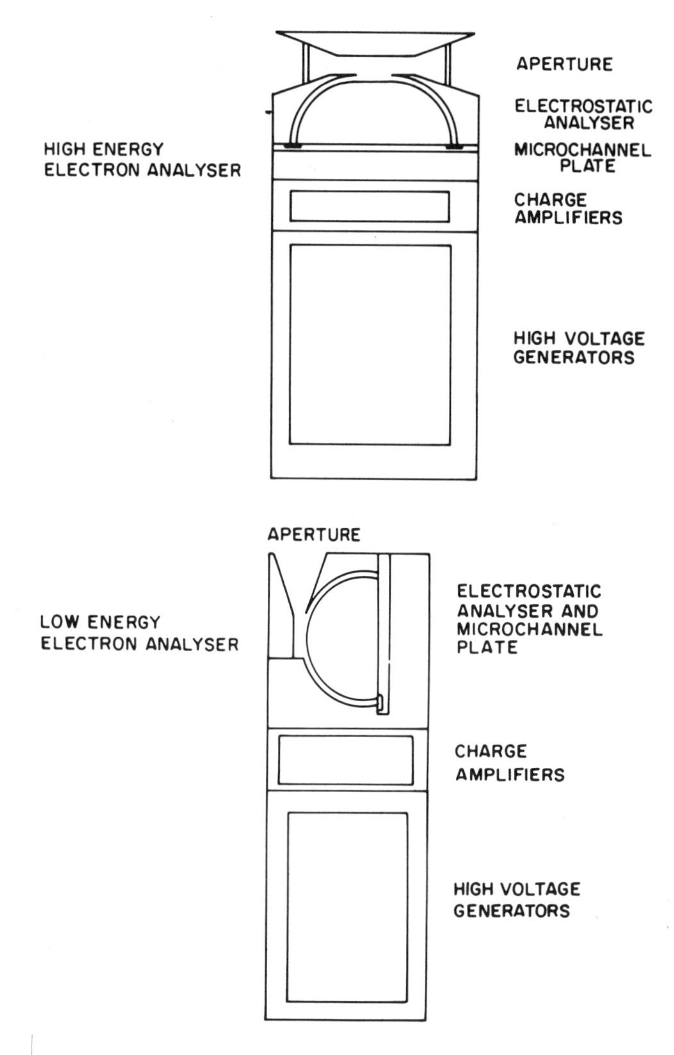

Fig. 11. HEEA (top) and LEEA (bottom) instruments.

Conclusions

Undoubtedly FTEs remain a priority in the understanding of solar wind-magnetospheric coupling. Much has been learned but as we have shown many problems still remain. Indeed we have not tried to cover all topics. For example: What is the scale size for plasma entry? What processes initiate FTE formation? How do they develop? We have shown examples where next-generation plasma instruments will provide useful data that will help in solving many of these outstanding problems. As an example, detailed plasma measurements will be able to distinguish between tearing mode and transient Petschek reconnection.

It is to be hoped that the instrumentation on the ESA Cluster mission will be of high enough quality to answer some of these questions. We have shown instrumentation that will meet the standards required. It is important that the instrumentation chosen for the Cluster mission (including the data processing unit) work as an integral unit to maximize the science return. In particular the electron and ion sensors should provide compatible data over the required parameter range. Beyond Cluster the outlook for new opportunities to study FTEs is bleak. One of the priorities for a new mission would be that of low-altitude/high-latitude measurements. To do this would require a multi-spacecraft mission with fast, three-dimensional, mass resolving plasma sensors.

In conclusion the study of FTEs has proven very fruitful. With the correctly instrumented spacecraft the Cluster mission may prove more so.

Acknowledgments. The instrumentation described is based on the Plasma Energy Angle and Composition Experiment (PEACE) proposal for the ESA Cluster mission. We thank the PEACE team. Our thanks also to Dave Hall and Mark Saunders for generously supplying their data; to the referee, Rick Elphic, for useful comments; and to Jill Johnson for helping prepare the final copy.

References

Coates, A. J., J. A. Bowles, R. A. Gowen, B. K. Hancock, A. D. Johnstone, and S. J. Kellock, The AMPTE UKS Three-dimensional ion experiment, IEEE Trans. Geosci. Remote Sensing, GE-23, pp. 287-292, 1985.

Cowley, S. W. H., The causes of convection in the Earth's magnetosphere: A review of developments during the IMS, Rev. Geophys. Space Phys., 20, pp. 531-565, 1982.

Daly, P. W., M. A. Saunders, R. P. Rijnbeek, N. Sckopke, and C. T. Russell, The distribution of reconnection geometry in flux transfer events using ion, plasma and magnetic data, J. Geophys. Res., 89, pp. 3843-3854, 1984.

Darbyshire, A. G., E. J. Gershuny, S. R. Jones, A. J. Morris, J. A. Thompson, G. A. Whitehurst, G. A. Wilson, and L. J. C. Woolliscroft, The UKS wave experiment, IEEE Trans. Geosci. Remote Sensing., GE-23, pp. 311-314, 1985.

Elphic, R. C., and C. T. Russell, ISEE-1 and -2 magnetometer observations of the magnetopause, in Magnetospheric Boundary Layers, Spec. Publ. ESA SP-148, edited by B. Battrick, pp. 43-50, European Space Agency, Noordwijk, The Netherlands, 1979.

Farrugia, C. J., R. C. Elphic, D. J. Southwood, and S. W. H. Cowley, Field and flow perturbations outside the reconnected field line region in flux transfer events: Theory, Planet. Space Sci., 35, pp. 227-240, 1987.

Farrugia, C. J., R. P. Rijnbeek, M. A. Saunders, D. J. Southwood, D. J. Rodgers, M. F. Smith, C. P. Chaloner, D. S. Hall, P. J. Christiansen, and L. J. C. Woolliscroft, A multi-instrument study of flux transfer event structure, J. Geophys. Res., in press, 1988.

Fritz, T. A., D. T. Young, W. C. Feldman, S. J. Bame, J. R. Cessna, D. N. Baker, B. Wilken, W. Studemann, P. Winterhoff, D. A. Bryant, D. S. Hall, J. F. Fennell, D. Chenette, N. Kartz, S. I. Imamoto, R. Koga, and F. Soreass, The mass composition instruments (AFGL-701-11), in CRESS/SPACERAD Experiment Descriptions, edited by M. S. Gussenhoven, E. G. Mullen, and R. C. Sagalyn, AFGL-TR-85-0017, 1985.

Greenly J. B., and B. U. O. Sonnerup, Tearing modes at the magnetopause, J. Geophys. Res., 86, pp. 1305-1312, 1981.

Haerendel G., G. Paschmann, N. Sckopke, H. Rosenbauer, and P. C. Hedgecock, The frontside boundary layer of the magnetosphere and the problem of reconnection, J. Geophys. Res., 83, pp. 3195-3216, 1978.

Lee, L. C., and Z. F. Fu, A theory of magnetic flux transfer at the Earth's magnetopause, Geophys. Res. Lett., 12, pp. 105-108, 1985.

Lockwood, M., M. F. Smith, C. J. Farrugia, and G. L. Siscoe, Ionospheric ion upwelling in the wake of Flux Transfer Events at the dayside Magnetopause, J. Geophys. Res., 93, pp. 5641-5654, 1988.

Mobius, E., G. Gloeckler, D. Hovestadt, F. M. Ipavich, B. Kloecker, M. Scholer, H. Arbinger, H. Hofner, E. Kunneth, P. Laeverenz, A. Luhn, E. O. Tums, and H. Waldleben, The time of flight spectrometer SULEICA for ions of the energy range 5-270 keV/charge on AMPTE IRM, IEEE Trans Geosci. Remote Sensing, GE-23, pp. 274-279, 1985.

Paschmann, G., G. Haerendel, I. Papamastorakis, N. Sckopke, S. J. Bame, J. T. Gosling, and C. T. Russell, Plasma and magnetic field characteristics of magnetic flux transfer events, J. Geophys. Res., 87, pp. 2159-2168, 1982.

Paschmann, G., I. Papamastorakis, W. Baumjohann, N. Sckopke, C. W. Carlson, B. U. O. Sonnerup, and H. Luhr, The magnetopause for large magnetic shear: AMPTE/IRM observations, J. Geophys. Res., 91, pp. 11,099-11,115, 1986.

Podgorny, I. M., E. M. Dubinin, and Yu. N. Potanin, The magnetic field on the magnetospheric boundary from laboratory simulation data, Geophys. Res. Lett., 5, pp. 207-210, 1978.

Podgorny, I. M., E. M. Dubinin, and Yu. N. Potanin, On the magnetic curl in front of the magnetosphere boundary, Geophys. Res. Lett., 7, pp. 247-250, 1980.

Quest, K. B., and F. V. Coriniti, Tearing at the dayside magnetopuse, J. Geophys. Res., 86, pp. 3,289-3,299, 1981.

Rijnbeek, R. P., S. W. H. Cowley, D. J. Southwood, and C. T. Russell, A survey of dayside flux transfer events observed by the ISEE 1 and 2 magnetometers, J. Geophys. Res., 89, pp. 786-800, 1984.

Rijnbeek, R. P., C. J. Farrugia, D. J. Southwood, M. W. Dunlop, W. A. C. Mier-Jedrzejowicz, C. P. Chaloner, D. S. Hall, and M. F. Smith, A magnetic boundary signature within flux transfer events, Planet. Space Sci., 35, pp. 871-878, 1987.

Russell, C. T., Reconnection at the Earth's magnetopause: Magnetic field observations and flux transfer events, in Magnetic Reconnection in Space and Laboratory Plasmas, Geophys. Monogr. Ser., vol. 30, edited by E. W. Hones, Jr., pp. 124-138, AGU, Washington, DC, 1984.

Russell, C. T., and R. C. Elphic, Initial ISEE magnetometer results: Magnetopause observations, Space Sci. Rev., 22, pp. 681-715, 1978.

Saflekos, N. A., J. L. Burch, M. Sugiura, D. A. Gurnett, and J. L. Horwitz, Observations of reconnected flux tubes within the mid-altitude cusp, J. Geophys. Res., submitted, 1988.

Saunders, M. A., C. T. Russell, and N. Sckopke, Flux transfer events: Scale size and interior structure, Geophys. Res. Lett., 11, pp. 131-134, 1984a.

Saunders, M. A., C. T. Russell, and N. Sckopke, A dual-satellite study of the spatial properties of FTEs, in Magnetic Reconnection in Space and Laboratory Plasmas, Geophys. Monogr. Ser., vol. 30, edited by E. W. Hones, Jr., pp. 145-152, AGU, Washington, DC, 1984b.

Saunders, M. A., C. J. Farrugia, M. W. Dunlop, W. A. C. Mier-Jedrzejowicz, M. F. Smith, and D. J. Rodgers, An unusual magnetopause encounter by AMPTE UKS, in Proc. 21st ESLAB Symposium, Norway, Spec. Publ. ESA SP-275, edited by B. Battrick, pp. 153-158, European Space Agency, Noordwijk, The Netherlands, 1987.

Saunders, M. A., C. J. Farrugia, M. W. Dunlop, M. F. Smith, and D. J. Rodgers, Observations of magnetopause fine structure by AMPTE UKS, J. Geophys. Res., submitted, 1988.

Shah, H. M., D. S. Hall, and C. P. Chaloner, The electron experiment on AMPTE UKS, IEEE Trans. Geosci. Remote Sensing, GE-23, pp. 293-299, 1985.

Smith, M. F. , D. J. Rodgers, R.P. Rijnbeek, D. J. Southwood, A. J. Coates, and A. D. Johnstone, Plasma and field observations with high time resolution in flux transfer events, in Solar Wind-Magnetospheric Coupling, edited by Y. Kamide and J.A. Slavin, pp. 321-329, Terra Scientific Publ. Co., Japan, 1986.

Smith, M. F., D. J. Rodgers, and M. A. Saunders, Ion flows in magnetospheric flux transfer events, in Proc. 21st ESLAB Symposium, Norway, Spec. Publ. ESA SP-275, edited by B. Battrick, pp. 153-158, European Space Agency, Noordwijk, The Netherlands, 1987.

Smith, M. F., J. D. Winningham, R. Heelis, and J. Slavin, Possible evidence for FTE signatures at ionospheric altitudes, J. Geophys. Res., to be submitted, 1988.

Sonnerup, B. U. O., Magnetic field reconnection at the magnetopause: An overview, in Magnetic Reconnection in Space and Laboratory Plasmas, Geophys. Monogr. Ser., vol. 30, edited by E. W. Hones, Jr., pp. 92-103, AGU, Washington, DC, 1984.

Sonnerup, B. U. O, On the stress balance in flux transfer events, J. Geophys. Res., 92, pp. 8,613-8,620, 1987.

Southwood, D. J., W. A. C. Mier-Jedrzejowicz, and C. T. Russell, The fluxgate magnetometer for the AMPTE UKS subsatellite, IEEE Trans. Geosci. Remote Sensing, GE-23, pp. 301-304, 1984.

Thomsen, M. F., J. A. Stansberry, S. J. Bame, S. A. Fuselier, and J. T. Gosling, Ion and electron velocity distributions within flux transfer events, J. Geophys. Res., 92, pp. 12,127-12,136, 1987.

Walén, C., On the theory of sunspots, Ark. Mat. Astron. Fys., 30A, pp. 1-87, 1944.

Winningham, D. J., and M. F. Smith, Low altitude observations of FTE's, in Physics of Space Plasmas, edited by C. T. Chang, submitted, 1988.

Young, D. T., S. J. Bame, M. F. Thomsen, R. H. Martin, J. L. Burch, J. A. Marshall, and B. Reinhard, A 2π field-of-view toroidal electrostatic analyser, Rev. Sci. Inst., submitted, 1988.

ION SPECTROMETERS FOR STUDYING THE INTERACTION OF THE SOLAR WIND WITH NON-MAGNETIC BODIES

Marcia Neugebauer

Jet Propulsion Laboratory, California Institute of Technology, Pasadena, CA 91109

Abstract. A good ion spectrometer for studying the interaction of the solar wind with non-magnetic bodies would have sufficient dynamic range to measure both ionospheric plasmas with temperatures of 300 K and accelerated pickup ions with energies of ≥300 keV, masses between 1 and ≥100 AMU, and charges of -1, +1, or +2; it would also have mass resolution sufficient to separate He^{++} from H_2^+ and N_2^+ from CO^+, a 4π ster field- of- view, and time resolution ≈ 0.2 s. Such an instrument does not now exist and probably never will. A combination of instruments is required to come even close to meeting these requirements. This paper is a survey of some of the techniques which have been used, which are being designed for future missions, or which might be tried to improve measurements of fast ions.

Requirements

The design of an ion spectrometer for studying the interaction of the solar wind with the atmosphere or ionosphere of a non-magnetic body is perhaps one of the greatest challenges in space plasma instrumentation. The principal problem is the large range of ion masses and distribution functions, varying from a dense, nearly stationary ionosphere, which may be very cold, to a tenuous distribution of energetic pickup ions in the upstream solar wind. Another major problem is the need to make complete measurements in as many as five dimensions (three dimensions in velocity space plus mass and charge) in a time less than the time over which there are significant changes in the plasma distribution functions.

Table 1 lists the ranges and resolutions required for ion observations in the vicinity of a short-period comet. Some of the entries require explanation:

The upper limit of ≥100 AMU is based on the observation of a series of mass peaks, shown in Figure 1, observed by the positive ion cluster composition analyzer (PICCA) on the Giotto mission to Comet Halley (Korth et al., 1986).

There are several factors that determine the mass resolution desired. First, one would like to obtain sufficient information to allow chemical modeling of the coma. At a minimum, that requires separation of the two most abundant ions in the inner coma, H_2O^+ and H_3O^+. As indicated by Figure 2, the identity of the ion species in the heavy ion peaks in Figure 1 remains ambiguous because the PICCA instrument was not a mass spectrometer; it measured energy/charge spectra, which can be unambiguously interpreted in terms of mass/charge only when the plasma is very cold and with all species flowing at the same velocity, so that peaks separated by 1 AMU/charge can be resolved. Huebner (1987) and Huebner and Boice (1989) have used the approximately 14 AMU/charge spacing of the mass peaks observed during some portions of the flyby to suggest that comets contain large amounts of polyoxymethylene; other interpretations of the PICCA data are certainly possible.

There are compelling reasons why we would like to separate different species which have approximately the same atomic or molecular weight. For example, chemical models would benefit from separation of CO^+ and N_2^+, which would require a mass resolution $m/\Delta m \geq 2800$. Mapping the abundance of solar wind plasma in the coma and understanding charge exchange processes between the solar wind and coma gases requires separation of He^{++} from H_2^+, which corresponds to a mass resolution ≥145. The high energy range spectrometer (HERS) of the Giotto ion mass spectrometer (IMS) was just barely able to distinguish these two species. Measurement of the isotopic composition of cometary carbon would require a mass resolution ≥2900 for the separation of $^{13}C^+$ from $^{12}CH^+$.

Table 2 summarizes the mass resolution requirements discussed above. In most fast ion mass spectrometer designs, mass resolution is a function of mass, with $m/\Delta m = 145$ at $m = 2$ AMU being much easier to achieve than a value of $m/\Delta m = 90$ at $m = 90$ AMU.

No measurements of ion charge have ever been made at a comet or in the region of interaction of the solar wind with other non-magnetic bodies. In interpreting plasma data, it has been assumed that all ions, except for solar wind helium, are positively and singly charged. This assumption should be experimentally verified, especially in light of the large flux

TABLE 1. Requirements for Ion Measurements at a Comet.

Parameter	Range	Resolution
m	1 to ≥100 AMU	See Table 2
q	-1, +1, +2	$\Delta q = 1$
n	10^5 to £10^{-4} cm^{-3}	$\Delta n/n \approx 0.1$
v	0.5 to 1500 km s^{-1}	$\Delta v/v \approx 0.1$
T	300 K to ≥300 keV	$\Delta T/T \approx 0.1$
θ, ϕ	4π ster	$\Delta\theta$, $\Delta\phi \approx 2°$
t	--	$\Delta t \approx 5$ s

of energetic particles observed in the vicinity of comets.

The ranges of densities, speeds, and temperatures listed in Table 1 all correspond to values detected in some part of the coma of Comet Halley. It is important to realize that the plasma can arrive from any direction. Far from the nucleus, the flow is generally supersonic and directed roughly antisunward. Close to the nucleus, however, the flow is outward from the nucleus. At intermediate ranges, the flow, especially that of picked-up cometary ions, can be subsonic, leading to anisotropic fluxes from almost every direction. The angular resolution of 2° given in Table 1 is predicated on a requirement to analyze plasma motions associated with upstream hydromagnetic waves.

The time resolution of ion measurements should be sufficient to resolve the internal structure of important surfaces in the interaction region. Neugebauer et al. (1987) found the thickness of the layer in which protons were thermalized at the Halley bow shock to be approximately 1500 km. Cravens (1988) has suggested that the Halley contact surface may have contained an "inner contact layer" of enhanced density whose thickness is predicted to be on the order of 100 km. Table 3 shows that the time required for such

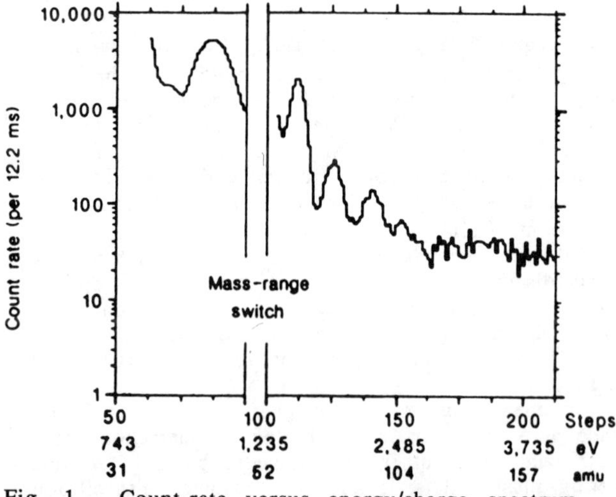

Fig. 1. Count-rate versus energy/charge spectrum measured at a distance of ~11,000 km from the nucleus of Comet Halley by the PICCA instrument (Korth et al., 1986).

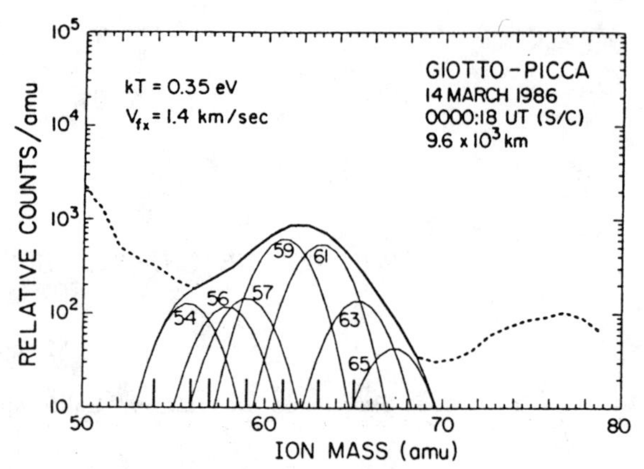

Fig. 2. Preliminary fit to one of the PICCA mass peaks (Mitchell et al., 1986).

structures to pass a spacecraft at the local Alfvén speed (assuming the spacecraft is essentially stationary relative to the comet) is of the order of 30 to 50 s. Therefore, time resolution of ~5 s should suffice for the study of these features. Time resolution of ~5 s would also allow study of the plasma component of the 10^{-2} Hz turbulence observed by the International Cometary Explorer (ICE) in the magnetic field at Comet Giacobini-Zinner (Tsurutani and Smith, 1986).

With some exceptions, parameter ranges and resolutions similar to those listed in Table 1 would be required for measurements near Venus, Mars, or Titan. The ion-neutral chemistry and the presence of very heavy ions is probably of less interest at those bodies than at comets, so the mass range could be considerably smaller. On the other hand, the time resolution requirement would be more stringent, principally because of the greater spacecraft speed relative to the body. For example, resolution of the plasma structure of a typical 20-km-wide magnetic flux rope (Elphic et al., 1980) in the ionosphere of Venus, from a spacecraft orbiting the planet at ~10 km s^{-1}, would require time resolution on the order of 0.2 s.

Instrumentation

The ranges of parameters can be daunting to an instrument designer, even though it is not necessary to plan for all possible combinations of all parameters. One can take advantage of an expected correlation between the highest masses, high densities, low temperatures, and low velocities, and the converse grouping of lower masses, low densities, high speeds,

TABLE 2. Mass Resolution Required for Separation of Ion Pairs

Ion Pair	$\Delta(m/q)$, AMU/chg	$m/\Delta m$
H_2O^+, H_3O^+	1	19
m/q = 89, 90	1	90
He^{++}, H_2^+	0.014	145
CO^+, N_2^+	0.010	2800
$^{13}C^+$, $^{12}CH^+$	0.0045	2900

TABLE 3. Typical Dimensions and Time Structure of Thin Cometary Features

Parameter	Proton Subshock	Inner Contact Layer
Thickness, km	1500	~100
Ion density, cm^{-3}	6	1500
Magnetic field, nT	8	20
Alfvén speed, km s^{-1}	50	2
Time, s	30	50

and high temperatures. Even with such a limitation to only part of the four-dimensional m-n-T-v space, an array of different types of sensors is required, with particle energy being the determining factor. Spacecraft to study atmospheric interactions of the solar wind generally carry one instrument for thermal and near thermal plasmas, another for super-thermal or fast ions, and a third for energetic particles. This paper concerns only the instrumentation for the middle of these three groups, the fast ions with energies in the range from ~100 eV to tens of keV. Fast ion instruments designed for studying the interaction of the solar wind with non-magnetic bodies have much in common with instruments optimized for magnetospheric observations in the same energy range, but with greater emphasis on mass range and mass resolution.

Almost all instruments for the analysis of fast ions use one of or some combination of the following four types of measurements:

(1) Response to an electric field, which gives a measure of the ion's kinetic energy per unit charge $E/q = mv^2/2q$.

(2) Response to a magnetic field, which gives a measure of the ion's rigidity $\rho = mv/q$.

(3) Time of flight between two points, which gives a measure of the ion's velocity = v.

(4) The response of a solid state detector, which gives a measure of the ion's kinetic energy = $mv^2/2$.

This list is not exhaustive. For example, some instruments have used either the secondary electron emission yield or the ratio of a current measurement to a particle counting rate to distinguish between protons and alpha particles. The four types of measurements listed above are the most frequently used parameters, however, and are the only ones discussed in the remainder of this paper.

Table 4 summarizes the different types of fast ion spectrometers developed to date. The list is in ap-

TABLE 4. Summary of Different Types of Fast Ion Spectrometers

Example	Filter	Image	Disperse	Measure	Detector Dim.	Detector Measure	Result	Ref.
Spherical CPA	ϕ, E/q	θ			1	θ	E/q, ϕ, θ	1, 2
Faraday cup	$(E/q)\cos^2\theta$	ϕ, θ			2	ϕ, θ	E/q, ϕ, θ	3
SPI	ϕ	θ	E/q		2	E/q, θ	E/q, ϕ, θ	
CPA + velocity filter	ϕ, θ, E/q, v				0		m/q, v, ϕ, (θ)	4, 5, 6
AMPTE/CHEM	ϕ, θ, E/q			v, E	0		m, q, v, ϕ, (θ)	7
Giotto/JPA/IIS	ϕ, θ, E/q			v	0		m/q, v, ϕ, θ	8
Giotto/NMS/M	ϕ, θ		E/q, ρ		1	m/q	m/q, v, (ϕ), (θ)	9
Giotto/IMS/HIS	ϕ, θ, E/q		ρ		1	m/q	m/q, v, ϕ, (θ)	10
Giotto/IMS/HERS	ϕ, $\rho\cos\theta$	$\theta/2$	$(E/q)\cos^2\theta$		2	m/q, θ	m/q, v, ϕ, θ	10
Galileo/PLS	ϕ, E/q, ρ	θ			1	θ	m/q, v, ϕ, θ	11
SOHO proposal	ϕ, θ			m/q	1	E/q	m/q, (ϕ), (θ)	12
Cluster proposal	θ, E/q	ϕ		v	1	ϕ	m/q, v, ϕ, θ	13

Nomenclature

- m = mass
- q = charge
- ϕ = azimuth angle
- θ = elevation angle
- v = speed
- E = $mv^2/2$
- ρ = mv/q
- ( ) = single fixed value
- CPA = curved plate analyzer

References

1. Bame et al., 1978a
2. Schwenn et al., 1975
3. Bridge et al., 1977
4. Balsiger et al., 1976
5. Coplan et al., 1978
6. Shelley et al., 1985
7. Gloeckler et al., 1983; 1985
8. Johnstone et al., 1986
9. Krankowsky et al., 1986a
10. Balsiger et al., 1986
11. Yeates et al., 1985
12. Gloeckler, 1988
13. Young et al., 1989

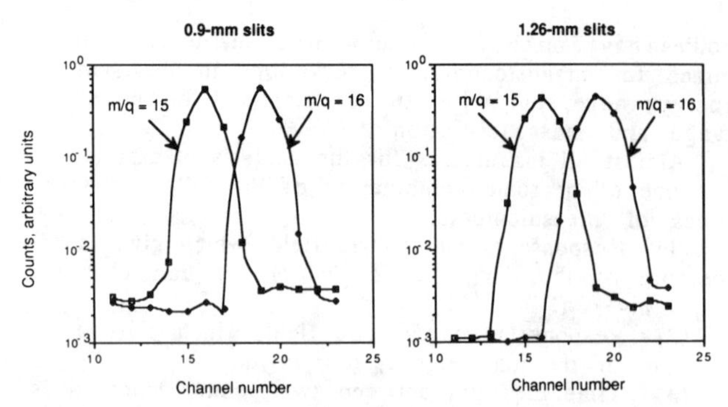

Fig. 3. Mass/charge calibration spectra for CH_3^+ and CH_4^+ ions for two different slit widths of the Giotto/IMS/HERS. Note that in both cases, the peak channel of each species is not significantly contaminated by the neighboring species, but that contamination of the next-to-peak channel is significantly increased for the wider slit width.

proximate chronological order, although the purely electrostatic spectrographic particle imager (SPI) (third in the list), made possible by the fairly recent development of two-dimensional detector arrays, is given early in the list with the other electrostatic analyzers.

In Table 4, the symbols ϕ and θ denote azimuth and elevation angles, respectively. An angle is considered to be a measure of azimuth if (1) it is the angle mapped out by spacecraft rotation, or (2), for a three-axis stabilized spacecraft, it corresponds to the angle defined by the cylindrical symmetry of the instrument.

The second column of Table 4 gives those parameters which are filtered by the instrument. By filtering, I mean that the arrangement of slits, electrodes, or whatever, is designed so that only those ions within a limited range of the filtered parameter are able to reach the detector. Mapping different regions of parameter space usually requires changing the filter. This can be done in several ways. The angular field of view can be varied by rotating the spacecraft, by having a set of several analyzers which look in different directions, and/or by electrostatic deflection of the ions before they reach the angular filter. Different ranges of E/q can be sampled by changing the voltages on electrodes within the filter. Some filters cannot be changed; the magnetic filter on the Giotto/IMS/HERS is an example. That instrument accelerated or decelerated ions by different amounts until they had the proper rigidity to pass through the fixed magnetic field. One of the problems with filtering is the amount of time required to vary the filter, or to vary other parameters so the ions can get through the filter. Each filtered parameter must be varied sequentially until all the desired combinations have been stepped through. The curved plate analyzer (CPA) combined with a velocity filter (crossed **E** and **B** fields) is an extreme example because each measurement cycle requires the sequential

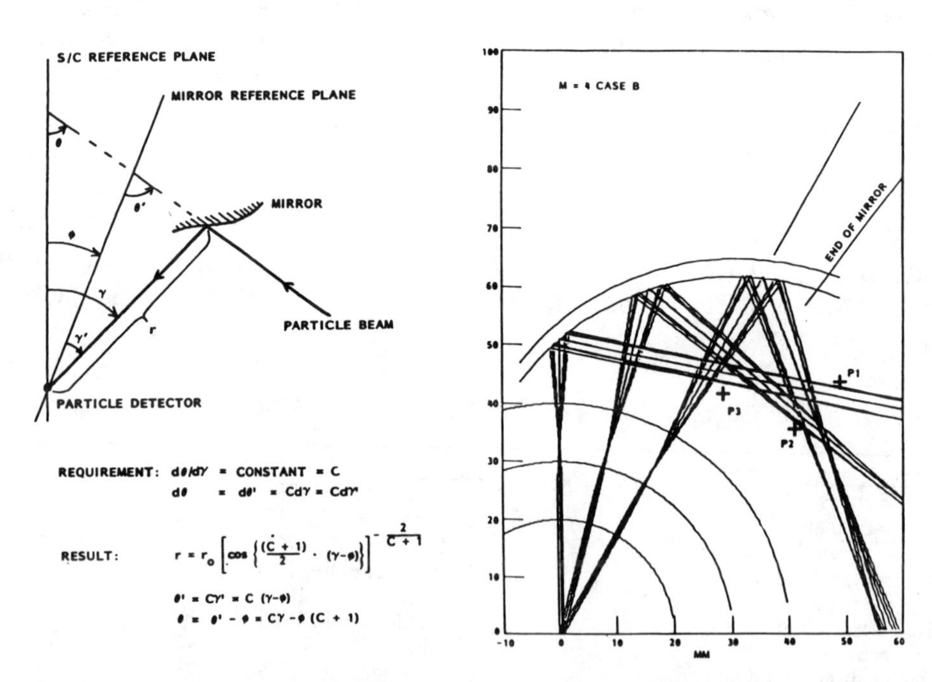

Fig. 4. (Left) Derivation of an algorithm for an ideal electrostatic mirror designed to give constant magnification for all angles of incidence. (Right) Ray tracing of ion reflection for an electrostatic mirror with a magnification factor of -2; although not indicated by this figure, finite penetration of the ions into the region between the two curved grids has been taken into account (Shelley and Simpson, 1985).

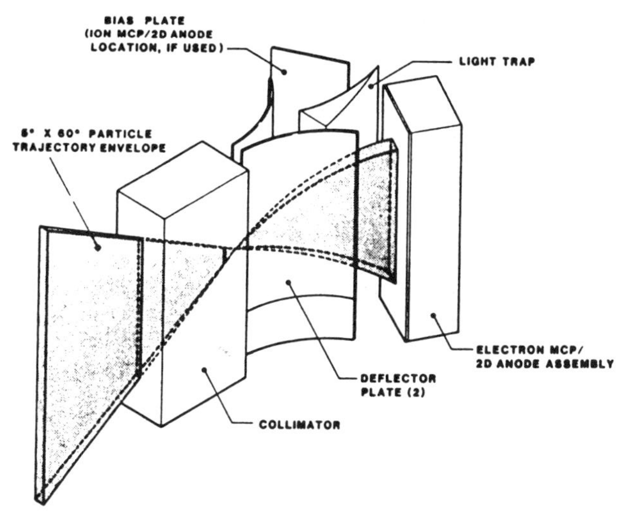

Fig. 5. Schematic of the design of an SPI, similar to the instrument selected for flight on the CRAF mission (J. Burch, Principal Investigator).

variation of four parameters. In practice, the time resolution of CPA-velocity filter combinations is usually improved by limiting the velocity observations to two dimensions, mapping only in azimuth and not in elevation. Filtering also limits the sensitivity or counting rate of an instrument because only a small fraction of parameter space is sampled at any time. Most filtering instruments are carefully designed to focus a range of ion energies and angles at the entrance slit or aperture onto the exit slit or aperture. Nonetheless, the experimenter has to decide the optimal trade between good mass, energy, and angular resolution on the one hand and high counting rates and good time resolution on the other. As an example, Figure 3 shows the effect on the mass resolution of the Giotto/IMS/HERS instrument of a pre-launch change in the slit widths to obtain a factor of 2 higher counting rate. Once the decision about instrument geometry has been made, it is fixed for the entire mission unless the instrument design is complicated by the addition of variable apertures or slits or other techniques.

The third column in Table 4 gives those parameters which are imaged by the instrument. In every case but one, the imaging is in the sense of imaging by a pinhole camera; the original angular distribution is maintained and can be measured by an array of detectors. Columns 6 and 7 give the dimensionality of the detector and the parameter measured in each dimension for each instrument type. One doesn't usually think of a Faraday cup as an imaging instrument, but in fact the collectors of these instruments can be segmented with the currents to each segment measured separately to obtain several two-dimensional angular "pixels." The exception to the pinhole approach was the focussing electrostatic mirror used in the Giotto/IMS/HERS instrument. Figure 4 shows calculated trajectories of ions through the mirror which linearly compresses the angular field of view by a factor of 2. The actual mirror, shown on the right, consisted of two nonconcentric circularly cylindrical grids which fairly accurately produced the performance of the ideal mirror shown on the left.

The fourth column of Table 4 gives those parameters which are dispersed by the instrument. These instruments used either electric fields to spread out a filtered beam of ions according to energy/charge or a magnetic field to spread out the beam in proportion to the rigidity of the ions. The dispersed beam can then be mapped with a detector array. Figures 5-7 illustrate three different uses of beam dispersal. Figure 5 shows the operation of the SPI electrostatic analyzer in which a fan-shaped (i.e., ϕ-filtered) beam is spread perpendicular to the plane of the fan to yield a two-dimensional map of E/q versus θ. Although such a detector is planned for use on the Comet Rendezvous Asteroid Flyby (CRAF) mission for the measurement of electrons over the energy range 0.1 eV to 30 keV, it would have equal applicability to analysis of the energy distribution of positive ions. Figure 6 illustrates the dispersion of an azimuth (ϕ) and rigidity filtered beam by the Giotto/IMS/HERS instrument to give a two-dimensional map of mass/charge versus elevation angle (θ). Finally,

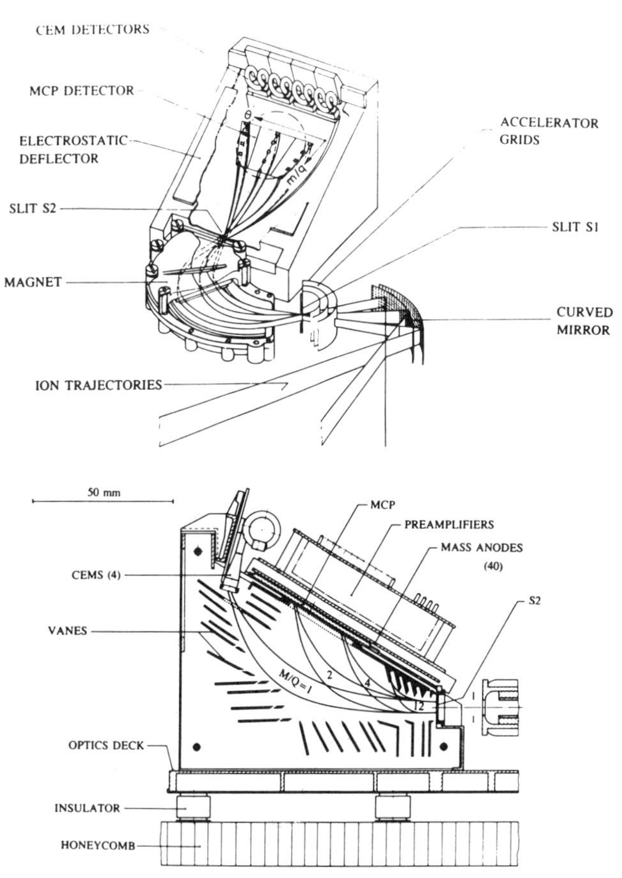

Fig. 6. Two schematic views of the Giotto/IMS/HERS (Balsiger et al., 1986).

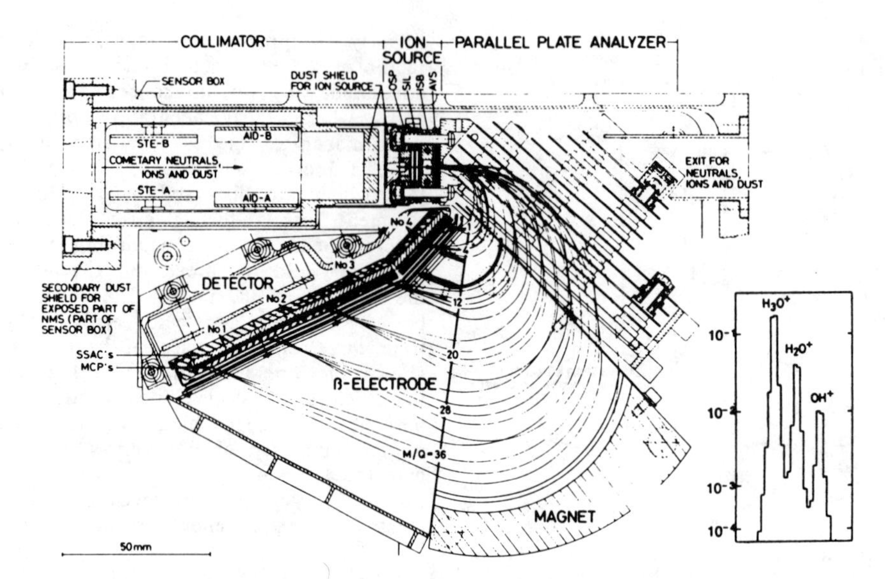

Fig. 7. (Left) Cross-section of the mass analyzer of the Giotto/NMS (Krankowsky et al., 1986a). Ion trajectories shown correspond to different entrance angles (± 4°) and energies (5 eV/ charge). (Right) Spectrum of 17-19 AMU/charge ions obtained by the Giotto/NMS mass analyzer at a distance of 2570 km from the nucleus of Comet Halley (Krankowsky et al., 1986b).

Figure 7 shows how the Giotto neutral mass spectrometer (NMS) used magnetic deflection (which could be thought of as dispersion in reverse) to focus ions dispersed in an electric field to yield mass dispersion over a finite, but limited, range of initial angles and energies. This instrument was "tuned" to velocities near the 68 km s^{-1} speed of the spacecraft past Comet Halley. Its resolution was very good, with $\Delta m/q$ = 0.15 AMU/charge, as illustrated in Figure 7 by the spectrum obtained 2570 km from the comet nucleus (Krankowsky et al., 1986b).

Column 5 in Table 4 indicates those instruments which measure either the ion energy through pulse-height analysis of solid state detector outputs or the ion speed through time-of-flight (although the latter measurement might equally well have been classified as dispersion in time). An important challenge of the energy measurement is the ~30 keV energy threshold of solid state detectors which requires the use of a high-voltage power supply to accelerate low-energy ions to this energy. To date, the time-of-flight technique has not achieved as high a mass resolution as has been achieved with magnetic fast ion spectrometers. Two of the sources of resolution limitation are path length differences of ions in the time-of-flight system and energy straggling of ions in the thin carbon foils which the ions must traverse to generate the "start" signal. One of the instruments described by Gloeckler (1988) overcomes the energy straggling problem by use of a particular electric field geometry in which time-of-flight is a direct measure of mass/charge, independent of velocity.

The design of the cometary matter analyzer (CoMA) selected for flight on CRAF may indicate a direction for future development of time-of-flight ion analyzers. A diagram of the CoMA instrument is shown in Figure 8. The principal objective of CoMA is to analyze secondary ions sputtered from collected dust samples, but the instrument also has gas and thermal-ion analysis modes. In its dust analysis mode, secondary ions generated by nanosecond pulses of primary ions are accelerated into a time-of-flight analyzer which achieves very high mass resolution because (1) it has a very long flight path of nearly 3 m, (2) it is an open system which does not require any foil traversal, and (3) it incorporates a "soft" electrostatic mirror which has the effect of energy/charge focussing and can compensate for up to ~30 eV of initial ion energy spread (This works because faster ions penetrate deeper into the mirror, thus taking a longer time to turn around than do the slower ions.). The expected mass resolution $m/\Delta m$ (where Δm = full width at half maximum of a mass peak) is >3000 at 13 AMU/charge and >13,000 at 350 AMU/charge. The CoMA time-of-flight spectrometer analyzes thermal coma ions by pulsed extraction, acceleration, and bunching of ions flowing through its source region.

Column 8 in Table 4, labeled "Result," lists the parameters which can be deduced from the various types of instruments. The first three instruments, which use only electrostatic analysis, cannot distinguish one ion species from another if their energy/charge spectra overlap, which is likely to happen when the solar wind picks up newly created ions with various masses at different locations and therefore at different speeds. Only instruments like the Active Magnetospheric Particle Tracer Explorers/charge-energy-mass (AMPTE/CHEM) spectrometer which measure the total energy in addition to the energy/charge can unambiguously

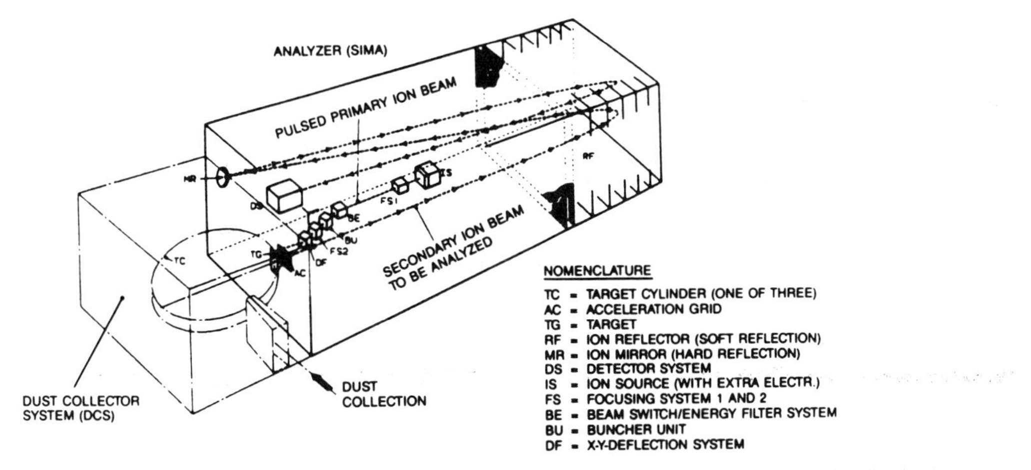

Fig. 8. Schematic of the time-of-flight analyzer of the CRAF CoMA experiment (J. Kissel, Principal Investigator).

determine the abundance of multiply charged ions. Both the neutral gas and ion mass spectrometer (NGIMS) and the CoMA instrument on CRAF will, however, be able to obtain the m/q spectrum of thermal ions for both positive and negative charge states. Most of the other fast ion spectrometers have been designed to obtain three-dimensional velocity distribution functions as a function of m/q, with emphasis increasingly shifting toward the best possible m/q resolution.

Table 4 shows that a large number of combinations of filtering, imaging, dispersing, and measuring mechanisms have been used in the design of fast ion spectrometers. No two rows in this table are the same.

Unfortunately, the word "fast" in the phrase "fast ion spectrometer" refers to the speed of the ions rather than to the speed of the measurement. There are several reasons why ion measurements have time resolutions which are usually orders of magnitude slower than the measurement of the magnetic field. First, there's the very large number of measurements required to completely characterize the distribution of ions; i.e., one must measure phase-space density for each bin in a four- or five-dimensional matrix (v, θ, ϕ, and m/q, or m and q separately), whereas only three numbers (B_x, B_y, and B_z) are needed to characterize the magnetic field. Spacecraft data rate often limits the rate at which ion spectra can be obtained. Another major constraint is the time required to step through all the stages of filtering required by the instrument design. Instrument sensitivity and the need to sum counts over the length of time required to get statistically significant data can also be an important factor for the detection of minor ion species. Finally, many experiments have been limited by the spacecraft spin period.

To the best of my knowledge, the fastest plasma measurements made to date were the four-dimensional (three components of velocity plus m/q) spectra obtained in 4 s by the high intensity spectrometer (HIS) part of the Giotto IMS (Balsiger et al., 1986). The circumstances of the HIS measurements were unusually favorable for fast measurements; however, the spacecraft velocity was much greater than the velocity of the ions being studied so that only a small part of phase space (i.e., limited energy range and angle of incidence for each ion species) had to be sampled and the plasma density and ion counting rates were unusually high. In a more general case, I believe the fastest ion spectral measurement was the two-dimensional energy/charge spectra obtained in 1.5 s by a set of electrostatic analyzers on the International Sun-Earth Explorer (ISEE) 1 and 2 (Bame et al., 1978b). A sometimes useful approach is to measure a single parameter, or only a very few parameters, at a very high rate at the same time that multidimensional measurements are being made by another part of the instrument. The apparent record holder in this regard is the total ion flux measurements made every 0.036 s by the Orbiting Geophysical Observatory 5 (OGO 5) solar wind experiment (Unti et al., 1973).

The Future

I believe the principal challenges for the next generation of fast ion mass spectrometers will be in achieving the highest possible mass resolution over the energy range of interest, in measuring ion charge states, and in obtaining three-dimensional velocity distributions on a reasonable time scale, often from a three-axis stabilized spacecraft. Another, related challenge will be the analysis of the mass and velocity distributions of fast neutral atoms and molecules. There are several papers at this conference which address either charge measurement, full-sky ion measurements from stationary instruments, or the measurement of neutral particles; I will try to complement rather than repeat those ideas.

One must realize that none of the instruments listed in Table 4 is ideal. It would clearly be possible to do a lot better if instrument mass and size were not important constraints. Then one could scale up the instrument size to obtain both high mass resolution, good sensitivity, and good time resolution. A large mass budget would also allow more sensors and thus more complete angular coverage without time-consuming scanning of the field of view. Unfortunately, mass constraints will probably not relax significantly for the next generation of planetary missions. So we must ask in what other areas can we look for progress. The use of toroidal electrostatic analyzers is one way to improve an instrument's geometric factor, and therefore its sensitivity and time resolution, without increasing its mass (Young et al., 1987; see also papers by Carlson, 1988, Young, 1989, and Young et al., 1989). In the remainder of this paper, I offer a few other (probably hare-brained) suggestions of things which might be fun, and possibly worthwhile, to consider.

First, it might be possible to extend some of the techniques invented for analyzing either thermal ions or the limited velocity range of fast, cold ions encountered in the inner coma of Comet Halley to the analysis of fast, hot ions. Specifically, it might be profitable to try to use a CPA together with pulsed electric fields to inject bursts of fast ions, all with approximately the same value of E/q, into a foil-less time-of-flight analyzer such as the one used for the CRAF CoMA experiment (Figure 8). There might be ways of adapting the type of high resolution mass spectrometer represented by the Giotto/NMS (Figure 7) to the detection of hot fast ions. One might think, for example, of preceding the Giotto/NMS by accelerating (or decelerating) grids followed by a velocity filter; such a kluge might be excessively heavy, however. Preacceleration combined with pulsed injection might be feasible. Or modulated preacceleration (similar to modulated Faraday cups) might be used to give an ac signal containing the mass spectrum, unless the dc level exceeded the sensor capacity or the modulated ghost peaks were too strong.

Many mass-analyzing fast ion spectrometers have curved plate electrostatic analyzers which act as E/q filters at the front end of the instrument; none of them uses ac-modulated grids (as in a Faraday cup) for this purpose. But because a modulated grid filters only a single parameter and has a large, wide-angle geometric factor, it might make a very appropriate first stage for a fast ion mass spectrometer for some applications. Small holes in the collector electrodes could pass ions through into either a magnetic or a time-of-flight mass spectrometer. One problem with this approach is the necessity to calculate the desired signal from the difference of two measurements, one at the top and one at the bottom of the modulation voltage pattern on the grid. The only other instrument listed in Table 4 which filters only a single parameter is the SPI, which might similarly serve as a fore-optics for magnetic or time-of-flight analysis to obtain m/q in addition to E/q.

It would also be possible to use a peak-tracking Faraday cup analyzer to measure the density, the vector velocity, and the temperature of ions at a rate of ≥10 measurements/s. The technique would be limited to analysis of a single peak in an energy/charge spectrum, such as protons in the solar wind. Use of a cylindrical band geometry would allow measurements for any azimuthal angle of incidence from a spinning spacecraft.

I'm sure there are other novel combinations of filtering, imaging, dispersing, and measuring, in addition to those shown in Table 4, which an imaginative person could use to design a better fast ion mass spectrometer. The rate of progress can be gauged by the fact that the last two entries in Table 4 were added during the course of this conference.

Acknowledgments. This paper presents the results of research at the Jet Propulsion Laboratory of the California Institute of Technology performed under contract from NASA.

References

Balsiger, H., P. Eberhardt, J. Geiss, A. Ghielmetti, H. P. Walker, D. T. Young, H. Loidl, and H. Rosenbauer, A satellite-borne ion mass spectrometer for the energy range 0 to 16 keV, *Space Sci. Instrum., 2,* 499, 1976.

Balsiger, H., et al., The Giotto ion mass spectrometer, *ESA SP-1077,* 129, 1986.

Bame, S. J., J. R. Asbridge, H. E. Felthauser, J. P. Glore, H. L. Hawk, and J. Chavez, ISEE-C solar wind plasma experiment, *IEEE Trans. Geosci. Elec., GE-16,* 160, 1978a.

Bame, S. J., J. R. Asbridge, H. E. Felthauser, J. P. Glore, G. Paschmann, P. Hemmerich, K. Lehmann, and H. Rosenbauer, ISEE-1 and ISEE-2 fast plasma experiment and the ISEE-1 solar wind experiment, *IEEE Trans. Geosci. Elec., GE-16,* 216, 1978b.

Bridge, H. S., J. W. Belcher, R. J. Butler, A. J. Lazarus, A. M. Mavretic, J. D. Sullivan, G. L. Siscoe, and V. M. Vasyliunas, The plasma experiment on the 1977 Voyager mission, *Space Sci. Rev., 21,* 259, 1977.

Carlson, C. W., Three-dimensional plasma measurement techniques, invited paper presented at the Yosemite 1988 Conference on Outstanding Problems in Solar System and Plasma Physics: Theory and Instrumentation, Yosemite National Park, California, February 2-5, 1988.

Coplan, M. A., K. W. Ogilvie, P. A. Bochsler, and J. Geiss, Ion composition experiment, *IEEE Trans. Geosci. Elec., GE-16,* 185, 1978.

Cravens, T, E., A magnetohydrodynamical model of the inner coma of Comet Halley, *J. Geophys. Res.,* in press, 1988.

Elphic, R. C., C. T. Russell, J. A. Slavin, and L. H. Brace, Observations of the dayside ionopause and ionosphere of Venus, *J. Geophys. Res., 85,* 7679, 1980.

Gloeckler, G., Measurements of the charge state and mass composition of hot plasmas, suprathermal ions and energetic particles using the time-of-flight vs. energy techniques, invited paper presented at the Yosemite 1988 Conference on Outstanding Problems in Solar System and Plasma Physics: Theory and Instrumetation, Yosemite National Park, California, February 2-5, 1988.

Gloeckler, G., J. Geiss, H. Balsiger, L. A. Fisk, F. Gliem, F. M. Ipavich, K. W. Ogilvie, W. Studemann, and B. Wilken, The ISPM solar-wind ion composition spectrometer, *ESA SP-1050*, 75, 1983.

Gloeckler, G., et al., The charge-energy-mass (CHEM) spectrometer for ~0.3 to 300 keV/e ions on AMPTE-CCE, *IEEE Trans. Geosci. Remote Sens., GE-23*, 234, 1985.

Huebner, W. F., First polymer in space identified in Comet Halley, *Science, 237*, 628, 1987.

Heubner, W. F., and D. C. Boice, Polymers and heavy ions in comet comae, these proceedings, 1989.

Johnstone, A. D., et al., The Giotto three-dimensional positive ion analyzer, *ESA SP-1077*, 15, 1986.

Korth, A., et al., Mass spectra of heavy ions near Comet Halley, *Nature, 321*, 335, 1986.

Krankowsky, D., et al., The Giotto neutral mass spectrometer, *ESA SP-1077*, 109, 1986a.

Krankowsky, D., et al., In situ gas and ion measurements at comet Halley, *Nature, 321*, 326, 1986b.

Mitchell, D. L., et al., Derivation of heavy (10-210 AMU) ion composition and flow parameters for the Giotto PICCA instrument, *ESA SP-250*, Vol. I, 203, 1986.

Neugebauer, M., F. M. Neubauer, H. Balsiger, S. A. Fuselier, B. E. Goldstein, R. Goldstein, F. Mariani, H. Rosenbauer, R. Schwenn, and E. G. Shelley, The variation of protons, alpha particles, and the magnetic field across the bow shock of Comet Halley, *Geophys. Res. Lett., 14*, 995, 1987.

Schwenn, R., H. Rosenbauer, and H. Miggenrieder, Das Plasmaexperiment auf Helios (E1), *Raumfahrtforschung, 19*, 226, 1975.

Shelley, E. G., A. Ghielmetti, E. Hertzberg, S. J. Battel, K. Altwegg-von Burg, and H. Balsiger, The AMPTE/CCE hot-plasma composition experiment (HPCE), *IEEE Trans. Geosci. Remote Sens., GE-23*, 241, 1985.

Shelley, E. G., and D. A. Simpson, Final report on Contract NASW-3572: A contract to develop a cometary ion mass spectrometer, Lockheed Palo Alto Research Laboratory Report LMSC/F018835, 1985.

Tsurutani, B. T., and E. J. Smith, Strong hydromagnetic turbulence associated with Comet Giacobini-Zinner, *Geophys. Res. Lett., 13*, 259, 1986.

Unti, T. W. J., M. Neugebauer, and B. E. Goldstein, Direct measurements of solar-wind fluctuations between 0.0048 and 13.3 Hz, *Ap. J., 180*, 591, 1973.

Yeates, C. M., T. V. Johnson, L. Colin, F. P. Fanale, L. Frank, and D. M. Hunten, Galileo: Exploration of Jupiter's system, *NASA SP-479*, 1985.

Young, D. T., A. G. Ghielmetti, E. G. Shelley, J. A. Marshall, J. L. Burch, and T. L. Booker, Experimental tests of a toroidal electrostatic analyzer, *Rev. Sci. Instrum., 58*, 501, 1987.

Young, D. T., In situ measurement techniques for multi-species, low density, high temperature (<60 keV) space plasmas, these proceedings, 1989.

Young, D. T., J. A. Marshall, J. L. Burch, S. J. Bame, and R. H. Martin, A 360° field-of-view toroidal ion composition analyser using time-of-flight, these proceedings, 1989.

CRITICAL PROBLEMS REQUIRING COORDINATED MEASUREMENTS OF LARGE-SCALE ELECTRIC FIELD AND AURORAL DISTRIBUTION

L. R. Lyons

Space Sciences Laboratory, The Aerospace Corporation
P.O. Box 92957, Los Angeles, CA 90009

O. de la Beaujardière

SRI International, 333 Ravenswood Ave., Menlo Park, CA 94025

Abstract. Important problems in magnetospheric and ionospheric physics involve the global electric field distribution and its relation to auroral arcs. If continuous observations of electric fields and auroras were available throughout the auroral zone and polar cap, it would be possible to determine the large-scale polar convection pattern for the first time, to identify the location of auroral arcs with respect to the convection pattern, and to determine the evolution of electric fields relative to that of auroras during substorms. This information, which cannot be obtained from presently available data, is required to fully understand the transfer of solar wind energy to the magnetosphere via the convection electric field, auroras, and substorms.

Introduction

Due to the sparsity of space missions, and the fact that satellites cannot provide continuous global measurements, it has become increasingly important to measure ionospheric and magnetospheric parameters from the ground. Here we argue that such measurements would be valuable for studying problems concerning the large-scale convection pattern, the generation of auroral arcs, and the causes of auroral substorms. Specifically we suggest that simultaneous measurements of the global distribution of electric fields and auroras within the ionosphere as a function time would yield significant information on the relation between auroral phenomena and magnetospheric phenomena. The measurements should be made throughout the auroral zone and polar cap region. In this paper, we consider only discrete auroral arcs, which are associated with field-aligned potential drops, and the diffuse aurora is not discussed.

The distribution of electric fields and aurora within the ionosphere could be obtained from the ground, so that the relation between auroral arc phenomena and electric field phenomena could be determined. Techniques for making the desired measurements are discussed in a companion paper (de la Beaujardière and Lyons, 1989). Here we discuss why such measurements would be valuable, and we describe specific questions that could be addressed with the measurements that would be difficult to answer with data available at present.

Association Between Auroras and Electric Fields

It is well established from both observations and theory that discrete auroral arcs are associated with structure in ionospheric electric fields **E**, where the divergence of **E** is negative. Such an electric field change lies along the afternoon-evening boundary between anti-sunward convection over the polar caps and sunward convection at lower latitudes, and the region of discrete aurora has been observed to be coincident with this boundary (Frank and Gurnett, 1971; Gurnett and Frank, 1973; Heelis et al., 1980). Also, measurements of ionospheric electric fields across individual arcs consistently show that each arc is coincident with a localized region of negative **E** divergence (e.g., Swift and Gurnett, 1973; Maynard et al., 1977; Evans et al.,1977; de la Beaujardière et al., 1977; Heelis et al., 1981; Marklund et al., 1983). The association between the electric field divergence and auroral arcs is a result of the converging ionospheric Pedersen currents driven by the electric fields. Ionospheric current continuity requires that converging Pedersen currents drive upward field-aligned currents $j_{\parallel}$ (e.g., Coroniti and Kennel, 1972). If the magnitude of $j_{\parallel}$ required for current continuity exceeds that which can be carried by magnetospheric electrons within the loss cone, then a field-aligned potential drop forms to increase $j_{\parallel}$ (Lyons, 1980,1981; Chiu and Cornwall, 1980).

An example of data showing the spatial association of discrete auroras with the convection reversal on the dusk side of the polar cap is shown in Figure 1. Data such as this also give information on the mapping of auroral field lines to the magnetosphere. Figure 1 shows energy-time spectrograms of electrons (0.17-33 keV) and ions (E/q from 0.09-3.9 keV/q) versus UT from the polar orbiting, spinning S3-3 satellite. The intensity is given by a grey scale in units of differential energy flux. In addition, the figure shows intensity coded strips for 235 keV electrons and >80 keV ions, the electric potential along the satellite trajectory, and the pitch angle of the particles measured as the satellite spins.

The spectrograms in Figure 1 allow us to identify regions of significant field-aligned potential drop $V_{\parallel}$ associated with discrete auroral arcs. This identification is based upon "inverted-V" profiles in the electron spectrograms (Frank and Ackerson, 1972), field-aligned up-flowing ions, field-aligned precipitating electrons, and enhanced electron loss cones (see Croley et al., 1978). In general, regions of $V_{\parallel} > 0.5$ kV can be identified; however, the spatial resolution of the measurements is not sufficient to resolve individual arcs having a latitudinal width < 20 km.

A clear region of discrete auroral arcs can be seen in Figure 1 between 27500s and 27600s UT. Comparison with the electric potential plot shows that this region lies at the minimum in the potential (i.e., at the afternoon convection reversal). This observation is consistent with

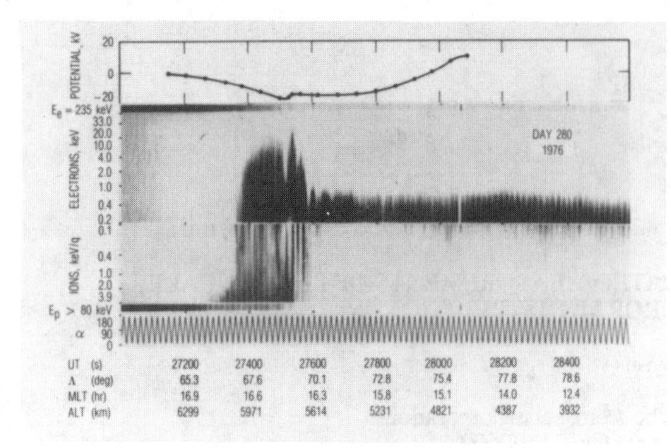

Fig. 1. Spectrogram of S3-3 plasma data and plot of electric potential along satellite trajectory for 27000 to 28600s UT on day 280, 1976 (October 6). The center panels show the energy flux for 0.2 to 33 keV electrons and for 0.1 to 3.9 keV/q ions versus time. Energy flux levels are encoded in a grey scale with darker shading representing higher flux. Grey scale bands at the top and bottom of the central spectrograms represent the intensities of 235 keV electrons and >80 keV protons, respectively. The pitch angle of the particle data is indicated by a line graph below the particle data. Time, invariant latitude, magnetic local time, and satellite altitude are annotated along the bottom of the figure.

Heelis et al. (1980). Poleward of the arcs, polar rain can be seen at electron energies <600 eV, and equatorward of the arcs, reduced fluxes can be seen in the downgoing loss cones at electron energies >5 keV. Assuming that polar rain results from solar particles entering the magnetosphere along open field lines and that reduced fluxes in the downgoing loss cone as well as in the upgoing loss cone indicates trapped particles on closed field lines, the observations in Figure 1 give information on the magnetospheric region responsible for forming the arcs. Specifically, they indicate that auroral arcs and the convection reversal lie approximately at the boundary between open and closed field lines. The association between the particle and electric field data in Figure 1 is expected for typical conditions, where convection is anti-sunward over the open polar cap region and sunward at lower latitudes.

There has generally been agreement that a simple two-cell convection pattern (e.g., Stern, 1977; Heppner, 1977), as is consistent with the observations in Figure 1, exists when the interplanetary magnetic field **B** (IMF) is southward, though a possible small extra cell has been proposed by Reiff and Burch (1985). However, convection patterns become more complicated when $|B_y| > |B_z|$ or $B_z > 0$. Under such conditions, satellite observations have been used to suggest that multi-cell convection patterns (Burke et al., 1979; Reiff and Burch, 1985) or significantly distorted two-cell patterns (Heppner and Maynard, 1987) might exist. To understand the transfer of solar wind energy to the magnetosphere, it is highly desirable to know how the convection pattern varies with the IMF. However, this problem cannot be unambiguously solved without measurements of the two-dimensional polar convection pattern.

An aspect of the above problem is whether there can be closed convection cells confined to the region of open polar cap field lines. Such convection cells have been proposed by Maezawa (1976), Crooker (1979), and Reiff and Burch (1985) to account for polar cap electric fields observed during northward IMF. These authors have suggested that the polar cap cells are associated with neutral points along open field lines. On the other hand, Lyons (1985) suggested that variations in the polar cap convection pattern as a function of the IMF could be explained by simply considering the connection between the geomagnetic and interplanetary magnetic field. Based on this proposal, two neutral points form along the separator between open and closed geomagnetic field lines, independent of the IMF orientation. Additional neutral points are not invoked. The proposal does not allow for closed convection cells on open polar cap field lines, and all polar cap equipotential contours cross the boundary between open and closed field lines. Other authors (e.g., Akasofu and Roederer, 1983; Kan and Burke, 1985) have suggested that the region of open polar cap field lines can split in two. In these models, sunward flow near the poles occurs on closed field lines, so that closed convection cells do not occur on open polar cap field lines.

Since the region of discrete aurora can be associated with the boundary between open and closed field lines, simultaneous measurements of the aurora and the two-dimensional convection pattern should be valuable in determining whether polar cap equipotentials cross the boundary. This question has recently been addressed using ion drift and auroral particle measurements from the Dynamics Explorer 2 satellite (Coley et al., 1987). These authors were able to study the direction of convection relative to the boundary between open and closed field lines. However, they concluded that a definitive answer concerning flow across the boundary could not be obtained from their satellite data, since flow trajectories could not be measured.

Because polar convection patterns are not well known, our understanding of the relation between the region of discrete auroral arcs and the convection reversal remains uncertain. Figure 2 shows a second example of S3-3 data from a traversal of the auroral zone. This example illustrates that the association between the aurora and the convection pattern can be significantly different from that seen in Figure 1.

The example in Figure 2 shows a region of intense discrete auroral arcs extending from $\sim 69^{\circ}$ to $\sim 72^{\circ}$ latitude. Poleward of the aurora, polar rain can be seen at electron energies <400 eV. Within portions of the arc region, and equatorward of the region, reduced fluxes can be seen in the downgoing loss cones at electron energies above a few keV. Thus in this example, as in the previous example, arcs apparently lie at the boundary between open and closed field lines. The region of arcs is broader in this example, and it extends equatorward onto field lines that can be identified as closed on the basis of the electron loss cone signature.

However, the potential plot in Figure 2 does not show a reversal in the large-scale electric field in the vicinty of the arcs and of the boundary between open and closed field lines. The electric field shows convection

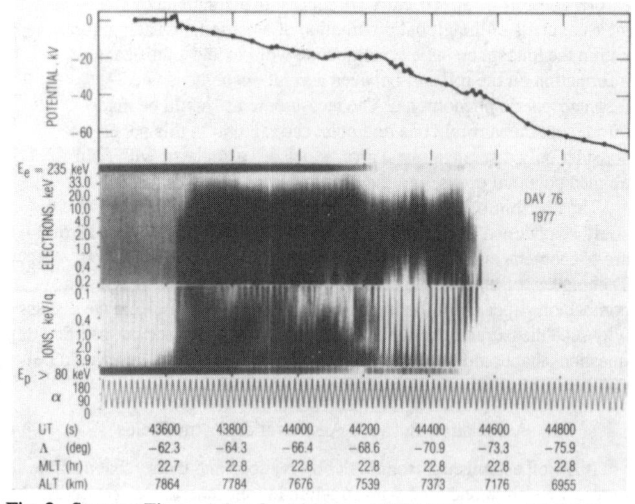

Fig. 2. Same as Figure 1, but for 43400 to 45000s UT on day 76, 1977 (March 17).

to be approximately uniform and in the midnight to dusk direction as expected from sunward convection from 62° to beyond 77° latitude. We do not currently have available simultaneous electric field and particle data from a sufficient number of S3-3 passes to do a statistical study. However, the lack of a convection reversal in the vicinity of the auroral zone and the open-closed field line boundary is not uncommon in our data. Another example, similar to that in Figure 2, has been presented by Mizera et al. (1981). They displayed a plot of electric field data with high spatial resolution along with electron and ion spectrograms and the potential along the satellite trajectory. While their data show no change in the large-scale electric field in the vicinity of the arcs, the data show large, highly variable electric field structures within the arc region. Thus data such as in Figure 2 are not inconsistent with the arcs being formed by regions of negative electric field divergence. However, the electric fields can be localized and not associated with a reversal in the large-scale convection pattern across the auroral zone.

The example in Figure 2 demonstrates that we have not yet determined a general relation between auroras and the large-scale convection pattern. Since the convection pattern varies with the IMF, it is possible that the differences in the convection patterns in Figures 1 and 2 are related to the IMF direction. It was southward during the time period shown in Figure 1, but IMF data are unavailable for the time periods of Figure 2 and the other similar examples we have available. Determining the relation between the region of discrete aurora and the convection pattern as a function of the IMF direction is an important problem that could be addressed with simultaneous measurements of auroras and the two-dimensional convection pattern. An aspect of this problem involves polar cap auroral arcs. In situations such as in Figure 2, the large-scale electric field should reverse somewhere within the polar cap. It would be interesting to determine whether such reversals are associated with polar cap arcs.

Substorms

Simultaneous measurements of the two-dimensional electric field pattern and the aurora as a function of time should also provide important new information concerning substorm processes. At the onset of a substorm expansion phase, quiet arcs become intense and active, beginning with the sudden brightening of a pre-existing quiet arc (Akasofu, 1964, 1977). Since arcs are associated with electric field structure, significant changes must occur in the electric fields in the vicinity of the auroral arcs during such breakups. The electric fields within the ionosphere and along auroral magnetic field lines dissipate energy and act as a load to generate processes that occur deep within the magnetosphere or along magnetospheric boundaries. Thus the auroral breakup identifies field lines along which substorm processes develop, and the associated electric field changes give information on electrodynamic processes occurring within the auroral generator during substorms.

With simultaneous electric field and aurora measurements, specific questions can be addressed that are important for understanding substorm processes. For example, using local measurements in the vicinity of an auroral breakup, we could ask:

1. Where is the breakup initiated with respect to the location of pre-existing discrete arcs and the diffuse aurora?

2. Where is the breakup initiated relative to electric field signatures such as the convection reversal, and the boundary between open and closed field lines?

3. How do ionospheric electric fields change when a breakup is initiated?

4. How do regions of breakup aurora evolve during the expansion phase relative to electric field signatures and the boundary between open and closed field lines?

The answers to questions 1 and 2 give information on the magnetospheric region where substorm processes are initiated. Since the diffuse aurora maps to the central plasma sheet and discrete arcs map to the source region for the plasma sheet boundary layer (Eastman et al., 1984; Lyons and Evans, 1984), question 1 considers the location of substorm onset within the plasma sheet. Question 2 relates the location of substorm initiation to the processes responsible for the generation of the large-scale magnetospheric electric fields. The answer to question 3 would give information on how magnetospheric electric fields change as a substorm is initiated and where these changes occur. Question 4 considers a longer time scale than do the other questions, and gives information on how electrodynamic parameters evolve during a substorm.

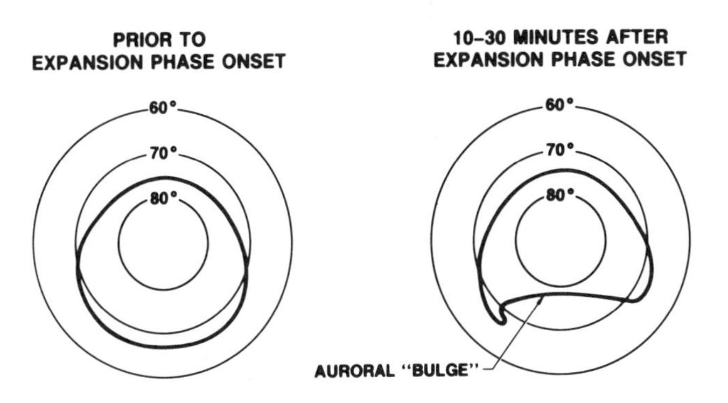

Fig. 3. Sketch of the poleward boundary of the auroral oval preceding, and during, a substorm expansion.

If measurements of the evolution of the large-scale convection pattern during substorms become available, the following additional questions can be asked:

1. Where is the region of auroral breakup located with respect to the large-scale convection pattern and how does this location vary with the IMF?

2. Does the large-scale electric field pattern, as well as the local electric field structure in the vicinity of a breakup, change at the onset of, or during, a substorm expansion phase?

3. If the large-scale electric fields change, are the changes in response to changes in the IMF, or are they solely related to substorm processes internal to the magnetosphere?

The answers to these questions would give information on how substorms are related to the generation of the large-scale convection electric field and the associated transfer of solar wind energy to the magnetosphere. They would also give information on the extent to which substorms involve global changes in magnetospheric fields versus more localized changes in the tail.

Evidence does exist that major changes occur in large-scale magnetospheric magnetic and electric fields during a substorm expansion; however, the nature of the changes is not well known. Figure 3, based upon auroral images (e.g., Akasofu, 1977; Craven and Frank, 1985, 1987), shows a sketch of changes in the poleward boundary of the auroral oval during a substorm. The poleward portion of the oval contains discrete auroras, at least along the post-noon to midnight portion. Thus significant electric field structure must exist in the vicinity of the poleward boundary of the oval. During the substorm expansion phase, a large "bulge" can form along the midnight portion of the oval. This bulge can protrude well into the pre-existing polar cap as shown in Figure 3. Such a change in the shape of the oval indicates that major changes must occur in the magnetospheric electric field distribution. To the best of our knowledge, it is not known what these changes are. If the poleward boundary of the aurora lies approximately along the boundary between open and closed field lines, then the bulge formation also indicates that significant, poorly understood, changes occur in the magnetic field geometry during substorms.

Subsidiary Models and Data

A quantitative model of the large-scale magnetospheric magnetic and electric fields would be desirable for interpreting the observations and for mapping phenomena observed in the ionosphere to magnetospheric regions. The model should be able to relate the large-scale structure of magnetospheric electric fields in polar regions to the boundary between open and closed magnetic field lines and to the IMF. The model should thus be able to map structure in the large-scale electric field distribution to specific regions in the magnetosphere. It must be able to accurately separate open from closed field lines, but it need not be able to accurately map field lines from the polar cap and auroral zone to specific positions in space. Such a model cculd be tested using measurements of the two-dimensional convection pattern. If the tests show the model gives reasonable results, it could then be used to map regions of aurora and auroral breakup to source regions in the magnetosphere. This would help identify the regions and processes responsible for the generation of auroral arcs and for substorm initiation.

Observations from polar orbiting satellites, such as those shown in Figures 1 and 2 and discussed earlier, would also be valuable. This data would help in relating auroras to electric field structures and the boundary between open and closed field lines. While the data could only be obtained for a limited number of points in space and time, it would be extremely useful in checking relations observed continuously from the ground and in checking the mappings from a model of the type described above. Ideally, the satellite data should include energetic (> 40 keV) particles, as well as lower-energy (0.1-20 keV) particles and electric fields. The energetic particle data are important because they identify field lines that thread the tail and magnetopause current sheets (Lyons and Evans, 1984; Lyons et al., 1987).

Conclusions

We have argued that important, unanswered questions related to magnetospheric energy transfer and dynamics could be addressed from ground-based measurements of the two-dimensional distribution of electric fields and auroras in polar regions. Specifically, answers to the questions we have discussed would give information on what magnetospheric regions and processes are responsible for arc formation and for substorm initiation and development. Information would also be obtained on magnetospheric electric and magnetic field changes associated with substorms and IMF variations. This information is required to fully understand the processes responsible for the transfer of solar wind energy to the magnetosphere via the convection electric field, auroras, and substorms.

Acknowledgments. This work was supported by NASA grant NAGW-853 to The Aerospace Corporation, The Aerospace Sponsored Research Program, and NSF cooperative agreement ATM-85-16436.

References

Akasofu, S.-I., The development of the auroral substorm, *Planet. Space Sci., 12*, 273, 1969.

Akasofu, S.-I., *Physics of Magnetospheric Substorms*, D. Reidel Publ. Co., Dordrecht, 1977.

Akasofu, S.-I., and M. Roederer, Polar cap arcs and open regions, *Planet. Space Sci., 31*, 193, 1983.

Burke, W. J., M. C. Kelley, R.C. Sagalyn, M. Smiddy, and S. T. Lai, Polar cap electric field structures with a northward interplanetary magnetic field, *Geophys. Res. Lett., 6*, 21, 1979.

Chiu, Y. T., and J. M. Cornwall, Electrostatic model of a quiet auroral arc, *J. Geophys. Res., 85*, 543, 1980.

Coley, W. R., R. A. Heelis, W. B. Hanson, P. H. Reiff, J. R. Sharber, and J. D. Winningham, Ionospheric convection signatures and magnetic field topology, *J. Geophys. Res., 92*, 12,352, 1987.

Coroniti, F. V., and C. F. Kennel, Polarization of the auroral electrojet, *J. Geophys. Res., 77*, 2835, 1972.

Craven, J. D., and L. A. Frank, The temporal evolution of a small auroral substorm as viewed from high altitudes with Dynamics Explorer 1, *Geophys. Res. Lett., 12*, 465, 1985.

Craven, J. D., and L. A. Frank, Latitudinal motions of the aurora during substorms, *J. Geophys. Res., 92*, 4565, 1987.

Crooker, N. U., Dayside merging and cusp geometry, *J. Geophys. Res., 84*, 951, 1979.

Croley, D. R., Jr., P. F. Mizera, and J. F. Fennell, Signature of a parallel electric field in ion and electron distributions in velocity space, *J. Geophys. Res., 83*, 2701, 1978.

de la Beaujardière, O., and L. R. Lyons, Instantaneous measurements of the global high-latitude convection pattern, this volume, 1989.

de la Beaujardière, O., R Vondrak, and M. Baron, Radar observations of electric fields and currents associated with auroral arcs, *J. Geophys. Res., 82*, 5051, 1977.

Eastman, T. E., L. A. Frank, W. K. Peterson, and W. Lennartsson, The plasma sheet boundary layer, *J. Geophys, Res., 89*, 1553, 1984.

Evans, D. S., N. C. Maynard, J. Troim, T. Jacobsen, and A Egeland, Auroral vector electric field and particle comparisons 2. Electrodynamics of an arc, *J. Geophys. Res., 82*, 2235, 1977.

Frank, L. A., and K. L. Ackerson, Local-time survey of plasma at low altitudes over the auroral zones, *J. Geophys. Res., 77*, 4116, 1972.

Frank, L. A., and D. A. Gurnett, Distributions of plasmas and electric fields over the auroral zones and polar caps, *J. Geophys. Res., 76*, 6829, 1971.

Gurnett, D. A., and L. A. Frank, Observed relationships between electric fields and auroral particle precipitation, *J. Geophys. Res., 78*, 145, 1973.

Heelis, R. A., J. D. Winningham, W. B. Hanson, and J. L. Burch, The relationships between high-latitude convection reversals and the energetic particle morphology observed by Atmosphere Explorer, *J. Geophys. Res., 85*, 3315, 1980.

Heelis, R. A., W. B. Hanson, and J. L. Burch, AE-C observations of electric fields around auroral arcs, in *Physics of Auroral Arc Formation*, edited by S.-I. Akasofu and J. R. Kan, Amer. Geophys. Union, Washington, D.C., 154, 1981.

Heppner, J.P., Empirical models of high-latitude electric fields, *J. Geophys. Res., 82*, 1115, 1977.

Heppner, J. P., and N. C. Maynard, Empirical high-latitude electric field models, *J. Geophys. Res., 92*, 4467, 1987.

Kan, J. R., and W. J. Burke, A theoretical model of polar cap arcs, *J. Geophys. Res., 81*, 4171,1985.

Lyons, L. R., Generation of large-scale regions of auroral currents, electric potentials, and precipitation by the divergence of the convection electric field, *J. Geophys. Res., 85*, 17, 1980.

Lyons, L. R., Discrete aurora as the direct result of an inferred, high-altitude generating potential distribution, *J. Geophys. Res., 86*, 1, 1981.

Lyons, L. R., A simple model for polar cap convection patterns and generation of auroras, *J. Geophys. Res., 90*, 1561, 1985.

Lyons, L. R., and D. S. Evans, An association between discrete aurora and energetic particle boundaries, *J. Geophys. Res., 89*, 2395, 1984.

Lyons, L. R., A. L. Vampola, and T. W. Speiser, Ion precipitation from the magnetopause current sheet, *J. Geophys. Res., 92*, 6147, 1987.

Maezawa, K., Magnetospheric convection induced by the positive and negative Z components of the interplanetary field: Quantitative analysis using polar cap magnetic records, *J. Geophys. Res.*, *81*, 2289, 1976.

Marklund, G., W Baumjohann, and I. Sandahl, Rocket and ground based study of an auroral breakup event, *Planet. Space Sci.*, *31*, 207,1983.

Maynard, N. C., D. S. Evans, B. Maehlum, and A. Egeland, Auroral vector electric field and particle comparisons, 1. Premidnight convection topology, *J. Geophys. Res.*, *82*, 2227, 1977.

Mizera, P. F., J. F. Fennell, D. R. Croley, Jr., A. L. Vampola, F. S. Mozer, R. B. Torbert, M. Temerin, R. Lysak, M. Hudson, C. A. Cattell, R. J. Johnson, R. D. Sharp, A. Ghielmetti, and P. M. Kintner, The aurora inferred from S3-3 particles and fields, *J. Geophys. Res.*, *86*, 2329, 1981.

Reiff, P. H., and J. L. Burch, IMF B_y-dependent plasma flow and Birkeland currents in the dayside magnetosphere: 2. A global model for northward and southward IMF, *J. Geophys. Res.*, *90*, 1595, 1985.

Stern, D. P., Large-scale electric fields in the earth's magnetosphere, *Rev. Geophys. Space Phys.*, *15*, 156, 1977.

Swift, D. W., and D. A. Gurnett, Direct comparison between satellite electric field measurements and the visual aurora, *J. Geophys. Res.*, *78*, 7306, 1973.

INSTANTANEOUS MEASUREMENTS OF THE GLOBAL HIGH-LATITUDE CONVECTION PATTERN

O. de la Beaujardière

Geoscience and Engineering Center, SRI International, Menlo Park, CA 94025

L. R. Lyons

Space Sciences Laboratory, The Aerospace Corporation, Los Angeles, CA 90009

Abstract. Knowledge of the ionospheric electric field is critically needed for the study of magnetospheric, ionospheric, and thermospheric phenomena. A companion paper by Lyons and de la Beaujardière argues for simultaneous, global measurements of the auroral precipitation and plasma convection. Here, we focus on how to obtain, with good temporal resolution, maps of the global electric field in the auroral zone and polar cap. We examine the advantages and drawbacks of several instruments from the perspective of obtaining continuous measurements over a large area. The best techniques appear to be the incoherent scatter and the high frequency (HF) backscatter radars. However, some questions still remain regarding the frequency of occurrence and spatial distribution of coherent HF echoes. The coverage from existing instruments is fairly dense in the longitudes from Scandinavia to central Canada, but the coverage is totally inadequate to infer a snapshot of the convection pattern. Possible locations for instruments that would need to be deployed are indicated.

Introduction

In a companion paper (Lyons and de la Beaujardière, 1989), we argue that important problems related to magnetospheric energy transfer and dynamics require measurements of the global distribution of electric fields. We emphasize the need for measuring the convection pattern as a function of time and for relating auroras to the electric field distribution.

The need for measuring the global convection pattern extends to ionospheric and atmospheric research, where our understanding and predictive capability are severely hampered by the lack of knowledge about the global convection pattern. For example, a critically important limitation in ionospheric and thermospheric modeling is the imprecise information on the high-latitude convection pattern (Sojka and Schunk, 1987).

To a first approximation, the magnetic field lines are equipotentials, and the magnetospheric electric field maps down to the ionosphere. Thus an extremely practical way of obtaining the magnetospheric electric field is to measure the convection pattern in the ionosphere.

The electric field, **E**, is generally obtained by measuring the ionospheric ion or electron velocity, **V**, and applying the relation

$$\mathbf{E} = -\mathbf{V} \times \mathbf{B}$$

where **B** is the magnetic field. Therefore, measuring the electric field is equivalent to measuring the plasma convection in the F region.

The distribution of the electrostatic potential across the auroral oval and polar cap can be inferred from the global convection pattern. This pattern would give the total potential drop across the polar cap, an important parameter because it provides an estimate of the total electromotive energy of the magnetospheric system (Reiff and Luhmann, 1986). Furthermore, the distribution of the potential across individual convection cells indicates convection pattern asymmetries and the relative importance of each convection cell.

Thus significant progress in understanding many aspects of the coupling between the solar wind, magnetosphere, ionosphere, and thermosphere requires that the high-latitude global convection pattern be reliably measured. However, at this time, the instantaneous convection pattern within the polar cap and the auroral oval is not known. Only average patterns are available, obtained from binning years of satellite or radar data. (By "instantaneous," we mean 10-20 min, i.e., over a time comparable to a low-altitude satellite transit time across the polar regions.) During the course of one day, individual ground-based stations can also measure the convection over a limited latitudinal range. However, such a pattern, obtained over 24 hours, is far from an instantaneous one.

Figure 1 illustrates how difficult it is to infer the high-latitude convection from currently available observations. The two convection patterns shown are from recent articles (Coley et al., 1987; Heppner and Maynard, 1987) obtained using observations from the same spacecraft. Discriminating criteria for orbit selection were not exactly the same, but these patterns correspond to the same time period and to the same interplanetary magnetic field (IMF) conditions. Yet the two patterns are strikingly dissimilar, dramatically illustrating the need for improved measurements. It is of fundamental importance that the available data be sufficiently extensive that the human interpretation factor be kept to a minimum.

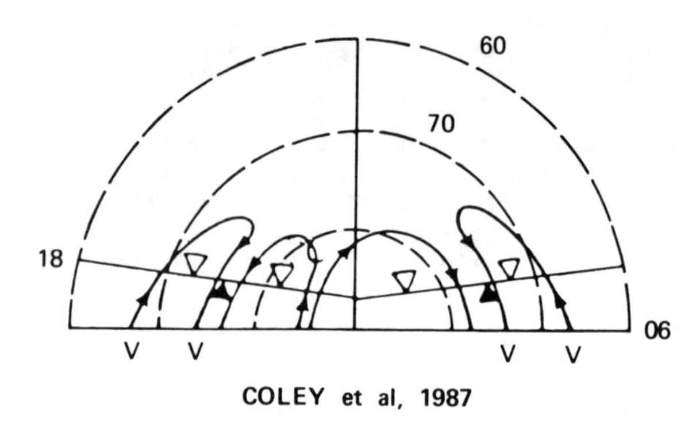

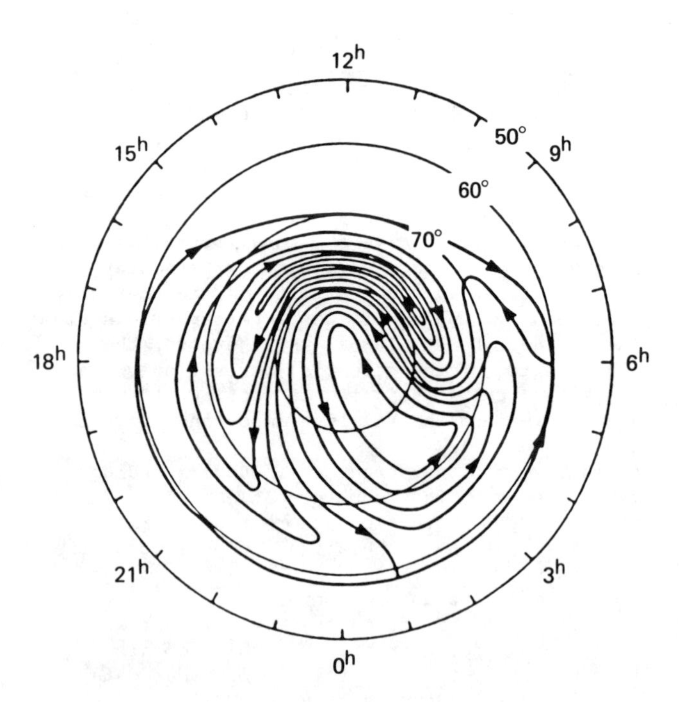

Fig. 1. Convection patterns, both inferred from the same Dynamics Explorer 2 satellite for the same IMF (Bz > 0 and By < 0).

In this paper, we identify various methods for measuring the plasma drift within the ionosphere, and we examine how they can be used in continuously monitoring the global convection electric field and the potential distribution across the polar cap. We point out the sparsity of measurements, particularly in the polar cap, and the need for more complete instrumentation. The advantages and limitations of each technique are presented, along with possible locations where instruments could be placed.

The challenge we address is this: Can we measure a snapshot of the convection pattern with sufficient accuracy to resolve some of the outstanding issues; and can we obtain another pattern a short time later, then another one, and so on, to allow progressive changes in the pattern to be seen as in a movie?

This paper focuses on plasma drift measurements. To obtain simultaneous and collocated auroral precipitation measurements would require the deployment of separate instruments, except in the case of incoherent scatter radars (ISRs).

Measurement Techniques

Instruments that have been used to measure or estimate electric fields include the following:

1. Incoherent scatter radars--three of them are at high latitudes: the European Incoherent Scatter (EISCAT), Millstone Hill, and Sondrestrom (Folkestad et al., 1983; Evans et al., 1979; Foster et al., 1986; Kelly, 1983; Wickwar et al., 1984).
2. Coherent backscatter radars at HF--Goose Bay and Halley Bay in the northern and southern hemispheres, respectively (Greenwald et al., 1985; Ruohoniemi et al., 1987).
3. Coherent backscatter radars at very high frequency (VHF)--Scandinavian Twin Auroral Radar Experiment (STARE), Sweden and Britain Radar Experiment (SABRE) (Greenwald et al., 1978; Nielsen and Schlegel, 1985), and Bistatic Auroral Radar System (BARS).
4. Magnetometers--estimate **E** indirectly from inferred currents (Heppner et al., 1971).

Figure 2 indicates the location of the existing radars and their minimum and maximum fields-of-view.

Other techniques are starting to emerge as potentially able to measure the electric field. They include:

1. HF digital sounders that measure Doppler (Strom et al., 1986)--the NOAA sounder and the Lowell University Digisonde.
2. Spaced receivers that cross-correlate ionospheric scintillation measurements (Rino and Livingston, 1982).

All of these techniques complement each other

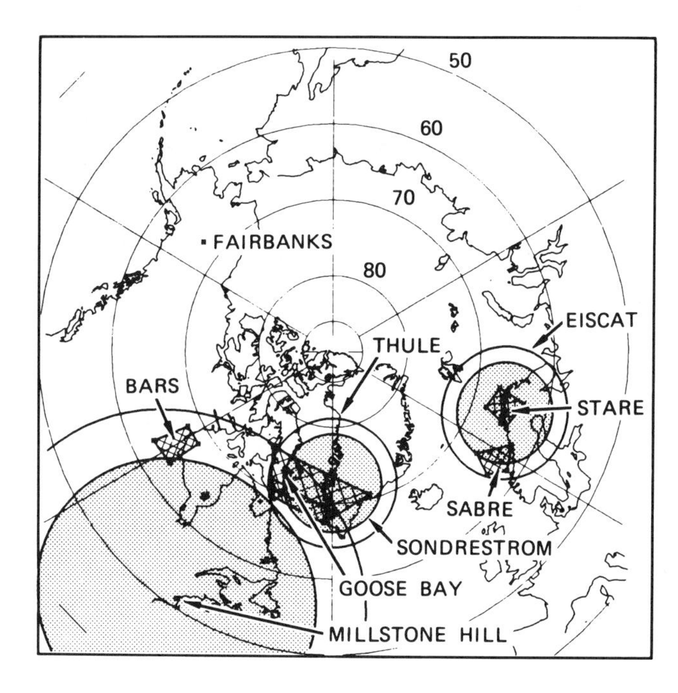

Fig. 2. Map in eccentric dipole coordinates showing nominal fields-of-view for backscatter radars and for incoherent scatter radars under solar minimum conditions (grey shaded areas) and solar maximum.

because they are adapted to study different processes. We now consider these techniques in more detail.

Incoherent Scatter Radars

ISRs can measure the ion drift directly from the Doppler shift of the Thomson-scattered return signal (Farley, 1971; Evans, 1969). Measurements can be made under all conditions. The only significant limitation is loss of signal at extremely low electron densities. ISRs can, therefore, provide continuous coverage for days at a time.

The radars' fields-of-view are quite large. They are circular with a diameter corresponding to a 17° and 50° latitude span during solar cycle maximum, for Sondrestrom and Millstone Hill, respectively (about 2.8 to 23 x 10^6 km^2); during solar minimum, these diameters are 13° and 40°, respectively. EISCAT's field-of-view is similar to Sondrestrom's. There are two reasons why Millstone Hill's field-of-view is so much larger than EISCAT's and Sondrestrom's. First, Millstone is located at the top of a hill, whereas the others are in a valley; second, it transmits at a lower frequency and thus can detect return signals from lower density plasmas because the return signal strength depends on the ratio between the Debye length and the transmitted wavelength.

Two line-of-sight velocity measurements are usually sufficient to compute the drift vector, because the component of the velocity perpendicular to **B** is much larger than that parallel to **B**. Measurements over a limited latitudinal extent (5°) can be made with a time resolution of about 2 min (Willis et al., 1986). Measurements over the whole field-of-view necessitate scanning the antenna through multiple positions, resulting in a time resolution of the order of 10 to 30 min. The latitudinal resolution varies from a few kilometers for nearly overhead measurement to about 200 km for measurements furthest from the radar.

The electron density profile is measured simultaneously with the ion drift, giving the distribution of auroras associated with the convection. Additionally, a wide variety of other ionospheric and thermospheric parameters can be obtained, most of them as a function of height and latitude. These parameters include: ion and electron temperatures, conductivities, energy distribution of the precipitating particles, current distribution perpendicular and parallel to the magnetic field, Joule heating rates, as well as neutral temperatures, densities, and winds. ISRs, therefore, provide a virtually complete set of physical parameters and are well suited for comprehensive research thrusts dealing with the magnetosphere, ionosphere, and upper atmosphere.

HF Backscatter Radars

These radars measure coherent echoes reflected from plasma irregularities when the wave vector is perpendicular to **B** (Greenwald et al., 1985). Perpendicularity is possible over a somewhat extended region because of refraction of the signal as it propagates through the ionosphere. This refraction allows these systems to be used for probing the polar cap. By electronically swinging the beam, return echoes are detected with an 80-s time resolution within a 55° azimuth sector that extends in latitude from 6° to 20°. Typically, F-region irregularities can be observed over a latitudinal width of 2° to 8° in one or more regions within this field of view (within about 0.15 to 0.7 x 10^6 km^2). Note that this latitude coverage is equivalent to that obtained from an ISR operated with similar temporal resolution.

To obtain the vector velocity, the returns from two radars whose viewing area overlaps can be combined, as will be done, in the case of the Goose Bay system, when a second system becomes operative in Quebec. In addition, sophisticated numerical techniques have been employed to deduce the plasma drift from single radar measurements. These techniques have been used for ISRs as well as for backscatter radars.

To measure the plasma drift, meter-scale irregularities must be present in the F-region; this limits the percentage of time a measurement can be made. The intensity of the backscattering irregularities is maximum under darkness, during solar cycle maximum, and when geomagnetic activity is high. However, very few data exist about the distribution over time and space of the F-region irregularities that are required. The problem of intermittent returns is somewhat alleviated because even though the probability of occurrence of an appreciable ionospheric echo may be fairly low, measurements can be made preferentially during active conditions (i.e., during conditions that often generate substantial scientific interest). However, measurements under quiet conditions are scarce. The strong point is that this instrument provides the plasma convection over a two-dimensional region with good temporal and spatial resolution (~10 to 40 km).

No statistical study has yet been conducted of the probability of significant F-region backscatter echoes in either the auroral zone or in the polar cap. In particular, the ability of HF backscatter radars to measure plasma drifts in the polar cap has not been demonstrated. It has been established that backscatter preferentially occurs in the poleward and equatorward edges of the auroral oval, (e.g., Bates et al., 1973; Moller, 1974; Nekrasov et al., 1982). Polar cap measurements of spread-F, which occurs when meter-scale irregularities are present, were conducted by Nekrasov et al. (1982) during solar cycle maximum. They found that during all seasons, the spread-F occurrence probability is less than 5% poleward of 70° invariant latitude. Furthermore, statistical studies have shown that, during solar minimum and in the polar cap, ionospheric scintillation (due to kilometer-scale size irregularities) occurs less than 5% of the time (Aarons et al., 1981). Because the strength of the irregularities decreases with size with a power law dependence (e.g., Fremouw et al., 1985), these results also imply that there may be a minimum in meter-size irregularities in the polar cap. Therefore, there is some evidence that the probability of occurrence of an appreciable backscatter echo could be very low in the polar cap (Tsunoda, 1988, and references therein).

VHF Backscatter Systems

Although these systems are similar to HF backscatter radars, a few important differences exist: VHF systems are sensitive to electrostatic plasma waves in the E-region only. VHF signals are not refracted, thereby precluding their use for polar cap measurements. The two-dimensional (0.2×10^6 km^2) area is smaller than that covered by HF systems, but the temporal resolution is excellent, about 20 s. As is the case with the HF system, appreciable echoes are received only a small fraction of the time: overall, about 25% in the case of SABRE (Jones et al., 1987).

The process for forming ionospheric irregularities requires that the electric field exceed 15 mV m^{-1} (Nielsen and Whitehead, 1983). Another problem is that the measured Doppler shift from which the velocities are obtained is that of the plasma waves present in the medium, not that of the ionospheric plasma motions. It has been shown that for sufficiently low electric fields this motion is nearly that of the electrons in the E-region. However, the measured Doppler shift reaches a saturation limit, so that when the electric field is above 35 mV m^{-1} the Doppler shift is not directly related to the electric field magnitude (Reinleitner and Nielsen, 1985; Nielsen and Schlegel, 1985). The direction of the electric field can be reliably estimated when the Doppler shift has reached the saturation limit. However, the rather narrow range of observable electric field magnitudes imposes a serious limitation on VHF systems for monitoring the global ionospheric convection.

Magnetometers

Magnetometers are probably the oldest systems to estimate ion drift velocities (Birkeland, 1908; Harang, 1946). The local electric field is inferred from magnetic field perturbations, which are due to horizontal and field-aligned currents. This E-field derivation is based on the knowledge of the Hall and Pedersen conductivities, as well as of the distribution of field-aligned currents. In mid and auroral latitude regions, when large gradients in conductivity are not present, magnetometers can reliably give the electric field direction. However, the electric field magnitude inferred contains a substantial error (unless Hall and Pedersen conductivities are known). In the polar cap, due to the effect of field-aligned currents, magnetometers do not provide reliable estimates of the electric field (Heppner et al., 1971; Hughes and Rostoker, 1977, 1979). Furthermore, no electric field can be derived if the conductivities in the E-region are too low to produce an appreciable current.

HF Sounders and Scintillation Spaced Receivers

These two techniques are not new instruments but are adaptations of traditional instruments to infer the plasma drift. HF sounders are ionosondes in which the Doppler shift of the return signal is measured over a narrow angle of arrival. Spaced receivers measure the motion of the ionospheric diffraction pattern by cross-correlating scintillation signals received from space-borne radio beacons (Rino and Livingston, 1982). Both techniques provide a point measurement close to overhead. They appear promising in that they are simple, unattended instruments that can operate continuously, but it has yet to be unambiguously demonstrated that the measured velocity is directly related to the F-region plasma drift. Studies are underway to determine their potential for plasma drift measurements.

To summarize, there are many techniques to measure the ionospheric plasma drift. Each has specific advantages and drawbacks that we have attempted to graphically illustrate on Figure 3. To indicate our estimate of the relative merit in each category we present a round symbol that ranges from best (full) to worst (empty). Figure 3 also lists the key to the various criteria we have used. Our ranking is based on how well the instrument can provide maps of the high-latitude convection during prolonged periods. For example, the criterion for the field-of-view is marked with a filled-in circle for ISRs, but with an empty circle for spaced receivers, HF sounders, and magnetometers because they only cover a very narrow region. In constructing this figure we have necessarily relied on our own judgement to a certain degree, and several ratings are based on incomplete knowledge, since the potential for some of the techniques listed is not yet well established. However, we feel this figure summarizes quite factually how well each instrument can measure the global convection pattern at high latitudes.

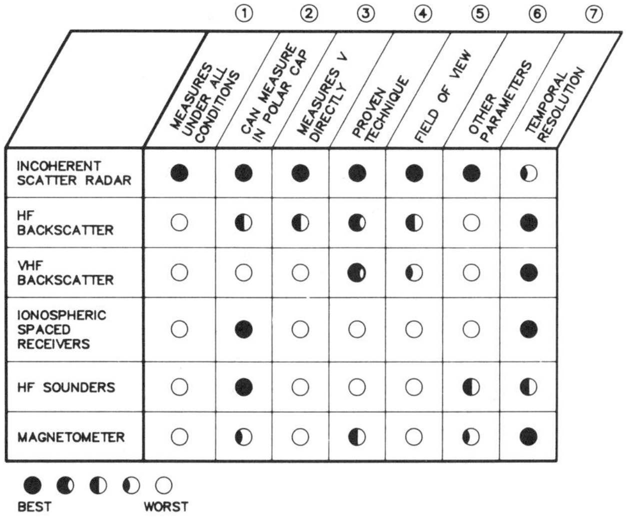

	① Measures under all conditions	② Can measure in polar cap	③ Measures V directly	④ Proven technique	⑤ Field of view	⑥ Other parameters	⑦ Temporal resolution
INCOHERENT SCATTER RADAR	●	●	●	●	●	●	◔
HF BACKSCATTER	○	◑	◑	◕	◑	○	●
VHF BACKSCATTER	○	○	○	◕	◔	○	●
IONOSPHERIC SPACED RECEIVERS	○	●	○	○	○	○	●
HF SOUNDERS	○	●	○	○	○	◑	◑
MAGNETOMETER	○	◔	○	◑	○	◔	●

● ◕ ◑ ◔ ○
BEST — WORST

KEY TO CRITERIA:

① Does the measurement of the ionospheric F-region plasma drift depend on specific conditions?

② Is measurement possible or valid in the polar cap?

③ Is the F-region plasma drift a parameter that can be directly and unambiguously obtained from the measurement?

④ Does unambiguous proof exist that the derived drift is a reliable measure of the actual F-region plasma drift?

⑤ What is the field of view of the instruments?

⑥ Does the technique simultaneously provide other physical parameters of interest for magnetosphere/ionosphere coupling studies?

⑦ How long does it take to make one measurement over the instrument's field of view?

Fig. 3. Ratings of the various techniques to measure the convection electric field.

Instrument Location

Our aim is to be able to obtain global maps of the convection in the polar ionosphere. At present, the number of instruments that can measure the electric field in the northern polar regions is very small, consisting of three ISRs and four backscatter systems (see Figure 2). In addition, about 40 magnetometers are in operation, a number considerably smaller than during the International Magnetospheric Study (IMS) period.

These instruments do provide extremely valuable data. The coverage they provide along the longitude from Scandinavia to central Canada is quite good. A number of studies are planned, or have been undertaken, to use simultaneous data acquired by these instruments during specific observation campaigns. For example, Heelis et al. (1983) and Richmond et al. (1988) inferred convection patterns by combining multiple instrument data taken during campaigns such as the Magnetosphere Ionosphere Thermosphere Radar Studies (MITHRAS) and the Global Ionosphere Study and Measurement of Substorms (GISMOS).

To achieve good coverage in the polar cap and auroral regions, it is clear from Figure 2 that several additional instruments need to be deployed. Because the different instruments complement each other, it is logical to assume that a variety of instruments will be required. However, it is too early to indicate precisely where instruments should be located.

Although feasibility studies, engineering studies, and cost estimates need to be done, some very preliminary ideas can be suggested to illustrate possibilities. An example is shown in Figure 4. To measure polar cap convection, it would be efficient to place an instrument close to the geomagnetic pole, so as to measure the convection all around. Thule is an excellent location for any instrument. A series of HF radar systems could be installed that would illuminate portions of arc segments. Building four systems similar to that at Goose Bay, arranged to look in four directions, might be necessary. A single ISR would illuminate all the area described by a circle of about 20° to 50° in diameter. This radar's field-of-view could be similar to that of Millstone Hill (one considerably larger than that of Sondrestrom or EISCAT)

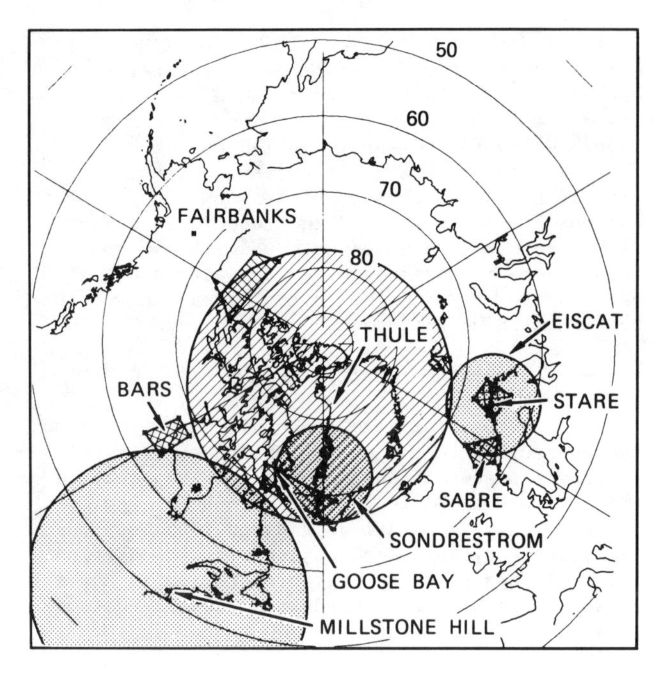

Fig. 4. Same as Figure 2, with, in addition, the fields-of-view of two proposed instruments: an ISR at Thule and an HF backscatter radar at Fairbanks.

because it might not have to be located in a valley and it would operate at a lower frequency. The Thule ISRs field-of-view would then cover a large portion of the polar cap.

If an HF backscatter system were also placed in Alaska, the electrostatic potential along a dawn/dusk meridian would be determined part of the time by the ensemble of radars from EISCAT to Alaska.

The already existing meridian chain of radars that extends from Sondrestrom, Goose Bay, Millstone Hill, and down to lower latitudes would be roughly perpendicular to this new meridian axis.

Conclusions

Even though considerable progress has been made in recent years, uncertainties concerning the global convection pattern have hindered our efforts to understand important aspects of the various coupling processes in the solar wind magnetosphere-ionosphere-thermosphere system. A companion paper stressed the importance of measuring simultaneously the plasma convection and the distribution of auroras.

In this paper, we have discussed several techniques for measuring electric fields in the polar cap and auroral regions and compared their respective merits. Figure 3 summarizes the results. Few techniques offer as many advantages as ISRs and backscatter systems. However, each offers a different set of advantages and disadvantages. Backscatter systems have a better time resolution, give instantaneous drift measurements with good spatial resolution over a two-dimensional region, and can be operated continuously. Nonetheless, it is not yet known what percentage of the time an appreciable backscatter signal can actually be measured, especially in the polar cap. These systems measure only one ionospheric parameter, the Doppler velocity, and would thus need to be supplemented by other instruments to measure ionospheric densities from which auroral precipitation can be inferred. The ISRs provide a large number of ionospheric and thermospheric parameters over a wide area and over a large altitude range. They can operate under all conditions. However, the temporal resolution for sampling their entire field-of-view is not very good. In conclusion, these instruments complement each other, with each addressing a different set of magnetosphere-ionosphere coupling processes. The ISRs provide the required continuous and global coverage, and the backscatter radars provide the high-time resolution for rapidly varying phenomena within a two-dimensional region.

Before any ambitious project to measure the global electric field distribution is implemented, two types of preliminary studies should be pursued. First, the exact capabilities and limitations of various instruments should be assessed, and the existing techniques should be improved and enhanced to allow determination of the optimum set of systems to be deployed. Second, the analysis of existing multiple instrument data sets should be vigorously pursued to allow development of the data handling and numerical analysis techniques essential to this endeavor.

We conclude that, although difficulties will need to be overcome, it will be technically possible to deploy instruments to make high resolution moving pictures that show the entire ionospheric image of magnetospheric circulation.

Acknowledgments. We are grateful for useful discussions with R. Tsunoda, R. Livingston, and R. Greenwald. This work was supported by NSF cooperative agreement ATM-85-16436, by AFOSR contract F4820-87-K-007, and by NASA grant NAGW-853.

References

Aarons, J., J. P. Mullen, H. E. Whitney, A. L. Johnson, and E. J. Weber, UHF scintillation activity over polar latitudes, Geophys. Res. Lett., 8, 277-280, 1981.

Bates, H. F., S.-I. Akasofu, D. S. Kimball, and J. C. Hodges, First results from the north polar auroral radar, J. Geophys. Res., 78, 3857, 1973.

Birkeland, K., On the cause of magnetic storms

and the origin of terrestrial magnetism, section 1 in The Norwegian Aurora Polaris Expedition 1902-1903, vol. 1, H. Aschehoug, Christiana, Norway, 1908.

Coley, W. R., R. A. Heelis, W. B. Hanson, P. H. Reiff, J. R. Sharber, and J. D. Winningham, Ionospheric convection signatures and magnetic field topology, J. Geophys. Res., 92, 12,352-12,364, 1987.

Evans, J. V., Theory and practice of ionosphere study by Thomson scatter radar, in Proceedings of the IEEE, 57, pp. 496-530, 1969.

Evans, J. V., J. M. Holt, and R. H. Wand, Millstone Hill incoherent scatter observations of auroral convection over $60° < \Lambda < 75°$, I. Observing and data reduction procedures, J. Geophys. Res., 84, 7059-7074, 1979.

Farley, D. T., Radio wave scattering from the ionosphere, Chapter 14 in Methods of Experimental Physics, 9B, edited by R. Lovberg and H. Greim, Academic Press, New York, 1971.

Folkestad, K., T. Hagfors, and S. Westerlund, EISCAT: An updated description of technical characteristics and operational capabilities, Radio Science, 18, 867, 1983.

Foster, J. C., J. M. Holt, R. G. Musgrove, and D. S. Evans, Ionospheric convection associated with discrete levels of particle precipitation, Geophys. Res. Lett., 13, 656-659, 1986.

Fremouw, E. J., J. A. Secan, and J. M. Lansinger, Spectral behavior of phase scintillation in the nighttime auroral region, Radio Science, 20, 923, 1985.

Greenwald, R. A., W. Weiss, E. Nielsen, and N. R. Thomson, STARE: A new radar auroral backscatter experiment in northern Scandinavia, Radio Science, 13, 1021, 1978.

Greenwald, R. A., K. B. Baker, R. A. Hutchins, and C. Hanuise, An HF phased-array radar for studying small-scale structure in the high-latitude ionosphere, Radio Science, 20, 63, 1985.

Harang, L., The mean field of disturbance of polar geomagnetic storms, J. Geophys. Res., 51, 353, 1946.

Heelis, R. A., J. C. Foster, O. de la Beaujardiere, and J. Holt, Multistation measurements of high-latitude ionospheric convection, J. Geophys. Res., 88, 10,111-10,121, 1983.

Heppner, J. P., and N. C. Maynard, Empirical high-latitude electric field models, J. Geophys. Res., 92, 4467-4489, 1987.

Heppner, J. P., J. D. Stolarik, and E. M. Wescott, Electric field measurements and the identification of currents causing magnetic disturbances on the polar cap, J. Geophys. Res., 76, 6028, 1971.

Hughes, T. J., and G. Rostoker, Current flow in the magnetosphere and ionosphere during periods of moderate activity, J. Geophys. Res., 82, 2271-2282, 1977.

Hughes, T. J., and G. Rostoker, A comprehensive model current system for high-latitude magnetic activity - I. The steady state system, Geophys. J.R. Astr. Soc., 58, 525-569, 1979.

Jones, T. B., J. A. Waldock, E. C. Thomas, C. P. Stewart, and T. R. Robinson, SABRE radar observations in the auroral ionosphere, in AGARD, Conference Proceedings, Propagation Effects on Military Systems in the High Latitude Region, pp. 6.4.1-6.4.15, 1987.

Kelly, J. D., Sondrestrom radar--initial results, Geophys. Res. Lett., 10, 1112-1115, 1983.

Lyons, L. R., and O. de la Beaujardière, Critical problems requiring coordinated measurements of large-scale electric field and auroral distribution, these proceedings, 1989.

Moller, H. G., Backscatter results from Lindau--II. The movement of curtains of intense irregularities in the polar F-layer, J. Atmos. Terr. Phys., 36, 1487, 1974.

Nekrasov, Y., A. V. Shirochkov, and I. A. Shumilov, Investigation of the irregular structure of the polar ionosphere using oblique incidence soundings, J. Atmos. Terr. Phys., 44, 769, 1982.

Nielsen, E., and J. D. Whitehead, Radar auroral observations and ionospheric electric fields, Adv. Space Res., 2, 131, 1983.

Nielsen, E., and K. Schlegel, Coherent radar Doppler measurements and their relationship to the ionosphere electron drift velocity, J. Geophys. Res., 90, 3498, 1985.

Reiff, P. H., and J. G. Luhmann, Solar wind control of the polar-cap voltage, in Solar Wind-Magnetosphere Coupling, edited by Y. Kamide and J. A. Slavin, pp. 453-476, Terra Scientific Publ. Co., Tokyo, 1986.

Reinleitner, L. A., and E. Nielsen, Self-consistent analysis of electron drift velocity measurements with the STARE/SABRE system, J. Geophys. Res., 90, 8479, 1985.

Richmond, A. D., et al., Mapping electrodynamic features of the high-latitude ionosphere from localized observations: Combined incoherent-scatter radar and magnetometer measurements for January 18-19, 1984, J. Geophys. Res., 93, 5760, 1988.

Rino, C. L., and R. C. Livingston, On the analysis and interpretation of spaced-receiver measurements of transionospheric radio waves, Radio Science., 17, 845-854, 1982.

Ruohoniemi, J. M., R. A. Greenwald, K. B. Baker, and J. P. Villain, Drift motions of small-scale irregularities in the high-latitude F region: An experimental comparison with plasma drift motions, J. Geophys. Res., 92, 4553-4564, 1987.

Sojka, J. J., and R. W. Schunk, Theoretical study of the high-latitude ionosphere's response to multicell convection patterns, J. Geophys. Res., 92, 8733-8744, 1987.

Strom, G. B., A. Brekke, O. Bratteng, and F. N.

Klokkervoll, Polar F-region plasma flow as observed by a spaced receiver system at Ny-Alesund, Svalbard, Annales Geophys., 4, 107-112, 1986.

Tsunoda, R. T., High-latitude F-region irregularities: A review and synthesis, Rev. Geophys., in press, 1988.

Wickwar, V. B., J. D. Kelly, O. de la Beaujardière, C. A. Leger, F. Steenstrup, and C. H. Dawson, Sondrestrom overview, Geophys. Res. Lett., 11, 883-886, 1984.

Willis, D. M., M. Lockwood, S. W. H. Cowley, A. P. van Eyken, B. J. I. Bromage, H. Rishbeth, P. R. Smith, and S. R. Crothers, A survey of simultaneous observations of the high-latitude and interplanetary magnetic field with EISCAT and AMPTE-UKS, J. Atmos. Terr. Phys., 48, 987, 1986.

SURFACE WAVES ON A GENERALIZED CURRENT SHEET

S. T. Suess and Z. E. Musielak*

Space Science Laboratory, NASA Marshall Space Flight Center, Huntsville, AL 35812

Abstract. We extend the theory for MHD surface waves in a non-isothermal layer of finite thickness in two different examples. The first is a tangential "discontinuity" in which the magnetic field undergoes a rotational change in direction parallel to the layer with no change in amplitude. The second is a sheet pinch across which the field undergoes a change in amplitude with no rotation or reversal. The field amplitude does not go through zero in either of these examples. Bound, normal mode surface wave solutions are found to be supported by both the sheet pinch and tangential discontinuities in which the field vector undergoes a limited rotation of less that 90°. However, field rotation does not always remove singularities in the wave equations and wave decay through mode conversion can then occur.

Introduction

Surface waves may occur when there is a change in physical properties over a small dimension compared to the wavelength. The waves propagate parallel to the surface and do not exist in the absence of the gradients. Many examples exist in nature, depending on the restoring force. A common example is a gravity wave on the surface of the ocean. In space physics, surface waves occur on interfaces sustained by a balance between magnetic and thermal pressure. Such interfaces are common and can often be taken as quasi-two-dimensional if the thickness is small compared to the curvature of the interface. Examples include the surface of magnetic flux tubes in the solar corona, the heliospheric current sheet and tangential discontinuities (i.e., continuously structured layers across which the total change in plasma and field quantities equals a tangential discontinuity) in the solar wind, and the magnetopause and plasma sheet in the Earth's magnetosphere. There is comparable, if not greater, interest in these waves in the laboratory plasma physics community because they can be used to heat magnetically confined plasmas.

* NRC-NAS Resident Research Associate

A large body of literature has been published on these waves in various contexts and under various assumptions. A brief, partial list of contributions in space science includes Davila (1985), Lee and Roberts (1986), Musielak and Suess (1988a,b), Hollweg (1982, 1987), Hopcraft and Smith (1986), Nenovski and Momchilov (1987), Edwin, et al. (1986), Chen and Hasegawa (1974a,b), and Roberts (1981). The main purpose of listing these references is to indicate the sustained level of interest in these waves and to give a few sources for further information.

A number of studies treat the interface in the limit of zero thickness. Using the linearized ideal MHD equations for an incompressible and non-isothermal sheet discontinuity lying in the x-y plane, in pressure balance, and with the ambient magnetic field lying in the y-direction, the wave equation for propagation in the y-direction reduces to (Preist, 1982)

$$D_a\left[V_s^2\frac{d^2U_{pz}}{dz^2}+D_sU_{pz}\right]=0 \tag{1}$$

where

$$\begin{aligned} D_a &= \omega^2-k^2V_a^2, \\ D_s &= \omega^2-k^2V_s^2, \\ V_s &= \sqrt{RT/\mu}, \\ V_a &= \sqrt{B^2/(4\pi\rho)}, \end{aligned}$$

k is the wave number in the y-direction, U_{pz} is the perturbation velocity normal to the layer, ω is the circular wave frequency, and R, B, T, and ρ are the gas constant and ambient magnetic field strength, temperature, and density, respectively.

Equation (1) describes surface waves when $D_s<0$ and unbound waves when $D_s\geq 0$, a case of no interest here. The surface waves are discrete magnetoacoustic modes and their solutions, of the form

$$U_{pz}\propto \exp\left[\left(\frac{-D_s}{V_s^2}\right)^{1/2}\mid z\mid\right] \tag{2}$$

are reviewed by Priest (1982).

The restrictions on this solution, that it be linear and non-dissipative, are potentially significant because sheet pinches are known to be resistively unstable and to exhibit turbulent behavior in the nonlinear regime. The growth time of resistive instabilities is typically the square root of the Alfvén time multiplied by the resistive decay time; this is much longer than the time scales of interest in the corona and interplanetary medium so it is not of any concern. Conversely, nonlinear interactions are likely to be important in both the corona and interplanetary medium given what we know about typical wave amplitudes. But, we ignore this complication for the present purpose of finding configurations in addition to a simple discontinuity for which surface waves do not undergo linear decay.

It is well known (e.g., Priest, 1982) that if the discontinuous interface is replaced by one with finite width with continuous density and magnetic field profiles, a normal mode solution no longer always exists. If the layer is a neutral sheet--across which the magnetic field reverses by its amplitude going through zero while the vector remains in the y-direction--the wave equation under conditions as above, but now permitting compressible motion, reduces to (Musielak and Suess, 1988b)

$$D_a U_{pz} + \frac{1}{\rho_o}\frac{d}{dz}\left[\rho_o \frac{D_a D_{as}}{D_f D_{sl}}\left(\frac{dU_{pz}}{dz}\right)\right] = 0 \tag{3}$$

where

$$\begin{aligned}
\vec{B}_o &= B_o(z)\hat{e}_y, \\
D_f &= \omega^2 - k^2 V_f^2, \\
D_{sl} &= \omega^2 - k^2 V_{sl}^2, \\
D_{as} &= D_a V_s^2 + \omega^2 V_a^2, \\
V_a &= B_o/\sqrt{4\pi\rho_o}, \\
V_f^2 &= max(V_s^2, V_a^2), \\
V_{sl}^2 &= min(V_s^2, V_a^2),
\end{aligned}$$

and $B_o(z)$ is an arbitrary function of z. Again applying the requirement that $D_a < 0$ outside the layer for surface waves--so that the solutions decay exponentially away from the layer--it is seen that there will always be a singularity in (3) where $D_a = 0$. The presence of this singularity implies that there is no nontrivial normal mode solution to (3). Detailed studies (Lee and Roberts, 1986) have shown that there is a continuous spectrum of solutions which decay through mode conversion with the associated energy flowing into local oscillations within the interface.

Such is not the case if the field does not go through zero in the layer, so that the layer represents a change in field strength alone--this is referred to here as a sheet pinch. One purpose of this paper is to show that the sheet pinch is able to sustain normal mode solutions and to solve the wave equation for one example.

A further generalization of the layer structure is necessary to reflect the properties of, for example, the structure within tangential discontinuities in the solar wind. This is that the magnetic field change direction through a vector rotation. The second purpose of this paper is to show that layers having magnetic field vector rotations of less than 90° are able to sustain normal mode solutions and to solve the wave equation for one example.

Sheet Pinch

The sheet pinch is an acceptable approximation to the boundary of magnetic flux tubes in the solar atmosphere and to the boundary of the plasma sheet in the Earth's magnetotail. We consider the sheet pinch in the low-β limit $V_a^2 >> V_s^2$, which leads to the wave equation (Musielak and Suess, 1988b)

$$\frac{1}{B_o^2}\frac{d}{dz}\left[B_o^2\left(\frac{dU_{pz}}{dz}\right)\right] = k_a^2 U_{pz} \tag{4}$$

where $k_a^2 = k^2 - \omega^2/V_a^2$, and, as before, $\vec{B}_o = B_o(z)\hat{e}_y$. Again, this is for propagation in the y-direction, or along the ambient magnetic field. In this case, we expect only two types of surface waves: purely transverse fast mode waves (identical to Alfvén waves) with $U_{pz} \neq 0$ and purely longitudinal slow mode waves with $U_{py} \neq 0$. There are no singularities in this equation since B_o^2 is never zero. If we assume the magnetic field increases monotonically with z from $B_o = B_{o1}$ at z_1 to $B_o = B_{o2}$ at z_2 and that between these two boundaries the field is given by

$$B_o^2 = A_1 + A_2 z$$

where A_1 and A_2 are constants given by

$$A_1 = \frac{B_{o1}^2 z_2 - B_{o2}^2 z_1}{z_2 - z_1}$$
$$A_2 = \frac{B_{o2}^2 - B_{o1}^2}{z_2 - z_1}$$

then we can solve (4) and for the associated density variation. The wave equation reduces to

$$(A_1 + A_2 z)\frac{d^2 U_{pz}}{dz^2} + A_2 \frac{dU_{pz}}{dz} - (A_1 + A_2 z)k_a^2 U_{pz} = 0 \tag{5}$$

where $k_a^2 > 0$ because we are solving for surface waves. For $k_a = const$ (which can be satisfied for a non-isothermal layer), this equation can be transformed into a modified Bessel's equation and its solution, proportional to the zeroth-order Macdonald function, K_o (a modified Bessel function (Watson, 1966)), given by

$$U_{pz} = C_1 K_o \left((A_1 + A_2 z) \frac{k_a}{A_2} \right) \tag{6}$$

where C_1 is a constant of integration. The function K_o decreases to zero as $z \to \infty$ and is smooth and well behaved so long as the argument does not go through zero (a plot is shown in connection with the next example). These waves are normal modes as the whole layer supports a collective oscillation with one frequency. The amplitude of the external, exponential solution is determined by requiring U_{pz} to be continuous at z_1 and z_2.

Field Rotation with Constant Field Strength

For our second example we consider a current sheet in the simple case when there are no gradients in density ($\rho_o = const$) and in the magnitude of the magnetic field ($|\vec{B}_o| = const$) across the layer. This is an extension of the calculations recently reported by Uberoi (1988). Letting Θ be the angle between the magnetic field on opposite sides of the layer, we look for solutions for $0^o \le \Theta \le 180^o$, again in the low-β limit and propagation parallel to the y-axis. The wave equation for this problem is (Musielak and Suess, 1988a)

$$\frac{d}{dz}\left[D_a(z) \left(\frac{dU_{pz}}{dz} \right) \right] = k_a^2 D_a(z) U_{pz} \tag{7}$$

where $D_a(z) = \omega^2 - k^2 V_a^2 \cos^2 \alpha(z)$ and the magnetic field variation in the layer is now defined by

$$\vec{B}_o = B_o(z)[\hat{e}_x \sin \alpha(z) + \hat{e}_y \cos \alpha(z)]. \tag{8}$$

$B_o(z)$ can be an arbitrary function of z, although it is taken to be constant in this example. $\alpha(z)$ is the angle between the direction of the magnetic field and the y-axis. Arbitrarily, we define $\alpha = 0$ for $z \ge d$. The geometry for this calculation is illustrated in the two panels in Figure 1 for the case that $\Theta = 180^o$. The top panel shows the field reversal layer with external quantities indicated by an subscript e. The bottom panel suggests the rotation of the magnetic field vector at an arbitrary location inside the layer.

For surface waves, $k_a^2 > 0$ and $D_a(z = d) < 0$. Therefore, D_a will always be zero somewhere in the layer for $\Theta_o \ge 90^o$ (see Musielak and Suess, 1988a, for further discussion). However, for $\Theta < 90^o$, there are values of ω for which $k_a^2 > 0$ while $D_a(z) < 0$ for all z, satisfying the requirements for collective motion of the layer. This type of behavior commonly occurs in the solar wind, for example on the boundaries of magnetic clouds (Suess, 1988; Klein and Burlaga, 1982).

We now show an example of a solution to (7) in the case that $\Theta < 90^o$, assuming that the wave frequencies ω are low enough to have the condition $D_a(z) < 0$ satisfied everywhere in the layer. First, defining

$$\eta = z/d$$
$$\sigma = \omega^2/(kV_a)^2,$$

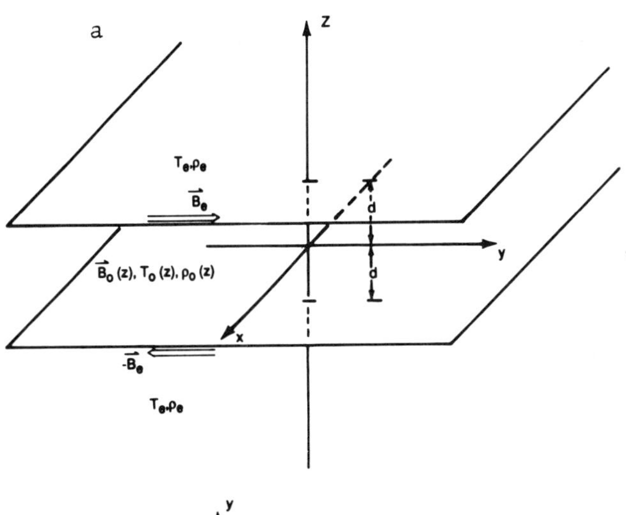

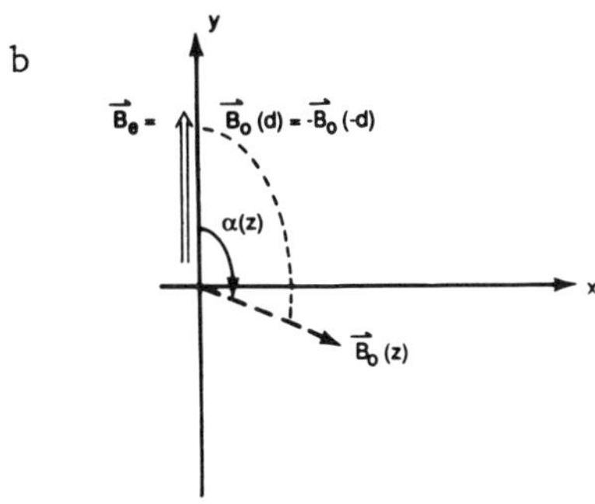

Fig. 1. Sketch of the current sheet layer structure for a magnetic field reversal via vector rotation, resulting in the wave equation (7). The top panel defines the coordinate system and the bottom panel suggests the rotation of the magnetic field vector at an arbitrary position within the layer.

we consider a simple situation in which the field rotation is given by

$$\vec{B}_o(z) = B_o[\hat{e}_x \sqrt{1-\eta} + \hat{e}_y \sqrt{\eta}] \tag{9}$$

with $B_o = const$. Then, wave equation (7) can be written in the form

$$(\sigma - \eta)\frac{d^2 U_{pz}}{d\eta^2} - \frac{dU_{pz}}{d\eta} - \lambda_a^2(\sigma - \eta) U_{pz} = 0, \tag{10}$$

where $\lambda_a^2 = k^2 d^2 (1 - \sigma)$. In this notation, surface waves can exist when $\sigma < 1$.

Equation (10) is, like equation (7), a modified Bessel's equation with its solution also being proportional to the zeroth-order Macdonald function (Watson, 1966), so that

$$U_{pz}(\eta) = C K_o((\sigma - \eta)\lambda_a) \tag{11}$$

where C is a constant of integration. The function K_o is well determined for $\lambda_a > 0$. Solution

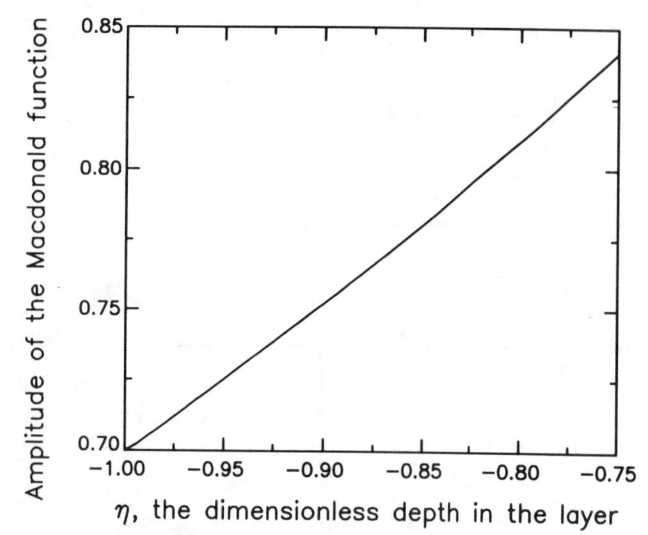

Fig. 2. Variation of the amplitude of a zeroth order Macdonald function across a layer extending only from $\eta = -1$ to $\eta = -0.75$. The argument of the function is $(\sigma - \eta)\lambda_a$ with $\sigma = 0.5$ and $\lambda_a = 0.443$.

(11) describes the amplitude of Alfvénic surface waves inside a current sheet with incomplete reversal of the magnetic field. In this case, the whole layer supports a collective motion with one frequency, thus the solution represents normal mode MHD surface waves.

If we choose $\sigma = 0.5$, $V_a = 10^7$ cm/s, $d = 10^8$ cm, and $k = 2\pi \times 10^{-9}$ cm^{-1} (numbers appropriate to a current sheet in the solar wind), and prescribe a 30° rotation in the field across the current sheet (such that $\Theta = -30^o$), then the resulting variation of the Macdonald function is shown in Figure 2. In this figure, η is constrained to lie between -1 and -0.75 by the limit on the total rotation angle of the field and the choice made in equation (9) for the way the field varies in the layer. The Macdonald function is, as stated, found to vary smoothly. The only irregularity that would be encountered is if the argument became zero, which is eliminated by the restriction on Θ and the above set of parameters.

Discussion

The purpose of this paper has been to describe configuration in which normal mode solutions for MHD surface waves exists that exhibit no decay through mode conversion. These configurations are sheet pinches and tangential "discontinuities" across which the magnetic field undergoes a rotation of less than 90°. On one hand, this means such waves will not contribute to local heating. On the other hand, these surface waves are ducted by the interface and can carry energy to large distance from their point of origin.

Acknowledgments. This research has been supported by the NASA Solar and Heliospheric Physics Branch and by the Space Plasma Physics Branch in the Office of Space Science and Applications. One of the authors (ZEM) held an NRC-NAS Resident Research Associateship while conducting this research.

References

Chen, L., and A. Hasegawa, A theory of long-period magnetic pulsations 1. Steady state excitation of field line resonance, *J. Geophys. Res.*, **79**, 1024, 1974a.

Chen, L., and A. Hasegawa, A theory of long-period magnetic pulsations 2. Impulse excitation of surface eigenmode, *J. Geophys. Res.*, **79**, 1033, 1974b.

Davila, J. M., A leaky magnetohydrodynamic waveguide model for the acceleration of high-speed streams in coronal holes, *Ap. J.*, **291**, 328, 1985.

Edwin, P. M., B. Roberts, and W. J. Hughes, Dispersive ducting of MHD waves in the plasma sheet: A source of Pi2 wave bursts, *Geophys. Res. Lett.*, **13**, 373, 1986.

Hollweg, J. V., Surface waves on solar wind tangential discontinuities, *J. Geophys. Res.*, **87**, 8065, 1982.

Hollweg, J. V., Resonance absorption of magnetohydrodynamic surface waves: Viscous effects, *Astrophys. J.*, **320**, 875, 1987.

Hopcraft, K. I., and P. R. Smith, Magnetohydrodynamic waves in a neutral sheet, *Planet. and Space Sci.*, **34**, 1253, 1986.

Klein, L. W., and L. F. Burlaga, Interplanetary magnetic clouds, *J. Geophys. Res.*, **87**, 613-624, 1982.

Lee, M. A., and B. Roberts, On the behavior of hydromagnetic surface waves, *Ap. J.*, **301**, 430, 1986.

Musielak, Z. E., and S. T. Suess, MHD bending waves in a current sheet, *Ap. J.*, **330**, 456-465, 1988a.

Musielak, Z. E., and S. T. Suess, MHD surface waves in high and low-beta plasmas. I. Normal mode solutions for low-beta sheet pinch structure, *J. Plasma Phys.*, submitted, 1988b.

Nenovski, P., and G. Momchilov, Model of polarization state of compressional surface MHD waves in the low-altitude cusp, *Planet. Space Sci.*, **35**, 1561, 1987.

Priest, E. R., *Solar Magnetohydrodynamics*, D. Reidel Publ. Co., Drodrecht, Holland, pp. 182-186, 1982.

Roberts, B., Wave propagation in a magnetically structured atmosphere I: Surface waves at a magnetic interface, *Solar Phys.*, **69**, 27, 1981.

Suess, S. T., Magnetic clouds and the pinch effect, *J. Geophys. Res.*, **93**, 5437-5445, 1988.

Uberoi, C., The Alfvén wave equation for magnetospheric plasmas, *J. Geophys. Res.*, **93**, 295, 1988.

Watson, G. N., *A Treatise on the Theory of Bessel Functions*, 2nd edition, Cambridge University Press, Cambridge, pp. 75-84, 1966.

THE ASPERA EXPERIMENT ON THE SOVIET PHOBOS SPACECRAFT

R. Lundin,[1] B. Hultqvist,[1] S. Olsen,[1] R. Pellinen,[2] I. Liede,[2] A. Zakharov,[3] E. Dubinin,[3] N. Pissarenko[3]

Abstract. ASPERA (Automatic Space Plasma Experiment with a Rotating Analyzer) is a three-dimensional plasma composition experiment for the Soviet mission to Mars and its moon Phobos in 1988-1989. Two spacecraft are targeted for a close flyby over the Phobos surface. ASPERA measures the composition of ions with energies 0.5 eV/e-25 keV/e and electrons with energies 1 eV-50 keV. The experiment utilizes a scanner platform to provide nearly complete coverage of the unit sphere on the three-axis stabilized Phobos spacecraft. The plasma analyzer comprises two spectrometer systems with a 360° field-of-view lying in a plane perpendicular to the plane of rotation (±90°). The 360° field-of-view is divided into 10 sectors for ions and 6 sectors for electrons, all sectors containing individual sensor elements. The unit sphere is covered after a 180° turn of the scanner platform.

Introduction

The Phobos project is a Soviet multi-disciplinary mission to the planet Mars and its moon Phobos. The launch is scheduled for July 7 and July 12, 1988, and the first rendevous with Phobos is expected to take place in the beginning of April 1989. If the first spacecraft is successfully completing its program the second spacecraft will possibly be rerouted for a Deimos flyby. The mission has three main objectives: to study the sun and interplanetary space (cruise phase), to study the Martian aeronomy and plasma environment (orbiting phase), and to study Phobos (Deimos). The ASPERA experiment is designed to address all three objectives.

Cruise Phase

During the approximately 7-month cruise phase the Phobos spacecraft (s/c) will perform solar-interplanetary studies of the solar wind. The ASPERA experiment is capable of measuring the plasma moments and composition of the solar wind with a time resolution of about 2 min (normal rate) or ≈20 s (event rate). An event triggering procedure , initiated by either internal (ASPERA) or by external means (e.g., the magnetometer) will enhance the data rate during, interplanetary shock crossings, for example.

The Phobos s/c is expected to provide several geotail crossings in September/October 1988. Attempts will be made to cover these tail crossings with maximum available data rate.

Mars

Important data are now available about the planet´s surface and its geological features, most of it obtained from the US Viking mission. However, many other fundamental fields of Martian science have remained relatively unexplored. For example, the Martian plasma environment is not as well understood as the more remotely located planets Jupiter and Saturn.

One of the most controversial questions, whether Mars has an intrinsic magnetic field, is still unresolved. Magnetic field measurements performed by the Soviet Mars 2, 3, and 5 orbiters (Dolginov et al., 1976) suggested a Martian magnetic dipole moment of the order $\approx 2.5 \times 10^{-4}$ M_E (M_E = Earth´s dipole moment) ; the general orientation of the dipole moment being opposite to that of the Earth. Russell et al. (1978a,b) criticized this interpretation and argued that, because of the bow shock location, the solar wind interacts directly with the Martian atmosphere and the dipole moment should consequently be less than $\approx 2.5 \times 10^{-5}$ M_E. On the other hand, Intriligator and Smith (1979) have concluded that the Martian atmospheric pressure is a factor of 4 too small to explain the location of the shock, indicating a magnetic moment of at least $\approx 10^{-4}$ M_E.

The Martian charged particle environment has remained comparatively little unexplored, the data again originating from the Soviet Mars missions (2, 3, 4, and 5). The Faraday cup data by Gringauz et al. (1976) were the first to confirm the existence of the bow shock and magnetopause, previously identified using the magnetometer. They also detected a region where the plasma density dropped below the measurement threshold, possibly disclosing the existence of a magnetospheric cavity. Alternatively the apparent drop in plasma density could have been due to the limited energy and angular coverage of the instruments. Based on an electrostatic analyzer system with a limited angular coverage on the Mars probes, Vaisberg et al. (1976) also argued that the smooth transition from the shocked solar wind plasma to the hot plasma "cushion" above the

---

[1]Swedish Institute of Space Physics, P.O. Box 812, S-981 28 Kiruna, Sweden.

[2]Finnish Meteorological Institute, P.O. Box 503, SF-00101 Helsinki, Finland.

[3]Space Research Institute of the Academia Nauk Profsojusnaja 88, 117810, Moscow, USSR.

ionosphere very much resembles the solar wind interaction with Venus, a planet without an intrinsic magnetic field.

The existence of a plasma tail near Mars was demonstrated by Vaisberg et al. (1986). They showed that the Martian magnetotail differed from the terrestrial magnetotail in the magnetotail boundary layer. Near Mars the boundary layer appeared to be dominated by a low-energy component of ionospheric origin. Vaisberg et al. (1986) also argued that the ionospheric component was dominated by O^+ ions. The plasma tail proper seemed to be dominated by a cold plasma component of ionospheric origin.

Ion composition measurements in the Earth´s plasma mantle by Lundin et al. (1982) demonstrated a substantial escape of O^+ ions. However, the percentage of O^+ was usually less than 10% in the terrestrial plasma mantle. This means that the relative loss of Martian ionospheric plasma may be much higher than that from the Earth´s ionosphere. The anticipated terrestrial net loss of about 3 kg s^{-1} ($\approx 2 \times 10^{26}$ ions s^{-1}) of the upper atmosphere via the magnetotail is insignificant on a cosmogonic time scale. However, a one order of magnitude higher loss is not. This means that Mars may have been exposed to a significant atmospheric evacuation on a cosmogonic time scale.

Phobos (Deimos)

The Martian satellites Deimos and Phobos, which by some scientists are believed to be captured asteroids, are very interesting celestial objects. Phobos, the innermost Martian satellite, shows traces of a spectacular evolution process. The Phobos surface is furrowed and contains many large craters, evidence of extensive meteorite bombardment. The mean density of Phobos is also only about 2 g cm^{-3}; i.e., it is likely to contain a significant fraction of volatile elements (e.g., H_2O and CO_2 ice).

Altogether, its small size ($\approx$27 by 15 km) and very low mass density indicates that it is in a state close to primordial. This is perhaps the most compelling reason for a close encounter and landing on Phobos. The Phobos spacecraft will be the first to take a very close look at a celestial object which is expected to bear traces from the very early time of the solar system formation.

The possible existence of volatile elements on Phobos and Deimos suggests that they may outgas at such a rate that they interact with the solar wind like a comet. Magnetic field observations by Bogdanov (1981) indicated that this was, in fact, the case for Deimos. However, no plasma measurements are yet available to prove, or disprove, this hypothesis. Thus, investigations of the plasma environment near the Martian satellites are important objectives for ASPERA.

The atmospheric loss process is a topic which raises much speculation. For instance, one may argue that atmospheric loss cannot occur as a result of a neutral gas outflow only, since such a gas will still be in a Kepler orbit around Mars similar to that of Phobos itself. There is a great probability that the same matter will eventually accrete on Phobos again. The most likely loss is instead due to ionization of the gaseous substances and a subsequent interaction with the solar wind. Once ionized, the gas will be strongly influenced by electric and magnetic forces. The process can be envisaged as one where the solar wind plasma carries away the ionized "atmosphere" in the anti-sunward direction.

In the cosmogonic theory by Alfvén (1954), matter in the solar nebula was strongly controlled by plasma-physical processes before the actual condensation into planetisimals occurred. However, Alfvén and Arrhenius (1976) also point out that fine grains with masses corresponding to 10^6 and up to 10^{12} of the molecular weight may,if electrically charged, be strongly controlled by magnetic and electric forces. Such a "dusty" plasma is then likely to be the source of larger grains which eventually accrete into larger bodies. To study the "dusty" plasma in the Phobos orbit is therefore another important objective for ASPERA. The very low gravitation field of Phobos leads to a strong interaction with the solar wind of even quite heavy ionized fragments.

The above mentioned topics are all related to processes expected to occur in the immediate vicinity of Phobos--its atmosphere and plasma environment. However, a substantial part of the mission is devoted to the surface properties of Phobos. ASPERA is expected to contribute in the remote sounding of the Phobos regolith layer by three independent means. The first is by studying the backscatter of solar wind particles impinging on the Phobos surface. ASPERA is in fact the only plasma instrument with mass resolution and a 360° field-of- view which can monitor the backscattering rate of various ion species during flyby. This will make it possible to study the ion implantation rate in the regolith, for example.

The second means of surface studies is by measuring the reflected Krypton beam. During flyby ASPERA will be switched into a mode which enables an analysis of approximately every third Kr^+ beam with a time resolution down to $\approx$50 ms. The idea is that both the backscatter spectrum and the decay of the ion pulse contains information about the surface characteristics.

The third means of surface studies is by investigating the indirect ionization of gaseous substances released by the laser beam impacts in the regolith. The neutral gas plumes possibly resulting from the laser firings may become ionized by a combined solar UV and solar wind interaction. ASPERA will be the instrument best suited to study the density and composition of these clouds of artificial ionosphere above Phobos.

ASPERA Scientific Objectives

ASPERA is a multi-purpose plasma experiment designed to address different objectives during the cruise phase, the planetary orbiting phase, and the Phobos flyby.

During the cruise phase ASPERA will be operated in a low data rate mode called the "interplanetary mode" (1.5 Mbits/5 day). This mode emphasizes measurements of the solar wind density, flow, and composition. During interplanetary shock crossings, identified by, for example, enhanced energetic electron fluxes, the instrument will automatically switch into an event mode with 8 times higher data rate that also covers the 30 min preceding the shock crossing. The typical duration of an interplanetary (IP) shock, $\approx$ 1 hour, will be covered using time resolution down to $\approx$20 s (see Table 4). This time resolution may not be sufficient to cover the ramp of an IP shock.

However, the capability of using two Phobos s/c will instead provide a two point measurement of the shock characteristics. Deep magnetotail encounters at a distance of ≈1000 R_E, identified, for example, by an enhanced O^+/H^+ density ratio, should also set the experiment into the high data rate event mode.

During Mars approach the two Phobos s/c will be inserted into an elliptic ≈3 day parking orbit around Mars (transfer orbit 1). The second s/c will stay in the transfer orbit 1 until the completion of the first Phobos flyby. Several orbit corrections within the first week around Mars will bring the first s/c into a circular orbit near Phobos. The s/c will reside in this orbit after completion of the flyby. A schematic picture of Mars with the two main s/c orbits is shown in Figure 1. In Figure 1 we have also included the Deimos orbit and the approximate scaling of the "quiet" Martian magnetosphere.

During the planetary orbiting phase the experiment will be operated in a high data rate mode (up to 4.8 Mbits/day). The objective near Mars is to study the solar wind interaction with the planetary atmosphere and ionosphere. This is a topic where previous missions to the red planet have provided very limited information. One intriguing question is the magnitude of the ionospheric plasma escape. ASPERA will be able to detect escape fluxes less than 10^4 ions cm^{-2} s^{-1}; i.e., several orders of magnitude lower than the expected escape rate (up to 10^8 ions cm^{-2} s^{-1}). Does Mars have a significant intrinsic magnetic field is another important question. If this cannot be resolved by magnetometers (because of the high perigee of the spacecraft) it can be possibly revealed from the existence of "trapped" energetic electrons near Mars. ASPERA will be the first particle experiment with full three-dimensional (3-D) coverage to study the plasma in the Martian magnetosphere.

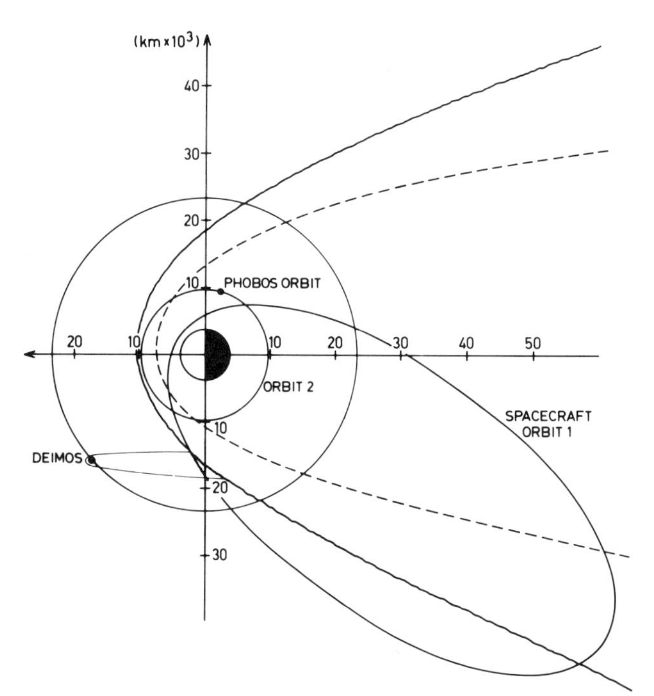

Fig. 1. Schematic picture of Mars with its two satellites Phobos and Deimos and the scale of the "quiet" Martian magnetosphere. The two main parking orbits (orbit 1 and orbit 2) are also depicted.

An important topic during the planetary orbiting phase is the solar wind interaction with Phobos and Deimos. Measurements from the Mars 5 spacecraft indicated that Deimos has a magnetic tail. Thus, Deimos may be surrounded by a thin atmosphere which makes the solar wind interaction cometary-like. The existence of heavy ions (e.g., O^+, O_2^+, H_2O^+) in the Deimos/Phobos magnetotail will be evidence for an atmospheric outgassing from the Martian satellites.

One theory suggests that Phobos and Deimos are fragments of a burst satellite (C-1 asteroid) that was much larger than the sum of the two satellites. If this is true, dusty fragments may still reside in orbit near Mars. ASPERA will therefore make a search for "dusty" plasma components, i.e., of ions with m/q $>10^2$. The identification of such ions can be made on the basis of data from the high resolution mass sensor (Figure 6b) and the expected velocity distribution of such heavy ions (ion pickup).

During the Phobos flyby, pictured in Figure 2, ASPERA will be set in a mode tailored for studying the solar wind interaction with Phobos. The instrument can determine if a thin atmosphere exists above the surface by at least two means: (1) the direct solar wind interaction with ions and neutrals (energy degradation, ionization, shock formation, etc.) during the approach and flyby, and (2) the escape of ionization products (ions and electrons) near Phobos (synchronous orbit and flyby). The good field-of-view (3-D) makes ASPERA the only onboard instrument which can completely study the deflection and backscattering of the solar wind near Phobos. Among other things this makes it possible to determine the deposition of solar wind particles into the regolith. The distribution and composition of backscattered ions and electrons can also be used to analyze the chemical properties of the Phobos regolith (e.g., backscatter spectrum of protons and electrons).

ASPERA is a support instrument for the active experiments on the spacecraft during flyby. The instrument will use about one-third of the measuring time studying backscattered Kr^+ ions from the ion gun (the DION experiment). The reflection rate of

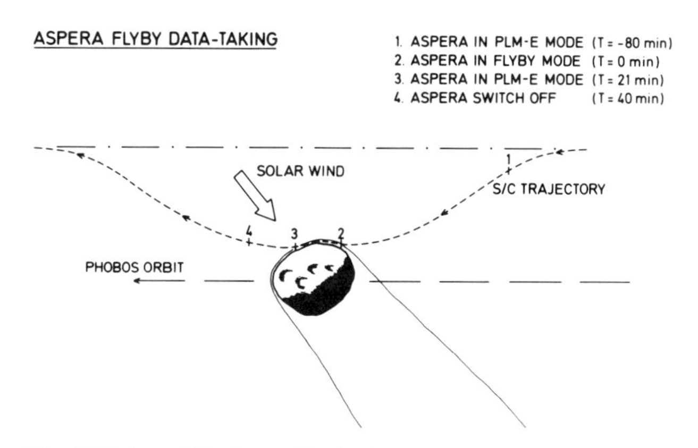

Fig. 2.Flyby of Phobos with the ASPERA flyby operations.

Kr^+ ions is expected to provide complementary information on the Phobos regolith properties.

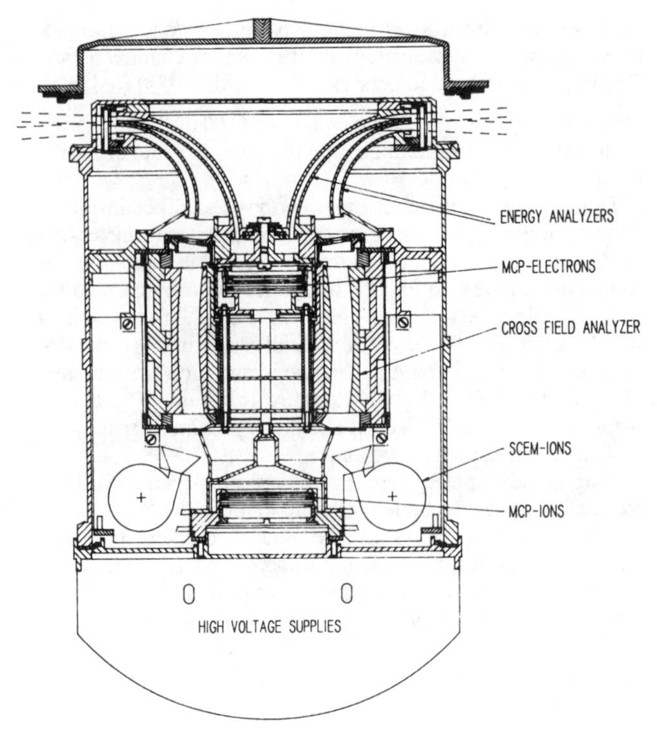

Fig. 3. Cross section of the ASPERA spectrometer unit.

The Spectrometer System

The electron spectrometer covers the energy range 1 eV-50 keV. It consists of a toroidal electrostatic energy per charge (E/q) analyzer and a microchannel plate (MCP) with six anodes as sensors. The six electron anode sectors provide an angle of acceptance of about 5° x 90°. This means an overlapping of about 30° between sensors. The main advantage with a toroidal analyzer is the stacking capability. This gives a similar azimuthal field-of-view for the ion and the electron spectrometers. However, because a toroidal system is defocussing in the azimuthal direction, it results in a spectrometer with a smaller geometric factor and a poorer angular resolution (particle trajectories crossing over). This explains the broad overlapping field-of-view of each electron sensor (Figure 7a). The calibrated response of the electron sensor (Table 1) shows that it is a good compromise, though. For instance, it enables the detection of solar wind fluxes ranging between $\approx 10^5$ and 10^{11} (cm^2 s sr $keV)^{-1}$ at 100 eV.

The ion mass spectrometer measures positive ions in the energy range 0.5 eV/q-25 keV/q. It comprises the inner part of the stacked toroidal E/q analyzer (Figure 3), placed in front of a curved cross-field mass analyzer (CFA). A split-beam technique is used to give both a low and a high mass resolution of the instrument. The single channel electron multiplier (SCEM) section of the instrument, denoted the moment sensors, has a sufficient mass resolution to resolve the "major" ion constituents (e.g., H^+, He^{++}, He^+, and O^+). The MCP section of the spectrometer, called the mass sensors, has a mass resolution that also enables separation of minor constituents (e.g., molecular ion species). Each one of the 10 moment sensors has a field-of-view of about 3° x 36°, altogether covering the 360° field-of-view in the plane perpendicular to the scanning plane (Figure 7b). The two mass sensors' fields-of-view (3° x 72°) are in the sunward and anti-sunward direction, respectively.

By means of a preacceleration voltage of -12 V, ions with very low energies ("cold" ions) will be accelerated into the analyzer system. Typical values for cold ion densities that can be measured by ASPERA are in the range 1-10^6 cm^{-3} for O_2^+ ions with a temperature of 1 eV. Cold plasma measurements are important near Mars and Phobos for studying the ionospheric plasma escape. In the cold plasma mode (0.5-500 eV) the increased resolution is utilized to enable the separation between atomic and molecular ion species.

The magnetic deflection system in the CFA represents a novel approach in closing the intrinsic field by the magnets. Since no yoke is used, only magnetic shielding by a high-mu metal is required. Despite a very low mass ($\approx$550 g), the system gives almost an order of magnitude lower stray field as compared to previously flown magnetic deflection systems. The magnetic stray field in the worst direction is no more than 1.0 nT at 1 m distance.

Figure 3 shows a cross section of the spectrometer with the different analyzer and sensor systems indicated. The sensor system has 10 SCEMs as moment sensors, a three-stage MCP for the multi-anode mass sensor, and a three-stage MCP for the multi-anode electron sensor. A block diagram of the experiment with the various functional systems is shown in Figure 4. A summary of the spectrometer characteristics and the mass and power requirements for ASPERA is given in Tables 1 and 2.

Sensor protection. The spectrometer unit is contained in a vacuum during the ground storage and launch phase. This will protect the sensors (particularly the MCPs) from contamination before launch. Once evacuated the sensor cover can only open in a vacuum and then after passing over a release mechanism during the first motor scan.

The Data Processing Unit

The ASPERA data processing unit (DPU) is built around a Texas 9989 16 bit cpu and 16 kbyte RAM memory. For reliability reasons the DPU has two redundant cpu and RAM memories. With one cpu and one RAM memory operating, the redundant cpu and RAM memory is powered off. Bootstrap loading and switching of the cpu and the RAM are controlled by a "watch-dog."

All experiment functions (high voltage settings, event triggering, sensor checking, motor scanning, power switching, etc.) are pre-programmed into the flight software. For instance, new high voltage settings for the energy and mass analyzer are computed within each time slice. The "flight program" is loaded from a 4 kbyte PROM. Updated settings of, for example, mass levels, event triggering conditions, and event duration, are

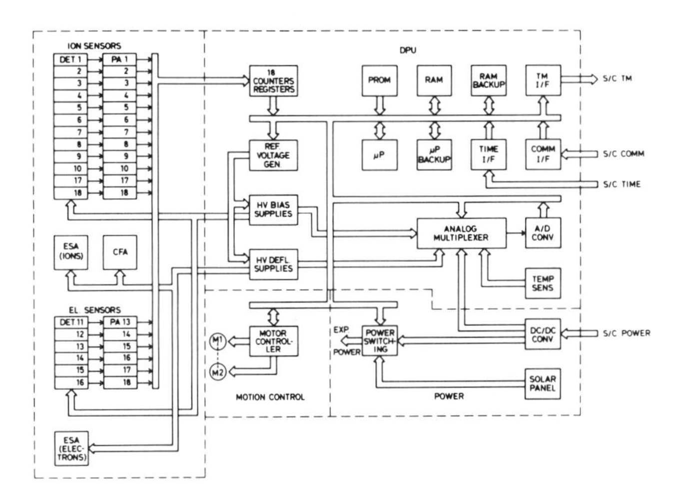

Fig. 4. Block diagram of the ASPERA experiment.

stored in the RAM. To avoid loss of memory the RAM is powered by a battery. The various systems in the DPU are described in the block diagram (Figure 4).

Data compression. To reduce the data rate, extensive onboard data analysis is performed. For instance, the first three moments of the distribution functions are computed by the microprocessor. The result is expressed as densities (N), mean flow velocities ($\langle \vec{v} \rangle$), and pressure tensors ($\tilde{P}$).

$$N = \int_0^\infty f(\vec{v})\, d^3v$$

$$\langle \vec{v} \rangle = \frac{1}{N} \int_0^\infty \vec{v} \cdot f(\vec{v})\, d^3v$$

$$\tilde{P} = \frac{1}{N} \int_0^\infty \vec{v} \cdot \vec{v} \cdot f(\vec{v})\, d^3v$$

TABLE 1. ASPERA Spectrometer Characteristics

Ions (10 + 2 sensors)	
Energy range	0.5 eV/e-25 keV/e
Energy resolution (ΔE/E)	0.10
Field-of-view per sensor	3° x 36°
Geom. factor Moment sensor	5-7 x 10^{-4} cm^2 sr
Geometric factor Mass sensor	2 x 10^{-4} cm^2 sr
Mass resolution (M/ΔM)	6 - 1
Electrons (6 sensors)	
Energy range	1.0 eV - 50 keV
Energy resolution (ΔE/E)	0.08
Field of view per sensor	5° x 90°
Geometric factor	4 x 10^{-4} cm^2 sr

TABLE 2. ASPERA Mass and Power Requirements

Total experiment mass	8.5 kg
Power consumption	8 W (average)
ASPERA solar panels	≈4 W (at 1 AU)
	≈2.5 W (at 1.5 AU)

From $\tilde{P}$ (the pressure tensor) the temperature can be deduced. The moments of the distribution function are computed for electrons and for ions of various m/q:s (at a minimum of six m/q).

Detailed ion composition data are obtained from the mass sensors. The data are accumulated in E/q and mass per charge (m/q) matrices. The basic E/q-m/q matrix consists of 32 energy levels and 128 m/q levels. Spectral information for electrons and ions is also obtained for up to six directions, but at a considerably lower rate (of the order tens of minutes) as compared to that for the plasma moments (of the order minutes).

Working Modes

ASPERA has seven different data output modes that are tailored to special interplanetary and planetary plasma regimes. For instance, during the cruise phase the instrument can be automatically triggered into an event mode by internal as well as external conditions. The event mode provides a data rate 8 times higher as compared to the normal mode.

To enable maximum data coverage near Mars, the experiment can be switched into a low power mode. ASPERA will then only require approximately 3 W from the satellite power for continuous operations.

The total telemetry output for ASPERA in various phases of the Phobos mission and for different modes of operation is given in Tables 3 and 4.

The Motor Scanner

The ASPERA scanner platform can turn the spectrometer unit 180° in about 2 min. One 180° turn provides full coverage of the

TABLE 3. ASPERA Telemetry Requirements

Phase	Data buffer (kbits)	Readout Time	Duration (days)
Cruise phase	1536	5 days	200
Transfer orbit 1	2560	3.3 days	12
Transfer orbit 2	2560	3 days	13
Circular orbit 1	4864	1 day	1
Synchronous orbit	4864	1 day	1
Flyby	768	1.5 hours	
Circular orbit 2	4864	1 day	200

TABLE 4. ASPERA Data Output for Different Modes of Operation

Mode	Data Block Size	Data Block Readout Time	Moments Time Resolution (min)				E/q Spectra Time Resol. (min)				E/q x m/q Matrices Int. Time
	(bytes)	(min)	N	$\langle v \rangle$	u^2	P	ions	el.	n_E	n_Ω	min (E x m)
IPM	480	34	2.1	8.5	8.5	-	68	68	8	3	34 (16 x 8)
IPE	3840	34	0.3	1.1	1.1	-	4.3	4.3	8	3	34 (16 x 64)
CAL	480	34	2.1	8.5	8.5	-	68	68	8	3	34 (4 x 32)
PLM	480	3.6	3.6	3.6	-	3.6	14	14	16	6	58 (32 x 64)
PLE	3840	3.6	1.8	1.8	-	1.8	3.6	3.6	32	6	29 (32 x 64)
FBM(*)	3840	3.6	13s	13s	13s	-	52s	52s	16	3	7 (16 x 64)
MON	3840	29	8.5	8.5	8.5	-	29	29	16	3	232(16 x 32)

*FBM also includes fast Kr^+ sampling on two energy levels with 50 ms time resolution at a 25% duty cycle.

Abbreviations:
- IPM = Interplanetary mode
- IPE = Interplanetary event mode
- CAL = Calibration mode
- PLM = Planetary mode
- PLE = Planetary event mode
- FBM = Flyby mode
- MON = Monitor mode (low power mode)

Moments = Plasma moments, N = density, $\langle v \rangle$ = mean drift velocity vector, u^2 = mean squared velocity, P = pressure tensor.

n_E = number of levels in the energy spectrum
n_Ω = number of directions (for energy spectra)

unit sphere by the plasma analyzer. However, due to the spacecraft shadow, about 35% (s/c 1) or 8% (s/c 2) of the unit sphere is obstructed. The spacecraft shadow is taken into account in the onboard computation of plasma moments. Motor scanning is utilized in two modes of operation (planetary mode and planetary event mode). In the other five modes the experiment is set in a fixed position with a 360° free field-of-view in one plane.

Power System

The experiment has redundant power converters producing stabilized voltages for the analog and digital electronics (5V, ±12 V). Furthermore, the experiment has its own solar battery as a backup in case of loss of power from the spacecraft. The two solar panels mounted inside the sensor cover and on a rear panel (see Figure 5) provide an additional 2.5 to 4 W to the 5 V power converter. Thus, the experiment is to some extent self-supporting. This is particularly valuable near Mars if the main power bus to ASPERA is switched off. Accordingly, the data processing unit may only be turned off during eclipses near Mars. A battery will prohibit loss of memory during eclipse. The sensors are also powered by redundant 3 kV power supplies. Odd sensors (1, 3, 5...) and even sensors (2, 4, 6...) are powered by different power supplies.

Calibrations and Tests

Vibration and electromagnetic tests were performed to qualification levels in Linköping, Sweden, at the SAAB vibration and EMC test facilities. The tests were successful and only minor items had to be modified from the qualification models to the flight units. The motor scanner was tested on a qualification model running continuously in thermal vacuum (-20 °C to +50 °C) for 2 months. No failure of the scan mechanism was detected. Thermal vacuum tests to qualification levels were performed on all flight units. The three ASPERA flight units were calibrated and tested at the IRF calibration facility in Kiruna. This facility can expose the spectrometers to a wide (≈10 cm diameter) homogeneous beam of electrons and ions of various species (e.g., H^+, H_2^+, He^+, O^+, Ar^{++}, Ar^+) within the energy range 200 eV/e-20 keV/e for ions and 500 eV-100 keV for electrons. A turntable with the calibration facility is used to determine the angular response of the spectrometers. The instrument performance was in general agreement with the design parameters. Figures 6a and 6b show examples of mass

Fig. 5. ASPERA general view with the sensor cover open.

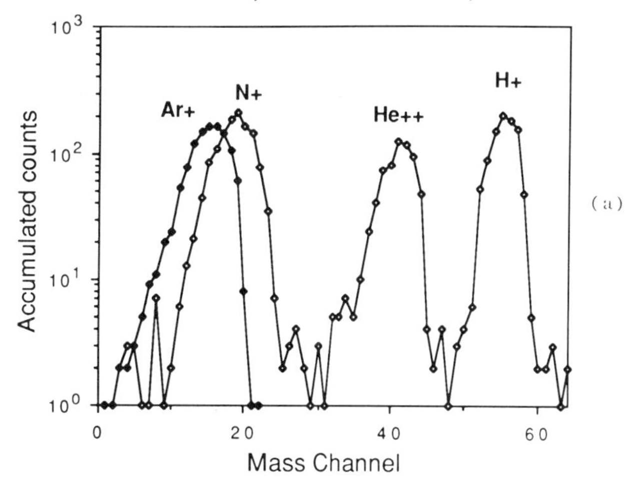

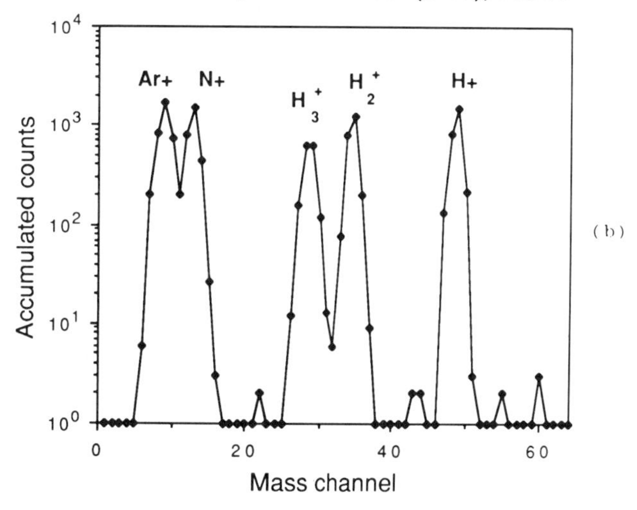

Fig. 6. Two examples of calibrated mass spectra for the ASPERA ion composition spectrometers. (a) Mass spectrum for the moment sensors at 10 keV/e. (b) Mass spectrum for the mass sensor at 605 eV/e.

spectra obtained from a moment sensor (S1) and mass (S17) sensor, respectively. Figures 7a and 7b give the calibrated azimuthal response for the electron sensor (7a) and ion sensors (7b), respectively.

Conclusions

The ASPERA experiment will enable quite detailed, almost three-dimensional, ion composition and plasma distribution measurements near the planet Mars and its satellite Phobos (possibly also Deimos). This will enable us to study a number of interesting scientific topics which are summarized below:

Cruise Phase

o Composition and flow of the solar wind, with special emphasis on interplanetary shock crossings
o Characteristics of the deep geotail (≈1000 R_E)

Martian Magnetosphere

o Scaling of the Martian magnetosphere compared with other planets in our solar system
o Possible existence of an intrinsic magnetic field
o Nature of the solar wind deflection boundary (magnetopause/ionopause)
o Characteristics of the solar wind interaction with the Martian upper ionosphere
o Magnitude and properties of the plasma escape from the Martian upper ionosphere

Phobos (Deimos)

o Accretion and loss of matter from a neutral celestial object (in general)
o Existence of an atmosphere/ionosphere above Phobos

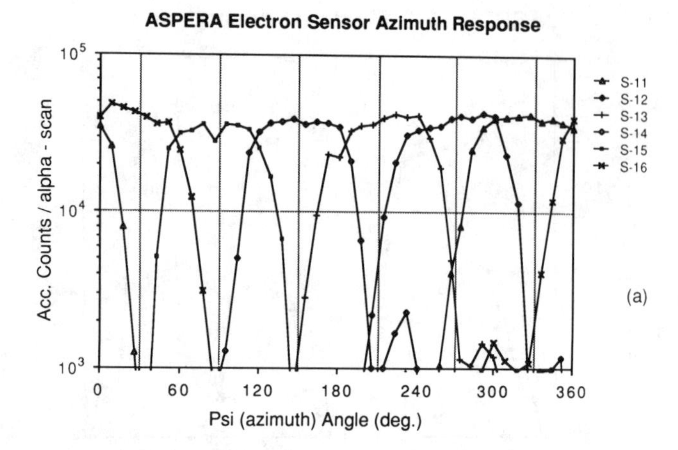

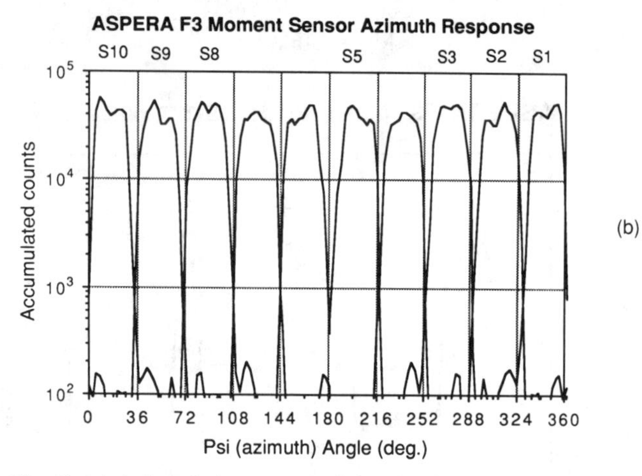

Fig. 7. (a) Azimuthal response of the six electron sensor heads (S11-S16). (b) Azimuthal response of the ten moment sensors (S1-S10).

o Solar wind interaction with Phobos

o Specific plasma-physical processes related to a supersonic plasma flow impinging on a celestial object

o Ion implantation on Phobos

o Abundance of ionized fragments (dusty plasma) along the Phobos orbit

o Surface/regolith properties of Phobos as determined from the backscattering characteristics of Kr^+ ions

o Presence of volatile elements in the Phobos surface layer from the outgassing and ionization resulting from the laser beam experiment

Acknowledgments. The ASPERA experiment was financed by grants from the Swedish Board for Space Activities and the Finnish Academy of Sciences.

References

Alfvén, H., On the Origin of the Solar System, Oxford Univ. Press, London, 1954.

Alfvén, H., and G. Arrhenius, Evolution of the Solar System, NASA SP-345, 1976.

Bogdanov, A. V.,Mars satellite Deimos interaction with the solar wind and its influence on flow around Mars, J. Geophys. Res., 86, 6926, 1981.

Dolginov, Sh. Sh., Ye. G. Yeroshenko, and L. N. Zhuzgov, The magnetic field of Mars according to data from the Mars 3 and Mars 5, J. Geophys. Res., 81, 3353-3362, 1976.

Gringauz, K.I., V.V. Bezrukikh, M.I. Vergin, and A.P. Rezimnov, On the electron and ion components of plasma in the antisolar part of near-Martian space, J. Geophys. Res., 81, 3349-3352, 1976.

Lundin, R., B. Hultqvist, N. Pissarenko, and A. Zakharov, The plasma mantle: Composition and other characteristics observed by means of the Prognoz-7 satellite, Space Sci. Rev., 31, 247, 1982.

Intriligator, D. S., and E. J. Smith, Mars in the solar wind, J. Geophys. Res., 84, 8427, 1979.

Russell, C.T., The magnetic field of Mars: Mars 3 evidence reexamined, Geophys. Res. Lett., 5, 81, 1978a.

Russell, C.T., The magnetic field of Mars: Mars 5 evidence reexamined, Geophys. Res. Lett., 5, 85, 1978b.

Vaisberg, O.L., Mars-Plasma e nvironment, in Physics of Solar Planetary Environment, vol. 2, edited by D.J. Williams, 845, AGU, Washington, D.C., 1976.

Vaisberg, O., and V. Smirnov, The Martian magnetotail, Adv. Space Res., 6, 301, 1986.

What are the important ring, moon, and dust interactions in planetary and cometary magnetosphere and ionosphere systems?

DUST-PLASMA INTERACTIONS IN PLANETARY RINGS

C. K. Goertz

Department of Physics and Astronomy, University of Iowa, Iowa City, IA 52242

Abstract. We discuss the electrostatic levitation of charged dust particles above Saturn's rings due to the enhanced surface electric field which exists in regions of enhanced plasma density. These density enhancements may be due to meteor impacts on the rings. Because the elevated dust moves relative to the plasma cloud a charge separation electric field in the azimuthal direction exists in the plasma cloud. This field causes a radial plasma motion. As the plasma moves more dust is elevated and the cloud leaves behind a radial dust trail - the spokes. The dust particles will settle down onto the ring after about half a Kepler period at a different radial distance thus transporting angular momentum in the radial direction. In addition the charged dust particles change their angular momentum due to electromagnetic forces. The net angular momentum transport can lead to significant mass transport in the ring and accounts nicely for the optical depth minimum at synchronous orbit. Small-scale optical depth variations are unstable to rapid growth. This instability may be able to account for the remarkable ring structure seen by Voyager. Finally we briefly discuss collective dust-plasma effects which may be important in the tenuous planetary rings.

Introduction

Voyager observations (Smith et al., 1981, 1982) of the rings of Saturn have revealed a great deal of unexpected structure and dynamics. It is obvious that the main rings of Saturn are not smooth discs but contain fairly large radial variations of the optical depth at scale lengths ranging from less than a few tens of kilometers to several 1000 km which are apparently not related to any known resonances and associated density waves. There is no universally accepted explanation of this structure but several speculations. In this paper I will discuss my own thoughts on this topic rather than give a review of all theories. I will try to show that at least some of the radial structure in the B ring is due to radial transport of submicron charged dust particles. The scattering properties of the dynamic spokes in Saturn's B ring suggest that they contain large amounts of such dust particles and I believe that an understanding of the ring structure must be based on a understanding of the spokes, especially their formation mechanism. The paper will therefore start with a description of the only detailed spoke formation mechanism published so far (Goertz and Morfill, 1983; Morfill and Goertz, 1983; Goertz, 1984). It will then discuss the mass transport induced by the radial displacement of the charged dust particles contained in a spoke and show that this mass transport should have carved out a significant density depletion near synchronous orbit over geological times (Goertz et al., 1986). An instability driven by this transport which leads to a very rapid growth of perturbations of the ring surface mass density (Goertz and Morfill, 1988) will then be described. The paper will conclude with a brief discussion of collective dust-plasma interactions which need to be considered in more detail in the future and a summary of outstanding questions. This paper is a very subjective review of our own work and makes no attempt to give adequate credit to competing theories. Most of these are very well represented in the literature and collected in an excellent book, Planetary Rings (Greenberg and Brahic, 1984).

The Formation of Spokes

The model of Goertz and Morfill (1988) is based on the assumption that spokes consist of electrostatically levitated submicron-sized dust particles and that the radial elongation of spokes is due to rapid radial motion of dense plasma clouds. Underneath these dense plasma clouds the ring charges to a surface potential, Φ_R, of the order of -6 V (see Figure 1). The potential drops to zero in the plasma through a Debye sheath. The surface electric field

$$E_{\perp} = \Phi_R/\lambda_D \tag{1}$$

may be strong enough to lift negatively charged dust particles of mass M_D if

$$E_{\perp} > \frac{M_D}{Q_D} g_{\perp} \tag{2}$$

where the dust charge is Q_D and the component of the acceleration due to gravity including the self-gravity of the ring is $g_{\perp}$. While the dust particle resides on the big ring particles, its average charge is very small because the surface charge density ($\sigma_q = E_{\perp}/4\pi$) times the area of a dust particle is small ($\sigma_q \pi a^2 << e$). In other words, it is very unlikely for a particular submicron-size particle to collect even one electron. This has been called in question by P. Goldreich (private communication, 1986) who claims that the charge on one dust particle fluctuates greatly between about $-4\pi\Phi_R a$ and $+4\pi\Phi_R a$ with only the average being small. This can be a large number ($\sim \pm 200e$). However, if a dust particle would, by chance, collect two or more electrons it would very rapidly discharge

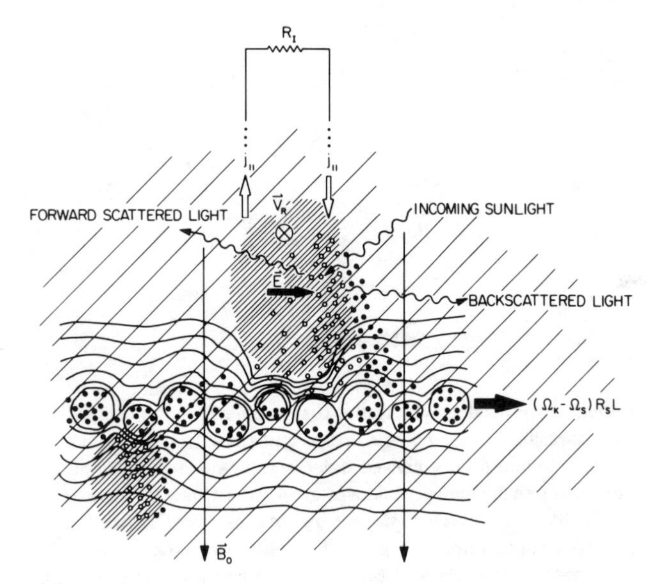

Fig. 1. A schematic sketch of the spoke formation mechanism discussed. The view is from Saturn through the ring along the ring plane. The magnetic field is downward. Dust particles with a large charge are drawn as open circles. Dust particles with a small charge are indicated by solid circles.

these electrons through conduction. Consider two neighboring dust particles one with an excess of negative charge and the other with an excess of positive charge. The electric field would be very large between them and the conduction current would rapidly neutralize the excess charges. The discharge time of a spherical dust grain embedded in a material of resistivity η is $\eta/4\pi$ (in cgs units). The charging time due to absorption of electrons or ions for dust particles in the ring plane is $T_c = 1/(4\pi a^2 F_i)$, where $F_i = F_e = nv_{th}$ is the ion flux to the ring. For a particle of $a = 0.5\ \mu$ a plasma density $n = 10^{-2}$ cm, and a plasma temperature of $T = 2$ eV, we obtain $T_c \simeq 3000$ s for a hydrogen plasma. Thus the resistivity must be larger than $3 \times 10^{14}\ \Omega$m for this charging time to be smaller than the discharge time by currents in the resistive medium. It seems unlikely that the resistivity of the ring (or the plasma in the ring plane) is that large. We thus believe that the surface charge density of the ring is uniform and does not fluctuate from one dust particle to the next as suggested by Goldreich. In that case the probability of a dust particle to have one excess electron is given by

$$P_e = \sigma_q \frac{4\pi a^2}{e} = \frac{\Phi_R}{\lambda_D} \frac{a^2}{e} \tag{3}$$

This is much smaller than 1 for particles satisfying equation (2). Once a dust grain acquires one excess electron the normal electrostatic force is larger than the gravitational force and the dust grain will be lifted out of the ring plane provided its radius is less than about 10^{-2} cm (see, e.g., Goertz, 1984). However, this does not guarantee that this dust grain will escape the Debye sheath because while being in the Debye sheath it is likely to acquire an ion and fall back onto the ring plane. The charging time T_c must be larger than the time to escape the Debye sheath for the grain not to fall back quickly. The time to escape the Debye sheath is

$$T_D \simeq \lambda_D (M_D/e\Phi_R)^{1/2} \tag{4}$$

Thus the net probability for a grain to be lifted above the Debye sheath can be written as

$$P = P_e e^{-T_D/T_c} \tag{5}$$

This probability has a peak at a grain radius given by

$$a_0^{-7/2} = \frac{4\pi}{7} n v_{thi} \lambda_D (\pi\rho/3\Phi_R)^{1/2} \tag{6}$$

where ρ is the mass density of the grain. The flux of escaping dust particles is given by

$$F = \int \frac{N_D(a)}{T_D(a)} P(a) da \tag{7}$$

where $N_D(a)$ is the dust column density in the ring plane. If the dust particles stay above the ring plane for a time T (roughly one half Kepler period or 5 hr) the optical depth of the elevated dust is (for continuous dust elevation)

$$\tau_D = T \int \frac{N(a)}{T_D(a)} P(a) \pi a^2 da \tag{8}$$

The optical depth of dust in the ring plane is m

$$m = \int N(a) \pi a^2 da \tag{9}$$

Most of the dust particles presumably reside on the big ring particles and form a thick regolith layer with $m > 1$. The dust particles orbiting Saturn between the big ring particles will be absorbed quickly if they have a velocity spread which is likely to be the case.

If we approximate the function $P(a)$ by a δ-function $P(a) = P(a_0)\delta(a - a_0)$ we find

$$\tau_D \approx m P(a_0) T/T_D \tag{10}$$

Morfill and Goertz (1983) have shown that if the plasma near the ring is only of ionospheric origin its density is small ($n = 10^{-2}$cm^{-3}). In that case $a_0 \approx 10^{-4}$ cm, $P(a_0) = 3 \times 10^{-6}$, $T_D = 100$ s, and we find that $\tau_D \sim m \times 5 \times 10^{-4}$. Such a small optical depth dust halo is difficult to detect.

However if n were increased to 300 cm^{-3} one obtains $a_0 = 5 \times 10^{-5}$ cm, $P(a_0) = 10^{-4}$, $T_D = 0.1$ s and thus $\tau_D = m \times 10^{-3} \times T$ where T is either the residence time of the dust grains above the ring plane or the duration during which the enhanced plasma density exists (whichever is smaller). Goertz and Morfill (1983) argue that enhanced plasma densities can be produced by meteor impacts on the ring. Such an enhanced density is, of course, a transient phenomenon and the time T needs to be calculated by considering the fate of the meteor impact produced plasma cloud. This involves the motion of the plasma cloud away from the point of impact rather than the absorption of the plasma by the ring.

After a dust grain escapes the Debye sheath its charge will increase with time because it is now in a region where the electron flux is much larger than the ion flux. The time scale for acquiring negative charge is now $T_c\sqrt{m_i/m_e}$ which for $a = 5 \times 10^{-4}$ cm, $T_e = 2$ eV and $n = 300$ cm^{-3} is only a few milliseconds. Thus the dust grains will all obtain their equilibrium charge Q_D, which depends on the dust density and the plasma density as discussed by Goertz and Ip (1984). The grains will move on nearly Keplerian orbits. The plasma cloud however corotates with Saturn. Thus there is an azimuthal current carried by the dust grains which causes an azimuthal charge separation field as indicated in Figure 1. The magnitude of this field has been calculated by Goertz and Morfill (1983) by balancing the current carried by the dust particles with the field-aligned current which eventually closes by a Pedersen current in Saturn's ionosphere

$$E_{\perp} = \frac{\tau_D}{\Sigma_p \pi a_0^2} Q_D(\Omega_K - \Omega_s) R_s L \qquad (11)$$

where Σ_p is the ionospheric conductance (~ 0.1 mho), Ω_K is the Keplerian rotation rate, Ω_s is Saturn's rotation rate, R_s is Saturn's radius, and $r = R_s L$ is the radial distance from the planet. The plasma cloud will $\vec{E} \times \vec{B}$ drift in the radial direction with a speed $V_R = cE_{\perp}/B$, where B ($B = B_0 L^{-3}$) is the magnetic field of Saturn at the distance r. The time for the plasma cloud of radius R to pass across a certain point is

$$T = R/V_R \qquad (12)$$

Since τ_D is proportional to T we can combine equations (10) through (12) to yield a radial velocity

$$V_R^2 = \left(\frac{mP(a_0)}{\pi T_D a_0^2}\right) \frac{cQ}{B} \frac{(\Omega_K - \Omega_s)}{\Sigma p} Rr \qquad (13)$$

Goertz and Morfill (1983) have shown that this velocity ranges from 0 exactly at synchronous orbit ($L_s = 1.866$) to several 10 km s^{-1} inside of $L = 1.8$ and outside of $L = 1.9$. They also show that the plasma cloud moves away from synchronous orbit if $Q_D < 0$ (as it should be). As the cloud moves away from synchronous orbit it leaves behind a radial trail of elevated dust particles, the spoke.

It has been pointed out that the radial velocity is zero at synchronous orbit where $\Omega_K = \Omega_s$ and it has been suggested that this would not allow spokes to extend across synchronous orbit as observed. However, only clouds whose radius is at least 1000 km will leave trails that are resolved by Voyager. Thus any meteor impact within $\pm$1000 km of synchronous orbit would produce clouds expanding both inward and outward from synchronous orbit. At a distance of 500 km from synchronous orbit the radial velocity is already 4 km s^{-1} sufficient to produce very long trails in short times.

This model also yields the right order of magnitude for the optical depth of a spoke 1000 km wide (equal to R). At $L \approx 1.9$ the radial velocity V_R is about 10 km s^{-1} and hence it takes about $T = 100$ s to pass a certain point. In this time the number of dust particles elevated will produce an optical depth of $\tau_D = m \times 10^{-1}$. If the dust grains completely cover the big ring particles ($m = 1$) the predicted optical depth is 0.1 in good agreement with observations. We also predict that the size of the dust grains in the spokes is 0.5 μ. This size is also indicated by the enhanced forward scattering of light by the spokes.

We close this section by noting that the dust elevated above the ring plane contained in the isolated spokes ($\tau \sim 0.1$) is larger than that contained in the dust halo ($\tau \sim 5 \times 10^{-4}$) if the spokes cover more than 0.5% of the ring area which they have done during the Voyager fly-by epoch.

Electromagnetically Induced Mass Transport in the Rings

The small dust particles elevated above the ring plane at a radial distance r are subject to significant electromagnetic forces which tend to force them into corotation with the planetary magnetic field. (Note that the dust in the ring plane has an exceedingly small charge to mass ratio and is thus not affected by electromagnetic forces.) Inside synchronous orbit the dust particles lose angular momentum to Saturn and outside of it they gain angular momentum. For $Q_D < 0$ the dust particles move toward synchronous orbit and will settle back onto the ring at a radial distance $r + \Delta r$. Since the specific angular momentum (angular momentum per mass) of the dust settling down at $r + \Delta r$ is different from the Keplerian specific angular momentum of the ring particles the ring material will experience a torque when the dust is absorbed. The magnitude of this torque is proportional to the flux of dust onto the ring material and the radial hopping distance, Δr. The ring material absorbing the dust will move onto a new circular Keplerian orbit corresponding to its new angular momentum. Averaged over many episodes of dust absorption we define a radial velocity of the ring material w which was given by Goertz et al. (1986, 1988)

$$w(r + \Delta r) = 2rF(r)\left(\frac{L_D}{L_K} - 1\right)/\sigma_R(r + \Delta r) \qquad (14)$$

where $F(r)$ is the flux of dust elevated at r. L_D is the dust specific angular momentum at $r + \Delta r$ and L_K is the Kepler angular momentum at $r + \Delta r$. The authors show that

$$2r\left(\frac{L_D}{L_K} - 1\right) = -\Delta r(1 - \epsilon) \qquad (15)$$

where

$$\epsilon = \frac{Q_D}{M_D} \frac{B}{c} \frac{1}{\Omega_K} \qquad (16)$$

We now make the assumption that the flux of dust elevated is proportional to the ring surface mass density, σ_R. This is quite reasonable since one would expect dust to be produced by grinding collisions between the big particles. Thus

$$F(r) = b\sigma_R(r) \qquad (17)$$

The continuity equation for the ring surface mass density can then be written as

$$\frac{\partial}{\partial t}\sigma_R(r) + \frac{1}{r}\frac{\partial}{\partial r}(rw(r)\sigma_R(r)) =$$

$$\frac{\partial^2}{\partial r^2}(D\sigma_R(r)) + (F(r - \Delta r) - F(r)) \qquad (18)$$

The first term on the right-hand side represents diffusion and the second the gardening effect of the dust. A similar equation can be written for the surface number density of the ring n_R

$$\frac{\partial}{\partial t}n_R(r) + \frac{1}{r}\frac{\partial}{\partial r}(rw(r)n_R(r)) = \frac{\partial^2}{\partial r^2}(Dn_R(r)) \quad (19)$$

The gardening effect does not change the number density. The surface number density is related to the surface mass density by

$$n = \frac{3}{4\pi}\frac{\sigma}{R_p^3\rho_p} \quad (20)$$

where R_p is the radius of the big ring particles whose mass density is ρ_p. The optical depth is given by

$$\tau = \pi R_p^2 n = 3\sigma/4R_p\rho_p \quad (21)$$

If we consider large scale density variations we neglect the difference between $\sigma_R(r - \Delta r)$ and σ_R and obtain the equation of Goertz et al. (1986). However, when considering small scale structures with scale lengths of the order of the hopping distance Δr we obtain the equations of Goertz et al. (1988). We should note that the radial transport discussed here is similar to the ballistic transport of Ip (1983), Lissauer (1984), and Durisen (1984). However, they consider only the radial displacement of uncharged debris produced by meteor impacts. In that case Δr is small and, most importantly, it can vary from positive to negative values depending on the velocity of the debris as it is ejected from the ring. Thus F(r) must be evaluated by integrating over all ejection velocities and the resulting angular momentum change is small because debris has almost the same probability of settling down at a larger radial distance than at a smaller one. For charged dust Δr is large and the sign of Δr depends only on the charge of the dust particles.

Let us now consider radial variations of σ_R and n_R whose scale lengths are much larger than Δr but smaller than r. In that case we write

$$w(r - \Delta r) = w(r) = -\Delta r(1-\epsilon)F(r)/\sigma(r)$$

$$F(r - \Delta r) - F(r) = -\Delta r \partial F(r)/\partial r \quad (22)$$

$$\frac{1}{r}\frac{\partial}{\partial r}r\sigma w \approx \frac{\partial}{\partial r}\sigma w$$

and obtain from equation (18)

$$\frac{\partial \sigma_R}{\partial t} + \frac{\partial}{\partial r}(\sigma_R w_{eff}) = \frac{\partial^2}{\partial r^2}(D\sigma) \quad (23)$$

where the effective radial mass transport velocity is given by

$$w_{eff} = -\frac{r}{2}(1 - \frac{\Omega_s}{\Omega_k})T_{DS}(\frac{Q_D B}{M_{DC}})^2\frac{<\sigma_D>}{<\sigma_R>} \quad (24)$$

This uses the relations (Goertz et al., 1986)

$$\Delta r = -(\frac{Q_D B}{M_{DC}})r(\Omega_K - \Omega_s)T_{DS}^2/2 \quad (25)$$

and (Goertz et al., 1988)

$$b = \frac{<\sigma_D>}{<\sigma_R>}\frac{1}{T_{DS}} \quad (26)$$

where T_{DS} is the average time a dust particle spends outside the Debye sheath (roughly 5 hr). We see that inside synchronous orbit the mass transport is inward and outside of it it is outward, independent of the dust particles sign of charge. The radial displacement, Δr, of a dust particle does depend on the sign of its charge. This, perhaps, puzzling fact is easily understood by realizing that the radial current carried by the radial displacement of the dust particles is always away from synchronous orbit no matter what the sign of Q_D is. The $\vec{j} \times \vec{B}$ force due to this current will decrease the angular momentum of the ring inside L_s and increase it outside. The change of ring angular momentum is taken up by the planet's ionosphere where the Pedersen currents exert the required $\vec{j} \times \vec{B}$ force.

The effective radial velocity, w_{eff}, can be quite large considering that $T_{DS} \sim 10^4$ s and $<\sigma_D>/<\sigma_R> \approx 2.5 \times 10^{-8}$. Goertz et al. (1986) show that in the B ring at $L = 1.8$, for example, the effective velocity is $7 \times 10^{-12}(Q_D/e)^2$cm s^{-1} which for $Q_D/e \approx 200$ becomes about 3×10^{-7}cm s^{-1}. The effective diffusive transport velocity for a typical ring viscosity of 20 cm^2 s^{-1} is only 2×10^{-9} cm s^{-1}. Clearly electromagnetic angular momentum transport is, at least, equal in importance to viscous transport. Since the electromagnetic transport is away from synchronous orbit one expects a minimum of σ_R there. This is, indeed, observed as shown in Figure 2.

For small scale length variations of σ_R we can make the ansatz

$$\sigma_R = \sigma_0(r) + \delta\sigma e^{-i(kr-\omega t)} \quad (27)$$

with $\sigma_0(r)$ the slowly varying density discussed above. Inserting this into equation (18) we obtain the dispersion relation (neglecting diffusion)

$$i\omega = b\{e^{ik\Delta r} - 1 + ik\Delta r(1-\epsilon)e^{ik\Delta r}\} \quad (28)$$

which indicates growth ($i\omega > 0$) for wave numbers in bands centered on $k\Delta r = \pm(n + 1/2)\pi$ with $n = 0, 1, 2, \ldots$ The growth rates are of the order of b and density perturbations with radial wavelengths $\lambda = \pm 2\Delta r(n+1/2)$ should grow to appreciable

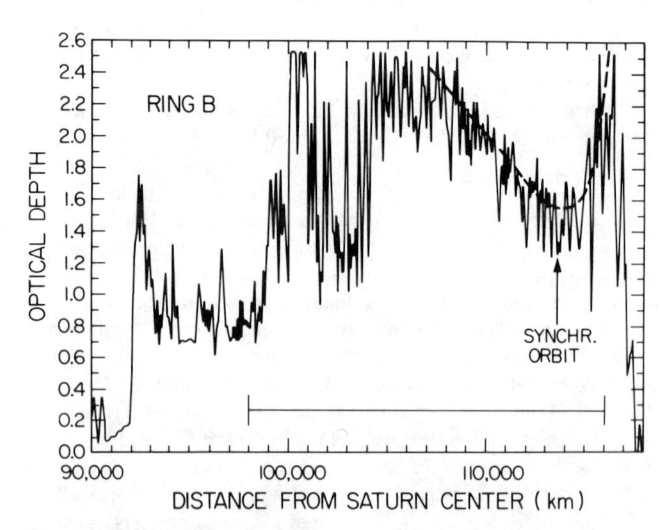

Fig. 2. The variation of optical depth with radial distance in Saturn's B ring. The decrease near synchronous orbit can be fit by a solution of equation (23) as indicated by the dashed line.

amplitudes in a relatively short time of order $b^{-1} \approx 4 \times 10^{11}$s or 10^4 years. Of course, diffusion will prevent very small wavelengths from growing because diffusive damping is proportional to $1/\lambda^2$.

In addition we must note that Δr is not a fixed number. For example, the charge on a dust particle fluctuates causing a fluctuation of Δr about a mean value. Goertz et al. (1988) have shown that the growth rate for $n \geq 1$ becomes, in fact, negative if the fluctuation of Δr exceeds 50%. In addition one should note that Saturn's magnetic field may not be constant over 10^4 years. If the field changes direction the sign of Δr changes. The duration of constant magnetic field since the last change may not have been long enough to allow perturbations to have grown to very large amplitudes. Note that a magnetic field turning would not effect w_{eff} because it is proportional to B^2. (I am grateful to Paul Cloutier for pointing this out to me.)

For $Q_D/e = 200$ we find that at $L = 1.8\Delta r \sim 50$ km. The unstable wavelength bands are thus centered at $\lambda =$ 400 km/(2n + 1). The $n = 0$ and $n = 1$ wavelengths are quite similar to the scale lengths of the optical depth variations observed in that region of Saturn's rings.

The model described above is crude in many respects. For example, the assumption (17) may not be correct. It is, perhaps, more reasonable to assume $F(r) = b\tau(r)$. More importantly on the right hand side of equation (18) we have assumed that all dust particles elevated at $r - \Delta r$ are absorbed at r. Clearly this requires that the ring optical depth is very large. In regions where $\tau < 1$ a significant fraction of the dust particles may traverse the ring plane and impinge on the rings at $r + \Delta r$ where, again, only a fraction are absorbed. This dependence of the absorbed flux of dust particles on the optical depth introduces a non- linearity into the equation which can only be treated numerically. We must also solve the set of equations, (18) through (21), rather than (18) alone. This, again, requires numerical solutions. We will report on the results of a numerical simulation of these processes in a future publication.

Collective Effects in Dusty Plasmas

When discussing the effects of charges on particles in planetary rings most authors have concentrated on the dynamics of individual particles subject to the overwhelmingly strong planetary gravitational field. For example, Mendis et al. (1984) have shown that significant perturbations of the Kepler orbits are only expected for submicron size particles. This is easy to understand when one compares the planetary gravitational force $F_G = M_D \Omega_K{}^2 r$ with the electromagnetic force in a corotating magnetosphere $F_E = Q_D B(\Omega_K - \Omega_p)r$. Since $Q_D = \Phi a$ the ratio F_G/F_E is exceedingly large even for micron size particles. Because the main planetary rings contain mainly particles whose radius a is of the order of centimeters or meters, electromagnetic effects are believed to play no direct role there. Only the minor rings of Saturn (E, F, G), the rings of Jupiter, and the spokes which contain large fractions of submicron particles are candidates for significant electromagnetic effects. However, there are many dynamically important phenomena (e.g., density waves, collisions, ring thickening) which are the result of interparticle forces such as self-gravity. When considering the interaction between ring particles one is dealing with small forces compared to the central planet's gravity. Long-range Coulomb forces between the charged particles are comparable or even stronger than the interparticle gravitational forces. Both forces scale as $1/r^2$ and the ratio of the interparticle gravitational force, F_g, to the Coulomb force, F_e is, to first order, independent of the particle density

$$\gamma = F_g/F_e = \left(\frac{M_D}{Q_D}\right)^2 G \times 10^{-21} \tag{29}$$

For a 1 cm particle with a surface potential of 4 V this ratio is $\gamma = 4 \times 10^{-4}$. However, two effects need to be considered which increase the ratio γ. The particles are imbedded in a plasma which screens out the interparticle Coulomb force if the distance between particles is larger than the plasma Debye length λ_D. On the other hand when the distance becomes much smaller than λ_D the average charge on a particle is reduced (Goertz and Ip, 1984). One can show that the electrostatic interparticle force maximizes when the quantity P introduced by Havnes et al. (1987) exceeds 3×10^{-9} where

$$P = Na(kT/e)/n_p \tag{30}$$

The number density of the particles is N, the plasma density is n_p, and the plasma temperature is T. For a more thorough discussion we refer to the recent paper by Goertz et al. (1988). The electrostatic force is repulsive, as expected.

Two possibly very exciting consequences of this electrostatic force immediately come to mind. It is obvious that electrostatic repulsion between particles may prevent gravitational collapse even when the random velocity of the particle is zero. In other words the electrostatic effect acts like a pressure. Previously it was believed that gravitational collapse can only be prevented by a finite value of the random velocity dispersion, $< c >$. However, these considerations indicate that even for $< c >= 0$ a ring may be gravitationally stable. Goertz et al. (1988) showed that this is the case in the Uranian rings, but not in the main rings of Saturn. A further consequence of these considerations is the possibility that when $< c >= 0$ the ring particles are actually collisionless, in which case very narrow rings may exist for long times. And, indeed, Michel (1986) has argued that if a mechanism for maintaining the Uranian ring particles on non-intersecting (i.e., collisionless) orbits could be found, the stability of the narrow Uranian rings may be understood. Coulomb repulsion between particles may prevent elastic collisions.

The second, even more intriguing possibility involves the concept of a Coulomb lattice. It has been shown (e.g., by Knorr, 1968, and May, 1967) that a two-dimensional plasma consisting of N_i particles of charge $Z_i e$ and N_j particles of charge $Z_j e$ obeys the following equation of state

$$pV = KT - \frac{1}{4} Z_i Z_j e^2 \tag{31}$$

Thus when $kT < \frac{1}{4} Z_i Z_j e^2$ the system collapses to a Coulomb lattice. Ikezi (1986) has recently suggested than an ensemble of particles which carry a charge $Q_D >> e$ imbedded in a charge-neutralizing fluid will form a Coulomb lattice provided that

$$\Gamma = \frac{Q_D^2}{bkT} \geq 170 \tag{32}$$

where b is the interparticle distance. Plasmas whose coupling constant Γ is large are called strongly coupled plasmas and have been subject to intense study. It is remarkable that in the Uranian ring where $a = 10$ cm, T (plasma) = 30 eV, and $b \sim 20$ cm the coupling constant is 20, quite close to the critical value. We thus expect that the Uranian rings may contain coherent well-ordered structures which are best described as Coulomb fluids or even Coulomb solids. So far this is only an inspired speculation which deserves further study.

Several points, in particular, need to be investigated:

(i) What is the average charge on a ring particle and how much does the charge vary due to statistical fluctuations? Is it sufficient to treat all particles of equal size or must the size distribution be taken into account?

(ii) What is the electrostatic force between the particles and how does it fluctuate?

(iii) What are the statistical properties of the particle-plasma ensemble? What is the equation of state of this strongly coupled classical plasma? Is the formation of a Coulomb lattice a real possibility?

(iv) What are the effects of Coulomb forces on the density wave dispersion relation? Can density waves be excited by perturbations of the particle orbits due to the planetary magnetic field or by perturbations of the particle charge due to the variation of solar EUV flux (photoelectron emission) along the orbits of the particles?

(v) Can instabilities be excited due to the motion of the charged particles relative to the corotating plasma?

The treatment of ring particles as a charged fluid is only now beginning. The possible rewards are very exciting.

Acknowledgments. This work was supported in part by NASA grants NAGW-970 and NAGW-871.

References

Durisen, R. H., Transport effects due to particle erosion mechanisms, in *Planetary Rings*, edited by R. Greenberg and A. Brahic, University of Arizona Press, p. 416, 1984.

Goertz, C. K., Formation of Saturn's spokes, *Adv. Space Res.*, *4*, 137, 1984.

Goertz, C. K., and G. Morfill, A model for the formation of spokes in Saturn's ring, *Icarus*, *53*, 219, 1983.

Goertz, C. K., and W.-H. Ip, Limitation of electrostatic charging of dust particles in a plasma, *Geophys. Res. Lett.*, *11*, 349, 1984.

Goertz, C. K., and G. E. Morfill, A new instability of Saturn's ring *Icarus*, in press, 1988.

Goertz, C. K., G. E. Morfill, W.-H. Ip, E. Grun, and O. Havnes, Electromagnetic angular momentum transport in Saturn's rings, *Nature*, *320*, 141, 1986.

Goertz, C. K., L. Shan, and O. Havnes, Electrostatic forces in planetary rings, *Geophys. Res. Lett.*, *15*, 84-87, 1988.

Greenberg, R., and A. Brahic (editors), *Planetary Rings*, Univ. of Ariz. Press, Tucson, Ariz., 1984.

Havnes, O., et al., Dust charges, cloud potential and instabilities in a dust cloud embedded in a plasma, *J. Geophys. Res.*, *92*, 2281, 1987.

Ikezi, H., Coulomb solid of small particles in plasmas, *Phys. Fluids*, *29*, 1764, 1986.

Ip, W.-H., Collisional interactions of ring particles: Ballistic transport process, *Icarus*, *54*, 253, 1983.

Knorr, G., The partition function of a two-dimensional plasma, *Phys. Lett.*, *28A*, 166, 1968.

Lissauer, J. J., Ballistic transport in Saturn's rings: An analytic theory, *Icarus*, *57*, 63, 1984.

May, R. M., Exact equation of state for a 2-dimensional plasma, *Phys. Lett.*, *25A*, 282, 1967.

Mendis, D. A., et al., Electrodynamic processes in the ring system of Saturn, in Saturn edited by T. Gehrels, Univ. of Ariz. Press, Tucson, Ariz., p. 546, 1984.

Michel, F. C., The collisionless rings of Uranus, *Geophys. Res. Lett.*, *13*, 442, 1986.

Morfill, G. E., and C. K. Goertz, Plasma clouds in Saturn's ring, *Icarus*, *55*, 111, 1983.

Smith, E. J., et al., Encounter with Saturn: Voyager 1 imaging results, *Science*, *212*, 163, 1981.

Smith, E. J., et al., A new look at the Saturn system: The Voyager 2 images, *Science*, *215*, 504, 1982.

COMETARY DUSTY GAS DYNAMICS

T.I. Gombosi and A. Körösmezey

Space Physics Research Laboratory, Department of Atmospheric, Oceanic and Space Sciences
The University of Michigan, Ann Arbor, MI 48109

Abstract. A well-developed dusty cometary atmosphere extends to distances over 4 orders of magnitude larger than the size of the nucleus. Similarly, the solar wind-dominated heliosphere extends to about 10^4 solar radii. The first part of this review explores the similarities and differences between the solar and cometary winds. The paper also presents initial results of a new time-dependent, two-dimensional dusty comet wind calculation.

Introduction

Our present, post-encounter understanding of cometary nuclei is based on Whipple's (1950) "dirty iceball" idea, which visualizes them as chunks of ice, rock, and dust with negligible surface gravity. Whipple's hypothesis quickly replaced the century-long series of "sandbank" models, wherein the nucleus was thought of as a diffuse cloud of small particles traveling together. As comets approach the Sun, water vapor and other volatile gases sublimate from the surface layers generating a rapidly expanding dusty atmosphere. The sublimated gas molecules (often called parent molecules) undergo collisions and various fast photochemical processes in the near nucleus region, thus producing a whole chain of daughter atoms and molecules. There is growing evidence that dust grain photochemistry, as well as gas-dust chemical reactions, also contribute to the maintenance of the observed atmospheric composition.

In the vicinity of the nucleus the gas and dust flows are strongly coupled: frequent gas-dust collisions accelerate small grains to velocities up to several hundreds of meters per second and inject them into the extensive cometary exosphere, where the gas and dust are decoupled. The expanding gas converts most of its original internal energy to bulk motion, while it also loses momentum and energy to the dust flow. At the same time the nucleus surface and the accelerating dust grains are heated by the attenuated and multiply scattered solar radiation. Most of the thermal radiation of the solid components is emitted in the 1-20 μm wavelength range, where several rotational and vibrational transients exist for the highly dipolar water molecules which have very large resonance cross sections. The resonant radiation is continuously absorbed and reemitted by the water molecules; in other words, it is trapped by the gas (Marconi and Mendis, 1986). A large fraction of the rotational/vibrational excitation energy is transformed into translational motion via molecular collisions, thus increasing the gas temperature. The higher gas temperature represents an increased source of internal energy, which eventually results in higher terminal velocities due to adiabatic expansion.

One of the most important factors influencing cometary dynamics is the "retarded" nature of gas and dust production. The radiation reaching the surface and supplying energy for sublimation must first penetrate an extensive, absorbing dusty atmosphere. Any change in the gas and dust production alters the optical characteristics of the atmosphere, thus causing a delayed (or "retarded") effect on the production rates themselves. This "retardation" makes inner coma modeling efforts complicated and time consuming.

Dust Production

The chemical composition and physical structure of the surface layers of a cometary nucleus are very important factors affecting the mass, momentum, and energy of the outflowing gas-dust mixture, as well as the relative abundances of various gas molecules. When the comet approaches the Sun, it absorbs an increasingly larger flux of solar radiation, and the vaporization rate of volatile molecules at the surface increases. Gravitational forces are very small; therefore, the vaporized gases leave the surface and form an expanding atmosphere. In this process the gas drags away some of those dust grains which have already been evacuated of their ice component (at least partially), but others may remain on the surface (or may fall back). In his original presentation of the icy-conglomerate model Whipple (1950) predicted that an inert layer of large dust particles, evacuated of the volatile component, would form an insulating crust on the surface (mantle). The thickness of the mantle varies with time because the continuous vaporization increases the thickness of the evacuated layer, and the "erosion" due to the drag of the outflowing gas decreases it. The development and thermal structure of such a mantle has extensively been discussed in the literature (Shulman, 1972; Mendis and Brin, 1977; Brin and Mendis, 1979; Horányi et al.,1984; Fanale and Salvail, 1984; Houpis et al.,1985). These models were able to predict several different mantle evolution patterns (for a detailed review, see Gombosi et al., 1986). The pre-encounter view of the mantle evolution process assumed that active periodic comets were covered with friable surface dust layers, so that a repetitive cycle appeared as the comet orbited the Sun. The prevailing view was that apart from the first approach to the vicinity of the Sun, the mantle thickness and the total gas production rate basically followed similar curves during subsequent revolutions. As a Halley-type comet approaches the Sun the mantle thickness increases up to a critical heliocentric distance, and then it starts to decrease. By the time the comet passes its perihelion most of the mantle is blown off and it keeps eroding further resulting in additional post-perihelion brightening. When the comet again leaves the vicinity of the Sun, a new mantle is developed; this new mantle is blown off during the next perihelion passage.

Elementary gas kinetic theory has been used to calculate the gas production rate since the early work of Delsemme and Swings (1952). It was widely assumed that the sublimated gas molecules leave the vicinity of the nucleus without collisions, so that the outflow velocity could be approximated reasonably well by the local sound velocity. At the same time the typical mean free path of the gas molecules near a cometary nucleus is on the order of 10 cm -10 m; therefore, a hydrodynamic rather than a kinetic approach is needed to calculate gas outflow rates. To resolve this problem, Gombosi et al. (1985) introduced a "reservoir outflow" model for gas production from the nucleus where the sublimating surface was replaced by a gas reservoir containing a stationary perfect gas. The surface of the reservoir was assumed to be covered by a thin layer of friable dust so that the gas could flow through it and which also "loaded" the discharging gas flow with dust grains. It was also assumed that the gas slowly diffused through the porous mantle and at every point was heated to the local mantle temperature. In this model the practically stationary gas at the top of the nucleus had a temperature identical to the mantle surface temperature and a pressure which was the same as the sublimation pressure. Gombosi et al. (1985) have considered the time-dependent dusty gas outflow from such a nucleus, assuming a realistic surface and sublimating temperatures.

The Comet Halley images revealed that most of the dust production (and supposedly the gas production, too) was concentrated on several active areas on the sunlit side, which covered only about 10% of the cometary surface. Careful analyses of ground-based observations also helped to identify active spots and line sources, which turn on and off fairly randomly. All this evidence points toward a much more complex picture of cometary gas and dust production than described by our present models; it seems to be increasingly probable that thermal stresses and other effects can cause rapidly developing openings (cracks) in the surface layers, which later slowly "heal" as a new dust mantle develops. Presently there are only a few initial attempts to model such localized, random active areas (cf. Kürth et al. 1986).

Coma Dynamics

It was recognized as early as the mid-1930's that gas outflow plays an important role in cometary dust production. In early treatments of the gas-dust interaction it was assumed that the gas drag coefficient was independent of the gas parameters and that the gas velocity was constant in the dust acceleration region. In the late 1960's this very naive picture was replaced by a two-component approach, which used free molecular approximation to describe the rarefied gas flow and neglected the random motion of the dust particles. In Probstein's (1968) dusty gas dynamic treatment (which later became the prototype of such calculations), the traditional gas energy conservation equation was replaced by a combined dust-gas energy integral. This approach was later considerably refined by a series of authors (Shulman, 1972; Hellmich and Keller, 1980; Gombosi et al., 1983, 1985; Marconi and Mendis, 1982, 1983, 1984, 1986; Kitamura, 1986), but it still represents the main method of dusty gas dynamics calculations.

Modeling efforts have shown that the spatial extent of the dust acceleration region (where dust particles accelerate to about 80% of their terminal velocity) is less than about 30 cometary radii. Gas particles typically spend less than 100 s in this region, which is not long enough for any significant change in the gross chemical composition of the gas. In a first approximation a single-fluid dusty gas hydrodynamical technique seems to be adequate for describing the overall dynamics of the gas-dust interaction.

In the early dusty gas dynamic calculations, attenuation of the solar radiation field by dust grains and gas particles was neglected. Based on a spherically symmetric dust density distribution produced by an isothermal nucleus, Hellmich (1979) developed a model to calculate the transfer of multiply scattered radiation in the inner coma and to determine the energy input to the nucleus. Surprisingly, Hellmich (1979) and later Weissman and Kieffer (1981) found that the net effect of the dust on the sublimation somewhat enhances the gas and dust production because the radiative flux scattered within the inner dust coma is partially trapped and this effect overcompensates the attenuation of direct solar radiation. Marconi and Mendis (1983) published an alternative model, which considered mainly the transfer of the solar UV radiation responsible for the major photolytic processes and treated the longer wavelength diffuse radiation field only superficially. It was also pointed out by Marconi and Mendis (1986) that several rotational and vibrational transitions exist for the highly dipolar water molecular in the 1-20 μm wavelength range, where most of the dust thermal radiation is emitted. In a collisionless gas the resonant radiation is continuously absorbed and reemitted by the water molecules; in other words, it is trapped by the gas. On the other hand, collisions play an important role in the inner coma and a large fraction of the rotational/vibrational excitation energy of water molecules can be transformed via collisions into translational energy, thus increasing the gas temperature. Marconi and Mendis (1986) have included a new approximate gas heating rate describing the effects of the dust thermal radiation and obtained a significant increase in the gas temperature and velocity profiles. The details of this complicated interaction process are not adequately understood, and the application of generalized transport equations may help us to have a better insight into this significant effect.

As the vaporized gases leave the surface they drag away some of those dust grains which have already been evacuated of their ice component. The gas drag force accelerates the dust particles to terminal velocities comparable with the gas flow velocity. The mass, momentum, and energy conservation equations of the single-fluid neutral gas are the following:

$$\frac{D\rho}{Dt} + \rho \, \nabla \cdot \mathbf{u} = 0 \tag{1}$$

$$\rho \frac{D\mathbf{u}}{Dt} + \nabla p - \rho \, \mathbf{g}_c = \mathbf{F}_{gd} \tag{2}$$

$$\frac{1}{\gamma - 1} \frac{Dp}{Dt} + \frac{\gamma}{\gamma - 1} p \, \nabla \cdot \mathbf{u} = Q_{ext} - Q_{gd} \tag{3}$$

where

$$\frac{D}{Dt} = \frac{\partial}{\partial t} + (\mathbf{u}\cdot\nabla) \quad (4)$$

is the convective derivative, ρ = mass density, p = gas pressure, $\mathbf{u}$ = gas velocity, $\mathbf{g}_c$ = acceleration due to the gravitational attraction of the comet, $\mathbf{F}_{gd}$ = gas to dust momentum transfer rate, Q_{gd} = gas to dust energy transfer rate, Q_{ext} = external heating rate. In the innermost coma where most of the gas-dust interaction takes place the radiation pressure effect can be neglected and the equation of motion of an individual dust grain becomes

$$\frac{D\mathbf{V}_a}{Dt} = \frac{3}{4a\rho_a} p\, C_D\, \mathbf{s}_a + \mathbf{g}_c \quad (5)$$

where $\mathbf{V}_a$ = dust particle velocity. The dimensionless gas-dust relative velocity, s_a, and the modified free molecular drag coefficient, C_D, are (Probstein, 1968)

$$\mathbf{s}_a = \frac{\mathbf{u} - \mathbf{V}_a}{\sqrt{2\frac{k}{m}T}} \quad (6)$$

$$C_D = \frac{2\sqrt{\pi}}{3}\sqrt{\frac{T_a}{T}} + \frac{2s_a^2+1}{s_a^2\sqrt{\pi}} e^{-s_a^2} + \frac{4s_a^4+4s_a^2-1}{2s_a^3}\operatorname{erf}(s_a) \quad (7)$$

where s_a = magnitude of the normalized gas-dust relative velocity vector, while T and T_a are the gas and dust temperatures, respectively. In the presence of an external radiation field the energy balance equation for a single dust particle is (Probstein, 1968)

$$C_a\frac{DT_a}{Dt} = \frac{3}{a\rho_a}\left[pC_H\sqrt{T} + \frac{1-A_{vis}}{4} I_{rad} - (1-A_{IR})\sigma T_a^4\right] \quad (8)$$

where C_a = dust specific heat, A_{vis} and A_{IR} are the visible and infrared dust albedos, respectively, while

$$C_H = \frac{\Gamma_a}{\gamma-1}\sqrt{2\frac{k}{m}}\left[2\gamma+2(\gamma-1)s_a^2 - \frac{(\gamma-1)\operatorname{erf}(s_a)}{s_a\Gamma_a} - (\gamma-1)\frac{T_a}{T}\right] \quad (9)$$

where

$$\Gamma_a = \sqrt{\pi}\, e^{-s_a^2} + \left(\frac{1}{2s_s} + s_a\right)\operatorname{erf}(s_a) \quad (10)$$

Finally, it is assumed that the dust particles do not undergo any further sublimation or fragmentation in the coma (there is recent indication that this assumption is probably violated to some extent); consequently, the dust size distribution function, f_a, must obey the following continuity equation:

$$\frac{Df_a}{Dt} + f_a \nabla\cdot\mathbf{V}_a = 0 \quad (11)$$

The gas to dust momentum and energy transfer rates can be obtained by integrating over all dust sizes

$$\mathbf{F}_{gd} = 2\pi p \int_0^{a_{max}} da\, C_D a\, f_a\, \mathbf{s}_a \quad (12)$$

$$Q_{gd} = 2\pi p \int_0^{a_{max}} da\, a\, f_a\left(C_D \nabla\cdot\mathbf{s}_a + 4\, C_H\sqrt{T}\right) \quad (13)$$

It should be noted that these integrals are dominated by the momentum and energy transfer to small particles. External gas heating is mainly caused by photochemical and radiative heating/cooling processes (cf. Gombosi et al., 1986), $Q_{ext} = Q_{phc} + Q_{IR}$. The main contribution to the photochemical heating rate comes from the photodissociation of water molecules (cf. Mendis et al., 1985)

$$Q_{phc} = Q_0 \frac{n}{d^2} e^{-\tau_{UV}} \quad (14)$$

where $n = \rho/m$, $Q_0 = 2.8\times10^{-17}$ erg cm^{-3} s^{-1}, d = heliocentric distance (AU), τ_{UV} = ultraviolet optical depth. The two main processes contributing to the infrared radiative heating/cooling term are the infrared radiation from the H_2O molecules (Shimizu, 1976; Crovisier, 1984) and the radiative trapping of the dust thermal radiation Marconi and Mendis (1986). The combined effect of these processes can be approximated as (cf. Marconi and Mendis, 1986; Gombosi et al., 1986):

$$Q_{IR} = Q_{emiss}\, n\, e^{-\tau_{UV}} + q_{abs} h_{IR}\sigma\left(1 - e^{-\tau_{IR}}\right)\int_0^{a_{max}} da\, a^2 f_a T_a^4 \quad (15)$$

where σ = Stefan-Boltzmann constant, τ_{IR} = infrared optical depth, h_{IR} = transforming efficiency of internal to translational energy ($h_{IR} \approx 0.5$), q_{abs}= relative width of absorbing bands with respect to the infrared spectrum emitted by dust ($q_{abs} \approx 0.001$), and

$$Q_{emiss} = \begin{cases} 4.4\times10^{-22}\, T^{3.35} & T<52K \\ 2.0\times10^{-20}\, T^{2.47} & T\geq 52K \end{cases} \quad (16)$$

It can be seen that for $\tau_{IR} << 1$ Q_{IR} gives back the well known Shimizu cooling, while in a dense coma ($\tau_{IR} >> 1$) it describes a strong heat source.

Spherically symmetric steady state approximation of the dusty gas mixture system yields a solar-wind type equation, which describes a transonic flow in the immediate vicinity of the nucleus

$$\frac{du}{dr} = -\frac{u}{1-M^2}\left(\frac{2}{r} - \frac{F_{gd}}{p} + \frac{(\gamma-1)(Q_{gd}-Q_{ext})}{\gamma p u}\right) \quad (17)$$

Here r denotes cometocentric distance, while M is the flow Mach number. The physical solution describes a reservoir outflow to a low pressure external medium (the external pressure is at least 10 orders of magnitude smaller than the pressure at the nucleus). This means that at large cometocentric distances the gas pressure must vanish while the flow velocity remains finite, i.e.,

$$\lim_{r \to \infty} M = \infty \tag{18}$$

At the surface the dust loading acts as a momentum sink, consequently the outflow is subsonic (in reservoir outflow problems unobstructed flows leave the reservoir with the local sound velocity; mass loaded flows always have subsolar outflow velocities)

$$M(r=R_n) < 1 \tag{19}$$

Equations (18) and (19) mean that the only physical solution to equation (17) is a transonic flow.

It is interesting to note that equation (17) is quite analogous to the classic solar wind equation. In the cometary case the gas to dust momentum transfer replaces the effect of solar gravity, otherwise the mathematical form of the equation is essentially unchanged. Figure 1 shows the various types of mathematically possible solutions, which are the same as the solutions of the solar wind equation.

The physical solution of equation (17) has a 0/0 type singularity at the sonic point, causing many numerical complications. Earlier models assumed a supersonic gas flow at large cometocentric distances and a subsonic flow close to the nucleus; in this case, there is one and only one transonic solution which passes smoothly through the sonic point. Transonic solutions cannot, in principle, be obtained numerically without additional assumptions. Generally, it has been assumed that the gas velocity (or Mach number) and its

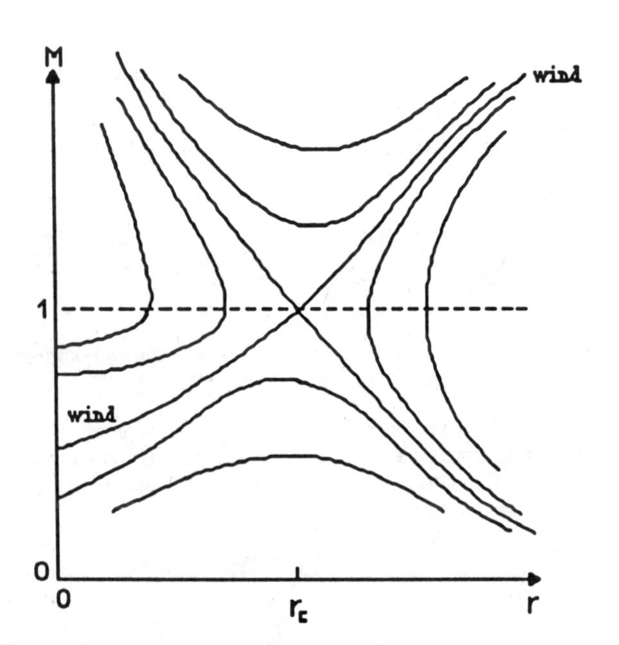

Fig. 1. Various types of solutions to equation (17). The only physical solution is the cometary wind, which starts subsonically at the surface, goes through a sonic point, and monotonously increases in the supersonic region.

first derivative behaved continuously at the sonic point. Following Probstein's original work (1968), practically all early transonic coma calculations used a type of "shooting method;" the initial flow velocity (or Mach number) value was "fine tuned" until a transonic solution was reached (Hellmich, 1979; Marconi and Mendis, 1982, 1983, 1984; Gombosi et al., 1983). In order to avoid the numerical difficulties of the earlier treatments and also to make it possible to describe dynamic phenomena in the coma, the first time-dependent (but still spherically symmetric) dusty gas dynamic model of the gas-dust interaction region was developed by Gombosi et al. (1985). The method was later combined with a three-dimensional, kinetic dust treatment in the outer coma (Gombosi and Horányi, 1986). In this model the gas dynamic equations are solved using a modified version of Godunov's first scheme which can naturally handle shocks and discontinuity surfaces. The model was able to describe such phenomena as dust halo formation in the inner coma, temporal evolution of dust and gas parameters following a comet outburst, etc. However, the numerical model can still use further improvements. In its first published form the model assumed a rather simple photochemistry, gas collisional cooling, neglected infrared radiation trapping (Marconi and Mendis, 1986), and assumed that the total radiative energy flux was approximately constant everywhere in the coma (an assumption that is more or less justified on the basis of earlier radiative transfer calculations (Keller, 1983; Marconi and Mendis, 1984)).

Time-Dependent, Axisymmetric Dusty Jet Models

The Comet Halley imaging experiments showed that cometary activity is concentrated to limited areas on the sunlit side of the nucleus, with most of the dust ejection coming from fairly localized jets. Spherically symmetric, steady state models were proven to be totally inadequate to describe cometary inner regions. Recently Kitamura (1986, 1987) has developed the first time-dependent, two-dimensional dusty gas dynamic code using one characteristic dust size, simple energetics, and a highly simplified chemistry. Kitamura's (1986, 1987) model has serious problems with the time-dependent treatment of the gas outflow, therefore he published only steady state results. Nevertheless, his first two-dimensional calculation represents an important step forward in describing dusty jets.

In this section we present preliminary results of a new, time-dependent, two-dimensional (axisymmetric) dusty jet model. We believe that these calculations will eventually help us to delineate the dominant physical processes governing the evolution of dusty cometary jets.

The model solves the coupled, time-dependent continuity, momentum, and energy equations for a dust-gas mixture (equations (1) through (11)). It is assumed that initially (t = 0) the near nucleus region is dust free and filled with low density gas. At t = 0 a localized dusty jet is generated on the nucleus surface (R_n = 6km). The adopted dust surface density distribution was similar to that of used by Kitamura (1986)

$$\rho(\Theta) = \rho_0 \left[(\alpha-1) \exp\left(-\frac{\Theta^2}{\Theta_0^2} \right) + 1 \right] \tag{20}$$

where α=10, Θ_0 = 10°, $\rho_0 = 2.71 \times 10^{-11}$ g cm^{-3}. The surface dust temperature was 350 K. For the sake of simplicity only one dust size is considered in the present model ($a = 0.65\ \mu m$). The dust to gas mass production ratio is 0.5, close to the observed Comet Halley value. A very low pressure external "vacuum cleaner" was placed at a distance of 100 km, which helped to ensure a supersonic flow in most of the integration region (the sonic point was located at about 100 m from the surface).

A 40×30 grid structure was employed in the integration region. There were 40 linearly spaced azimuthal and 30 logarithmically spaced radial grids extending from 6 to 100 km. The time-dependent, coupled multidimensional partial differential system was solved with a second order upwind biased Godunov-type numerical scheme developed at the University of Michigan.

Figures 2 and 3 show snapshots of the gas and dust densities. The snapshots present two-dimensional equidensity curves at t = 20s, 40's, 60's, 100's, 200 s and 300 s after onset. Inspection of Figures 2 and 3 reveals a very interesting feature of the localized axisymmetric jet. Initially the newly ejected gas and dust expands radially with relatively little horizontal broadening. The main reason for this almost entirely radial outflow pattern is that due to the ~10° spatial extent of the source region at the nucleus (the active "spot" on the surface has a radius of about 1 km) the surface radial pressure gradient is much larger than the azimuthal gradient. As the gas leaves the nucleus-coma interface region the azimuthal pressure gradient becomes comparable to the radial one and the gas flow starts to expand in the azimuthal direction as well. This effect can be observed at t = 40s. In this phase the gas expansion goes faster than r^2 (especially near the outer edge of the jet); therefore, two interesting things happen. First of all the jet cone expands (at t = 60's the jet cone is about 45° wide), and second the dust grains also attain a significant azimuthal velocity component. A natural consequence of this azimuthal velocity component is that a large fraction of the dust particles is "swept" away from the region near the axis of symmetry, thus resulting in a dust density depletion above the active area.

By about t = 100 s the azimuthal gas expansion is stopped by the background gas. It has already been demonstrated by Kitamura's calculations (1986, 1987) that this effect is quite sensitive to the background gas production rate (gas production from the inactive part of the nucleus). This means that beyond about 50 km or so there is no significant azimuthal gas velocity and consequently the dust particles also lose most of their azimuthal velocity component. The late time (few hundred seconds) distributions, which are in a good qualitative agreement with Kitamura's steady state results, show that most of the dust is concentrated on a region surrounding the surface of a cone with a half-opening angle of about 50°. The dust density is quite low inside this cone.

Our conclusion is that the structure of a dusty jet resulting from an axisymmetric active region is cone-like. The opening angle of the dust cone is largely determined by the ratio of the gas pressures inside the active surface area and the inactive surface background. This jet structure is significantly different from the ones predicted by earlier spherically symmetric calculations (cf. Mendis et al., 1985; Gombosi et al., 1986);

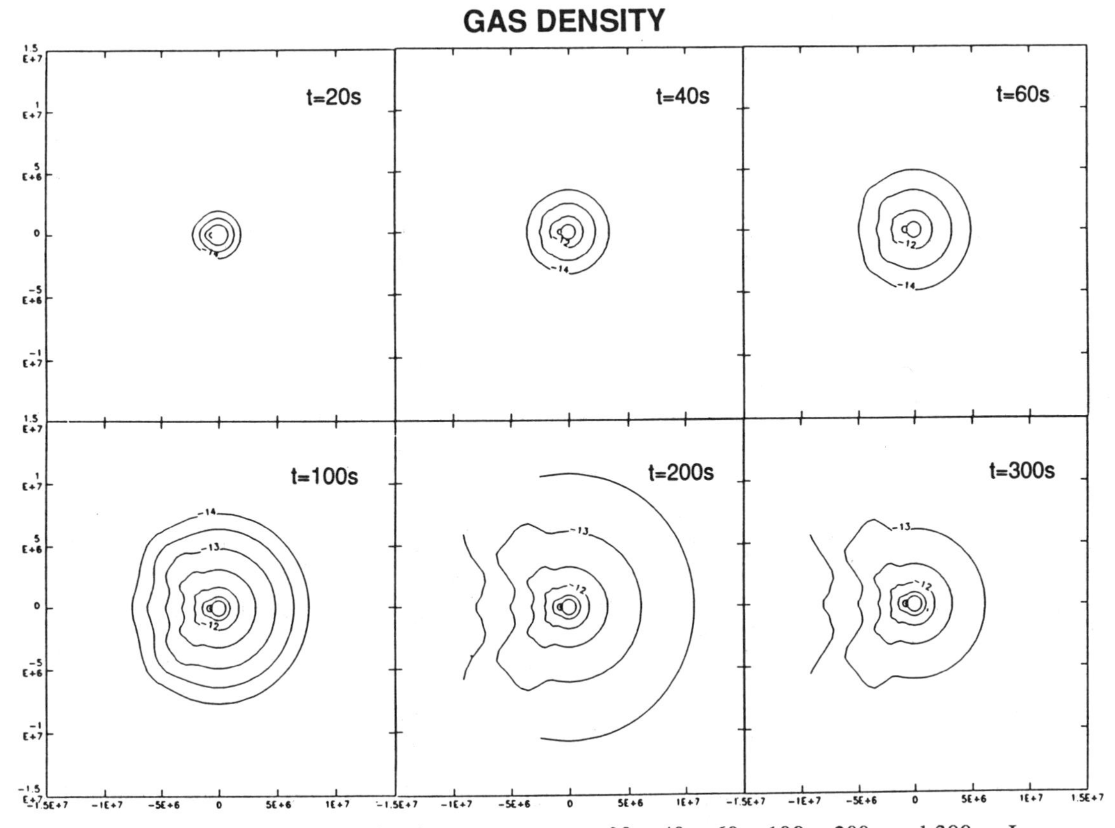

Fig. 2. Snapshots of gas isodensity contours at t = 20 s, 40 s, 60 s, 100 s, 200 s and 300 s. In each panel the axis of symmetry is a horizontal line going through the center.

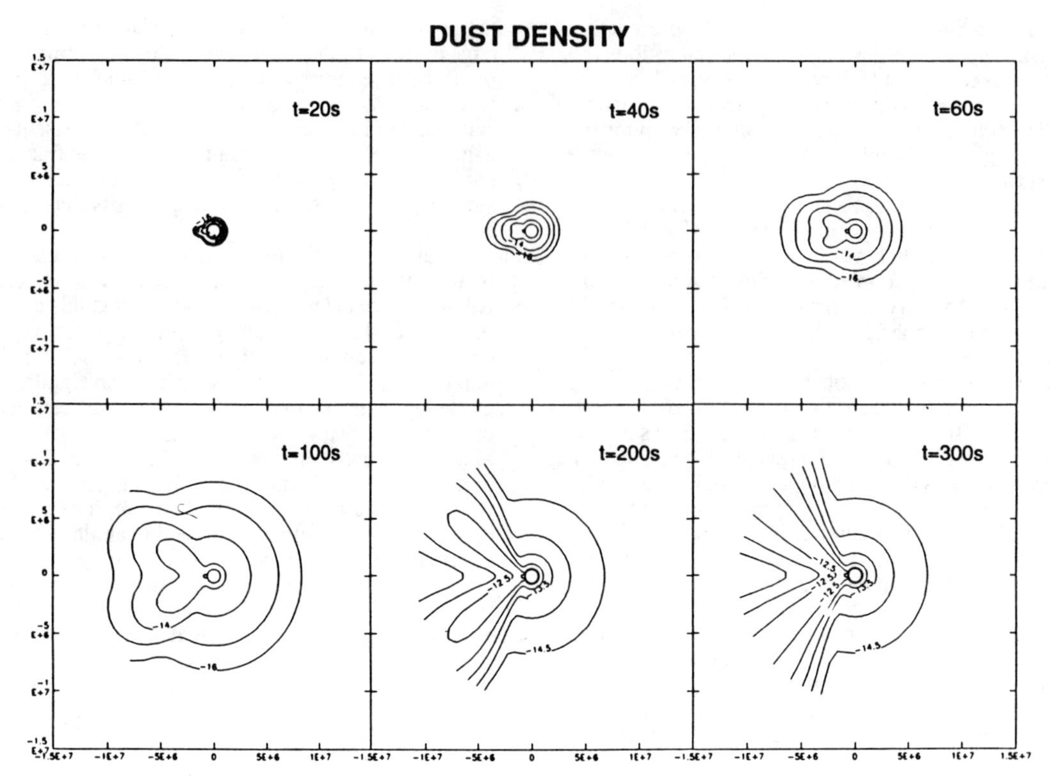

Fig. 3. Snapshots of dust isodensity contours at t = 20 s, 40 s, 60 s, 100 s, 200 s and 300 s. In each panel the axis of symmetry is a horizontal line going through the center.

therefore, it is clear that our earlier picture of the dust acceleration region has to be considerably revised.

Acknowledgments. The authors are indebted to Bram van Leer for his advice in developing the second-order upwind biased Godunov-type numerical scheme to solve the coupled, time-dependent multidimensional partial differential equation system. This work was supported by NSF grant AST-8605994 and NASA grant NGR 23-005-015. Acknowledgments are also made to the University of Michigan and to the National Center for Atmospheric Research sponsored by NSF, for the computing time used in this research.

References

Brin, G.D., and D.A. Mendis, Dust release and mantle development in comets, Ap. J., 229, 402, 1979.

Crovisier, J., The water molecule in comets: Fluorescence mechanisms and thermodynamics of the inner coma, Astron. and Astrophys., 130, 361, 1984.

Delsemme, A.H., and P. Swings, Hydrates de gaz dans les noyaux cometaires et les grains interstellaires, Ann. Astrophys., 15, 1, 1952.

Divine, N., and R.L. Newburn, Modeling Halley before and after the encounters, Astron. Astrophys., in press, 1987.

Fanale, F.P., and J.R. Salvail, An idealized short-period comet model: Surface insolation, H_2O flux,dust flux and mantle evolution, Icarus, 60, 476, 1984.

Gombosi T.I., and M. Horányi, Time-dependent modeling of dust halo formation at comets, Ap. J., 311, 491, 1986.

Gombosi, T. I., K. Szegö, B.E. Gribov, R.Z. Sagdeev, V.D. Shapiro, V.I. Shevchenko, and T.E. Cravens, Gas dynamic calculations of dust terminal velocities with realistic dust size distributions, Cometary Exploration, edited by T. I. Gombosi, KFKI Press, Budapest, Hungary, Vol 1., p. 99, 1983.

Gombosi, T.I., T.E. Cravens, and A.F. Nagy, Time-dependent dusty gas dynamical flow near cometary nuclei, Ap. J., 293, 328, 1985.

Gombosi, T.I., A.F. Nagy, and T.E. Cravens, Dust and neutral gas modeling of the inner atmospheres of comets, Rev. Geophys., 24, 667, 1986.

Hellmich, R., Anisotrope Mehrfachstreunung in der Staubkoma und ihr Einfluss auf die Sublimationsrate des Kometenkerns, Ph.D. thesis, Univ. Göttingen, FRG, 1979.

Hellmich, R., and H.U. Keller, On the dust production rates of comets, in Solid Particles in the Solar System, edited by J. Halliday and B.A. McIntosh, p. 255, D. Reidel, Hingham, Mass., 1980.

Horányi M., T.I. Gombosi, T.E. Cravens, A. Körösmezey, K. Kecskeméty, A.F. Nagy, and K. Szegö, The friable sponge model of a cometary nucleus, Ap. J., 278, 449, 1984.

Houpis H.L.F., W.-H. Ip, and D.A. Mendis, The chemical differentiation of the cometary nucleus: The process and its consequences, Ap. J., 295, 654, 1985.

Keller, H.U., Dust and gas models in the coma, in Cometary Exploration, edited by T.I. Gombosi, Vol. 1, p. 119, Hungarian Academy of Sciences, Budapest, 1983.

Kitamura, Y., Axisymmetric dusty gas jet in the inner coma of a comet, Icarus, 66, 241, 1986.

Kitamura, Y., Axisymmetric dusty gas jet in the inner coma of a comet II. The case of isolated jets, Icarus, 72, 555, 1987.

Kürth, E., D. Möhlmann, B. Giese, and F. Tauber, Thermal stresses and dust dynamics on comets, in Proc. 20th ESLAB Symposium on the Exploration of Halley's Comet, edited by B. Battrik, E.J. Rolfe, and R. Reinhard, ESA SP-250, Vol. 2, p. 385, 1986.

Marconi, M.L., and D.A. Mendis, The photochemical heating of the cometary atmosphere, Ap. J., 260, 386, 1982.

Marconi, M. L., and D.A. Mendis, The atmosphere of a dirty cometary nucleus, A two-phase multi-fluid model, Ap. J., 273, 381, 1983.

Marconi, M. L., and D.A. Mendis, The effects of the diffuse radiation fields due to multiple scattering and thermal reradiation by dust on the dynamics and thermodynamics of a dusty cometary atmosphere, Ap. J., 287, 445, 1984.

Marconi, M.L., and D.A. Mendis, IR heating of Comet Halley's atmosphere, Earth, Moon and Planets, 36, 249, 1986.

Mendis, D.A., and G.D. Brin, On the monochromatic brightness variations of comets II, The core-mantle model, Moon and Planets, 17, 359, 1977.

Mendis, D.A., Houpis, H.L.F., and Marconi, M.L., The Physics of Comets, Fundamentals of Cosmic Physics, 10, 1, 1985.

Probstein R.F., The dusty gas dynamics of comet heads, in Problems of Hydrodynamics and Continuum Mechanics, Soc. Industr. Appl. Math., p. 568, 1968.

Shimizu, M., Neutral temperature of cometary atmospheres, in The Study of Comets, edited by B. Donn, M. Munna, W. Jackson, and R. Harrington, NASA S-393, p. 363, 1976.

Shulman, L.M., Dinamika Kometnykh Atmospher - Neutral'nyi Gaz, Naukova Dumka, Kiev, 1972.

Weissman, P.R. and Kieffer, H.H., Thermal modeling of cometary nuclei, Icarus, 47, 302, 1981.

Whipple, F.L., A comet model I., The acceleration of comet Encke, Ap. J., 111, 375, 1950.

THREE-DIMENSIONAL PLASMA MEASUREMENTS FROM THREE-AXIS STABILIZED SPACECRAFT

S. J. Bame, R. H. Martin, and D. J. McComas

Space Plasma Physics Group, Los Alamos National Laboratory, Los Alamos, NM 87545

J. L. Burch, J. A. Marshall, and D. T. Young

Southwest Research Institute, San Antonio, TX 78284

Abstract. Future planetary missions require that comprehensive three-dimensional measurements of electrons and mass-resolved ions be made from three-axis stabilized spacecraft. In order to make these measurements without requiring expensive and resource intensive platforms to scan space mechanically, we are developing various systems that are designed to scan space electrostatically. These systems also make it possible to circumvent the significant shadowing that would be present even with a scan platform, caused by necessary spacecraft appendages such as communications antennas and a power source (RTG or solar cell panels). The systems, which are axially symmetric, select particles arriving from 360° in azimuth along conical surfaces whose polar (or elevation) angles, referenced to the instrument symmetry axes, are determined by applying suitable deflection voltages to shaped deflectors. Particles thus selected in polar angle pass into spherically or toroidally-shaped electrostatic analyzers. After analysis, the 360° outputs of the analyzers are divided into discrete angular swaths to provide azimuthal angle resolution. In the case of electrons, the analyzed particles can be detected directly; in the case of ions, the particles in each swath can be counted directly, or further analyzed with time-of-flight or magnetic analyzers to obtain the velocity distributions of the separated major ion constituents. We present computer simulations of particle paths through the various analyzers of this type and show results from laboratory calibrations of prototypes.

Introduction

One of the fundamental realizations brought about by the planetary exploration program of the past two decades is the wide range of complex plasma physics phenomena that occurs in the vicinity of the terrestrial and Jovian planets, as well as comets. All of the solar system bodies which have been probed seem to exhibit strong and unique plasma and electrodynamic phenomena. However, in many cases the reconnaissance and exploratory missions which have been flown have left these phenomena only partially investigated and not completely understood. Two important limitations on the detailed understanding of plasma phenomena at other planets have been the incomplete coverage of plasma velocity distributions provided by previous missions and the absence of unambiguous composition measurements.

To obtain a quantitative understanding of the various plasma physical phenomena occurring in the vicinity of a planet or other body, it is important to have accurate measurements of the fluid moments, including the density, bulk velocity, temperature, and heat flux of each of the plasma components. These measurements must resolve the entire velocity distributions of electrons and ions with adequate temporal, angular, mass, and energy resolution. Full angular coverage is particularly important when the distributions are anisotropic in the spacecraft frame, either due to an actual pitch-angle anisotropy or to a significant flow velocity. Unfortunately, obtaining the requisite full velocity-space coverage is difficult because planetary missions typically use three-axis stabilized spacecraft in order to facilitate imaging experiments, and to maintain good telemetry links with Earth. Thus, one cannot take advantage of spacecraft spin to obtain full angular coverage of the plasma distribution. An additional problem exists even if a single scan platform is provided for plasma instrumentation. Necessary spacecraft appendages such as communications antennas and power sources (RTGs or solar cell panels), and the spacecraft itself block significant portions of velocity-space.

Accurate measurement of plasma moments requires knowledge of the ion composition, the importance of which has only recently begun to be appreciated. This appreciation has been fueled by spacecraft observations of significant populations of heavy ions, with clear terrestrial origins, within the Earth's magnetosphere (see Young, 1983, for a review), and even more so by the discovery of major heavy-ion components in the plasmas of the Jovian and Saturnian magnetospheres (cf. Lazarus and McNutt, 1983). Voyager observations at Jupiter and Saturn suggest that the various satellites and rings within those magnetospheres may be the dominant sources of magnetospheric plasma. As a consequence, it is now clear that a knowledge of plasma composition is a first-order requirement for a proper understanding of the physics of various planetary magnetospheres.

The importance of understanding planetary plasma phenomena is now well recognized, and missions such as Galileo and Comet Rendezvous and Asteroid Flyby (CRAF) include comprehensive plasma and field instrumentation in their payloads. Two new missions now in the planning stages are the Mars Aeronomy Observer and Cassini, which is comprised of a Saturn Orbiter and Titan Probe. Fairly complete plasma instrumentation has also been recommended for these missions. However, the study reports for both missions assume that the needed instruments have already been developed for previous planetary or solar-terrestrial missions. Unfortunately, this is not the case. Previous plasma instruments on planetary missions have provided neither unambiguous mass analysis nor full velocity-space coverage. Furthermore, since for communication and imaging reasons planetary spacecraft are usually three-axis stabilized, plasma instrument of the present generation that make two-dimensional measurements, provide three-dimensional coverage of accessible portions of space only if mechanical scanning is available with scan platforms or separate spinning spacecraft sections. Even with such platforms, as mentioned above significant portions of velocity-space are blocked by spacecraft appendages.

Provision of a rotating platform introduces reliability problems, increases the cost of the spacecraft, uses limited spacecraft resources such as attitude control gas, and in any event does not solve the full accessibility to space problem caused by obstructions. Additionally, because of the need to conserve spacecraft resources, the measurement time resolution using a scan platform is necessarily constrained by the speed and number of operating cycles that can be allotted it.

The CRAF spacecraft presents a good example of the problems that arise when mechanical scanning is necessary to convert two-dimensional measurements into three-dimensional plasma distributions. CRAF, like Cassini, will use the Mariner Mark II spacecraft, which is three-axis stabilized with scan platforms. The proposed instrument for providing composition-resolved three-dimensional plasma measurements has a two-dimensional acceptance fan ($5° \times 160°$) and will need to rely on single-axis mechanical scanning projected to be provided by the CRAF low precision scan platform in order to obtain three-dimensional coverage. However, because of fuel limitations, scanning will be limited to a small fraction of the spacecraft operational time (less than 25%), a serious degradation of the measurement capability. Additionally, other experiments on the platform have pointing requirements that are incompatible with those of the plasma experiment, further reducing measurement capability.

Taking the lessons learned from CRAF and generalizing to future missions such as Cassini and Mars Aeronomy Observer as well as potential missions such as Lunar Polar Orbiter, Mercury Magnetospheric Orbiter, etc., it is important to examine how fully three-dimensional plasma velocity distributions might be obtained using fixed-attitude spacecraft. An optimum solution must avoid substantial blockage of velocity-space by the spacecraft and its appendages and should not require a scan platform. Search for an answer to this problem has been a major impetus for the work reported here that was carried out at the Los Alamos National Laboratory and the Southwest Research Institute (SwRI). As discussed in the following sections, mechanical scanning can be eliminated by the use of electrostatic scanning. Several candidate systems are presented. The problem of solid angle obscuration by the spacecraft and appendages can be overcome by appropriately locating two sensor heads on opposite sides of the spacecraft.

Measurement Objectives and Approach

As noted previously, the plasmas found within all of the planetary magnetospheres sampled to date are highly complex in makeup and in velocity distribution. Moreover, they are highly variable both spatially and temporally. For example, within the magnetosphere of Saturn plasma densities range from less than 10^{-2} up to $\sim$10 cm^{-3}, flow speeds range from $\sim$10 to $\sim$200 km s^{-1}, and both ion and electron temperatures range from several electron volts to $\sim$1 keV (e.g., Sittler et al., 1983; Richardson, 1986). Comparable or greater variations are found within other magnetospheres.

In addition, in many situations of great interest more than one particle population is present. For example, both very cold (several electron volts) and very hot (several hundred electron volts) electron and/or ion components commonly coexist in the same region, and several different elemental or molecular species may be present (for example, H, N, O, H_2O, CO, etc.). Each elemental species may also occupy more than one charge state (for example, O^+, O^{++}). Further, velocity distributions may be highly non-Maxwellian, and strong anisotropies may be present. Often strong streaming along the local magnetic field is observed. The plasma flow may be supersonic and therefore appear as a beam (typically transverse to the field within the magnetospheres of the outer planets), or subsonic and therefore more nearly isotropic in the spacecraft frame.

Measurement objectives during upcoming planetary missions should therefore be directed toward resolving and distinguishing these complexities with a sensitive instrument placed on a three-axis stabilized spacecraft, and if possible these observations should be made without requiring a scan platform and in such a way as to avoid obscuration of significant portions of space by such mission-essential elements as the spacecraft antennas and the spacecraft itself. The plasma measurements should be fully three-dimensional or cover the pertinent portions of velocity space required for the specific scientific problems of interest with adequate angular resolution to resolve moderately narrow beams independent of their flow direction or the spacecraft orientation. The instrument should be sensitive enough to measure densities of major ion species as low as 10^{-2} cm^{-3} on a temporal scale of several tens of seconds or less, and should have an energy range extending from near 1 eV to ~30 keV where energetic particle solid state detectors can take over. Energy resolution should be adequate for both cold supersonic and hot subsonic flows. Major ion species should be separated under all conditions; their separation should not be dependent on fortuitous combinations of ion flow speed and direction. To implement these objectives, the SwRI/Los Alamos joint program has two major goals: (1) to explore and develop instrument concepts aimed at providing fully three-dimensional ion and electron distribution measurements on three-axis stabilized spacecraft that might have large, blocking appendages, without requiring mechanical scan platforms or spinning sections, and (2) to develop an appropriate mass analysis system to be used in conjunction with the three-dimensional ion instrument to resolve all important ion species. Principal emphasis in this paper is given to goal (1). A parallel development of time-of-flight (TOF) mass analysis is being carried out; some results of this development have been reported by Young et al. (1989).

Our approach to the field-of-view (FOV) problem associated with fixed-attitude spacecraft has been to develop the candidate electrostatic deflection systems described in the following sections. These systems scan a large fraction of the unit sphere with programmed deflection and analyzer voltages. The voltages required, that are no higher than those that have been successfully used in some previous space flight plasma experiments, are shielded from the external environment by grounded enclosures. Combining a FOV scanning system with a high throughput, lightweight, high-resolution TOF system will provide high time resolution and M/q composition-resolved measurements of the three-dimensional distributions of the plasmas which surround planetary objects of interest.

Elevation Analyzers Under Development

In this section we describe five "elevation analyzer" concepts that achieve desired fully steerable FOVs. These elevation analyzers are in various stages of development at SwRI and Los Alamos. Each concept has been demonstrated to be viable, but some require further characterization with computer ray tracing studies and verification measurements on laboratory prototypes. Names given to the five candidate systems described below, in the order of presentation, are Beacon-C, Beacon-D, fast ion mass spectrometer (FIMS), gridded truncated hemisphere (GTH), and CRAF non-scanning platform (NSP).

Beacon-C Elevation Analyzer for Ion Composition Measurements

The Beacon-C elevation analyzer is one of two versions of a system called "Beacon" under development at Los Alamos. With a single Beacon sensor on a fixed-attitude spacecraft, particles can be resolved in energy, azimuth, and polar angle over all or most of a 2π str hemisphere. Fully three-dimensional measurements over the entire 4π str unit sphere can be implemented by placing diametrically opposed Beacon sensor heads on opposite sides of a spacecraft. A single set of power supplies and data processing and control logics can be used to operate both sensor heads. The Beacon-C is an ion composition version of Beacon designed to measure separately, but concurrently, the velocity distributions of the major ion species to be found in a planetary or cometary magnetosphere. Another version, Beacon-D, which is described in the next section, is configured to measure the velocity distributions of magnetospheric and solar wind electrons. It can also be employed to measure both electrons and ions (without composition) with a single detector system on missions with severely constrained spacecraft resources. Both the -C and -D versions make use of a Beacon system of rotationally symmetric electrodes shown in Figure 1, to deflect parti-

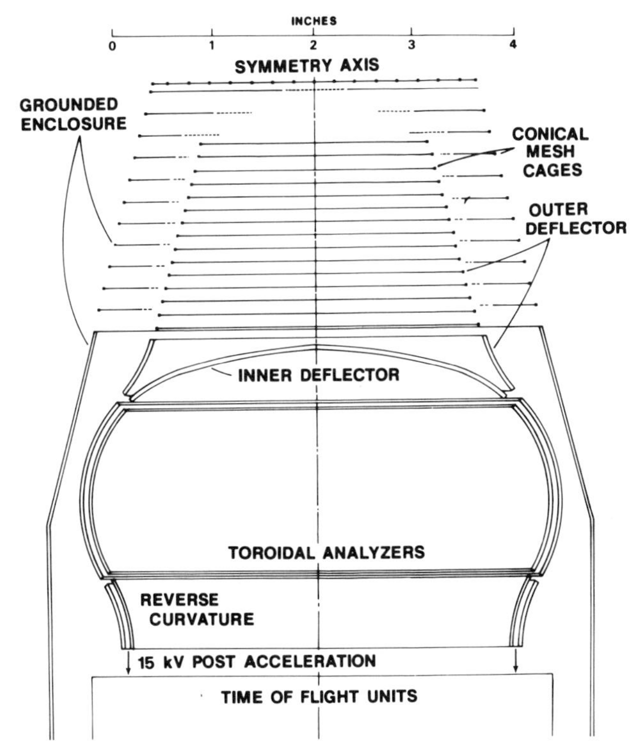

Fig. 1. The Los Alamos Beacon-C ion composition elevation analyzer, shown in a schematic cross-sectional view.

cles arriving along a conical surface into a toroidal electrostatic analyzer for E/q analysis. The acceptance cone angle, equivalent to the polar or elevation angle as referenced to the symmetry axis, is selected by applying appropriate voltages to the deflector electrodes. Particles arriving from 360° in azimuth around the conical surface are accepted and can be resolved in azimuth by physically dividing the 360° range. This division is implemented in different ways for Beacon-C and Beacon-D as described below.

Figure 1 shows a schematic, cross-sectional view of Beacon-C. This instrument employs the specially shaped inner and outer deflector electrodes of Beacon to deflect incoming ions into a two-stage toroidal analyzer. The deflector shapes were empirically determined using a 2 1/2-dimensional ray tracing computer program. This program, which traces only those rays that pass through the symmetry axes of rotationally symmetric systems, was utilized to maximize the deflection sensitivity vs. voltage of Beacon-C while permitting the desired range of polar angle acceptance to be achieved. The inner deflector, as shown, can be entirely solid, while the outer deflector is made up of a solid section, where bending is strongest, followed by a conical section of wire or mesh that is largely transparent to incoming ions.

The purpose of the second reverse curvature stage of the compound toroidal analyzer is to extend the bending angle of the analyzer to provide adequate energy resolution and UV rejection and at the same time to bring the analyzed ions out along a cylindrically shaped surface centered on and parallel to the instrument symmetry axis. For mass per charge separation the ions are then post-accelerated by −15 kV and injected into a circular array of TOF analyzers such as the array with 22.5° azimuthal resolution described by Young et al. (1989). Mass resolution might also be provided by an appropriate magnetic analyzer system, or a combination of TOF and magnetic analyzers, as used for the CRAF NSP instrument described in a later section.

Positive voltages are used on the deflectors and analyzer of the Beacon to avoid outward acceleration of the photo- and secondary electrons that will be produced on the instrument surfaces in space. The instrument is encased in a grounded outer container, as shown in Figure 1 to shield the instrument exterior from the interior voltages which may extend as high as 7 kV. (Such voltages have been successfully used in past space experiments and are not expected to cause any problems.)

Ray tracing with the 2 1/2-dimensional computer program has verified and characterized the capability of the Beacon deflection system. The program calculates the electrostatic configuration of an axially symmetric system and then incrementally traces particles launched with given initial conditions through the system along a plane which contains the symmetry axis. Further characterization of the Beacon response is in progress using a program which tracks particles through the three-dimensional representation of the instrument without constraining them to any particular plane.

Results using the 2 1/2-dimensional program are given in Figure 2. The program calculates rays one at time; a number of them have been superimposed in the figure. With an appropriate positive potential applied to the outside analyzer plate, ions launched from deep within the analyzer follow central paths out through the analyzer entrance and the deflectors. For electrostatic systems the rays are, of course, completely reversible. The rays shown were traced for a selection of voltages on the deflectors. They exhibit a pattern extending from ~1° to ~78° in polar or elevation angle. (This pattern prompted the name "Beacon" for this elevation analyzer.) Rays at polar angles extending from 46° down to 1°, near the sensor equator, were obtained with outer deflector voltages extending from 0 V to +7000 V in 700 V increments. Rays from 46° up to 78°, 12° from a polar angle of 90°, were obtained with the same voltages applied to the inner deflector. In either case the opposite deflector is held at 0 V. For these calculations, the ion launch energy was 34.0 keV/q corresponding to an incoming energy of 35.5 keV/q. (The energy difference is ~1/2 the analyzer voltage.)

Figure 3 is a plot of ion deflection angle vs. deflection voltage applied to the inner and outer deflectors. The slope of the curve is a measure of deflection sensitivity, which increases from high to low polar angle. This is understandable from inspection of Figure 2 which shows that larger portions of the trajectories of

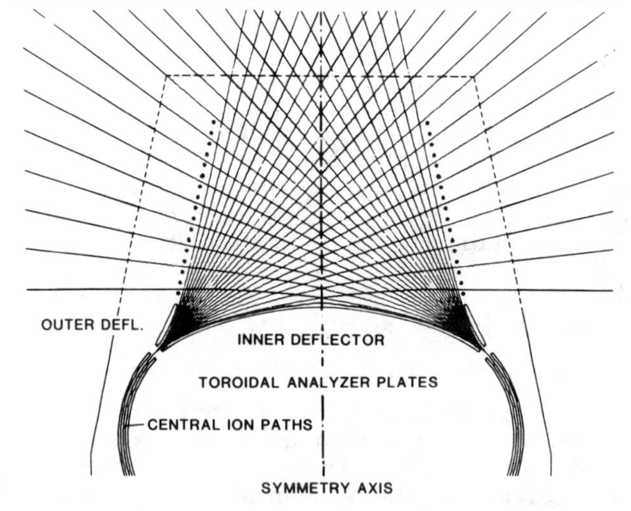

Fig. 2. Beacon coverage of the hemisphere. 35 keV ions with polar angles extending from ~0° to ~78° are deflected into the analyzer with voltages extending from 0 to 7000 V.

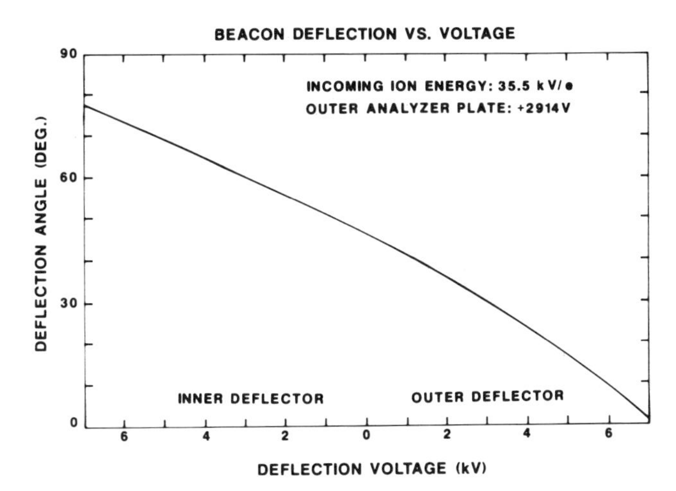

Fig. 3. Beacon deflection angle vs. voltage. Voltage is applied to one deflector at a time holding the other at 0 V.

ions deflected from low polar angles lie in regions in which the field strength is high and appropriately directed in comparison to the trajectories of ions deflected from high polar angles.

The ray tracing program has also been used to verify that the reverse curvature analyzer shown in Figure 1 performs as desired. A particle launched downward from near the entrance of the top toroidal section successfully passed into the lower section and exited with a trajectory parallel to the symmetry axis. Further simulations have shown that the Beacon can be configured to accept particles from an entire hemisphere. The version of Figure 1 accepts particles up to ~78° polar angle, which allows coverage of ~98% of the unit sphere with two sensor heads. Giving the outer deflector a more nearly cylindrical shape, coverage up to 100% has been calculated. The penalty for this small amount of additional spatial coverage is either a reduction of the maximum energy covered or a requirement for higher deflection voltages. For most applications the particle distributions to be encountered are broad enough that 98% coverage is quite adequate.

In order to verify that the Beacon operates as simulated and to further characterize its full three-dimensional behavior, a proof-of-principle test model was constructed at Los Alamos and taken to SwRI for preliminary tests in the SwRI calibration facility. A photograph of the test model is shown in Figure 4. It is constructed of aluminum alloy with Kel-F insulators in such a way that it can be assembled in both the Beacon-C and Beacon-D configurations. The figure shows it in the -C configuration without the external grounded container. The dome-shaped electrode in the center is the inner deflector; the outer deflector, visible above the inner, is extended by the wire cage continuation of the solid part of the deflector. The wire is 0.5 mm diameter BeCu which is relatively thick in comparison to the 1.5 mm analyzer plate spacing. Some of the Beacon response characteristics at low polar angle may have been affected by the thickness of the wire. We plan to do further tests with smaller wire or mesh.

At SwRI the calibration tests were conducted by placing the Beacon-C in the calibration system beam, mounted to permit rotation in both the azimuthal and polar angle directions. A broad parallel beam of 1120 eV N_2^+ ions was used for all the tests and the analyzed ions were detected in an imaging microchannel plate (MCP) system which had adjustable spacing behind the test model. Figure 5 shows 12 MCP images of the spatial distribution of the output ions from Beacon-C as displayed on a storage oscilloscope screen. For this series the deflectors were at 0 V, the instrument was set so that the beam polar angle was 44°, and the analyzer voltage was tuned to 84 V. As the figure shows, the analyzed beam at this polar angle is bifurcated at close and distant spacings but has a broad, well-defined crossover or focussed distribution at intermediate spacings extending from ~8 to ~21 mm. This bifurcation, which is under further study with fully three-dimensional simulations, is thought to result from transmission of those

Fig. 4. Photograph of the Beacon proof-of-principle test model, with the outer grounded enclosure removed.

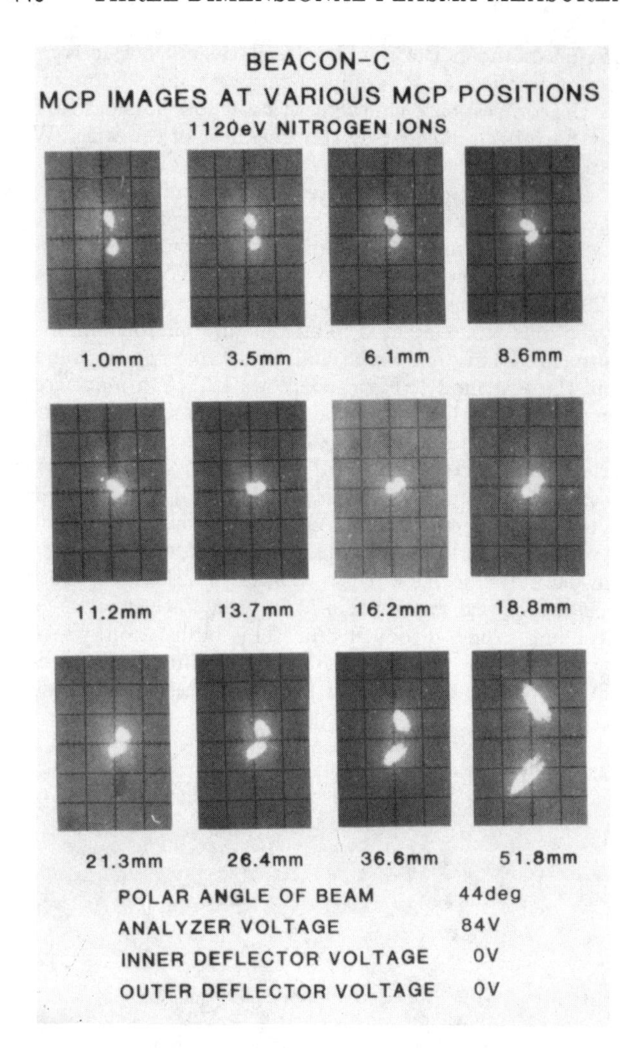

Fig. 5. Photographs of the spatial distributions of ions striking the imaging MCP taken at various MCP positions behind Beacon-C with the beam at a polar angle of 44° and deflectors at 0 V.

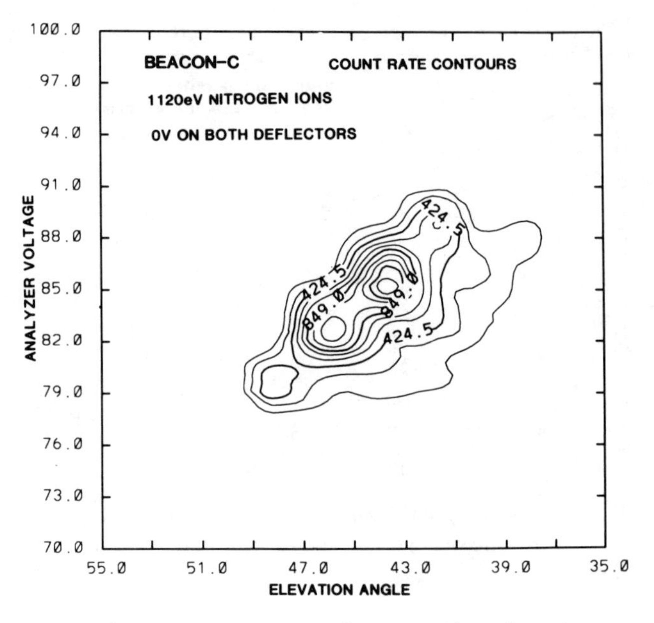

Fig. 6. Count rate contours taken over the relevant ranges of analyzer voltage and polar angle with both deflectors at 0 V.

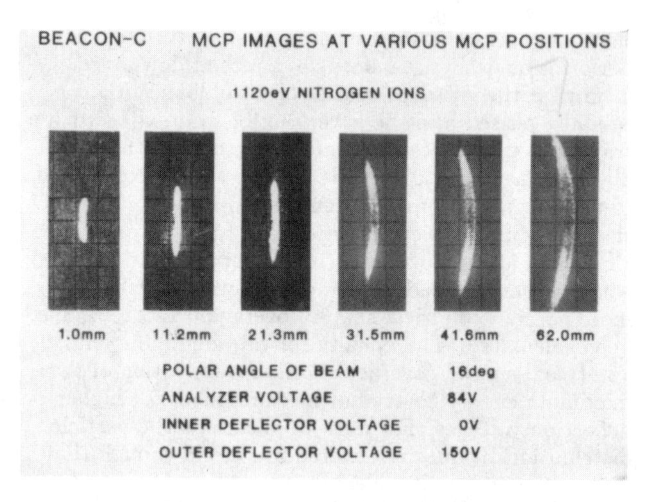

Fig. 7. Photographs of the spatial distributions of ions striking the imaging MCP at various spacings with the beam at a polar angle of 16°.

ions in the broad parallel beam which pass on either side of the region around the symmetry axis. Any of these intermediate focussed distributions would be suitable for 15 kV post acceleration into a TOF system such as that described by Young et al. (1989).

Figure 6 shows count rate contours taken over a field composed of the relevant ranges of analyzer voltage and polar angle. Both deflectors were held at 0 V for this set of measurements. The contours show evidence of structure in the response peak which is probably caused by grid wire shadows; this will be investigated further using finer grid wires.

Significant responses were found for Beacon-C over a range of polar angles extending from 0° to 79°. At angles above 79° the outer deflector wire cage interferes with the incoming ions. Figure 7 shows photographs of MCP images for a beam at a polar angle of 16°, obtained with 84 V on the analyzer, 0 V on the inner deflector, and 150 V on the outer deflector. For this angle the beam is somewhat broader than for the previous case, but the image is still quite adequate between ~8 and ~20 mm to feed into a TOF system. Count rate contours over the relevant ranges of analyzer voltage and polar angle for 150 V on the outer deflector are shown in Figure 8. Multiple response peaks are not as

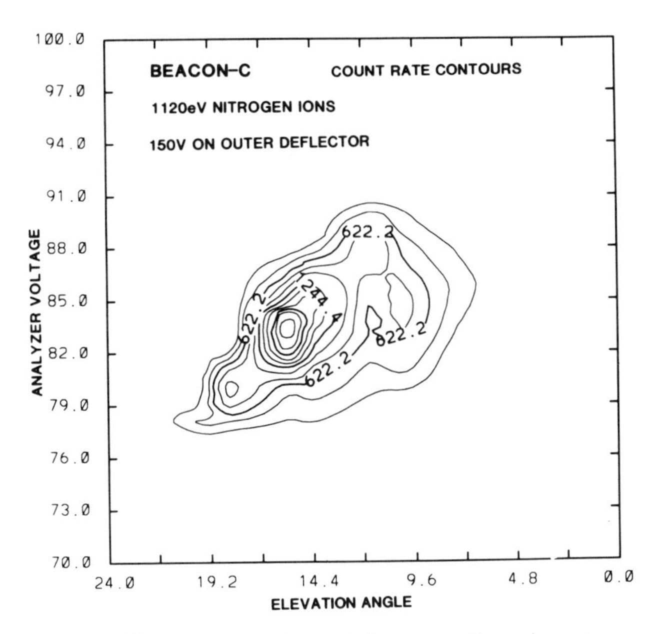

Fig. 8. Count rate contours taken over the relevant ranges of analyzer voltage and polar angle with the outer deflector at 150 V.

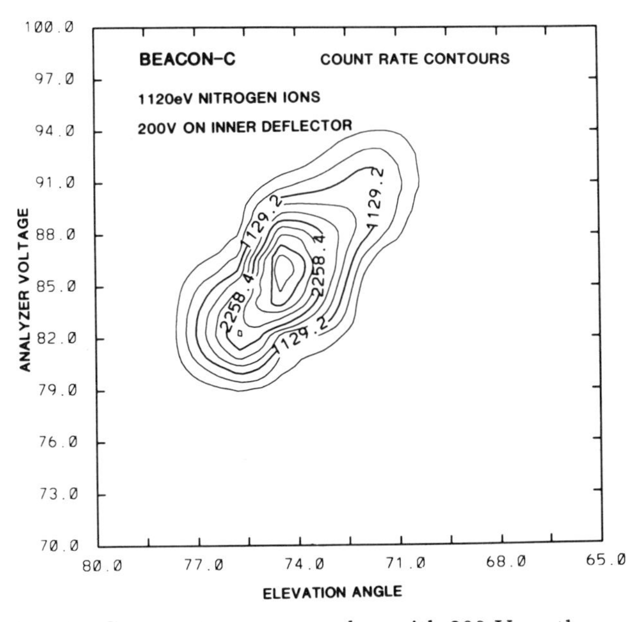

Fig. 9. Count rate contours taken with 200 V on the inner deflector, resulting in high polar angle response.

prominent here as for the previous case. Count rate contours for 200 V on the inner deflector are shown in Figure 9, and again there seems to be only one principal peak, positioned at ~74° polar angle and 85 V analyzer voltage. Here the beam does not go through the wire grid.

As a final test of the Beacon-C capabilities, the analyzer voltage was set at 84 V and polar angle sweeps were made for deflector voltages set at 0, 50, 100, 150, and 200 V, first on one deflector and then on the other. The results, given in Figure 10, show clean separation of the polar angle deflection ranges, particularly at high polar angles. At low polar angles, fine structure in the response peaks, as mentioned above, appears to be real and repeatable. Again, we believe this structure is probably due to the relatively large deflector wire size in comparison to the analyzer plate spacing. Future tests using finer deflector grid wire or mesh should help resolve this issue and reduce the amplitude of the fine structure, if indeed the wire size is the source of this response.

Beacon-D Elevation Analyzer for Electron Distribution Measurements

The Beacon-D analyzer shown in Figure 11 is intended for measurements of the velocity distributions of electrons in planetary magnetospheres and in the solar wind. It has deflection optics identical to those of Beacon-C, but instead of utilizing a compound toroidal analyzer, it uses a single toroid with a 133° bending angle. Analyzed electrons are post-accelerated, e.g., by +200 V, into an exit aperture housing where they strike a secondary emitter and create secondary electrons which are attracted into an MCP with its front face at +400 V. Since the azimuthal angle distributions of the

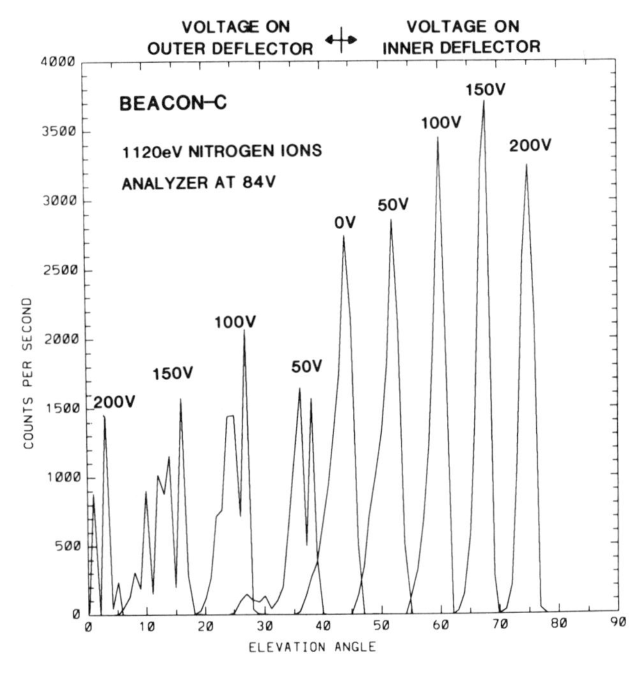

Fig. 10. Polar angle response of Beacon-C, measured at SwRI, for the various deflector voltages on the inner and outer deflectors shown above each peak.

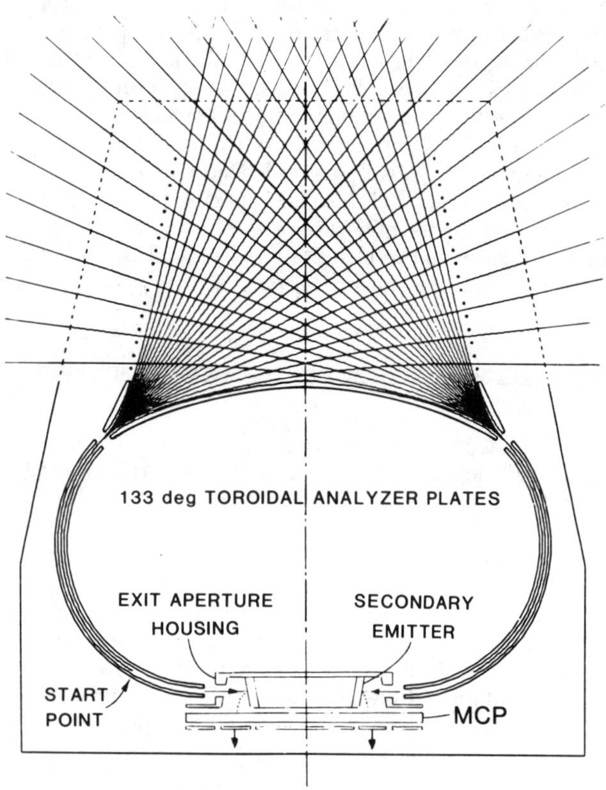

Fig. 11. Beacon-D configuration for detecting magnetospheric and solar wind electrons. The deflection optics are identical to those of Beacon-C.

electrons entering the sensor from outside are retained after deflection and electrostatic analysis, azimuthal angle resolution can be obtained by spatially resolving the electrons striking the MCP by employing a divided anode. For example, division of the anode into 16 parts would give 22.5° resolution.

As mentioned earlier, this configuration could be used to detect both electrons and ions for missions where both need to be measured, but spacecraft resources do not allow for flying independent electron and ion detectors. Referring again to Figure 11, ions as well as electrons can be analyzed if a second analyzer supply is provided. No additional deflector voltage supplies are required since the existing set serves to deflect both electrons and ions. To detect both electrons and ions after analysis, a programmable secondary emitter supply must be added to change the voltage configuration of the exit aperture housing/secondary emitter suitably. Computer simulations show this scheme to be viable; we plan to demonstrate it with a breadboard model as well. However, if resources permit, it is of course preferable to measure electrons and ions with separate instruments, to allow concurrent measurements.

At SwRI the Beacon-D configuration was tested in the same manner as Beacon-C with similar results. Polar angle sweeps for deflector voltages of 0, 50, 100, 150, and 200 V on the two deflectors, shown in Figure 12, demonstrate that the polar angle response of Beacon-D is similar to that of Beacon-C.

Fast Ion Mass Spectrometer (FIMS) Elevation Analyzer for Ion Composition Measurements

At SwRI, work has proceeded over the past several years on a class of gridded hemisphere elevation analyzers. This work was motivated by the need for three-dimensional, composition-resolved plasma measurements on CRAF, coupled with a lack of sufficient availability of mechanical scanning because of fuel limitations. Three instrument concepts that have been investigated are described here, beginning with FIMS. A schematic cross-sectional sketch of the FIMS is presented in Figure 13. This system, an outgrowth of an elevation analyzer system described by Wilhelm (1985), consists of three parts, which for convenience have been called an "elevation" analyzer, an "energy" analyzer, and an "azimuth" analyzer. The system is designed to cover a hemisphere, like Beacon, and would require two sensor heads with opposed viewing axes to cover the 4π str unit sphere. For ion mass measurements, the "azimuth" analyzer could directly feed into a cylindrical array of TOF analyzers as described by Young et al. (1989).

The elevation analyzer consists of a hemispherical system with the outer hemisphere made of grid or

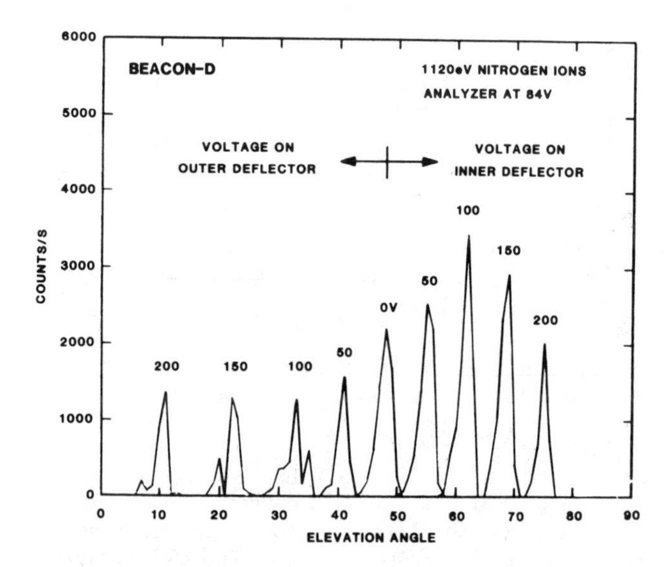

Fig. 12. Polar angle response of Beacon-D, measured for the deflector voltages shown above the response peaks.

mesh material. The inner elevation hemisphere consists of concentric conducting rings electrically isolated from each other. By programming the analyzer voltage onto first the bottom unshaded #1 ring, then the first two rings, the bottom three rings, etc., ions with polar angles beginning near 90° and extending down to near 0° can be accepted and further energy analyzed in the central truncated hemispherical analyzer. Particles then cross over from the energy analyzer into the azimuth analyzer which then brings them out in a cylindrical shell parallel to and centered on the instrument symmetry axis. By dividing the output of the azimuth analyzer into swaths, the azimuthal distribution of incoming ions, which is mirrored at the output, can be resolved, while polar angle resolution is achieved, as described, by programming the analyzer voltage on the rings of the elevation analyzer.

Figure 14 shows the results of a FIMS angular response measurement at SwRI. The instrument was placed in a 1 keV N_2^+ beam in the SwRI calibration chamber and the polar angle response was measured with 365 V on the various combinations of conducting rings. It is apparent that the FIMS can clearly resolve ion distributions in polar angle, as well as azimuth, thus facilitating three-dimensional determinations. By adding TOF units at the exit of the azimuth analyzer, the measurements of all major ion species can be made simultaneously.

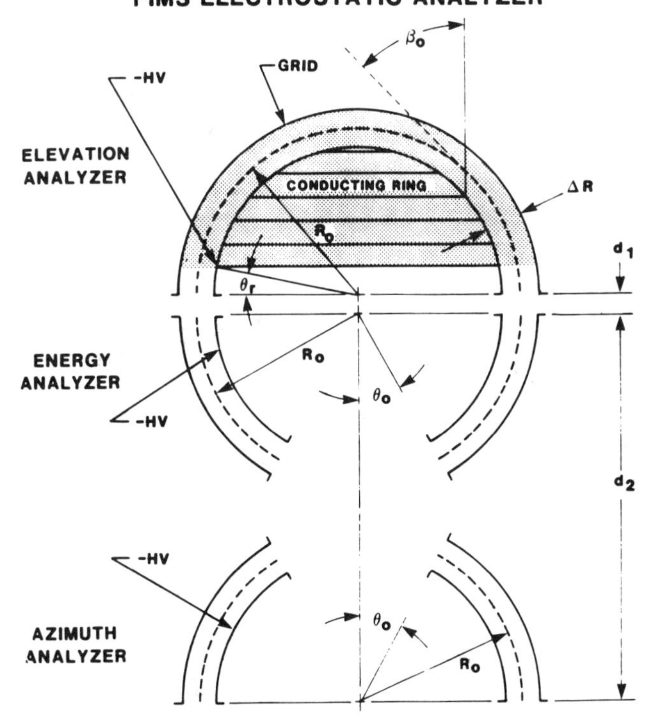

Fig. 13. Cross-sectional schematic of the FIMS electrostatic analyzer system.

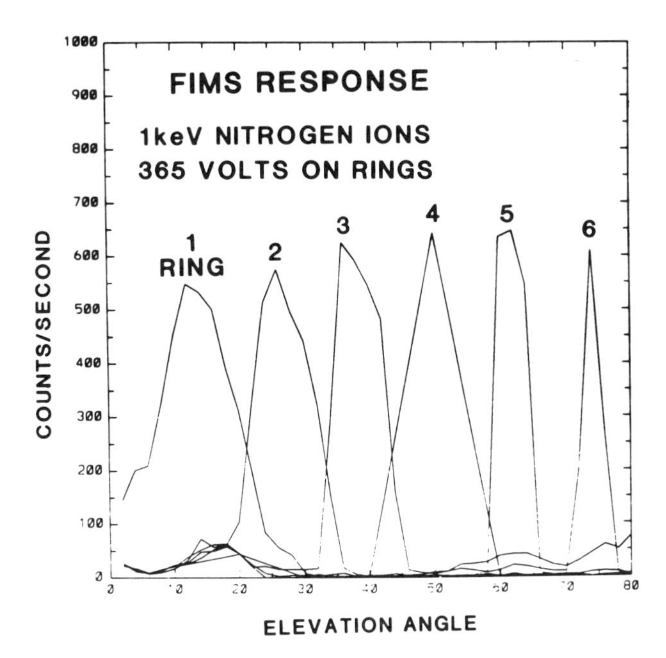

Fig. 14. FIMS polar angle response. The colatitude elevation or polar angle is used in this plot.

Gridded Truncated Hemisphere (GTH) Elevation Analyzer

Another member of the class of gridded hemisphere elevation analyzers under development at SwRI is the GTH, shown sketched in Figure 15. The inner deflection plate consists of a hemisphere that is truncated and closed with a flat surface at a latitude of 60°; i.e., it is a 360° by 60° hemispherical section, capped by a flat plate parallel to the equatorial plane of the section. The outer deflection plates are of similar geometry, but the flat plate is raised above the point of truncation. Both are made of 85% transmission conducting mesh or grid. In essence, the GTH consists of a parallel plate deflector followed by a spherical section deflector. For selecting particles from high polar (elevation) angle directions, as illustrated in Figure 15a, only the spherical section deflector is used, with no deflection occurring in the parallel plate deflector. To select particles from low polar angles, as in Figure 15b, the hemispherical analyzer is set at a constant deflection voltage, while the voltage of the parallel plate deflector is varied. The polar angle, α, is defined in Figure 16.

The reason two deflector systems are combined for the GTH, rather than using a full hemisphere, is that laboratory tests and ray tracing results have shown that if a full hemisphere is used the angular resolution (as measured by $dV/d\alpha$, where dV is the change in deflection voltage and $d\alpha$ is the resulting change in polar angle) becomes very coarse for polar angles less than about 30°. Figure 15a shows how the angle for 1 keV ions is controlled by varying the deflection voltage on the hemispherical section. In Figure 15b we show how

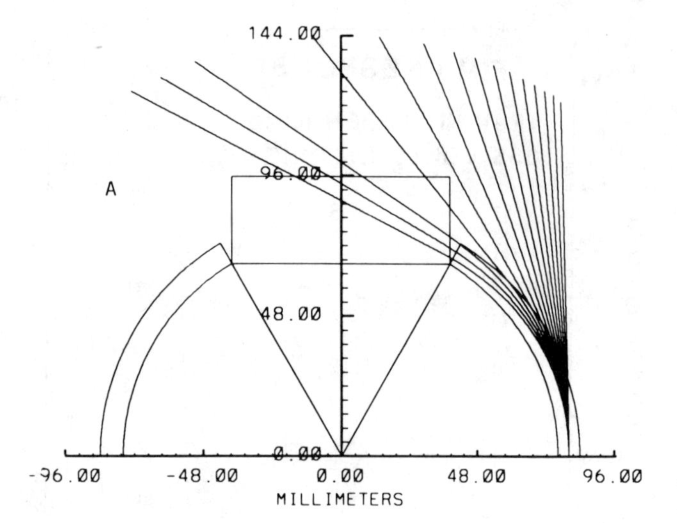

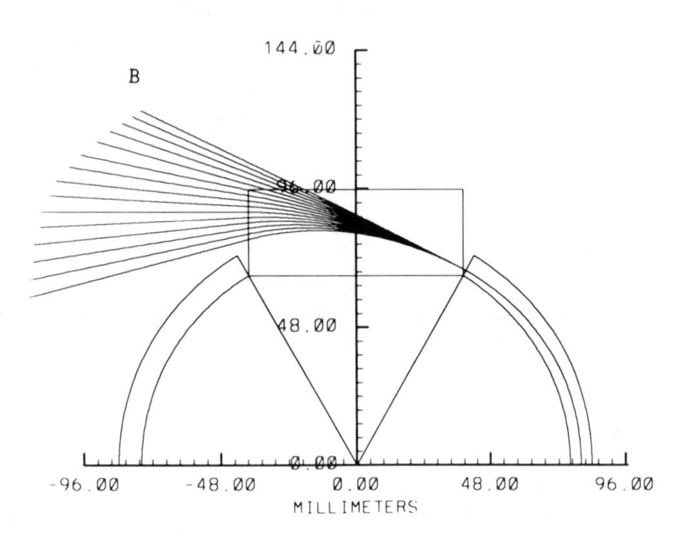

Fig. 15. (a) Cross-sectional view of the ion optics of the GTH elevation analyzer. Ions are shown entering from high polar angles through the outer walls of the parallel plate and curved plate deflectors which are made of screen or mesh. (b) Ions shown entering from low polar angles.

polar angles less than 30° are selected, with good angular resolution, by changing the voltage on the parallel plate deflector while holding the hemispherical section voltage constant. As shown in Figure 15b, good angular resolution is obtained even to polar angles beyond 0°, i.e., the instrument equatorial plane, with the circular parallel plate deflection system.

With the type of analyzer shown in Figure 15, it is necessary to have additional energy filtering because a large matrix of energies and elevation angles can enter the elevation analyzer at different locations along the entrance grid and pass through the analyzer exit. For this reason, the GTH deflector system is fitted onto a second hemispherical-section analyzer for energy analysis. Similarly to the FIMS, shown in Figure 13, these first two stages are referred to for simplicity as the elevation analyzer and the energy analyzer, respectively. Again like the FIMS, a third analyzer, which is referred to as the azimuth analyzer, is added to provide a relatively large annular output aperture which can be segmented to obtain good azimuthal angle resolution. The complete GTH stack is shown in Figure 16, which also shows the computed tracks of two ions passing through the entire system. The azimuth analyzer also provides a nearly parallel output beam, which can then be mass-analyzed with high resolution using either a magnetic or a TOF mass analyzer, not shown in Figure 16.

CRAF Non-Scanning Platform (NSP) Elevation Analyzer for Ion Composition Measurements

A recent addition to the SwRI candidate elevation analyzers for flight on the CRAF spacecraft, called CRAF NSP, would not require placement on the low resolution scan platform. Shown schematically in Fig-

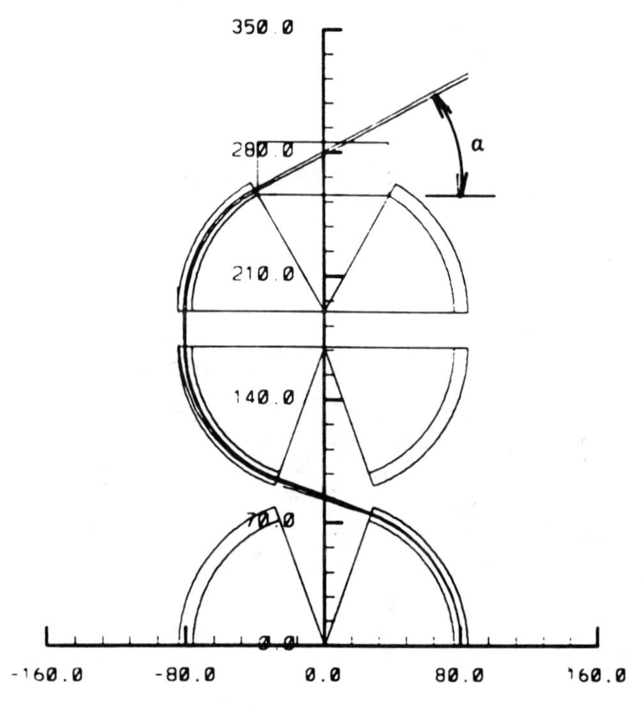

Fig. 16. View of the GTH stack, assembled with elevation, energy, and azimuth analyzers. Computed tracks of two ions are shown passing through the entire system.

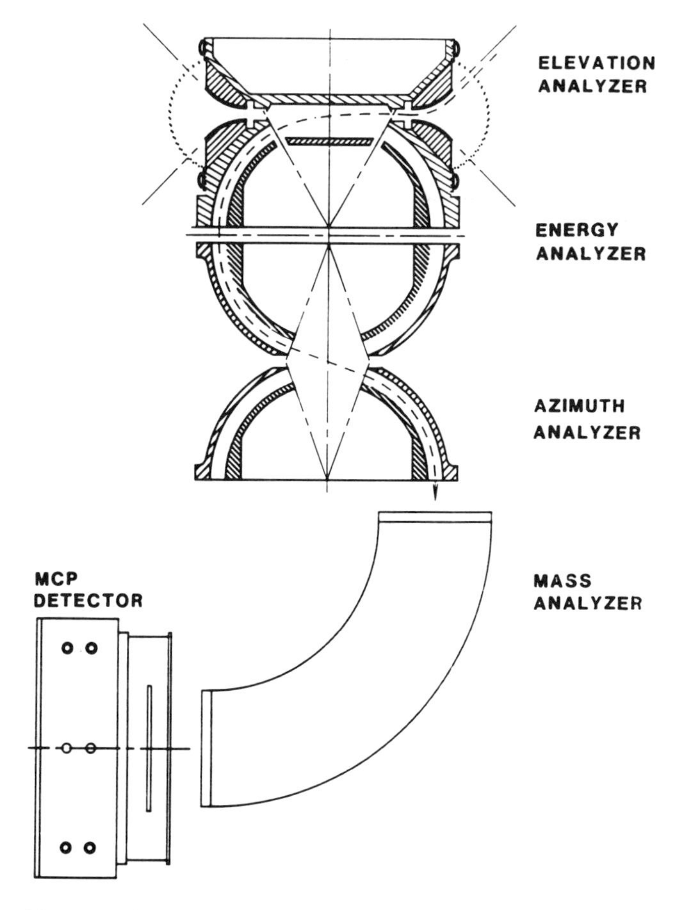

Fig. 17. Cross-sectional view of the CRAF NSP detector.

ure 17, it is composed of three truncated hemispherical analyzer sections in a stack with an additional deflection system on top. The top two truncated hemispheres comprise an energy analyzer which is followed by an azimuth analyzer. The upper elevation analyzer is no longer a hemisphere, but instead is composed of two curved deflector plates with rotational symmetry. The deflectors are enclosed in a grounded toroidal section of screen to isolate the deflection voltages from the spacecraft exterior. A distinct advantage of this geometry is that, like the Beacon geometry, particles entering into the deflector stage from outside pass through the screen almost normally, while with the FIMS and the GTH they pass through almost tangentially so that transmission is substantially reduced by the screen.

The energy analyzer for the CRAF NSP is composed of two near-hemispheres fed by the elevation analyzer. Referring again to Figure 17, ions from the energy analyzer cross over into the azimuth analyzer which separates them again to provide a nearly parallel output beam for mass analysis. In the present example, analyzed ions passed into a magnetic analyzer for mass analysis. Laboratory measurements with a prototype show that this system provides high-resolution separation of singly charged molecular ions of nitrogen and oxygen.

A TOF mass analyzer array like that described by Young et al. (1989) could also be used to mass analyze the ions passing out of the azimuth analyzer, or a combination of TOF analyzers and magnetic analyzers could be used.

Tests and ray tracing results show that the practical polar angle range that the CRAF NSP can cover is about ±45° from the instrument equator. This range constitutes more than 70% of the full 4π str unit sphere and may provide sufficient spatial coverage for the CRAF mission.

Summary

The Los Alamos National Laboratory and the Southwest Research Institute are cooperating in a program to develop instruments capable of making three-dimensional measurements of space plasmas without requiring access to a spacecraft scan platform or spinning section. Such instruments could greatly reduce spacecraft requirements, simplify the acquisition of plasma data on planetary missions, and substantially increase the useful operating time of a plasma experiment. Importantly, the obscuration problem caused by spacecraft appendages such as antennas, RTGs, other instruments, and by the spacecraft itself can be circumvented by placing two 2π sensor heads on opposite sides of the spacecraft.

Five different instrument concepts have been explored. Instruments based on all five of these have the ability to select plasma particles from all pertinent directions in space by using electrostatic deflection techniques. In a continuing joint program these instruments will be investigated and tested further with the goal of developing one or more of them to a state of readiness for an appropriate planetary mission.

Acknowledgments. The authors wish to thank Bruce Barraclough, Jack Gosling, and Michelle Thomsen at Los Alamos who have generously contributed their time and enthusiasm in support of this joint SwRI/Los Alamos development program. We also wish to thank Tom Booker and David Strain at SwRI for their assistance with the calibrations of all the instruments described in this paper. Work at Los Alamos has been under the auspices of the U.S. Department of Energy. At SwRI this work has been supported by Internal Research Project 15-9455.

References

Lazarus, A. J., and R. L. McNutt, Jr., Low-energy plasma ion observations in Saturn's magnetosphere, J. Geophys. Res., 88, 8831, 1983.

Richardson, J. D., Thermal ions at Saturn: Plasma parameters and implications, J. Geophys. Res., 91, 1381, 1986.

Sittler, E. C., Jr., K. W. Ogilvie, and J. D. Scudder, Survey of low-energy plasma electrons in Saturn's magnetosphere: Voyagers 1 and 2, J. Geophys. Res., 88, 8847, 1983.

Wilhelm, K., W. Studemann, and W. Riedler, Observations of the electron spectrometer and magnetometer (Experiment 1 ES 019) onboard Spacelab 1 in response to electron accelerator operations, Earth-Orient. Applic. Space Technol., 5, 47, 1985.

Young, D. T., Near-equatorial magnetospheric particles from ~1 eV to ~1 MeV, Rev. Geophys. Space Phys., 21, 402, 1983.

Young, D. T., J. A. Marshall, J. L. Burch, S. J. Bame, and R. H. Martin, A 360° field-of-view toroidal ion composition analyzer using time-of-flight, these proceedings, 1989.

POLYMERS IN COMET COMAE

W. F. Huebner and D. C. Boice

Southwest Research Institute, San Antonio, TX 78284

Abstract. Heavy ions have been detected in the coma of Comet Halley with various instruments on the Giotto and Vega spacecraft. Ions of sulfur and sulfur compounds are one class of species that can account for some masses observed in the ion mass spectrometer, but other heavy mass components are still not identified. The ionized decay products of a polymer, polyoxymethylene (POM), have been identified from the PICCA mass peaks at 45, 61, 75, 91, and 105 AMU. POM appears to be associated with the dust component of the coma. We suggest that polymers are a source for many unidentified species and that POM, in particular, is responsible for the observed extended source of CO in the coma. There are indications that at least one other polymer, containing CN, is associated with the dust and may be responsible for some of the unidentified mass spectra. We present results from our chemical coma model and suggest polymers and their ionic fragments that show potential for detection in existing spectra or future in situ measurements.

Introduction

To identify a space polymer it is desirable to (1) explain the sequence of processes that leads to its formation, (2) obtain its optical or mass spectrum, and (3) correlate its properties with other observed properties of the environment. All three criteria have been met in the case of the identification of polymerized formaldehyde $(H_2CO)_n$ (polyoxymethylene or POM). A sequence for the formation of POM in interstellar clouds has been presented (Huebner et al., 1987a,b), ions of decay products of POM have been identified (Huebner, 1987) in the mass spectrum from the positive ion cluster composition analyzer (PICCA) (Korth et al., 1986a,b; Mitchell et al. 1986, 1987), and comet data from several instruments are consistent with the properties of POM. Of particular relevance are (1) its affinity for silicate and carbon surfaces that can bond submicron-sized grains into larger dust particles which can disintegrate on heating as evidenced by the enhancement in the ratio of small to large particles with increasing cometocentric distance (McDonnell et al., 1986), (2) the light trapping by porous layers of dust particles held together by POM whiskers to result in a low albedo of the comet nucleus (Keller et al., 1986), and (3) the radio detection (Snyder et al., 1988) and possible infrared detection (Combes et al., 1986; Knacke et al., 1986) of formaldehyde for which POM may be the mother substance in the coma. Some of these observations are difficult to explain without the existence of POM. Here we present evidence that this polymer may also be the extended source of CO in the coma of the comet (Eberhardt et al., 1987) and that the presence of POM and another polymer containing CN is also consistent with the data from the particle impact analyzer (PIA) instrument on Giotto and the PUMA instruments on Vega 1 and Vega 2. In particular we draw attention to the need for measurements with ion and neutral mass spectrometers, especially in the mass range from 50 to 60 AMU, in future comet missions. In order to improve our understanding of polymers in space, basic molecular data for ionization, charge exchange, dissociation, protonization, and other end-capping processes for polymers in the gas phase are needed.

POM as an Extended Source for CO

Analyzing the data from the Giotto neutral mass spectrometer (NMS) in the mass channel corresponding to 28 AMU (CO and N_2), Eberhardt et al. (1987) found that the abundance times distance squared increases with cometocentric distance out to about 10^4 km and leveled off to a constant value at about 2×10^4 km. Eberhardt et al. concluded that the ratio of the number density of CO to H_2O is about 0.07 at 1000 km (the CO coming directly from the nucleus) and is about 0.24 at 20,000 km, the CO partially contributed by an extended source associated with the dust in the coma. They also showed that the production of CO from the dust reaches a maximum at about 10^4 km.

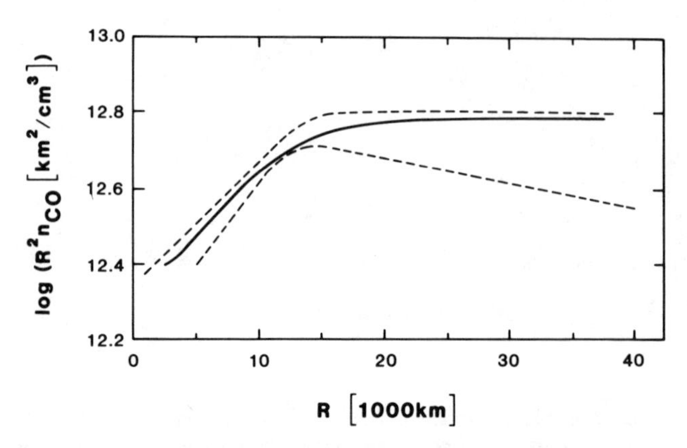

Fig. 1. Comparison of data analyzed by Eberhardt et al. (1987) (between dashed lines) with model calculations (solid line) using POM on dust as the extended source of CO.

We apply our comet coma model to the analysis of this problem. Since no data for ionization or dissociation for polymers in the gas phase exist, we estimate data for these processes based on monomer formaldehyde and other, similar molecules. We assume that the rate coefficient for photodissociation of POM is 10^{-4} s^{-1}, for photoionization 10^{-5} s^{-1}, and for electron dissociative recombination 2.5 x 10^{-7} cm^3 s^{-1}. Electron collisional dissociation and ionization and other processes such as charge exchange and end-capping by protons and other species have not been included, although they should be important. Our model is therefore incomplete. However, detailed chemical reactions for formaldehyde (which is a decay product of POM) and for water (which is a main constituent of the comet nucleus) have been considered. For an assumed release rate for POM_5 from dust particles of 3 x 10^{-4} s^{-1}, an abundance of 4% POM_5 on the dust and 5% CO and 91% H_2O in the ice component, we find that CO/H_2O is 6% at 1000 km and 24% at 20,000 km from the nucleus. The symbol POM_5 represents an oligomer with five units of formaldehyde. Less POM is needed for longer chains. The above composition represents a simplified approximation to Halley. Nevertheless, our result agrees well with the analysis of the NMS spectrum by Eberhardt et al. Our preliminary model indicates that POM is a likely candidate for the extended source of CO (see Figure 1). The large affinity of POM in the solid phase for silicates, and therefore for association with dust grains, further strengthens this conclusion. In view of the crude estimates for the rate coefficients, improvements such as distribution functions for POM chain lengths and size, temperature, and velocity distributions for the dust particles are not justified at this time.

We have already reported (Huebner et al., 1987a,b) on the comparison of the ratios of the observed ionized decay products of POM in the PICCA spectrum with the results from our computer model. The agreement we obtained was within a factor of 4, which we consider good, considering the uncertainty in the rate coefficients. There was no significant difference in the results using either POM_4 or POM_5 in the initial composition. However, POM_4 cannot be nearly as abundant as POM_5, because it would overpopulate the mass channels at about 120 AMU. Whether the dominant POM chain length is 5, 6, or many more H_2CO units long cannot be predicted at this time.

In light of the above results, the analysis of the data from the ion mass spectrometer (IMS) on the Giotto spacecraft by Allen et al. (1987) and Wegmann et al. (1987a,b) will need to be reviewed for mass channels 12 through 16, corresponding to the ions of C, CH, CH_2, CH_3, and CH_4. POM_5 contributes significantly to some of these species.

Other Polymers as Extended Sources for Coma Radicals and Ions

It has been suggested that the CN component in comet spectra also originates from the dust (A'Hearn et al., 1986). Indeed, from the analysis of the PIA instrument on the Giotto spacecraft one can speculate that POM is released from dust and disintegrates at cometocentric distances out about 10^4 km, resulting in the relatively high occurrence of H-, C-, and O-bearing parents. Another polymer containing N may be released from dust at even larger distances, resulting in an increase of H-, C-, N-, and O-bearing particles (Clark et al., 1986; see Figure 3.4-1). To identify this polymer will be more difficult than the identification of POM. However, the PICCA instrument reveals a small bump in the mass spectra at about 53 AMU (D. L. Mitchell, private communication, 1988) that we suspect may be evidence for a polymer containing CN. The high-intensity spectrometer (HIS) of the IMS also indicates a signal in this mass range, which has been tentatively identified as Fe^+ (Balsiger et al., 1986). Gringauz et al. (1986) have also tentatively identified a rather broad peak in this mass range as Fe^+. Unfortunately the PICCA instrument switched mass ranges at 50 AMU and the mass range of the HIS instrument is limited to 57 AMU. Ion mass spectrometers on future missions should concentrate on the 50 to 60 AMU mass range.

From chemical considerations one can expect a substitution of sulfur for oxygen in POM. However, since the cosmic abundance of S to O is only 1:40, such a substitution will occur only occasionally, resulting in a co-polymer. If only one S is substituted for O in POM, the mass spectrum of the products is shifted by 16 AMU to higher masses, or effectively by 2 AMU in an alternating sequence of 14 and 16 AMU. This shift of 2 AMU to higher mass is still consistent with the width of the POM spectrum from the PICCA. Low-mass decay products of such a co-polymer would result in CH_2-S (46 AMU), O-CH_2-S (62 AMU), and the ions associated with these molecules and the corresponding protonated species at 47 and 63 AMU. The energy of a C-S bond is significantly smaller than that for a C-O bond. Thus the C-S bond is the weak link in the chain, thereby increasing the relative abundance of these species.

Another polymer with similar structure to POM is polyethylene. The energy of the C-C bond is similar to that of the C-O bond. This polymer could contribute to C_2, C_3, but particularly to the species C_2H_4, C_3H_6, C_4H_8, etc. and their simple and protonated ions. It should be noted that C_2H_4 has the same mass as CO and N_2 and could influence the extended source analysis of Eberhardt.

Conclusions

The heavy ions of the decay products of POM, its sulfur co-polymer, and of polyethylene contribute to mass channels for which observations from the HIS instrument are higher than model calculations have been able to account for (see Wegmann et al. 1987a,b). Specifically HCO^+ (29 AMU), H_2CO^+ (30 AMU), H_3COH^+ (32 AMU), $C_3H_6^+$ (42 AMU), $C_3H_7^+$ and H_2COCH^+ (43 AMU), $H_2COCH_2^+$ (44 AMU), $H_3COCH_2^+$ and HCS^+ (45 AMU), $C_4H_7^+$ (55 AMU), and, finally, $C_4H_8^+$ (56 AMU). Mass channel 45, which contains HCS^+, shows a much larger decrease in intensity than other high-mass channels from 1500 to 6000 km cometocentric distance, consistent with the low energy for the C-S bond.

Polymers may play an important role as extended sources for cometary radicals and ions. POM is a likely candidate for the extended production of CO in Comet Halley. Other polymers that may be extended sources of species in comet comae are hydrogen cyanide polymers and polyethylene. The design of ion and neutral mass spectrometers for future comet missions should include the higher mass range (50 to 200 AMU) to observe possible correlated decay products of polymers, especially the ranges between 50 and 60 AMU and 100 to 120 AMU for detection of a suspected hydrogen cyanide polymer.

Acknowledgments. This work was supported by the NASA Planetary Atmospheres Program.

References

A'Hearn, M. F., S. Hoban, P. V. Birch, C. Bowers, R. Martin, and D. A. Klinglesmith, III, Cyanogen jets in comet Halley, Nature, 324, 649, 1986.

Allen, M., M. Delitsky, W. Huntress, Y. Yung, W.-H. Ip, R. Schwenn, H. Rosenbauer, E. Shelley, H. Balsiger, and J. Geiss, Evidence for methane and ammonia in the coma of comet P/Halley, Astron. Astrophys., 187, 502, 1987.

Balsiger, et al., Ion composition and dynamics at comet Halley, Nature, 321, 330, 1986.

Clark, B., L. W. Mason, and J. Kissel, Systematics of the "CHON" and other light-element particle populations in comet Halley, in 20th ESLAB Symposium on the Exploration of Halley's Comet, ESA SP-250, Vol. III, edited by B. Battrick, E. J. Rolfe, and R. Reinhard, p. 353, Noordwijk, The Netherlands, 1986.

Combes, M., et al., Detection of parent molecules in comet Halley from the IKS-Vega experiment, in 20th ESLAB Symposium on the Exploration of Halley's Comet, ESA SP-250, Vol. I, edited by B. Battrick, E. J. Rolfe, and R. Reinhard, p. 353, Noordwijk, The Netherlands, 1986.

Eberhardt, P., et al., The CO and N abundance in comet P/Halley, Astron. Astrophys., 187, 481, 1987.

Gringauz, K. I., et al., First in situ plasma and neutral gas measurements at comet Halley, Nature, 321, 282, 1986.

Huebner, W. F., First polymer in space identified in comet Halley, Science, 237, 628, 1987.

Huebner, W. F., D. C. Boice, and C. M. Sharp, Polyoxymethylene in comet Halley, Ap. J. Lett., 320, L149, 1987a.

Huebner, W. F., D. C. Boice, C. M. Sharp, A. Korth, R. P. Lin, D. L. Mitchell, and H. Rème, Evidence for first polymer in comet Halley: Polyoxymethylene, in Symposium on the Diversity and Similarity of Comets, ESA SP-278, edited by E. J. Rolfe and B. Battrick, p. 163, 1987b.

Keller, H. U., et al., First Halley Multicolour Camera imaging results from Giotto, Nature, 321, 320, 1986.

Knacke, R. F., T. Y. Brooke, and R. R. Joyce, Observations of 3.2-3.6 micron emission features in comet Halley, Ap. J. Lett., 310, L49, 1986.

Korth, A., et al., Mass spectra of heavy ions near comet Halley, Nature, 321, 335, 1986a.

Korth, A., et al., Radial variations of flow parameters and composition of cold heavy ions within 50,000 km of Halley's nucleus, in 20th ESLAB Symposium on the Exploration of Halley's Comet, ESA SP-250, Vol. I, edited by B. Battrick, E. J. Rolfe, and R. Reinhard, p. 199, Noordwijk, The Netherlands, 1986b.

McDonnell, J. A. M., et al., Dust density and mass distribution near comet Halley from the Giotto observations, Nature, 321, 338, 1986.

Mitchell, D. L., et al., Derivation of heavy (10-210 AMU) ion composition and flow parameters for the Giotto PICCA instrument, in 20th ESLAB Symposium on the Exploration of Halley's Comet, ESA SP-250, Vol. I, edited by B. Battrick, E. J. Rolfe, and R. Reinhard, p. 203, Noordwijk, The Netherlands, 1986.

Mitchell, D. L., R. P. Lin, K. A. Anderson, C. W. Carlson, D. W. Curtis, A. Korth, H. Rème, J. A. Sauvaud, C. d'Uston, and D. A. Mendis, Evidence for chain molecules enriched in carbon, hydrogen, and oxygen in comet Halley, Science, 237, 626, 1987.

Snyder, L. E., P. Palmer, and I. de Pater, Radio

detection of formaldehyde emission from comet Halley, Astron. J., submitted, 1988.
Wegmann, R., H. U. Schmidt, W. F. Huebner, and D. C. Boice, Cometary MHD and chemistry: Application to Halley, in Symposium on the Diversity and Similarity of Comets, ESA SP-278, edited by E. J. Rolfe and B. Battrick, p. 277, 1987a.
Wegmann, R., H. U. Schmidt, W. F. Huebner, and D. C. Boice, Cometary MHD and chemistry, Astron. Astrophys., 187, 339, 1987b.

CHARGED DUST IN THE EARTH'S MAGNETOSPHERE

Mihaly Horanyi

Supercomputer Computations Research Institute, Florida State University, Tallahassee, FL 32306.

Abstract. Computer simulations were carried out on the spatial distribution of small Al_2O_3 particles dumped into the Earth's magnetosphere during solid rocket propellant burns. In addition to the standard gravitational and light pressure forces, we have taken into account the electrodynamic forces as the particle will be electrostatically charged because it is immersed in the plasma and radiative environment of the Earth. We will conclude that the lifetime of a grain in the magnetosphere is not sensitive to the electrodynamic forces but the number of grains lost to the solar wind will dramatically increase at the expense of the flux lost by colliding with the Earth.

Introduction

In a recent paper by Horanyi, et al. (1988)— hereafter referred to as (HHM, 1988) we investigated the electrostatic charging processes of small grains in the Earth's magnetosphere. We arrived at the conclusion that the charge on the grain will be modest; therefore, only small grains will be affected by the electrodynamic forces. For gains with radii of $r_g \lesssim 0.1\ \mu$ the electrodynamic forces are comparable with the light-pressure force and become important in shaping the trajectory of these grains. In the present study we only consider $0.1\ \mu$ grains as the production rate of Al_2O_3 spherules has its maximum yield around this range (Mueller and Kessler, 1985). In the present study we will discuss the computer simulation results on the evolution of a dust ring around geosynchronous orbit. In the next paragraph we briefly summarize the dynamics and charging processes and refer to (HHM, 1988) for more details.

Dynamics and Charging

In the Earth-centered inertial frame, the motion of a charged dust grain of mass m is governed by the equation

$$m\ddot{\bar{r}} = \bar{F}_G + \bar{F}_{LP} + \bar{F}_L \qquad (1)$$

where the three terms included on the right-hand side are the gravitational, light pressure, and Lorentz forces. In the region around the geosynchronous orbit, we may neglect the neutral gas and plasma (Coulomb) drags on the dust (Peale, 1966). Furthermore, we regard the interparticle distances to be larger than the Debye shielding length λ, so that interparticle Coulomb interactions are also negligible.

The Lorentz force:

$$\bar{F}_L = \frac{Q(t)}{m}\left(\bar{E} + \frac{\dot{\bar{r}}}{c} \times \bar{B}\right) \qquad (2)$$

where the electrical field $\bar{E}$ has two sources—the cross tail field $\bar{E}_{CT}$ and the corotational electrical field. Assuming rigid corotation and a simple aligned centered dipole magnetic field $\bar{B}$, $\bar{E}$ becomes

$$\bar{E} = \bar{E}_{CT} - \frac{1}{c}\left(\bar{\Omega}_p \times \bar{r}\right) \times \bar{B} \qquad (3)$$

The cross tail electrical field depends on the geomagnetic activity level for which in this study we assume average conditions.

The charge on the grain can be obtained by solving the current balance equation

$$\frac{dQ}{dt} = \sum_i I_i \qquad (4)$$

where I_i represents the various charging currents. In the present study we included all the important charging currents: (1) electron and protron thermal currents, (2) secondary electron currents by electron and proton impact, (3) backscattered electron current, (4) photoelectron current.

The optical and physical parameters of these Al_2O_3 sphercules are not too well known and we will therefore consider two essentially extreme cases. In one case (type 1), we will assume the grain to have the scattering efficiency Q_{pr} of conducting magnetite with $\rho \approx 3.2\ g\,cm^{-3}$, whereas in the other case (type 2) we will assume it to have the scattering efficiency of dielectric olivine with

$\rho \approx 2.2\ g\,cm^{-3}$. Other material constants of these two types of grains (such as the photoelectron and secondary emission yields) are also different.

The relevant expressions for the various charging currents as well as the list of the material constants for type 1 and type 2 grains are listed in HHM (1988). As the charge on the grain is not only a function of the local plasma parameters but also of the previous charging history, the equation of the motion (equation (1)) must be integrated simultaneously with the current balance equation (equation (9)). In order to follow the evolution of the trajectory and the charge, we need a plasma model of the Earth's environment.

The Plasma Environment

In order to evaluate the various currents, we need to know the parameters of the plasma in which the grains are immersed. To describe the plasma environment of the Earth, we have followed Hill and Whipple (1985) with the modification that we included the local time-dependence of the plasmapause. Inside the plasmapause we assume thermal equilibrium between the ions and the electrons, with their density given by the empirical formula

$$n(L) = 10^{(15-L)/3.5}\mathrm{cm}^{-3} \tag{5}$$

where L denotes the magnetic shell, whose equatorial radius is L Earth radii. The temperature is likewise given by the empirical relation

$$kT(L) = 0.09293L^{2.7073}\ eV \tag{6}$$

when the temperature of the plasma exceeds 1 eV, the plasma is considered to have two components, a hot one with T given in equation (6) and $n = 1\ \mathrm{cm}^{-3}$, and a cold one with $kT = 1\ eV$ and density 1 cm^{-3} less than that given by (5).

The boundary of the plasmapause for average geomagnetic conditions is given by

$$L_{pp} = 7.6R_E\left[\frac{\sqrt{1+\sin\Psi}-1}{\sin\Psi}\right] \tag{7}$$

where Ψ is the local hour angle (Lyons and Williams, 1984). Between the plasmapause and the magnetopause, we used the analytic expression fitted for the measured energy distribution of electrons and ions (Garrett and DeForest, 1979). The plasmapause extends beyond the 18 hours LT geosynchronous radius, but it leaves no obvious signature on the above satellite data. Perhaps this is because the detectors had a higher-energy threshold, and it is for this reason that we kept the local time dependence of the plasmapause. The bi-Maxwellian energy distribution, which is different for electrons and ions, depends on the daily geomagnetic activity index A_p which is kept constant at its average value of 120. The magnetospheric plasma density is of the order of 1 cm^{-3}, with characteristic energies of about 2 keV for electrons and about 5 keV for protons. The magnetopause which delineates the magnetospheric plasma, is described by a paraboloidal shape whose apex points toward the Sun. The subsolar point of the magnetopause is at 10.5 R_E and opens up to 14.5 R_E at the terminators.

Beyond the magnetopause, we assume solar wind conditions with $n = 5\ \mathrm{cm}^{-3}$ and $kT \approx 10$ eV.

Discussion

We have followed the evolution of a dust cloud (1000 particles) which at $T = 0$ formed a uniform density ring between 0.8–1.2 times the geosynchronous orbit in the equatorial plane. Initially, all the particles move with Kepler speed on circular orbits. With these conditions, if one could switch off the light pressure and Lorentz forces, the ring would be maintained forever. As a test of our code, we checked the evolution without light pressure and Lorentz forces, and up to about a year the ring is maintained without noticeable distortion. As the grains move in the various plasma regimes, the charge varies accordingly. Inside the plasmapause we have a few volts negative potential (the potential in this case is only the charge divided by the radius of the grain as we assume low dust density, this also justifies that the dust-dust collisions are entirely negligible) smoothly varying with altitude. In the magnetosphere the potential is positive. It can reach up to 10 volts, and is a function of local time only. In the solar wind we have a few volts positive potential. Because of the finite capacitance of the small grains, accompanied by low densities in the magnetosphere and solar wind, the charge variation along the trajectory of a grain is smooth except when the grain enters the plasmapause where, because of the high densities, the charge will change abruptly. The time scales as well as the equilibrium values of the potential in the various plasma regimes are discussed in HHM (1988).

In Figure 1 we show the spatial distribution of the grains. Figure 1(a) shows the initial distribution and Figures 1(b)–(c) is the distribution after 0.75 and 1.5 days, for the type 1 grains. Figures 1(d)–(e) show the distribution after 3.5 and 4.5 days for type 2 grains. These distributions would be similar if one would neglect the Lorentz forces. The noticeable difference is a rotation of the outer edge of the dust cloud, which neglecting the Lorentz forces, would point toward 6 hours LT. As the charges inside the plasmasphere are much smaller than in the magnetosphere, the corotational electric fields become less important than the cross tail field. As a result, the perturbing force will be approximately the sum of the

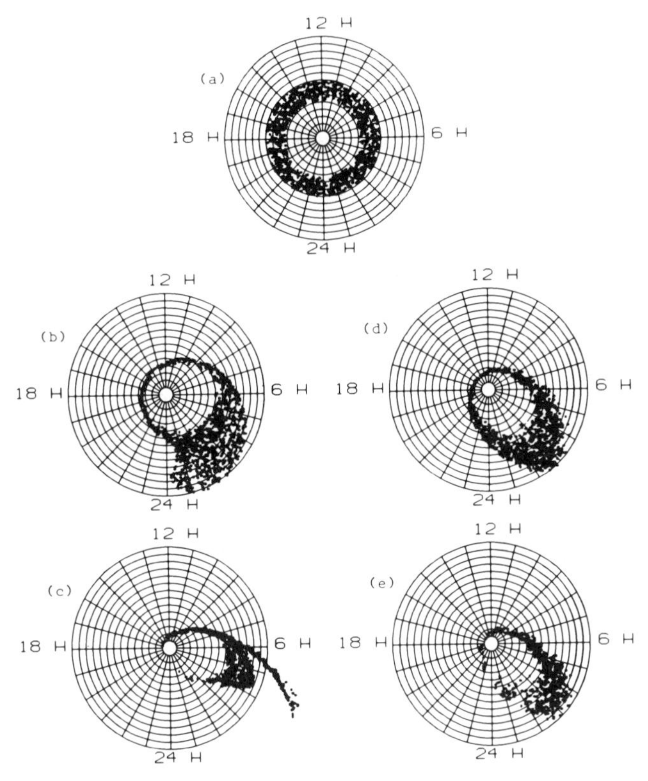

Fig. 1. Spatial distribution of grains in the equatorial plane. (a) Initial distribution. (b)–(c) Distribution after 0.75 and 1.5 days, respectively, for the type 1 grains. (d)–(e) Distribution after 3.5 and 4.5 days accordingly for type 2 grains. The radial unit is 1 R_E.

vectors of the light pressure force and the electrostatic force due to the cross tail field, which points from 6H toward 18H local time. The net perturbing force points toward approximately 21H instead of 24H local time, as was the case neglecting the Lorentz force.

In Figure 2 we show the time dependence of the number of particles, still orbiting the Earth within 20 R_E (n_p), the number of particles collided with the Earth (n_e), and the number of particles left beyond 20 R_E (n_s), which we consider as lost to the solar wind. In Figures 2(a)–(b), we show the time dependence of the various groups of particles n_p, n_e and n_s for type 1 and type 2 grains, for the simulations where the Lorentz forces were included and in Figures 2(c)–(d) where the Lorentz forces were neglected again for type 1 and type 2 accordingly.

For type 1 grains, we lose all of the 1000 initial particles within approximately 2.6 days regardless of the Lorentz forces. In the case where the Lorentz forces are neglected, the only particles lost to the solar wind are those which, at $T = 0$, are far enough from the Earth toward the Sun. By the time they reach 24H local time these particles have gained enough energy from the solar radiation pressure that their total energy becomes or exceeds zero. If we do not neglect the Lorentz force there will be another way to gain energy. The average charge on the particle between 6H and 18H is somewhat larger than between 18H and 6H local time. That will result in a slight gain in energy at each revolution, because the grain has a larger charge when moving in the direction of the cross tail field than when moving against it. As a result, the loss to the solar wind increases from 8% to 32% of the original number of the particles.

As the light scattering efficiency is much smaller for type 2 grains (Q_{pr} is approximately 1.5 for type 1 and approximately 0.3 for type 2 grains, when $R_g = 0.1\ \mu$), it takes longer to clean the magnetosphere of type 2 grains. For both cases, with or without the Lorentz force, it takes approximately 6 days to lose all the particles. The number of particles lost to the solar wind increases from zero to 16% of the original amount.

Conclusions

The central aim of our present study is to assess the role of electromagnetic forces that are experienced by fine dust particles injected into the Earth's magnetosphere. In this study we used a simple model of the particles and field environments of the Earth and confined our attention to particle orbits in the equatorial plane only. We

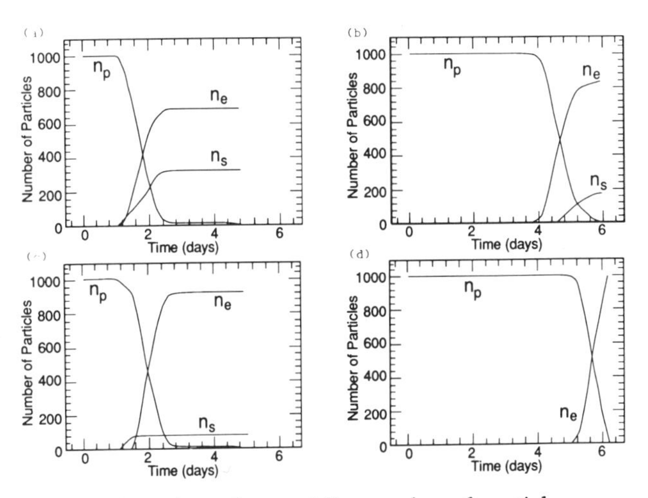

Fig. 2. Time dependence of the number of particles, orbiting the Earth within 20 R_E, n_p, the number of particles collided with the Earth, n_e, and the number of particles lost to the solar wind, n_s. (a)–(b) Case for type 1 and type 2 grains, respectively. (c)–(d) Case for type 1 and type 2 grains, respectively, when the electrodynamical forces were neglected.

have considered all the important charging currents on the grains. The electric charge acquired by the grains is rather modest. It varies typically from a few volts (negative) inside the plasmasphere to about 10 volts (positive) in the magnetosphere. For small grains the electromagnetic forces are important in the shaping of their trajectories, but the time of residence for the grains is not significantly changed. Most important is the change in the flux of particles lost to the solar wind; for the type 1 grains, this flux is increased from 8% to 32% (of the original input) and from zero to 16% for type 2 particles due to the electrostatic forces.

We have neglected the effects of the tilted dipol as well as the fluctuations in the geomagnetic activity index and electrostatic charge. These issues will be addressed later on and might significantly change some of our conclusions.

Acknowledgments. We would like to acknowledge support from the U.S. Department of Energy through contract DE-FC05-85ER250000 and the Florida State University Computing Center who donated part of the computation time.

References

Garrett, H.B. and DeForest, S.E., *J. Geophys. Res.*, *84*, 2083, 1979.

Hill, J.R. and Whipple, E.C., *J. Spacecraft and Rockets*, *22*, 245, 1985.

M. Horanyi, H.L.F. Houpis and D.A Mendis, to appear in *Astronomy and Space Sci.*, 1988.

Lyons, L.R. and Williams, D.J., *Quantitative Aspects of Magnetospheric Physics*, D. Reidel Pub. Co., p. 79, 1984.

Mueller, A.C. and Kessler, D.J., *Adv. Space Res.*, *5* (2), 1985.

Peale, S.J., *J. Geophys. Res.*, *71*, 911, 1966.

AUTHOR INDEX

LIST OF PARTICIPANTS

S.-I. Akasofu
Geophysical Institute
University of Alaska
903 Koyukuk Avenue, North
Fairbanks, AK 99775

Roger Anderson
Department of Physics and Astronomy
University of Iowa
Iowa City, IA 52242

Fran Bagenal
HAO/NCAR
P.O. Box 3000
Boulder, CO 80307

S. J. Bame
Mail Stop D438
Los Alamos National Laboratory
P.O. Box 1663
Los Alamos, NM 87545

P. Bochsler
Physikalisches Institut
University of Bern
Sidlerstrasse 5, CH-3012
Bern, Switzerland

P. L. Bornmann
R/E/SE
NOAA/SEL
Boulder, CO 80303

M. E. Bruner
Lockheed Palo Alto Research Laboratory
Department 91-20, Bldg. 255
3251 Hanover Street
Palo Alto, CA 94304

J. L. Burch
Southwest Research Institute
P.O. Drawer 28510
San Antonio, TX 78284

Richard C. Canfield
Institute for Astronomy
University of Hawaii
2680 Woodlawn Drive
Honolulu, HI 96822

Charles W. Carlson
Space Sciences Laboratory
University of California at Berkeley
Berkeley, CA 94720

D. P. Cauffman
Lockheed Missiles and Space Company
Department 91-60, Bldg. 256
3251 Hanover Street
Palo Alto, CA 94304

J. T. Clarke
Code 681
NASA/Goddard Space Flight Center
Greenbelt, MD 20771

W. A. Coles
University of California/San Diego
C-014
La Jolla, CA 92093

John E. Connerney
Code 695
NASA/Goddard Space Flight Center
Greenbelt, MD 20771

C. J. Crannell
Code 682
NASA/Goddard Space Flight Center
Greenbelt, MD 20771

T. E. Cravens
Space Physics Research Laboratory
Department of Atmospheric, Oceanic, and Space Sciences
The University of Michigan
Ann Arbor, MI 48109

C. C. Curtis
Department of Physics, Bldg. 81
The University of Arizona
Tucson, AZ 85721

Steven Curtis
Code 695
NASA/Goddard Space Flight Center
Greenbelt, MD 20771

R. B. Dahlburg
Laboratory for Computational Physics and Fluid Dynamics
Naval Research Laboratory
Code 4440
Washington, D.C. 20375

O. de la Beaujardière
Geoscience and Engineering Center
SRI International
333 Ravenswood
Menlo Park, CA 94025

G. A. Dulk
Department of Astrophysical, Planetary, and Atmospheric Scienes
Campus Box 301
University of Colorado
Boulder, CO 80309

R. C. Elphic
Institute of Geophysics and Planetary Physics
University of California at Los Angeles
Los Angeles, CA 90024

George H. Fisher
Institute for Astronomy
University of Hawaii
2680 Woodlawn Drive
Honolulu, HI 96822

Ernest G. Fontheim
Space Physics Research Laboratory
The University of Michigan
Ann Arbor, MI 48109

H. A. Garcia
Space Environment Laboratory, NOAA
Boulder, CO 80303

Dale E. Gary
Solar Astronomy 264-33
California Institute of Technology
Pasadena, CA 91125

George Gloeckler
Department of Physics and Astronomy
University of Maryland
College Park, MD 20742

C. K. Goertz
Department of Physics and Astronomy,
The University of Iowa
Iowa City, IA 52242

Martin V. Goldman
Department of Astrophysical, Planetary,
and Atmospheric Scienes
Campus Box 301
University of Colorado
Boulder, CO 80309

Melvyn L. Goldstein
Code 692
Laboratory for Extraterrestrial Physics
NASA/Goddard Space Flight Center
Greenbelt, MD 20771

Raymond Goldstein
Mail Stop 169-506
Jet Propulsion Laboratory
4800 Oak Grove Drive
Pasadena, CA 91109

T. I. Gombosi
Space Physics Research Laboratory
Department of Atmospheric, Oceanic,
and Space Sciences
The University of Michigan
Ann Arbor, MI 48109

John Harmon
Arecibo Observatory
P.O. Box 995
Arecibo, PR 00613

Ernest Hildner
Space Environment Laboratory, R/E/SE
325 Broadway
Boulder, CO 80303

Mihaly Horanyi
Supercomputer Computations Research
Institute
Florida State University
Tallahassee, FL 32306

K. C. Hsieh
Department of Physics, Bldg. 81
The University of Arizona
Tucson, AZ 85721

W. F. Huebner
Southwest Research Institute
P.O. Drawer 28510
San Antonio, TX 78284

B. V. Jackson
Code 014
University of California/San Diego
La Jolla, CA 92093

J. R. Kan
Geophysical Institute
University of Alaska, Fairbanks
Fairbanks, AK 99775

J. T. Karpen
Code 4175-K
Naval Research Laboratory
Washington, D.C. 20375

E. P. Keath
Applied Physics Laboratory
The Johns Hopkins University
Johns Hopkins Road
Laurel, MD 20707

Paul M. Kintner
School of Electrical Engineering
Cornell University
5147 Upson
Ithaca, NY 14853

Craig Kletzing
Center for Astrophysics and Space
Sciences, Code C-011
University of California/San Diego
La Jolla, CA 92093

James A. Klimchuk
Center for Space Science and
Astrophysics, ERL 300
Stanford University
Stanford, CA 94305

John L. Kohl
Harvard-Smithsonian Astrophysical
Observatory
60 Garden Street
Cambridge, MA 02138

S. M. Krimigis
Applied Physics Laboratory
The Johns Hopkins University
John Hopkins Road
Laurel, MD 20707

Martin A. Lee
EOS-SERC
University of New Hampshire
Durham, NH 03824

R. P. Lin
Space Sciences Laboratory
University of California at Berkeley
Berkeley, CA 94720

B. C. Low
High Altitude Observatory
National Center for Atmospheric
Research
P.O. Box 3000
Boulder, CO 80307

R. Lundin
Institut for Rymdfysik
Swedish Institute of Space Physics
P.O. Box 812
S-981, 28 Kiruna, Sweden

L. R. Lyons
Space Sciences Laboratory
Mail Stop M2-260
The Aerospace Corporation
P.O. Box 92957
Los Angeles, CA 90009

J. A. Marshall
Southwest Research Institute
P.O. Drawer 28510
San Antonio, TX 78284

B. H. Mauk
Applied Physics Laboratory
The Johns Hopkins University
Johns Hopkins Road
Laurel, MD 20707

A. N. McClymont
Institute for Astronomy
University of Hawaii
2680 Woodlawn Drive
Honolulu, HI 96822

D. J. McComas
MS-D438
Los Alamos National Laboratory
P.O. Box 1663
Los Alamos, NM 87545

Billy M. McCormas
Lockheed Missiles and Space Company
3251 Hanover Street
Palo Alto, CA 94304

R. W. McEntire
Applied Physics Laboratory
Johns Hopkins University
Johns Hopkins Road
Laurel, MD 20707

James P. McFadden
Space Sciences Laboratory
University of California at Berkeley
Berkeley, CA 94720

Carl E. McIlwain
Center for Astrophysics and Space
Sciences, Code C-011
University of California/San Diego
La Jolla, CA 92093

Ralph L. McNutt, Jr.
Center for Space Research
Massachusetts Institute of Technology
77 Massachusetts Avenue
Cambridge, MA 02139

Yi Mei
University of California at Los Angeles
405 Hilgard Avenue
Los Angeles, CA 90024

Muhammad Adel Miah
Department of Physics and Astronomy
Louisiana State University
Baton Rouge, LA 70803

J. A. Miller
Code 665
NASA/Goddard Space Flight Center
Greenbelt, MD 20771

D. G. Mitchell
Applied Physics Laboratory
Johns Hopkins University
Johns Hopkins Road
Laurel, MD 20707

Ron L. Moore
Code ES52
NASA/Marshall Space Flight Center
Huntsville, AL 35812

Marcia Neugebauer
Mail Station 264-664
Jet Propulsion Laboratory
4800 Oak Grove Drive
Pasadena, CA 91109

J. D. Nichols
Center for Astrophysics and Space Sciences, Code C-011
University of California/San Diego
La Jolla, CA 92093

Francesco Paresce
Space Telescope Science Inst.
3700 San Martin Drive
Baltimore, MD 21218

Joachim Raeder
University of California at Los Angeles
405 Hilgard Avenue
Los Angeles, CA 90024

Reuben Ramaty
Code 660
Laboratory for High Energy Astrophysics
NASA/Goddard Space Flight Center
Greenbelt, MD 20771

John D. Richardson
Center for Space Research, 37-655
Massachusetts Institute of Technology
77 Massachusetts Avenue
Cambridge, MA 02139

E. C. Roelof
Applied Physics Laboratory
Johns Hopkins University
Johns Hopkins Road
Laurel, MD 20707

Robert Rosner
Astronomy and Physics
University of Chicago
5640 South Ellis Avenue
Chicago, IL 60637

Ilan Roth
Space Sciences Department
University of California at Berkeley
Berkeley, CA 94720

R. Schmidt
Space Science Department of ESA/ESTEC
Postbus 199
2200 AG, Noordwijk, The Netherlands

Michael Schulz
Mail Stop M2-259
The Aerospace Corporation
P.O. Box 92957
Los Angeles, CA 90009

G. W. Simon
Air Force Geophysics Laboratory/ National Solar Observatory/ Sacramento Peak
Sunspot, NM 88349

M. F. Smith
Southwest Research Institute
P.O. Drawer 28510
San Antonio, TX 78284

R. S. Steinolfson
Institute for Fusion Studies
The University of Texas at Austin
Austin, TX 78712

Nobie H. Stone
Code ES53
NASA/Marshall Space Flight Center
Huntsville, AL 35812

Peter A. Sturrock
Center for Space Science and Astrophysics, ERL 306
Stanford University
Stanford, CA 94305

S. T. Suess
Code ES52
NASA/Marshall Space Flight Center
Huntsville, AL 35812

Richard M. Thorne
Department of Atmospheric Science
University of California at Los Angeles
Los Angeles, CA 90024

Alan Title
S2-10 Solar Physics
Lockheed Palo Alto Research Laboratories
3251 Hanover Street
Palo Alto, CA 94304

Roy Torbert
Research Institute, Room D4A
The University of Alabama in Huntsville
Huntsville, AL 35899

J. Hunter Waite, Jr.
Code ES53
NASA/Marshall Space Flight Center
Huntsville, AL 35812

Raymond J. Walker
Institute of Geophysics and Planetary Physics
University of California at Los Angeles
Los Angeles, CA 90024

C. C. Wu
Department of Physics
University of California at Los Angeles
Los Angeles, CA 90024

Tyan Yeh
Cooperative Institute for Research in Environmental Sciences
University of Colorado
Boulder, CO 80309

D. T. Young
Southwest Research Institute
P.O. Drawer 28510
San Antonio, TX 78284